# ENGINEERING THERMODYNAMICS

## THROUGH EXAMPLES



**Y V C Rao**

Department of Chemical Engineering

IIT Kanpur

**Universities Press**

**Universities Press (India) Private Limited**

*Registered Office*
3-6-747/1/A & 3-6-754/1, Himayatnagar,
Hyderabad 500 029 (A.P.), India
e-mail: info@universitiespress.com

*Distributed by*
Orient Blackswan Private Limited

*Registered Office*
3-6-752, Himayatnagar, Hyderabad 500 029 (A.P.), India

*Other Offices*
Bangalore / Bhopal / Bhubaneshwar / Chennai / Ernakulam / Guwahati
Hyderabad / Jaipur / Kolkata / Lucknow / Mumbai / New Delhi / Patna



First published 2003
Reprinted 2005, 2008, 2009, 2011

ISBN: 978 81 7371 423 8



*Typeset by*
S.K. Typesetters

Printed in India at
Multi Fold
Chennai 600 085

***Published by***
Universities Press (India) Private Limited
3-6-747/1/A & 3-6-754/1, Himayatnagar
Hyderabad 500 029 (A.P), India

*Dedicated to*

Ramana and Sawjanya

# The Author

Prof. Y.V.C. Rao obtained his B.Tech degree in Chemical Engineering from Andhra University, Vishakapatnam and his M.Tech and Ph.D degrees in Chemical Engineering from the IIT–Kanpur. Prof. Rao has been teaching thermodynamics at undergraduate and postgraduate levels since 1973. He has published several research papers in national and international journals in the areas of thermodynamics, molecular energy transfer and shock waves and presented several research papers in national and international conferences.

Prof. Rao was a Fulbright visiting research scholar at the Columbia University, New York. He is a Fellow of the Institution of Engineers and a Fellow of the Indian Institute of Chemical Engineers. In addition to this book, he has published four other books on Classsical Thermodynamics, Statistical Thermodynamics and Chemical Engineering Thermodynamics.

# Contents

# 1

# Basic Concepts

### Example 1.1

List some of the main activities of an engineer and how are they related to a study of thermodynamics ?

*Solution* :

The main activities of an engineer are concerned with :

1. The design and development of new processes or the improvement of the existing processes.
2. To use the available resources namely space, material, energy and time in an optimal way. That is, the processes are to be designed such that they require minimum energy, space and material to perform the required task.

### Example 1.2

An engineer who is entrusted with the task of producing methanol according to the reaction

$$CO\ (gas) + 2H_2\ (gas) \rightarrow CH_3OH\ (liquid)$$

would like to know the answers to a few questions, before he undertakes the design and development of a process. List some of the questions which can be answered by a study of thermodynamics.

*Solution* :

Application of thermodynamic principles would provide solutions to the following questions.

1. Whether the anticipated reaction is feasible or not at the specified temperature and pressure.
2. If the reaction is feasible, to what extent the raw materials can be converted into products or what is the degree of conversion ?

3. Is it possible to alter the degree of conversion by changing the pressure and temperature at which the reaction is taking place ?
4. Is it possible to increase the degree of conversion by controlling the ratio of reactants while they are fed to the reactor ?
5. How much energy is required to change the state of the reactants to the desired pressure and temperature ?
6. How much energy is required to separate the desired product from the byproducts and reactants ?
7. How much energy is required to carry out the reaction ?

**Example 1.3**

Thermodynamics can be studied by adopting either a macroscopic or a microscopic approach. Distinguish between macroscopic and microsocopic approaches.

*Solution* :

Thermodynamics can be studied by adopting two different approaches, namely, macroscopic and microscopic. In the macroscopic approach, the state of matter is described by specifying only a small number of variables and these variables can be measured or estimated. For example, the state of a gas enclosed in a container can be described by specifying the pressure ($P$), temperature ($T$) and volume ($V$), all of which can be easily measured. In the macroscopic approach the structure of matter under consideration is not taken into account. That is, the state of a gas, liquid or solid can be described by specifying the variables like $P$, $T$ and $V$.

All matter consists of a large number of microscopic particles called atoms and molecules. In the microscopic approach the structure of matter or the state of agregation, like gas, liquid or solid is taken into account. The atoms or molecules, constituting the matter, move at random with independent velocities and there exist intermolecular attractive and repulsive forces. In the microscopic approach, the state of matter is described by specifying the position and velocity vectors of each of the constituting atoms or molecules and the intermolecular forces that exist among them. Thus, in the microscopic approach a large number of variables are required to describe the state of matter and these variables cannot be directly measured. A knowledge of the structure of matter is essential in analyzing the behavior of the matter.

The macroscopic approach is adopted in the study of classical thermodynamics whereas the microscopic approach is adopted in the study of statistical thermodynamics.

**Example 1.4**

For a thorough understanding of any science it is essential to have a good knowledge of the units of measurement. Every system of units adopts a few base units. What are the base units in the International System of units.?

*Solution* :

The base units adopted in the International System of units (or SI units) are :

Kilogram (kg) for mass; metre (m) for length ; second (s) for time ; Kelvin (K) for temperature; mole (mol) for quantity of matter ; ampere (A) for electric current and candela (cd) for luminous intensity.

**Example 1.5**

Define the base units (*a*) metre (*b*) second (*c*) Kelvin and (*d*) mole.

*Solution* :

1. Metre (m) : In 1960 the General Conference on Weights and Measures (Conference Generale des poids et Measures, abbreviated as CGPM) defined the metre as the length equal to 1,650, 763.73 wavelengths, in vacuum of the radiative transition between $2p_{10}$ and $5d_5$ in krypton - 86.
2. Second (s) : The CGPM 1967 defined the second as the duration of 9, 192, 631, 770 periods of the radiation corresponding to the transition between the two hyperfine levels of the ground state of cesium - 133 atom.
3. Kelvin (K) : The CGPM 1967 defined Kelvin as equal to 1/273.16 of the thermodynamic temperature of the triple point of water (0.01°C).

The relation between the Kelvin scale of temperature and the Celsius temperature scale is given by

$$K = {}^{o}C + 273.15$$

4. Mole (mol) : The CGPM 1971 defined the mole as the amount of substance which contains as many elementary entities as there are atoms in 0.012 kg of carbon - 12.

The number of elementary entities in one mole is equal to the Avogadro's number $6.022\ 045 \times 10^{23}$. When the unit mole is used it is necessary to specify the elementary entities as atoms, molecules, ions, electrons etc.

**Example 1.6**

What are the principal features of the SI units ?

*Solution* :

All the physical quantities are classified into two groups as primary quantities and secondary quantities. The primary quantities are measured in terms of the base units and the secondary quantities are measured in terms of derived units. The principal features of the SI units are :

1. Minimum number of base units (only seven) are used to cover the field of science and engineering.
2. Based on the definitions or physical laws, the derived units are obtained in terms of base units. The derived units do not contain any arbitrary constants.
3. The familiar derived units have been given internationally accepted names.
4. Multiples and submultiples of units have been recommended to facilitate the use of SI units to cover a wide range of values.

**Example 1.7**

Derive the dimensions of the derived units–force, pressure, energy and power and name their SI units. Also provide the relations between the base units and derived units.

*Solution* :

The Newton's second law of motion gives force ($F$) as the product of mass ($m$) and acceleration ($a$). That is

$$F = ma = m\frac{d^2L}{dt^2}$$

where $L$ is distance and $t$ is time.

$$\text{Units of } F = \text{kg}\frac{\text{m}}{\text{s}^2} = \text{kg m s}^{-2} = \text{N (newton)}$$

Force has the dimensions $MLT^{-2}$

The SI unit of force is named as newton (abbreviated as N) in honor of the Scientist Newton. One newton is the force required to produce an acceleration of 1 $\text{m/s}^2$ in a body of mass 1 kg.

Pressure ($P$) is defined as the force per unit area. That is

$$P = F/A$$

$$\text{Units of P} = \frac{\text{kg m s}^{-2}}{\text{m}^2} = \text{kg m}^{-1}\text{ s}^{-2} = \text{Pa (pascal)}$$

Pressure has the dimensions $ML^{-1}T^{-2}$

The SI unit of pressure is pascal (abbreviated as Pa) and Pa = N / $\text{m}^2$

If a force $F$ acting on a body moves it through a distance $L$ in the direction of the applied force, then the work done ($W$) (or the energy transferred to the body) is given by

$$W = F.L$$

$$\text{Units of energy} = \text{kg m s}^{-2}\ \text{m} = \text{kg m}^2\ \text{s}^{-2} = \text{J (joule)}$$

$$\text{Dimensions of energy} = ML^2\,T^{-2}$$

The SI unit of energy is joule (abbreviated as J) and J = N m

The power is the rate at which energy is consumed or delivered per unit time. That is

$$\text{Power} = \frac{\text{Energy}}{\text{Time}}$$

$$\text{Units of power} = \frac{\text{kg m}^2\text{s}^{-2}}{\text{s}} = \text{kg m}^2\ \text{s}^{-3} = \text{W (watt)}$$

$$\text{Dimensions of power} = ML^2\,T^{-3}$$

The SI unit of power is watt (abbreviated as W) and W = J/s

***Note*** *The SI units newton, pascal, joule and watt which are named after the famous scientists are abbreviated using the capital letters* N, Pa, J *and* W. *All other units are abbreviated using the lower case letters.*

**Example 1.8**

What are the accepted prefixes in SI units to express large and small quantities ?

*Solution* :

The internationally accepted SI prefixes are given below.

| Factor | Prefix | Symbol | Factor | Prefix | Symbol |
|---|---|---|---|---|---|
| 10 | deca | da | $10^{-1}$ | deci | d |
| $10^2$ | hecto | h | $10^{-2}$ | centi | c |
| $10^3$ | kilo | k | $10^{-3}$ | milli | m |
| $10^6$ | mega | M | $10^{-6}$ | micro | μ |
| $10^9$ | giga | G | $10^{-9}$ | nano | n |
| $10^{12}$ | tera | T | $10^{-12}$ | pico | p |
| $10^{15}$ | peta | P | $10^{-15}$ | femto | f |
| $10^{18}$ | exa | E | $10^{-18}$ | atto | a |

**Example 1.9**

What conventions are recommended by CGPM to express the SI units and quantities?

*Solution :*

The following conventions are recommended by CGPM to express the SI units and quantities,

1. The symbol of a prefix should be combined with a single unit symbol forming a new symbol. For example 1000 × Pa = kPa.
2. Compound prefixes should not be used to express a quantity. For example $10^{-9}$ m cannot be written as 1 m μm (milli micro metre), but should be expressed as 1 nm (nanometre).
3. The SI base unit for mass is kg which already contains the SI prefix k (for kilo). Therefore, $10^{-6}$ kg can be expressed as 1 mg and not as 1 μkg.
4. Except at the end of a sentence, a period (.) should not be used after a symbol of an SI unit.
5. Capital letters should not be used is SI units while writing in full, except at the beginning of a sentence. They should be written as pascal, newton, joule etc.
6. The unit symbols should not be used in plural form. It is incorrect to write 150 Js for 150 joules but it should be written as 150 J.
7. The unit symbol should be placed after the numerical value, leaving a space in between the numerical value and the unit symbol.
8. Care should be taken to avoid confusion between the SI prefix milli (m) and the base unit metre (m) while expressing a quantity. For example,

   mN is used for millinewton, where as m N is used to express (metre × newton). Similarly ms stands for millisecond while m s indicates (metre × second).
9. When a unit, is obtained by multiplication or division of two or more units, they can be indicated as shown below:

   newton × metre = N m

   metre ÷ $(\text{second})^2$ = $m/s^2$ or m $s^{-2}$
10. The SI prefixes should be selected such that the numerical value lies between 0.1 and 1000. For example, 101 325 Pa can be written as 101.325 kPa or 0.101 325 MPa; 5500 Å (= 5500 × $10^{-8}$ cm) can be written as 550 nm.

However, when the values are presented is a tabular form, for the same quantity, it is advisable to use the same SI prefix throughout even if some of the numerical values lie outside the range 0.1 to 1000.

11. While writing a quantity with several digits, to facilitate reading, they should be separated into groups of three, counting from the decimal point towards right and left, These groups should be separated by a small gap and should not include a comma.

### Example 1.10

If a long glass tube closed at one end is completely evacuated and the open end is immersed in a pool of mercury, keeping the glass tube upright, the mercury rises in the glass tube. At one standard atmospheric pressure, the mercury rises to a height ($h$) of 760 mm. If the acceleration due to gravity ($g$) is 9.806 65 m/s$^2$ and the density ($\rho$) of mercury is $1.359\ 51 \times 10^4$ kg/m$^3$, what is the value of the standard atmospheric pressure in SI units ?

*Solution* :

The pressure exerted by the column of mercury, which is balanced by the standard atmospheric pressure is given by

$$P = \rho\, hg = 1.359\ 51 \times 10^4 \left(\frac{\text{kg}}{\text{m}^3}\right) \times 760\ (\text{mm}) \times \frac{1}{1000}\left(\frac{\text{m}}{\text{mm}}\right) \times 9.806\ 65 \left(\frac{\text{m}}{\text{s}^2}\right)$$

$$= 101\ 325.01\ \text{kg}\left(\frac{\text{m}}{\text{s}^2}\right)\left(\frac{1}{\text{m}^2}\right)$$

$$= 101\ 325.01\ \text{Pa} = 101.\ 325\ \text{kPa}$$

***Note***

1. *The SI unit of pressure (Pa) is quite small in magnitude and hence for convenience it is customary to express pressure in kilopascal (kPa) or megapascal (MPa).*
2. *There are certain units which are outside the SI and are recognised because of their practical convenience. They are 1 litre (l) = $1 \times 10^3$ cm$^3$; 1 bar = $10^5$ Pa = 100 kPa.*
3. *The pressure exerted by 1 millimetre column of mercury is named as Torr. Usually subatmospheric pressure or pressure below atmospheric pressure is expressed in Torr.*

1 standard atmospheric pressure =101.325 kPa = 1.013 25 bar = 760 Torr

1 Torr = 133.322 Pa

### Example 1.11

Express the following quantities as indicated.

(*a*) 300,000,000 W in kW and in MW

(*b*) 10 kPa in bar and in Torr

(*c*) 20 $dm^3$ in litre and in $m^3$

(*d*) 2Å in μm and in nm

(*e*) 1 $g/cm^3$ in $kg/m^3$

(*f*) 0.15 J in kJ and in mJ

*Solution* :

(*a*) 300,00,000 W = 300,000 kW = 300 MW

$$(b)\ 10\text{ kPa} = \frac{10\times10^3}{10^5} = 0.1\text{ bar}$$

$$10\text{ kPa} = \frac{10\times10^3}{133.322}\left(\frac{\text{Pa}}{\text{Pa/Torr}}\right) = 75.006\text{ Torr}$$

$$(c)\ 20\text{ dm}^3 = 20\text{ dm}^3\times\left(10\frac{\text{cm}}{\text{dm}}\right)^3 = 20\times10^3\text{ cm}^3 = 20\ l$$

$$20\text{ dm}^3 = 20\text{ dm}^3\times\left(10^{-1}\frac{\text{m}}{\text{dm}}\right)^3 = 0.02\text{ m}^3$$

$$(d)\ 2\text{ Å} = 2(\text{Å})\times10^{-8}\left(\frac{\text{cm}}{\text{Å}}\right)\times\frac{1}{100}\left(\frac{\text{m}}{\text{cm}}\right) = 2\times10^{-10}\text{ m}$$

$$= 2\times10^{-4}\ \mu\text{m} = 0.2\text{ nm}$$

$$(e)\ 1\frac{\text{g}}{\text{cm}^3} = 1\left(\frac{\text{g}}{\text{cm}^3}\right)\times\frac{1}{1000}\left(\frac{\text{kg}}{\text{g}}\right)\times\left(100\frac{\text{cm}}{\text{m}}\right)^3 = 1\times10^3\text{ kg/m}^3$$

(*f*) 0.15 J = 0.000 15 kJ = 150 mJ

**Example 1.12**

According to Newton's universal law of gravitation, the force ($F$) between two bodies of masses $m_1$ and $m_2$ which are separated by a distance $R$ is given by

$$F = G\frac{m_1 m_2}{R^2}$$

where $G = 6.672\times10^{-11}\text{ N m}^2\text{ kg}^{-2}$, is the universal gravitational constant.

(*a*) Given that the radius and mass of earth are 6371 km and $5.9659 \times 10^{24}$ kg respectively, estimate the acceleration due to gravity on earth.

(*b*) Given that the radius and mass of moon are 1738 km and $7.35 \times 10^{22}$ kg respectively, estimate the weight of an astronaut of mass 70 kg on moon.

(*c*) Determine the ratio of the acceleration due to gravity on earth to the acceleration due to gravity on moon.

*Solution* :

(*a*) Let $m$ = mass of a body.

Gravitational force acting on the body, $F = \dfrac{Gm\,M_e}{R_e^2} = mg$

or
$$g = \frac{GM_e}{R_e^2} = \frac{6.672\times10^{-11}\times5.9659\times10^{24}}{(6371\times10^3)^2} = 9.8066 \text{ m/s}^2$$

(*b*) Weight of the astronaut = Force exerted by the astronaut as a result of gravitational force field

$$F = \frac{GmM_m}{R_m^2} = \frac{6.672\times10^{-11}\times70\times7.35\times10^{22}}{(1738\times10^3)^2} = 113.64 \text{ N}$$

(*c*) $g_{\text{earth}} = \dfrac{GM_e}{R_e^2}$; $g_{\text{moon}} = \dfrac{GM_m}{R_m^2}$

$$\frac{g_{\text{earth}}}{g_{\text{moon}}} = \frac{M_e}{M_m}\left(\frac{R_m}{R_e}\right)^2 = \frac{5.9659\times10^{24}}{7.35\times10^{22}}\times\left(\frac{1738}{6371}\right)^2 = 6.04$$

**Example 1.13**

An experiment is designed to correlate the volumetric flow rate of fluid through a pipe of uniform cross section with the pressure drop across a specified length of the pipe. In this experiments a U– tube manometer with mercury ($\rho$ = $1.3595 \times 10^4$ kg/m$^3$) is used to measure the pressure drop of a fluid, the density of which is 800 kg/m$^3$. At a particular flow rate, the difference in the levels of mercury in the two limbs is 20 cm. Determine the pressure drop.

*Solution* :

The mercury manometer is shown in Fig.E 1.13. At the surface S–S, pressure is the same. Hence pressure exerted by the fluid in the left limb AS is identical to the pressure exerted by the fluid in the right limb BS. That is

$$P_A + \rho_f\, g\,(h + l) = P_B + \rho_f\, g\,(l) + \rho_m\, gh$$

or
$$P_A - P_B = gh\,(\rho_m - \rho_f) = 9.81 \times 0.2 \times (1.3595 \times 10^4 - 800)$$
$$= 25.104 \text{ kPa}$$

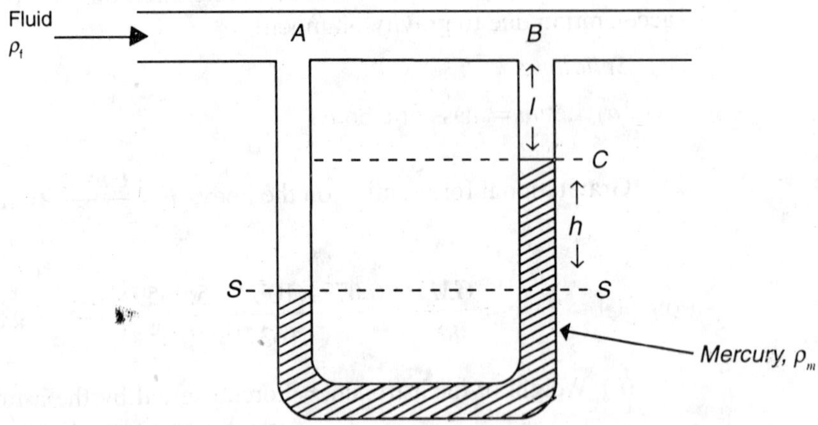


Fig.E 1.13. Sketch for Example 1.13.

**Example 1.14**

The atmospheric pressure is due to the air surrounding the earth. Assuming that the air behaves like an ideal gas ($P = \rho\, RT$) where $\rho$ is molar density, mol / m$^3$ and the atmospheric air has a uniform temperature $T$, derive a relation to estimate the pressure as a function of altitude.

*Solution* :

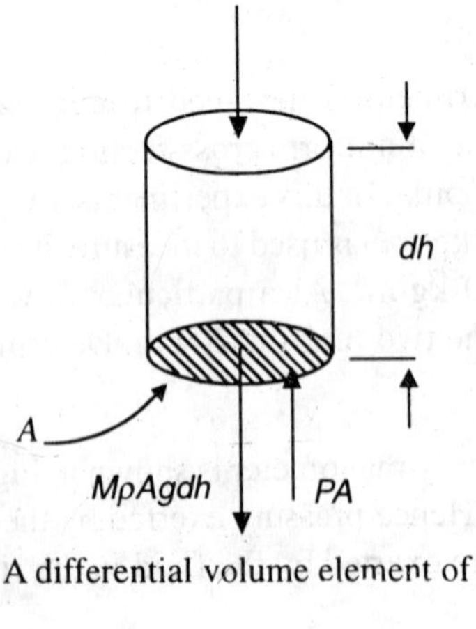


Fig.E 1.14. A differential volume element of the atmospheric air.

Consider a differential volume element of the atmospheric air as shown in Fig.E 1.14 and apply the force balance to obtain

$(P + dP)\,A + M\rho\,A\,gdh = PA$

where $M$ is the molar mass of air

or $$-dP = gM\rho dh = \frac{gMP}{RT}dh$$

or $$\int_0^h dh = \frac{-RT}{Mg}\int_{P_0}^{P}\frac{dP}{P}$$

or $$h = \frac{-RT}{Mg}\ln(P/P_0)$$

or $$P = P_0 \exp\left(\frac{-Mgh}{RT}\right)$$

**Example 1.15**

What is molar mass ? Assuming that atmospheric air is a mixture of nitrogen and oxygen in the mole ratio 79:21 estimate the molar mass of air.

*Solution* :

The mass of one mole of a substance is called molar mass.

Basis : 1 mole of air.

$N_2$ present in air = 0.79 mol and $O_2$ present in air = 0.21 mol

Molar mass of Nitrogen = $28 \times 10^{-3}$ kg ;

Molar mass of Oxygen = $32 \times 10^{-3}$ kg

Molar mass of air = $0.79 \times 28 \times 10^{-3} + 0.21 \times 32 \times 10^{-3}$

$= 28.84 \times 10^{-3}$ kg

***Note*** *The molar mass of air is usually taken as $28.97 \times 10^{-3}$ kg because of the presence of small quantities of $CO_2$ and argon.*

**Example 1.16**

Define a thermodynamic system and state its characteristics.

*Solution* :

A definite quantity of matter (which is enclosed by a boundary that separates it from the rest of the universe) on which we focus our attention for thermodynamic analysis is called a thermodynamic system.

The important characteristics of a system are:

1. A system contains a definite quantity of matter . This quantity does not undergo any change as the system undergoes a process.Therefore, a system is sometimes called as **Control mass.**
2. The boundary enclosing the system may be real (for example the glass wall of a thermos flask acts as a boundary for enclosing coffee (system) contained in the flask) or imaginary (for example one may focus his attention on air contained in $1m \times 1m \times 1m$ space of a room).
3. The system boundary may be rigid or may change its shape, as well as in size during a given process. That is the volume of a system may undergo a change.
4. The system may exchange energy either in the form of work or heat or both, with its surroundings or with other systems. If a system does not exchange energy, in the form of work as well as heat, with its surroundings it is called an **isolated system**.
5. A system may be very simple, like a certain quantity of coffee held in a thermos flask or it may be complex like a huge petrochemical plant.
6. The choice of a system is not unique and it depends on the analyst.

***Note*** *All matter external to the system is called surroundings, The combination of a system and its surroundings is called universe.*

**Example 1.17**

What are the essential features of a thermodynamic property.?

*Solution* :

A thermodynamic property is a characteristic which can be used to describe the state of a system. The essential features of a property are:

1. A thermodynamic property should have a definite value when a system is in a particular state.
2. The change in the value of the property should not depend on the path followed by the system in reaching the specified state. In other words, the

change in the property depends on the initial and final states of the system only. That is the change in the property $Z$ can be written as

$$\int_1^2 dZ = \Delta Z = Z_2 - Z_1.$$

That is a thermodynamic property is a point function or state function and it is not a path fuction. In other words, its differential is exact.

*Note* *If* $Z = Z(x,y)$, *then*

$$dZ = \left(\frac{\partial Z}{\partial x}\right)_y dx + \left(\frac{\partial Z}{\partial y}\right)_x dy = Mdx + Ndy$$

$dZ$ is an exact differential if

$$\left(\frac{\partial M}{\partial y}\right)_x = \left(\frac{\partial N}{\partial x}\right)_y$$

**Example 1.18**

For a particular thermodynamic system, the following expression has been suggested for the differential of pressure ($P$).

$$dP = \frac{RdT}{v-b} - \left\{\frac{RT}{(v-b)^2} - \frac{2a}{v^3}\right\} dv$$

Where $T$ is temperature, $v$ is volume, $a$, $b$ and $R$ are constants.

Judge whether the given differential is exact or not. If it is exact find the relation between $P$, $v$ and $T$.

*Solution* :

The given relation tells that $P = P(T,v)$. Then

$$dP = \left(\frac{\partial P}{\partial T}\right)_v dT + \left(\frac{\partial P}{\partial v}\right)_T dv = MdT + Ndv \qquad (A)$$

Where $M = (\partial P / \partial T)_v$ and $N = \left(\frac{\partial P}{\partial v}\right)_T$

The given differential for pressure is

$$dP = \frac{RdT}{v-b} - \left\{\frac{RT}{(v-b)^2} - \frac{2a}{v^3}\right\} dv \qquad (B)$$

Comparing Eqns. (*A*) and (*B*), we get

$$M = \frac{R}{v-b} \text{ and } N = \frac{-RT}{(v-b)^2} + \frac{2a}{v^3}$$

The given differential will be exact if and only if

$$\left(\frac{\partial M}{\partial v}\right)_T = \left(\frac{\partial N}{\partial T}\right)_v \qquad (C)$$

Therefore, let us check whether Eqn. (*C*) is satisfied or not.

$$\left(\frac{\partial M}{\partial v}\right)_T = \frac{-R}{(v-b)^2} \text{ and } \left(\frac{\partial N}{\partial T}\right)_v = \frac{-R}{(v-b)^2}$$

Eqn.(*C*) is satisfied. Hence, the given expression is an exact differential.

To determine the explicit relation $P = P(T,v)$, we can partially integrate $M$ with respect to $T$ and obtain

$$P = \int (MdT)_v = \int \left[\frac{RdT}{v-b}\right]_v = \frac{RT}{v-b} + f(v) \qquad (D)$$

Where $f(v)$ is an integration constant, which is a function of $v$ only. To determine the value of $f(v)$, we can evaluate $\left(\frac{\partial P}{\partial v}\right)_T$ from Eqn. (*D*) and compare it with the given value of $N$

$$\left(\frac{\partial P}{\partial v}\right)_T = \frac{-RT}{(v-b)^2} + f'(v) = N = \frac{-RT}{(v-b)^2} + \frac{2a}{v^3}$$

or $$f'(v) = \frac{2a}{v^3} \text{ or } f(v) = \frac{-a}{v^2} + C \qquad (E)$$

Where $C$ is a constant which is independent of both $T$ and $v$. Now, substitute for $f(v)$ from Eqn. (*E*) in Eqn. (*D*) to obtain.

$$P = \frac{RT}{v-b} - \frac{a}{v^2} + C$$

**Example 1.19**

Determine the explicit relation $P = P(T, v)$ for Example 1.18 by partially integrating $N = \left(\frac{\partial p}{\partial v}\right)_T$ with respect to $v$.

*Solution* :

$$M = \frac{R}{v-b} \text{ and } N = \frac{-RT}{(v-b)^2} + \frac{2a}{v^3}$$

$$P = \int [Ndv]_T = \int \left[ \left\{ \frac{-RT}{(v-b)^2} + \frac{2a}{v^3} \right\} dv \right]_T = \frac{RT}{v-b} - \frac{a}{v^2} + f(T) \qquad (A)$$

Where $f(T)$ is an integration constant, which is a function of $T$ only. To determine the value of $f(T)$, evaluate $\left(\frac{\partial P}{\partial T}\right)_v$ from Eqn. ($A$) and compare it with the given $M$.

$$\left(\frac{\partial P}{\partial T}\right)_v = \frac{R}{v-b} + f'(T) = \frac{R}{v-b}$$

or $$f'(T) = 0 \text{ or } f(T) = C \qquad (D)$$

Where $C$ is a constant which is independent of $T$ and $v$.

Substitute the value of $f(T)$ from Eqn. ($D$) in Eqn. ($A$) to obtain

$$P = \frac{RT}{v-b} - \frac{a}{v^2} + c$$

**Example 1.20**

Rework Example 1.18 by simultaneously integrating both the terms to obtain the relation $P = P(T, v)$

*Solution* :

The given relation is

$$dP = \frac{RdT}{v-b} - \left\{ \frac{RT}{(v-b)^2} - \frac{2a}{v^3} \right\} dv \qquad \text{(See Example 1.18)}$$

Integration of the above relation gives

$$P = \frac{RT}{v-b} + \left\{ \frac{RT}{v-b} - \frac{a}{v^2} \right\} + C \qquad (A)$$

Where $C$ is an integration constant.

In Eqn.($A$), it is necessary to drop one of the terms which is repeated. In this case the term $\dfrac{RT}{v-b}$ is repeated. By dropping one of the terms $\dfrac{RT}{v-b}$ in Eqn. ($A$) we get

$$P = \frac{RT}{v-b} - \frac{a}{v^2} + C$$

**Example 1.21**

Verify whether

$$dZ = (4xy + 4y^2)\, dx + (2x^2 + 8xy)dy$$

is an exact differential or not. If exact find $Z = Z(x, y)$.

*Solution :*

$$dZ = (4xy + 4y^2)\, dx + (2x^2 + 8xy)\, dy = Mdx + Ndy$$

where $M = 4xy + 4\, y^2$

and $N = 2x^2 + 8xy$

If $dZ$ is an exact differential it must satisfy the condition

$$\left(\frac{\partial M}{\partial y}\right)_x = \left(\frac{\partial N}{\partial x}\right)_y \qquad (A)$$

Therefore, let us check whether the above condition is satisfied or not.

$$\left(\frac{\partial M}{\partial y}\right)_x = \frac{\partial}{\partial y}(4xy + 4y^2) = 4x + 8y$$

$$\left(\frac{\partial N}{\partial x}\right)_y = \frac{\partial}{\partial x}(2x^2 + 8xy) = 4x + 8y$$

Therefore Eqn. ($A$) is satisfied. Hence $dZ$ is an exact differential

To determine $Z = Z(x,y)$, one can partially integrate $M$ w.r.t.$x$. or $N$ w.r.t.$y$.

Integrating $M$ partially w.r.t.$x$. we get

$$Z = \int Mdx = \int (4xy + 4y^2)\, dx = 2x^2y + 4xy^2 + f(y) \qquad (B)$$

where $f(y)$ is a function of $y$ only.

Integrating $N$ partially w.r.t. $y$, we get

$$Z = \int Ndy = \int (2x^2 + 8\, xy)\, dy = 2x^2y + 4xy^2 + f(x) \qquad (C)$$

where $f(x)$ is a function of $x$ only,

From Eqn. ($B$) we get

$$\left(\frac{\partial Z}{\partial y}\right)_x = 2x^2 + 8xy + f'(y) = N = 2x^2 + 8xy$$

or

$$f'(y) = 0 \text{ or } f(y) = C$$

Where $C$ is a constant which is independent of $x$ and $y$. Hence

$$Z = 2x^2y + 4xy^2 + C \qquad (D)$$

Alternatively, form Eqn. ($C$) we get

$$\left(\frac{\partial Z}{\partial x}\right)_y = 4xy + 4y^2 + f'(x) = M = 4xy + 4y^2$$

or

$$f'(x) = 0 \text{ or } f(x) = C$$

where $C$ is a constant which is independent of $x$ and $y$. Hence

$$Z = 2x^2y + 4xy^2 + C.$$

**Example 1.22**

Distinguish between extensive and intensive properties with the help of an example.

*Solution* :

The properties which depend on the mass of a system are called extensive properties, where as the properties which are independent of the mass of a system are called intensive properties. At *NTP* (normal temperature and pressure i.e. at 0°C and 1 standard atmospheric pressure) one mole of a gas occupies a volume of 22.4 litres. Consider two systems **A** and **B** such that system **A** contains one mole of oxygen at *NTP* and system **B** contains 10 moles of oxygen at *NTP*. If one measures the volumes occupied by the systems **A** and **B**, it will be found that system **B** occupies a volume of 224 litres which is equal to ten times the volume occupied by system **A**, That is the volume occupied by a system depends on the mass of the system and hence volume is an extensive thermodynamic property. Measurements of temperature and pressure show that systems **A** and **B** will have the same temperature (0°C) and pressure (1 standard atmospheric pressure). That is, the properties $P$ and $T$ do not depend on the mass of the system and they are called intensive properties.

***Note***

1. *The ratio of an extensive property to the mass (or the property per unit mass) is called the specific property. For example*

*Specific volume,* $v = \dfrac{\text{Total volume}}{\text{Total mass}} = \dfrac{V}{m} = \text{m}^3/\text{kg}$

2. *The ratio of an extensive property to the mole number (or the property per mole) is called the molar property. For example,*

*Molar volume,* $v = \dfrac{\text{Total volume}}{\text{Mole number}} = \dfrac{V}{n} = \text{m}^3/\text{mol}$

*For simplicity of notation we use* v *to denote, both molar volume and specific volume. The units* m³ / kg or m³ / mol *indicates whether it is specific volume or molar volume.*

**Example 1.23**

Classify the following properties. (*a*) pressure (*b*) temperature (*c*) volume (*d*) internal energy (*e*) volume per mole (*f*) mass (*g*) enthalpy per unit mass.

*Solution :*

Extensive property : (*c*); (*d*); (*f*)

Intensive property : (*a*); (*b*)

Specific property : (*g*)

Molar property : (*e*)

**Example 1.24**

Assuming that the molar volume of a system depends on pressure and temperature, derive the following relation.

$$\frac{dv}{v} = \beta\, dT - \kappa dP$$

Where $\beta = \dfrac{1}{v}\left(\dfrac{\partial v}{\partial T}\right)_P$ = coefficient of volume expansion

$\kappa = -\dfrac{1}{v}\left(\dfrac{\partial v}{\partial P}\right)_T$ = Isothermal compressibility

*Solution :*

$$v = v\,(T, P)$$

$$dv = \left(\frac{\partial v}{\partial T}\right)_P dT + \left(\frac{\partial v}{\partial P}\right)_T dP$$

or
$$\frac{dv}{v} = \frac{1}{v}\left(\frac{\partial v}{\partial T}\right)_P dT + \frac{1}{v}\left(\frac{\partial v}{\partial P}\right)_T dP$$
$$= \beta\, dT - \kappa\, dP$$

**Example 1.25**

Show that $\left(\frac{\partial \beta}{\partial P}\right)_T = -\left(\frac{\partial \kappa}{\partial T}\right)_P$

*Solution* :

Consider $v = v\,(T, P)$. Then

$$\frac{dv}{v} = \beta\, dT - \kappa dP \quad \text{[See Example 1.14]}$$

or
$$dv = \beta\, vdT - \kappa vdP$$

Since volume is a property, its differential must be exact. Therefore, the following condition must be satisfied.

$$\left[\frac{\partial}{\partial P}(\beta v)\right]_T = -\left[\frac{\partial}{\partial T}(-\kappa v)\right]_P$$

or
$$\left(\frac{\partial \beta}{\partial P}\right)_T = -\left(\frac{\partial \kappa}{\partial T}\right)_P$$

**Example 1.26**

Explain the meaning of the terms

(*a*) kinetic energy

(*b*) potential energy

(*c*) mechanical energy.

*Solution* :

*(a) Kinetic energy.* The energy possessed by a body by virtue of its motion is called the kinetic energy. If a body of mass of $m$ moves with a velocity $\mathbf{V}$ the kinetic energy (KE) possessed by the body is given by

$$\text{KE} = \frac{1}{2} m\mathbf{V}^2$$

*(b) Potential energy.* The energy possessed by a body by virtue of its location or configuration is called potential energy. If a body of mass $m$ is at an elevation

of $h$ above the datum plane, the potential energy (PE) possessed by the body is given by

$$PE = mgh$$

where $g$ is the accleration due to gravity.

A compressed spring or a stretched rubber band also possesses potential energy.

*(c) Mechanical energy.* The sum of the potential energy and Kinetic energy of a body is called the mechanical energy of the body.

***Note***

1. *The kinetic energy and potential energy possessed by a body can be estimated in terms of macroscopically measurable quantities (**V** and h) and hence they are called macroscopic modes of energy.*
2. *The kinetic energy and potential energy can be completely converted from one form to the other. These are the most useful forms of energy.*
3. *The principle of conservation of energy for mechanical systems (proposed by Leibnitz) states that the sum of the kinetic energy and potential energy (mechanical energy) remains constant during the motion of a body.*

**Example 1.27**

An elevator having a mass of 2000 kg is moving upright and after reaching a height of 80 m above the ground level the cable holding the elevator breaks. The elevator falls freely to the ground level where it hits a strong spring and brought to rest at the position of the maximum spring compression by a locking arrangement. Assume that there is no friction between the elevator and its shaft. Calculate

(*a*) The potential energy of the elevator at the moment when the cable is about to break.

(*b*) The work done on the elevator in raising it to the height of 80 m above the ground level.

(*c*) The velocity and the kinetic energy of the elevator when it is about to touch the spring at the bottom.

*Solution :*

(*a*) Potential energy, PE = $mgh$ = 2000 × 9.81 × 80 = 1569.6 kJ

(*b*) Minimum force required to move the elevator, $F = mg$

Work done, $W = Fs = mgh$ = 1569.6 kJ

(*c*) When the cable is about to break, PE = 1569.6 kJ; Kinetic energy, KE = 0

$$PE + KE = 1569.6 \text{ kJ}$$

When the elevator falls freely under gravity and when it is about to touch the spring PE = 0. Since total energy remains constant

$$KE = 1569.6 \text{ kJ}$$

$$KE = \frac{m\mathbf{V}^2}{2}$$

or $$\mathbf{V} = \sqrt{\frac{2KE}{m}} = \sqrt{\frac{2 \times 1569.6 \times 10^3}{2000}} = 39.62 \text{ m/s}$$

**Example 1.28**

It is required to eject a projectile of mass 2 kg from a rocket, so that it reaches to a maximum height of 300 m. Calculate the velocity with which the projectile should leave the rocket. The projectile after reaching the maximum height falls freely and on the return path it hits a tower after travelling a distance of 250 m. Calculate the velocity with which the projectile hits the tower.

*Solution* :

When the projectile reaches the maximum height, its Kinetic energy KE = 0 and potential energy, PE = $mgh = 2 \times 9.81 \times 300 = 5886$ J

$$PE + KE = 5886 \text{ J}$$

While the projectile is freely falling, its total energy (PE + KE) = 5886 J

At a height of 50 m above the ground, that is after travelling 250 m during the return path,

$$PE = mgh = 2 \times 9.81 \times 50 = 981 \text{ J}$$

Therefore $$KE = 5886 - 981 = 4905 \text{ J}$$

Velocity of the projectile when it hits the tower, $\mathbf{V} = \sqrt{\frac{2KE}{m}}$

or $$\mathbf{V} = \sqrt{\frac{2 \times 4905}{2}} = 70.04 \text{ m / s}$$

**Example 1.29**

A diatomic molecule can be modelled as a system of two particles connected to each other by a spring. It is well known that the gas molecules are not at rest and they move at random with independent velocities. Consider 1 $cm^3$ sample of nitrogen gas at NTP held in a rigid container resting on a table top. Considering the table top as the datum plane does the gas possess potential energy, kinetic energy, internal energy? Explain the origin of the internal energy for the gas.

*Solution* :

At NTP, the number of molecules in 1 $cm^3$ is approximately $10^{20}$. The sample of gas is resting on the table top. At a macroscopic level, the gas sample is not moving and hence its KE is equal to zero. Considering the table top as the datum plane, the PE of the gas sample is also equal to zero. Thus the gas does not possess any macroscopically observable modes of energy.

At a microscopic level, the gas molecules move at random. If $V_{xi}$, $V_{yi}$ and $V_{Zi}$ denote the velocity vectors of a molecule $i$ in $x,y$ and $z$ directions, the translational kinetic energy possessed by a molecule is $m\,(V_{xi}^2 + V_{yi}^2 + V_{zi}^2)\,/\,2$, where $m$ is the mass of a molecule. Consider the line joining the two nuclei of nitrogen atoms as the $X$–axis and place the centre of mass of the molecule at the origin of the coordinate axes. Then the molecule rotates about $Y$ and $Z$ axis and possesses rotational kinetic energy. The rotational kinetic energy for rotation about each axis is given by $I\,\omega^2\,/\,2$, where $I$ is the moment of inertia of the molecule about the centre of mass and $\omega$ is the angular velocity. In addition to the translational and rotational motion of the molecule as a whole, the atoms of the molecule vibrate about their equilibrium positions. Then the molecule posseses certain vibrational energy which is associated with the vibrational motion. At a microscopic level, it is not possible to measure the translational, rotational and vibrational energies possessed by each molecule. These modes of energy are called microscopic modes of energy. Similarly any sample of gas possesses electronic energy, nuclear energy, etc. in addition to the above modes of energy. The sum of the energies associated with all these microscopic modes is called the internal energy, Therefore, the sample possesses internal energy which cannot be measured at a macroscopic level.

**Note** *The energy possessed by a system due to its location or configuration is called potential energy. The energy possessed by a system by virtue of its motion is called kinetic energy. A macroscopic system possesses* PE, KE *and internal energy (U). Then the total energy (E) of the system is given by*

$$E = \text{KE} + \text{PE} + U$$

**Example 1.30**

Distinguish between steady state and equilibrium with the help of an example.

*Solution* :

Suppose one end of a copper rod is kept in contact with condensing steam at 100°C while the other end of the rod is in contact with melting ice at 0°C. Then energy flows through the copper rod in the form of heat from the condensing steam to the melting ice. At time $t = 0$, that is at the instance when the copper rod is suddenly brought into simultaneous contact with the condensing steam and melting ice, the temperature of the rod is uniform along its length. Measurement of temperature at a specified location of the copper rod reveals that the temperature gradually changes with time initially and attains a constant value after some time. There will be no further change in the temperature of the rod (at a specified location), however long one may observe. This observation will be true at any location of the rod and the temperature of the rod continuously varies from 100°C at the condensing steam end to 0°C at the melting ice end. Then the copper rod is said to be in a steady state. Steady state implies that the properties (like temperature in this case) do not change with respect to time. In other words, the time derivatives of properties at any given location (say $dT / dt$) is equal to zero in steady state.

Now, isolate the copper rod (system) from the condensing steam and melting ice baths and measure the temperature at any specified location of the rod. The temperature gradually changes again and attains a definite value which will not undergo any further change. Measurement of temperature at several locations along the length of the rod indicates that the temperature is uniform through out the length of the rod. Then the system is said to be in a state of equilibrium. In a state of equilibrium, the system has no tendency to undergo any further change. Equilibrium is a concept associated with the absence of any tendency for further change on a macroscopic level. The tendency to change occurs due to the presence of driving forces (like temperature difference). Thus equilibrium implies the absence of driving forces. In a state of equilibrium, the thermodynamic properties of a system have unique values.

### Example 1.31

What criteria a system should satisfy for it to be in a state of thermodynamic equilibrium?

*Solution* :

For a system to be in a state of thermodynamic equilibrium it should simultaneously satisfy the criteria for thermal equilibrium, mechanical equilibrium and chemical equilibrium.

### Example 1.32

Explain the criteria for (*a*) thermal equilibrium (*b*) mechanical equilibrium and (*c*) chemical equilibrium.

*Solution* :

(*a*) If there is no further change in the thermal condition (temperature) of a system when it is isolated, the system is said to be in a state of **thermal equilibrium.** In a state of thermal equilibrium, there is no temperature gradient within the system. Since a system in thermal equilibrium has uniform temperature, a unique value can be specified for the temperature of the system. If two systems are in thermal equilibrium with each other, the temperatures of both the systems are identical. In other words, the criterion for thermal equilibrium is given by

$$T_A = T_B$$

where $T_A$ and $T_B$ denote the temperatures of the systems **A** and **B,** respectively.

(*b*) If there is no imbalance of forces in a system when it is isolated it is said to be in a state of **mechanical equilibrium.** In a state of mechanical equilibrium, the system has a uniform pressure and a unique value of pressure can be specified. If two systems are in a state of mechanical equilibrium with each other, the pressures of both systems are identical. That is, the criterion for mechanical equilibrium is given by

$$P_A = P_B$$

where $P_A$ and $P_B$ denote the pressures of systems **A** and **B**, respectively.

(*c*) If the composition of a system does not undergo any change due to diffusion, mass transfer or chemical reaction and if there is no tendency for a change in its chemical composition when the system is isolated, it is said to be in a state of **Chemical equilibrium.** The difference in the chemical potential (which is defined in a later chapter) causes diffusion, mass transfer or chemical reaction. In a state of chemical equilibrium the chemical potential is uniform through out the system. Suppose two systems **A** and **B** each consisting of *N* components are in a state of chemical equilibrium. Then the criterion of chemical equilibrium can be stated as

$$\mu^A{}_i = \mu^B{}_i \quad \text{for } i = 1, 2........N$$

where $\mu_i^A$ and $\mu_i^B$ denote the chemical potentials of component *i* in systems **A** and **B**, respectively.

**Example 1.33**

Suppose a rigid and insulated vessel is filled with a mixture of liquid water and water vapor at 100°C and standard atmospheric pressure. Considering the contents of the vessel as a system, is it in a state of equilibrium ? Can we say that the molecules which were originally in liquid phase continue to exist in the same phase at all times ?

*Solution* :

The system satisfies the criteria of thermal, mechanical and chemical equilibrium. Hence, the system is in a state of thermodynamic equilibrium. At a macroscopic level the amounts of liquid and vapor phases remain constant and the system appears to be static. However, at a microscopic level the system is not at rest. Some liquid water evaporates and some water vapor condenses. That is the molecules move from liquid to vapor and vice–versa, but at such a rate that the rate of evaporation is exactly balanced by the rate of condensation. Hence the molecules which were originally present in the liquid phase do not continue to exist in the same phase at all times.

**Example 1.34**

An equimolar mixture of $H_2$ and $O_2$ is contained in a rigid vessel at 1 bar and 25°C. No detectable change in its composition has been observed over a period of 30 days. Since there is no change in the composition of the mixture can we say that the mixture is in a state of chemical equilibrium ? If not, specify why ?

*Solution* :

The mixture of $H_2$ and $O_2$ is not in a state of chemical equilibrium, though there is no detectable change in its composition over a long period of time. The rate of reaction between $H_2$ and $O_2$ to form water vapor at 25°C is very small and hence there is no change in the composition of the mixture. The mixture has a tendency to undergo a chemical reaction. If a small amount of energy (in the form of an electric spark or by introducing a lighted match stick) is added to the mixture to create a small disturbance, the mixture reacts violently to form water vapor. Hence the system is not in a state of chemical equilibrium.

**Example 1.35**

Classify the states of equilibrium with the help of a mechanical analogue.

*Solution* :

Consider a sphere resting at the bottom of a spherical bowl as shown in Fig.E 1.35 (*a*). If the sphere is disturbed it will settle into the original state, however large the disturbance may be. This state of equilibrium, in which the system (here the sphere) returns to its original state even if it is subjected to large disturbances is called stable equlibrium. In thermodynamics we are mostly interested in stable equilibrium states. The sphere shown in Fig.E 1.35 (*b*) returns to its original state if it is subjected to small disturbances but settles in a different equilibrium state if the disturbance exceeds a certain magnitude. Such a state of equilibrium is called metastable. The sphere shown in Fig.E 1.35 (*c*) settles into a different state when it is subjected to a small disturbance and this state is called unstable state of equilibrium.

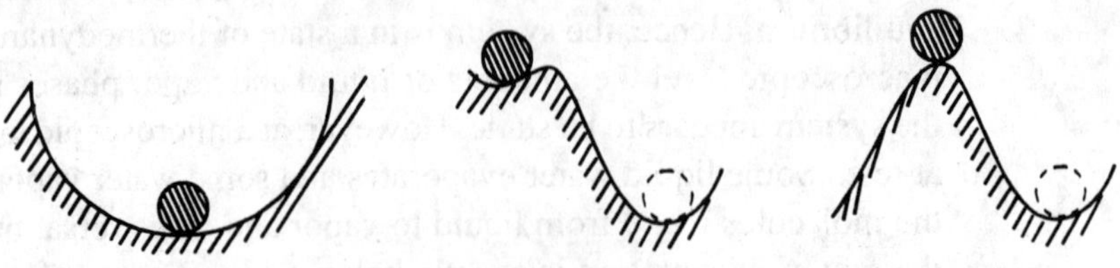

(a) (b) (c)

Fig.E 1.35. States of equilibrium (*a*) stable (*b*) metastable (*c*) unstable.

**Example 1.36**

With careful cooling it is possible to cool water to a temperature (say 10°C) below its normal freezing point. The cold water produced in this manner is usually called super cooled water. Is the super cooled water in a state of equilibrium?

*Solution* :

Consider the super cooled water as a system and let us apply the criteria of equilibrium to judge whether it is in a state of equilibrium or not. The system has a uniform temperature and hence it satisfies the criterion of thermal equilibrium. The system also satisfies the criterion of mechanical equilibrium since the pressure is uniform through out the system. The system contains pure water with uniform temperature and pressure. Hence the chemical potential is uniform through out the system. Thus it satisfies the criterion of chemical equilibrium also. Since the system simultaneously satisfies the criterion of thermal, mechanical and chemical equilibrium, it is in a state of thermodynamic equilibrium. If the system is subjected to a minor disturbance either by adding a small amount of dust or by shaking the container, the supercooled liquid immediately freezes into ice. Hence, the system is in a state of metastable equilibrium.

**Example 1.37**

What is a process ?

*Solution* :

When a system changes from a specified initial equilibrium state to another equilibrium state, the path followed by the system in reaching the final equilibrium state from the given initial state is called the process.

**Example 1.38**

Explain the meaning of Quasi–Static process.

*Solution* :

A process which takes place very slowly and with infinitesimal driving force is called a quasi–static process. Suppose a gas at pressure $P$ is enclosed in a piston cylinder assembly. If the piston is restrained by an external force such that the external pressure on the piston is $(P - dP)$ where $dP > 0$, then the gas can expand slowly. Such an expansion occurs at a slow rate and the system appears to be almost static. Then the process followed by the gas is quasi–static.

**Example 1.39**

What is meant by a reversible process ? Explain it with the help of an example.

*Solution* :

A process which proceeds with no driving forces is a reversible process. As an example consider a gas which is enclosed in a frictionless piston cylinder assembly, as shown in Fig.E1.39.

Suppose the piston is loaded with a certain mass of fine sand particles. If the sand particles are moved one after another on to adjacent platforms, the piston moves up slowly and the gas expands.

During the expansion process, the system (here, the gas) does work on the surroundings and the sand particles are raised to different elevations as shown in Fig.E1.39 (*b*). If the sand particles one after another are restored on to the piston from the adjacent plaforms, the piston moves down slowly and the gas gets compressed. During the compression process the work done on the system is exactly equal to the work done by the system during the expansion process. Thus the process can be carried out in either direction. By reversing the direction of the process both the system and surroundings are restored to their respective initial states. Then the process undergone by the system is reversible. During a reversible process the system is always in a state of equilibrium with its surroundings.

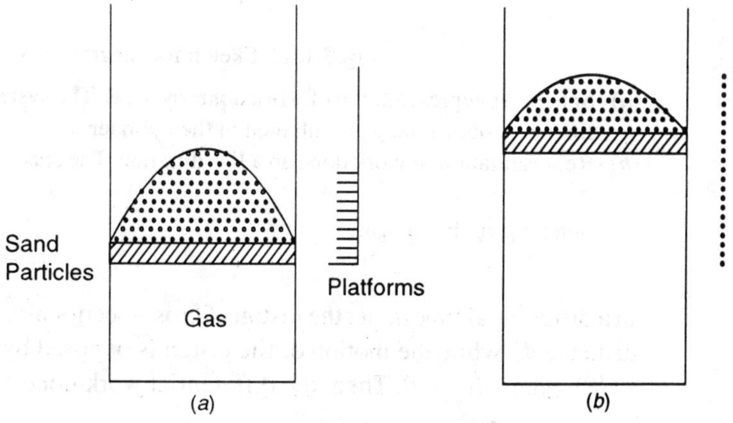


Fig.E 1.39. Reversible Process.

**Note** *A reversible process is a quasi- static process, but a quasi-static process need not be a reversible process.*

### Example 1.40

State the mechanics definition of work done by a force. Apply the mechanics definition of work and show that the work done by a gas is given by $\int_1^2 Pdv$

*Solution* :

In mechanics work done by a force is defined as the product of the force and the displacement in the direction of the applied force. That is the differential work done ($dw$) is given by

$$dw = F.ds$$

where $F$ is the applied force and $ds$ is the differential displacement.

Consider a certain amount of gas contained in a cylinder piston assembly as shown in Fig.E 1.40. At time $t$ let $P$ be the pressure of the gas inside the cylinder.

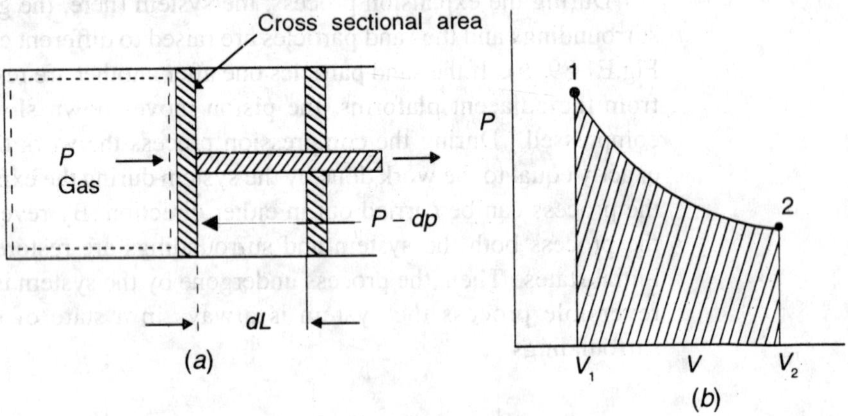


Fig.E 1.40. Sketch for diagram 1.40.

(*a*) Schematic representation of work done by a gas. The system (enclosed by the dotted line) constitutes the gas contained in the cylinder.

(*b*) Representation of work done an a *P-V* diagram. The shaded area represents the work done $\int_1^2 PdV$ by the gas.

In a differential time $dt$, let the piston of cross–sectional area $A$ move a differential distance $dL$ while the motion of the piston is opposed by an external pressure ($P - dP$) where $dP \to 0$. Then, the differential work done by the gas is given by

$$dW = F.ds = (PA).dL = P\,(AdL) = Pdv$$

or
$$W = \int_1^2 Pdv$$

It may be noted that the external pressure $(P - dP)$ is always infinitesimally smaller than the gas pressure. Hence the expansion process can be reversed at any time by increasing the external pressure infinitesimally more than the gas pressure. That is, if the external pressure is $(P + dP)$, the gas undergoes a compression process. During the expansion/ compression of the gas the forces on the moving boundary are balanced and hence the process is reversible. The process can be represented on a $P - V$ diagram as shown in Fig.E 1.40.

### Example 1.41

Suppose a piston cylinder assembly initially contains a gas at pressure $P_1$ and occupies a volume $V_1$ and it is allowed to expand reversibly till it attains a pressure $P_2$. Is it possible to calculate the work done by the gas from a knowledge of its initial and final conditions ? Explain whether work done by a gas is a property of the gas or not.

*Solution* :

From the given data the initial $(P_1, V_1)$ and final $(P_2, V_2)$ states of the gas are known. The work done by a gas is given by

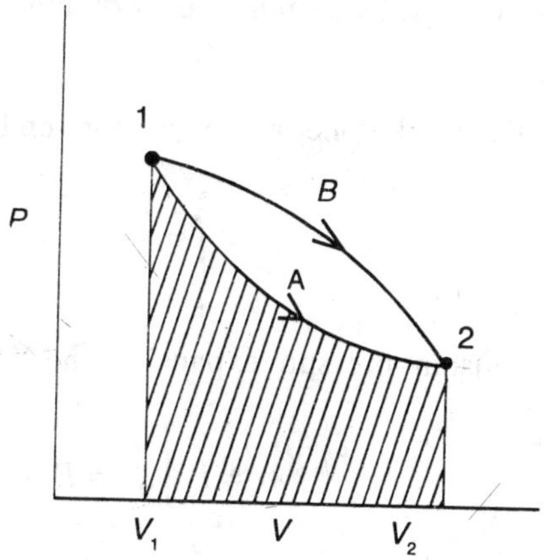


Fig.E 1.41. Representation of two different reversible processes between the specified initial and final states of the gas. The area under the curve 1A2 represents the work done by the gas, if it follows the path A. The area under the curve 1B2 represents the work done by the gas if it expands along the path B.

$$W = \int_1^2 P dv$$

To integrate the above relation, one should know the relation between the pressure ($P$) and volume ($V$) of the gas during expansion/compression process. That is the path followed by the gas must be known to evaluate the work done. Fig.E 1.41 shows two different reversible path 1A2 and 1B2 connecting the given initial and final states of the gas. Since $\int Pdv$ represents the area under the curve the work done during path 1A2 is different from the work done during the path 1B2. In other words, the work done depends on the path followed by a system. Hence, it is not possible to evaluate the work done purely from a knowledge of the initial and final states of a system. The work done by a system is a path function and hence it is not a property of a system. Work is an interaction, that is energy transfer, between a system and its surroundings. Hence work is energy in transit.

**Example 1.42**

A piston–cylinder assembly contains one mole of an ideal gas at $(P_1, v_1, T_1)$. If the gas expands reversibly such that ($a$) the final volume is $V_2$ while the pressure is held constant ($b$) the final pressure is $P_2$ while the temperature is held constant and ($c$) the final pressure is $P_2$ while $Pv^{\gamma}$ = constant during expansion. Here $\gamma$ is the ratio of heat capacity at constant pressure to the heat capacity at constant volume. Determine the work done by the gas.

*Solution* :

($a$) The work done by the gas is given by

$$W = \int_1^2 P dv$$

For constant pressure expansion, the above relation reduces to

$$W = P \int_1^2 dv = P_1 (v_2 - v_1)$$

($b$) We know that for an ideal gas $Pv = RT$. If the temperature of the gas is held constant, we get

$$P_1 v_1 = P_2 v_2 = RT_I$$

Then the work done by the gas is given by

$$W = \int_1^2 Pdv = \int_1^2 \frac{RT}{v} dv = RT_1 \int_1^2 \frac{dv}{v} = RT_1 \ln \frac{v_2}{v_1}$$

(*c*) The path followed by the gas during the expansion process is

$$Pv^\gamma = \text{constant or } P_1 v_1^\gamma = P_2 v_2^\gamma = K = Pv^\gamma$$

or

$$P = Kv^{-\gamma}$$

The work done by the gas is given by

$$W = \int_1^2 Pdv = \int_1^2 Kv^{-\gamma} dv = \frac{K(v_2^{1-\gamma} - v_1^{1-\gamma})}{1-\gamma} = \frac{Kv_1^{1-\gamma} - Kv_2^{1-\gamma}}{\gamma - 1}$$

$$= \frac{P_1 v_1^\gamma v_2^{1-\gamma} - P_2 v_2^\gamma v_2^{1-\gamma}}{\gamma - 1} = \frac{P_1 v_1 - P_2 v_2}{\gamma - 1} = \frac{R(T_1 - T_2)}{\gamma - 1}$$

**Example 1.43**

It is desired to compress one mole of air (ideal gas with $\gamma = 1.4$) from 1 bar and 27°C to 10 bar and 27°C. For this purpose compare the work to be done for the following processes :

(*a*) Isothermal compression.

(*b*) constant volume heating followed by a constant pressure compression and

(*c*) adiabatic compression ($PV^\gamma$ = Constant) followed by constant volume cooling.

*Solution :*

(*a*) The isothermal compression process is shown as 1A2 on the $P$–$v$ diagram in Fig.E 1.43

The work done for the isothermal process 1A2 is given by

$$W_{1A2} = \int_1^2 Pdv = \int_1^2 RT \frac{dv}{v} = RT \ln \frac{v_2}{v_1} = RT \ln \frac{P_1}{P_2}$$

$$= 8.314 \times 300 \times \ln \frac{1}{10}$$

$$= -5.7431 \text{ kJ}$$

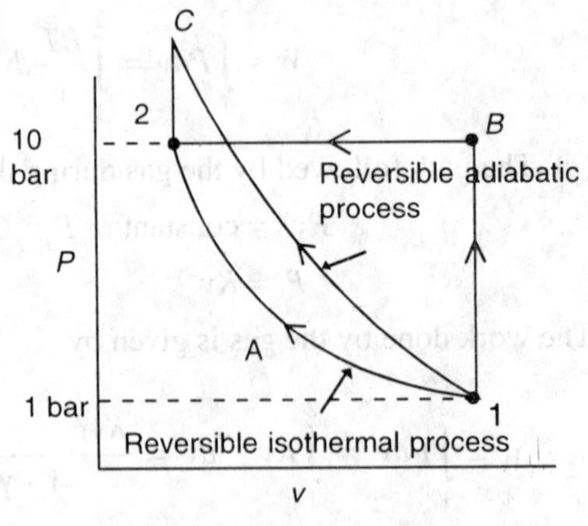


Fig.E 1.43. The isothermal compression process is shown as 1A2.

The path 1B2 shows the constant volume heating followed by constant pressure compression. The path 1C2 shows adiabatic compression followed by a constant volume cooling.

(*b*) The work done along the path 1B2 is given by

$$W_{1B2} = W_{1B} + W_{B2} = \int_1^B Pdv + \int_B^2 Pdv$$

$$= 0 + P_2(v_2 - v_B) = P_2(v_2 - v_1) = P_2 v_2\left(1 - \frac{v_1}{v_2}\right)$$

$$= P_2 v_2\left(1 - \frac{P_2}{P_1}\right) = RT_2\left(1 - \frac{P_2}{P_1}\right) = RT_1\left(1 - \frac{P_2}{P_1}\right)$$

$$= 8.314 \times 300\left(1 - \frac{10}{1}\right) = -\ 22.4478 \text{ kJ}$$

(since $T_1 = T_2$ and $Pv = RT$)

The work done along the path 1C2 is given by

$$W_{1C2} = W_{1C} + W_{C2} = \int_1^C Pdv + \int_C^2 Pdv$$

$$= \frac{P_1 v_1 - P_C v_C}{\gamma - 1} + 0 = \frac{P_1 v_1 - P_C v_2}{\gamma - 1}$$

$$P_1 v_1^{\gamma} = P_c v_c^{\gamma} = P_c v_2^{\gamma} \qquad \text{(since } V_C = V_2\text{)}$$

or
$$P_C = P_1 \left(\frac{v_1}{v_2}\right)^{\gamma}$$

$$\frac{P_1 v_1}{T_1} = \frac{P_2 v_2}{T_2} \text{ or } \frac{v_1}{v_2} = \frac{P_2}{P_1} \qquad \text{(since } T_1 = T_2\text{)}$$

Therefore,
$$W_{1C2} = \frac{P_1 v_1 - P_1 \left(\frac{v_1}{v_2}\right)^{\gamma} v_2}{\gamma - 1} = \frac{P_1 V_1 - P_1 \left(\frac{P_2}{P_1}\right)^{\gamma} v_1 \left(\frac{P_1}{P_2}\right)}{\gamma - 1}$$

$$= \frac{P_1 v_1 \left[1 - \left(\frac{P_2}{P_1}\right)^{\gamma - 1}\right]}{\gamma - 1} = \frac{RT_1 \left[1 - \left(\frac{P_2}{P_1}\right)^{\gamma - 1}\right]}{\gamma - 1}$$

$$= \frac{8.314 \times 300 [1 - 10^{0.4}]}{0.4} = -9.4274 \text{ kJ}$$

***Note*** *The negative work done by the gas indicates that work is to be done on the gas by the surroundings.*

### Example 1.44

A piston–cylinder assembly contains air (ideal gas with $\gamma = 1.4$) at 200 kPa and occupies a volume of 0.01 $m^3$. The piston is attached to one end of a spring and the other end of the spring is fixed to a wall. The force exerted by the spring on the piston is proportional to the decrease in the length of the spring from its natural length. The ambient atmospheric pressure is 100 kPa. Now, the air in the cylinder is heated till the volume is doubled and at this instant it is found that the pressure of the air in the cylinder is 500 kPa. Calculate the work done by the gas.

*Solution* :

We know that the force exerted by the spring on the piston is given by

$$F_S = KX = K \frac{(V - V_0)}{A}$$

Where $K$ = spring constant

$A$ = cross–sectional area of piston

$V_0$ = Volume of air if the spring is at its natural length, that is the spring is neither compressed nor elongated.

$V$ = Volume of air if the spring is compressed by a distance $X$,

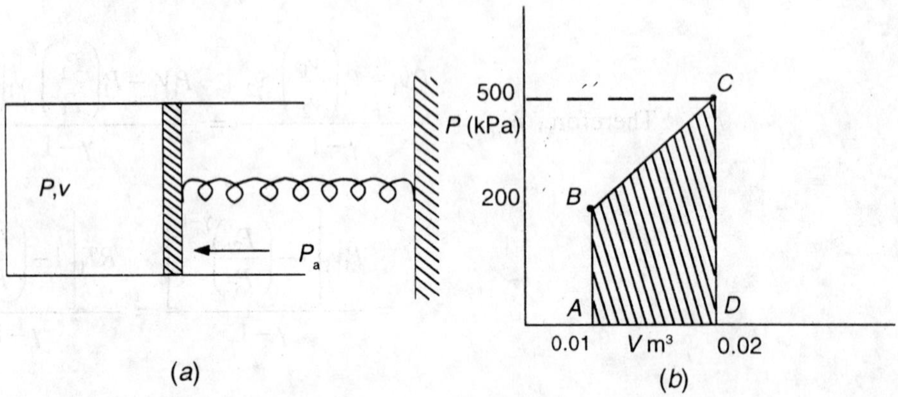


Fig.E 1.44 (*a*) Sketch for Example 1.44.
(*b*) *P-v* diagram in which shaded area represents the work done.

Force acting on the piston due to ambient atmosphere is given by

$$F_a = P_a A$$

Force balance on the piston when the air volume is $V$ is given by

$$PA = P_a A + \frac{K(V - V_0)}{A}$$

or
$$P = P_a + \frac{K}{A^2}(V - V_0) \qquad (A)$$

Force balance on the piston in the initial state gives

$$P_1 A = P_a A + K\left(\frac{V_1 - V_0}{A}\right) \qquad (B)$$

Force balance on the piston in the final state gives

$$P_2 A = P_a A + K\left(\frac{V_2 - V_0}{A}\right) \qquad (C)$$

The work done by the air is given by

$$W = \int F.dx = \int PA.d X = \int PdV$$

$$= \int_1^2 \left\{P_a + \frac{K}{A^2}(V - V_0\right\}dV = P_a(V_2 - V_1) + \frac{K}{2A^2}\left\{(V_2 - V_0)^2 - (V_1 - V_0)^2\right\}$$

$$= P_a(V_2 - V_1) + \frac{K}{2A^2}\{(V_2 - V_0) + (V_1 - V_0)\}\{(V_2 - V_0) - (V_1 - V_0)\}$$

$$= P_a(V_2 - V_1) + \frac{K}{2A^2}(V_2 + V_1 - 2V_0)(V_2 - V_1)$$

$$= \left\{P_a + \frac{K}{2A^2}(V_2 + V_1 - 2V_0)\right\}(V_2 - V_1) \qquad (D)$$

From Eqns. *(B)* and *(C)*, we get

$$\frac{P_1 + P_2}{2} = P_a + \frac{K}{2A^2}(V_2 + V_1 - 2V_0) \qquad (E)$$

Substituting Eqn.(*E*) in Eqn. (*D*), we obtain

$$W = \frac{(P_1 + P_2)}{2}(V_2 - V_1)$$

$$= \frac{(200 \times 10^3 + 500 \times 10^3)}{2} \times (0.02 - 0.01) = 3.5 \text{ kJ}$$

The shaded area of the trapezium ABCD in Fig.E 1.44 (*b*) represents the work done by air. It can be c' [illegible] the area of the trapezium ABCD is given by

$$\frac{(P_1 + P_2)}{2}(V_2 - V_1)$$

**Example 1.45**

A balloon which is initially collapsed and flat is slowly filled with hydrogen at 100 kPa so as to form it into a sphere of radius 1m. Determine the work done by the gas in the balloon during the filling process.

*Solution :*

During the filling process, the gas pressure remains constant at 100 kPa. Therefore, the work done during the constant pressure process is given by

$$W = \int_1^2 PdV = P(V_2 - V_1) = P \times \frac{4}{3}\pi(r_2^3 - r_1^3) = \frac{P4\pi r_2^3}{3}$$

$$= \frac{100 \times 10^3 \times 4\pi \times 1^3}{3} = 418.88 \text{ kJ}$$

(Since $r_1 = 0$ when the balloon is collapsed and flat)

**Example 1.46**

One mole of a Van der Waals gas undergoes a reversible and isothermal expansion from the initial volume $V_1$ to the final volume $V_2$. Determine the work done by the gas.

*Solution* :

The Van der Waals equation of state is given by

$$\left(P + \frac{a}{v^2}\right)(v - b) = RT \text{ or } P = \frac{RT}{v-b} - \frac{a}{v^2}$$

The work done by the gas is given by

$$\text{W} = \int Pdv = \int_1^2 \left(\frac{RT}{v-b} - \frac{a}{v^2}\right) dv = RT \ln\left\{\frac{v_2 - b}{v_1 - b}\right\} - a\left(\frac{1}{v_1} - \frac{1}{v_2}\right)$$

**Example 1.47**

One mole of air (ideal gas with $\gamma = 1.4$) at pressure $P_1$ and temperature $T_1$ is compressed at constant volume till its pressure is doubled. Then it is allowed to expand reversibly and isothermally to the original pressure and finally restored to the original temperature by cooling at constant pressure. Sketch the path followed by the gas, on a $P$–$V$ diagram and calculate the net work done by the gas.

*Solution* :

Fig.E 1.47 shows the path followed by the gas on $P$–$v$ diagram.

The net work done by the gas is given by $W = W_{12} + W_{23} + W_{31}$

$$= \int_1^2 Pdv + \int_2^3 Pdv + \int_3^1 Pdv$$

$$= 0 + RT_2 \ln\frac{v_3}{v_2} + P_3(v_1 - v_3)$$

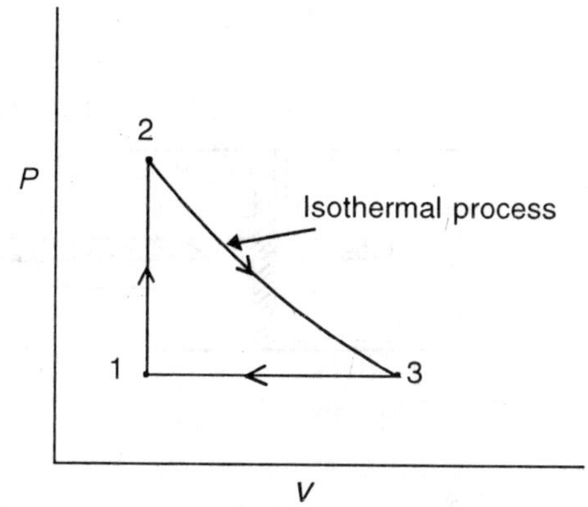


Fig.E 1.47. Path followed by the gas on $P$–$v$ diagram. 1–2 is a constant volume process, 2–3 is isothermal process and 3–1 is a constant pressure process.

For an ideal gas $\frac{P_1 v_1}{T_1} = \frac{P_2 v_2}{T_2}$. For the path 1– 2, $v_1 = v_2$. Hence,

$$T_1 \frac{P_2}{P_1} = 2T_1 \quad (\text{Since } P_2 = 2P_1)$$

For the path 1–3, pressure is constant. Hence $P_1 = P_3$. Therefore

$$\frac{P_1 v_1}{T_1} = \frac{P_1 v_1}{T_1} = \frac{P_3 v_3}{T_3} \quad \text{or} \quad v_3 = \frac{T_3}{T_1} \cdot \frac{P_1}{P_3} v_1 = \frac{T_2 v_1}{T_1} = 2v_1$$

(Since 2–3 is a constant temperature process.)

Then the net work done by the gas is given by

$$W = RT_2 \ln \frac{v_3}{v_2} + P_3(v_1 - v_3) = R\,2T_1 \ln \frac{2v_1}{v_1} + P_1(v_1 - 2v_1)$$

$$= 2RT_1 \ln 2 - 2P_1 v_1 = 2RT_1 \ln 2 - 2\,RT_1 = 2RT_1(\ln 2 - 1)$$

**Example 1.48**

Is it possible to view the expansion of a gas in a piston–cylinder assembly as the one and the only one effect external to the system (the gas contained in the cylinder) as the raising of a mass against the gravitational force? If so devise a means to view such an effect.

*Solution* :

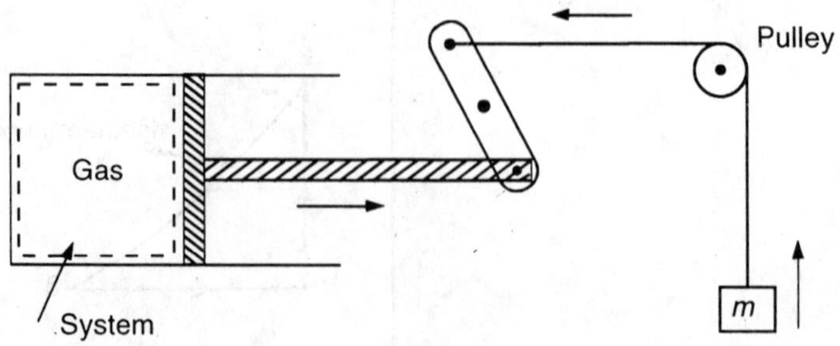


Fig.E 1.48 To visualise the work done by the gas enclosed in a piston cylinder assembly as equivalent to raising of a mass in a gravitational field.

Yes, It is possible to modify the surroundings, such that the one and the only one effect (sole effect) external to the system is the raising of a mass against the gravitational force as shown in Fig.E 1.48.

***Note*** *One can generalize and state that all mechanical work done by a system is equivalent to raising of a mass, through a certain distance, in a gravitational field.*

**Example 1.49**

Suppose the terminals of a storage battery are connected to an electrical resistor through a switch. When the switch is closed electrical current flows through the resistor, the resistor becomes warm and the charge of the storage battery decreases. Consider the storage battery as a system and apply the mechanics definition of work to identify whether work is done by the system or not. Is it possible to modify the surroundings such that the sole effect external to the system is equivalent to raising of a mass ? If so devise a method.

*Solution* :

In mechanics work done is defined as the product of the force and the displacement in the direction of applied force. When the terminals of a storage battery are connected to an electrical resistor, current flows through the resistor, thereby decreasing the charge of the battery. In the interaction between the battery and the resistor, no force has moved through any distance and hence the work done by the battery from the mechanics point of view, is zero. We know that an electric current is a manifestation of the flow of electrons which are driven by an electrical potential difference. The electrons are crossing the boundary of the system (battery) and some work is done in moving the electrons. However, the mechanics definition of work fails to identify the work done.

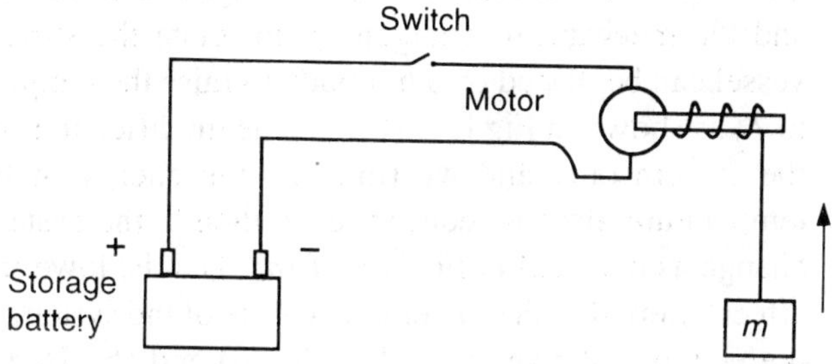


Fig.E. 1.49. Visualization of battery interaction as equivalent to raising of a mass.

It is possible to modify the surroundings such that the sole effect external to the system is equivalent to raising of a mass as shown is Fig.E 1.49. The resistor can be replaced by an electric motor which draws the same amount of current as the resistor. One can wind a string around the shaft of the motor and suspend a mass through this string. When the switch is closed the motor runs and winds the string which in turn raises the mass.

**Example 1.50**

State the thermodynamic definition of work.

*Solution* :

In thermodynamics work done by a system on its surroundings is defined as that interaction whose sole effect, external to the system, is equivalent to raising of a mass through a certain distance.

**Example 1.51**

Consider a certain amount of liquid contained in a vessel. A paddle wheel or stirrer is inserted in the liquid and the stirrer is rotated, for some time, by lowering a mass as shown in Fig.E 1.51. While the mass is lowered work is done on the liquid. Is it possible to define the work done on the system as the interaction whose sole effect, external to the system, is equivalent to lowering of a mass ? Explain.

*Solution* :

It is not correct to state the thermodynamic definition of work done on the system by its surroundings as the interaction, whose sole effect, external to the system is equivalent to lowering of a mass. The system (liquid) is initially at some state, say at temperature $T_1$. After performing work on the system, by rotating the stirrer, the system reaches a different state–say at temperature $T_2$. Now, leaving the system untouched, the surroundings can be modified as follows. The stirrer and the mechanical arrangement to rotate the stirrer can be removed and the vessel can be placed on a hot body to raise the temperature of the liquid from $T_1$ to $T_2$ as shown in Fig.E 1.51 (*b*). The modifications have been made external to the system only and we find that the energy of the hot body (possibly its temperature also) is reduced, even though the system has undergone the same change as depicted in Fig.E 1.51 (*a*). That is, lowering of a mass is not the sole effect external to the system. The state of the system can be changed by either of the methods shown in Fig.E. 1.51 (*a*) and (*b*). Hence in thermodynamics, the work done on the system by its surroundings is not defined in terms of lowering of a mass.

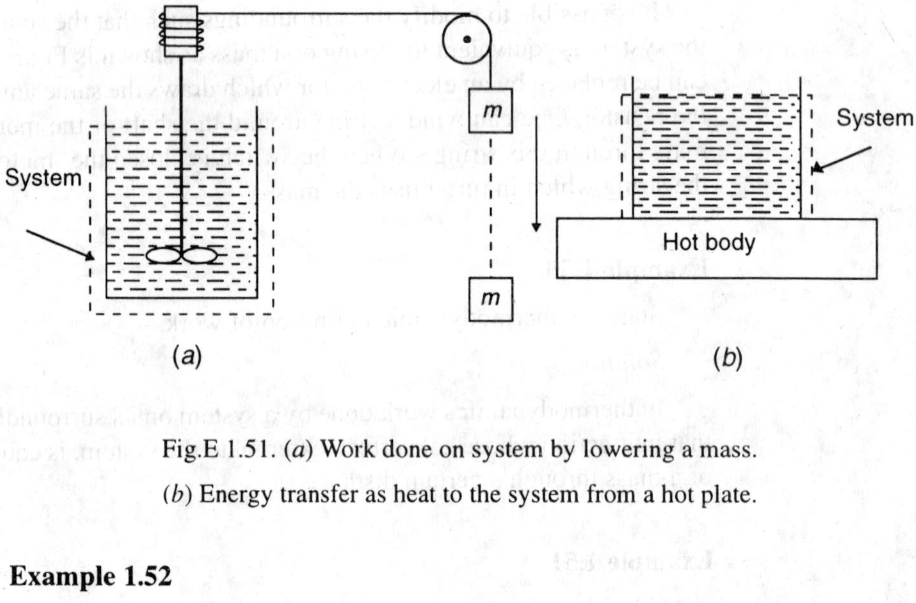


Fig.E 1.51. (*a*) Work done on system by lowering a mass.
(*b*) Energy transfer as heat to the system from a hot plate.

**Example 1.52**

The work associated with the boundary movement (expansion / compression work) is given by $\int PdV$. Is there a situation where

$$W \neq \int PdV$$

*Solution* :

The work done during a reversible process is given by

$$W = \int PdV$$

If the process is irreversible, the work done is not equal to $\int PdV$. Suppose a vessel is divided into two compartments 1 and 2. The compartment 1 of volume

$V_1$ contains a gas at pressure $P_1$ and the compartment 2 is evacuated. If the partition separating the compartments is removed, the gas expands and occupies the entire volume of the vessel. Let $P_2$ and $V_2$ denote the final pressure and volume of the gas. The pressure and volume of the gas have changed. To calculate $\int PdV$, a knowledge of initial and final values of pressure and volume are not enough. It is necessary to know the variation of pressure and volume during the expansion of the gas. To obtain some information regarding the intermediate values of pressure and volume, the expansion process can be carried out in steps as shown in Fig.E 1.52. Suppose several movable partitions A, B, C, D, E are placed in compartment 2.

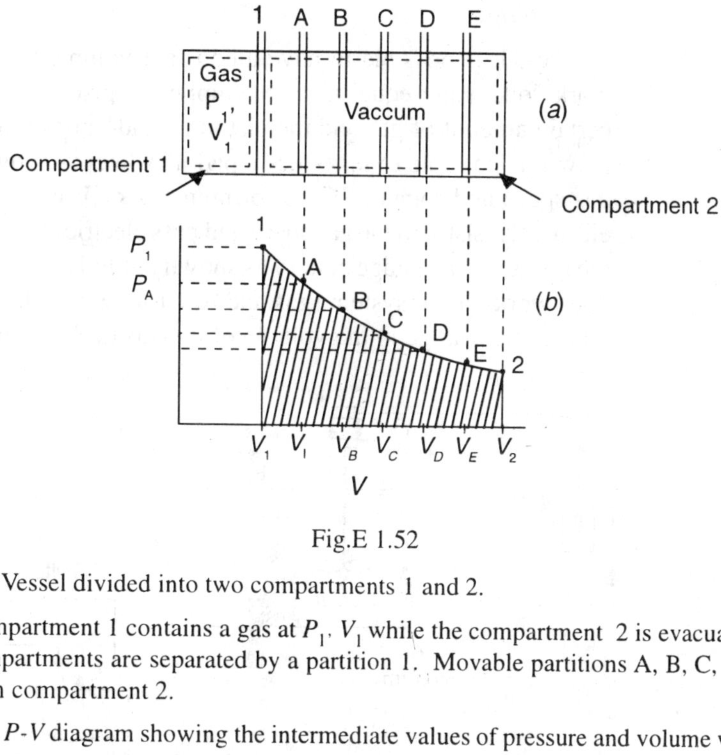


Fig.E 1.52

(*a*) Vessel divided into two compartments 1 and 2.

Compartment 1 contains a gas at $P_1$, $V_1$ while the compartment 2 is evacuated. The two compartments are separated by a partition 1. Movable partitions A, B, C, D, E are placed in compartment 2.

(*b*) *P-V* diagram showing the intermediate values of pressure and volume when the partitions A, B, C, D, E are removed one after another.

If the partition 1 is removed, the gas in compartment 1 expands and moves upto partition A, at which instant the gas pressure is $P_A$ and its volume is $V_A$. If the partition A is removed, the gas expands and its state is given by $P_B$ and $V_B$. Similarly one can remove the partitions B, C, D and E and measure the pressure and volume of the gas after removing each partition. A plot of *P* versus *V* then will appear as shown in Fig.E 1.52 (*b*). It is possible to evaluate the area under the curve 1 ABCDE 2 (Shaded area in Fig.E 1.52 (*b*)) to determine $\int_1^2 PdV$. However, this value does not represent the work done by the gas. In reality, the gas has not done any work since no force has been moved. Such an expansion is called *free expansion*, that is an expansion which is not restrained by any force. During the expansion of a gas, the work done is equal to zero. In this case or in a situation which is not reversible, the work done $W$ is not equal to $\int PdV$.

**Example 1.53**

Is there is situation where $dV = 0$ and yet $dW \neq 0$ ?

*Solution* :

Yes. It is possible to have a constant volume ($dV = 0$) process and yet the work done is not equal to zero. Suppose a piston cylinder assembly contains a certain amount of gas and the piston is held in position with the help of latches. Now it is possible to introduce a paddle wheel as shown in Fig.E 1.53 (*a*) in the gas space and rotate it by performing work. It is also possible to introduce an electrical resistor in the gas space and pass electric current through it by connecting the resistor to a storage battery as shown in Fig.E 1.53 (*b*). In this case, the battery does work on the system (gas and resistor). Since the piston is held in position, there is no change in the volume of the gas ($dV = 0$) and yet $dW \neq 0$.

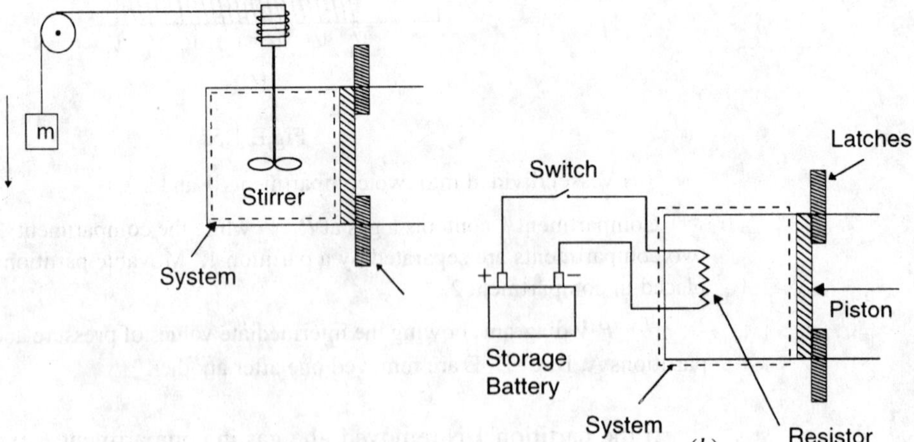


(*a*) (*b*)

Fig.E 1.53. Illustratation of process to show that $dV = 0$ and yet $dW \neq 0$.

(*a*) paddle wheel (or shaft) work done on system

(*b*) Electrical work done on system.

**Example 1.54**

One mode of work is associated with the movement of the boundary of the system and it is usually called expansion/compression work. Are there any other modes of work interaction between a system and its surrounding ?. If so, specify some of them.

*Solution* :

There are several modes of work interaction between a system and its surroundings. Some of the common modes of work interaction are

- Compression / expansion (boundary movement work)
- Extension of a solid rod (elastic work)
- Spring work
- Stretching of a liquid film
- Work transmitted through a shaft or shaft work
- Electrical work
- Polarization
- Magnetization

**Example 1.55**

Derive a relation to estimate the work done if an elastic rod is subjected to tension.

*Solution* :

Consider an elastic rod of length $L$ and cross–sectional area $A$, which is fixed at one end as shown in Fig.E 1.55 as the system.

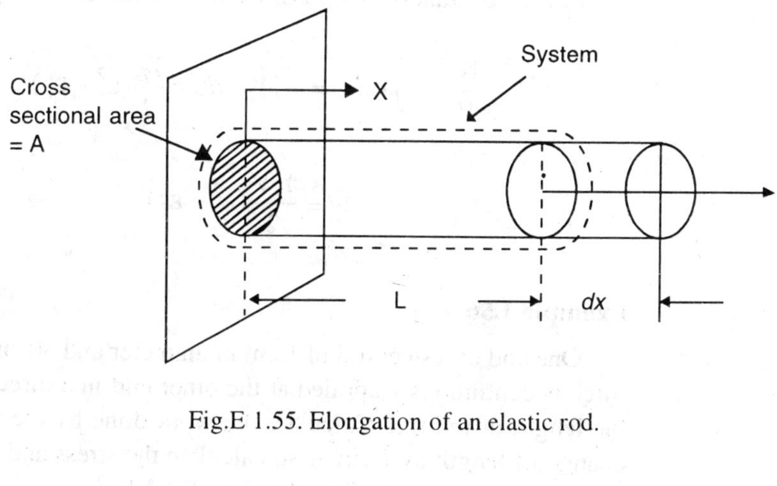


Fig.E 1.55. Elongation of an elastic rod.

The elastic rod is fixed at $X = 0$ and a force $F$ is applied at the other end. If $\sigma$ is the normal stress acting at the other end of the rod, then the force $F$ acting on the rod is given by

$$F = \sigma A$$

As a result of the applied force, if the rod is elongated by $dx$, the work done by the elastic rod is given by

$$dW = -F.dx = -\sigma A dx \text{ or } \frac{dW}{V} = -\sigma \frac{dx}{L} - \sigma d\varepsilon$$

where $\varepsilon = \dfrac{dx}{L}$ = elongation per unit length = strain

$V = AL$ = Volume of the rod

A negative sign is introduced in the above relation to take into account that work is to be done on the rod if $dx$ is positive. The work done per unit volume ($W/V$) of the rod is given by

$$w = \frac{W}{V} = -\int \sigma d\varepsilon$$

To integrate the above relation it is essential to have a knowledge of the relation between stress ($\sigma$) and strain ($\varepsilon$). If the elastic rod is isothermally elongated the stress and strain are related by the Youngs isothermal modulus ($Y_T$) which is given by

$$Y_T = \frac{\sigma}{\varepsilon}$$

If the elongation of the rod occurs in the Hooke's law region

$Y_T$ is a constant or $\sigma = Y_T \varepsilon$. Then the work done is given by

$$\frac{W}{V} = -\int_1^2 \sigma d\varepsilon = -\int_1^2 Y_T \varepsilon d\varepsilon = \frac{Y_T}{2}(\varepsilon_1^2 - \varepsilon_2^2)$$

or $$W = \frac{Y_T V}{2}(\varepsilon_1^2 - \varepsilon_2^2)$$

**Example 1.56**

One end of a steel rod of 1 cm in diameter and 50 cm in length is fixed and a force is continuously applied at the other end in a direction which is parallel to the length of the rod. Calculate the work done by the force on the steel rod to change its length by 1 cm. Also calculate the stress and force applied on the rod. The Young's modulus of steel is $2 \times 10^{11}$ N/m$^2$.

*Solution* :

The work done by the steel rod is given by

$$W = \frac{Y_T}{2} V(\varepsilon_1^2 - \varepsilon_2^2) \qquad (A)$$

Initial volume of the steel rod $= V = \dfrac{\pi d^2 L}{4} = \dfrac{\pi(0.01)^2 \times 0.5}{4}$

$= 3.927 \times 10^{-5}\ \text{m}^3$

Youngs modulus of steel $= Y_T = 2 \times 10^{11}\ \text{N/m}^2$

Initial strain $= \varepsilon_1 = 0$

Final Strain $= \varepsilon_2 = \dfrac{x}{L} = \dfrac{1}{50} = 0.02$

Substituting these values in Eqn. (A), we get

$$W = -2 \times 10^{11} \times \frac{3.927 \times 10^{-5}}{2} \times (0.02)^2 = -1.5708\ \text{kJ}$$

Stress $= \sigma = Y_T\,\varepsilon = 2 \times 10^{11} \times 0.02 = 4 \times 10^9\ \text{N/m}^2$

Force $= F = \sigma A = 4 \times 10^9 \times \dfrac{\pi(0.01)^2}{4} = 3.1416 \times 10^5\ \text{N}$

**Example 1.57**

One end of steel rod of 1 cm in diameter and 1 m in length is fixed to the ceiling of a roof and the other end is connected to a hook to which a mass of 10000 kg is suspended. If the Young's modulus of steel is $2 \times 10^{11}\ \text{N/m}^2$ calculate the elongation of the rod and the work done on the steel rod.

*Solution* :

Force acting on the steel rod $= F = mg = 10000 \times 9.8 = 98\ \text{kN}$

Stress $= \sigma = \dfrac{F}{A} = \dfrac{4F}{\pi d^2} = \dfrac{4 \times 98 \times 10^3}{\pi(0.01)^2} = 1.2478 \times 10^9\ \text{N/m}^2$

Strain $= \varepsilon = \dfrac{\sigma}{Y_T} = \dfrac{1.2478 \times 10^9}{2 \times 10^{11}} = 6.2389 \times 10^{-3}$

Elongation $= x = \varepsilon L = 6.2389 \times 10^{-3} \times 1 = 6.24\ \text{mm}$

$$\text{Work done on the steel rod} = -W = \frac{Y_T V}{2}(\varepsilon_2^2 - \varepsilon_1^2) \qquad (A)$$

$$\text{Volume of the rod} = V = \frac{\pi d^2 L}{4} = \frac{\pi (0.01)^2 \times 1}{4} = 7.854 \times 10^{-5}\,\text{m}^3$$

$$\text{Therefore } -W = \frac{2 \times 10^{11} \times 7.854 \times 10^{-5}}{2} \times (6.2389 \times 10^{-3})^2 = 305.71 \text{ J}$$

**Example 1.58**

Derive a relation to calculate the work done by a spring when it is subjected to compression or extension.

*Solution* :

To change the length of a spring either by compression or extension, it is necessary to exert a force which causes a displacement $X$. The displacement $X$ of a spring is given by $(L - L_0)$ where $L$ is the length or the spring under compression / extension and $L_0$ is its natural length or the spring length when it is neither under compression nor extension. The force exerted by the spring or the restoring force is in a direction to restore the spring to its equilibrium position. The restoring force $F_s$ is given by

$$F_S = -KX = -K\,(L - L_0)$$

The negative sign indicates that the force is always opposite in direction from the displacement of its free end.

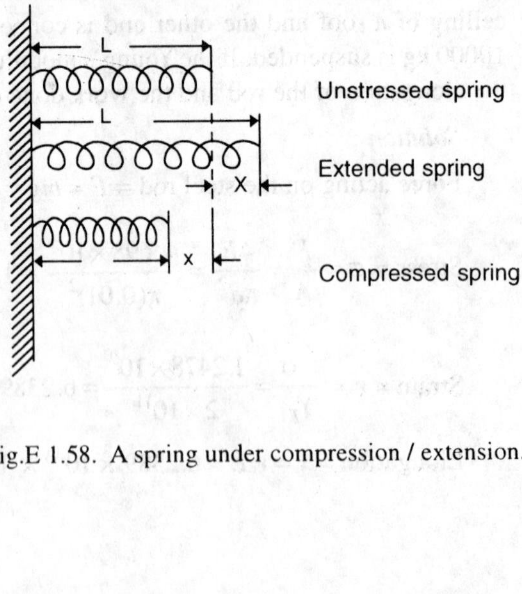


Fig.E 1.58. A spring under compression / extension.

Fig.E 1.58 shows a spring under compression and extention. The work done by the spring is given by

$$W = \int_{X_1}^{X_2} F_s.dX = -\int_{X_1}^{X_2} KX.dX = \frac{-1}{2}K(X_2^2 - X_1^2)$$

### Example 1.59

It is required to apply a force of 500 N to the free end of a spring to stretch it by 5 cm from its relaxed length. (*a*) Calculate the spring constant and the work done on the spring. (*b*) Now the spring is connected to a block and it is allowed to return slowly such that the final length of the spring is 1 cm more than its natural length. Calculate the work done by the spring.

*Solution* :

(*a*) Restoring force,

$$F = -500 \text{ N}$$

Displacement, $X = 0.05$ m

Spring constant, $K = \frac{-F}{X} = \frac{500}{0.05} = 50 \text{ kN/m}$

$$= -62.5 \text{ J}$$

Work done by the spring , $W = \frac{-KX^2}{2} = -\frac{50\times10^3\times(0.05)^2}{2}$

Work done on the spring ($-W$) = 62.5 J

(*b*) Work done by the spring, $W = \frac{-1}{2}K(X_2^2 - X_1^2)$

$$= \frac{-1}{2}\times 50\times10^3\{(0.01)^2 - (0.05)^2\} = 60 \text{ J}$$

### Example 1.60

Suppose a liquid film is held between a wire frame, one side of which can be moved with the help of a sliding wire as shown in Fig.E 1.60. The interfacial tension or surface tension which is the force per unit length normal to a line on the surface is given by $\tau$. Obtain a relation to estimate the work done on the liquid film when the surface area of the film is stretched.

*Solution* :

The force $F$ applied on the liquid film by an external agent is given by.

$$F = 2\,L\,\tau$$

The factor 2 is included because two film surfaces act on the wire. If the sliding wire is to be displaced by a distance $dX$, the external agent must do work on the liquid film.

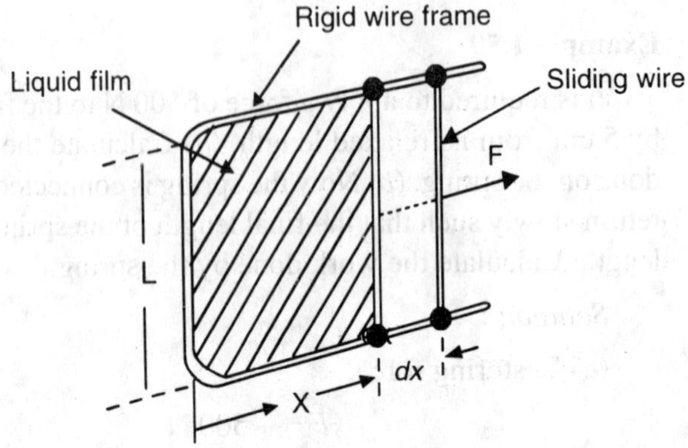


Fig. E 1.60. Liquid film held between a rigid wire frame.

Then the work done by the liquid film which is the negative of the work done by the external agent is given by

$$W = -\int_{X_1}^{X_2} F dX = -\int_{X_1}^{X_2} 2L\tau dX$$

Associated with the displacement $dX$ there is a change in the surface area of the liquid film. The change in the surface area ($dA$) of the liquid film is given by

$$dA = 2\,LdX$$

Then the work done by the liquid film can be rewritten as

$$W = -\int_{X_1}^{X_2} 2L\tau dX = -\int_{A_1}^{A_2} \tau dA = -\,\tau(A_2 - A_1)$$

**Example 1.61**

An aqueous soap solution is maintained in the form of a thin film, on a wire frame as shown in Fig.E 1.61. If the surface tension of the liquid is 72.75 mN / m calculate the work done when the movable wire is displaced by 1 cm in the direction indicated.

*Solution* :

The work done, $W = -\tau\ (A_2 - A_1)$

$$A_2 - A_1 = 2L\,(X_2 - X_1) = 2 \times 0.02\,(0.03 - 0.02) = 4 \times 10^{-4}\ \text{m}^2$$

Therefore $W = -72.75 \times 10^{-3} \times 4 \times 10^{-4} = -29.1 \times 10^{-6}$ J

The negative work indicates that work is done on the liquid film.

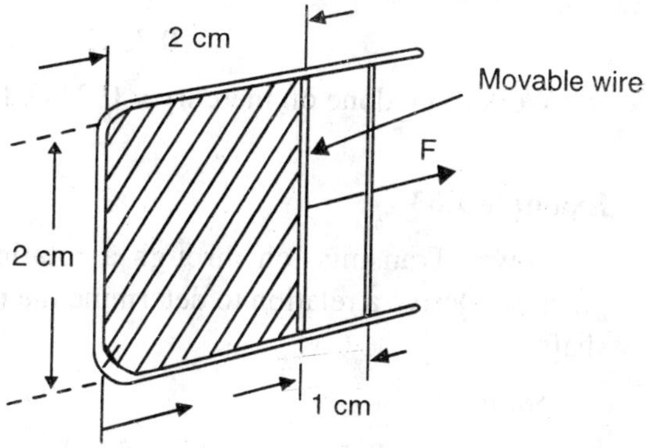


FigE. 1.61. Sketch for Example 1.61.

**Example 1.62**

Calculate the work to be done on 1 cm$^3$ of mercury (which is available as a spherical drop) to disperse it into fine droplets of 4 μm diameter. The surface tension of mercury is 487 mN / m.

*Solution* :

Applying the principle of conservation of mass, we get

$$\rho V_1 = N\rho V_2$$

or

$$N = \frac{V_1}{V_2}$$

where $V_1$ = volume of the initial drop

$V_2$ = volume of the final droplet = (4/3) $\pi r_2^3$

$$N = \frac{V_1}{V_2} = \frac{1 \times 10^{-6}}{\frac{4}{3}\pi(2 \times 10^{-6})^3} = 3 \times 10^{10}$$

Total surface area of the droplets, $A_2 = N\,(4\,\pi\,r_2^2)$

$$= 3 \times 10^{10} \times 4\pi\,(2 \times 10^{-6})^2 = 1.508\ \text{m}^2$$

$$V_1 = \frac{4}{3}\pi r_1^3 \text{ or } r_1 = \left(\frac{3V_1}{4\pi}\right)^{1/3} = \left(\frac{3\times1\times10^{-6}}{4\pi}\right)^{1/3} = 6.235\times10^{-3}\ \text{m}^3$$

Initial surface area of mercury, $A_1 = 4\,\pi\,r_1^{\,2} = 4\pi \times (6.235 \times 10^{-3})^2$

$$= 4.885 \times 10^{-4}\ \text{m}^2$$

Work done, $W = -\,\tau\,(A_2 - A_1) = -\,487 \times 10^{-3}(1.508 - 4.885 \times 10^{-4})$

$$= -\,0.7342\ \text{J}$$

Work to be done on mercury = 0.7342 J

**Example 1.63**

Power Transmission through a rotating shaft is a common engineering practice. Derive a relation to determine the power transmitted through a rotating shaft.

*Solution* :

Suppose a shaft of radius $r$ is fixed at one end and a force $F_t$ is tangentially applied at the other end as shown in Fig.E 1.63.

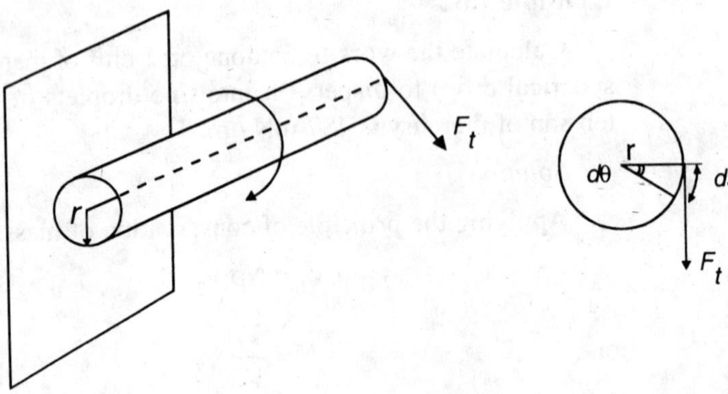


Fig.E 1.63. Torsion of a shaft.

The torque applied to the shaft = $\tau = F_t r$

The application of the torque results in angular displacement $d\theta$ of the point at which the torque is applied.

The linear displacement, $ds = rd\,\theta$

Work done, $dW = F.ds = F_t rd\,\theta = \tau\ d\theta$

$$W = \int_{\theta_1}^{\theta_2} \tau d\theta$$

Power transmitted, $P = \frac{dW}{dt} = \tau \frac{d\theta}{dt} = \tau\omega$

where $\omega$ is the angular velocity (in radians) of the shaft.

**Example 1.64**

Mixing and stirring of fluids is usually done with the help of a paddle wheel or stirrer. A paddle wheel consists of a shaft to which paddles or blades are attached. When a torque is applied to the shaft, the fluid forces on the paddles balance the applied torque and the shaft rotates at constant speed. Determine the power transmitted through a shaft to a liquid if the applied torque is 2 Nm and the shaft rotates at a rate of 600 revolutions per minute.

*Solution* :

The power transmitted through the shaft, $P = \tau\ \omega$

Angular velocity, $\omega\ = 2\pi n$

where $n$ = number of revolutions per second.

Therefor, $P = \tau\ \omega\ = \tau\ 2\ \pi\ n = 2 \times 2\pi \times 600/60 = 125.66$ W

**Example 1.65**

It is required to raise a body of mass 100 kg to a height of 60 m in the earth's gravitational field, where the acceleration due to gravity is 9.81 m/s$^2$. Calculate the work to be done on the body.

*Solution* :

In the earth's gravitational field, the force acting on a body of mass $m$ is given by $F = mg$. To raise the body, the external agent must apply a force $F$ in a direction which is opposite to the direction of acceleration due to gravity. Then the work done by the external agent is given by

$$W = \int F.ds = \int_0^z mgdz$$

$$= mgz = 100 \times 9.81 \times 60 = 58.86 \text{ kJ}$$

**Example 1.66**

The acceleration due to gravity as a function of elevation above the sea level is given by $g = 9.81 - 3.32 \times 10^{-6}\ z$ where $g$ is in m/s$^2$ and $z$ is in meters. A satellite with a mass of 300 kg is put in an orbit at an elevation of 400 km above the earths surface. Calculate the work done on the satellite.

*Solution* :

$$\text{Work done on the satellite, } W = \int_0^z mgdz = \int_0^z 300\,(9.81 - 3.32 \times 10^{-6}\, z)\, dz$$

$$= \int_0^z (2943 - 9.96 \times 10^{-4} z)\, dz = 2943z - 9.96 \times 10^{-4}\, z^2/2$$

$$2943 \times 400 \times 10^3 - \frac{9.96 \times 10^{-4}}{2} \times (400 \times 10^3)^2 = 1.0976 \text{ GJ}$$

**Example 1.67**

Derive a relation to estimate the work to be done on a body of mass $m$ in accelerating it from an initial velocity $\mathbf{V}_1$ to a final velocity $\mathbf{V}_2$.

*Solution* :

When a body of mass $m$ is acted upon by a force $F$, its acceleration $a$ is related to the force $F$ by the Newton's second law of motion. The Force $F$ acting on the body is given by

$$F = ma = m\, d\mathbf{V}/dt$$

Then the work done on the body is given by

$$W = \int F.ds = \int m\,(d\mathbf{V}/dt)\, ds = \int m\,(ds/dt)\, d\mathbf{V} = \int_{\mathbf{V}_1}^{\mathbf{V}_2} m\,\mathbf{V} d\mathbf{V} = m\, \frac{\mathbf{V}_2^2 - \mathbf{V}_1^2}{2}$$

**Example 1.68**

120 kJ work was done on a car in accelerating is from an initial speed of 10 km/h to 100 km/h. (*a*) Determine the mass of the car. (*b*) What could be the final speed of the car if the work done on the car is 200 kJ ?

*Solution* :

$$(a) \text{ Initial velocity of the car, } \mathbf{V}_1 = \frac{10 \times 10^3}{3600} = 2.778 \text{ m/s}$$

$$\text{Final velocity of the car, } \mathbf{V}_2 = \frac{100 \times 10^3}{3600} = 27.778 \text{ m/s}$$

Work done, $$W = \frac{1}{2} m (\mathbf{V}_2^2 - \mathbf{V}_1^2)$$

or $$120 \times 10^3 = \frac{1}{2} m \{(27.778)^2 - (2.778)^2\} = 381.95 \text{ m}$$

or $$m = 314.18 \text{ kg}$$

(*b*) Work done, $W = \frac{1}{2} m (\mathbf{V}_2^2 - \mathbf{V}_1^2)$

or $$200 \times 10^3 = \frac{314.18}{2} \{\mathbf{V}_2^2 - (2.778)^2\}$$

or $$\mathbf{V}_2 = 35.79 \text{ m/s} = 128.8 \text{ km/h}$$

**Example 1.69**

Suppose the terminals of a storage battery are connected to an electrical resistor through a switch as shown in Fig.E 1.69. When the switch is closed, the battery discharges doing work. Derive an expression to calculate the work done by the battery.

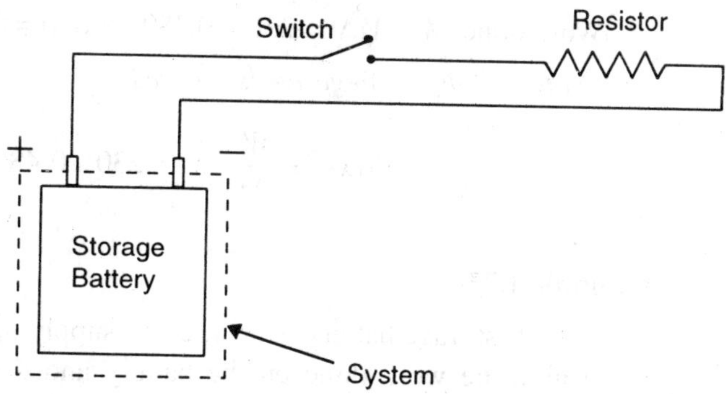


Fig.E 1.69. Discharging of a storage of battery through a resistance.

*Solution :*

Consider the storage battery as the system. The electric current flowing through the resistor is a manifestation of the flow of electrons. Work is required to displace the electric charge across the boundary of the system. We may recall that the electrical potential at a point is defined as the work done on a unit charge in moving it, which is at infinite distance, to that point. The electric current is defined as the rate of change of charge. That is electric current $i$ is given by

$$i = \frac{dQ_e}{dt} \quad \text{or} \quad dQ_e = idt$$

Where $Q_e$ = electric charge

If the differential charge $dQ_e$ is transferred during a differential time $dt$ with a potential difference $V$ then the work done is given by

$$dW = -VdQ_e \quad \text{or} \quad W = -\int VdQ_e = -\int Vidt$$

If the charge leaves the system $dQ_e$ is negative and the battery does work on the surroundings. On the other hand if $dQ_e$ is positive, that is during the charging of a battery, work has to be done on the system. To take this into account a negative sign is introduced in the above expression.

**Example 1.70**

An electric potential difference of 230 V is maintained across a 470 Ω resistor for a period of 10 minutes. Determine the electrical work done and the power.

*Solution :*

Current $i = \frac{V}{R} = \frac{230}{470} = 0.4894$ A

(where $R$ is electrical resistance)

Work done, $W = Vi\Delta t = 230 \times 0.4894 \times 600 = 67.537$ kJ

*(Note that $dQ_e$ is negative is this case)*

$$\text{Power} = \frac{W}{\Delta t} = Vi = 230 \times 0.4894 = 112.56 \text{ W}$$

**Example 1.71**

A 12 V storage battery is charged by supplying a current of 2A for 3 hours. Determine the work done on the battery and the quantity of electric charge transferred.

*Solution :*

Work done, $W = -Vi\Delta t = -12 \times 2 \times 3 \times 3600 = -259.2$ kJ

Work done on storage battery = 259.2 kJ

Electric charge transferred to battery, $Q_e = i\ \Delta t = 2 \times 3 \times 3600 = 21600$ coulombs

**Example 1.72**

Potential energy can be stored in a spring by stretching it. Similarly one can store energy as potential energy in an electric field and the device used for this purpose is called a capacitor. In the present electronic age the capacitors play a vital role. Develop an expression to calculate the work done in charging a capacitor.

*Solution* :

A parallel plate capacitor shown in Fig.E 1.72 consists of two parallel conducting plates of area $A$ separated by a distance $d$. When the capacitor is charged, its plates have equal but opposite charges $+Q_e$ and $-Q_e$. Usually, the charge of a capacitor is represented by $Q_e$. Since the plates are conductors, they are equipotential surfaces-that is all points on a plate are at the same electric potential. The potential difference $V$ between the plates is given by

$$V = Ed \tag{A}$$

where $E$ is the electric field.

The charge $Q_e$ and the potential difference $V$ for a capacitor are proportional to each other. That is

$$Q_e = CV \tag{B}$$

where $C$ is the capacitance of the capacitor.

From electrostatics we know that

$$Q_e = \varepsilon_0 EA \tag{C}$$

Where $\varepsilon_0$ is the permittivity constant and is given by

$\varepsilon_0 = 8.85 \times 10^{-12}$ c$^2$/N m$^2$ = $8.85 \times 10^{-12}$ Faraday/m

From Eqns. ($A$) to ($C$) ; we get

$$C = \varepsilon_0 \frac{A}{d} \tag{D}$$

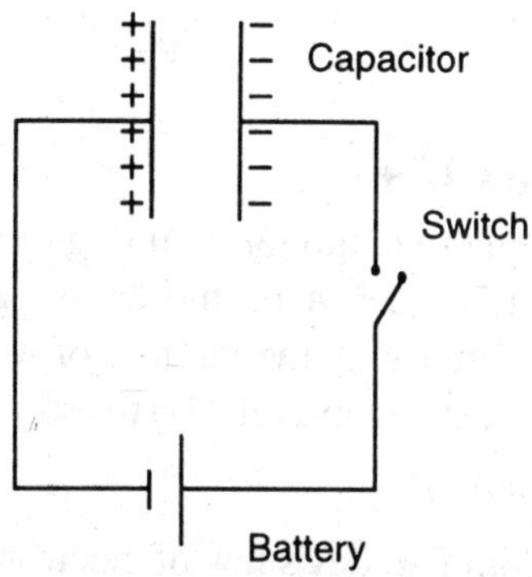


Fig.E 1.72. Charging a capacitor.

To charge a capacitor work must be done by an external agent. Starting with an uncharged capacitor, electrons are to be removed from one plate and transferred to the other plate. The electric field which builds up in the space between the plates is in a direction to oppose further transfer. Therefore, as charge accumulates on the capacitor larger amount of work has to be done for further transfer of electrons. The differential work $dW$ done in transferring a differential charge $dQ_e$ through a potential difference $V$ is given by

$$dW = -VdQ_e = -\frac{Q_e}{C}dQ_e$$

or
$$W = -\frac{1}{C}\int Q_e dQ_e = \frac{Q_e^2}{2C} = \frac{VQ_e}{2}$$

(Since work is done on the capacitor in charging it.)

**Example 1.73**

A parallel plate capacitor is charged to 230 V. Given that the capacitance of the capacitor is 10 $\mu F$, calculate the work done in charging the capacitor.

*Solution :*

Work done in charging a capacitor is given by

$$W = \int VdQ_e = \frac{1}{C}\int Q_e dQ_e = \frac{Q_e^2}{2C} = \frac{VQ_e}{2}$$

The charge transferred $Q_e$ in given by

$$Q_e = CV = 10 \times 10^{-6} \times 230 = 23 \times 10^{-4}\ \text{C}$$

$$W = \frac{VQ_e}{2} = \frac{230 \times 23 \times 10^{-4}}{2} = 0.2645\ \text{J}$$

**Example 1.74**

Suppose a thin toroidal ring of material is covered with winding as shown in Fig.E 1.74. Let $A$ be the cross–sectional area of the substance, $L$ the mean circumference, $N$ the number of windings and $i$ the current passing through the winding, Develop a relation to estimate the work done in magnetising the material.

*Solution* :

From Faraday's law of electromagnetic induction the induced electromotive force $E$ in the winding (or back *emf*) is given by

$$E = -NA\left(\frac{dB}{dt}\right)$$

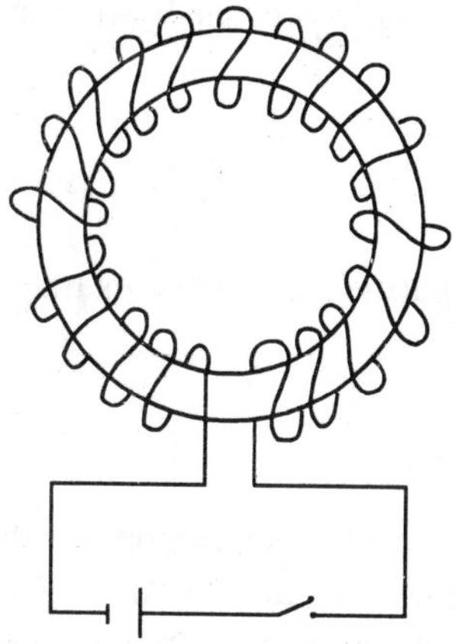

Fig.E 1.74. Magnetic work done on a toroidal ring.

where $B$ is the magnetic field. During the differential time $dt$ if the differential quantity of charge $dQ_e$ is transferred in the circuit, then the work done is given by

$$dW = EdQ_e = -NA\frac{dB}{dt}dQ_e = -NA\left(\frac{dQ_e}{dt}\right)dB = -NAidB \qquad (A)$$

In a toroidal winding the magnetic field intensity ($\boldsymbol{H}$) is given by

$$\boldsymbol{H} = \frac{Ni}{L} \qquad (B)$$

Substituting Eqn. $(B)$ in Eqn. $(A)$, we get

$$dW = -AL\boldsymbol{H}dB = -V\boldsymbol{H}dB \qquad (C)$$

where $V = AL$ = volume of the material.

The action of a magnetic field on the substance is to produce magnetization (or magnetic moment) $\boldsymbol{M}$. The total effect of the current in the winding is to establish the magnetic field and magnetization of matter within the field. The effect is called the magnetic induction $B$ and is given by

$$B = \mu_o(\boldsymbol{H} + \boldsymbol{M}) \qquad (D)$$

where $\boldsymbol{M}$ is the magnetization per unit volume and $\mu_0$ is the permeability of free space.

From Eqns. ($C$) and ($D$), we get

$$dW = -\mu_0 V\boldsymbol{H}d\boldsymbol{H} - \mu_0 V\boldsymbol{H}d\boldsymbol{M} \qquad (E)$$

The first term on the RHS of Eqn.($E$) represents the work done in increasing the magnetic field in the volume $V$ of the empty space and the second term represents the work done in increasing the magnetization of the material by the amount $dM$. Therefore, the work done in magnetization (of a material) is given by

$$dW = -\mu_0 VHdM \quad \text{or} \quad \frac{dW}{V} = -\mu_0 HdM \qquad (F)$$

The negative sign indicates that work is done on the system (material to be magnetized).

**Example 1.75**

Curie's law for paramagnetic substances is given by $M = CH/T$ where $C$ is Curies constant. Calculate the work done per unit volume of a paramagetic substance in changing the magnetization from $M_1$ to $M_2$.

*Solution :*

The work done for magnetization is given by

$$dW = \mu_0 VHdM$$

or
$$\frac{dW}{V} = \mu_0 HdM$$

or
$$\frac{W}{V} = -\mu_0 \int_{M_1}^{M_2} HdM = \mu_0 \int_{M_1}^{M_2} \frac{TM}{C} dM = -\frac{\mu_0 T}{2C}(M_2^2 - M_1^2)$$

Therefore, work to be done on the substance $= \frac{\mu_0 T}{2C}(M_2^2 - M_1^2)$

**Example 1.76**

In mechanics work done ($dW$) is given by the relation $dW = F.ds$ where $F$ is the force which is an intensive parameter and $ds$ is the displacement, which is an extensive parameter. Is it possible to generalize the above relation so that it is applicable to all modes of work interactions? If so,give a summary of the relevent forces and displacements.

*Solution :*

The work done for boundary movement (compression /expansion) is given by $dW = PdV$, where $P$ is an intensive parameter and $dV$ is an extensive parameter. The elastic work done is given by $dW / V = -\sigma\, d\varepsilon$ where the stress $\sigma$ is an intensive parameter and the stain ($\varepsilon$) is an extensive parameter. Similarly for all modes of work the work done by a system can be generalized as

$$dW = F_k.dX_k$$

where $F_k$ = generalized force, which is an intensive parameter $dX_k$ = generalized displacement, which is an extensive parameter . The extensive and intensive parameters associated with several modes of work interactions are summarized below.

**Generalized Work Interactions**

| Work mode | Generalized force (intensive) | Generalized displacement (Extensive) | Equation for work ($dW =$ ) |
|---|---|---|---|
| Boundary motion (Compression/Expansion) | Pressure ($P$) | Volume ($dV$) | $PdV$ |
| Elastic (Extension of solid) | Stress ($\sigma$) | Strain ($d\,\varepsilon$) | $-\sigma d\varepsilon$ |
| Spring | Force ($F$) | Displacement ($dX$) | $FdX$ |
| Surface (Liquid film) | Surface tension $\tau$ | Surface area ($dA$) | $-\tau dA$ |
| Shaft work (Torsion) | Torque ($\tau$) | Angular displacement ($d\theta$) | $\tau\, d\theta$ |
| Electrical (Reversible cell) | Voltage (V) | Charge ($dQ_e$) | $-\text{V}dQ_e$ |
| Capacitor (Electric polarization) | Potential difference ($V$) | Charge($dQ_e$) | $-VdQ_e$ |
| Magnetization (Magnetic polarization) | Magnetic field ($\boldsymbol{H}$) | Magnetization($d\boldsymbol{M}$) | $-\mu_o \boldsymbol{H}d\boldsymbol{M}$ |

**Example 1.77**

Name the different modes by which the energy of a system can be changed.

*Solution* :

Consider a gas enclosed in a piston –cylinder assembly as our system. The energy of the gas can be increased/decreased by compressing/expanding the gas under adiabatic conditions. That is, the energy of the system can be altered through work interaction. The energy of the gas can also be altered by cooling or heating. That is the energy of a system can be altered through heat interaction also. Suppose an elastic rod is considered as our system. The system's energy can be changed through elastic work as well as through heat interaction. If we consider an elastic and magnetic substance as our system, its energy can be altered by two different types of work interactions– elastic work and magnetization work in addition to heat interaction. Now we can generalize to state that the energy of a system can be altered by two modes. They are (1) work interaction and (2) heat interaction.

**Example 1.78**

Suppose a rigid vessel containing a gas is placed in thermal contact with a hot body. The temperature of the gas and hence the energy of the gas increases. Considering the gas as our system, identifty the interaction between the system and the surroundings from (*a*) macroscopic and (*b*) microscopic points of view.

*Solution* :

(*a*) Since the vessel is rigid, its volume does not undergo any change. Hence the boundary motion (compression/expansion) work is not associated with the system. At a macroscopic level, no work interaction is responsible for the change in the energy of the system. Since there are only two modes, work and heat, by which the system can change its energy, the above interaction can be called heat interaction.

(*b*) At a microscope level, the atoms or molecules of the container wall vibrate vigorously since the container wall heats up when it is placed in thermal contact with a hot body. The gas consists of a large number of atoms/molecules which move with random velocities and the average velocity is proportional to $\sqrt{T}$. Some of the gas atoms/molecules come into collision with the atoms of the container and during the collision energy transfers from the atoms of the container to the gas atoms/molecules. Because of intermolecular collision the energy is transferred to the other atoms/molecules of the gas also, resulting in an increase in the energy of the system. During a collision, the kinetic energy of one atom/molecule is transferred to another atom/molecule. That is one atom/molecule does work on another atom/molecule. Therefore, at a microscope level, the energy transfer occurs through work interaction only. However, at a macroscopic level, we fail to observe any work interaction.

**Example 1.79**

Explain the meaning of 'Heat interaction' or 'Heat'

*Solution* :

The mode of energy transfer which cannot be accounted as work from a macroscopic point of view is called heat interaction or simply heat. The heat interaction occurs by virtue of temperature difference across the boundary. The heat interaction depends on the path followed by a system and hence it is a path function and not a point function. Heat is energy in transit. It is not a property of the system and its differential is not exact.

**Example 1.80**

The two modes of energy transfer are work and heat. Does the mode of energy transfer depend on the choice of a system? Support your answer with the help of an example.

*Solution* :

The mode of energy transfer depends on the choice of a system. Consider a vessel containing some water, the temperature of which is raised by switching on an immersion heater (placed in water) connected to a storage battery. Suppose we consider the water alone as our system and the rest of the universe as surroundings, then the interaction between the system and its surroundings is called heat, because the energy transfer occurs by virtue of temperature difference between the immersion heater (or heating coil) and water. On the other hand, if we choose the water and the immersion heater as our system, the interaction between the system and its surroundings is called work, because the energy transfer is due to the discharging of the storage battery which can be visualised as the raising of a mass. Therefore, the work and heat interactions depend on the choice of a system.

**Example 1.81**

Consider the water contained in a container as our system. Suppose the water is initially at temperature $T_1$ and it is allowed to interact with its surroundings. As a result of energy transfer suppose the water reaches a temperature $T_2$, where $T_2 > T_1$. Is it possible to identify the interaction as work or heat by which the system has changed its energy?

*Solution* :

It is not possible to name the interaction as work or heat from a knowledge of the initial and final conditions of the system. The mode of energy transfer (heat or work) can be identified only while the system is interacting with its surroundings. It is possible that the container is placed on a hot body for some time to raise the temperature of the water. In such a case, the interaction is called heat. It is also possible that a paddle wheel is placed inside the water and it is rotated for sometime to raise the temperature of water. In such a case, the interaction is called work (paddle wheel work or shaft work).

**Example 1.82**

Which of the following characteristics are true for work ?

(*a*) It is a path function.

(*b*) It is not a property of a system.

(*c*) Its differential is not exact

(*d*) It is not stored in a system

(*e*) It is energy in transit.

(*f*) It can be identified only during the course of interaction.

*Solution :*

All these characteristics are true for work.

**Example 1.83**

Which of the characteristics given in Example 1.82 are true for heat ?

*Solution :*

All the characteristics are true for heat.

# 2

# Zeroth Law of Thermodynamics

### Example 2.1

How do we characterize a thermodynamic system?

*Solution* :

A thermodynamic system is characterized without any ambiguity by specifying certain properties. A thermodynamic system contains a definite quantity of matter and matter possesses energy in its various forms. Therefore, a thermodynamic system is completely described by specifying some of the following quantities.

- The quantity as well as the composition of matter in terms of mole numbers of each constituent.
- The measurable properties like volume, pressure and temperature of system.
- The energy of the system

### Example 2.2

Consider one mole of an ideal gas enclosed in a piston-cylinder assembly as a system. The system can be characterized by specifying temperature ($T$), pressure ($P$), Volume ($v$) etc. Are the properties $T$, $P$ and $v$ are independent of each other?

*Solution* :

All the three properties $P$, $T$ and $v$ are not independent. There exists a relationship of the form $f(P, v, T) = 0$ among the three properties. For an ideal gas, the relationship is given by $Pv = RT$ where $v$ is the molar volume and $R$ is the universal gas constant. Only two of the three properties $P$, $v$ and $T$ can be independently varied.

**Example 2.3**

Consider a compressible substance, say an ideal gas, enclosed in a piston-cylinder assembly as the system. Is it possible to independently vary the energy and volume of the system ?

*Solution* :

For a compressible substance, energy and volume can be independently varied. The energy of a system can be varied by heat and work interactions. Suppose the piston is held in position and the cylinder can be brought into thermal contact with a hot body, transferring energy as heat to the system. Thus the volume of the system can be held constant while its energy can be changed. It is also possible to change the volume of the system by compression /expansion while the system is held at constant temperature by immersing it in a constant temperature bath. For an ideal gas, the energy depends on temperature only. Hence, the energy of the system is held constant while the volume changes. Therefore, the energy and volume of a compressible substance can be independently varied.

**Example 2.4**

Is it possible to vary the pressure while energy and volume of a compressible substance are held constant?

*Solution* :

It is not possible to vary pressure while the energy and volume of a compressible substance are held constant. We know that the pressure of a gas can be changed by (*a*) compression/expansion (*b*) heating / cooling and (*c*) vigorous stirring. If the gas is compressed / expanded, the volume undergoes a change even though the energy can be held at a constant value by immersing the system in a constant temperature bath. If the gas is heated / cooled in a rigid container, the energy of the system changes even though the volume is held constant. If the gas enclosed in a rigid vessel is vigorously stirred its energy increases, even though the volume is held constant. Thus we find that it is not possible to change the pressure of a gas while its energy and volume are held constant.

**Example 2.5**

What does the state postulate specify and explain its meaning.

*Solution* :

The State postulate states that for a given thermodynamic system, the number of independently variable thermodynamic properties is equal to the number of reversible work modes, associated with the system, plus one.

The energy of a system can be altered by work and heat interactions. A thermodynamic system may exchange its energy by more than one work interaction. A compressible substance has only one work interaction which is due to boundary movement (compression/expansion). An elastic rod can exchange its energy through one work interaction which is the elastic work. However, an elastic and magnetic substance can exchange its energy through two work interactions - namely elastic work and magnetic work. In addition to work interactions, a thermodynamic system can exchange its energy through heat interaction also. Therefore, the number of ways in which the energy of a thermodynamic system, can be changed is equal to the number of work modes associated with the system plus one. One thermodynamic property is associated with every work mode. The property $v$ is associated with compression / expansion work mode, the property strain ($\varepsilon$) is associated with elastic work mode and the property magnetization ($\boldsymbol{M}$) is associated with the magnetization work mode. Hence the number of thermodynamic properties which can be independently varied is equal to the number of work modes plus one. If a given thermodynamic system is capable of exchanging its energy through $n$ work modes, then the number of independently variable thermodynamic properties of the system is equal to $(n + 1)$.

**Example 2.6**

How many thermodynamic properties of a system are required to describe its state ?

*Solution* :

If a thermodynamic system has $n$ independent variable thermodynamic properties, the state of such a system is completely described, without any ambiguity, by specifying $n$ independent thermodynamic properties. If the independently variable thermodynamic properties of a system are held constant, no other property of that system can be varied independently.

**Example 2.7**

Is there any limitation on the nature (say extensive or intensive) of the properties to be used in describing the state of a thermodynamic system ?

*Solution* :

There is no restriction on the nature of the thermodynamic properties that can be used to describe the state of a thermodynamic system. One can use only extensive properties or only intensive properties or a combination of extensive and intensive properties as long as they are independent of each other.

**Example 2.8**

How many thermodynamic properties are to be specified for complete description of a thermodynamic system consisting of (*a*) an incompressible substance (*b*) an elastic substance and (*c*) an elastic and magnetic substance?

*Solution* :

(*a*) Two (*b*) Two (*c*) Three

**Example 2.9**

For a simple compressible substance which of the following sets of thermodynamic properties can be used to describe the state of the system ?

(*a*) Energy ($u$) and volume ($v$)

(*b*) Energy ($u$) and Temperature ($T$)

(*c*) Energy ($u$) and density ($\rho$)

(*d*) Pressure ($P$) and volume ($v$)

(*e*) Volume ($v$) and density ($\rho$)

(*f*) Pressure ($P$) and temperature ($T$)

(g) Temperature ($T$) and volume ($v$)

*Solution* :

The state of a simple compressible substance can be described by any of the following sets of properties,

$(u,v)$; $(u,\rho)$; $(P,v)$; $(P,T)$;$(T,v)$

**Note** *If one does not insist that the energy of a system should remain constant, then one can specify the pressure instead of energy (u) as one of the independent properties to specify the state of a thermodynamic system.*

**Example 2.10**

Why is it not possible to use energy ($u$) and temperature ($T$) or volume ($v$) and density ($\rho$) as a set of thermodynamic properties to specify the state of a simple compressible substance ?

*Solution* :

Energy ($u$) and temperature ($T$) are not independently variable properties of a thermodynamic system because there exists a relation between these two properties.

Similarly volume ($v$) and density ($\rho$) are not independent of each other. Hence one cannot use the sets $(u,T)$ and $(v, \rho)$ to specify the state of a simple compressible substance.

**Example 2.11**

You may recall (from physical chemistry course) that the Gibbs phase rule can be used to determine the number to degrees of freedom (or the number of properties needed for complete description) of a system. Apply the Gibbs phase rule and determine the degrees of freedom of a simple compressible substance (like nitrogen gas) existing in gas phase.

*Solution* :

According to the Gibbs phase rule, the number of degree of freedom ($F$) of a system is given by

$$F = C + 2 - P$$

where $C$ = number of independent components

$P$ = number of phases

For a simple compressible substance (like nitrogen gas) existing in gas phase, $C = 1$ and $P = 1$. Hence

$$F = C + 2 - P = 1 + 2 - 1 = 2$$

Therefore, there are two degrees of freedom. In other words, two thermodynamic properties can be independently varied.

**Example 2.12**

Is the state postulate equivalent to Gibbs phase rule? What is the difference between these two?

*Solution* :

The state postulate and the Gibbs phase rule are independent. The state postulate gives the number of thermodynamic properties needed for a complete description of the state of a system. These properties can be extensive, intensive or a combination of extensive and intensive. The Gibbs phase rule gives the degrees of freedom (or the number of independent thermodynamic properties) of a system. These properties are intensive properties only. They cannot be extensive properties.

**Example 2.13**

State the zeroth law of thermodynamics.

*Solution* :

The zeroth law of thermodynamics states that if each of the two given systems are in thermal equilibrium with a third system then the two given systems are in thermal equilibrium with each other.

**Example 2.14**

There are four laws of thermodynamics namely zeroth, first, second and third law of thermodynamics. Usually one counts starting with 1. Why the counting of the laws of thermodynamic started with zero ?

*Solution* :

The first law of thermodynamics was proposed by J.P Joule in 1851 which is followed by the second and third laws of thermodynamics. During the development of the laws of thermodynamics the content or the principle involved in the zeroth law of thermodynamics was assumed to be true without ever stating it so, However, for a logical development of the science it is necessary to state the zeroth law before presenting the other laws of thermodynamics. This was realized by R.H. Fowler in 1931. Since the other laws were formulated before 1931, counting them as first, second and third, it was necessary to call the postulate or law proposed by Fowler as the zeroth law of thermodynamics.

**Example 2.15**

Explain how it is possible to compare the thermal state (degree of hotness) of several bodies by making use of the zeroth law of thermodynamics.

*Solution* :

A particular body called a thermometer can be brought into thermal contact with another body, say body A and allow these two bodies to reach a state of thermal equilibrium. In that state, some measurable characteristic of the thermometer - called thermometric property can be recorded. Then the thermometer in turn can be placed in thermal contact with a second body, say body B, till thermal equilibrium is established between them. Once they reach a state of thermal equilibrium one can measure the thermometric property again. If the two thermometric property readings—one measured while in contact with body A and the other measured while in contact with Body B - are identical, then the two bodies A and B are in thermal equilibrium with each other. Knowing the thermal state of one of the bodies A or B, the thermal state of the other body can be determined.

**Example 2.16**

Name a few measurements or quantities which can be used as thermometric properties.

*Solution* :

The following measurements or quantities can be conveniently used as thermometric properties in order to quantify the temperature.

- Length of mercury column in a capillary tube.
- Electrical resistance of a wire
- Pressure of a fixed quantity of gas confined in a constant volume bulb.
- Volume of a fixed quantity of gas held at constant pressure.
- emf generated at a junction of two dissimilar metal wires.

**Example 2.17**

Explain the two fixed points on which the Celsius temperature scale (earlier it was called centigrade scale) is based.

*Solution* :

The Celsius scale is based on two easily reproducible fixed points. They are the ice point and the steam point. The ice point is the equilibrium temperature of the ice with water at standard atmospheric pressure and it is assigned a value of 0°C.

The steam point is the equilibrium temperature of pure water with its own vapour at standard atmospheric pressure and it is assigned a value of 100°C.

**Example 2.18**

At the tenth conference Generale des poids et Measures (General conference of weights and measures CGPM) held in 1954, the Celsius scale in redefined in terms of a single fixed point. What is that single fixed point?

*Solution* :

In 1954, the CGPM redefined the Celsius scale in terms of a single fixed point. The single fixed point is the triple point of water (the state in which all the three states-solid, liquid and vapor of water coexist in equilibrium) which is assigned a value of 0.01°C.

**Example 2.19**

Consider a mercury in glass and alcohol in glass thermometers. Both the thermometers are calibrated using the two standard fixed points - ice point and steam point. Suppose one measures the temperature of a body, which is approximately at 50°C, using these two thermometers will the readings be identical? Explain the reason for the discrepancy.

*Solution* :

Though both the thermometers are calibrated using the same two fixed points, the values of temperature as read by both these thermometers may not be identical at any other temperature. Assuming that the variation of the length of the liquid column is linear with temperature the scales are marked on both the thermometers. It is not necessary that the variation of the volume (or the length of the liquid column) of both liquids to be linear.

**Example 2.20**

What is the motivation is devising an ideal gas temperature scale ?

*Solution* :

It is known that all materials do not change their thermometric properties linearly with temperature. Therefore, different thermometers, made of different thermometric materials, show different values of the temperature of a body except at the points at which they are calibrated. Hence to have reliable and reproducible values of temperatures it is necessary to devise a temperature scale which does not depend on the nature of the thermometric material. This was the basis for the establishment of an ideal gas temperature scale.

**Example 2.21**

What is an ideal gas ?

*Solution* :

An ideal gas obeys the relation $Pv = RT$, where $v$ is the molar volume and $R$ is the universal gas constant. The value of $R$ is 8.314 J/mol K. This equation is only an approximation to the actual behavior of gases. All gases behave like ideal gas at sufficiently low pressure (in the limit $P \to 0$). The internal energy of an ideal gas is a function of temperature only.

**Example 2.22**

Describe a constant volume gas thermometer and explain how it can be used to measure the temperature of a body.

*Solution* :

A schematic diagram of a constant volume thermometer is shown in Fig.E 2.22 (*a*). It consists of a constant volume gas bulb which is connected to a mercury column through a mercury reservoir and flexible tube.

Helium or some other gas is filled in the gas bulb and the limb AB, which has the mercury column is raised or lowered such that the mercury in the reservoir stands at the reference mark *R*. Then the volume of the gas held in the bulb corresponds to the volume of gas bulb and the volume of the connecting limb up to the reference mark *R*. The gas bulb is immersed in the fluid, the temperature of which is to be measured, such that the entire gas bulb is in thermal contact with the fluid. Once thermal equilibrium is established between the fluid and the gas, the limb AB (mercury column) is raised or lowered such that the mercury in the reservoir stands at the reference mark *R*. Thus the gas is held at constant volume. Then the pressure (*P*) of the gas is measured from the height of the mercury column, above the reference mark *R* in the limb AB. A similar measurement is made by immersing the gas bulb in a bath which is maintained at the triple point of water.

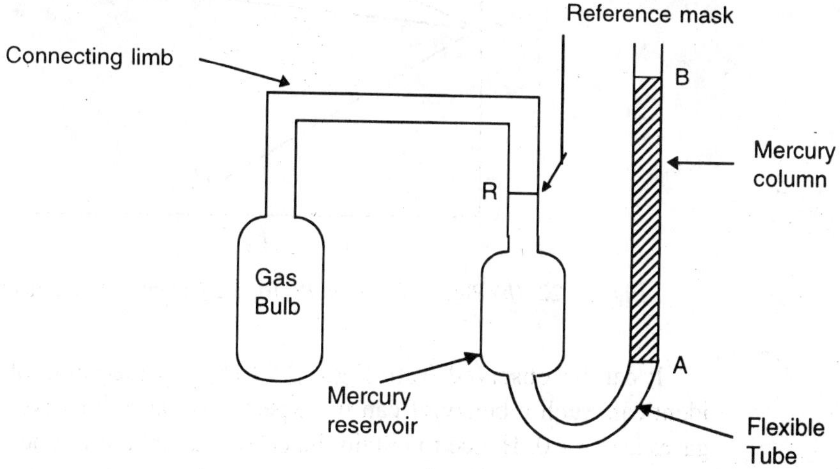


Fig.E 2.22 (*a*). Schematic diagram of a constant volume gas thermometer.

Let $P_{tp}$ denote the pressure of the gas when the gas bulb is in thermal equilibrium with the bath at the triple point temperature $T_{tp}$.

For an ideal gas at constant volume $P \propto T$. Therefore,

$$\frac{T}{T_{tp}} = \frac{P}{P_{tp}} \quad \text{or} \quad T = T_{tp}\frac{P}{P_{tp}} \qquad (A)$$

From a knowledge of $T_{tp}$, one can estimate the temperature of the fluid from the above relation. The triple point temperature of water has been assigned a value of 273.16 K.

Therefore, Eqn (*A*) can be rewritten as

$$T = 273.16\frac{P}{P_{tp}} \qquad (B)$$

Suppose a series of such measurements are made with different quantities and different types of gases in the constant volume gas bulb. Then the measured pressure (*P*) at the system temperature (*T*) as well as the pressures ($P_{tp}$) at the triple point temperature change depending on the quantity of gas in the gas bulb. A plot of *T* calculated from Eqn. (*B*) as a function of $P_{tp}$ appears as shown in Fig.E 2.22 (*b*).

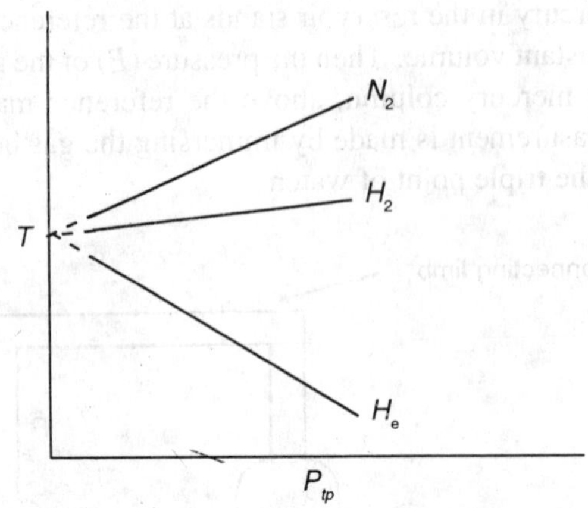


Fig.E 2.22. (*b*) Plot of $T$ versus $P_{tp}$ for a constant volume gas thermometer.

It can be observed from Fig.E 2.22 (*b*) that the value of $T$ as $P_{tp} \rightarrow 0$ is identical. Such a behavior can be expected because all gases behave like ideal gases as $P \rightarrow 0$. Hence to obtain the correct temperature one should ensure that $P_{tp}$ is as low as possible. Therefore to obtain the actual temperature of the system, Eqn. (*B*) is modified as

$$T = 273.16 \lim_{P_{tp} \to 0} \left( \frac{P}{P_{tp}} \right)$$

**Example 2.23**

It is known that the variation of electrical resistance of a wire with temperature is not linear. Is it possible to use the electrical resistance as a thermometric property ?

*Solution* :

If the variation of a thermometric property is not linear with temperature the functional dependence of the property on the temperature should be known for it to be used as a thermometric property.

**Example 2.24**

Suppose one devises a constant pressure ideal gas thermometer for the measurement of temperature. If the measured volumes of the gas at temperature $T$ and at the triple point of water are $V$ and $V_{tp}$ respectively, determine the relation to estimate $T$.

*Solution* :

For an ideal gas $PV = nRT$

where $V$ is the total volume, $n$ is the mole number of the gas. At constant pressure, the above relation gives $T \propto V$.

Therefore, $\frac{T}{T_{tp}} = \frac{V}{V_{tp}}$ or

$$T = T_{tp} \frac{V}{V_{tp}} = 273.16 \frac{V}{V_{tp}}$$

The above relation will be exact only when the gas behaves like an ideal gas or in the limit as $P_{tp} \to 0$. Hence, one can estimate the temperature $T$ from the relation

$$T = 273.16 \underset{P_{tp} \to 0}{Lt} \left( \frac{V}{V_{tp}} \right)$$

**Example 2.25**

What are the practical difficulties associated with the use of ideal gas thermometer?

*Solution* :

The ideal gas thermometer can be used at temperatures where the behavior of the gases tend to the behavior of ideal gas. It is known that at low temperatures the behavior of gases deviate from the ideal gas behavior. More over, the lowest temperature (4.2 K) can be achieved by using helium gas at its normal boiling point. This sets a limit on the lowest temperature that can be measured. In addition to this great skill and experience are required to operate an ideal gas thermometer since it is necessary to apply several corrections to obtain accurate values of temperature.

**Example 2.26**

What is IPTS-68 and name some of the primary fixed points on which it is based.

*Solution* :

In view of the practical difficulties associated with the use of ideal gas thermometer, the international committee on weights and measures have adopted the International Practical Temperature Scale (IPTS-68) in 1968 based on a number of easily reproducible fixed points. Some of the primary fixed points on which IPTS-68 is based are given below :

| **Primary fixed point** | **Temperature (°C)** |
|---|---|
| Triple point of hydrogen | – 259.34 |
| Normal boiling point of hydrogen | – 252.87 |
| Triple point of oxygen | – 218.789 |
| Normal boiling point of oxygen | – 182.962 |
| Triple point of water | 0.01 |
| Normal freezing point of silver | 961.93 |
| Normal freezing point of gold | 1064.43 |

# 3

# Properties of Simple Compressible Fluids

### Example 3.1

What is meant by a pure substance ? Can we treat air as a pure substance ?

*Solution* :

In general a pure substance is one which consists of a single chemical species like $H_2O$, He, $N_2$, $O_2$, etc. Quite often in engineering analysis one deals with systems of fixed composition like air. Air is a mixture of 79 % nitrogen and 21% oxygen (by volume). If a certain quantity of air, enclosed in a piston-cylinder assembly, undergoes cooling/ heating or compression/expansion processes, its composition does not undergo any change. In such cases it can be treated as a pure substance. However, if the air is cooled to sufficiently low temperatures where it can be present in both liquid and vapor phases, then it cannot be considered as a pure substance, because the composition of the liquid will be quite different from that of the vapor.

### Example 3.2

Explain the meaning of the term 'phase'

*Solution* :

A system which is uniform throughout both in chemical composition and physical state, is called a phase. A pure substance can exist in three different states of physical aggregation. For example, water can be found as solid (ice), liquid and gas (water, vapor or steam). Solids can be found in different allotropic forms. For example, sulfur can be found in rhombic form (lemon yellow in color) and in monoclinic form (orange in color). The properties of these allotropic forms differ from each other. Hence solid sulfur can be found in two different phases. Similarly solid carbon can exist in two different phases, namely- graphite and diamond.

**Example 3.3**

It is known that pure substances can exists in different phase. Is it possible to find mixtures also in different phases ? Give one example.

*Solution* :

Mixtures also can be found in different phases. A mixture of alcohol and water can exist in liquid as well as vapor phases. However, the chemical composition of the vapor phase is usually different from that of the liquid phase.

**Example 3.4**

What is meant by compressible substance ?

*Solution* :

All substances– solids, liquids and gases–change their specific /molar volume when they are subjected to a change in temperature and pressure. Depending on the range of the temperature, applied pressure and on the nature of the substance the change in the volume may be significant or insignificant. If the change in the volume is negligible the substance is called incompressible. Usually liquids and solids are treated as incompressible substances. On the other hand, if the change in the volume is quite significant the substance is called compressible.

**Example 3.5**

Define isothermal compressibility ($\kappa$), isentropic compressibility ($\kappa_s$), and coefficient of volume expansion. ($\beta$).

*Solution* :

The isothermal compressibility ($\kappa$) is defined as the rate of decrease of volume with pressure at constant temperature and per unit volume. That is.

$$\kappa = -\frac{1}{V}\left(\frac{\partial V}{\partial P}\right)_T$$

The isentropic compressibility ($\kappa_s$) is defined as the rate of decrease of volume, per unit volume, with pressure at constant entropy. That is

$$\kappa_s = -\frac{1}{V}\left(\frac{\partial V}{\partial P}\right)_s$$

The coeffecient of volume expansion ($\beta$) is defined as the rate of change of volume, per unit volume, with temperature at constant pressure. That is

$$\beta = \frac{1}{V}\left(\frac{\partial V}{\partial T}\right)_P$$

### Example 3.6

Explain the meanings of the terms 'simple substance' 'and simple compressible substance'.

*Solution* :

A substance which can exchange energy through only work mode is called a simple substance. If the work mode through which a simple substance can exchange energy is the boundary movement (or compression/expansion) then it is called a simple compressible substance. In other words, a simple compressible substance can exchange energy with its surroundings though $Pdv$ work mode in addition to the energy exchange through heat interaction.

### Example 3.7

Describe the commonly employed methods by which the thermodynamic data of substances are presented.

*Solution* :

The thermodynamic properties which are in wide use are pressure ($P$), molar/specific volume ($v$), temperature ($T$), enthalpy ($h$) and entropy ($s$) .[ The terms enthalpy and entropy are defined later]. The widely used and most convenient methods of representing the thermodynamic data are:

(*a*) Graphical representation or presenting the data in the form of diagrams or charts.

(*b*) Tabular representation or presenting the data in the form of tables.

(*c*) Mathematical representation or presenting the data in the form of equations.

### Example 3.8

Name the widely used thermodynamic diagrams for a pure substance.

*Solution* :

The most widely used thermodynamic diagrams for a pure substance are:

Pressure versus volume ($P - v$) diagram.

Pressure versus temperature ( $P - T$) diagram.

Temperature versus volume ($T - v$) diagram.

Temperature versus entropy ($T$ – s) diagram.

Enthalpy versus entropy ($h - s$) diagram. or Mollier diagram.

Pressure versus enthalpy. ($P - h$) diagram.

### Example 3.9

Explain the terms – 'saturated liquid' and 'saturated vapor' :

*Solution* :

A liquid which is in equilibrium with its own vapor at a specified temperature / pressure is called a saturated liquid. That is a liquid at its boiling point at specified pressure is a saturated liquid. A vapor which is in equlibrium with its own liquid at a specified temperature / pressure is called saturated vapor. That is a vapor at its condensing temperature at specified pressure is a saturated vapor. It is known that at standard atmospheric pressure water boils at 100°C generating its own vapor. Hence, at standard atmospheric pressure, water at 100°C is a saturated liquid. The steam or water vapor which is in equilibrium with boiling water at 100°C (at standard atmospheric pressure) is at its condensing temperature and hence it is saturated vapor.

### Example 3.10

How many thermodynamic degrees of freedom are there for a pure substance existing as (*a*) saturated liquid and (*b*) saturated vapor ?

*Solution* :

(*a*) According to the Gibbs phase rule, the thermodynamic degrees of freedom ($F$) are given by

$$F = C + 2 - P$$

Where $C$ is the number of components and $P$ is the number of phases. For a pure substance $C = 1$. If the substance is existing as saturated liquid, it is in equilibrium with its own vapor ( the quantity of vapor may be negligible). Hence $P = 2$. Application of the phase rule gives $F = C + 2 - P = 1 + 2 - 2 = 1$. Hence only one intensive property can be specified for the complete description of the state of the system. Therefore, one can specify either pressure ($P$) or temperature ($T$). For example, the system is saturated liquid at standard atmosphere pressure ($P = 101.325$ kPa). Then the temperature at which the saturated liquid is existing is 100°C. On the other hand if the temperature of the saturated liquid is specified as 100°C then the system pressure is one standard atmosphere.

(*b*)Saturated vapor implies that it is in equilibrium with its own liquid. Hence $C = 1$ and $P = 2$. Therefore, application of the phase rule gives $F = C + 2 - P = 1 + 2 - 2 = 1$. Hence, one can specify either temperature ($T$) or pressure ($P$).

### Example 3.11

Explain the terms 'saturation temperature' and 'saturation pressure'.

*Solution* :

The temperature at which a liquid boils (or in equilibrium with its own vapor) at a specified pressure is called saturation temperature. The pressure, at a specified temperature, at which the liquid boils (or in equilibrium with its own vapor) is called saturation pressure. We know that at standard atmospheric pressure ($P$ = 101.325 kPa), water boils at 100°C. Therefore, at $P$ = 101.325 kPa, the saturation temperature of water is 100°C. At 100°C, the saturation presure of water is 101.325 kPa. Sometimes, the saturation presure is called as vapor pressure.

### Example 3.12

What is quality or dryness fraction ?

*Solution* :

A mixture of saturated liquid and saturated vapor is described by specifying its quality ($X$) or dryness faction . The quality is defined as the ratio of the mass of saturated vapor to the total mass of the mixture. That is,

$$\text{quality} = \frac{\text{mass of saturated vapor}}{\text{mass of mixture}}$$

or

$$X = \frac{m_g}{m} = \frac{m_g}{m_f + m_g}$$

where $m$ = mass of mixture

$m_f$ = mass of saturated liquid

$m_g$ = mass of saturated vapor

### Example 3.13

Express the specific / molar volume of a mixture in terms of the quality of the mixture.

*Solution* :

Since volume is an extensive property, it is additive. That is, the total volume of a mixture is the sum of the volumes of saturated liquid and saturated vapor. Let $v_f$ and $v_g$ denote the specific / molar volume of saturated liquid and saturated vapor, respectively. Then the total volume ($V$) of a mixture of mass ($m$) is given by

$$V = mv = m_f v_f + m_g v_g \qquad (A)$$

or

$$v = \frac{m_f v_f}{m} + \frac{m_g v_g}{m} = (1 - X)\, v_f + X v_g$$

or $$v = v_f + X(v_g - v_f) = v_f + Xv_{fg}$$

Where $v$ = specific volume of mixture

$v_{fg} = v_g - v_f$ = Change in specific volume due to vaporization.

Eqn. (*A*) can be rewritten as

$$mv = (m_f + m_g)\, v = m_f v_f + m_g v_g$$

or $$\frac{m_f}{m_g} = \frac{v_g - v}{v - v_f} \qquad (B)$$

Eqn. (*B*) is commonly known as the lever rule. Similar relations can be used to describe any extensive property (like internal energy *u*, enthalpy *h*, entropy *s*) of a mixture in terms of its quality, *X*.

**Example 3.14**

What is triple point ? For a pure substance how many degrees of freedom are there at the triple point?

*Solution* :

The state at which all the three phases – solid, liquid and vapor coexist in equilibrium is called the triple point. At the triple point , $C = 1$ ; $P = 3$ and application of the phase rule gives $F = C + 2 - P = 1 + 2 - 3 = 0$. Hence, the number of degrees of freedom at the triple point is equal to zero. In other words the triple point is invariant ($F = 0$). One cannot arbitrarily assign either temperature or pressure for the triple point. The triple point exists at a definite pressure and temperature. For example, the triple point of water is $P$ = 0.611kPa and $t$ = 0.01°C.

**Example 3.15**

What is critical point ?

*Solution* :

Critical point represents the highest temperature and pressure above which a liquid cannot exist. At the critical point the liquid and vapor are indistinguishable. That is, the properties of saturated liquid and saturated vapor are identical at the critical point, The temprature and pressure at the critical point are called critical temperature ($T_c$) and critical pressure ($P_c$). If a vapor or gas is to be condensed (or liquified) at constant temperature by the application of pressure, its temperature must be below the critical temperature. Similarly, if a gas is to be liquified by cooling at constant pressure its pressure must be below the critical pressure.

### Example 3.16

Suppose energy is continuously transferred as heat at constant pressure to a solid till it completely converts into vapor. Represent the path followed by the system (solid) on a temperature versus volume diagram.

*Solution* :

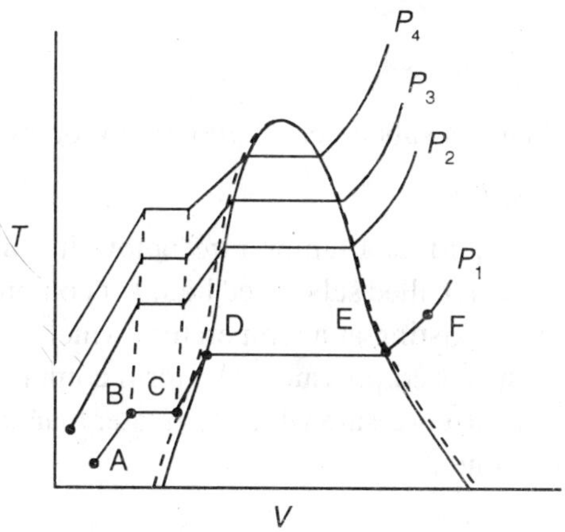


Fig.E 3.16. Temperature–volume diagram at four different pressures $P_4>P_3>P_2>P_1$.

If energy is transferred as heat at constant pressure $P_1$ to a solid at state A, its temperature and specific volume increase and follows the path A B as shown in Fig.E 3.16. At state B, the solid starts melting while the temperature remains constant, till it reaches state C where it completely exists as liquid. The segment BC represents the fusion process. On further heating, the specific volume and temperature of the liquid increase and the system follows and path CD. At state D the liquid starts vaporizing. With continued heating, more and more of liquid vaporizes while its temperature remains constant and the path followed by the system is represented by DE in Fig.E 3.16. The segent DE represents the vaporization process. At state D, the liquid is available as saturated liquid. At state E all the liquid is completely vaporized. The vapor at state E is called saturated vapor. On further heating the vapor follows the path E F. If similar experiments are made at higher pressures $P_2$ *to* $P_4$ $(P_4 > P_3 > P_2 > P_1)$ and the results are represented on the same diagram, they will appear as shown in Fig.E 3.16.

### Example 3.17

What is meant by superheated vapor ?

*Solution* :

A vapor existing at a temperature above the saturation temperature at a specified pressure is called superheated vapor. For example steam at 101.325 kPa and 150°C is a superheated vapor, since at the specified pressure $P = 101.325$ kPa the saturation temperature of water is 100°C and the vapor is available at 150°C which is above the saturation temperature.

### Example 3.18

What is subcooled liquid or compressed liquid ?

*Solution* :

A liquid at a temprature below its saturation temperature, at a specified pressure, is called subcooled liquid or compressed liquid. Subcooled liquid implies that it is existing at a temperature which is less than the saturation temperature, at the specified pressure. The term compressed liquid implies that the liquid is existing at a pressure which is greater than the saturation pressure, at the specified temperature.

### Example 3.19

Sketch the pressure–volume diagram for a pure substance and indicate the various regions on it.

*Solution* :

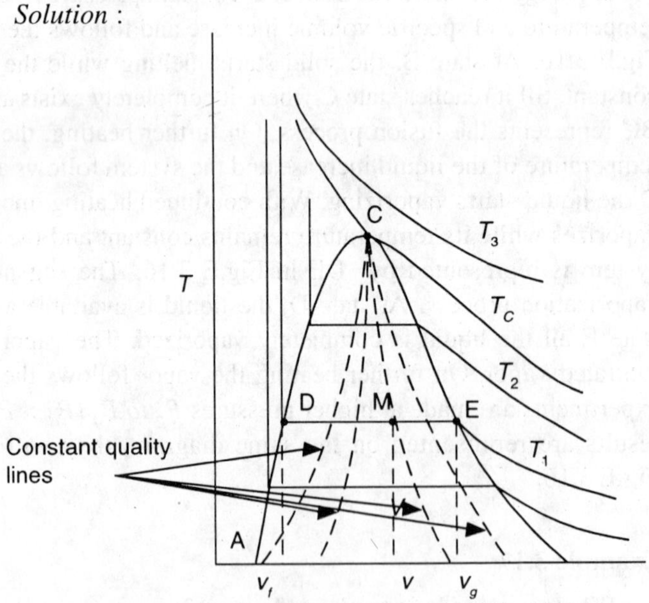


Fig.E 3.19. Pressure versus volume diagram for a pure substance.

The curve AB represents the saturated liquid. Saturated vapor is represented by the curve BC. The point C denotes the critical point. The isotherms are shown by the curves $T_1$ to $T_3$ ($T_3 > T_c > T_2 > T_1$). The regions denoted by $L$ and $V$ correspond to subcooled liquid and superheated vapor, respectively. The region enclosed by the dome ACB denotes the two phase region where mixtures of saturated liquid and saturated vapor are present. The constant quality lines are indicated by dashed lines and these constant quality lines merge at the critical point. The points D and E represent the saturated liquid and saturated vapor which are in equilibrium with each other. The point M, lying on the line segment DE, in the two phase region, denotes a mixture of quality $X$.

**Example 3.20**

Suppose a constant volume vessel is initially filled with a small quantity of saturated liquid and the remaining volume is filled with saturated vapor at a given temperature and the vessel is heated. Trace the path followed by the system (material contained in the vessel) on a $P$–$v$ diagram. Show the path on the same $P$–$v$ diagram had the vessel been filled with a large quantity of saturated liquid and the remaining volume with saturated vapor.

*Solution* :

The volume of the vessel as well as the mass of material in the vessel remain constant during heating. Hence, the specific volume of the material remains constant. In other words, the system follows a constant volume process. If the vessel is filled with a small quantity of saturated liquid, the quality ($X$) of the initial mixture is large. That is the initial state is close to the saturated vapor curve as shown by A in Fig.E 3.20. As energy is transferred in the form of heat to the system, its temeperature and pressure increase at constant volume. The path followed by the system is as shown by ABCD. Finally the system ends up in the superheated vapor state 1 (point D).

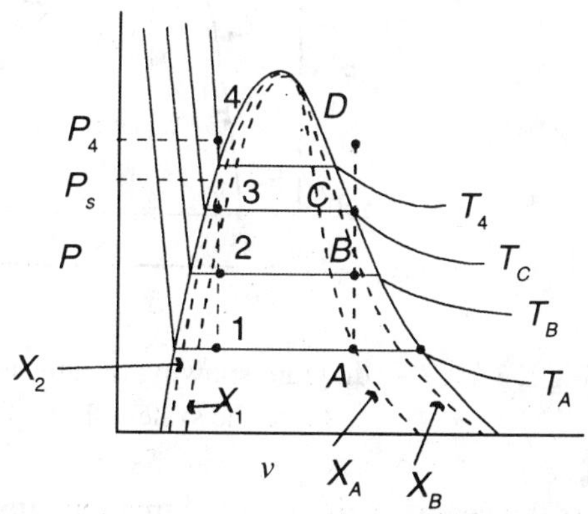


Fig.E 3.20. $P-v$ diagram showing constant volume heating process.

At state B, the quality of the mixture is $X_B$ which is greater than $X_A$. At state C, the substance is saturated vapor. That is during heating, the liquid vaporizes increasing its quality from $X_A$ to $X_B$ and at state C, $X_c = 1$. With further heating the system reaches state D.

If the vessel is initially filled with a large quantity of saturated liquid, the quality of the mixture is low ($X_1$) and the system is in state 1. During the constant volume heating, the system follows the path 1234 as shown in Fig.E 3.20. With continued transfer of energy as heat at constant volume, the liquid expands, its quality decreases ($X_2 < X_1$) and occupies the entire vessel at state 3 where it is available as saturated liquid ($X_3 = 0$). With further heating, the system reaches state 4, where it is available as subcooled liquid or compressed liquid. The isotherm passing through state 4 is shown as $T_4$. The saturation pressure corresponding to $T_4$ is shown as $P_s$. The system is at pressure $P_4$ which is greater than $P_s$ and hence the system at state 4 exists as a compressed liquid.

**Example 3.21**

Suppose a closed and rigid vessel is initially filled with a mixture of saturated liquid water and saturated water vapor at 100 kPa. On transferring energy as heat, the water is found to pass through the critical point. Determine the ratio of the volume of saturated vapor to the volume of the saturated liquid with which the vessel is initially filled. At 100 kPa, $v_f = 0.001\ 043\ 4\ m^3/kg$ and $v_g = 1.694\ m^3/kg$. The critical volume of water is $0.003\ 17\ m^3/kg$.

*Solution* :

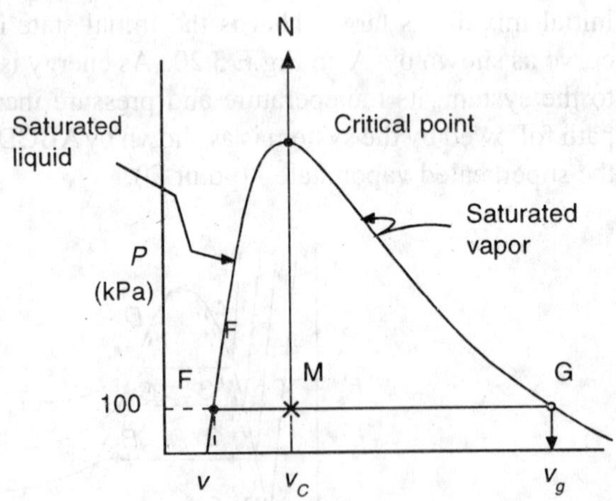


Fig.E 3.21. *P–v* diagram showing the constant volume heating process which takes the system through critical point.

As the vessel is rigid, its volume remains constant. Since the vessel is closed and the total mass of mixture remains constant the system follows a constant volume process. It is given that the system passes through the critical state on heating. Therefore, the specific volume at the critical point is equal to the specific volume of the mixture. That is

$$v_c = v_M = Xv_g + (1 - X)\, v_f \qquad (A)$$

Where $v_M$ is the specific volume of the mixture. The path follwed by the system is shown as M N on a $P$–$v$ diagram in Fig.E 3.21. The point M denotes the initial mixture at 100 kPa and the points F and G denote the saturated liquid and saturated vapor, respectively at 100 kPa. Substituting the given values in Eqn. ($A$), we get

$$0.003\,17 = 1.694\,X + (1 - X)\,0.001\,043\,4$$

or

$$X = \frac{0.003\,17 - 0.001\,043\,4}{1.694 - 0.001\,043\,4} = 1.256 \times 10^{-3}$$

$$\frac{\text{Volume of saturated vapor}}{\text{Volume of saturated liquid}} = \frac{m_g v_g}{m_f v_f} = \frac{(m_g / m)v_g}{(m_f / m)v_f} = \frac{Xv_g}{(1 - X)v_f}$$

$$= \frac{1.256 \times 10^{-3} \times 1.694}{(1 - 1.256 \times 10^{-3})0.001\,043\,4} = 2.04$$

**Example 3.22**

The pressure–volume–temperature relationship for a substance can be conveniently shown on a $P$ –$v$– $T$ diagram. Sketch the $P$– $v$– $T$ diagram for a substance which contracts on freezing.

*Solution* :

Fig.E 3.22 shows the $P$ – $v$– $T$ diagram for a substance which contracts on freezing. Many substances follow the trend shown in the figure. However, there are some exceptions like water which expands on freezing.

The solid, liquid and vapor phases appear as surfaces. These single phase regions (marked as S, L and V) are separated from the two phase regions (S + L), (S + V) and (L + V) by surfaces which represent a two phase mixture. For example, the solid and liquid regions are separated by a solid–liquid surface. Similarly, the liquid and vapor regions are separated by a liquid–vapor surface. Any state denoted by a point lying on a line separating the single phase region from a two phase region is called saturated. It can be observed from Fig.E 3.22 that the isotherm shows a discontinuity at the saturated state . For example the isotherm 1234 is discontinuous at 2 and 3. The liquid existing at state 2 is called saturated liquid while the vapor at state 3 is called saturated vapor.

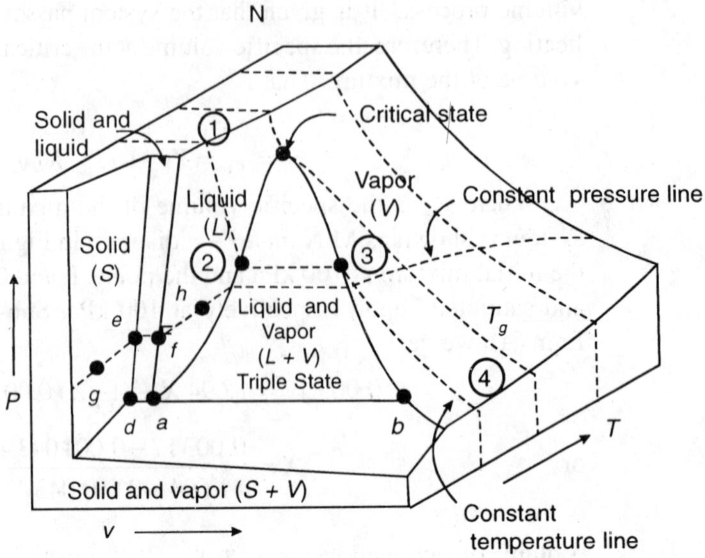


Fig.E 3.22. $P-v-T$ diagram for a substance which contracts on freezing.

### Example 3.23

Sketch a phase ($P$ versus $T$) diagram for water and indicate the solid, liquid and vapor regions on it. Explain how this diagram differs from the phase diagram of other substances.

*Solution* :

The phase diagram for water is shown in Fig.E 3.23 in which the regions of solid, liquid and vapor are indicated. The single phase regions are separated from each other by saturation lines. For example, the solid and liquid regions are separated by the fusion curve $T2$. Along the fusion curve both solid and liquid phases coexist in equilibrum. It can be observed that the slope of the fusion curve is negative.

That is,

$$\left(\frac{\partial P}{\partial T}\right) < 0$$

which shows that the melting point of ice decreases with increasing pressure. The sublimation curve 1T separates the solid region from the vapor region, while the vaporization curve $TC$ separates the liquid region from the vapor region. Along the sublimation curve the solid and vapor phases coexist in equilibrium while the liquid and vapor phases coexist in equilibirium along the vaporization curve. The phase diagram shown in Fig.E 3.23 represents the phase behavior of substances (like water, bismuth and antimory) which contract on melting, that is, the specific volume of the liquid is smaller than the specific volume of the solid. The state T represents the triple point where all the three phases–solid, liquid and vapor– coexist in equilibrium. The state C represents the critical point where the properties of liquid and vapor are identical and they cannot be distinguished from one another. For all other substances the slope of the fusion curve will be positive (as shown by T 2′ on the diagram) which indicates that the melting point increases with increasing pressure.

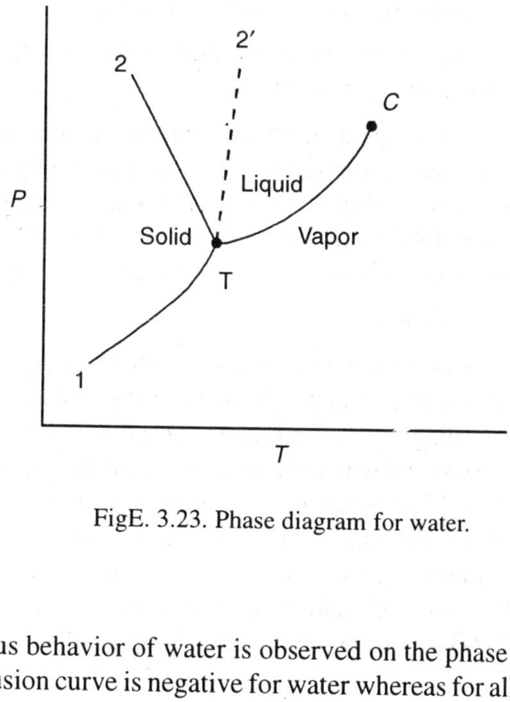


FigE. 3.23. Phase diagram for water.

**Example 3.24**

An anomolous behavior of water is observed on the phase diagram. That is, the slope of the fusion curve is negative for water whereas for all other substances the fusion curve has a positive slope. Give two examples where this anomolous behavior is found to be advantageous.

*Solution.* :

When water freezes into ice, its specific volume increases or density decreases. Hence ice floats on water. In cold countries the water in the lakes and sea starts freezing and the ice thus formed floats on water. The floating ice acts as an insulating material and does not expose the water at the bottom layers to the cold atmosphere thus preventing further freezing of water at the bottom layers. Therefore, the acquatic life can survive in the water. The melting point of ice decreases with increasing pressure. This phenomenon is taken to advantage by ice skaters. In the cold countries as well as in some parts of north India, like Simla, Jammu and Kashmir the water in the lakes freeze into ice at temperature less than 0°C. At atmospheric pressure, ice does not melt below 0°C. The ice skates are usually tapered at the bottom and under normal conditions of use, the area of contact between the skates and ice is very low. When a person stands on the skates, he/she exerts enormous pressure on the ice. Then the ice below the skates starts melting because of the applied pressure. The melting ice provides a thin layer of water between the ice surface and skates which acts as a lubricant which is turn helps in smooth gliding.

## Example 3.25

Suppose a closed and rigid vessel is initially filled with a

(*a*) large quantity of saturated liquid water and the remaining volume is filled with saturated water vapor.

(*b*) small quantity of saturated liquid water and the remaining volume is filled with saturated water vapor and energy is transferred as heat to the vessel. Since the volume of the vessel and the quantity of water remain constant, the water follows a constant volume process. Sketch the constant volume paths followed by water, in both the above cases, on a phase diagram.

*Solution* :

(*a*) If the vessel is filled with a large quantity of saturated liquid water and heated at constant volume the path followed by water is shown as 1234 on a *P–v* diagram in Fig.E 3.20. At states 1 and 2, both the liquid and vapor phases coexist and hence these states lie on the vaporization curve on a phase diagram. State 3 represents saturated liquid and hence this state also lies on the vaporization curve. The state 4 represents liquid and hence it lies in the liquid zone. Since the compressibility of liquid is very low, the rise in pressure with temperature will be very high and hence the slope of 3-4 is large. The constant volume path 1-2-3-4 is shown in Fig.E 3.25 on a phase diagram.

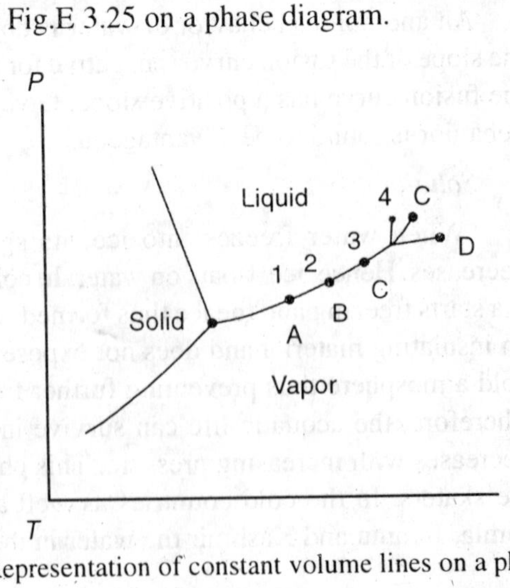


Fig.E 3.25. Representation of constant volume lines on a phase diagram.

(*b*) When the vessel is filled with a small quantity of saturated liquid water and heated at constant volume, the path followed by water is shown as ABCD on a $P$–$v$ diagram in Fig.E 3.20. The states A, B, and C, lie on the vaporization curve while the state D lies in the vapor region. The constant volume path followed by water is shown as ABCD on a phase diagram in Fig.E 3.25.

**Example 3.26**

Represent the following processes on a phase diagram.

(*a*) Constant pressure heating of liquid water to produce superheated vapor.

(*b*) Isothermal vaporization of liquid water to produce superheated steam.

(*c*) Liquid water is converted to superheated steam without observing the phase transition. That is the vaporization of liquid cannot be detected during phase change.

*Solution* :

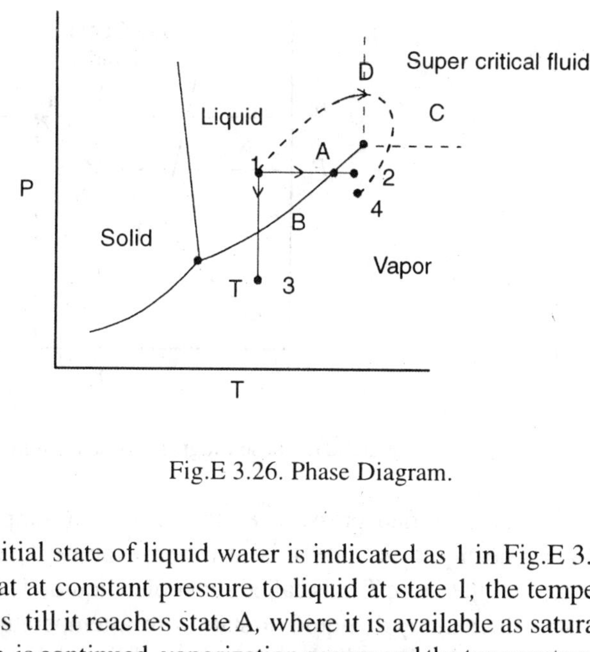


Fig.E 3.26. Phase Diagram.

(*a*) The initial state of liquid water is indicated as 1 in Fig.E 3.26. If energy is added as heat at constant pressure to liquid at state 1, the temperature of the liquid increases till it reaches state A, where it is available as saturated liquid. If energy addition is continued, vaporization occurs and the temperature and pressure remain constant till all the liquid is converted into saturated steam. The state A represents both the saturated liquid and saturated vapor. With continued addition of energy water reaches state 2, which is superheated vapor.

(*b*) If liquid water at state 1 is maintained at constant temperature and the pressure is reduced, it follows the path 1B3. State B represents saturated liquid as well as saturated vapor and state 3 denotes superheated vapor.

*(c)* If liquid water at state 1 is subjected to a process path indicated by 1D4 as shown in Fig.E 3.26, it is not possible to observe the phase change at any stage during the course of the process. The process path passes through the supercritical fluid region. It is always possible to convert a liquid into vapor or vice -versa without observing phase transition, by taking the fluid through the supercritical fluid region.

**Example 3.27**

Sketch the phase diagram for a substance that expands on freezing and indicate the regions of subcooled liquid and superheated vapor.

*Solution* :

The phase diagram for a substance like water which expands on freezing in shown in Fig.E 3.27. The liquid existing at state 1 in Fig.E 3.27 is called subcooled liquid. The liquid at state 1 is characterized by pressure $P_1$ and temperature $T_1$. At

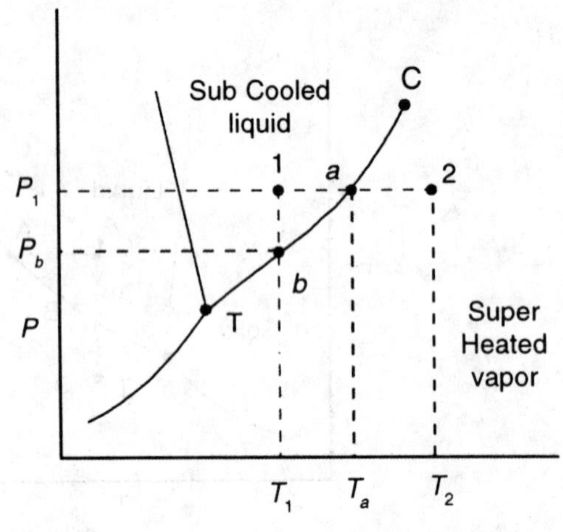


Fig.E 3.27. Phase diagram for a substance that expands on freezing.

the specified pressure $P_1$ the saturation temperature of the liquid is $T_a$ .Since the liquid at state 1 is at a temperature $T_1$ which is less than the saturation temperature $T_a$ it is called subcooled liquid. This state can be achieved by cooling the saturated liquid at state *a* to a temperature $T_1 < T_a$. The liquid at state 1 can also be called as compressed liquid since at the specified temperature $T_1$ the liquid exists at a presure $P_l$ which is greater than the saturation pressure $P_b$. This state can be achieved by compressing the saturated liquid at state *b* to a pressure $P_1 > P_b$. Similarly, the vapor at state 2 is at a temperature $T_2$ which is greater than the saturation temperature $T_a$ at the specified pressure $P_1$. Hence the vapor at state 2 is called super heated vapor. Usually the term superheating is used if the temperature of a vapor is increased at constant pressure. For a pure substance exsisting in two phases liquid and vapor–the saturation pressure is usually referred as vapor pressure. The temperature at which a given liquid boils at standard atmospheric pressure (101.325 kPa) is called the normal boiling point.

**Example 3.28**

Name a few drawbacks associated with the graphical representation of thermodynamic data.

*Solution.* :

The drawbacks associated with the graphical representation of thermodynamic data are:

1. The values of the properties cannot be read accurately.
2. Sometimes it is necessary to interpolate the data leading to loss of accuracy, since the data is usually presented at some selected values of the independent variables only.
3. Quite often it is necessary to have a knowledge of the partial derivatives of the thermodynamic properties and such information is not readily available on thermodynamic diagrams. The estimation of the partial derivatives of thermodynamic properties from diagrams is quite difficult and invariably the estimated values can be used as approximate values only and are not accurate.
4. The number of properties that can be presented on a diagrams is limited. Any attempt to increase the number of properties renders the diagram unreadable and leads to confusion.

**Example 3.29**

Suggest an alternative method of representing the thermodynamic data, which is likely to overcome some of the drawbacks associated with the graphical representation.

OR

List the advantages of representing the thermodynamic data in a tabular form compared to the graphical representation.

*Solution* :

An alternative method of representing the thermodynamic data of pure substances is to present the data in tabular form. The tabular representation of data overcomes some of the drawbacks associated with the graphical representation of the same data. The advantages associated with the tabular representation of data are given below .

1. The values can be read accurately.
2. The data can be listed in fine increments of the independent variables.
3. There is no limitation on the number of properties that can be presented in a tabular form.

**Example 3.30**

Describe the most convenient and widely used format for the tabular represeniation of thermodynamic data of saturated steam.

*Solution* :

The important thermodynamic properties like $P$, $v$, $T$, $u$, $h$ and $s$ for saturated liquid and saturated vapor are presented in a tabular form. Such tables are available for water, ammonia, sulfur dioxide, carbon dioxide, freons etc. which are commonly used as working fluids in heat engines and refrigerators. In general, the format of all these tables is similar. Hence, it is enough if a representative table for one fluid is described. Since steam is extensively used in power plants and in several industries and an exact equation of state which is applicable over wide ranges of temperature and pressure is not available, the steam tables are described below.

If a single component simple system exists in two phases, say liquid and vapor, then the system (the substance under consideration) has only one degree of freedom. That is, only one independent thermodynamic property ( either $P$ or $T$), can be specified for the complete specifiction of the state of the system. If the temperature $T$ is chosen as the independent variable , then the system has definite values for all other thermodynamic properties. The most convenient and widely used format of the saturated steam (Temperature) table is as given in table for Example 3.30 (page 93).

**Example 3.31**

What is the reference state chosen for tabulating the thermodynamic properties of water in the steam tables ?

*Solution* :

The selection of the reference state for the purpose of tabulating the thermodynamic properties of a substance is quite arbitrary and it is governed by convenience only and not by any law. The selected reference states differ from substance to substance. By convention, the saturated liquid water at its triple point (0.01°C and 0.6113 kPa) is chosen as the reference state while tabulating the thermodynamic properties of steam. The specific (molar) internal energy ($u$)

## Table for Example 3.30.

| Temperature $t$(°C) | Pressure $P$ (kPa) | Specific volume ($m^3$/kg) | | Specific enthalpy (kJ/kg) | | Specific entropy (kJ/kg K) | |
|---|---|---|---|---|---|---|---|
| | | Sat. liquid $v_f$ | Sat. Vapor $v_g$ | Sat. liquid $h_f$ | Sat. vapor $h_g$ | Sat. liquid $s_f$ | Sat. vapor $s_g$ |
| 0 | 0.6108 | 0.001 000 2 | 206.3 | –0.04 | 2501.6 | –0.0002 | 9.1577 |
| 0.01 | 0.6113 | 0.001 000 2 | 206.15 | 0.000 61 | 2501.4 | 0.0000 | 9.1573 |
| – | – | – | – | – | – | – | – |
| – | – | – | – | – | – | – | – |
| 100 | 101.33 | 0.001 043 7 | 1.673 | 419.06 | 2676.0 | 1.3069 | 7.3554 |
| – | – | – | – | – | – | – | – |
| 200 | 1554.9 | 0.001 156 5 | 0.1272 | 852.37 | 2790.9 | 2.3307 | 6.4278 |
| 374.15 | 22120.0 | 0.003 17 | 0.003 17 | 2107.4 | 2107.4 | 4.4429 | 4.4429 |

## Table for Example 3.33

| Pressure $P$ (kPa) | Temperature $t$(°C) | Specific volume ($m^3$/kg) | | Specific enthalpy (kJ/kg) | | Specific entropy (kJ/kg K) | |
|---|---|---|---|---|---|---|---|
| | | Sat. liquid $v_f$ | Sat. Vapor $v_g$ | Sat. liquid $h_f$ | Sat. vapor $h_g$ | Sat. liquid $s_f$ | Sat. vapor $s_g$ |
| 0.6113 | 0.01 | 0.001 000 2 | 206.15 | 0.000 61 | 2501.4 | 0.0000 | 9.1573 |
| 1 | 6.9828 | 0.001 000 1 | 129.20 | 29.34 | 2514.4 | 0.1060 | 8.9767 |
| 5 | 32.898 | 0.001 005 2 | 28.19 | 137.77 | 2561.6 | 0.4763 | 8.3960 |
| 10 | 45.833 | 0.001 010 2 | 14.67 | 191.83 | 2584.8 | 0.6493 | 8.1511 |
| – | – | – | – | – | – | – | – |
| 100 | 99.632 | 0.001 043 4 | 1.694 | 417.54 | 2675.4 | 1.3027 | 7.3598 |
| – | – | – | – | – | – | – | – |
| 200 | 120.23 | 0.001 060 8 | 0.8854 | 504.70 | 2706.3 | 1.5301 | 7.1268 |
| – | – | – | – | – | – | – | – |
| 22120 | 374.15 | 0.003 17 | 0.003 17 | 2107.4 | 2107.4 | 4.4429 | 4.4429 |

and the specific (molar) entropy ($s$) of saturated liquid water at the triple point have been arbitrarily assigned a value of zero in the steam tables,

**Example 3.32**

State the ranges of temperature and pressure as well as the states for which the thermodynamic properties of saturated steam are tabulated in the steam tables.

*Solution* :

All the steam tables present the thermodynamic properties of saturated liquid and saturated vapor starting from the triple point (0.01°C and 0.6113 kPa) to the critical point (374.15°C and 22.12 MPa). Since the properties of water at 0°C are quite often needed in the thermodynamic calculation, the properties of saturated liquid and saturated vapor at this temperature are also tabulated in the steam tables.

**Example 3.33**

Describe the format of the saturated steam tables if one chooses to use the pressure as the independent variable for the specification of the state of the saturated liquid or saturated vapor.

*Solution* :

If the pressure $P$ is chosen as the independent variable for the description of a two phase single component system ( liquid and vapor phases coexisting in a state of equilibrium), then the system has definite values for all other thermodynamic properties. The widely used format of the saturated steam (pressure) tables is given on page 93. The data presented in the saturated steam (temperature) and saturated steam (pressure) tables is the same but the format differs from each other.

**Example 3.34**

In thermal power plants and in process industries one deals with superheated steam, that is steam at a temperature which is above the saturation temperature. Describe the most widely used format for presenting the thermodyanmic properties of superheated steam.

*Solution* :

The superheated steam exists in only one phase. Hence two independent variables-pressure ($P$) and temperature ($T$)- are to be specified for complete description of the state of the superheated steam. The most convenient and widely used format for presenting the thermodynamic properties of superheated steam is given below.

| Pressure P(bar) (Saturation temperature °C) | | Temperature $t$(°C) | | |
|---|---|---|---|---|
| | | 100 | 200 | 300 |
| 0.2 | $v$(m³/kg) | 8.585 | 10.507 | 13.219 |
| (60.09) | $h$(kJ/kg) | 2686.3 | 2879.2 | 3076.4 |
| | $s$(kJ/kg K) | 8.1261 | 8.5839 | 8.9618 |
| 1 | $v$ | 1.6960 | 2.1720 | 2.6390 |
| (99.63) | $h$ | 2676.2 | 2875.4 | 3074.5 |
| | $s$ | 7.3618 | 7.8349 | 8.2166 |
| 2 | $v$ | – | 1.0804 | 1.3162 |
| (120.23) | $h$ | – | 2870.5 | 3072.1 |
| | $s$ | – | 7.5072 | 7.8937 |
| – | – | – | – | – |
| 20 | $v$ | – | – | 0.1255 |
| (212.37) | $h$ | – | – | 3025.0 |
| | $s$ | – | – | 6.7696 |

**Example 3.35**

Freon–12 (dichloro difluoro methane $CCl_2F_2$ ) and ammonia are used as refrigerants. Specify the reference state which is normally adopted for tabulating the thermodynamic properties of these substances.

*Solution* :

By convention the saturated liquid at–40°C is chosen as the reference state for tabulating the thermodynamic properties of refrigerants. The internal energy ($u_f$) and entropy ($s_f$) of saturated liquid at –40°C have been arbitrarily assigned a value of zero.

**Example 3.36**

Does the arbitrary selection of the reference state for tabulating the thermodynamic properties of substance affect the thermodynamic analysis of processes ?

*Solution* :

The choice of reference state does not affect the thermodynamic analysis of processes since in thermodynamics one is concerned with the changes in the thermodynamic properties of the substances while they undergo a process.

**Example 3.37**

What is the saturation temperature of water at 0.5 bar pressure ? What is the saturation pressure of water at 130°C ?

*Solution* :

From the saturated steam tables (pressure) given in the Appendix–l one can read the saturation temperature as 81.345°C at 0.5 bar pressure. From the saturated steam (Temperature) tables given in Appendix –2, one can read the saturation pressure as 2.7013 bar.

**Example 3.38**

The domestic pressure cookers maintain a pressure of 2 bar while food is cooked in it. Determine the temperature at which water boils in a domestic pressure cooker.

*Solution* :

The boiling point of water in a domestic pressure cooker is equal to the saturation temperture at 2 bar. From the saturated steam tables (Pressure) given in the Appendix– 1, one can read the saturation temperature of water at 2 bar pressure as 120.23°C.

**Example 3.39**

Determine the specific volume of wet steam of quality 0.7 at 200°C.

*Solution* :

Read the following values from the saturated steam (Temperature) tables given in Appendix –2.

$$v_f = 0.001\ 156\ 5\ \text{m}^3/\text{kg}\ ;\ v_g = 0.1272\ \text{m}^3/\text{kg}$$

The specific volume of a two phase mixture can be calculated from the relation $v = Xv_g + (1-X)\,v_f$ where $X$ = quality of mixture.

$v = 0.7\ (0.1272) + 0.3\ (0.001\ 156\ 5) = 0.0894\ \text{m}^3/\text{kg}$

**Example 3.40**

Determine the quality of wet steam at 200°C with a specific volume of 0.095 m³/kg.

*Solution* :

From steam tables (Appendix–2) for steam at 200°C we get $v_f = 0.001\ 156\ 5$ m³/kg and $v_g = 0.1272$ m³/ kg

we know that $v = Xv_g + (1-X)\,v_f$

Subsitituing the values of $v$, $v_g$ and $v_f$ in the above relation we get.

$$0.095 = X\,(0.1272) + (1 - X)\,0.001\ 156\ 5$$

or $$0.095 - 0.001\ 156\ 5 = (0.1272 - 0.001\ 156\ 5)X$$

or $$X = \frac{0.095 - 0.001\,156\,5}{0.1272 - 0.001\,156\,5} = 0.7445$$

Therefore, the quality ($X$) of the wet steam is 0.7445.

**Example 3.41**

Wet steam at 6 bar has a specific volume of 0.29 m³/kg. Determine the specific volume if wet steam of the same quality is maintained at 3 bar.

*Solution* :

Read the following data from the steam tables given in Appendix –1. Saturated steam at 6 bar : $v_f = 0.001\ 100\ 9$ m³/kg ; $v_g = 0.3155$ m³/kg. Saturated steam at 3 bar : $v_f = 0.001\ 073\ 5$ m³/kg ; $v_g = 0.6056$ m³/kg. we know that

$$v = Xv_g + (1-X)\,v_f \qquad (A)$$

Eqn ($A$) can be rearranged as

$$X = \frac{v - v_f}{v_g - v_f} \qquad (B)$$

Substituting the values of $v$, $v_f$ and $v_g$ at 6 bar in Eqn. (*B*), we get

$$X = \frac{0.29 - 0.001\,100\,9}{0.3155 - 0.001\,100\,9} = 0.9188$$

The specific volume of wet steam of quality 0.9188 at 3 bar is calculated from Eqn. (*A*) by substituting $X = 0.9188$ and the values of $v_f$ and $v_g$ at 3 bar.

$$v = 0.9188\ (0.6056) + (1 - 0.9188)\ 0.001\ 073\ 5 = 0.5565\ \text{m}^3/\text{kg}$$

**Example 3.42**

A vessel of volume 2m³ contains wet steam of quality 0.8 at 210°C. Determine the masses of liquid and vapor present in the vessel.

*Solution* :

The following data can be read from the steam tables (see Appendix – 2) at 210°C

$v_f = 0.001\ 172\ 6\ \text{m}^3/\text{kg}$ ; $v_g = 0.1042\ \text{m}^3/\text{kg}$

The specific volume of wet steam of quality (*X*) 0.8 is given by

$v = Xv_g + (1 - X)\ v_f = 0.8\ (0.1042) + 0.2\ (0.001\ 172\ 6) = 0.0836\ \text{m}^3/\text{kg}$

mass of steam ($m$) in the vesse 1 $= \dfrac{V}{v} = \dfrac{2}{0.0836} = 23.923$ kg

where $V$ = volume of the vessel

Mass of vapor ($m_g$) $= mX = 0.8 \times 23.923 = 19.138$ kg

Mass of liquid $m_f = (1 - X)\ m = 0.2 \times 23.923 = 4.785$ kg

**Example 3.43**

Determine the volume occupied by 1 kg of steam at the following conditions (*a*) $P = 0.4$ bar : $t = 300$°C (b) $P = 4$ bar ; $t = 300$°C and (c) $P = 10$ bar ; $t = 500$°C

*Solution* :

The following values can be directly read from the superheated steam tables given in Appendix – 3

(*a*) $P = 0.4$ bar and $t = 300$°C : $v = 6.6065\ \text{m}^3/\text{kg}$

(*b*) $P = 4$ bar and t = 300°C : $v = 0.6549\ \text{m}^3/\text{kg}$

(*c*) $P = 10$ bar and $t = 500$°C : $v = 0.3540\ \text{m}^3/\text{kg}$

### Example 3.44

Determine the specific volume of steam at 25 bar and 330°C.

*Solution* :

The steam tables given in Appendix – 3 does not have enteries at 25 bar and 330°C. If the data is not tabulated in close intervals, it is recommended to estimate the required data by making use of linear interpolation. First, let us calculate the specific volume at 25 bar pressure at temperatures 300°C and 400°C as indicated below. Now read the following data from steam tables ( see Appendix – 3). $P = 24$ bar and $t = 300$°C : $v = 0.103\,36$ m$^3$/kg

$P = 26$ bar and $t = 300$°C : $v = 0.094\,83$ m$^3$/kg

Therefore, the specific volume $(v_1)$ at $P = 25$ bar and $t_1 = 300$°C can be calculated as.

$$v_1 = \frac{0.103\,36 + 0.094\,83}{2} = 0.099\,095 \text{ m}^3\text{/kg}$$

Similarly. we read the following values also

$P = 24$ bar and $t = 400$°C : $v = 0.125\,22$ m$^3$/kg

$P = 26$ bar and $t = 400$°C : $v = 0.115\,26$ m$^3$/kg

Therefore, the specific volume $(v_2)$ at $P = 25$ and $t_2 = 400$°C is given by.

$$v_2 = \frac{0.125\,22 + 0.115\,26}{2} = 0.120\,24 \text{ m}^3\text{/kg}$$

Having determined the specific volumes at 300°C and 400°C at a pressure of 25 bar, we can again use the method of linear interpolation to determine the specific volume at $P = 25$ bar and $t = 330$°C as shown below.

The specific volume $(v)$ at temperature $t$ is given by

$$v = v_1 + \left(\frac{v_2 - v_1}{t_2 - t_1}\right)(t - t_1)$$

$$0.099\,095 + \frac{(0.120\,24 - 0.099\,095)}{(400 - 300)}(330 - 300)$$

$$= 0.105\,439 \text{ m}^3\text{/kg}$$

### Example 3.45

List some of the drawbacks associated with the tabular representation of thermodynamic properties of a substance.

*Solution* :

Some of the drawbacks associated with the tabular representation of thermodynamic properties are given below.

1. The data cannot be presented in fine increments of the independent variables and for all the substances as the data is likely to occupy several volumes.
2. In thermodynamic analysis quite often one is interested in the values of the partial derivatives of the thermodynamic properties. Accurate values of these partial derivatives cannot be obtained from the data if it is presented in tabular form.
3. Since the data is usually presented in finite increments of the independent variables, some times it is necessary to interpolate the data. The accuracy of the interpolated values is usually poor.

**Example 3.46**

What is an equation of state ?

*Solution* :

A mathematical expression of the type $f(P,v,T) = 0$ which describe the $P-v-T$ behavior of substances is called an equation of state. Such a relation is called the mechanical equation of state. For a single component simple system there are three equations of state. They are (a) Mechanical equation of state (b) Thermal equation of state and (c) Chemical equation of state. It may be noted that all the equations of state of a substance are not independent and there exists a relationship among them. In general an equation which expresses the intensive parameter as a function of the extensive parameters is called an equation of state. A detailed discussion of the equations of state is beyond the scope of this book and hence we focus our attention on mechanical equation of state only.

**Example 3.47**

State the Boyle's law and Charles law.

*Solution* :

Boyle's law states that the molar volume ($v$) of a gas varies inversely with pressure ($P$) when tempertaure ($T$) is held constant. That is,

$$v \propto \frac{1}{P} \quad \text{or} \quad Pv = \text{constant} \qquad \text{(at constant Temperature).}$$

Charles law states that ($a$) the molar volume of a gas is directly proportional to temperature when the pressure is held constant. That is $v \propto T$ (at constant pressure) and (*b*) the pressure exerted by a gas is directly proportional to temperature when the volume is held constant. That is $P \propto T$ (at constant volume).

### Example 3.48

What is an ideal gas equation of state ? What assumptions are made in deriving the ideal gas equation of state from the kinetic theory point of view and how does one justify these assumptions ?

*Solution* :

The ideal gas equation of state is given by $Pv = RT$ or $PV = nRT$ where $v$ = molar volume, $V$ = total volume, $n$ = number of moles and $R$ = universal gas constant which is equal to 8.314 J /mol K or 8.314 kJ/ kmol K. Based on the following assumptions, the ideal gas equation of state can be derived through the kinetic theory of gases.

1. The volume occupied by the gas molecules is negligible compared to the volume available for free motion.
2. The molecules are independent of each other and do not exert any attractive forces.

At normal temperature and pressure the number of molecules per cubic metre are approximately $2.4 \times 10^{25}$. Hence, the volume available per molecule is approximately $4 \times 10^{-24}$ m$^3$. The diameter of a molecule is approximately $2 \times 10^{-10}$ m and hence the volume of each molecule is approximately $4 \times 10^{-27}$ m$^3$. Hence the volume occupied by the molecule is about 1/1000 of the volume available for free motion . Therefore, it is reasonable to assume that the molecules do not occupy any volume. The volume available per molecule is $4 \times 10^{-24}$ m$^3$ or the distance between two adjacent molecules is approximately $10^{-8}$ m whereas the diameter of the molecule is $2 \times 10^{-10}$ m. Thus, the distance between two molecules is 50 times the diameter of a molecule. Hence it is reasonable to assume that a molecule does not exert any attractive force on other molecules.

### Example 3.49

Is the ideal gas equation of state applicable to all gases and at all temperatures and pressures ? Does it represent the actual $P-v-T$ behavior of gases ?

*Solution* :

The ideal gas equation of state is applicable only at low densities where the assumptions made in deriving it, through the kinectic theory of gases, can be justified. The ideal gas law is only an approximation or idealization to the behavior of gases. It is applicable at low pressures and high temperatures where the density is sufficiently low. In general the $P-v-T$ behavior of real gases differs from that predicted by the ideal gas law.

**Example 3.50**

It is desired to store 28 kg of nitrogen at 14 MPa pressure and 27°C in a cylinder. Assuming that nitrogen behaves like and ideal gas determine the size of the cylinder.

*Solution* :

The ideal gas equation of state is given by.

$$PV = nRT$$

Hence $$V = \frac{nRT}{P}$$

$$\text{Quantity of nitrogen} = \frac{28}{28 \times 10^{-3}} = 10^3 \text{ mol}$$

(molar mass of nitrogen. = $28 \times 10^{-3}$ kg/mol)

Therefore volume of the cylinder,

$$V = \frac{nRT}{P} = \frac{10^3 \times 8.314 \times 300}{14 \times 10^6} = 0.1782 \text{ m}^3$$

**Example 3.51**

A 0.1 m³ cylinder is charged with ethane ( $C_2H_6$ ) to a pressure of 120 bar at 27°C. Assuming that ethane behaves like an ideal gas, estimate the mass of ethane in the cylinder.

*Solution* :

Molar mass of ethane ( $C_2H_6$ ) = $30 \times 10^{-3}$ kg/ mol

$P = 120$ bar $= 120 \times 10^5$ Pa ; $T = 273 + 27 = 300$ K ; $v = 0.1$ m³.

We know that for an ideal gas $PV = nRT$ or

$$n = \frac{PV}{RT} = \frac{120 \times 10^5 \times 0.1}{8.314 \times 300} = 481.12 \text{ mol}$$

Mass of ethane in the cylinder = $481.12 \times 30 \times 10^{-3} = 14.434$ kg

**Example 3.52**

(*a*) Suppose it is desired to charge 15 kg of ethane at 300 K into a cylinder of volume 0.08 m$^3$, determine the pressure to which the cylinder is to be charged.

(*b*) If 20 kg of ethane is charged into a cylinder of capacity 0.15 m$^3$ and the cylinder is held at a pressure of 15 MPa, determine the temperature of the gas in the cylinder. Assume that ethane behaves like an ideal gas.

*Solution* :

(*a*) Given data: $T = 300$ K ; $V = 0.08$ m$^3$

$$\text{Quantity of ethane} = 15 \text{ kg} = \frac{15}{30 \times 10^{-3}} = 500 \text{ mol}$$

(Molar mass of ethane , $C_2H_6 = 30 \times 10^{-3}$ kg /mol)

We know that for an ideal gas $PV = nRT$

$$P = \frac{nRT}{V} = \frac{500 \times 8.314 \times 300}{0.08}$$

$$= 15.589 \text{ MPa}$$

Therefore, pressure at which the cylinder is to be charged is 15.589 Mpa

$$(b) \text{ Quantity of ethane} = 20\text{kg} = \frac{20}{30 \times 10^{-3}} = 666.67 \text{ mol}$$

$$P = 15 \text{ MPa} ; V = 0.15 \text{ m}^3$$

We know that $PV = nRT$

or $$T = \frac{PV}{nR} = \frac{15 \times 10^6 \times 0.15}{666.67 \times 8.314} = 405.94 \text{ K}$$

Therefore, temperature of ethane in the cylinder = 405.94 K = 132.79°C

**Example 3.53**

The Van der Waals equation of state is a modification of the ideal gas law. Name the factors which were taken into account by Van der Waal in modifying the equation of state.

*Solution* :

The ideal gas law is only an approximation or idealization to the $P-v-T$ behavior of gases and it is applicable at low densities only. The $P-v-T$ behavior of real gases differ from that predicted by the ideal gas law. In 1873 Van der Waals proposed an equation of state which takes into account

1. The unavailable volume for molecular motion, that is the excluded volume and

2. The attractive forces between the molecules which leads to a reduction in pressure exerted by the gas.

while modifying the ideal gas equation of state.

**Example 3.54**

Explain the terms excluded volume and reduction in pressure which were incorporated in the ideal gas equation of state leading to Van der Waals equation of state.

*Solution* :

While deriving the ideal gas law, in the kinetic theory of gases, it is assumed that the molecules do not occupy any volume. In real gases, such asumption is not valid since the molecules do occupy some volume. The Van der Waals equation of state takes into account the volume occupied by the molecules as explained below. The gas molecules are considered to be hard spheres of radius $r$. Then the centre to centre distance between two molecules at the instant of collision is $2r$. Therefore, a molecule cannot reach another molocule such that its centre lies within a distance of $2r$ from the center of the other molecule. That is, the centre of a molecule is excluded from a spherical region, called the sphere of exclusion, of radius $2r$ from other molecules a shown in Fig.E 3.54. Then, the total unavailable volume or the excluded volume $b$ for molecular motion of $N_0$ molecules (Avagadros number) is given by

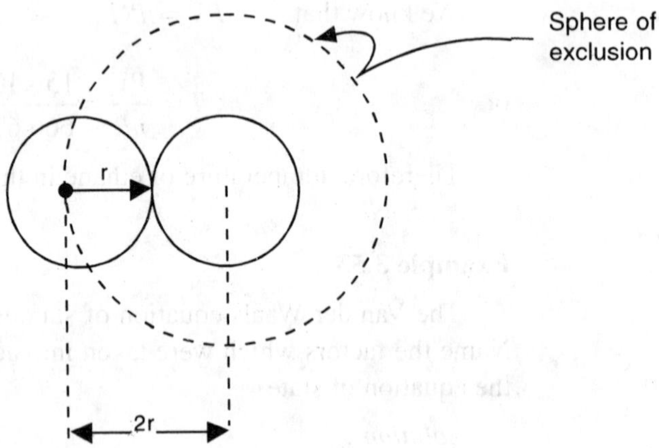


Fig.E 3.54. Sphere of exculsion.

$$b = \frac{N_0}{2} \times \frac{4}{3}\pi(2r)^3 = \frac{16}{3} N_0 \pi r^3$$

In the above equation, the factor 2 is included to avoid counting each pair twice. The actual volume occupied by $N_0$ molecules is equal to $\frac{4}{3} N_0 \pi r^3$. Thus, the excluded volume $b$ is four times the actual volume occupied by the molecules. Therefore, the free volume available for molecular motion is $(v - b)$.

In the kinetic theory of gases, the ideal gas equation of state is derived with the assumption that there are no intermolecular forces of attraction between the molecules. In real gases, such an assumption is not valid since there are forces of attraction and repulsion between any pair of molecules.The molecules in the interior of the gas are attracted equally in all directions by the surrounding molecules and hence they do not experience any net attractive and repulsive forces. However, the molecules adjacent to the container wall experience a net inward force, since they are attracted by the surrounding molecules on one side only. Therefore,a molecule adjacent to the container wall exerts less force on the wall. Thus, the observed pressure will be less than the pressure the gas would have exerted in absence of the attractive forces. The reduction in pressure is proportional to the number of molecules per unit volume $(N_0/v)$ in the layer adjacent to the wall and to the number of molecules per unit volume in the next layer, which is attracting. Therefore, the reduction in pressure $(P' - P)$ is given by.

$$P' - P = a'\left(\frac{N_0}{v}\right)^2 \quad \text{or} \quad P' = P + \frac{a}{v^2}$$

where $P'$ is the pressure the gas would have exerted in absence of the attractive forces. $P$ is the observed pressure, $a'$ is a constant and $a$ is another constant. If the corrections for excluded volume and attractive forces are incorporated in the ideal gas law, one obtains

$$\left(P + \frac{a}{v^2}\right)(v - b) = RT$$

which is known as the Van der Waals equation of state. The constants $a$ and $b$ appearing in the above equation are called the Van der Waals constants and they depend on the nature of the gas.

**Example 3.55**

Sketch the isotherms for a Van der Waals gas on pressure versus volume diagram and explain the significance of the values of volume at $T < T_C$ ; $T = T_C$ and $T > T_C$ where $T_C$ is the critical temprature.

*Solution* :

The Van der Waals equation of state is cubic in volume and in general it gives three values for the volume $(v)$ at a give temperature and pressure. However, for some range of values of $P$ and $T$ it gives only one real value and in some other range of values of $P$ and $T$, it gives three real values for volume. The isotherms of a Van der Waals gas are shown in Fig.E 3.55. It can be observed from Fig.E 3. 55 that at $T > T_C$, there is only one real value for volume. However, at $T < T_C$ it gives three real values $v_f, v_i$ *and* $v_g$ for volume. It can also be observed from Fig.E 3.55 that at $T < T_C$, the segment 1 2 3 4 5 is not in agreement with the experimental observation. The practically realizable variation of volume is indicated by the dotted line 1-3-5. In segment 2 3 4, the Van der Waals equation shows that the volume increases with increasing pressure at constant temperature which is not realistic. It can be shown, based on the stability criteria (which is beyond the scope of this book) that the system (gas) is unstable in the region 1 2 3 4 5 and the system should actually follow the dotted straight line path 1-3-5 in this region. Hence, the volume $v_i$ which is predicted by the Van der Waals equation of state is inadmissible. Of the three real values the smallest value $v_f$ represents the volume of the liquid phase and the largest value $v_g$ represents the volume of the gas phase.

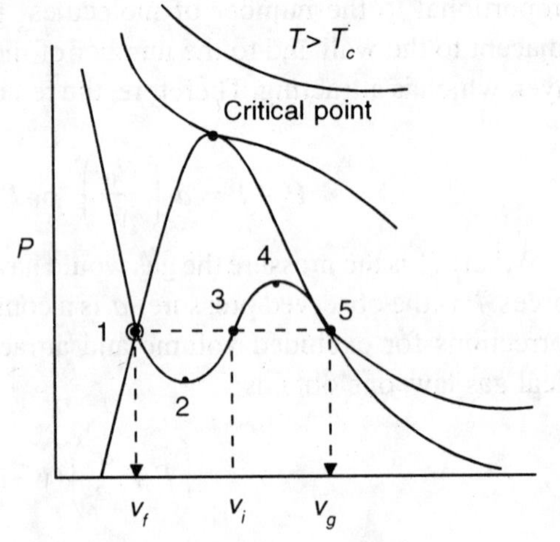


Fig.E 3.55. Isotherms of a Van der Waals gas on $P$–$v$ diagram.

**Example 3.56**

A cylinder of 0.01 m$^3$ volume is filled with 0.727 kg of n-octane ( $C_8H_{18}$) at 427.85 K. Assuming that n – octane obeys the Van der Waals equation of state, calculate the pressure of the gas in the cylinder. The Van der Waals constants $a$ and $b$ for n – octane are 3.789 Pa (m$^3$ / mol)$^2$ and $2.37 \times 10^{-4}$ m$^3$/mol, respectively.

*Solution* :

Molar mass of n – octane ($C_8H_{18}$) = $114 \times 10^{-3}$ kg / mol

Quantity of n – octane is the cylinder,

$$= 0.727 \text{ kg} = \frac{0.727}{114 \times 10^{-3}} = 6.377 \text{ mol}$$

Molar volume of n–octane in the cylinder, $v = \dfrac{0.1}{6.377} = 15.68 \times 10^{-3} \text{ m}^3/\text{mol}$

The Van der Waals equation of state is given by,

$$\left(P + \frac{a}{v^2}\right)(v - b) = RT \text{ or } P = \frac{RT}{v - b} - \frac{a}{v^2}$$

$$P = \frac{8.314 \times 427.85}{15.68 \times 10^{-3} - 2.37 \times 10^{-4}} - \frac{3.789}{(15.68 \times 10^{-3})^2} = 2.149 \text{ MPa}$$

**Example 3.57**

If a cylinder of volume $0.1 \text{m}^3$ is filled with 1.373 kg of ammonia at 1.95 MPa, determine the temperature at which ammonia exist in the cylinder. Assume that ammonia obeys the Van der Waals equation of state. The Van der Waals constants $a$ and $b$ for ammonia are $422.546 \times 10^{-3}$ Pa $(\text{m}^3/\text{ mol})^2$ and $37 \times 10^{-6}$ $\text{m}^3$ / mol, respectively.

*Solution* :

Molar mass of ammonia ($NH_3$) = $17 \times 10^{-3}$ kg /mol

Quantity of ammonia = $1.373 \text{ kg} = \dfrac{1.373}{17 \times 10^{-3}} = 80.765 \text{ mol}$

Molar volume of ammonia, $v = \dfrac{0.1}{80.765} = 1.238 \times 10^{-3} \text{ m}^3/\text{mol}$

We know that the Van der Waals equation of state is given by

$$\left(P + \frac{a}{v^2}\right)(v - b) = RT$$

substituting the values of $P$, $v$, $a$ and $b$ in the above equation, we get

$$\left(1.95 \times 10^6 + \frac{422.546 \times 10^{-3}}{(1.238 \times 10^{-3})^2}\right)(1.238 \times 10^{-3} - 37 \times 10^{-6}) = 8.314T$$

or $\qquad T = 321.5$ K

**Example 3.58**

Develop iterative scheme for the calculation of volume from a knowledge of $P$ and $T$ for the Van der Waals equation of state.

*Solution* :

Given the value of volume the calculation of $P$ or $T$ for a Van der Waals equation of state is straight forward. However, the calculation of volume from a knowledge of $P$ and $T$ is difficult since the Van der Waals equation of state is cubic in volume.

To facilitate the calculation of volume from a knowledge of $P$ and $T$ the Van der Waals equation of state can be rearranged as

$$v = \frac{RT}{(P + a/v^2)} + b$$

For iterative calculations, the above equation can be written as

$$v_{i+1} = \frac{RT}{(P + a/v_i^2)} + b$$

where $v_i$ = value of the molar volume in $i$th iteration.

$v_{i+1}$ = value of the molar volume in $(i + 1)$th iteration.

To initate the iterative calculation, when $i$ = o, $v_0$ can be taken as equal to that predicted by the ideal gas law if one is interested in calculating the gas volume. On the other hand, if one is interested in estimating the liquid volume $v_0$ can be taken as equal to $b$. Subsitute the value of $v_0$ and the known values of $P$ and $T$ to obtain the value of $v_1$ (the value of volume in the first iteration). Then compare $v_1$ with $v_0$ and check whether they are approximately equal or not. If the difference is large, continue the iterative calculation till the values converge or the difference $(v_{i+1} - v_i)$ is less than a preset value.

**Example 3.59**

Assuming that ammonia obeys the Van der Waals equation of state, calculate the molar volumes of ammonia vapor and liquid ammonia at 1.95 MPa and 321.55 K. The Van der Waals constants $a$ and $b$ for ammonia are $422.546 \times 10^{-3}$ Pa $(m^3/mol)^2$ and $37 \times 10^{-6}$ $m^3/mol$, respectively.

*Solution* :

For iterative calculations, the Van der Waals equation of state can be rewritten as

$$v_{i+1} = \frac{RT}{P + a/v_i^2} + b$$

$$v_0 = \frac{RT}{P} = \frac{8.314 \times 321.55}{1.95 \times 10^6} = 1.3709 \times 10^{-3} \text{ m}^3/\text{mol}$$

$$v_1 = \frac{RT}{P + a/v_0^2} + b$$

$$= \frac{8.314 \times 321.55}{\left[1.95 \times 10^6 + \dfrac{422.546 \times 10^{-3}}{(1.3709 \times 10^{-3})^2}\right]} + 37 \times 10^{-6}$$

$$= 1.2662 \times 10^{-3} \text{ m}^3/\text{mol}$$

Since the value of $v_1$ is much different from the value of $v_0$ we shall continue the iterative calculations.

$$v_2 = \frac{RT}{(P + a/v_1^2)} + b = \frac{8.314 \times 321.55}{\left\{1.95 \times 10^6 + \dfrac{422.546 \times 10^{-3}}{(1.2662 \times 10^{-3})^2}\right\}} + 37 \times 10^{-6}$$

$$= 1.2447 \times 10^{-3} \text{ m}^3/\text{mol}$$

The calculated values of successive iterations are as given below :

| **Iteration No.** | $v_i \times 10^3$ **m³/mol** | **Iteration No.** | $v_i \times 10^3$ **m³/mol** |
|---|---|---|---|
| 1. | 1.2662 | 5 | 1.2383 |
| 2. | 1.2447 | 6 | 1.2382 |
| 3. | 1.2397 | 7 | 1.2382 |
| 4. | 1.2386 | | |

The calculated values of volume in the sixth and seventh iterations agree with each other. Hence the molar volume of ammonia vapor is $1.2382 \times 10^{-3}$ m³/ mol.

To calculate the molar volume of liquid ammonia, we take $v_0 = b = 37 \times 10^{-6}$ m³/ mol. The iterative calcualtions can be performed. The calculated values of liquid volume for successive iterations are as given below.

| **Iteration No.** | $v_i \times 10^6$ **m³/mol** | **Iteration No.** | $v_i \times 10^6$ **m³/mol** |
|---|---|---|---|
| 1. | 45.607 | 5 | 55.419 |
| 2. | 50.035 | 6 | 56.160 |
| 3. | 52.658 | 7 | 56.668 |
| 4. | 54.322 | | |

The difference in the values of volume in sixth and seventh iterations is $0.508 \times 10^{-6}$ m$^3$/mol, which is small enough. Hence iterative calculations can be terminated and the molar volume of liquid ammonia at these conditions is 56.668 $\times 10^{-6}$ m$^3$/mol.

**Example 3.60**

Suppose the experimental $P-v-T$ data for a substance is known and it is desired to obtain the Van der Waals constants for that substance. Indicate a method of estimating the Van der Waals constants from the experimental $P-v-T$ data

*Solution :*

The Van der Waals equation of state is given by

$$(P + a/v^2)(v-b) = RT$$

This equation can be rearranged as

$$P = \frac{RT}{v-b} - \frac{a}{v^2}$$

The above equation suggests that a plot of $P$ versus $T$ at constant volume gives a straight the with a slope of $\left(\frac{R}{v-b}\right)$ and an intercept $(-a/v^2)$. Hence for a known value of $v$ plot $P$ versus $T$ and determine the slope and intercept of the resulting straight line. Then, one can evaluate the constants $a$ and $b$ from a knowledge of slope and intercept.

**Example 3.61**

Specify the conditions which must be satisfied by every equation of state.

*Solution :*

Every equation of state must satisfy the following criteria.

1. The slope of an isotherm on a $P$ versus $v$ diagram must be negative. That is $\left(\frac{\partial P}{\partial v}\right)_T < 0$. In other words the isothermal compressibility.

$$\kappa = -\frac{1}{v}\left(\frac{\partial v}{\partial P}\right)_T > 0$$

2. It should reduce to the ideal gas law, $Pv = RT$ as $P \to 0$
3. At the critical point, the critical isotherm must exhibit a point of inflexion. Stated mathematically.

$$\left(\frac{\partial P}{\partial v}\right)_{T_C} = 0 \text{ and } \left(\frac{\partial^2 P}{\partial v^2}\right)_{T_C} = 0$$

where $T_C$ is the critical temperature.

**Example 3.62**

The constants of any two parameter equation of state can be determined from a knowledge of the critical constants – critical temperature $T_C$ and critical pressure $P_C$. Develop the necessary relations for estimating in Van der Waals constants in terms of $P_C$ and $T_C$.

*Solution* :

It is known that every equation of state must satisfy the following condition. The critical isotherm must exhibit a point of inflexian at the critical point. That is, it should satisfy the conditions

$$\left(\frac{\partial P}{\partial v}\right)_{T_C} = 0 \text{ and} \qquad (A)$$

$$\left(\frac{\partial^2 P}{\partial v^2}\right)_{T_C} = 0 \qquad (B)$$

The Van der Waals equation of state can be written as. $P = \frac{RT}{v-b} - \frac{a}{v^2}$

Therefore $$\left(\frac{\partial P}{\partial v}\right)_{T_C} = -\frac{RT_c}{(v_C - b)^2} + \frac{2a}{{v_C}^3} = 0 \qquad (C)$$

or $$\frac{RT_c}{(v_C - b)^2} = \frac{2a}{{v_C}^3} \qquad (D)$$

$$\left(\frac{\partial^2 P}{\partial v^2}\right)_{T_C} = \frac{2RT_c}{(v_C - b)^3} - \frac{6a}{{v_C}^4} = 0 \qquad (E)$$

or $$\frac{2RT_c}{(v_C - b)^3} = \frac{6a}{v_C^4} \qquad (F)$$

At the critical point, the Van der Waals equation of state is written as

$$P_C = \frac{RT_C}{v_C - b} - \frac{a}{{v_C}^2}$$

Where $v_C$ is the critical volume.

Dividing Eqn (*F*) with Eqn (*D*) we get

$$\frac{2}{v_C - b} = \frac{3}{v_C} \text{ or } v_C = 3b \qquad (H)$$

Substitute the value of $v_C$ in Eqns (*D*) and (*G*) to obtain

$$T_C = \frac{8a}{27Rb} \qquad (I)$$

$$P_C = \frac{RT_c}{2b} - \frac{a}{9b^2} = \frac{a}{27b^2} \qquad (J)$$

Therefore, one can estimate the values of $P_C$, $T_C$ and $v_C$ from a knowledge of the Van der Walls constants, by making use of the Eqns. (*H*) – (*J*). Alternatively, one can determine the constants *a* and *b* from a knowledge of $P_C$ and $T_C$. Eqn (*I*) can be rearranged as

$$a = \frac{27RT_Cb}{8} \qquad (K)$$

Similarly Eqn (*J*) can be re-arranged as $a = 27P_Cb^2$ (*L*)

From Eqns (*K*) and (*L*), we get

$$b = \frac{RT_C}{8P_C} \qquad (M)$$

From Eqns (*K*) and (*M*) we get

$$a = \frac{27R^2T_C^{\;2}}{64P_C} \qquad (N)$$

These Eqns. (*M*) and (*N*) express the Van der Waals constants in terms of the critical constant $T_C$ the $P_C$.

**Example 3.63**

The Van der Waals constants *a* and *b* for ammonia are $422.546 \times 10^{-3}$ Pa $(m^3 / mol)^2$ and $37 \times 10^{-6}$ $m^3$/mol, respectively. Determine the values of $P_C$ and $T_C$ for ammonia.

*Solution* :

We know that $P_C = \frac{a}{27b^2}$ and $T_C = \frac{8a}{27Rb}$. Therefore,

$$P_C = \frac{422.546 \times 10^{-3}}{27 \times (37 \times 10^{-6})^2} = 11.432 \text{ MPa}$$

$$T_C = \frac{8 \times 422.546 \times 10^{-3}}{27 \times 8.314 \times 37 \times 10^{-6}} = 406.99 \text{ K}$$

**Example 3.64**

Estimate the Van der Waals constants for ethane ($C_2H_6$) and Freon −12 (Dichloro difluoro methame $CCl_2F_2$ ) assuming that the critical constants of the substances are known.

*Solution* :

The critical constants of some selected substances are presented in Appendix – 4. From Appendix – 4, we read the following data

| **Substance** | $T_C$ **(K)** | $P_C$ **(bar)** |
|---|---|---|
| Ethane ($C_2H_6$) | 305.43 | 48.84 |
| Feron–12 ($CCl_2F_2$) | 384.70 | 40.12 |

We know that, for a Van der Waals gas, the constants are given by

$$a = \frac{27R^2T_C^2}{64P_C} \text{ and } b = \frac{RT_C}{8P_C}$$

Substituting the values of $T_C$ and $P_C$ in the above equations we get

$$a\,(C_2H_6) = \frac{27 \times (8.314 \times 305.43)^2}{64 \times 48.84 \times 10^5} = 0.557 \text{ Pa(m}^3\text{/mol)}^2$$

$$b(C_2H_6) = \frac{8.314 \times 305.43}{8 \times 48.84 \times 10^5} = 64.99 \times 10^{-6} \text{ m}^3\text{/mol}$$

$$a\,(CCL_2F_2) = \frac{27 \times (8.314 \times 384.70)^2}{64 \times 40.12 \times 10^5} = 1.0757 \text{ Pa(m}^3\text{/mol)}^2$$

$$b(CCl_2/F_2) = \frac{8.314 \times 384.70}{8 \times 40.12 \times 10^5} = 99.651 \times 10^{-6} \text{ m}^3\text{/mol}$$

**Example 3.65**

The Berthelot equation of state is given by

$$\left(P+\frac{a}{Tv^2}\right)(v-b) = RT$$

where $a$ and $b$ are characteristic of the gas. Develop necessary relations to determine the parameters $a$ and $b$ in terms of the critical constants $T_C$ and $P_C$.

*Solution :*

The Berthelot equation of state is rewritten as.

$$P=\frac{RT}{v-b}-\frac{a}{Tv^2}$$

We know that the critical isotherm must show a point of inflexion at the critical point. Therefore,

$$\left(\frac{\partial P}{\partial v}\right)_{T_C} = -\frac{RT_C}{(v_C-b)^2}+\frac{2a}{T_C v_C^{\,3}}=0 \text{ or } \frac{RT_c}{(v_C-b)^2}=\frac{2a}{T_C v_C^{\,3}} \qquad (A)$$

$$\left(\frac{\partial P}{\partial v^2}\right)_{T_C} = \frac{2RT_C}{(v_C-b)^3}-\frac{6a}{T_C v_C^{\,4}}=0 \text{ or } \frac{2RT_C}{(v_C-b)^3}=\frac{6a}{T_C v_C^{\,4}} \qquad (B)$$

At the critical point the equation of state gives

$$P_C=\frac{RT_C}{v_C-b}-\frac{a}{T_C v_C^{\,2}} \qquad (C)$$

Dividing Eqn (*B*) with Eqn (*A*) we get

$$\frac{2}{v_C-b}=\frac{3}{v_C} \text{ or } v_C=3b \qquad (D)$$

substitute Eqn (*D*) in Eqn (*A*) to get

$$\frac{RT_C}{4b^2}=\frac{2a}{27T_C b^3} \text{ or } a=\frac{27RT_C^{\,2}b}{8} \qquad (E)$$

Substitute Eqn. (*D*) in Eqn (*C*) to get

$$P_C=\frac{RT_C}{2b}-\frac{27RT_C^2 b}{8T_C(3b)^2} \text{ or } b=\frac{RT_C}{8P_C} \qquad (F)$$

Substituting Eqn (*F*) in Eqn (*E*) we obtain

$$a = \frac{27R^2T_C^3}{64P_C} \qquad (G)$$

The parameters $a$ and $b$ can be calculated from Eqns. ($F$) and ($G$) with a knowledge of the critical constant $T_C$ and $P_C$.

**Example 3.66**

The Dieterici equation of state is given by

$$P(v-b)exp\left(\frac{a}{RTv}\right) = RT$$

where $a$ and $b$ are characteristic of the gas. Determine the relations to estimate $a$ and $b$ in terms of the critical constants $T_C$ and $P_C$.

*Solution* :

The Dieterci equation of state is given by.

$$P = (v-b)\, exp\left(\frac{a}{RTv}\right) = RT$$

or $$\ln P + \ln(v-b) + \frac{a}{RTv} = \ln\,(RT)$$

$$\frac{dP}{P} + \frac{dv}{v-b} - \frac{adv}{RTv^2} - \frac{adT}{RT^2v} = \frac{dT}{T}$$

$$\left(\frac{\partial P}{\partial v}\right)_{T_C} = \frac{P_C a}{RT_C v_C{}^2} - \frac{P_C}{v_C - b} = 0$$

or $$(v_C - b) = \frac{RT_C v_C^2}{a} \qquad (A)$$

$$\left(\frac{\partial^2 P}{\partial v^2}\right)_{T_C} = P_C\left[\frac{-2a}{RT_C v_C^3} + \frac{1}{(v_C-b)^2}\right] = 0$$

or $$(v_C - b)^2 = \frac{RT_C v_C^3}{2a} \qquad (B)$$

Dividing Eqn ($B$) with Eqn ($A$) we get

$$v_C = 2b \qquad (C)$$

From Eqns ($A$) and ($C$) we get

$$a = 4RbT_C \tag{D}$$

At the critical point, the equation of state gives

$$P_C(v_C - b)\, exp\left(\frac{a}{RT_C v_C}\right) = RT_C$$

or

$$P_C = \frac{RT_C}{v_C - b}\, exp\left(\frac{-a}{RT_C v_C}\right) \tag{E}$$

Substituting Eqns ($C$) and ($D$) in Eqn ($E$), we get

$$P_C = \frac{RT_C}{b}\, exp\,(-2) \quad \text{or}$$

$$b = \frac{RT_C}{P_C}\, exp\,(-2)$$

or

$$b = 0.135\,335\,\frac{RT_C}{P_C} \quad \text{and} \tag{F}$$

$$a = 4RbT_C = 0.541\,341\,\frac{R^2 T_C^{\,2}}{P_C} \tag{G}$$

The values of constants $a$ and $b$ can be estimated by making use of the Eqns. ($F$) and ($G$).

**Example 3.67**

Derive relations for estimating the coefficient of thermal expansion $\beta$ and the coefficient of isothermal compressibility $\kappa$ for ($a$) an ideal gas and ($b$) a Van der Waals gas.

*Solution* :

($a$) The ideal gas equation of state is given by

$$Pv = RT \quad \text{or} \quad v = \frac{RT}{P}$$

Therefore

$$\left(\frac{\partial v}{\partial T}\right)_P = \frac{R}{P} \quad \text{and} \quad \left(\frac{\partial v}{\partial P}\right)_T = \frac{-RT}{P^2}$$

Hence $$\beta = \frac{1}{v}\left(\frac{\partial v}{\partial T}\right)_P = \frac{R}{Pv} = \frac{1}{T}$$

$$\kappa = -\frac{1}{v}\left(\frac{\partial v}{\partial P}\right)_T = \frac{RT}{P^2 v} = \frac{1}{P}$$

(*b*) The Van der Waals equation of state is given by

$$P = \frac{RT}{v-b} - \frac{a}{v^2}$$

Therefore,

$$dP = \frac{RdT}{v-b} - \frac{RTdv}{(v-b)^2} + \frac{2adv}{v^3} \qquad (A)$$

or $$\frac{RdT}{v-b} = \left[\frac{RT}{(v-b)^2} - \frac{2a}{v^3}\right] dv \text{ at constant } P$$

or $$\left(\frac{\partial v}{\partial T}\right)_P = \frac{R}{v-b} \frac{1}{\left[\frac{RT}{(v-b)^2} - \frac{2a}{v^3}\right]} = \frac{Rv^3(v-b)}{RTv^3 - 2a(v-b)^2}$$

Hence $$\beta = \frac{1}{v}\left(\frac{\partial v}{\partial T}\right)_P = \frac{Rv^2(v-b)}{RTv^3 - 2a(v-b)^2}$$

From Eqn. (*A*), we get

$$dP = -\left[\frac{RT}{(v-b)^2} - \frac{2a}{v^3}\right] dv \qquad \text{(at constant } T)$$

or $$\left(\frac{\partial v}{\partial P}\right)_T = -\frac{v^3(v-b)^2}{RTv^3 - 2a(v-b)^2}$$

Hence $$\kappa = \frac{-1}{v}\left(\frac{\partial v}{\partial P}\right)_T = \frac{v^2(v-b)^2}{RTv^3 - 2a(v-b)^2}$$

**Example 3.68**

State the Redlich – Kwong (RK) equation of state.

*Solution* :

The Redlich – Kwong (RK) equation of state is in wide use and is given by.

$$P = \frac{RT}{(v-b)} - \frac{a}{v(v+b)}$$

where $a$ and $b$ are characteristic of a gas. The parameters $a$ and $b$ can be related to the critical constants $T_C$ and $P_C$ by making use of the fact that the critical isotherm shows a point of inflexion at the critical point.

**Example 3.69**

State the Peng – Robinson (PR) equation of state.

*Solution* :

The Peng – Robinson (PR) equation of state is widely used for predicting the therodynamic properties of hydrocarbons. The PR equation of state is given by

$$P = \frac{RT}{v-b} - \frac{a}{v(v+b)+b(v-b)}$$

where the parameters $a$ and $b$ are characteristic of a gas.

**Example 3.70**

State the Virial equation of state and explain the meaning of the constants appearing in that equation of state.

*Solution* :

In 1901 Kamerlingh Onnes suggested the virial equation of state which is expressed as a power series in reciprocal volumes. The Leiden form ( it is called to since it originated from the University of Leiden) or the pressure explicit form of the virial equation of state is given by

$$\frac{Pv}{RT} = 1 + \frac{B}{v} + \frac{C}{v^2} + \frac{D}{v^3} \quad \ldots\ldots\ldots\ldots$$

where $B$, $C$, $D$, etc. are known as second virial coefficient, third virial coefficient, fourth virial coefficient etc. The term virial comes from the Latin word vis meaning force. Thus the virial coefficients take into account the interaction forces between the molecules. The virial coefficients express the deviation from ideal gas behavior and they depend on the nature the gas and temperature only. The second virial coefficient takes into account two body interactions and the third virial coefficient takes into account three body interactions etc. The advantage of the virial equation of state is that it can be made to represent the experimental $P - v - T$ data more accurately by increasing the number of terms in the power series depending on the complexity of the substance.

**Example 3.71**

Is there another form of the virial equation of state ? If so state the other form of the virial equation of sate.

*Solution* :

The virial equation of state is sometimes expressed as a power series in the pressure as

$$\frac{Pv}{RT} = 1 + B'P + C'P^2 + D'P^3 +$$

where $B'$, $C'$, $D'$ etc. are called virial coefficients which depend on the nature of the substance and temperature only. This equation is called the Berlin form or the volume explicit virial equation of state.

**Example 3.72**

Are the virial coefficient $B'$, $C'$, $D'$ of the Berlin form or volume explicit form related to the virial coefficient $B,C,D$ of the Leiden form or the pressure explicit form ? If so derive the necessary relations.

*Solution* :

The virial coefficients $B'$, $C'$, $D'$ are related to the virial coefficients $B$, $C$, $D$ of the pressure explicit virial equation of state.

The pressure explicit virial equation is given by

$$\frac{Pv}{RT} = 1 + \frac{B}{v} + \frac{C}{v^2} + \frac{D}{v^3} + \ldots\ldots\ldots.$$

$$P = \frac{RT}{v} + \frac{BRT}{v^2} + \frac{CRT}{v^3} + \frac{DRT}{v^4} + \ldots\ldots\ldots. \quad (A)$$

The volume explicit virial equation of state is given by

$$\frac{Pv}{RT} = 1 + B'P + C'P^2 + D'P^3 + \ldots\ldots\ldots.$$

or

$$P = \frac{RT}{v} + \frac{B'PRT}{v} + \frac{C'P^2RT}{v} + \frac{D'P^3RT}{v} + \ldots\ldots\ldots. \quad (B)$$

Substitute Eqn $(A)$ in Eqn $(B)$ to get

$$P = \frac{RT}{v} + \frac{B'RT}{v}\left[\frac{RT}{v} + \frac{BRT}{v^2} + \frac{CRT}{v^3} + \frac{DRT}{v^4} + \ldots\ldots\ldots\right]$$

$$+ \frac{C'RT}{v}\left[\frac{RT}{v} + \frac{BRT}{v^2} + \frac{CRT}{v^3} + \frac{DRT}{v^4} + \ldots\ldots\ldots\ldots\right]^2$$

$$+ \frac{D'RT}{v}\left[\frac{RT}{v} + \frac{BRT}{v^2} + \frac{CRT}{v^3} + \frac{DRT}{v^4} + \ldots\ldots\ldots\right]^3 \quad (C)$$

Now, we find that both Eqns ($A$) and ($C$) are expressed as power series is reciprocal volumes. Comparing the like terms we get the following results.

$$\text{Terms of } \frac{1}{v^2} \text{ gives : } \frac{BRT}{v^2} = \frac{B'(RT)^2}{v^2} \text{ or } B' = \frac{B}{RT}$$

$$\text{Terms of } \frac{1}{v^3} \text{ gives : } \frac{CRT}{v^3} = \frac{B'B(RT)^2}{v^3} + \frac{C'(RT)^3}{v^3}$$

or $$C = B'BRT + C'(RT)^2 = B^2 + C'(RT)^2$$

or $$C = \frac{C - B^2}{(RT)^2}$$

$$\text{Terms of } \left(\frac{1}{v^4}\right) \text{ gives :}$$

$$\frac{DRT}{v^4} = \frac{B'C(RT)^2}{v^4} + \frac{2BC'(RT)^3}{v^4} + \frac{D'(RT)^4}{v^4}$$

$$= \frac{BCRT}{v^4} + \frac{2B(C - B^2)RT}{v^4} + \frac{D'(RT)^4}{v^4}$$

or $$D' = \frac{D - 3BC + 2B^3}{(RT)^3}$$

**Example 3.73**

Determine the molar volume of n – butane at 500 K and 8 MPa by making use of ($a$) the ideal gas law and ($b$) the virial equation of state. The virial coefficients of n – butane are given by

$B = -0.265 \times 10^{-3}\ \text{m}^3/\text{mol}$ and $C = 0.3025 \times 10^{-7}\ \text{m}^6/\text{mol}^2$

*Solution* :

(*a*) The ideal gas equation of state gives

$$v = \frac{RT}{P} = \frac{8.314 \times 500}{8 \times 10^6} = 0.5196 \times 10^{-3}\ \text{m}^3/\text{mol}$$

(*b*) The virial equation of state is given by

$$\frac{Pv}{RT} = 1 + \frac{B}{v} + \frac{C}{v^2} \qquad (A)$$

This virial equation of state is pressure explicit and hence it is easier to calculate $P$ if $v$ is given. On the otherhand to calculate $v$ from a knowledge of $P$ it is necessary to solve the equation by trial and error. Let us assume $v = 0.219 \times 10^{-3}$ m³/mol and check whether equation ($A$) is satisfied or not.

L.H.S of Eqn ($A$) gives : $\dfrac{Pv}{RT} = \dfrac{8 \times 10^6 \times 0.219 \times 10^{-3}}{8.314 \times 500} = 0.4215$

R.H.S of Eqn ($A$) gives $= 1 + \dfrac{B}{v} + \dfrac{C}{v^2}$

$$= 1 - \frac{0.265 \times 10^{-3}}{0.219 \times 10^{-3}} + \frac{0.3025 \times 10^{-7}}{(0.219 \times 10^{-3})^2} = 0.4206$$

The L. H. S and R . H . S of Eqn. ($A$) differs in the third decimal place only and hence we can say that Eqn. ($A$) is satisfied.

Therefore, $v = 0.219 \times 10^{-3}\ \text{m}^3/\text{mol}$

**Example 3.74**

Define the terms reduced temperature, reduced pressure and reduced volume.

*Solution* :

The reduced temperature of a substance is defined as the ratio of the temperature of the substance to its critical temperature. That is,

Reduced temperature $T_r = T/T_C$.

Similarly reduced pressure $P_r = P/P_C$

Reduced volume $v_r = v/v_C$

**Example 3.75**

Show that for all substances which obey the Van der Waals equation of state.

$$\frac{P_C v_C}{RT_C} = 0.375$$

*Solution* :

We know that for a Van der waals gas

$$v_C = 3b \text{ and } b = \frac{RT_C}{8P_C}$$

Therefore $$b = \frac{v_C}{3} = \frac{RT_C}{8P_C}$$

or $$\frac{P_C v_C}{RT_C} = \frac{3}{8} = 0.375$$

**Example 3.76**

Express the Van der Waals equation of state in terms of reduced parameters

or

Deduce the reduced equation of state.

*Solution* :

We know that the Van der Waals equation of state is given by

$$(P + a/v^2)(v - b) = RT \qquad (A)$$

and the Van der Waals constants are given by

$$a = \frac{27R^2T_C^2}{64P_C} \qquad (B)$$

and $$b = \frac{RT_C}{8P_C} \qquad (C)$$

Substituting Eqns. (*B*) and (*C*) in Eqn (*A*), we get

$$P + \left(\frac{27R^2T_C^2}{64P_C v^2}\right)\left(v - \frac{RT_C}{8P_C}\right) = RT \qquad (D)$$

Expressing *P*, *v* and *T* in terms of the reduced parameters Eqn. (*D*) becomes

$$\left(P_r P_C + \frac{27R^2T_C^2}{64P_C v_C^2 v_r^2}\right)\left(v_C v_r - \frac{RT_C}{8P_C}\right) = RT_C T_r$$

or $$\left(\frac{P_C v_C}{RT_C}\right)\left[P_r + \frac{27}{64v_r^2}\left(\frac{RT_C}{P_C v_C}\right)^2\right]\left[v_r - \frac{1}{8}\left(\frac{RT_C}{P_C v_C}\right)\right] = T_r \qquad (E)$$

We also know that for a Van der Waals gas

$$v_C = 3b \qquad (F)$$

From Eqns (*C*) and (*F*), we get

$$b = \frac{v_C}{3} = \frac{RT_C}{8P_C} \text{ or } \frac{P_C v_C}{RT_C} = \frac{3}{8} \qquad (G)$$

Substituting Eqn. (*G*) in Eqn (*E*) we get,

$$\frac{3}{8}(P_r + 3/v_r^2)(v_r - \frac{1}{3}) = T_r$$

or $$(P_r + 3/v_r^2)(3v_r - 1) = 8T_r \qquad (H)$$

The Eqn (*H*) contains only reduced parameters and is called the reduced equation of state.

## Example 3.77

State the law of corresponding states.

*Solution* :

The Van der Waals equation when expressed in reduced parameters gives $(P_r + 3/v_r^2)(3v_r - 1) = 8T_r$. This equations does not contain any constants which are characteristic of individual gases. This equation is true for all fluids which obey the Van der Waals equation of state and leads to the law of two parameter corresponding states. The law of corresponding states tells that all fluids when considered at the same values of reduced temperature and reduced pressure will have the same value of reduced volume. That is all fluids having the same value of $T_r$ and $P_r$ are said to be in corresponding states.

## Example 3.78

Define the compressibility factor and explain its siginificance.

*Solution* :

The compressibility factor Z is defined as the ratio of the actual volume to the volume predicted by the ideal gas law at the same temperature and pressure. That is,

$$Z = \frac{\text{Actual volume}}{\text{Volume predicted by the ideal gas}}$$

or $$Z = \frac{v}{RT/P} = \frac{Pv}{RT}$$

For an ideal gas Z = 1at all temperatures and pressures. For a real gas Z can be different from unity. The value of Z indictates the extent of deviation from the ideal gas law.

**Example 3.79**

Express the Van der Waals equation of state in terms of the compressibility factor, reduced temperature and reduced pressure.

*Solution* :

The Van der Waals equation of state is given by

$$(P + a/v^2)(v - b) = RT \tag{A}$$

and the parameters $a$ and $b$ are given by

$$a = \frac{27R^2T_C^2}{64P_C} \tag{B}$$

$$b = \frac{RT_C}{8P_C} \tag{C}$$

We also know that $$v = \frac{ZRT}{P} \tag{D}$$

substituting Eqns ($B$) – ($D$) in Eqn ($A$) and expressing it in terms of reduced parameters, we get

$$\left(P_rP_C + \frac{27R^2T_C^2P^2}{64P_CZ^2R^2T^2}\right)\left(\frac{ZRT}{P} - \frac{RT_C}{8P_C}\right)$$

or $$\frac{P_r}{T_r}\left(1 + \frac{27}{64}\frac{P_r}{T_r^2Z^2}\right)\left(\frac{ZT_r}{P_r} - \frac{1}{8}\right) = 1$$

or $$\left(Z + \frac{27P_r}{64ZT_r^2}\right)\left(1 - \frac{P_r}{8ZT_r}\right) = 1 \tag{E}$$

Eqn ($E$) expresses the compressibility factor $Z$ for a Van der Waals gas in terms of reduced temperature and reduced pressure.

**Example 3.80**

What conclusions can be drawn from the generalized reduced equation of state which is given below

$$\left(Z + \frac{27P_r}{64ZT_r^2}\right)\left(1 - \frac{P_r}{8ZT_r}\right) = 1_1$$

*Solution :*

The given reduced equation of state is of the form $Z = Z(T_r, P_r)$ and tells that all fluids when considered at the same values of $T_r$ and $P_r$ will have the same compressibility factor.

**Note** *The Van der Waals equation of state predicts that the critical compressibility factor $Z_C$ which is defined as*

$$Z_c = \frac{P_c v_c}{RT_c}$$

*is 0.375 for all fluids. A close look at the values of $Z_C$ given in Appendix–4 shows that $Z_C$ varies from 0.23 to 0.31. Therefore, the Van der Waals equation of state is not very accurate for predicting the P–v–T behavior of fluids. However, it leads to an important conclusion that $Z = Z(T_r P_r)$. The generalized compressibility chart has been developed based on this conclusion.*

**Example 3.81**

There are two broad regions in the generalized compressibility chart. In one region $Z < 1$ and in the other $Z > 1$. Name the important factors which are likely to play a dominant role in these regions.

*Solution :*

When $Z < 1$, the pressure exerted by the gas is $P = \frac{ZRT}{V}$. This value of pressure is less than the pressure exerted by an ideal gas $\left(P = \frac{RT}{V}\right)$. That is the actual gas pressure is less than the ideal gas pressure. In other words, there is a reduction in pressure and this can be caused by intermolecular attractive forces. Therefore, in the region where $Z < 1$, the intermoleculor forces of attraction play a dominant role.

In the region where $Z > 1$, the volume occupied by the gas is $\left(v = \frac{ZRT}{P}\right)$ This value of volume occupied by the actual gas is greater than the volume occupied by the ideal gas $\left(v = \frac{RT}{P}\right)$. That is, the actual volume occupied by the gas is the sum of the volume occupied by the molecules and the volume available for free molecular motion (which is the ideal gas volume). Hence in the region in which $Z > 1$, the volume occupied by the molecules plays a dominant role.

**Example 3.82**

Is it necessary to incorporate any additional correction while using the generalized compressibility chart ?

*Solution* :

It has been observed that the gases— hydrogen, helium and neon do not correlate very well on a generalized compressibility chart. This difficulty can be overcome, at temperatures above 50 K, by redefining the reduced pressure and reduced temperature for these three gases as given below.

$$P_r = \frac{P}{P_C + 8} \text{ and } T_r = \frac{T}{T_C + 8}$$

where $P$ is in atmospheres and $T$ is in kelvin.

**Example 3.83**

Calculate the molar volume of methane ($CH_4$) at 209.77 K and 69.62 bar, using the generalized compressibility chart.

*Solution* :

The generalized compressibility chart is presented in Appendix – 5 which can be used in analyzing this problem. The critical constants for methane are read from Appendix – 4

$$T_C = 190.70 \text{ K and } P_C = 46.41 \text{ bar}$$

$$T_r = \frac{T}{T_C} = \frac{209.77}{190.70} = 1.1$$

$$P_r = \frac{P}{P_C} = \frac{69.62}{46.41} = 1.5$$

At $T_r = 1.1$ and $P_r = 1.5$, read the value of $Z = 0.48$ from the compressibility chart given in Appendix – 5 as shown in Fig.E 3.83.

Then $$v = \frac{ZRT}{P} = \frac{0.48 \times 8.314 \times 209.77}{69.62 \times 10^5}$$

$$= 0.1202 \times 10^{-3} \text{ m}^3/\text{mol}$$

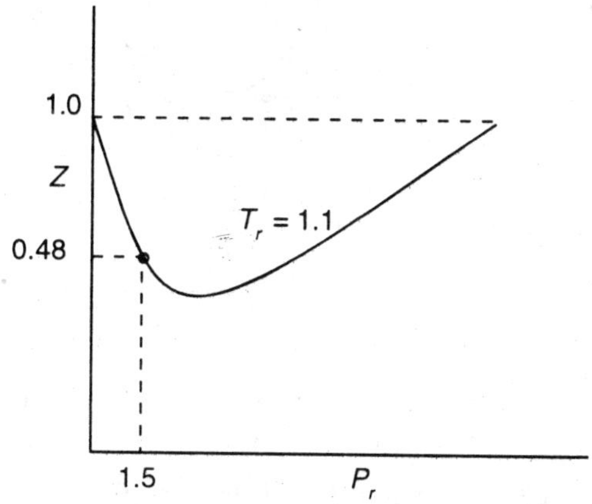


Fig.E 3.83. Sketch for Example 3.83.

**Example 3.84**

Calculate the specific volume of steam at 500°C and 300 bar using (*a*) the ideal gas law (*b*) generalized compressibility chart and (*c*) steam tables.

*Solution* :

(*a*) We know that the ideal gas equation of state gives

$$v = \frac{RT}{P} = \frac{8.314 \times 773.15}{300 \times 10^5} = 0.2143 \times 10^{-3} \text{ m}^3/\text{mol}$$

$$v = \frac{0.2143 \times 10^{-3}}{18 \times 10^{-3}} = 0.0119 \text{ m}^3/\text{kg}$$

(Molar mass of water = $18 \times 10^{-3}$ kg/mol)

(*b*) The critical constants for water are (see Appendix – 4) given by

$$T_C = 647.30 \text{ K and } P_c = 221.20 \text{ bar}$$

Therefore, $$T_r = \frac{T}{T_C} = \frac{773.15}{647.30} = 1.19$$

$$P_r = \frac{P}{P_C} = \frac{300}{221.2} = 1.36$$

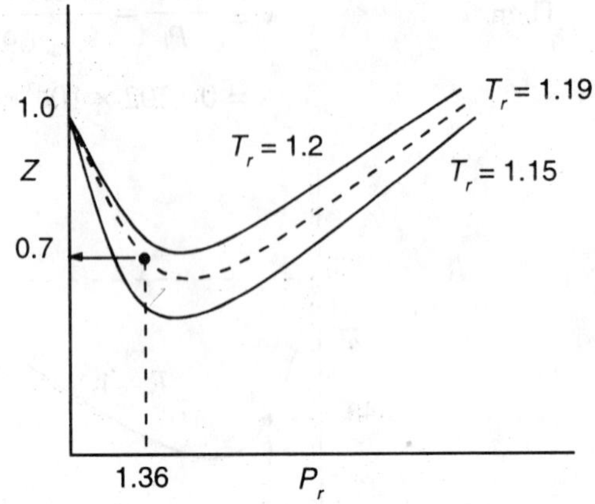


Fig.E 3.84. Sketch for Example 3.84.

At $T_r = 1.19$ and $P_r = 1.36$, read the value of $Z = 0.7$ as shown in Fig.E 3.84.

$$\text{There } v = \frac{ZRT}{P} = \frac{0.7 \times 8.314 \times 773.15}{300 \times 10_5} = 0.145 \times 10^{-3} \text{ m}^3 / \text{mol}$$

or
$$v = \frac{0.145 \times 10^{-3}}{18 \times 10^{-3}} = 0.0081 \text{ m}^3 / \text{kg}$$

(*c*) From the superheated steam tables ( see Appendix –3 ) at $t = 500$°C and $P = 300$ bar read $v = 0.008\ 681$ m³/ kg

**Example 3.85**

One kmol of methane is stored in a rigid vessel of volume 0.6 m³ at 20°C. Determine the pressure developed by the gas by making use of the compressibility chart.

*Solution* :

The critical constants of methane ( see Appendix – 4 ) are

$$T_C = 190.70 \text{ K and } P_C = 46.41 \text{ bar}$$

$$T_r = \frac{T_r}{T_C} = \frac{293.15}{190.70} = 1.537$$

To calculate the pressure developed by the gas, we should first determine the value of the compressibility factor $Z$, which requires a knowledge of $T_r$ and $P_r$. However, $P_r$ cannot be calculated which requires a knowledge of $P$, the property we are supposed to estimate. Therefore, the following method may be adopted to evaluate $Z$ and $P$.

$$P_r = P / P_C \text{ or } P = P_C P_r = 46.41 \times 10^5 P_r$$

$$Z = \frac{Pv}{nRT} = \frac{46.41 \times 10^5 P_r (0.6)}{1 \times 10^3 \times 8.314 \times 293.15} = 1.1425 P_r \qquad (A)$$

Eqn. (*A*) shows that the gas satisfies the relation $Z = 1.1425\ P_r$. Hence, the state of the gas must lie on a straight line passing through the origin and having a slope of 1.1425 on *Z* vessus $P_r$ chart. More over, the state of the gas must satisfy the condition that $T_r = 1.537$. Therefore, draw a straight line passing through the origin and with slope 1.1425 on the compressibility chart and locate the intersection of this line with $T_r = 1.537$ as shown in Fig.E 3.85.

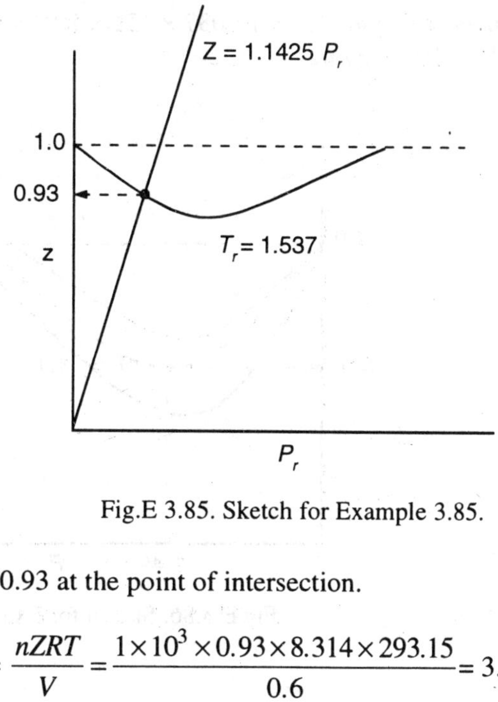


Fig.E 3.85. Sketch for Example 3.85.

Read Z = 0.93 at the point of intersection.

$$\text{Then } P = \frac{nZRT}{V} = \frac{1 \times 10^3 \times 0.93 \times 8.314 \times 293.15}{0.6} = 3.778 \text{ MPa}$$

**Example 3.86**

Freon – 12 for charging domestic refrigerators is usually sold in small cylinders of volume 3 litres. Determine the mass of Freon – 12 ($CCl_2F_2$) contained in one such cylinder at 10 MPa and 450 K.

*Solution* :

The critical constants of Freon –12 ($CCl_2F_2$) are

$$T_C = 384.7 \text{ K} ; P_C = 40.12 \text{ bar}$$

$$T_r = \frac{T}{T_C} = \frac{450}{384.7} = 1.17$$

$$P_r = \frac{P}{P_C} = \frac{10 \times 10^6}{40.12 \times 10^5} = 2.49$$

At $T_r = 1.17$ and $P_r = 2.49$ read $Z = 0.5$ from the compressibility chart as shown in Fig.E 3.86.

Then $$n = \frac{PV}{ZRT} = \frac{10 \times 10^6 \times 3 \times 10^{-3}}{0.5 \times 8.314 \times 450} = 16.037 \text{ mol}$$

or mass of Freon –12 = 16.037 × 121 × $10^{-3}$ = 1.941 kg ( molar mass of $CCl_2F_2$ = 121 × $10^{-3}$ kg / mol).

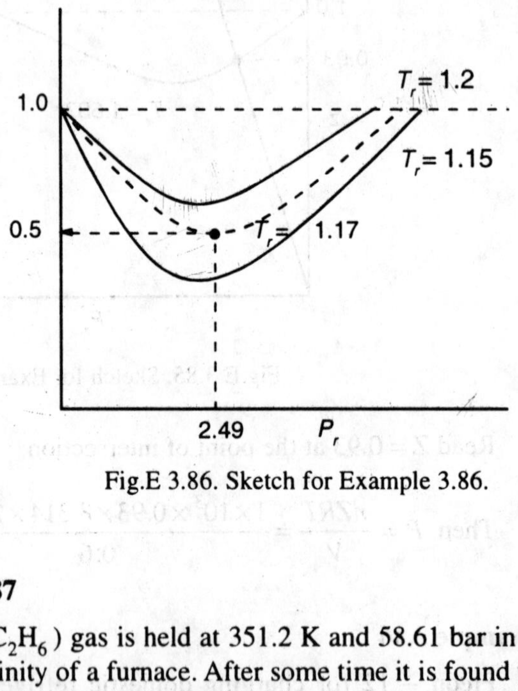


Fig.E 3.86. Sketch for Example 3.86.

**Example 3.87**

Ethane ($C_2H_6$ ) gas is held at 351.2 K and 58.61 bar in a rigid cylinder and left in the vicinity of a furnace. After some time it is found that the gas pressure in the cylinder reached a value of 68.38 bar. Determine the temperature of the gas in the cylinder.

*Solution* :

The critical constants of $C_2H_6$ are

$$T_C = 305\ 43 \text{ K} ; P_C = 48.84 \text{ bar}$$

Initial values of the reduced parameters are

$$T_{r1} = \frac{T_1}{T_C} = \frac{351.2}{305.43} = 1.15$$

$$P_{r1} = \frac{P_1}{P_C} = \frac{58.61}{48.84} = 1.2$$

From the compressibility chart read $Z_1 = 0.59$ at $P_{rl} = 1.2$ and $T_{r1} = 1.15$. Then

$$v_1 = \frac{Z_1 R T_1}{P_1} = \frac{0.59 \times 8.314 \times 351.2}{58.61 \times 10^5} = 0.2939 \times 10^{-3}\ \text{m}^3/\text{mol}$$

Since the gas is contained in a rigid cylinder, the quantity as well as volume of the cylinder remain unchanged. Therefore, the molar volume ($v_2$) in the final state is the same as the initial molar volume ($v_1$).

$$v_2 = v_l = 0.2939\ x\ 10^{-3}\ \text{m}^3/\text{mol}$$

Final values of the reduced parameters are

$$P_{r_2} = \frac{P_2}{P_C} = \frac{68.38}{48.84} = 1.4$$

$$T_{r2} = \frac{T_2}{T_C} \text{ or } T_2 = T_C T_{r_2} = 305.43\ T_{r2}$$

$$Z_2 = \frac{P_2 v_2}{R T_2} = \frac{68.38 \times 10^5 \times 0.2939 \times 10^{-3}}{8.314 \times 305.43 T_{r2}} = \frac{0.7914}{T_{r2}}$$

or $$Z_2 T_{r_2} = 0.7914 \qquad (A)$$

The compressibility chart is a plot of $Z$ versus $P_r$ with $T_r$ as a parameter and hence Eqn. ($A$) cannot be directly plotted on the compressibility chart to evaluate the values of $Z_2$ and $T_{r_2}$. Hence a trial and error procedure as indicated below can be adopted to determine $Z_2$ and $T_{r_2}$.

We know that the state of the gas should lie on the line $P_{r2} = 1.4$. Therefore, we read the following values at $P_{r2} = 1.4$

$$T_{r_2} = 1.18\ ;\ Z_2 = 0.665\ ;\ Z_2\ T_{r_2} = 0.7729$$

$$T_{r_2} = 1.2\ ;\ Z_2 = 0.685\ ;\ Z_2\ T_{r_2} = 0.822$$

Since the required value of $Z_2\ T_{r_2}$ must be equal to 0.7914, $T_{r_2}$ lies in between 1.18 and 1.2. We read at $P_{r2} = 1.4$

$$T_{r_2} = 1.19 \; ; Z_2 = 0.665 \; ; Z_2 \, T_{r_2} = 0.7914$$

Hence the final value of $T_{r_2} = 1.19$

or $$T_2 = 305.43 \, T_{r2} = 305.43 \times 1.19 = 363.46 \text{ K}$$

Therefore, final temperature of gas in the cylinder = 363.46 K

**Example 3.88**

Propane ( $C_3H_8$ ) initially at 388.4 K and 63.855 bar is heated at constant pressure till the temperature reaches 425.4 K. Calculate the change in the specific volume ($m^3$/kg) using (*a*) the ideal gas law and (*b*) the compressibility chart

*Solution* :

(*a*) Molar mass of propane ( $C_3H_8$) = 44 × $10^{-3}$ kg/mol.

Initial molar volume, $v_1 = \dfrac{RT_1}{P_1}$

Final molar volume, $v_2 = \dfrac{RT_2}{P_2} = \dfrac{RT_2}{P_1}$ (Since $P_2 = P_1$)

$$\text{Change in molar volume} = \frac{R}{P_1}(T_2 - T_1) = \frac{8.314(425.4 - 388.4)}{63.855 \times 10^5}$$

$$= 4.8174 \times 10^{-5} \text{ m}^3/\text{mol}$$

$$\text{Change in specific volume} = \frac{4.8174 \times 10^{-5}}{44 \times 10^{-3}} = 1.0949 \times 10^{-3} \text{ m}^3/\text{kg}$$

(*b*) The critical constants (see Appendix – 4) of propane are $T_C = 369.9$ K : $P_C = 42.57$ bar.

$$T_{r1} = \frac{T_1}{T_C} = \frac{388.4}{369.9} = 1.05$$

$$P_{r1} = \frac{P_1}{P_C} = \frac{63.855}{42.57} = 1.5$$

From the compressiblity chart read $Z_1 = 0.32$ at $T_{r1} = 1.05$ and $P_{r1} = 1.5$.

$$P_{r2} = P_{r1} = 1.5$$

(Since the gas in maintained at constant pressure)

$$T_{r_2} = \frac{T_2}{T_C} = \frac{425.4}{369.9} = 1.15$$

Now read $Z_2 = 0.59$ at $P_{r2} = 1.5$ and $T_{r_2} = 1.15$ from the compressibility chart.

$$v_1 = \frac{Z_1RT_1}{P_1} \text{ and } v_2 = \frac{Z_2RT_2}{P_2}$$

( $P_2 = P_1$ as $P$ is held constant )

$$\text{Change in molar volume, } (v_2 - v_1) = \frac{R}{P_1}(Z_2T_2 - Z_1T_1)$$

$$= \frac{8.314}{63.855 \times 10^5}(0.59 \times 425.4 - 0.32 \times 388.4)$$

$$= 1.6496 \times 10^{-4} \text{ m}^3/\text{mol}$$

$$\text{Change in specific volume} = \frac{1.6496 \times 10^{-4}}{44 \times 10^{-3}} = 3.851 \times 10^{-3} \text{ m}^3/\text{kg}$$

### Example 3.89

A dewar flask is filled to one half of its volume with liquid nitrogen at its normal boiling point 77 K. The molar volume of liquid nitrogen at these conditions is $34.7 \times 10^{-6}$ m$^3$/mol. The flask is sealed and left in a corridor. After some time if the temperature of the flask rises to 328.1 K, determine the pressure developed by the nitrogen in the flask.

*Solution* :

Let $V$ = volume of dewar flask

$$\text{Quantity of liquid nitrogen} = \frac{V/2}{34.7 \times 10^{-6}} \text{ mol}$$

A small quantity of nitrogen vapor occupies the volume $V/2$, which may be neglected.

$$\text{Final molar volume of nitrogen} = \frac{V}{V/(2 \times 34.7 \times 10^{-6})} = 69.4 \times 10^{-6} \text{ m}^3/\text{mol}$$

The critical constants of nitrogen (see Appendix – 4) are $T_C = 126.2$ K and $P_C = 33.94$ bar. In the final state

$$T_{r_2} = \frac{T_2}{T_C} = \frac{328.1}{126.2} = 2.6$$

$$P_{r_2} = \frac{P_2}{P_C} \text{ or } P_2 = P_C P_{r_2} = 33.94 \times 10^5 P_{r_2}$$

$$Z_2 = \frac{P_2 v_2}{RT_2} = \frac{33.94 \times 10^5 (P_{r2}) 69.4 \times 10^{-6}}{8.314 \times 328.1}$$

$$= 0.0863\ P_{r_2} \qquad (A)$$

Eqn $(A)$ represents a straight line passing through the origin and having a slope of 0.0863 on the compressibility chart.

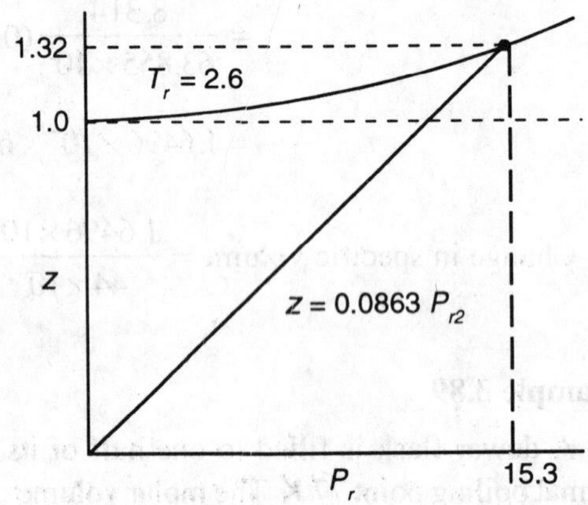


Fig.E 3.89. Sketch for Example for 3.89.

Find the intersection of the line $Z = 0.0863\ P_{r_2}$ with the curve $T_r = 2.6$ on the compressibility chart as shown in FigE.3.89 and read $Z_2 = 1.32$ and $P_{r_2} = 15.3$.

Then $P_2 = 33.94 \times 10^5\ P_{r_2} = 33.94 \times 10^5 \times 15.3 = 51.928$ MPa

Hence, the pressure developed in the flask is 51.928 MPa

**Example 3.90**

The industrial gas cylinders are color coded to assure that the gases are filled in the correct cylinders only. Accidentally a nitrogen cylinder is filled with ethylene ($C_2H_4$) at 80 bar and 311.4 K. Had the cylinder been filled with nitrogen under identical conditions, it contains 7 kg nitrogen. Estimate the mass of ethylene filled in the cylinder.

*Solution* :

The critical constants (see Appendix – 4) of nitrogen and ethylene are

$$T_{C_{N_2}} = 126.2 \text{ K};\ P_{C_{N_2}} = 33.94 \text{ bar}$$

$$T_{C_{C_2H_4}} = 283.1 \text{K};\ P_{C_{C_2H_4}} 51.17 \text{ bar}$$

$$T_{r_{N_2}} = \frac{T}{T_{c_{N_2}}} = \frac{311.4}{126.2} = 2.4675$$

$$P_{r_{N_2}} = \frac{P}{P_{c_{N_2}}} = \frac{80}{33.94} = 2.3571$$

Read $Z_{N_2} = 0.98$ at $P_{r_{N_2}} = 2.3571$ and $T_{r_{N_2}} = 2.4675$ from the compressibility chart.

$$P_{r_{C_2H_4}} = \frac{P}{P_{c_{C_2H_4}}} = \frac{80}{51.17} = 1.5634$$

$$T_{r_{C_2H_4}} = \frac{T}{T_{c_{C_2H_4}}} = \frac{311.4}{283.1} = 1.1$$

Read $Z_{C_2H_4} = 0.47$ at $P_{r_{C_2H_4}} = 1.5634$ and $T_{r_{C_2H_4}} = 1.1$ from the compressibility chart. Let $V$ = volume of cylinder

$$n_{N_2} = \frac{PV}{Z_{N_2}RT}; \quad n_{C_2H_4} = \frac{PV}{Z_{C_2H_4}RT}$$

$$\frac{n_{C_2H_4}}{n_{N_2}} = \frac{Z_{N_2}}{Z_{C_2H_4}} = \frac{0.98}{0.47} = 2.0851$$

Molar mass of nitrogen = $28 \times 10^{-3}$ kg/mol. Molar mass of $C_2H_4 = 28 \times 10^{-3}$ kg/ mol. Since the molar masses of $N_2$ and $C_2H_4$ are the same, we get

$$\frac{\text{Mass of } C_2H_4}{\text{Mass of } N_2} = \frac{n_{C_2H_4}}{n_{N_2}} = 2.0851$$

Hence the mass of ethylene in the cylinder = $2.0851 \times 7 = 14.596$ kg.

**Example 3.91**

What accuracy can be expected in the predicted values of $P$, $v$ and $T$ if one uses the generalized compressiblity chart ?

*Solution :*

The generalized compressibility chart predicts the volumetric data within 5 percent error for non–polar gases and slightly polar gases. However, the deviations are very large for highly polar gases and hence this chart cannot be used for polar gases. It can be observed from the compressibility chart (see Appendix – 5) that the critical compressibility factor $Z_C$, that is $Z$ at $P_r = 1$ and $T_r = 1$ is 0.3 whereas the experimental values range from 0.23 to 0.31. This indicates that the generalized compressiblity chart is necessarily not accurate in predicting the volumetric data. It can be used to obtain approximate values only and never as a substitute for experimental *P–v–T* data. More over the three gases hydrogen, helium and neon do not correlate very well on the compressibility chart. To determine the compressibility factor of these three gases , it is necessary to redefine the reduced temperature and reduced pressure by using the following relations

$$T = \frac{T}{T_c + 8} \text{ and } P_r = \frac{P}{P_c + 8}$$

Where $T$ is in kelvin $P$ is in atmospheres.

**Example 3.92**

Define the acentric factor (ω).

*Solution* :

The acentric factor (ω). is defined as

$$\omega = -\log_{10} P_r^s|_{Tr=0.7} - 1$$

where $P_r^s$ is the reduced saturation pressure. The acentric factor is defined such that its value is equal to zero for the spherical molecules argon, krypton and xenon. The acentric factor is a measure of deviation of the intermolecular potential of the molecule from that of a spherical molecule..

**Example 3.93**

State the three parameter law of corresponding states.

*Solution* :

The three parameter law of corresponding states tells that all fluid having the same values of $P_r$, $T_r$ and ω will have the same value of the compressibility factor $Z$, and they are said to be in corresponding states. In other words $Z = Z(T_r, P_r, \omega)$

**Note** ***Pitzer and co workers have developed the generalized compressibility charts based on the three parameter law of corresponding states. The discussion on the modified compressibility chart is not included in here as it is beyond the scope of the present book.***

# 4

# First Law Of Thermodynamics and its Applications

**Example 4.1**

What is meant by cyclic change.?

*Solution* :

If a thermodynamic system undergoes a process or a combination of processes, as a result of which the system is restored to its initial state then the system is said to have undergone a cyclic change.

**Example 4.2**

Describe the classic paddle wheel experiment performed by Joule ? What conclusion was drawn based on the experimental observations ?

*Solution* :

Joule carried out a series of experiments during 1843 and 1848. Of the several experiments conducted by Joule, the paddle wheel experiment described below is considered as a classic. In the paddle wheel experiment a known quantity of liquid is taken in a rigid and well insulted container. The container was provided with a thermometer and a paddle wheel. The initial temperature of the liquid was measured. The paddle wheel was made to rotate by lowering the mass $m$ which is tied to a string and the string was wound on the paddle wheel as shown in Fig.E 4.2. While lowering the mass $m$ through a distance $Z$, work was done on the liquid (which is the system) and the amount of work done on the system was measured. Then the system has reached a different state and the temperature of the liquid was noted. Then the insulation surrounding the container was removed and the container was placed in a water bath. The water bath was provided with another thermometer. A known quantity of water at a known temperature was taken in the water bath. The container was placed in the water bath till the temperature of the liquid in the container was restored to the initial value. Then the rise in the temperature of the water bath was measured. Knowing the change in the temperature of the water bath, the amount of energy transferred as heat from the system to the surroundings (the water bath) was estimated. In this experiment the system, the liquid contained in the container, was restored to its initial state. In other words, the system has undergone a cyclic change. As a result of this experiment Joule found that the work done on the system is equal to the energy transferred as heat from the system.

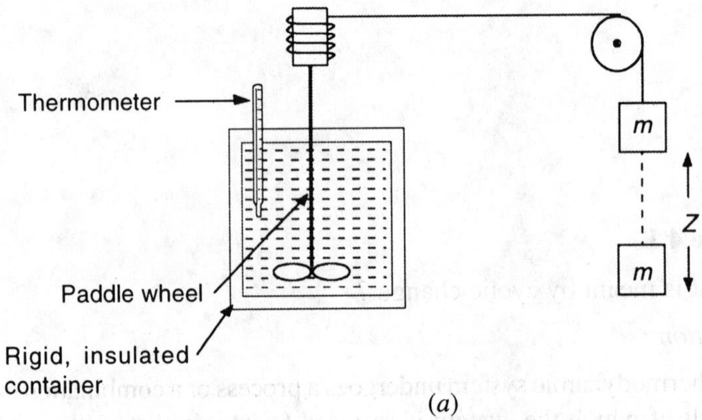


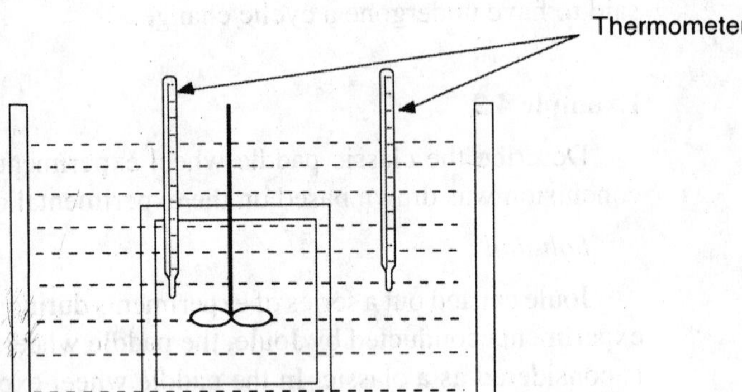


Fig.E 4.2. Schematic arrangement of paddle wheel experiment.

Joule carried out several experiments involving different types of work interaction. Based on these experimental results Joule found that the net work done on the system is exactly equal to the net heat interaction, irrespective of the type of work interaction, the rate at which work was done and the method employed for transferring the energy in the form of heat from the system. That is

$$\underset{\text{From system}}{\oint dQ} = \underset{\text{on system}}{\oint dW}$$

or

$$\underset{\text{From system}}{\oint dQ} - \underset{\text{on system}}{\oint dW} = 0$$

where $\oint$ denotes the integral over a cycle.

**Example 4.3**

Specify the most widely used sign convention for work and heat interaction.

*Solution* :

The sign convention for work and heat interactions is quite arbitrary. The widely used sign convention is shown in Fig.E 4.3. The work done by a system on the surroundings is treated as a positive quantity. Similarly, the energy transferred as heat to a system from the surroundings is also treated as a positive quantity.

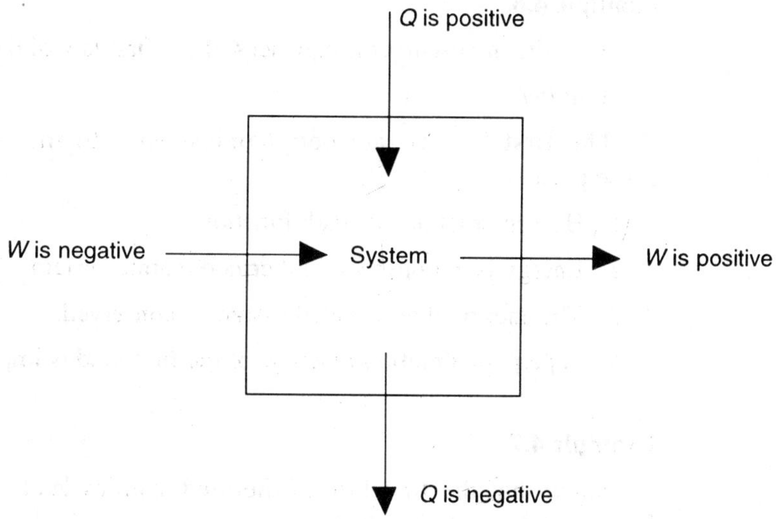


Fig.E 4.3. Sign convention for work and heat interaction.

**Example 4.4**

State the first law of thermodynamics.

*Solution*:

The first law of thermodynamics states that whenever a system undergoes a cyclic change, however complex the cycle may be, the algebraic sum of the work interactions is equal to the algebraic sum of the heat interactions. That is

$$\oint dQ \quad - \quad \oint dW = 0$$

From system On system

By introducing the sign convention for work and heat interactions,the above equation reduces to

$$-\oint dQ - \oint(-dW) = \oint dQ - \oint dW = 0$$

or $$\oint(dQ - dW) = 0$$

**Example 4.5**

Is the first law of thermodynamics applicable to irreversible processes ?

*Solution*:

The first law of thermodynamics is applicable to all types of processes– reversible as well as irreversible.

**Example 4.6**

State the important consequences of the first law of thermodynamics.

*Solution* :

The first law of thermodynamics leads to the following important consequences.

1. Heat interaction is a path function.
2. Energy is a property of a thermodynamic system.
3. The energy of an isolated system is conserved.
4. A perpetual motion machine of the first kind is impossible.

**Example 4.7**

Show that the first law of thermodynamics leads to the fact that heat interaction is a path function.

*Solution* :

Suppose a system follows the path 1-a -2 in reaching the final state 2 starting from the initial state 1. Then the system can be restored to the initial state 1 either by the path 2b1 of 2c1 as shown in Fig.E 4.7. Then the combination of the processes leads to two different cycles, namely 1a 2b1 and 1a2c1. Applying the first law of thermodynamics to the cycles 1a2b1 and 1a2c1, we get

$$\int_{1a2} dQ + \int_{2b1} dQ - \int_{1a2} dW - \int_{2b1} dW = 0 \tag{A}$$

$$\int_{1a2} dQ + \int_{2c1} dQ - \int_{1a2} dW - \int_{2c1} dW = 0 \tag{B}$$

Subtracting Eqn. (*B*) from Eqn. (*A*), we get

$$\int_{2b1} dQ - \int_{2c1} dQ - \int_{2a1} dW - \int_{2c1} dW = 0 \tag{C}$$

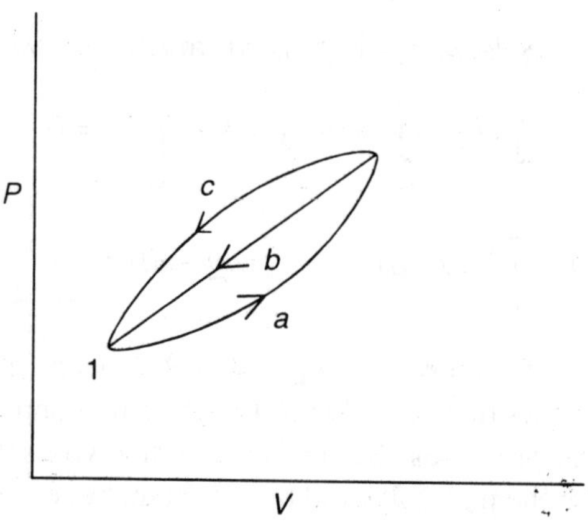


Fig.E 4.7. Two reversible cycles1-a-2-b-1 and 1-a-2-c-1 on *P* versus *V* diagram.

We know that work interaction is a path function and $\int dW$ represents the area under the *P–V* curve. Hence

$$\int_{2b1} dW - \int_{2c1} dW \neq 0 \tag{D}$$

From Eqns.(*C*) and (*D*), we get

$$\int_{2b1} dQ - \int_{2c1} dQ \neq 0 \quad \text{or} \quad \int_{2b1} dQ \neq \int_{2c1} dQ$$

That is, the heat interaction along the path 2b1 is different from the heat interaction along the path 2c1. Thus, heat interaction depends on the path followed by a system and heat interaction is a path function. It is not a point function and hence its differential is not exact. Heat interaction is not a property of a thermodynamic system.

**Example 4.8**

Show that the first law of thermodynamics leads to the consequence that energy is a property of a thermodynamic system.

*Solution* :

Consider the two reversible cycles 1a2b1 and 1a2c1 shown in Fig.E 4.7. Application of the first law of thermodynamic to both the cycles 1a2b1 and 1a2c1 we get

$$\int_{1a2} dQ + \int_{2b1} dQ - \int_{1a2} dW - \int_{2b1} dW = 0 \qquad (A)$$

$$\int_{1a2} dQ + \int_{2c1} dQ - \int_{1a2} dW - \int_{2c1} dW = 0 \qquad (B)$$

Subtracting Eqn.(*B*) from Eqn.(*A*), we get

$$\int_{2b1} dQ - \int_{2c1} dQ - \int_{2b1} dW + \int_{2c1} dW = 0$$

or $$\int_{2b1} (dQ - dW) = -\int_{2c1} (dQ - dW) \qquad (C)$$

We know that $\int dQ$ and $\int dW$ depend on the path followed by a system. Eqn.(*C*) shows that the value of $\int (dQ - dW)$ connecting the states 1 and 2 is the same for the processes 2b1 and 2c1. In other words the quantity $(dQ - dW)$ does not depend on the path followed by a system, but depends only on the initial and final states of a system. Hence, the quantity $(dQ - dW)$ is an exact differential. It can be said that it is the differential of a property of a system. This property is called the energy $E$. Then the differential change in the energy of a system is given by

$$dE = dQ - dW \qquad (D)$$

That is, there exits a property called the energy of a system and its differential change is related to the work and heat interactions as given by Eqn.(*D*). We also know that the energy of a system is the sum of macroscopic and microscopic modes of energy and is given by

$$E = \text{KE} + \text{PE} + U \qquad (E)$$

where KE = Kinetic energy

PE = Potential energy

$U$ = Internal energy (the sum of all microscopic modes of energy)

Comparing Eqns. (*D*) and (*E*), we get

$$dE = d(\text{KE}) + d(\text{PE}) + dU = dQ - dW$$

**Example 4.9**

Show that the first law of thermodynamics leads to the result that the energy of an isolated system is conserved.

*Solution* :

A system which does not exchange energy with its surroundings through work and heat interactions is called an isolated system. That is for an isolated thermodynamic system $dW = 0$ and $dQ = 0$.

The first law of thermodynamics gives

$$dE = dQ - dW$$

Hence, for an isolated system, the first law of thermodynamics reduces to $dE = 0$ or $E_2 = E_1$. In other words, the energy of an isolated thermodynamic system remains constant. This fact is quite often stated as the principle of conservation of energy.

**Example 4.10**

Show that the first law of thermodynamics implies that a Perpetual Motion Machine of the First Kind (PMMFK) is impossible.

*Solution* :

We know that the macroscopic modes of energy can be converted from one form to the other and work can be obtained. However, the microscopic modes of energy cannot be readily converted into macroscopic modes of energy. An important application of thermodynamics is to devise means of converting the microscopic modes of energy into the macroscopic modes of energy. For this purpose heat engines which work cyclically are devised. The first law of thermodynamics when applied to a cyclic process gives

$$\oint (dQ - dW) = 0 \quad \text{or} \quad \oint dQ = \oint dW$$

or
$$Q = W$$

where $Q$ is the net heat interaction and $W$ is the net work delivered.

An imaginary device which would deliver work continuously without absorbing energy as heat is called a Perpetual Motion Machine of the First Kind (PMMFK). A perpetual motion machine of the first kind has to operate on a cycle to deliver work continuously. If the device does not absorb any energy as heat in a cycle, then $\int dQ = Q = 0$. Then the first law of thermodynamics tells that for such a device $\oint dW = W = 0$. Thus, the first law of thermodynamics implies that it is impossible to devise a PMMFK. A PMMFK violates the first law of thermodynamics.

### Example 4.11

Give an example of an irreversible constant volume process and a reversible constant volume process.

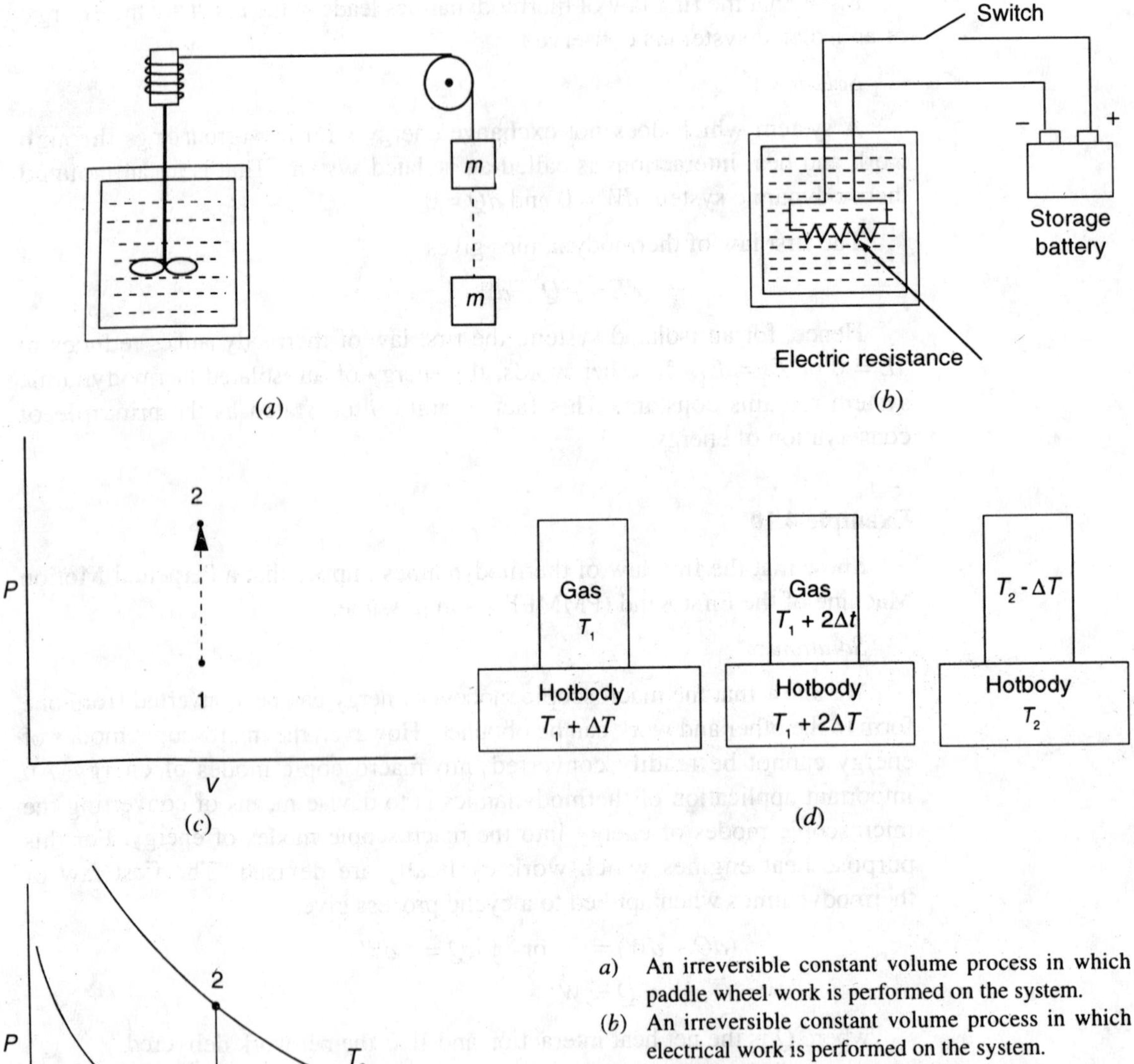


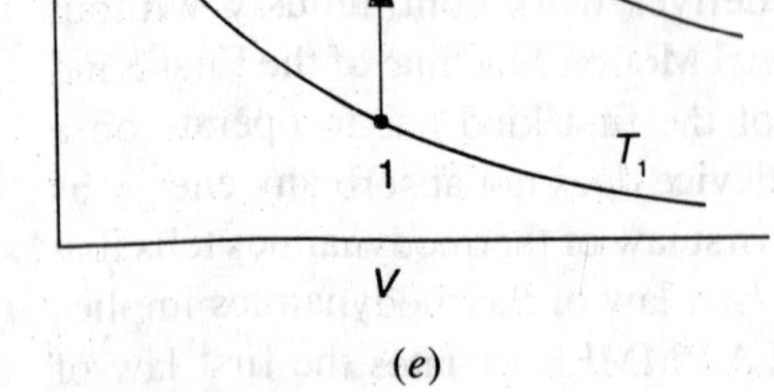


*a)* An irreversible constant volume process in which paddle wheel work is performed on the system.

*(b)* An irreversible constant volume process in which electrical work is performed on the system.

*(c)* *P-V* diagram for an irreversible constant volume process.

*(d)* Schematic arrangement showing a reversible constant volume process.

*(e)* *P-V* diagram for a reversible constant volume process.

Fig.E 4.11

*Solution* :

A process undergone by a system in which the volume of the system remains constant is called a constant volume process or an isochoric process. That is for a constant volume process, $V =$ constant or $dV = 0$.

Consider the liquid contained in a rigid and insulated vessel shown in Fig.E 4.11 (*a*), (*b*) as our system. In Fig.E 4.11 (*a*) work is done on the system by rotating the paddle wheel and in Fig.E 4.11 (*b*) work is done on the system by the storage battery. By performing work on the system the system temperature can be raised. This process can take place in one direction only but cannot proceed in the reverse direction. That is, we never observe in practice that the paddle wheel which is initially at rest wll rotate or the storage battery gets charged while the liquid cools. Therefore, the processes undergone by the system in Fig.E 4.11 (*a*) and (*b*) are irreversible. The irreversible process followed by a system cannot be traced on a thermodynamic diagram. We can represent the initial and final states of the system only. Usually an irreversible process is denoted by a dotted line connecting the initial and final states of the system as shown in Fig.E 4.11 (*c*).

Consider the gas held in a rigid vessel as the system. Let the initial temperature of the gas be $T_1$. The vessel can be placed in thermal contact with a hot body at $(T_1 + \Delta T)$ where $\Delta T \to 0$. Then energy transfers as heat from the hot body to the gas. Since $\Delta T \to 0$, the process can take place in either direction. Once the system attains the temperature $(T_1 + \Delta T)$ it can be placed in thermal contact with another hot body at $(T_1 + 2\Delta T)$ till the system attains the temperature $(T_1 + 2\ \Delta T)$. The system can be placed in thermal contact with a series of hot bodies at $(T_1 + 3\Delta T)...T_2$ till the system attains the temperature $T_2$ as shown in Fig.E 4.11 (*d*). Thus the system can be made to undergo a reversible constant volume process. In Fig.E 4.11 (*e*) the reversible constant volume process is shown by the solid line 1–2 on a $P$–$V$ diagram.

## Example 4.12

What conclusions can be drawn from the first law of thermodynamics for a constant volume process ?

*Solution* :

We know that for a constant volume process, $V =$ constant or $dV = 0$. Then the reversible work done is given by $dW = PdV = 0$.

The first law of thermodynamics gives

$$dE = d(\text{KE}) + d(\text{PE}) + dU = dQ - dW$$

Quite often in engineering thermodynamics the systems under consideration are such that there is no change in kinetic energy and potential energy, That is $d(\text{KE}) = d(\text{PE}) = 0$. Then the first law of thermodynamics reduces to

$$dU = dQ - dW \qquad (A)$$

For a reversible constant volume process, Eqn. ($A$), reduces to

$$dU = dQ \text{ or } U_2 - U_1 = Q$$

That is the change in the internal energy of a system is equal to the heat interaction in a reversible constant volume process.

Now consider the irreversible constant volume process shown in Fig.E 4.11 ($a$) and ($b$). In these irreversible constant volume processes, the system is insulated and hence $dQ = 0$. However, either paddle wheel work or electrical work is done on the system. Application of the first law of thermodynamics to such systems, gives

$$dU = -dW \text{ or } U_2 - U_1 = -W$$

Where ($-W$) represents the work done on the system by the surroundings. That is, the increase in the internal energy of the system is equal to the work done on the system.

***Note*** *In an irreversible constant volume process the system does not perform work on the surroundings at the expense of its internal energy.*

**Example 4.13**

Define the specific heat at constant volume and express it in terms of internal energy.

*Solution* :

The specific heat of a substance at constant volume is usually defined as the amount of energy required in the form of heat to change the temperature of a unit mass of the substance by one degree. That is, the specific heat at constant volume, $C_v$ is given by

$$C_v = \left(\frac{dq}{dT}\right)_v$$

where $q$ is the heat interaction per unit mass.

We know that for a constant volume process, the first law of thermodynamics reduces to

$$dU = dQ \text{ or } du = dq$$

Where $du$ is the change in the specific interal energy or internal energy per unit mass. Thus we find that when a system undergoes a constant volume process the heat interaction across the boundary is equal to the change in the property $u$ of the system. Then Eqn. ($A$) can be rewritten as

$$C_v = \left(\frac{\partial u}{\partial T}\right)_v \qquad (B)$$

We have also noticed that the temperature of a system can be increased at constant volume by performing either paddle wheel work or electrical work on it. The amount of work done on a system at constant volume is equal to the increase is its interal energy. Hence it is more appropriate to define $C_v$ according to Eqn.(*B*) which expresses the property $C_v$ of a system in terms of the property change ($du$).

**Example 4.14**

A rigid and insulated tank of volume 100 litres contains nitrogen at 300 K and 1 bar. An electrical resistor placed inside the tank is connected to a battery and current flows through the resistor till the gas raises to a temperature of 350 K. Assume that nitrogen is an ideal gas. The molar heat capacity at constant volume ($C_v$) of nitrogen is $5R/2$ where $R$ is the universal gas constant. Determine the work done by the battery and the final pressure of the gas in the tank.

*Solution* :

We know that for an ideal gas $PV = nRT$. Therefore,

$$n = \frac{PV}{RT} = \frac{1\times10^5\times100\times10^{-3}}{8.314\times300} = 4.0093 \text{ mol}$$

we know that $\quad C_V = \left(\frac{\partial u}{\partial T}\right)_v$ or $du = C_v\, dT$

or $\quad \Delta u = u_2 - u_1 = C_v (T_2 - T_l)$

$$\Delta U = (U_2 - U_2) = nC_v (T_2 - T_1)$$

$$= 4.0093 \times \frac{5}{2} \times 8.314 \times (350 - 300) = 4.1667 \text{ kJ}$$

The first law of thermodynamics gives

$$U_2 - U_1 = -W \text{ or } W = -4.1667 \text{ kJ}$$

Therefore, work done by the battery on the gas = 4.1667 kJ

$$P_1V = nRT_1 \text{ and } P_2V = nRT_2$$

or $$\frac{P_2}{P_1} = \frac{T_2}{T_1}$$

or $$P_2 = P_2 = \frac{T_2P_1}{T_1} = \frac{350\times1\times10^5}{300} = 1.1667 \text{ bar}$$

### Example 4.15

A system is taken through a series of processes as a result of which it is restored to the initial state. The work and heat interactions for some of the processes are measured and they are as given below.

| Process | $W$ (kJ) | $Q$ (kJ) | $\Delta U$ (kJ) |
|---|---|---|---|
| 1–2 | 100 | 100 | – |
| 2–3 | – | –150 | 200 |
| 3–4 | –250 | – | – |
| 4–1 | 300 | – | 50 |

Complete the table and determine the net work done and net heat interaction.

*Solution* :

We know that the first law of thermodynamics gives $\Delta U = Q - W$. Application of the first law of thermodynamics to all the processes 1–2, 2–3, 3–4, and 4–1 gives us

$$U_2 - U_1 = Q_{12} - W_{12}$$

*or* $$U_2 - U_1 = 200 - 100 = 100 \text{ kJ}$$

$$U_3 - U_2 = Q_{23} - W_{23}$$

or $$200 = -150 - W_{23} \text{ or } W_{23} = -350\text{kJ}$$

$$U_1 - U_4 = Q_{41} - W_{41}$$

or $$50 = Q_{41} - 300 \text{ or } Q_{41} = 350\text{kJ}$$

For the entire cycle $\Delta U = 0$

$$(U_2 - U_1) + (U_3 - U_2) + (U_4 - U_3) + (U_1 - U_4) = 0$$

or $$100 + 200 + (U_4 - U_3) + 50 = 0$$

or $$U_4 - U_3 = -350\text{kJ}$$

$$U_4 - U_3 = Q_{34} - W_{34}$$

or $$-350 = Q_{34} - (-250) \text{ or } Q_{34} = -600 \text{ kJ}$$

Net work done, $W = W_{12} + W_{23} + W_{34} + W_{41}$

$$= 100 + (-350) + (-250) + 300 = -200 \text{ kJ}$$

That is 200 kJ of work has to be done by the surroundings on the system, in one complete cycle.

The first law of thermodynamics gives $\oint dQ = \oint dW$

Therefore, net heat interactive = $\oint dQ = W = -200$ kJ

That is 200 kJ of energy is transferred as heat from the system to the surroundings in one complete cycle.

**Example 4.16**

A system undergoes a process 1- 2 in which it absorbs 200 kJ energy as heat while it does 100 kJ work. Then it follows the path 2 - 3 in which it rejects 50 kJ energy as heat when 80 kJ work is done on it. If it is required to restore the system to state 1 through an adiabatic path, calculate the work and heat interactions along the adiabatic path. Also calculate the net work and heat interactions.

*Solution* :

Application of the first law of thermodynamics to the process 1–2 gives

$$U_2 - U_1 = Q_{12} - W_{12} = 200 - 100 = 100 \text{ kJ}$$

Similarly $\quad U_3 - U_2 = Q_{23} - W_{23} = -50 - (-80) = 30 \text{ kJ}$

For the complete cycle $\Delta U = 0$

or $\quad (U_2 - U_1) + (U_3 - U_2) + (U_1 - U_3) = 0$

or $\quad 100 + 30 + (U_1 - U_3) = 0$ or $U_1 - U_3 = -130$ kJ

The process 3–1 is desired to be adiabatic. Therefore, $Q_{31} = 0$

The first law of thermodynamics for the process 3–1, gives

$U_1 - U_3 = Q_{31} - W_{31}$ or $-130 = 0 - W_{31}$ or $W_{31} = 130$ kJ

Therefore, work done during the adiabatic process = 130 kJ

Net work, $W = \oint dW = W_{12} + W_{23} + W_{31} = 100 + (-80) + 130 = 150$ kJ

We know that for a cyclic process $\oint dQ = \oint dW = 150$ kJ

Therefore, net heat interaction = 150 kJ

**Example 4.17**

The specific heat at constant volume for a particular substance is given by the relation

$$C_v = a + bT + cT^2$$

where $a$,$b$ and $c$ are constants. If it is desired to change the temperature of the substance from $T_1$ to $T_2$ at constant volume, develop the relation to calculate the heat interacton.

*Solution* :

We know that for a constant volume process $dq = du$ and

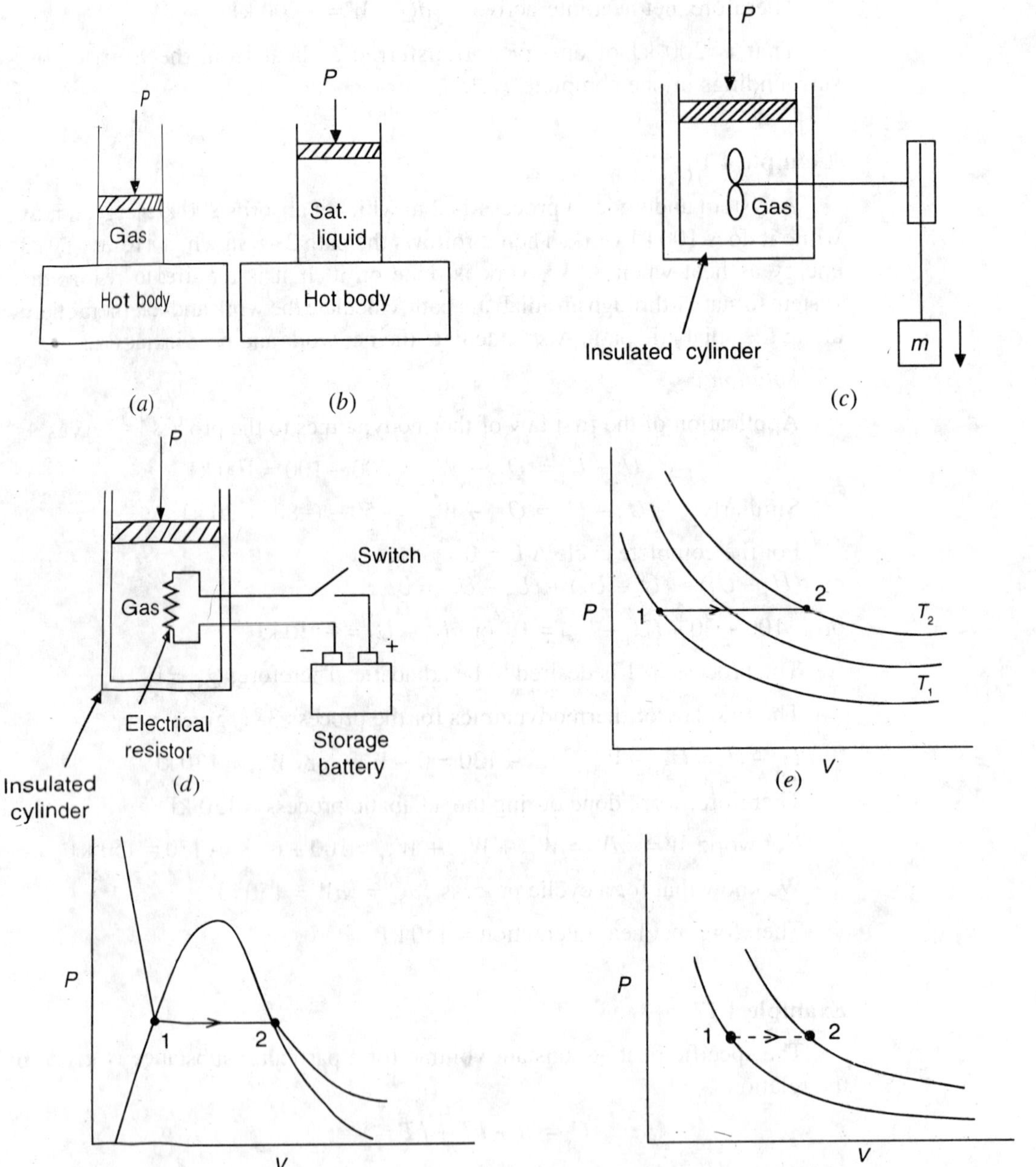


Fig.E 4.18

(*a*) Constant pressure expansion of a gas in a piston-cylinder assembly.

(*b*) Constant pressure vaporization of saturated liquid.

(*c*) Constant pressure expansion of a gas in an insulated cylinder in which paddle wheel work is performed on the gas.

(*d*) Constant pressure expansion of a gas in an insulated cylinder in which electrical work is performed on the gas.

(*e*) Reversible constant pressure process on a *P-V* diagram followed by the gas shown in (*a*).

(*f*) *P-V* diagram showing the vaporization process shown in (*b*).

(g) *P-V* diagram showing the irreversible constant pressure process shown in (*c*) and (*d*).

$$C_v = \left(\frac{\partial u}{\partial T}\right)_v$$

Therefore $$q = \int du = \int \left(\frac{\partial u}{\partial T}\right)_v dT = \int C_v dT = \int_{T_1}^{T_2} (a + bT + cT^2) dT$$

$$= a(T_2 - T_1) + \frac{b}{2}(T_2^2 - T_1^2) + \frac{c}{3}(T_2^3 - T_1^3)$$

**Example 4.18**

Quite a few industrially important processes are carried out under isobaric conditions. Give some examples of reversible isobaric processes and irreversible isobaric processes.

*Solution* :

If a system undergoes a process in which the pressure is held constant then the process is called a constant pressure process or isobaric processes. A few constant pressure processes are schematically represented in Fig.E 4.18.

Suppose a gas enclosed in a frictionless piston–cylinder assembly shown in Fig.E 4.18 (*a*) is considered as the system. The ambient pressure *P* acts on the piston. If the cylinder is placed in contact with a hot body, the temperature of which is infinitesimally greater than that of the gas, energy flows as heat into the system and the gas expands. Since the temperature difference between the hot body and the system is infinitesimal, the process can proceed in either direction. That is the system undergoes a reversible isobaric process. Another example of a reversible constant pressure process is the vaporization of saturated liquid held in a frictionless piston–cylinder assembly shown in Fig.E 4.18 (*b*). During the vaporization process the temperature as well as the pressure of the system (fluid held in the cylinder) remain constant. These reversible processes can be represented in a thermodynamic diagram. Fig.E 4.18 (*e*) shows the constant pressure process followed by the gas shown in Fig.E 4.18 (*a*). The constant pressure process followed by the fluid during vaporization is shown in Fig.E 4.18 (*f*).

Suppose a gas is enclosed in a frictionless piston–cylinder assembly and the piston as well as the cylinder are insulated as shown in Fig.E 4.18 (*c*) and (*d*). Consider the gas enclosed in the cylinder as the system. Work can be performed on the system by the surroundings. In Fig.E 4.18 (*c*) paddle wheel work is done on the system and in Fig.E 4.18 (*d*) electrical work is done on the system. While the gas undergoes expansion, the ambient pressure acting on the piston remains constant. Since it is not possible to rotate a paddle wheel which is initially stationary or to charge a battery by cooling the gas, the process followed by the gas shown in Fig.E 4.18 (*c*) and (*d*) are irreversible. The irreversible constant pressure process followed by the gas is shown in Fig.E 4.18 (*g*).

**Example 4.19**

Define enthalpy.

*Solution* :

The enthalpy $H$ is defined as $H = U + PV$. The specific enthalpy (molar enthalpy) is given by

$$h = u + Pv.$$

**Example 4.20**

What is meant by energy analysis and in how many ways the energy analysis can be carried out?

*Solution* :

The first law of thermodynamics gives

$$dE = d(\text{KE}) + d(\text{PE}) + dU = Q - W$$

The quantity $dE$ or $[d\,(\text{KE}) + d\,(\text{PE}) + dU]$ represents the net change in the energy of the system while $dQ$ and $dW$ represent the energy transfer across the boundary of the system in the form of heat and work, respectively. The first law of thermodynamics in essence tells that the net change in the energy of the system is exactly equal to the net energy transfer across the boundary of the system. In other words, the first law of thermodynamics tells

Accumulation of energy = Inflow of energy – outflow of energy.

The application of the first law of thermodynamics to systems or processes concentrates on energy flows and such an analysis is called energy analysis. Basically there are two types of energy analysis. They are (*a*) Control–mass analysis and (*b*) control–volume analysis.

**Example 4.21**

Explain the essential steps involved in the control–mass analysis.

*Solution* :

The control–mass analysis is usually used to analyze non–flow processes. The essential steps in the control–mass analysis are :

1. The system or the control-mass on which attention is focussed must be clearly identified since the work and heat interactions depend on the choice of the control-mass.

2. Represent the process followed by the system on a thermodynamic diagram. This step helps in identifying the states of the system and in the analysis of the process.

3. Follow the energy transfer across the boundary of the system while applying the principle of conservation of energy.
4. The $P$–$v$–$T$ behavior of the substance which is available through charts, tables or equations of state should be incorporated while relating the energy changes of the system.

### Example 4.22

Apply the first law of thermodynamics to a constant pressure process and deduce relations to estimate heat and work interactions.

*Solution* :

Consider a simple compressible substance (say a gas) enclosed in a frictionless piston--cylinder assembly as the system or the control mass as shown in Fig.E 4.22. Let $P_1$, $v_1$ and $T_1$ denote the initial state of the system. Let the piston be loaded with a fixed mass such that the gas is held at constant pressure. If energy transfers as heat reversibly into the system, then the gas undergoes a reversible constant pressure process.

The process followed by the control–mass is shown on a $P - V$ diagram in Fig.E 4.22 ($b$). The work done by the system or the control–mass is given by

$$W = \int_1^2 PdV = P(V_2 - V_1) \qquad \text{Since } P_1 = P_2 = P$$

Applying the first law of thermodynamics to the control--mass, we get

$$dU = dQ - dW = dQ - PdV$$

$$= dQ - (PdV + VdP) \qquad \text{Since } dP = 0$$

or $$dU = dQ - d(PV)$$

or $$dQ = dU + d(PV) = d(U + PV) = dH$$

or $$Q = H_2 - H_1$$

Therefore, in a reversible constant pressure process, the heat interaction across the boundary of the system is exactly equal to the change in the enthalpy (a property) of the system.

Now let us consider an irreversible constant pressure process shown in Fig.E 4.22 ($c$) and ($d$). In Fig.E 4.22 ($c$) and ($d$) the control–mass is the gas enclosed in an insulated cylinder. The piston is loaded with a fixed mass and exerts a constant pressure on the gas. In Fig.E 4.22 ($c$) paddle–wheel work is performed on the gas and in Fig.E 4.22 ($d$) electrical work is done on the gas.

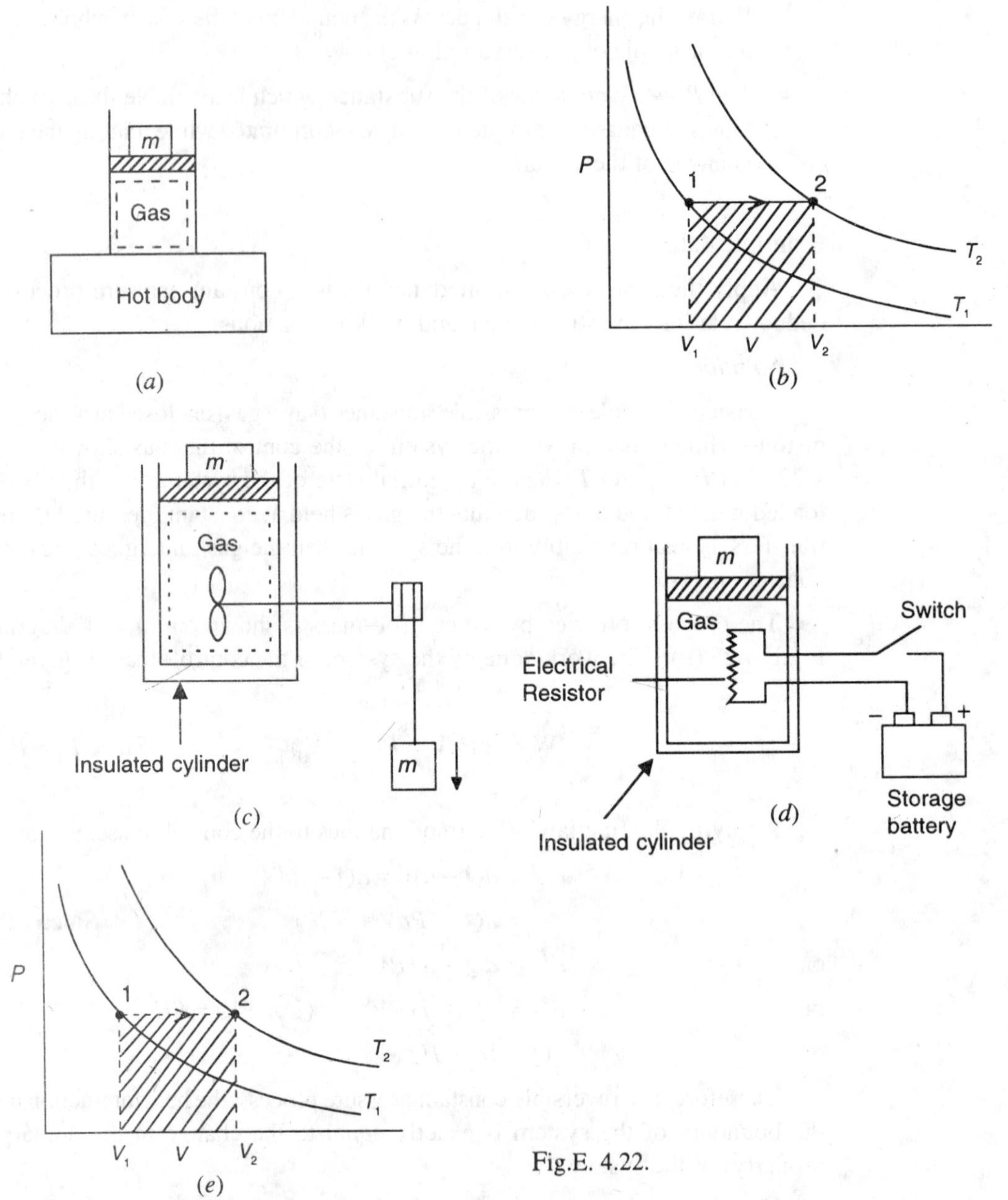


Fig.E. 4.22.

(*a*) Schematic representation of a reversible constant pressure process. The region enclosed by the dotted line is the system or control mass.

(*b*) $P - V$ diagram showing the constant pressure process 1 – 2. The shaded area represents the work done $W = \int_1^2 PdV$ by the control–mass.

(*c*) Irreversible constant pressure process in which paddle wheel work is performed on the control-mass.

(*d*) Irreversible constant pressure process in which electrical work is performed on the control mass.

(*e*) $P - V$ diagram showing the process path.

The cylinder and the piston are insulated and hence $dQ = 0$. Across the boundary of the control–mass there are two modes of work interaction. They are the ($PdV$) work associated with the motion of the boundary and ($dW_s$) the shaft work (either the paddle wheel work or the electrical work) done by the surroundings on the system. That is

$$dW = PdV + dW_s$$

Applying the first law of thermodynamics to the control–mass, we get

$$dU = dQ - dW = 0 - PdV - dW_s$$

or $$-dW_s = dU + PdV$$

$$= dU + d(PV) \qquad \text{Since } VdP = 0$$

or $$-dW_s = d(U + PV) = dH$$

or $$-W_s = H_2 - H_1$$

Therefore, the increase in the enthalpy of the system is equal to the shaft work done on the system. It may be noted here that $W_s$ is always a negative quantity since shaft work is done by the surroundings on the system.

## Example 4.23

Define specific heat at constant pressure and express it in terms of a property change.

*Solution* :

If a system changes its state at constant pressure, it is accompanied by a temperature change (except in the case of constant pressure phase change). The specific heat of a substance at constant pressure is defined as the energy required (in the form of heat) to change the temperature of a unit mass of the substance by one degree. That is, the specific heat at constant pressure $C_P$ is given by

$$C_P = \left(\frac{dq}{dT}\right)_P$$

where $q$ is the heat interaction per unit mass.

We know that the heat interaction during a constant pressure process is equal to the change in the enthalpy or $dq = dh$. Therefore,

$$C_P = \left(\frac{dq}{dT}\right)_P = \left(\frac{\partial h}{\partial T}\right)_P$$

Thus $C_P$ which is a property of the substance is expressed in terms of another property change.

**Example 4.24**

Calculate the internal energy of dry saturated steam at 100°C from the data given in the steam tables.

*Solution :*

We read the following values from the steam tables (see Appendix – 2).

$t = 100°C$, $P = 1.0133$ bar ; $v_g = 1.673\ m^3/kg$ and $h_g = 2676$ kJ/kg.

We know that $h = u + Pv$

or $u = h - Pv = 2676 \times 10^3 - 1.0133 \times 10^5 \times 1.673 = 2506.475$ kJ/kg

**Example 4.25**

Calculate the internal energy of the wet steam of quality 0.8 at 1 bar.

*Solution :*

Read the following values from the steam tables (Appendix – 1).

$P = 1$ bar : $v_f = 0.001\ 043\ 4\ m^3/kg$ ; $v_g = 1.694\ m^3/kg$

$h_f = 417.54$ kJ/kg ; $h_g = 2675.4$ kJ/kg

$v = Xv_g + (1 - X)\, v_f = 0.8 \times 1.694 + 0.2\ (0.001\ 043\ 4) = 1.3554\ m^3/kg$

$h = Xh_g + (1 - X)\, h_f = 0.8 \times 2675.4 + 0.2\ (417.54) = 2223.828$ kJ/kg

We know that $u = h - Pv$. Therefore

$$u = 2223.828 \times 10^3 - 1 \times 10^5 \times 1.3554$$
$$= 2088.288 \text{ kJ/kg}$$

**Example 4.26**

Calculate the internal energy of saturated liquid water at 100°C.

*Solution :*

We read the following values from the tables (Appendix–2).

$t = 100°C$; $P = 1.0133$ bar ; $v_t = 0.001\ 043\ 7\ m^3/kg$ ; $h_f = 419.06$ kJ/kg

We know that $u = h - Pv$

or $$u = 419.06 \times 10^3 - 1.0133 \times 10^5 \times 0.001\ 043\ 7$$
$$= 418.95 \text{ kJ/kg}$$

Thus we find that for saturated liquid at 100°C : $h_t = 419.06$ kJ/kg and $u_f = 418.95$ kJ/kg,

Therefore, for saturated liquid $u_f \cong h_t$ at low to ordinary pressures.

**Example 4.27**

Estimate the latent heat of vaporization or the enthalpy of vaporization for the water at 100°C, 200°C, 300°C and critical point,

*Solution* :

Read the following values from the steam tables (Appendix–2).

$$t = 100°C;\ h_f = 419.06 \text{ kJ/kg};\ h_g = 2676.0 \text{ kJ/kg}$$

$$t = 200°C;\ h_f = 852.37 \text{ kJ/kg};\ h_g = 2790.9 \text{ kJ/kg}$$

$$t = 300°C;\ h_f = 1345.0 \text{ kJ/kg};\ h_g = 2751.0 \text{ kJ/kg}.$$

Critical point ($t = 374.15°C$ and, $P = 221.20$ bar)

$$h_f = 2107.4 \text{ kJ/kg};\ h_g = 2107.4 \text{ kJ/kg}$$

The latent heat of vaporization is the quantity of energy required as heat for converting saturated liquid into saturated vapor at constant temperature and pressure,by unit mass of the substance. That is the latent heat of vaporization or the enthalpy of vaporization ($h_{fg}$) is given by

$$h_{fg} = h_g - h_f$$

$$h_{fg} \text{ (at 100°C)} = 2676.0 - 419.06 = 2256.94 \text{ kJ/kg}$$

$$h_{fg} \text{ (at 200°C)} = 2790.9 - 852.37 = 1938.53 \text{ kJ/kg}$$

$$h_{fg} \text{ (at 300°C)} = 2751.0 - 1345.0 = 1406.0 \text{ kJ/kg}$$

$$h_{fg} \text{ (at critical point)} = 2107.4 - 2107.4 = 0 \text{ kJ/kg}$$

***Note*** *The enthalpy of vaporization of a substances decreases with increasing temperature or pressure and it is equal to zero at the critical point.*

**Example 4.28**

A piston–cylinder assembly contains 0.1 kg wet steam of quality 0.7 at 1 bar. If 200 kJ energy is transferred as heat at constant pressure, determine the final state of the steam and the work done by the steam.

*Solution* :

We read the following values from the steam tables (see Appendix – 1):

$P = 1$ bar : $t = 99.632°C$ ; $v_f = 0.001\ 043\ 4 \text{ m}^3\text{/kg}$ ; $v_g = 1.694 \text{ m}^3\text{/kg}$

$$h_f = 417.54 \text{ kJ/kg};\ h_g = 2675.4 \text{ kJ/kg}$$

$$v_1 = Xv_g + (1 - X)v_f = 0.7 \times 1.694 + 0.3 \times 0.001\ 043\ 4 = 1.1861 \text{ m}^3\text{/kg}$$

$$h_1 = Xh_g + (1 - X)h_f = 0.7 \times 2675.4 + 0.3 \times 417.54 = 2123.304 \text{ kJ/kg}$$

We know that for a constant pressure process $q = h_2 - h_l$. Therefore

$$h_2 = h_1 + q = 2123.304 + 10 \times 200 = 4123.304 \text{ kJ/kg}$$

*Note* *Energy transferred is 200 kJ for 0.1 kg or (10 × 200) kJ/kg. Since $h_2 > h_g$, the steam exists in the superheated state at 1 bar. From superheated steam tables (see Appendix – 3) we read the following values.*

| **P = 1 bar :** | **t = 700°C** | **t = 800°C** |
|---|---|---|
| $v$ | 4.4900 $m^3$/kg | 4.9520 $m^3$/kg |
| $h$ | 3928.2 kJ/kg | 4158.3 kJ/kg |

The value $h_2 = 4123.304$ kJ/kg lies in between 700°C and 800°C. Therefore, we interpolate to determine the temperature of steam.

$$4123.304 = 3928.2 + \frac{(4158.3 - 3928.2)\Delta t}{800 - 700}$$

or $$\Delta t = 84.79 \text{ or } t = 700 + 84.79 = 784.79°\text{C}$$

The process followed by the steam is shown in Fig.E 4.28.

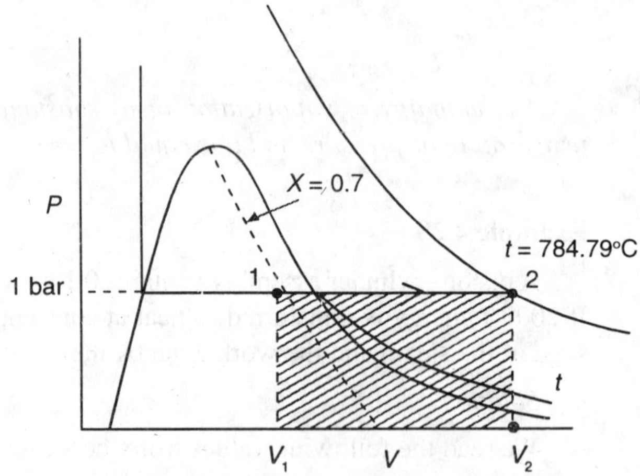


Fig.E 4.28. *P-V* diagram showing the path followed by steam. The shaded area represents the work done by steam.

Now we calculate the specific volume of steam at 784.79°C as

$$v_2 = 4.4900 + \frac{(4.9520 - 4.4900) \times 84.79}{800 - 700} = 4.8817 \text{ m}^3/\text{kg}$$

Work done $W = \int PdV = P(v_2 - v_1) = 1 \times 10^5 (4.8817 - 1.1861) = 369.56$ kJ/kg

Work done by steam (0.1 kg) = 36.956 kJ

**Example 4.29**

One kg of ice at 0°C is completely melted into water at 0°C at 1 bar pressure. Given that the latent heat of fusion of water is 333.44 kJ/kg and the densities of water and ice at 0°C are 999.85 kg/m³ and 916.23 kg/m³, respectively, calculate the work done and energy transferred as heat,

*Solution* :

Specific volume of ice $v_i = 1/916.23 = 1.0914 \times 10^{-3}$ m³/kg

Specific volume of water $v_f = 1/999.85 = 1.0002 \times 10^{-3}$ m³/kg

The phase transition occurs at constant pressure. Therefore, work done is given by

$$W = \int Pdv = P(v_f - v_i) = 1 \times 10^5 (1.0002 - 1.0914) \times 10^{-3} = -9.38 \text{ J}$$

$$q = h_f - h_i = h_{if} = 333.44 \text{ kJ}$$

9.38 J work was done by the surroundings on water and 333.44 kJ energy is transferred form surroundings.

**Example 4.30**

A rigid and sealed tank of volume 1m³ is initially filled with dry saturated steam at 200°C and left in the open. After a while the temperature of the steam is reduced to 150°C. Determine the final conditions of steam in the tank and the amount of energy transferred as heat.

*Solution* :

From steam tables we find at $t = 200$°C

$$P = 15.549 \text{ bar} ; v_g = 0.1272 \text{ m}^3/\text{kg} ; h_g = 2790.9 \text{ kJ/kg}$$

$$u_g = h_g - Pv_g = 2790.9 \times 10^3 - 15.549 \times 10^5 \times 0.1272$$

$$= 2593.12 \text{ kJ/kg}$$

Since the tank is rigid and sealed, the mass of steam and the total volume remain constant. Therefore, the specific volume ($v = V/m$) of steam also remains constant. Hence $v_2 = v_1 = v_g = 0.1272$ m³/kg.

For steam at 150°C, we read the following values from the steam tables

$t = 150$°C ; $P = 4.76$ bar; $v_f = 0.001\ 090\ 8$ m³/kg;

$v_g = 0.3924$ m³/kg ; $h_f = 632.15$ kJ/kg ; $h_g = 2745.4$ kJ/kg

The final value of the specific volume $v_2$ lies in between $v_f$ and $v_g$ at 150°C. Hence the steam exists as wet steam of quality $X_2$,

$$v_2 = v_1 = 0.1272 = X_2 v_g + (1 - X_2) v_f$$

or $0.1272 = 0.3924\, X_2 + 0.001\,090\,8\,(1 - X_2)$

$$X_2 = \frac{0.1272 - 0.001\,090\,8}{0.3924 - 0.001\,090\,8} = 0.3223$$

$$h_2 = X_2 h_g + (1 - X_2)$$

$$h_t = 0.3223 \times 2745.4 + (1 - 0.3223)\ 632.15$$

$$= 1313.25 \text{ kJ/kg}$$

$$u_2 = h_2 - P_2 v_2 = 1313.25 \times 10^3 - 4.76 \times 10^5 \times 0.1272$$

$$= 1252.703 \text{ kJ/kg}$$

For a constant volume process we know that $q = u_2 - u_1$

or $q = 1252.703 - 2593.12 = -1340.417$ kJ/kg

Mass of steam in the tank $m = \dfrac{V}{v} = \dfrac{1}{0.1272} = 7.862$ kg

Energy Transferred as heat $Q = mq = 7.862 \times (-1340.417) \times 10^3 = -10.538$ kJ

Hence, energy is transferred from steam to the surrounding.

**Example 4.31**

It is required to raise the temperature of 1 kmol of ethylene at constant pressure from 300 K to 400 K. Determine the amount of energy to be transferred as heat to ethylene. The molar heat capacity of ethlene is given by

$$C_P = 4.196 + 154.565 \times 10^{-3}\, T - 81.076 \times 10^{-6}\, T_2 + 16.813 \times 10^{-9}\, T^3$$

Where $C_p$ is in J/mol K and $T$ is in kelvin.

*Solution* :

We know that for a constant pressure process $q = h_2 - h_1$ and

$$C_P = \left(\frac{\partial h}{\partial T}\right)_P$$

Therefore, $$q = \int dh = \int_{T_1}^{T_2} \left(\frac{\partial h}{\partial T}\right)_P dT = \int_{T_1}^{T_2} CpdT$$

$$= \int_{T_1}^{T_2} (a + bT + cT^2 + dT^3)dT$$

or $$q = a(T_2 - T_1) + \frac{b}{2}\left(T_2^2 - T_1^2\right) + \frac{C}{3}\left(T_2^3 - T_1^3\right) + \frac{d}{4}\left(T_2^4 - T_1^4\right) \quad (A)$$

Substituting the values of $T_1, T_2$, $a, b, c$ and $d$ in Eqn. ($A$), we get

$$q = 4.196(400 - 300) + \frac{154.565 \times 10^{-3}}{2}(400^2 - 300^2)$$

$$-\frac{81.076 \times 10^{-6}}{3}(400^3 - 300^3) + \frac{16.813 \times 10^{-9}}{4}(400^4 - 300^4)$$

or $$q = 419.6 + 772.8 - 999.9 + 70.6 = 263.1 \text{ J/mol}$$

Therefore, the amount energy to be transferred as heat to raise the temperature of 1 kmol of ethylene = 263.1 kJ

**Example 4.32**

Give an example of an isothermal process.

*Solution* :

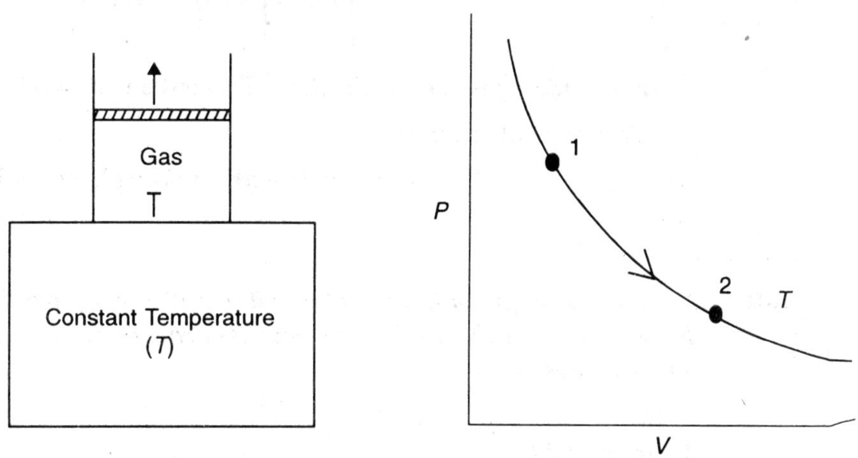


($a$) ($b$)

Fig.E 4.32

($a$) Schematic arrangement to show isothermal expansion of a gas

($b$) Isothermal process 1–2 on a pressure versus volume diagram.

A process in which the temperature of a system remains constant is called an isothermal process or constant temperature process. Consider a gas enclosed in a frictionless piston–cylinder assembly shown in Fig.E 4.32 (*a*) as a system. If the piston-cylinder assembly is placed in thermal contact with a constant temperature bath and if the pressure of the gas is slowly and continuously decreased, the gas undergoes expansion while its temperature is held constant. Since, the process is reversible it can be traced on a thermodynamic diagrams shown in Fig.E 4.32 (*b*).

### Example 4.33

Suppose an ideal gas which is initially at $P_1$, $v_1$, $T$ undergoes isothermal expansion till it reaches the final state $P_2, v_2$, $T$ estimate the work and heat interactions. For an ideal gas $u = u(T)$ only.

*Solution* :

The first law of thermodynamics gives

$$du = dq - dW \tag{A}$$

The work done by the gas is given by $W = \int PdV$.

We know that for an ideal gas $Pv = RT$ or $P = RT/v$. Therefore,

$$W = \int_1^2 Pdv = \int_1^2 RT\frac{dv}{v} = RT\ln\frac{v_2}{v_1}$$

For an ideal gas $du = C_v dT$. Since $T$ = constant $du = 0$.

Then Eqn. (*A*) reduces to

$$du = 0 = dq - dW \text{ or } dq = dW \text{ or } q = W = RT \ln (v_2/v_1)$$

**Note** *If an ideal gas undergoes a reversible isothermal expansion, the work done by the gas is exactly equal to the amount of energy transferred as heat from the surroundings.*

### Example 4.34

It is desired to compress isothermally one kmol of butane ($C_4H_{10}$) at 300 K from the initial volume of 0.5 m$^3$/mol to the final volume of 0.1 m$^3$/mol. Determine the amount of work to be done on the gas and the amount of energy to be removed as heat from the gas. The critical constants of butane are $T_C = 425.2$ K and $P_C =$

37.97 bar. Assume that butane behaves like a Van der Waals gas. For a Van der Waals gas $du = C_v dt + (a/v^3)dv$.

*Solution* :

The Van der Waals equation of state is given by

$$(P + a/v^2)(v - b) = RT \quad \text{or} \quad P = \frac{RT}{v-b} - \frac{a}{v^2}$$

$$\text{Workdone, } W = \int_1^2 P dv = \int_1^2 \left( \frac{RT}{v-b} - \frac{a}{v^2} \right) dv$$

$$= RT \ln \frac{v_2 - b}{v_1 - b} - a\left( \frac{1}{v_1} - \frac{1}{v_2} \right) \quad (A)$$

The Van der Waals constant are given by

$$a = \frac{27R^2T_c^2}{64P_c} = \frac{27 \times (8.314 \times 425.2)^2}{64 \times 37.97 \times 10^5}$$

$$= 1.3885 \text{ Pa (m}^3\text{/mol)}^2$$

$$b = \frac{RT_c}{8P_c} = \frac{8.314 \times 425.2}{8 \times 37.97 \times 10^5}$$

$$= 0.1164 \times 10^{-3} \text{ m}^3\text{/ mol}$$

Subsituting the values of $v_1$, $v_2$, $a$, $b$ and $T$ in Eqn. $(A)$, we get

$$W = RT \ln \frac{v_2 - b}{v_1 - b} - a\left( \frac{1}{v_1} - \frac{1}{v_2} \right)$$

$$= 8.314 \times 300 \ln \left( \frac{0.1 - 0.1164 \times 10^{-3}}{0.5 - 0.1164 \times 10^{-3}} \right)$$

$$-1.3885 \left( \frac{1}{0.5} - \frac{1}{0.1} \right) = -4.0055 \text{ kJ/mol}$$

$$du = C_v\, dT + \frac{a}{v^2} dv = \frac{a}{v^2} dv \qquad \text{since } T = \text{constant}$$

or $$u_2 - u_1 = \int_1^2 \frac{a}{v^2} dv = a\left(\frac{1}{v_1} - \frac{1}{v_2}\right)$$

$$= 1.3885\left(\frac{1}{0.5} - \frac{1}{0.1}\right) = -11.108 \text{ J/mol}$$

The first law of thermodynamics gives

$$u_2 - u_1 = q - w$$

or $$-11.108 = q - (-4005.5) \text{ or } q = 4016.61 \text{ J/mol}$$

Work to be done on one kmol of gas = 4005.5 kJ

Energy to be transferred as heat to the gas = 4016.61 kJ

**Example 4.35**

Give some examples of irreversible adiabatic processes.

*Solution* :

A process in which there is no energy transfer across the boundaries of the system, that is when the system is enclosed by an insulated wall, is called adiabatic. For an adiabatic process $Q = 0$. A few irreversible adiabatic processes are shown in Fig.E 4.35.

In Fig.E 4.35 (*a*) a fluid is taken in an insulated container and paddle wheel work is performed on the system (fluid). In Fig.E 4.35 (*b*) the system is again a fluid held in an insulated container and electrical work is done on the system. In Fig.E 4.35 (*c*) an insulated tank is divided into two compartments by a removable partition. One compartment contains a fluid at a given $T$ and $P$ while the second compartment contains a different fluid or the same fluid at different $T$ and $P$. If the partition is removed, the fluids mix with each other. Considering both the fluids as the system, the system undergoes an irreversible adiabatic process. In Fig.E 4.35 (*d*) an insulated tank is divided into two compartments by a removable partition. One compartment is filled with a gas at a given pressure while the other compartment is evacuated. If the partition is removed, the gas expands and occupies the entire tank. The expansion of the gas is unrestrained because no force is opposing the expansion of the gas. Such an expansion is called free expansion. If the gas is considered as the system, the system undergoes an irreversible adiabatic process.

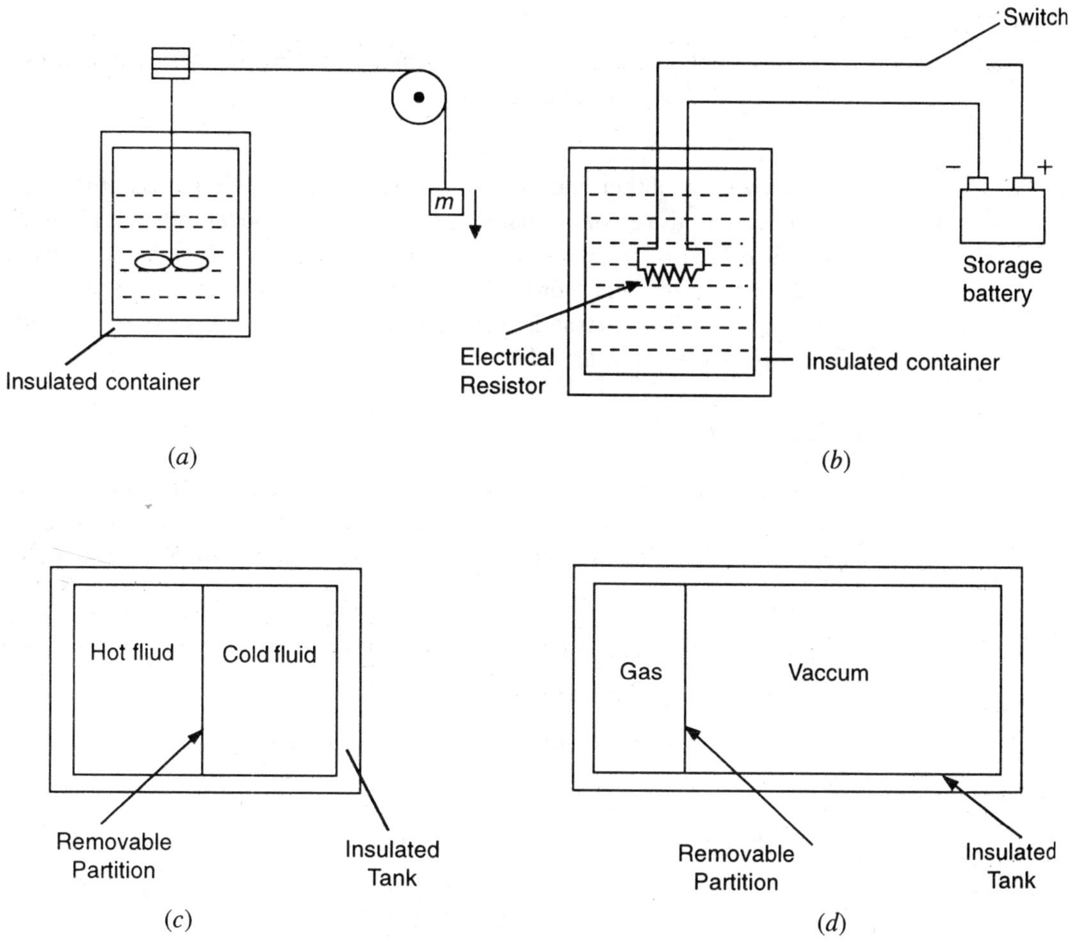


Fig.E 4.35. Schematic representation of irreversible adiabatic processes involving
(*a*) paddle wheel work
(*b*) electrical work
(*c*) mixing of fluids at different states of temperature
(*d*) free expansion of gas.

## Example 4.36

Give an example of a reversible adiabatic process.

*Solution* :

Suppose a gas is held in a frictionless piston– cylinder assembly. Let the piston and cylinder be insulated. If the gas is allowed to expand or compress slowly by adjusting the opposing pressure such that the gas pressure is always equal to the opposing pressure then the expansion or compression process is reversible, Such a process is a reversible adiabatic process.

**Example 4.37**

What conclusions can be drawn from the first law of thermodynamics for a system undergoing adiabatic process?

*Solution* :

The first law of thermodynamics gives $du = dq - dW$. For an adiabatic process $dq = 0$. Hence, if a system undergoes an adiabatic process, then $du = -dW$ or $W = -\Delta u$. That is, the work done by the system is equal to the decrease in its internal energy. In other words, the system does work at the expense of its own internal energy. This is true irrespective of whether the process is reversible or irreversible since $du = dq - dW$ is always true.

**Example 4.38**

Estimate the work done by an ideal gas, if it undergoes a reversible adiabatic expansion from $P_1$, $v_1$, $T_1$ to $P_2$, $v_2$, $T_2$. It is known that the internal energy of an ideal gas is a function of temperature only. That is, for an ideal gas $u = u(T)$ only.

*Solution* :

For an ideal gas $Pv = RT$ and $u = u(T)$ only.

Then $du = C_v dT$ for an ideal gas.

We know that for an adiabatic process $du = -dW$

or
$$W = -\Delta u = u_1 - u_2 = C_v (T_1 - T_2) \qquad (A)$$

For an ideal gas, we know that $C_P - C_v = R$

$$\gamma = \frac{C_P}{C_v} = \frac{C_v + R}{C_v} = \frac{R}{C_v} \quad \text{or} \quad C_v = \frac{R}{\gamma - 1}$$

where $\gamma$ is the ratio of the specific heats $C_P$ and $C_v$.

Substituting for $C_v$ is Eqn. (*A*), we get

$$W = C_v (T_1 - T_2) = \frac{R}{\gamma - 1}(T_1 - T_2)$$

$$= \frac{RT_1 - RT_2}{\gamma - 1} = \frac{P_1 v_1 - P_2 v_2}{\gamma - 1}$$

**Example 4.39**

If an ideal gas undergoes a reversible adiabatic process, trace the paths of the process on temperature versus volume diagram, pressure versus volume diagram and temperature versus pressure diagram.

*Solution* :

Application of the first law of thermodynamics to an adiabatic process gives

$$du = -\,dW$$

If the process is reversible $dW = Pdv$

For an ideal gas $du = C_v dT$ and $C_p - C_v = R$

$$\gamma = \frac{C_P}{C_v} = \frac{C_v + R}{C_v} = 1 + \frac{R}{C_v} \quad \text{or} \quad C_v = \frac{R}{\gamma - 1}$$

Therefore, $$Pdv = \frac{-RdT}{\gamma - 1} \quad \text{or} \quad \frac{RTdv}{v} = \frac{-RdT}{\gamma - 1}$$

or $$\frac{dT}{T} = -(\gamma - 1)\frac{dv}{v} \quad \text{or} \quad \frac{T_2}{T_1} = \left(\frac{v_1}{v_2}\right)^{\gamma - 1} \qquad (A)$$

or $Tv^{\gamma - 1} = \text{constant}$

or $\ln T + (\gamma - 1)\ln v = 0$ or $\ln T = (1 - \gamma)\ln v$

Hence a plot of $\ln T$ versus $\ln v$ yields a straight line with slope $(1 - \gamma)$ as shown in Fig.E 4.39 (*a*).

For an ideal gas $P_1 v_1 = RT_1$ and $P_2 v_2 = RT_2$.

Therefore $$\frac{T_2}{T_1} = \frac{P_2 v_2}{P_1 v_1} \qquad (B)$$

Substituting Eqn.(*B*) in Eqn. (*A*), we get

$$\frac{P_2 v_2}{P_1 v_1} = \left(\frac{v_1}{v_2}\right)^{\gamma - 1} \quad \text{or} \quad P_1 v_1^{\gamma} = P_2 v_2^{\gamma} \qquad (C)$$

or $Pv^{\gamma} = \text{constant}$ or $\ln P + \gamma \ln v = 0$

or $$\ln P = -\gamma \ln v$$

Thus a plot of $\ln(P)$ versus $\ln(v)$ yields a straight line with slope $-\gamma$ as shown in Fig.E 4.39 (*b*)

Eqn.(*C*) can be rewritten as

$$\frac{v_1}{v_2} = \left(\frac{P_2}{P_1}\right)^{1/\gamma} \qquad (D)$$

Substituting Eqn.(*D*) in Eqn.(*A*) gives

$$\frac{T_2}{T_1} = \left(\frac{v_1}{v_2}\right)^{\gamma-1} = \left(\frac{P_2}{P_1}\right)^{\frac{\gamma-1}{\gamma}}$$

or $$\frac{T_2}{T_1} = \left(\frac{P_2}{P_1}\right)^{\left(\frac{\gamma-1}{\gamma}\right)} \quad \text{or } TP^{\left(\frac{1-\gamma}{\gamma}\right)} = \text{constant}$$

or $$\ln T + \frac{(1-\gamma)}{\gamma}\ln P = 0 \text{ or } \ln T = \left(\frac{\gamma-1}{\gamma}\right)\ln P$$

or $$\ln P = \left(\frac{\gamma}{\gamma-1}\right)\ln T$$

Hence a plot of ln $P$ versus ln $T$ yields a straight line with slope $(\gamma/\gamma-1)$ as shown in Fig.E 4.39 (*c*).

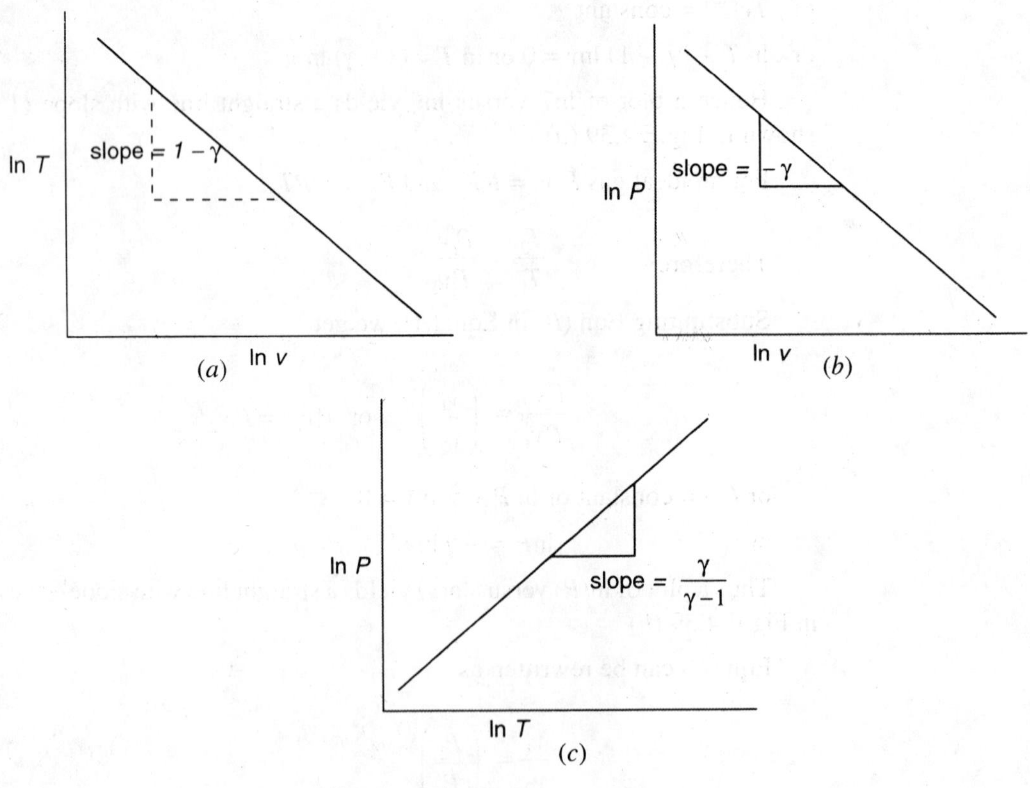


Fig.E 4.39. Reversible adiabatic paths for an ideal gas on
(*a*) ln $T$ versus ln $v$ diagram (*b*) ln $P$ versus ln $v$ diagram and (c) ln $P$ versus ln $T$ diagram

**Example 4.40**

In a Diesel cycle the atmospheric air at 300 K and 1 bar is compressed reversibly and adiabatically till the volume of the air reduces to 1/16 of the original value. Calculate the temperature and pressure of the air at the end of the compression and the amount of work done in compressing the gas. Treat air as an ideal gas with $\gamma = 1.4$

*Solution* :

$$\frac{\text{Final volume}}{\text{Initial volume}} = \frac{v_2}{v_1} = \frac{1}{16}$$

For reversible adiabatic compression of an ideal gas

$$\frac{T_2}{T_1} = \left(\frac{v_1}{v_2}\right)^{\gamma-1} \text{ or } \frac{T_2}{300} = (16)^{0.4} = 3.0314$$

or $T_2 = 909.4$ K

$$\frac{P_2}{P_1} = \left(\frac{v_1}{v_2}\right)^{\gamma} \text{ or } \frac{P_2}{1} = (16)^{1.4} = 48.503$$

or $P_2 = 48.503$ bar

$$\text{Work done, } W = \frac{R(T_1 - T_2)}{\gamma - 1} = \frac{8.314(300 - 909.4)}{0.4}$$

$$= -12.666 \text{ kJ/mol}$$

Therefore 12.666 kJ/mol work is to be done on the air.

**Example 4.41**

One mol of an ideal gas with $\gamma = 1.4$ initially at 300 K and 1 bar is compressed reversibly and adiabatically to 6 bar and then it is cooled at constant pressure to the original temperature. The gas is then restored to the initial state through an isothermal process. Calculate the net work and heat interaction.

*Solution* :

The path followed by the gas is shown in Fig.E 4.41.

Path 1–2 is reversible and adiabatic. Therefore,

$$T_2 = T_1\left(\frac{P_2}{P_1}\right)^{\frac{\gamma-1}{\gamma}} = 300(6)^{\frac{0.4}{1.4}} = 500.55 \text{ K}$$

$q_{12} = 0$ since it is an adiabatic process.

Path 2–3 is at constant pressure. Therefore, $q_{23} = h_3 - h_2 = C_p (T_3 - T_2)$.

We know that $C_v = R/(\gamma - 1)$ for an ideal gas. Hence

$$C_p = \gamma C_v = \frac{R\gamma}{\gamma - 1} = \frac{1.4 \times 8.314}{0.4} = 29.099 \text{ J/mol K}$$

$q_{23} = C_p (T_3 - T_2) = 29.099\,(300 - 500.55) = -5835.8$ J/mol

($T_3 = T_1$, because 3–1 is an isothermal process)

Path 3–1 represents isothermal process. For an ideal gas $du = C_v\,dT$.

Hence $du = 0 = dq - dW$ or $0 = q_{31} - W_{31}$ or $q_{31} = W_{31}$

$$\text{Work done } W_{31} = \int_3^1 P dv = RT_1 \ln \frac{v_1}{v_3}$$

$$P_1 v_1 = P_3 v_3 \text{ or } v_1/v_3 = P_3/P_1 = 6$$

Therefore, $q_{31} = W_{31} = RT_1 \ln v_1/v_3 = 8.314 \times 300 \ln 6 = 4469$ J/mol

Net heat interaction, $q = q_{12} + q_{23} + q_{31} = 0 + (-5835.8) + 4469 = -1366.8$ J/mol

The gas undergoes the cycle 1–2–3–1. For a cyclic process, the first law of thermodynamics gives $\oint dq =$ or $\oint dw$ or $q = W$.

Hence, net work done, $W = -1366.8$ J/mol

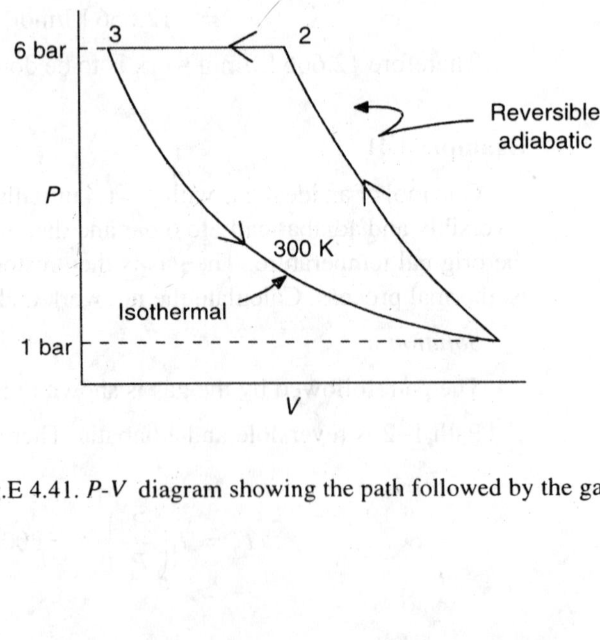


Fig.E 4.41. *P-V* diagram showing the path followed by the gas.

### Example 4.42

An ideal gas is heated at constant volume until its tempertature is doubled and then cooled at constant pressure to the original temperature. Finally the gas is allowed to expand isothermally to the initial state. Derive a relation to estimate the net work done.

*Solution* :

The process paths of the gas are shown on a $P - v$ diagram in Fig.E 4.42.

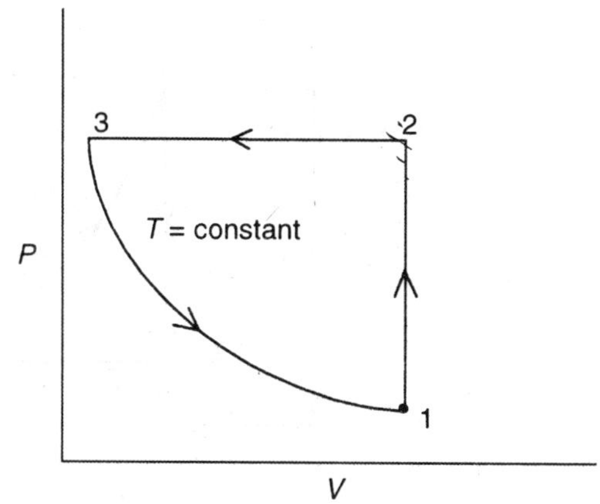


Fig.E 4.42. $P - v$ diagram for Example 4.42.

Path 1–2 represents a constant volume process. Hence $W_{12} = 0$

Path 2–3 represents a constant pressure process. Hence $W_{23} = P_2(v_3 - v_2)$

or
$$W_{23} = P_2v_3 - P_2v_2 = P_3v_3 - P_2v_2$$
$$= R(T_3 - T_2) = R(T_1 - T_2) = R(T_1 - 2T_1) = -RT_1$$

(Since $P_2 = P_3$ ; $T_3 = T_1$ and $T_2 = 2T_1$)

Path 3–1 represents an isothermal process. Hence,

$$W_{31} = \int Pdv = RT_1 \ln (v_1/v_3)$$

$$W_{31} = RT_1 \ln \frac{v_2}{v_3} = RT_1 \ln \frac{T_2}{T_3} = RT_1 \ln \frac{2T_1}{T_1} = RT_1 \ln 2$$

(Since $v_1 = v_2$ ; $P_2 = P_3$ ; $\frac{P_2v_2}{T_2} = \frac{P_3v_3}{T_3}$

Net work = $W = W_{12} + W_{23} + W_{31} = 0 + (-RT_1) + RT_1 \ln 2 = RT_1 (\ln 2 - 1)$

or $W = -0.3069\, RT$.

**Example 4.43**

One mole of an ideal gas initially at $P_1$ and $T_1$ is compressed reversibly and adiabatically till the pressure is $P_2$ and then it is cooled at constant volume to the initial pressure. Finally the gas is restored to the iniitial state through an isobaric process, calculate the work done by the gas,

*Solution :*

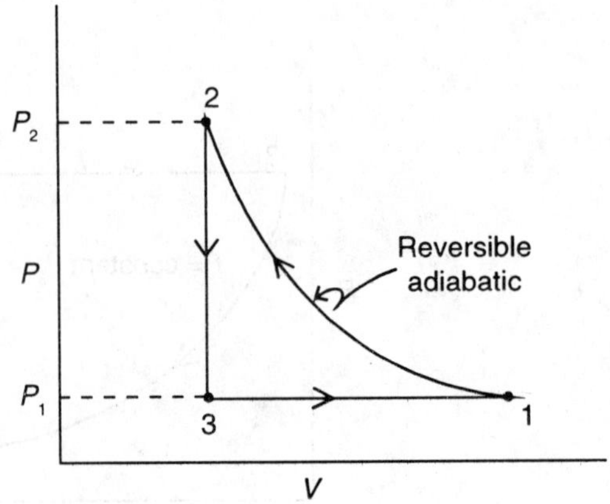


Fig.E 4.43. Sketch for Example 4.43.

The various processes followed by the gas are shown in Fig.E 4.43.

1–2 is a reversible adiabatic path. Therefore,

$$W_{12} = u_1 - u_2 = C_v(T_1 - T_2) = \frac{R}{\gamma - 1}(T_1 - T_2)$$

2–3 is a constant volume process. Therefore, $W_{23} = 0$

3–1 is a constant pressure process, Therefore, $W_{31} = P_1 (v_1 - v_3)$

or $W_{31} = R(T_1 - T_3)$

Since 2–3 is a constant volume process,

$$\frac{P_2}{T_2} = \frac{P_3}{T_3} = \frac{P_1}{T_3}$$

$$\text{or} \quad T_3 = \frac{P_1 T_2}{P_2}$$

Since 1–2 is a reversible adiabatic process

$$\frac{T_2}{T_1} = \left(\frac{P_2}{P_1}\right)^{\frac{\gamma-1}{\gamma}}$$

or $$T_2 = T_1(P_2 / P_1)^{\frac{\gamma-1}{\gamma}}$$

Therefore $$T_3 = \frac{P_1}{P_2} T_1 \left(\frac{P_2}{P_1}\right)^{\frac{\gamma-1}{\gamma}} = T_1(P_2 / P_1)^{-1/\gamma}$$

or $$\frac{T_3}{T_1} = (P_2 / P_1)^{-1/\gamma}$$

Network, $W = W_{12} + W_{23} + W_{31}$

$$= \frac{R}{\gamma - 1}(T_1 - T_2) + 0 + R(T_1 - T_3)$$

$$W = \frac{RT_1}{\gamma - 1}\left(1 - \frac{T_2}{T_1}\right) + RT_1\left(1 - \frac{T_3}{T_1}\right)$$

$$= \frac{RT_1}{\gamma - 1}\left[1 - (P_2 / P_1)^{\frac{\gamma-1}{\gamma}}\right] + RT_1\left[1 - (P_2 / P_1)^{-1/g}\right]$$

or $$W = RT_l\left[\frac{1 - (P_2 / P_1)^{\frac{\gamma-1}{\gamma}}}{\gamma - 1} + 1 - (P_2 / P_1)^{-1/\gamma}\right]$$

**Example 4.44**

It is desired to compress one mole of air (an ideal gas with $\gamma = 1.4$) from the initial state 300 K and 1 bar to the final state 300 K and 10 bar. There are several possibilities by which the air can be taken from the initial state to the final state. Some of the possible paths are:

(*a*) Isothermal compression.

(*b*) Cooling at constant pressure followed by heating at constant volume.

(*c*) Adiabatic compression followed by cooling at constant volume.

(*d*) Heating at constant volume followed by cooling at constant pressure.

Since there are several ways in which the task can be achieved, it is necessary to choose the most economical process. Therefore, calculate the net work required for all the above processes.

*Solution* :

The various possible processes by which the air can be compressed are shown the Fig.E 4.44.

(*a*) Isothermal compression path is shown as 1–2 in Fig.E 4.44.

$$W_{12} = \int_1^2 P dv = RT_1 \ln \frac{v_2}{v_1} = RT_1 \ln \frac{P_1}{P_2}$$

$$= 8.314 \times 300 \ln \frac{1}{10} = -5743.11 \text{ J/mol}$$

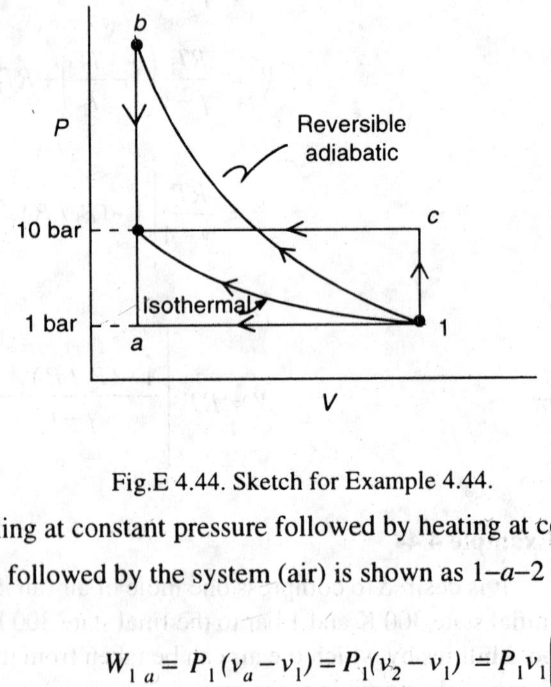


Fig.E 4.44. Sketch for Example 4.44.

(*b*) Cooling at constant pressure followed by heating at constant volume.

The path followed by the system (air) is shown as 1–*a*–2 in Fig.E 4.44.

$$W_{1a} = P_1(v_a - v_1) = P_1(v_2 - v_1) = P_1 v_1 \left(\frac{v_2}{v_1} - 1\right)$$

$$= P_1 v_1 \left(\frac{P_1}{P_2} - 1\right)$$

$$= RT_1\left(\frac{P_1}{P_2}-1\right)=8.314\times 300\left(\frac{1}{10}-1\right)$$

$$= -\ 2244.78 \text{ J/mol}$$

$$W_{a2} = 0$$

Therefore, $W_{1a2} = W_{1a} + W_{a2} = -\ 2244.78$ J/mol

(*c*) Adiabatic compression followed by cooling at constant volume. The path followed by air is shown as 1–*b*–2 in Fig.E 4.44.

$$W_{1b} = u_1 - u_b = C_v(T_1 - T_b) = \frac{R}{\gamma-1}(T_1 - T_b)$$

$$\frac{T_b}{T_1} = \left(\frac{v_1}{v_b}\right)^{\gamma-1} = \left(\frac{v_1}{v_2}\right)^{\gamma-1} = \left(\frac{P_2}{P_1}\right)^{\gamma-1}$$

or $T_b = T_1 (P_2 / P_1)^{\gamma-1}$

or $$T_b = 300\times(10)^{0.4} = 753.57 \text{ K}$$

$$W_{1b} = \frac{R}{\gamma-1}(T_1 - T_b) = \frac{8.314}{0.4}(300-753.57)$$

$$= 9427.45 \text{ J / mol}$$

$$W_{b2} = 0$$

Therefore $W_{1b2} = W_{1b} + W_{b2} = -\ 9427.45$ J/mol

(*d*) Heating at constant volume followed by cooling at constant pressure. The path followed by the air is shown as 1–*c*–2 in FigE. 4.44

$$W_{1c} = 0$$

$$W_{c2} = P_2(v_2 - v_c) = P_2(v_2 - v_1)$$

$$= P_2v_2\left(1-\frac{v_1}{v_2}\right) = P_2v_2\left(1-\frac{P_2}{P_1}\right)$$

$$= RT_2\left(1-\frac{P_2}{P_1}\right) = RT_1\left(1-\frac{P_2}{P_1}\right)$$

$$= 8.314\times 300(1-10) = -22.448 \text{ kJ / mol}$$

$$W_{1c2} = W_{1c} + W_{c2} = -22.448 \text{ kJ/mol}$$

**Example 4.45**

The work to be done in compressing an ideal gas from a given initial state to a given final state depends on the path followed. In Example 4.44, the minimum work required is found for the process in which the gas is cooled at constant pressure followed by heating at constant volume. What are the practical difficulties associated with this process?

*Solution* :

Refer Example 4.44.

The constant pressure cooling process is shown by path 1–*a* in Fig.E 4.44. The heat interaction among the path 1–*a* is given by

$$q_{1a} = h_a - h_1 = C_P(T_a - T_1) = \frac{R\gamma}{\gamma - 1}(T_a - T_1)$$

$$= \frac{RT_1\gamma}{\gamma - 1}\left(\frac{T_a}{T_1} - 1\right)$$

$$\frac{P_1V_1}{T_1} = \frac{P_aV_a}{T_a} \quad \text{or} \quad \frac{T_a}{T_1} = \frac{v_a}{v_1} = \frac{v_2}{v_1} = \frac{P_1}{P_2}$$

or

$$T_a = T_1 \cdot \frac{P_1}{P_2} = 300 \times \frac{1}{10} = 30 \text{ K}$$

$$q_{1a} = \frac{RT_1\gamma}{\gamma - 1}\left(\frac{T_a}{T_1} - 1\right)$$

$$= \frac{8.314 \times 300 \times 1.4}{0.4}\left(\frac{1}{10} - 1\right)$$

$$= -7856.73 \text{ J / mol}$$

The air which is initially at 300 K is to be cooled to 30 K and 7856.73 J/mol of energy is to be removed at this low temperature. This necessitates the use of a refrigerating device. More over it is difficult to achieve this low temperature. Air is treated as an ideal gas (which does not liquefy) in the problem. In practice air starts liquefying at about 150 K. Hence this process cannot be adopted in practice.

**Example 4.46**

Isothermal compression of a gas (of all feasible processes) requires the least amount of work input. However, it is difficult to carry out a process at constant temperature. Suggest an alterenative scheme, for compressing a gas, which requires the work input which is marginally more than that required by an isothermal process.

*Solution* :

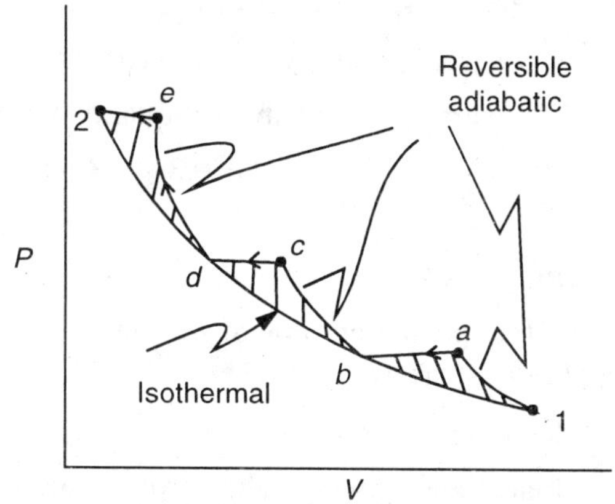


Fig.E 4.46. *P–v* diagram showing a combination of reversible adiabatic processes and isobaric processes which can be closely approximated as an isothermal process.

The gas which is initially at state 1 can be compressed reversibly and adiabatically to state *a* followed by isobaric cooling to state *b*. Then it can be again compressed reversibly and adiabatically to state *c* followed by constant pressure cooling to state *d*. The sequence of steps can be continued as shown in Fig.E 4.46 to attain the final state 2. The shaded area in the figure represents the additional work input compared to an isothermal process. The Fig.E 4.46 shows three stages of compression. By increasing the number of stages of compression, that is by reducing the pressure ratio in each stage of compression, it is possible to closely follow the isothermal compression. In practice, multistage compressors are used to achieve higher compression ratios.

**Note** *A reversible adiabatic process is an isentropic process. That is the entropy of the system remains constant when it undergoes a reversible and adiabatic process or s = constant where s is the specific entropy of the substance. This fact will be shown while discussing the second law of thermodynamics.*

**Example 4.47**

Calculate the specific entropy of wet steam of quality 0.8 at 1 bar pressure.

*Solution* :

Read the following values from the steam tables (Appendix–1) at 1 bar pressure.

$s_f = 1.3027$ kJ/kg K ; $s_g = 7.3598$ kJ/kg K

where $s_f$ = specific entropy of saturated liquid, and

$s_g$ = specific entropy of saturated vapor.

The entropy of a mixture of quality $X$ is given by $s = Xs_g + (1 - X)\, s_f$

$$= 0.8 \times 7.3598 + 0.2 \times 1.3027 = 6.1484 \text{ kJ/kg K}$$

**Example 4.48**

If the specific entropy of wet steam at 100°C is 6.50 kJ/kg K, determine the quality of steam.

*Solution* :

Read the following values from steam tables at 100°C.

$s_f = 1.3069$ kJ/kg K ; $s_g = 7.3554$ kJ/kg K

We know that $s = Xs_g + (1 - X)\, s_f$

Therefore,

$$6.50 = 7.3554\, X + (1 - X)\, 1.3069$$

or

$$X = \frac{6.50 - 1.3069}{7.3554 - 1.3069} = 0.8586$$

Hence, the quality of wet steam = 0.8586

**Example 4.49**

A piston–cylinder assembly contains 0.1kg of steam at 4 bar and 300°C. If it is allowed to expand reversibly and adiabatically to a pressure of 1 bar, calculate the work done by steam.

*Solution* :

Read the following values from superheated steam tables at $P = 4$ bar and $t = 300$°C.

$v_1 = 0.6549$ m$^3$/kg ; $h_1 = 3067.2$ kJ/kg ; $s_1 = 7.5675$ kJ/kg K

We know that for a reversible adiabatic expansion $s$ = constant.

Therefore, $s_1 = s_2 = 7.5675$ kJ/kg K

At 1 bar, we find from the saturated steam tables $s_g = 7.3598$ kJ/kg K. Since $s_2 > s_g$, the final state of steam is superheated at 1 bar. From the superheated steam tables we find the following values at 1 bar.

| | $t = 100°C$ | $t = 200°C$ |
|---|---|---|
| $v$ | 1.6960 $m^3/kg$ | 2.1720 $m^3/kg$ |
| $h$ | 2676.2 kJ/kg | 2875.4 kJ/kg |
| $s$ | 7.3618 kJ/kg K | 7.8349 kJ/kg K |

Since $s_2 = 7.5675$ kJ/kg K lies in between the values of $s$ at 100°C and 200°C, we determine the temperature of the steam by interpolation as given below.

$$s_2 = 7.5675 = 7.3618 + \frac{(7.8349 - 7.3618)}{200 - 100} \Delta t$$

or

$$\Delta t = 94.84$$

Therefore, the final temperature $t$ of steam = 100 + 94.84 = 194.84°C

$$h_2 \text{ (at 194.84°C)} = 2676.2 + \frac{(2875.4 - 2676.2)}{100} 94.84$$

$$= 2865.12 \text{ kJ/kg}$$

$$v_2 \text{(at 194.84°C)} = 1.6960 + \frac{(2.1720 - 1.6960)}{100} 94.84$$

$$= 2.1474 \text{ m}^3\text{/kg}$$

$$u_2 \text{(at 194.84°C)} = h_2 - P_2 v_2 = 2865.12 \times 10^3 - 1 \times 10^5 \times 2.1474$$

$$= 2650.38 \text{ kJ/kg K}$$

$u_1 = h_1 - P_1 v_1 = 3067.2 \times 10^3 - 4 \times 10^5 \times 0.6549 = 2805.24$ kJ/kg K

For an adiabatic process, $W = -\Delta u = u_1 - u_2$

$= 2805.24 - 2650.38 = 154.86$ kJ/kg

Therefore, work done by 0.1 kg steam = 15.486 kJ

**Example 4.50**

If superheated steam at 10 bar and 300°C is allowed to expand isentropically to a pressure of 1 bar in a piston –cylinder assembly, calculate the amount of work that can be obtained per kg of steam.

*Solution* :

For steam at 10 bar and 300°C

$v_1 = 0.2580$ $m^3/kg$ ; $h_1 = 3052.1$ kJ/kg ; $s_1 = 7.1251$ kJ/kg K

$u_1 = h_1 - P_1 v_1 = 3052.1 \times 10^3 - 10 \times 10^5 \times 0.2580 = 2794.1$ kJ/kg

For isentropic expansion $s_2 = s_1 = 7.1251$ kJ/kg K

At 1 bar, the following values are read from saturated steam tables.

$v_f = 0.001\,043\,4$ m$^3$/kg ; $v_g = 1.694$ m$^3$/kg

$h_f = 417.54$ kJ/kg ; $h_g = 2675.4$ kJ/kg

$s_f = 1.3027$ kJ/kg K ; $s_g = 7.3598$ kJ/kg K

We find $s_f = 1.3027 < s_2 = 7.1251 < s_g = 7.3598$ kJ/kg K

Therefore, the final state is wet steam at 1 bar. Then we find the quality of steam as

$$s_2 = X_2\, s_g + (1 - X_2)\, s_f$$

or
$$X_2 = \frac{s_2 - s_f}{s_g - s_f} = \frac{7.1251 - 1.3027}{7.3598 - 1.3027} = 0.9613$$

$$h_2 = X_2\, h_g + (1 - X_2)\, h_f = 0.9613 \times 2675.4 + (1 - 0.9613)\, 417.54$$
$$= 2588.02 \text{ kJ/kg}$$

$$v_2 = X_2\, v_g + (1 - X_2)\, v_f = 0.9613 \times 1.694 + (1 - 0.9613)\, 0.001\,043\,4$$
$$= 1.6285 \text{ m}^3/\text{kg}$$

$$u_2 = h_2 - P_2 v_2 = 2588.02 \times 10^3 - 1 \times 10^5 \times 1.6285 = 2425.17 \text{ kJ/kg}$$

For reversible adiabatic process $W = -\Delta u = u_1 - u_2$

or $W = 2794.1 - 2425.17 = 368.93$ kJ/kg

**Example 4.51**

What is a polytropic process and how does it reduce to other known processes ?

*Solution* :

If an ideal gas undergoes a reversible adiabatic process the path followed by the gas is given by $Pv^{\gamma}$ = constant and the work done is given by.

$$W = \frac{P_1 v_1 - P_2 v_2}{\gamma - 1}$$

These relations are quite convenient in engineering calculations and they can be easily extended to other processess also. In general, if a gas is exchanging energy, with its surroundings, in the form of work and heat, the process occurs is such a way that $Pv^n$ = constant, where $n$ is a constant. A process which follows the path $Pv^n$ = constant is called a polytropic process. The work done in a polytropic process is given by

$$W = \frac{P_1v_1 - P_2v_2}{n-1}$$

If $n$, the index of expansion is equal to $\gamma$, the process reduces to a reversible adiabatic or isentropic process. Similarly, the polytropic process reduces to other known processes as listed below.

If $n = 0$ ;Constant –pressure process

$n = 1$ ; Constant –temperature process

$n = \gamma$ ; Reversible–adiabatic process

$n = \infty$ ; Constant–volume process

These processes are shown in Fig.E 4.51 on $P$ versus $v$ diagram. The lower right quardrant in the figure (considering the initial state as the origin) shows expansion proceses and the upper left quadrant shows the compression processes.

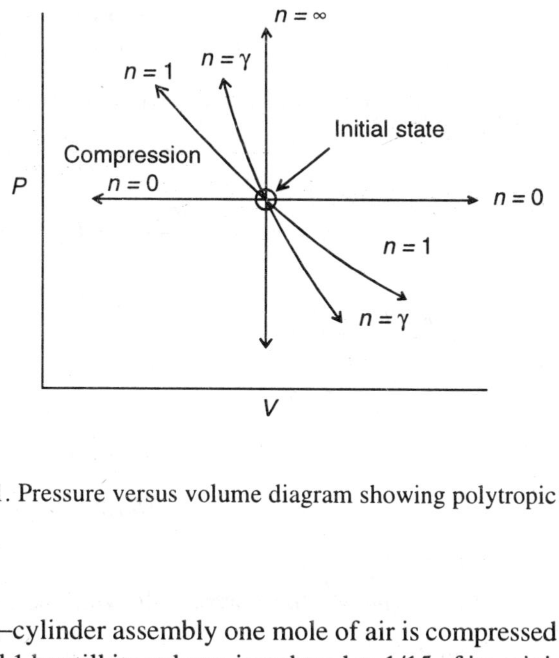


Fig.E 4.51. Pressure versus volume diagram showing polytropic processes.

### Example 4.52

In a piston–cylinder assembly one mole of air is compressed from the initial state 300 K and 1 bar till its volume is reduced to 1/15 of its original volume. The compression process can be approximated as $Pv^{1.25}$ = constant. Determine the (*a*) pressure and temperature of the gas at the end of the compression process (*b*) work done on the gas and (*c*) the energy transferred as heat. The air may be treated as an ideal gas.

*Solution* :

(*a*) The polytropic process follows the relation $Pv^{1.25}$ = constant.

or $$\frac{P_2}{P_1} = \left(\frac{v_1}{v_2}\right)^{1.25} \quad \text{or} \quad P_2 = P_1\left(\frac{v_1}{v_2}\right)^{1.25}$$

or $$P_2 = 1\times10^5(15)^{1.25} = 29.52 \text{ bar}$$

For an ideal gas $$\frac{P_1v_1}{T_1} = \frac{P_2v_2}{T_2} \quad \text{or} \quad T_2 = \frac{P_2v_2T_1}{P_1v_1}$$

or $$T_2 = \frac{29.52\times10^5}{1\times10^5}\left(\frac{1}{15}\right)\times300 = 590.4 \text{ K}$$

(*b*) Work done $$W = \frac{P_1v_1 - P_2v_2}{n-1} = \frac{R(T_1-T_2)}{n-1}$$

$$= \frac{8.314(300-590.4)}{1.25-1} = -9657.5 \text{ J/mol}$$

(*c*) The first law of thermodynamics gives

$$du = dq - dw \quad \text{or} \quad u_2 - u_1 = q - w$$

or $$C_v(T_2-T_1) = q - w \quad \text{or} \quad \frac{R}{\gamma-1}(T_2-T_1) = q - w$$

or $$q = \frac{R\,(T_2-T_1)}{\gamma-1} + w$$

$$= \frac{8.314(590.4-300)}{1.4-1} + (-9657.5)$$

$$= -3621.5 \text{ J/mol}$$

**Example 4.53**

Sketch the Mollier diagram showing the essential features.

*Solution* :

The Mollier diagram (enthalpy versus entropy diagram) was introduced into engineering thermodynamics by Richard Mollier and hence it was named after him. The processes commonly encountered in engineering thermodynamics are—constant volume, constant pressure, constant temperature, constant entropy etc. All these processes can be conveniently analyzed by using the Mollier diagram. The Mollier diagram shows the two phase region and the vapor region. It contains constant pressure lines, constant temperature lines and also the constant quality lines in the two phase region. A schematic of a Mollier diagram is shown in Fig.E 4.53.

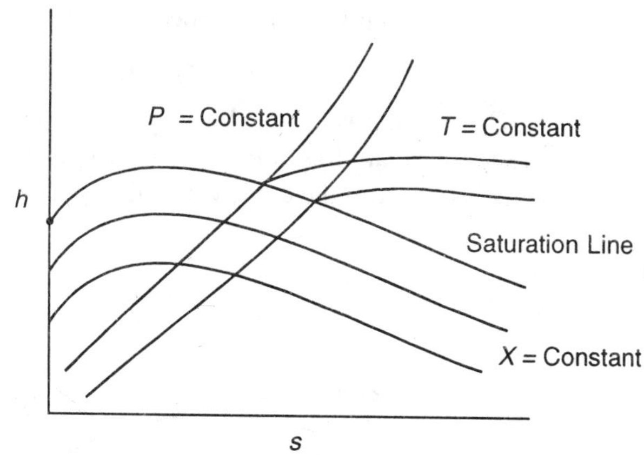


Fig.E 4.53. Schematic of a Mollier diagram.

It may be noted that

1. The constant temperature lines and constant pressure lines coincide in the two phase region.
2. The constant temperature and constant pressure lines are straight in the two phase region.
3. The constant temperature lines are usually curved in the superheat region but at increased temperatures they tend to become horizontal as the gas behavior approaches that of an ideal gas [$h = h\ (T)$ only].
4. The constant pressure lines are usually curved.
5. In general or in most of the diagrams, the constant volume lines may not be shown.
6. The saturation line corresponds to quality ($X$) is equal to one.
7. The constant quality lines are usually marked as constant dryness fraction lines and they are approximately parallel to the saturation line.

### Example 4.54

A piston– cylinder assembly contains 0.1 kg of superheated steam at 150°C and 1 bar pressure. If 23 kJ energy is removed from the steam while the pressure is held constant, determine the final state of steam. Use the Mollier diagram to solve the problem.

*Solution* :

Energy interaction at constant pressure = – 23 kJ

or $q = -230$ kJ/kg

For a constant pressure process

$$q = h_2 - h_1$$

The schematic of the Mollier diagram is shown in Fig.E 4.54. The initial state of the steam lies at the intersection of $t = 150$°C line and $P = 1$ bar and it is shown as 1. From the chart we read $h_1 = 2780$ kJ/kg

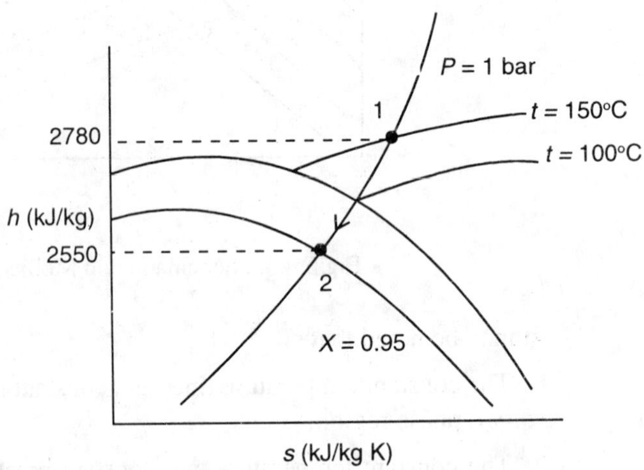


Fig.E 4.54. Sketch for Example 4.54.

Then $q = h_2 - h_1$

or $-230 = h_2 - 2780$ or $h_2 = 2550$ kJ/kg

The final state of steam should have $h = 2550$ kJ/kg and be at a pressure of 1 bar. Hence follow the constant pressure line of 1 bar till $h = 2550$ kJ/kg and the final state is represented as 2 in the figure. At the intersection of $h = 2550$ kJ/kg and $P = 1$bar, we find $X_2 = 0.95$. Hence the final state is wet steam of quality 0.95 at 1 bar pressure.

**Example 4.55**

A piston-cylinder assembly contains 0.1 kg superheated steam at 5 bar and 400°C. If the steam is allowed to expand reversibly and adiabatically at what pressure it will be available as dry saturated steam? Use the Mollier diagram to answer this question.

*Solution* :

A schematic of the Mollier diagram is shown in Fig.E 4.55. The initial state of the steam is shown as 1 at the intersection of the lines $P = 5$ bar and $t = 400°C$. We know that $s_2 = s_1$ for a reversible and adiabatic expansion. Hence follow the vertical line ($s$ = constant) from point 1 till it intersects the saturation curve (shown as 2). At the point 2, read $P = 0.3$ bar.

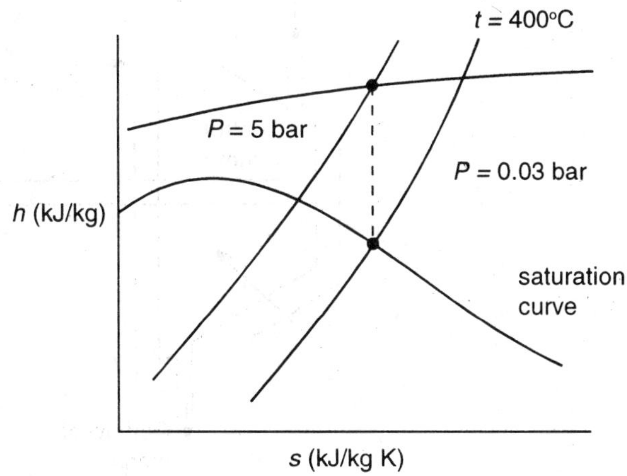


Fig.E 4.55. Sketch for Example 4.55.

## Example 4.56

What is a control volume? Why is it necessary to introduce control volume in the energy analysis of processes ?

*Solution* :

The first law of thermodynamics is the outcome of experimental observations on systems (that is a definite quantity of matter) or control masses. In engineering practice one deals with turbines, compressors, pumps, expanders, heat exchangers, reactors etc. where matter flows in and out of the equipment. In such cases it is extremely difficult to identify a definite quantity of matter and to follow the course of each such element of matter to perform the energy analysis. Therefore, it is necessary to have a different formulation of energy analysis, To analyze such flow processes the concept of control volume is introduced.

A control volume is a region in space bounded by a closed envelope on which attention is focussed for the energy analysis. The boundary of the control volume many coincide with the physical boundary of the equipment or it may be in part a well defined physical boundary or all of it may be an imaginary envelope.

The envelope which encloses the control volume is called a control surface. The control volume is shown in Fig.E 4.56. The control volume need not be fixed in size as well as in shape. It may not be fixed in position also. However, in most of the applications we deal with control volumes which are fixed in size and shape as well as fixed in position relative to the observer.

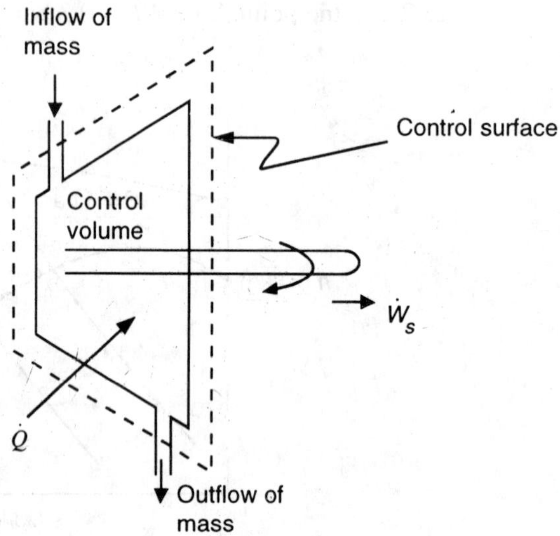


Fig.E 4.56. Sketch for Example 4.56.

**Example 4.57**

Compare and contrast the properties of a control mass and a control volume.

*Solution* :

| Control mass | Control volume |
|---|---|
| 1. Refers to a definite quanity of matter on which attention is focused. | Refers to a defined region of space on which attention is focused. |
| 2. Bounded by a closed boundary, which may be real or imaginary. | Enclosed by a control surface which may be real or imaginary. |
| 3. Boundary need not be constant in shape and size. | Control volume need not be constant in shape and size. |
| 4. Matter does not cross the boundaries of a control mass. | Matter continuously flows in and out of the control volume |
| 5. Heat and work interactions are present across the system boundary. | Control volume can exchange heat and work through control surface. |

**Example 4.58**

Express the law of conservation of mass for a control volume.

*Solution* :

Consider the control volume shown in Fig.E 4.58.

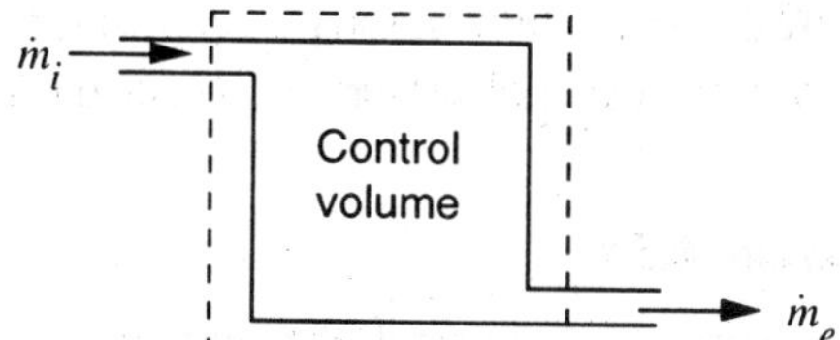


Fig.E 4.58. Sketch for Example 4.58.

Let

$dm_i$ = mass entering the control volume in a time interval $dt$

$dm_e$ = mass leaving the control volume in time interval $dt$

$m(t)$ = mass inside the control volume at time $t$

$m(t + dt)$ = mass inside the control volume at time $(t + dt)$

The law of conservation of mass gives

$$m(t + dt) = m(t) + dm_i - dm_e$$

or $$m(t + dt) - m(t) = dm_i - dm_e$$

or $$\frac{m(t+dt) - m(t)}{dt} = \frac{dm_i}{dt} - \frac{dm_e}{dt}$$

in the limit $dt \to 0$, we get

$$\frac{dm}{dt} = \dot{m}_i - \dot{m}_e \qquad (A)$$

$\dot{m}_i = \dfrac{dm_i}{dt}$ = rate of mass entering the control volume

$\dot{m}_e = \frac{d\dot{m}_c}{dt}$ = rate of mass leaving the control volume

If the control volume has several inlets and outlets, the law of conservation of mass given by Eqn. (*A*) can be written as

$$\frac{dm}{dt} = \sum_i \dot{m}_i - \sum_e \dot{m}_e$$

Stated in words, the law of conservation of mass is given by

(Rate of accumulation of mass inside the control volume) = (Rate of mass entering the control volume) – (Rate of mass leaving the control volume.)

**Example 4.59**

Steam at 5 bar and 400°C enters a control volume through a pipe of 10 cm diameter at a velocity 5 m/s and leaves the control volume at 4 bar and 375°C through a pipe of 5 cm dia. The steam flows through the pipe in a steady state. Determine (*a*) the flow rate of steam in kg/s and the (*b*) exit velocity in m/s.

*Solution* :

(*a*) If the flow is in a steady state, there is no accumulation of mass inside the control volume. Then, the law of conservation of mass of mass reduces to

$$\dot{m}_i = \dot{m}_e$$

We know that $\dot{m} = \rho A\mathbf{V} = A\mathbf{V}/v$

where $\rho$ = density of the substance

$A$ = cross– sectional area of the pipe

$\mathbf{V}$ = velocity

From the steam tables, we read $v = 0.6172\ m^3/kg$ at 5 bar and 400°C

Then, $$\dot{m}_i = \rho_i A\mathbf{V}_i = \frac{A\mathbf{V}_i}{v}$$

$$= \pi\frac{(0.1)^2}{4} \times \frac{5}{0.6172} = 0.0636\ \text{kg/s}$$

Therefore the mass flow rate of steam = 0.0636 kg/s

(*b*) $\dot{m}_e = \dot{m}_i = 0.0636$ kg/s

$$\dot{m}_c = \rho_e A_e \mathbf{V}_e = \frac{A_e \mathbf{V}_e}{v_e}$$

From the steam tables at 4 bar we find $v = 0.6549$ m$^3$/kg at 300°C and 0.7725 m$^3$/kg at 400°C. We interpolate between these two values to find $v$ at 375°C,

$$v_e = 0.6549 + \frac{(0.7725 - 0.6549)75}{100}$$

$$= 0.7438 \text{ m}^3/\text{kg}$$

$$\dot{m}_e = \frac{A_e \mathbf{V}_e}{v_e}$$

or $$\mathbf{V}_e = \frac{\dot{m}_e v_e}{A_e} = \frac{0.0636 \times 0.7438}{\pi (0.05)^2/4} = 24.09 \text{ m/s}$$

Therefore, the exit velocity of steam = 24.09 m/s

**Example 4.60**

Starting from the known form of the first law of thermodynamics for a control mass, derive the general form of the first law of thermodynamics for a control volume.

*Solution* :

Consider the control volume shown in Fig.E 4.60.

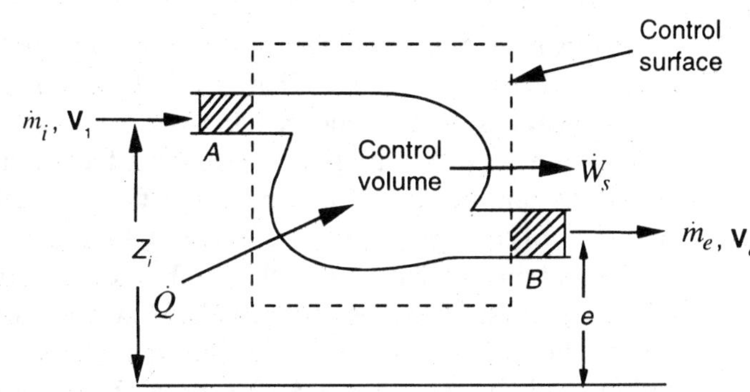


Fig.E. 4.60. Schematic of a Control-volume for energy analysis.

Let

$\dot{m}_i$ = mass flow rate into the control volume

$\dot{m}_e$ = mass flow rate out of the control volume

$\mathbf{V}_i$ = flow velocity at the inlet of the control volume

$\mathbf{V}_e$ = flow velocity at the exit of the control volume

$Z_i$ = elevation of the inlet of the control volume

$Z_e$ = elevation of the exit of the control volume

$P_i$ = pressure at the inlet of the control volume

$P_e$ = pressure at the exit of the control volume

$v_i$ = specific volume at the inlet of the control volume

$v_e$ = specific volume at the exit of the control volume

$Q$ = rate of energy flow as heat into the control volume

$W_s$ = rate of shaft work done by the control volume

$m(t)$ = mass inside the control volume at time $t$

$m(t + dt)$ = mass inside the control volume at time $(t + dt)$

$e_i = (u_i + \mathbf{v}_i^2/2 + gZ_i)$ = specific energy of matter at the inlet of the control volume

$e_e = (u_e + \mathbf{v}_i^2/2 + gZ_e)$ = specific energy of matter at the exit of the control volume

$e(t)$ = specific energy of matter inside the control volume at time $t$

$e(t + dt)$ = specific energy of matter inside the control volume at time $(t + dt)$

To apply the first law of thermodynamics for a control mass we should choose the control mass or system, That is we should identify a definite quantity of matter and focus our attention on it. In other words we should choose the boundary of a system such that it always contains a fixed mass and the boundary may change its configuration. The continuous flow of matter can be considered as a sequence of a large number of events taking place and each event takes a differential time $dt$. Now let us choose the control mass (system) at time $t$ as the sum of the mass inside the control volume at time $t$ and the mass which is about to enter the control volume in the differential time $dt$ which is occupying the region marked as A in Fig.E 4.60. That is the boundary of the system at time $t$ constitutes the control volume and the region A.

Mass of system at time $t = m(t) + m_i\, dt$

During the differential time $dt$, the mass in the region A enters the control volume and some mass leaves the control volume and occupies region B. Therefore, the system at time $(t + dt)$ is defined as the sum of the mass inside the control volume at time $(t + dt)$ and the mass in the region B. In other words, the system boundary at time $(t + dt)$ includes the control volume and the region B.

Then mass of system at time $(t + dt) = m(t + dt) + \dot{m}_e\, dt$.

Since the mass of the system remains constant, it is necessary that

$$m(t) + \dot{m}_i\, dt = m(t + dt) + \dot{m}_e\, dt$$

Energy of the system at time $t = E_1 = m(t)\, e(t) + \dot{m}_i\, e_i\, dt = E(t) + \dot{m}_i\, e_i dt$

Energy of system at time$(t + dt) = E_2 = m(t + dt)\, e(t + dt) + \dot{m}_e\, e_e dt$

$$= E(t + dt) + \dot{m}_e\, e_e dt$$

where $E(t)$ = energy of the matter in the control volume at time $t$

$E(t + dt)$ = energy of the matter in the control volume at time $(t + dt)$

To accomodate the mass in region A $(= \dot{m}_i dt)$ in the control volume, the mass inside the control volume has to be compressed such that its volume decreases by the amount $\dot{m}_i\, v_i\, dt$. This is accomplished by the pressure $P_i$ acting on the material at the inlet of the control volume. Therefore, the surroundings are doing work on the system.

Work done by system while mass enter the control volume $= -P_i v_i \dot{m}_i dt$

During the differential time $dt$ the mass $\dot{m}_e\, dt$ leaves the control volume at the exit. By similar reasoning, we can obtain the work done by the system while mass leaves the control volume.

Work done by system while mass leaves the control volume $= P_e v_e \dot{m}_e dt$

The shaft work delivered during the differential time $dt = \dot{W}_s dt$

Therefore, the total work done $W$ is given by

$$W = \dot{W}_s\, dt + P_e v_e\, \dot{m}_e dt - P_i v_i\, \dot{m}_i\, dt$$

The energy transfered as heat $Q$ during the differential time $dt$ is given by

$$Q = \dot{Q}\, dt$$

The first law of thermodynamics is given by

$$E_2 - E_1 = Q - W$$

Substituting for $E_2$, $E_1$, $Q$ and $W$ in the above equation, we get

$$\{E(t + dt) + \dot{m}_e\, e_e dt\} - \{E(t) + \dot{m}_i\, e_i dt\}$$

$$= \dot{Q}\, dt - \{\dot{W}_s\, dt + P_e v_e \dot{m}_e dt - P_i v_i\, \dot{m}_i dt\}$$

or $\dot{m}_e(e_e + P_e v_e) - \dot{m}_i(e_i + P_i v_i) = \dot{Q} - \dot{W}_s - \dfrac{E(t+dt) - E(t)}{dt}$

In the limit $dt \to 0$, we get

$$\dot{m}_e\left(u_e + \frac{\mathbf{V}_e^2}{2} + gZ_e + P_e v_e\right) - \dot{m}_i\left(u_i + \frac{\mathbf{V}_i^2}{2} + gZ_i + P_i v_i\right) = Q - \dot{W}_s - \frac{dE}{dt}$$

or $$\dot{m}_e\left(h_e + \frac{\mathbf{V}_e^2}{2} + gZ_e\right) - \dot{m}_i\left(h_i + \frac{\mathbf{V}_i^2}{2} + gZ_i\right) = \dot{Q} - \dot{W}_s - \frac{dE}{dt}$$

Which is the required first law of thermodynamics for a control volume.

In the above expression $h_i = u_i + P_i v_i$ = specific enthalpy of the substance at the inlet to the control volume

$h_e = u_e + P_e v_e$ = specific enthalpy of the substance at the exit of the control volume

The first law of thermodynamic for a control volume (given above) can be rewritten as

$$\frac{dE}{dt} = \left[\dot{m}_i\left(h_i + \frac{\mathbf{V}_i^2}{2} + gZ_i\right) + \dot{Q}\right] - \left[\dot{m}_e\left(h_e + \frac{\mathbf{V}_e^2}{2} + gZ_e\right) + \dot{W}_s\right]$$

or, stated in words

(Rate of energy accumulation in the control volume) = (Rate of energy flow into the control volume) – (Rate of energy flow out of the control volume)

**Example 4.61**

What conditions are implied if the flow is termed as steady state steady flow ?

*Solution* :

If the control volume is operating in steady state conditions, then there are no changes inside the control volume with time. That is

$$\frac{dE}{dt} = 0 \text{ and } \frac{dm}{dt} = 0$$

The steady flow situation arises when $\dot{m}_i$ and $\dot{m}_e$ do not change with time. In the case of steady state, steady flow there is no accumulation of mass and energy inside the control volume. That is

$$\dot{m}_i = \dot{m}_e = \dot{m} \quad \text{and} \quad \frac{dE}{dt} = 0$$

In the case of steady state, steady flow the rate of energy transfer as heat $\dot{Q}$ and work $\dot{W}_s$ are constant, That is

$$\dot{Q} = \text{constant and } \dot{W}_s = \text{constant}$$

**Example 4.62**

Many of the engineering devices like turbines, compressors, pumps, etc. operate at steady state steady flow conditions. An analysis of steady state flow processes is useful in designing and evaluating the performance of such devices. Reduce the general expression for the first law of thermodynamics for a control volume so that it can be directly used in the energy analysis of steady state, steady flow processes,

*Solution* :

The general expression for the first law of thermodynamics for a control volume is given by

$$\dot{m}_e \left(h_e + \frac{\mathbf{V}_e^2}{2} + gZ_e\right) + \dot{m}_i\left(h_i + \frac{\mathbf{V}_i^2}{2} + gZ_i\right)$$

$$= \dot{Q} - \dot{W}_s - \frac{dE}{dt}$$

If the flow is characterized as steady state, steady flow then

$$\dot{m}_i = \dot{m}_e = \dot{m} \text{ and ; } dE/dt = 0.$$

Hence, the above expression reduces to

$$\left(h_e + \frac{\mathbf{V}_e^2}{2} + gZ_e\right) - \left(h_i + \frac{\mathbf{V}_i^2}{2} + gZ_i\right) = \frac{\dot{Q} - \dot{W}_s}{\dot{m}}$$

**Example 4.63**

In a thermal power plant operating in steady state an adiabatic steam turbine receives 1 kg/s of superheated steam at 3 MPa and 400°C. The steam enters the turbine with a velocity of 10 m/s at an elevation of 10 m above the ground level. The steam leaves the turbine at 0.1 bar with 10% moisture content. The velocity of the steam at the exit is 30 m/s and the exit is at an elevation of 4 m above the ground level.Calculate the power output of the turbine.

*Solution :*

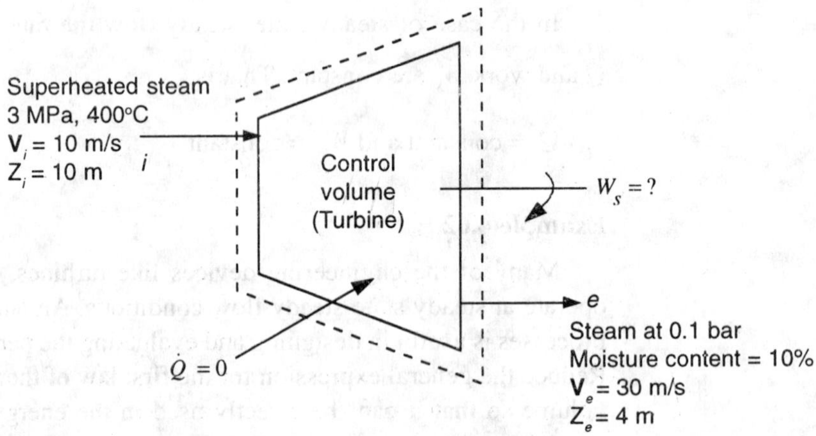


Fig.E 4.63. Sketch for Example 4.63.

Choose the turbine as the control volume as shown in Fig.E 4.63.

$\dot{m} = 1$ kg/s ; $\mathbf{V}_i = 10$ m/s ; $\mathbf{V}_e = 30$ m/s

$Z_i = 10$ m ; $Z_e = 4$ m

$X_e = 0.9$ (Since 10% moisture content is specified)

$\dot{Q} = 0$ (Since the turbine is adiabatic),

Read the following values from the steam tables.

At $P = 3$ MPa and $t = 400$°C ; $h_i = 3232.5$ kJ/kg

At $P = 0.1$ bar ; $h_f = 191.83$ kJ/kg ; $h_g = 2584.8$ kJ/kg

The first law of thermodynamics for a control volume operating in steady state steady flow is given by

$$(h_e + \frac{\mathbf{V}_e^2}{2} + gZe) - (h_i + \frac{\mathbf{V}_i^2}{2} + gZ_i) = \frac{\dot{Q} - \dot{W}_s}{\dot{m}}$$

This equation can be rearranged as

$$(h_e - h_i) + (\frac{\mathbf{V}_e^2}{2} - \frac{\mathbf{V}_i^2}{2}) + (gZ_e - gZ_i) = \frac{\dot{Q} - \dot{W}_s}{\dot{m}} \quad (A)$$

$h_e = X_e h_g + (1 - X_e)\, h_f = 0.9 \times 2584.8 + 0.1 \times 191.83 = 2345.5$ kJ/kg

Substituting the values of $h_e$, $h_i$, $\mathbf{V}_e$, $\mathbf{V}_i$, $Z_e$, $Z_i$, $Q$ and in Eqn ($A$), we get

$$(2345.5 - 3232.5) \times 10^3 + \frac{(30^2 - 10^2)}{2} + 9.81(4 - 10) = -\dot{W}_s$$

or $\quad -887 \times 10^3 + 400 - 58.86 = -\dot{W}_s$

or $\quad \dot{W}_s = 886.66$ kW

Power output of the turbine = 886.66 kW

**Example 4.64**

In Example 4.63 what percentage error will be introduced by ignoring the changes in ($a$) kinetic energy and ($b$) potential energy terms?

*Solution :*

($a$) Refer Example 4.63. In that example we find

$$(1/2)\,(\mathbf{V}^2_e - \mathbf{V}^2_i) = 400 \text{ J/s}$$

$$g(Z_e - Z_i) = -58.86 \text{ J/s}$$

$$(h_e - h_i) = -887 \times 10^3 \text{ J/s}$$

$$\dot{W}_s = 886.66 \text{ kW}$$

$$-\dot{W}_s = (h_e - h_i) + \frac{(\mathbf{V}_2^e - \mathbf{V}_1^e)}{2} + g\,(Z_e - Z_i)$$

If the changes in kinetic energy in ignored,

$$-\dot{W}_s = -887 \times 10^3 - 58.86 = -887.06 \text{ kW}$$

or $\quad \dot{W}_s = 887.06$ kW

Therefore, percentage error due to neglecting the kinetic energy terms

$$= \frac{(887.06 - 886.66)}{886.66} \times 100 = 0.045\,\%$$

($b$) $\quad -\dot{W}_s = -887 \times 10^3 + 400 = -886.6$ kW

or $\quad \dot{W}_s = 886.6$ kW

Percentage error due to neglecting the potential energy terms

$$= \frac{(886.66 - 886.6)}{886.66} \times 100 = 0.0068\,\%$$

***Note*** *The changes in the kinetic energy and potential energy terms are negligible compared to the change in the enthalpy. The omission of the kinetic energy and potential energy terms in the analysis of turbines introduces an error of the order of 0.05%. Hence, it is legitimate to ignore the changes in the kinetic energy and potential energy while analyzing the performance of turbines.*

**Example 4.65**

Apply the first law of thermodynamics for a control volume and derive a simple relation to analyze the performance of an adiabatic turbine operating in a steady state.

*Solution* :

The first law of thermodynamics for a control volume operating in steady state is given by

$$\left(h_e + \frac{\mathbf{V}_e^2}{2} + gZ_e\right) - \left(h_i + \frac{\mathbf{V}_i^2}{2} + gZ_i\right) = \frac{\dot{Q} - \dot{W}_s}{\dot{m}} \qquad (A)$$

Adiabatic turbine $\Rightarrow \dot{Q} = 0$

Ignore changes in the kinetic energy and potential energy terms. That is 1/2 $(\mathbf{V}^2_e - \mathbf{V}^2_i) = 0$ and $g\,(Z_e - Z_i) = 0$. Then Eqn.(*A*) reduces to $\dot{W}_s / \dot{m} = h_i - h_e$

Thus, in an adiabatic turbine operating in a steady state the work done, per unit mass of the fluid flowing through it, is equal to the decrease in the specific enthalpy of the fluid.

**Example 4.66**

An adiabatic turbine operating in a steady state is supplied with 10 kg/s superheated steam at 30 bar and 350°C. If the steam leaves the turbine at 5 kPa with 15% moisture content, estimate the power output of the turbine. Ignore KE and PE change and use the Mollier diagram to slove the problem.

*Solution :*

The first law of thermodynamics for an adiabatic control volume operating in steady state is given by

$$\dot{W}_s = \dot{m}\,(h_i - h_e)$$

when changes in KE and PE are ignored. The process followed by the steam in the turbine is shown as *i–e* on the Mollier diagram in Fig.E 4.66.

Locate the inlet state *i* and the exit state *e* of the steam as shown in Fig.E 4.66 and read $h_i$ = 3110 kJ/kg and $h_e$ = 2200 kJ/kg.

Then $\dot{W}_s = \dot{m}\ (h_i - h_e) = 10\ (3110 - 2200) \times 10^3 = 9.1$ MW

Power output of the turbine = 9.1 MW.

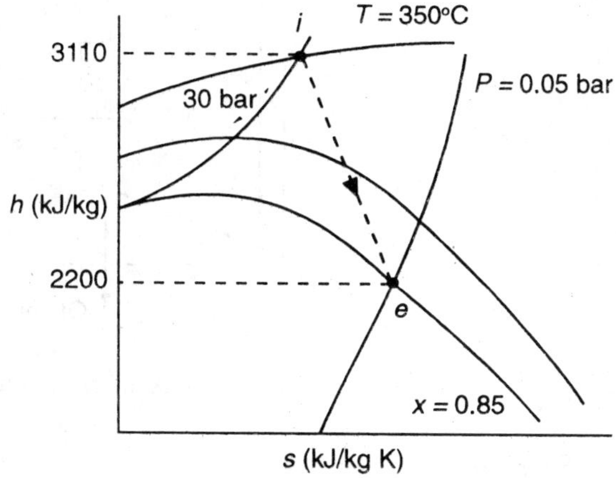


Fig.E 4.66. Sketch for Example 4.66.

### Example 4.67

Compressors and pumps are devices in which work is done on a fluid which results in an increase in the pressure of the fluid. Usually compressors are used for gases and pumps are used for liquids. In these devices the fluid flows in and out. Apply the first law of thermodynamics for a control volume and develop an expression to estimate the work input to these devices.

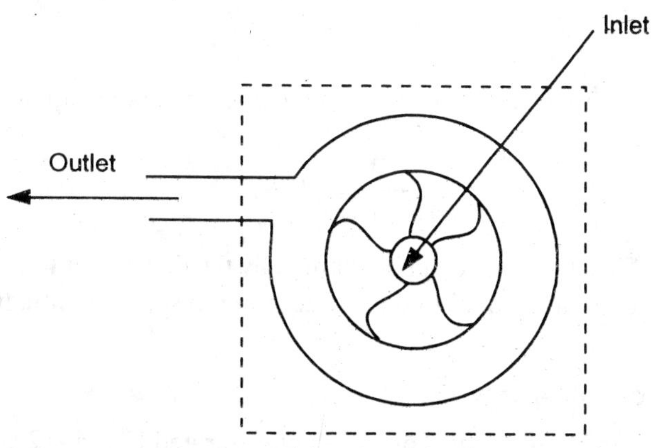


Fig.E 4.67. (*a*) Centrifugal flow compressor/pump.

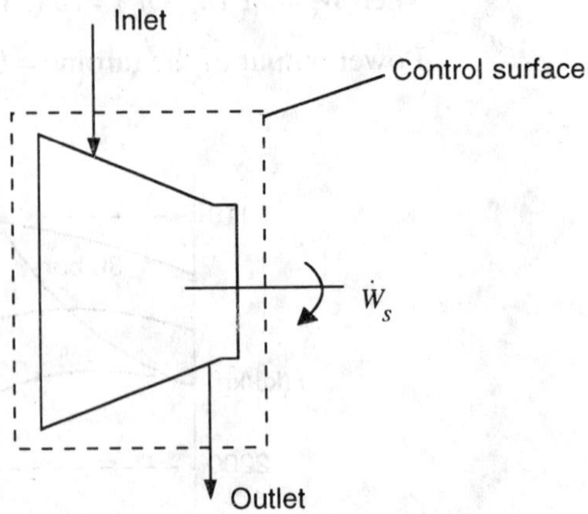


Fig.E 4.67. (*b*) Axial flow compressor

*Solution :*

Schematics of cetrifugal flow compressor/pump and an axial flow compressor are shown in Fig.E 4.67. In these devices the potential and kinetic energy changes can be ignored. The first law of thermodynamics for a control volume operating in steady state is given by

$$\left(h_e + \frac{\mathbf{V}_e^2}{2} + gZ_e\right) - \left(h_i + \frac{\mathbf{V}_i^2}{2} + gZ_i\right) = \frac{\dot{Q} - \dot{W}_s}{\dot{m}}$$

Ignoring the KE and PE changes, we get

$$h_e - h_i = \frac{\dot{Q} - \dot{W}_s}{\dot{m}}$$

Further, if the device is adiabatic, the above equation reduces to

$$\frac{-\dot{W}_s}{\dot{m}} = h_e - h_i$$

That is, the power consumed by the device per unit mass of the fluid flowing through it is equal to the increase in the specific enthalpy of the fluid.

**Example 4.68**

In an air–conditioning plant, saturated Freon–12 at –20°C with a quality of 0.8 enters an adiabatic compressor and leaves as saturated vapor at 40°C. If the flow rate of Freon–12 through the compressor is one kg/s, estimate the power input to the compressor.

*Solution* :

The thermodynamics properties of Freon –12 are

| $t$(°C) | $h_f$(kJ/kg) | $h_g$ (kJ/kg) |
|---|---|---|
| –20 | 17.9517 | 178.9017 |
| 40 | 75.1134 | 203.1063 |

$$h_i = X_i h_g + (1 - X_i) h_f = 0.8 \times 178.9017 + 0.2 \times 17.9517$$

or $h_i = 146.7117$ kJ/kg and $h_e = 203.1063$ kJ/kg

We know that for an adiabatic compressor

$$-\dot{W}_s/\dot{m} = h_e - h_i \text{ or } -\dot{W}_s = 203.1063 - 146.7117$$
$$= 56.395 \text{ kW}$$

Hence, power input to the compressor = 56.395 kW

**Example 4.69**

In an adiabatic compressor operating in steady state, the fluid enters at $P_1, v_1$ and leaves at $P_2$ and $v_2$. Show that the work done on the compressor is given by $W = -\int_1^2 Pdv$. Sketch the $P$–$v$ diagram and represent the work done on the compressor. Also compare the workdone on the compressor with the work done had the gas been compressed in a cylinder–piston assembly.

*Solution* :

The first law of thermodynamics for an adiabatic control volume operating in steady state is given by

$$h_e - h_i = -\dot{W}_s/\dot{m} \text{ or } dh = -dW_s \text{ (considering } \dot{m} = 1)$$

or $$du + Pdv + vdP = -dW_s \qquad (A)$$

Considering the mass undergoing the compression as our system, we can write the first law of thermodynamics as

$$du = dq - dW \text{ or } du = -dW = -Pdv \qquad (B)$$

(Since $dq = 0$ for an adiabatic process)

Substituting Eqn.($B$) in Eqn.($A$), we get

$$-d\dot{W}_s = du + Pdv + vdP = -Pdv + Pdv + vdP = vdP$$

or $$d\dot{W}_s = -vdP \text{ or } \dot{W}_s = -\int_i^e vdP \qquad (C)$$

That is, the work done in a flow process is given by the area under the curve when $v$ is plotted against $P$. The compression process in a flow device is shown in Fig.E 4.69 (*a*) and the shaded area represents the work done on the compressor.

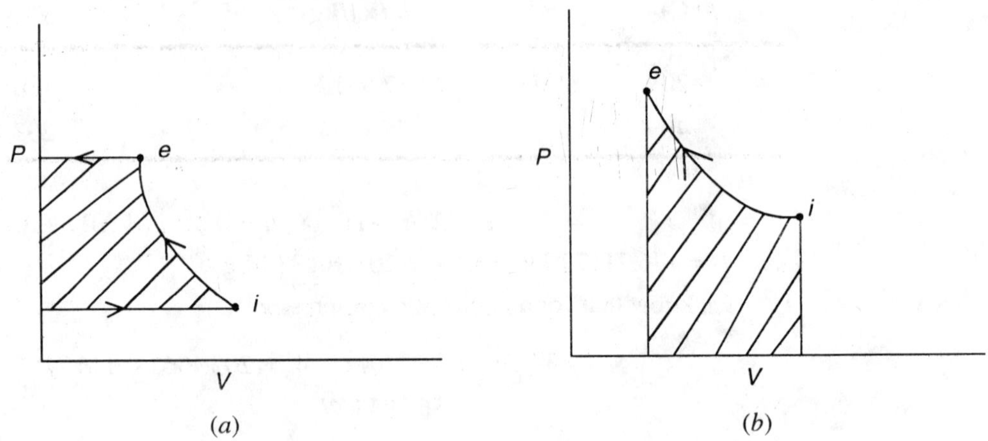


Fig.E 4.69.

(*a*) *P-v* diagram for compression in a flow process

(*b*) *P-v* diagram for compression in a piston-cylinder assembly or non flow process. The shaded area represents the work done.

In Fig.E 4.69 (*b*) the shaded area represents the work done in the case of compression in a piston–cylinder assembly. In other words, the flow work is given by $-\int vdP$ and the non flow work is given by $\int Pdv$.

**Example 4.70**

In a thermal power plant a feed pump receives saturated liquid water at 30°C and delivers it to a boiler at 30 bar. Assuming that liquid water is incompressible and the pump is adiabatic, estimate the work input to the pump per kg of liquid.

*Solution* :

We know that the work done by an adiabatic pump is given by

$$W_s = -\int_i^e vdP$$

If liquid water is treated as incompressible $v$ = constant. From the steam tables we obtain the following data for saturated liquid water at 30°C.

$$v_i = v_f = 0.001\ 004\ 3\ \text{m}^3/\text{kg}\ ;\ P_i = 0.042\ 41\ \text{bar}$$

Then $$W_s = -\int_{P_i}^{P_e} v dP = -v_i(P_e - P_i)$$

or $$-W_s = v_i(P_e - P_i) = 0.001\,004\,3\,(30 - 0.042\,41) \times 10^5$$
$$= 3.009 \text{ kJ/kg}$$

**Example 4.71**

In many steady state flow processes a need arises to increase the velocity of the fluid, A device which increases the velocity and hence the kinetic energy of a fluid at the expense of energy (enthalpy) is called a nozzle. Starting from the general form of the first law of thermodynamics for a control volume, derive a relation to evaluate the performance of a nozzle.

Solution :

A schematic of a nozzle is shown in Fig.E 4.71.

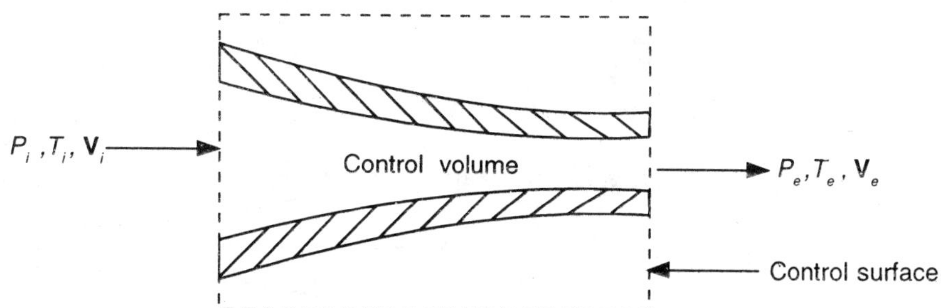


Fig.E 4.71. Schematic of a nozzle.

The general form of the first law of thermodynamics for a control volume is given by

$$\dot{m}_e\left(h_e + \frac{\mathbf{V}_e^2}{2} + gZ_e\right) - \dot{m}_i\left(h_i + \frac{\mathbf{V}_i^2}{2} + gZ_i\right) = \dot{Q} - \dot{W}_s - \frac{dE}{dt} \qquad (A)$$

Consider an adiabatic nozzle operating in steady state steady flow. Then

$\dot{Q} = 0$ (Adiabatic nozzle)

$dE/dt = 0$ (No accumulation of energy since it is operating in steady state).

$\dot{m}_e = \dot{m}_i$ (steady flow)

$\dot{W}_s = 0$ (Nozzle does not produce any shaft work)

Then Eqn.(*A*) reduces to

$$\left(h_e + \frac{\mathbf{V}_e^2}{2} + gZ_e\right) - \left(h_i + \frac{\mathbf{V}_i^2}{2} + gZ_i\right) = 0 \qquad (B)$$

The change in potential energy $g(Z_e - Z_i)$ can be ignored. However, the change in kinetic energy cannot be neglected since the primary objective of the nozzle is to increase the flow velocity. Then Eqn.(*B*) reduces to

$$(\mathbf{V}_e^2 - \mathbf{V}_i^2) = 2\,(h_i - h_e)$$

**Example 4.72**

Dry saturated steam at 5 bar enters an adiabatic nozzle at a velocity of 2 m/s and leaves as dry saturated steam at 2 bar. Calculate the exit velocity of the steam.

*Solution* :

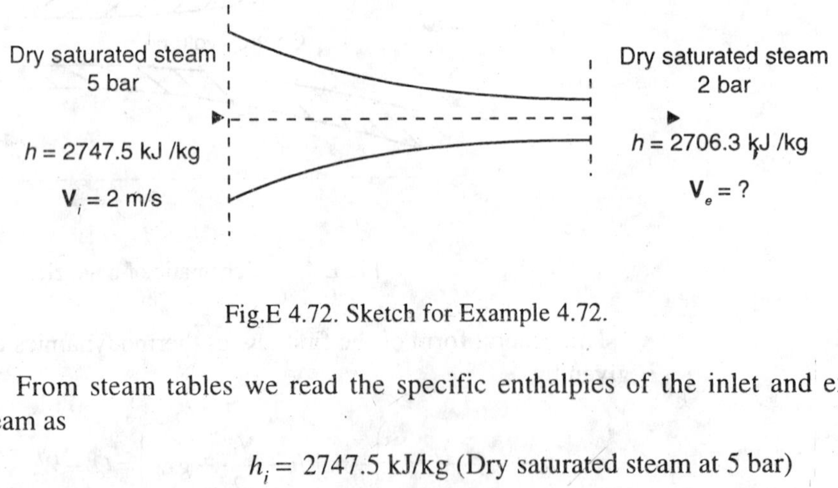


Fig.E 4.72. Sketch for Example 4.72.

From steam tables we read the specific enthalpies of the inlet and exit steam as

$h_i = 2747.5$ kJ/kg (Dry saturated steam at 5 bar)

$h_e = 2706.3$ kJ/kg (Dry saturated steam at 2 bar)

We know that for an adiabatic nozzle

$$\mathbf{V}_e^2 - \mathbf{V}_i^2 = 2\,(h_i - h_e)$$

or $\mathbf{V}_e^2 - 4 = 2(2747.5 - 2706.3) \times 10^3$

or $\mathbf{V}_e = 287.06$ m/s

### Example 4.73

Suppose an ideal gas at $P_i$ and $T_i$ enters a reversible and adiabatic nozzle with negligible velocity and leaves at pressure $P_e$. Derive an expression to estimate the exit velocity.

*Solution* :

We know that for an adiabatic nozzle

$$\mathbf{V}_e^2 - \mathbf{V}_i^2 = 2\ (h_i - h_e) \qquad (A)$$

For an ideal gas $h = h(T)$ only. Therefore

$$h_i - h_e = C_p (T_i - T_e) \qquad (B)$$

For an ideal gas undergoing reversible and adiabatic expansion

$$\frac{T_e}{T_i} = \left(\frac{P_e}{P_i}\right)^{\frac{\gamma-1}{\gamma}} \text{ or } T_e = T_i\left(\frac{P_e}{P_i}\right)^{\frac{\gamma-1}{\gamma}} \qquad (C)$$

Substituting Eqns ($B$) and ($C$) in Eqn.($A$) we get

$$\mathbf{V}_e^{\,2} - \mathbf{V}_i^{\,2} = 2C_pT_i\left[1-\left(\frac{P_e}{P_i}\right)^{\frac{\gamma-1}{\gamma}}\right]$$

If the gas enters the nozzle with negligible velocity, $\mathbf{V}_i = 0$. Then

$$\mathbf{V}_e = \sqrt{2C_pT_i\left[1-\left(\frac{P_e}{P_i}\right)^{\frac{\gamma-1}{\gamma}}\right]}$$

### Example 4.74

Air (ideal gas with $\gamma = 1.4$) at 4 bar and 350 K enters a reversible and adiabatic nozzle with negligible velocity and leaves at 1 bar. Calculate the velocity and temperature of the air leaving the nozzle. The molar mass of air is $29 \times 10^{-3}$ kg/mol.

*Solution* :

We know that $T_e = T_i\left(\frac{P_e}{P_i}\right)^{\frac{\gamma-1}{\gamma}}$

$$\text{or } T_e = 350\left(\frac{1}{4}\right)^{\frac{0.4}{1.4}} = 235.5 \text{ K}$$

$$C_P = \frac{R\gamma}{\gamma - 1} = \frac{8.314 \times 1.4}{0.4} = 29.1 \text{ J/mol K}$$

$$\text{or } C_P = \frac{29.1}{29 \times 10^{-3}} = 1.0034 \text{ kJ / kg K}$$

$$\mathbf{V}_e^2 - \mathbf{V}_i^2 = 2(h_i - h_e) = 2C_p(T_i - T_e)$$

$$\mathbf{V}_e = \sqrt{2C_P(T_i - T_e)}$$

$$= \sqrt{2 \times 1.0034 \times 10^3 (350 - 235.5)}$$

$$= 479.35 \text{ m/s}$$

**Example 4.75**

One of the most important steady flow devices of engineering interest is a heat exchanger. The heat exchangers are used to heat certain fluids while cooling some other fluids in a process industry. In steam power plants, the heat exchangers are used as condensers. A thermodynamic analysis of heat exchangers in quite important. Apply the first law of thermodynamics for a control volume for a heat exchanger operating in steady state and obtain a relation to evaluate its performance.

*Solution* :

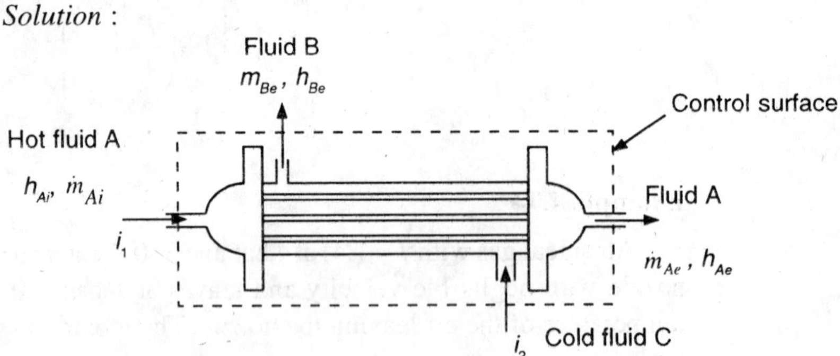


Fig.E 4.75. Schematic of a heat exchanger.

A schematic of a heat exchanger is shown in Fig.E 4.75. The hot fluid enters the heat exchanger at the inlet $i_1$ and leaves at exit $e_1$, while the cold fluid enters the device at inlet $i_2$ and leaves at exit $e_2$. The first law of thermodynamics for the control volume with multiple inlets and exits in given by

$$\sum \dot{m}_e \left(h_e + \frac{\mathbf{V}_e^2}{2} + gZ_e\right) - \sum \dot{m}_i \left(h_i + \frac{\mathbf{V}_i^2}{2} + gZ_i\right) = \dot{Q} - \dot{W}_s - \frac{dE}{dt} \qquad (A)$$

For steady state operation $\frac{dE}{dt} = 0$.

It is legitimate to ignore KE and PE changes. A heat exchanger does not deliver any shaft work and hence $\dot{W}_s = 0$. Moreover, if the heat exchanger is adiabatic, $\dot{Q} = 0$. Then Eqn.(*A*) reduces to

$$\Sigma \dot{m}_e h_e - \Sigma \dot{m}_i h_i = 0$$

or $\dot{m}_{Ae} h_{Ae} + \dot{m}_{Be} h_{Be} = \dot{m}_{Ai} h_{Ai} + \dot{m}_{Bi} h_{Bi}$

or $\dot{m}_{Ae} h_{Ae} - \dot{m}_{Ai} h_{Ai} = \dot{m}_{Bi} h_{Bi} - \dot{m}_{Bi} h_{Be}$ (*B*)

Since the fluids do not mix in the heat exchanger, we have

$$\dot{m}_{Ae} = \dot{m}_{Ai} = \dot{m}_A \text{ and } \dot{m}_{Be} = \dot{m}_{Bi} = \dot{m}_B$$

Therefore, Eqn (*B*) reduces to

$$\dot{m}_A (h_{Ae} - h_{Ai}) = \dot{m}_B (h_{Bi} - h_{Be})$$

**Example 4.76**

Another important type of equipment is an adiabatic mixer. It is quite often used in thermal power plants where steam is mixed with liquid water to preheat the water before it is fed to a boiler and this equipment is usually called open feed water heater. In air conditioning plants chilled air coming from different units is fed to a common duct.

These devices have several inlets but the mixture leaves through a single exit only. Consider an open feed water heater shown in Fig.E 4.76 and apply the first law of thermodynamics for a control volume to this device.

*Solution* :

The first law of thermodynamics for a control volume in steady state operation is given by.

$$\sum \dot{m}_e \left(h_e + \frac{\mathbf{V}_e^2}{2} + gZ_e\right) - \sum \dot{m}_i \left(h_i + \frac{\mathbf{V}_i^2}{2} + gZ_i\right) = \dot{Q} - \dot{W}_s \qquad (A)$$

In an adiabatic open feed water heater there is no shaft work ($\dot{W}_s = 0$) and $\dot{Q} = 0$. One can ignore changes in KE and PE. Then applying Eqn. (*A*) to the control volume shown in Fig.E 4.76, we get

$$\dot{m}_3 h_3 - \dot{m}_1 h_1 - \dot{m}_2 h_2 = 0$$

or
$$\dot{m}_3 h_3 = \dot{m}_1 h_1 + \dot{m}_2 h_2 \tag{B}$$

The law of conservation of mass gives

$$\dot{m}_3 = \dot{m}_1 + \dot{m}_2$$

Therefore, Eqn.(*B*) can be rewritten as

$$(\dot{m}_1 + \dot{m}_2) h_3 = \dot{m}_1 h_1 + \dot{m}_2 h_2$$

or
$$h_3 = \frac{\dot{m}_i h_i + \dot{m}_2 h_2}{\dot{m}_i + \dot{m}_2}$$

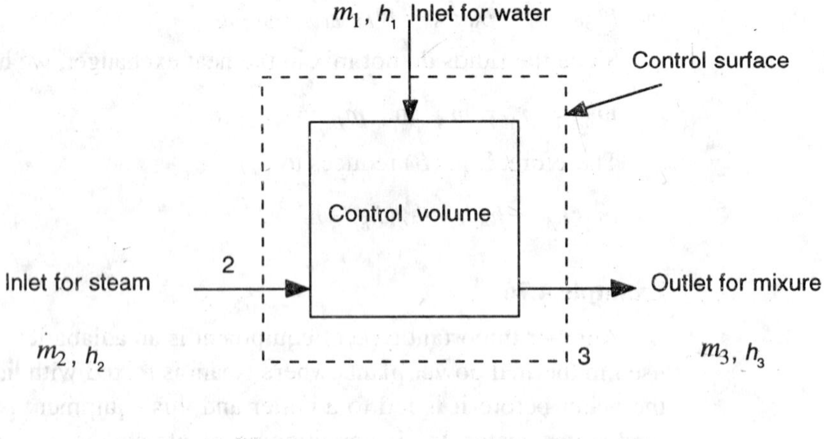


FigE. 4.76. Schematic of an open feed water heater.

**Example 4.77**

In a steam power plant dry saturated steam at 10 bar is mixed with liquid water at 10 bar and 30°C such that the resulting liquid is saturated liquid at 10 bar. Determine the ratio in which the steam and liquid water are fed to the open feed water heater.

*Solution* :

See Fig.E 4.76 for a schematic of the open feed water heater. We know that

$$\dot{m}_1 + \dot{m}_2 = \dot{m}_3 \text{ and} \tag{A}$$

$$(\dot{m}_1 + \dot{m}_2) h_3 = \dot{m}_1 h_1 + \dot{m}_2 h_2 \tag{B}$$

From steam tables, we obtain

$h_1$ (liquid water at 10 bar, 30°C) = 125.66 kJ / kg

(Enthalpy of compressed liquid water at 30°C) ≅ (Enthalpy of saturated liquid water at 30°C)

$h_2$ (Dry saturated steam at 10 bar) = 2776.2 kJ / kg

$h_3$ (Saturated liquid at 10 bar) = 762.61 kJ / kg

Let $\dot{m}_1$ = 1 kg /s then Eqn. (*B*) gives

$$(1 + \dot{m}_2)\, 762.61 = 125.66 + \dot{m}_2\,(2776.2)$$

or
$$\dot{m}_2 = 0.3163 \text{ kg/s}$$

Therefore, $\dot{m}_2/\dot{m}_1 = 0.3163$

Hence 0.3163 kg of dry saturated steam at 10 bar is mixed with every kg of liquid water at 10 bar and 30°C.

**Example 4.78**

Define the Joule–Thomson (or Joule–Kelvin) coefficient.

*Solution* :

The Joule–Thomson (or Joule–Kelvin) coefficient is defined as

$$\mu_{JT} = (\partial T / \partial P)_h$$

That is the slope of an isenthalpic curve is called the Joule–Thomson coefficient.

**Example 4.79**

Describe the porous plug experiment of Joule and Thomson and what conclusion can be drawn from the results of porous plug experiment ?

*Solution* :

The porous plug experiment was designed by Joule and Thomson (or Lord Kelvin) in 1853 to measure the temperature change in a fluid flowing in a steady state through a porous plug. A schematic of the apparatus used by Joule and Thomson is shown in Fig.E 4.79. It consists of a long insulated and horizontal pipe in which a porous plug is placed. A gas at known pressure $P_i$ and $T_i$ enters the pipe and leaves at pressure $P_e$ and $T_e$. The upstream and down stream pressures are maintained at constant values.

Consider the control volume shown in Fig.E 4.79. The first law of thermodynamics for a control volume gives

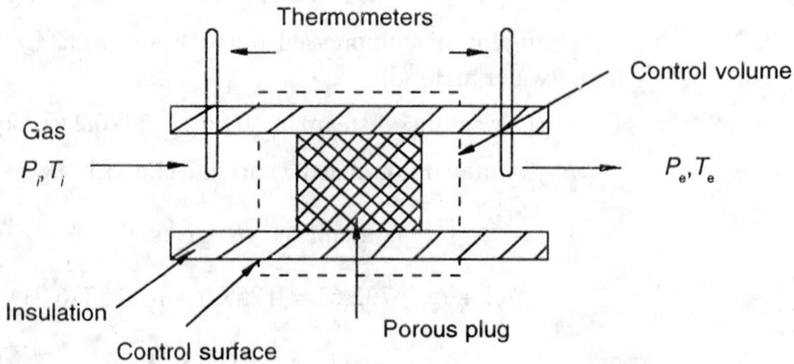


Fig.E 4.79. Schematic of porous plug apparatus.

$$\dot{m}_e\left(h_e+\frac{\mathbf{V}_e^2}{2}+gZ_e\right)-\dot{m}_i\left(h_i+\frac{\mathbf{V}_i^2}{2}+gZ_i\right)=\dot{Q}-\dot{W}_s-\frac{dE}{dt} \qquad (A)$$

Steady state steady flow : $\frac{dE}{dt}=0$ ; $\dot{m}_i=\dot{m}_e$

Insulated pipe : $\dot{Q}=0$ No shaft work is involved : $\dot{W}_s=0$

Horizontal pipe : $Z_e=Z_i$

When the fluid flows through the porous plug from high pressure to low pressure, it is associated with a change in the flow velocity. However, the change in the flow velocity is negligible. Therefore, one can assume that $\mathbf{V}_e \equiv \mathbf{V}_i$ Then Eqn. (*A*) reduces to

$$h_e = h_i$$

Therefore, whenever a fluid expands from a region of high pressure to a region of low pressure through a porous plug or a partially opened valve or some obstruction, without exchanging energy as heat and work with the surroundings, the enthalpy of the fluid remains constant. The process described above is called throttling. Hence throttling process is an isenthalpic process. Joule and Thomson measured the temperatue of the exit fluid at various downstream pressures while the upstream conditions are held constant.

**Example 4.80**

What is an inversion point and what is an inversion curve ?

***Solution*** :

The Joule–Thomson coefficient $\mu_{JT}$ can be positive, negative or zero. Usually the value of $\mu_{JT}$ changes with pressure and temperature. The point at which $\mu_{JT}=$

0 is called an inversion point. That is at the inversion point (specificed by pressure and temperature) the Joule–Thomson coefficient changes its sign from negative to positive or positive to negative. The locus of all the inversion points is called the inversion curve.

### Example 4.81

Sketch the isenthalpic curves for a gas and show the inversion curve. What is the significance of Joule–Thomson coefficient ?

*Solution* :

In the Joule –Thomson experiment, if one measures the exit temperature of the gas at different exit pressures, for a given values of inlet pressure and temperature and plots the results on temperature versus pressure diagram, he gets an isenthalpic curve. Since inlet conditions, $P_i$ and $T_i$ are held constant the enthalpy of the gas for all the measured conditions ($P_e$ and $T_e$) is constant. If the experiment is repeated with several inlet conditions, one obtains several isenthalpic curves as shown in Fig.E 4.81.

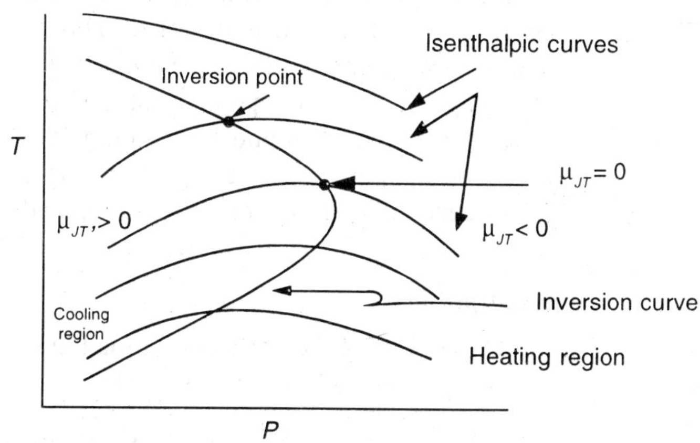


Fig.E 4.81. Isenthalpic curves of a real gas. $T$ and $P$ represent the down stream conditions.

From Fig.E 4.81 we find that $\mu_{JT} > 0$ to the left of the inversion curve. If

$$\mu_{JT} = \left(\frac{\partial T}{\partial P}\right)_h > 0$$

there will be a decrease in temperature of the gas when it is throttled ($\Delta P < 0$). Therefore, the region of $\mu_{JT} > 0$ represents the region of cooling. That is, if a real gas is subjected to a throttling process at the conditions falling in the region of $\mu_{JT} > 0$, the gas will be cooled. $\mu_{JT} < 0$ to the right of the inversion curve and this region denotes the region of heating. That is, if a real gas is throttled such that its state lies to the right of the inversion curve, it gets heated.

### Example 4.82

How does a knowledge of the inversion temperatures of a gas help an engineer?

*Solution* :

For almost all gases at ordinary temperatures and pressures, $\mu_{JT} > 0$ and the maximum inversion temperature is above room temperature. However, for gases like hydrogen, helium and neon the maximum inversion temperature is much below the room temperature. For hydrogen, the maximum inversion temperature is 200 K and for helium the maximum inversion temperature is 24 K. If hydrogen is throttled at temperatures greater than 200 K, the temperature of the gas increases. To liquefy hydrogen by throttling, it is necessary to cool hydrogen to a temperature below 200 K, before it is throttled. This is usually accomplished by cooling hydrogen with liquid nitrogen. Similarly, in the production of liquid helium by throttling, the initial temperature of helium should be below 24 K. Hence it is usually cooled with liquid hydrogen prior to throttling. A knowledge of the inversion temperatures and inversion curves of real gases is of considerable importance in the design of refrigeration and liquefaction equipment.

### Example 4.83

Describe the Linde process for the liquefaction of gases.

*Solution* :

When a real gas at high pressure is throttled to a low pressure, its temperature decreases. Linde made use of this principle for the liquefaction of gases. A schematic of the Linde process is shown in Fig.E 4.83. The gas to be cooled is compressed to a high pressure in a multistage compressor with intercooling. Then the high pressure gas flows through a counter current heat exchanger in which it is cooled by the cold gas leaving the separator. The cooled gas is throttled to a low pressure which results in the liquefaction of a part of the gas. Then the mixture enters a separator where the gas is separated from the liquid and the cold gas is recirculated through a heat exchanger in which the incoming gas is cooled.

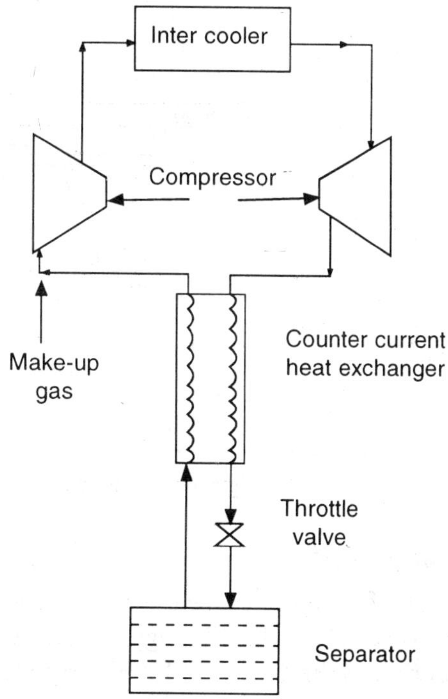


Fig.E 4.83. Schematic of Linde process.

**Example 4.84**

What is a throttling calorimeter and how is it used ?

*Solution* :

A throttling calorimeter is a device which is used to measure the quality of wet steam. This calorimeter is based on the principle of throttling. A part of the wet steam flowing through a pipe is throttled to a low pressure, usually atmospheric pressure, in a throttling calorimeter as shown in Fig.E 4.84.

After throttling the steam becomes superheated. The temperature and pressure of the steam in the calorimeter are recorded. Knowing the values of $P$ and $T$ of steam in the calorimeter one can determine its enthalpy. Since throttling is an isenthalpic process, the enthalpy of steam in the main line is equal to the enthalpy of steam in the calorimeter. Knowing the enthalpy and pressure in the main line one can determine the quality of steam by making use of the relation.

$$h_e = h_i = X_i h_g + (1 - X_i) h_f$$

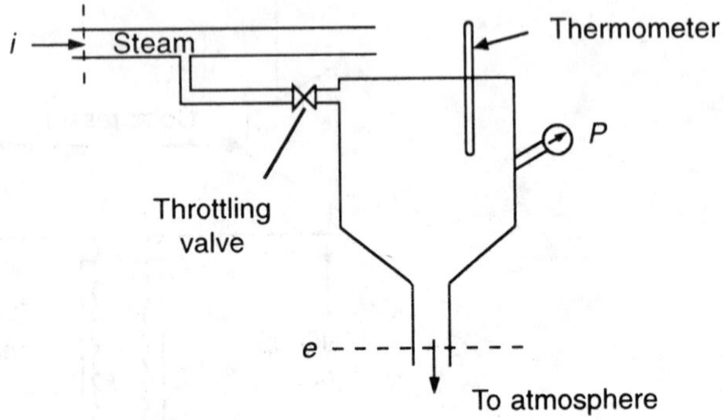


Fig.E 4.84. Schematic of throttling calorimeter.

**Example 4.85**

A throttling calorimeter is used to measure the quality of wet steam at 20 bar. The steam is throttled through a calorimeter, The pressure and temperature in the calorimeter are 1 bar and 130°C, respectively. Calculate the quality of steam.

*Solution* :

$h_e$(at 1 bar and 130°C) = 2735.96 kJ/kg

(obtained by interpolating between 100°C and 200°C)

At 20 bar: $h_f = 908.59$ kJ/kg ; $h_g = 2797.2$ kJ/kg

$$h_e = h_i = X_i h_g + (1 - X_i) h_f$$

or $2735.96 = 2797.2\, X_i + (1 - X_i)\, 908.59$

or $X_i = \dfrac{2735.96 - 908.59}{2797.2 - 908.59} = 0.9675$

Therefore, the quality of wet steam = 0.9675

**Example 4.86**

Rework Example 4.85 using the Mollier diagram.

*Solution* :

The Mollier diagram is shown in Fig.E 4.86. The steam in the calorimeter is at 1 bar and 130°C. Hence, locate the point of intersection *e* of the $P = 1$ bar line and the $t = 130$°C. Since enthalpy is constant during throttling, draw a horizontal line from point *e* and find the point of intersection *i* of this line with $P = 20$ bar line as shown in Fig.E 4.86. Then read the value $X = 0.9675$ (interpolation is needed between 0.96 and 0.97).

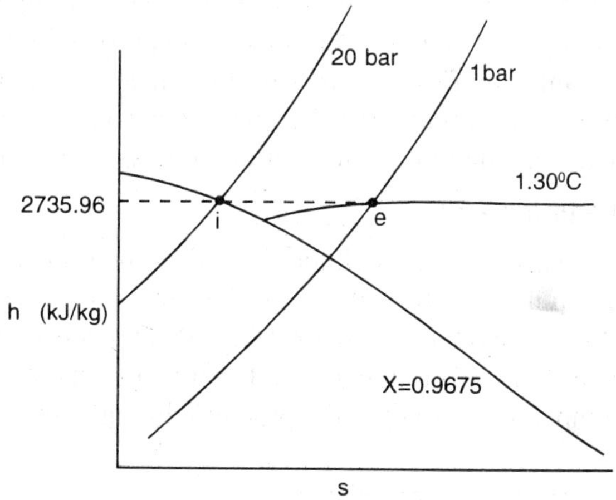


Fig.E 4.86. Sketch for Example 4.86.

**Example 4.87**

Explain why the measurement of $P$ and $T$ does not provide information about the quality of wet steam whereas measurement of $P$ and $T$ after throtting enables one to determine the quality of steam.

*Solution* :

Wet steam contains both the liquid and vapor phases. In a single component system when both the liquid and vapor phases coexist in equiibrium, there is only one degree of freedom. That is $P$ and $T$ are not independent of each other. If $P$ is specified its temperature will be the saturation temperature at that pressure. The system may be present as saturated liquid only or saturated vapor only or any arbitrary mixture of saturated liquid and saturated vapor. Hence it is not possible to determine the quality of wet steam by measuring its pressure and temperature. On the other hand, if the wet steam is throttled such that the final state is superheated steam, the state of the superheated steam (single phase) can be completely determined by measuring two independent variables $P$ and $T$.In particular, the enthalpy of the steam has a definite value. Since enthalpy is an extensive property, it depends on the amounts of liquid and vapor present in the mixture. Therefore, a knowledge of enthalpy and either saturation pressure or saturation temperature will enable one to determine the quality of wet steam.

**Example 4.88**

What are the characteristics of a transient flow process ?

*Solution* :

Many processes of interest to an engineer involve unsteady flow. During the startup or shut down of turbines, compressors and while charging and discharging of tanks the flow is transient. In transient flow processes, the properties of the fluid change with time. The mass flow rates into and out of the device are not the same and there is accumulation of mass as well as energy in the device.

**Example 4.89**

Suppose a gas cylinder initially containing a gas of mass $m_0$ is connected to a supply line through which the same gas at temperature $T$ and pressure $P$ is flowing and the cylinder is filled with the gas till the gas pressure in the cylinder rises to $P$. It is required to determine the final mass of gas in the cylinder and its temperature. Analyse the problem through control mass analysis.

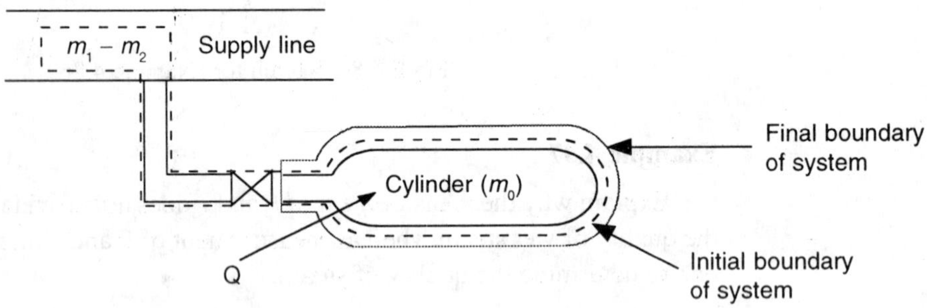


Fig.E 4.89. Control-mass for charging of a cylinder.

*Solution* :

Let $m_0$ = initial mass of gas in the cylinder

$m_f$ = final mass of gas in the cylinder

$P$ = pressure of gas in the supply main

$v$ = specific volume of gas in the supply main

$u$ = specific internal energy of gas in the supply main

$u_0$ = initial specific internal energy of gas in the cylinder

$u_f$ = final specific internal energy of gas in the cylinder

$Q$ = amount of energy transferred as heat into the cylinder

For control mass analysis one has to clearly identify the control mass or system. That is the mass of the system should remain constant all the time, but the configuration of the boundary enclosing the system may change. Therefore, we choose the system such that it contains the same mass of gas $m_f$. The system boundary is shown in the Fig.E 4.89. Initially the system contains $m_0$ mass of gas in the cylinder and $(m_f - m_0)$ mass of gas in the supply main. As the process proceeds. the mass of gas $(m_f - m_0)$ enters the cylinder and the system boundary finally coincides with the physical boundary of the cylinder. That is the boundary configuration undergoes a change.

The energy of the system before the start of the filling operation $E_1$ is the sum of the energy of the gas of mass $m_0$ in the cylinder and the energy of the mass of the gas $(m_f - m_0)$ in the supply line. That is

$$E_1 = m_0 u_0 + (m_f - m_0)\, u$$

Energy of the system at the end of the filling process, $E_2$ is given by

$$E_2 = m_f\, u_f$$

For the mass of gas $(m_f - m_0)$ to enter the cylinder, the system boundary has to be collapsed reducing its volume by $(m_f - m_0)v$. This is done by the pressure $P$ of the gas in the supply line. That is the surroundings does work on the system in pushing the gas into the cylinder. Therefore, the work done $W$ by the system is

$$W = -(m_f - m_0)\, v$$

(Note that work is done on the system and hence a negative sign is introduced)

The first law of thermodynamics gives

$$E_2 - E_1 = Q - W$$

or $m_f u_f - [m_0 u_0 + (m_t - m_0)u] = Q - [-(m_t - m_0)\, Pv]$

or $m_f u_f - m_0 u_0 = Q + (m_f - m_0)\,(u + Pv) = Q + (m_f - m_0)\, h$

or $$(m_f - m_0)\, h = m_f u_f - m_0 u_0 - Q \qquad (A)$$

where $h = u + Pv$ = specific enthalpy of gas in the supply line.

**Example 4.90**

Rework Example 4.89 through control volume analysis.

*Solution* :

The charging of a cylinder is a transient flow process and it can be most conveniently analyzed through control – volume approach. Choose the cylinder as the control–volume as shown in Fig.E 4.90.

The general form of the first law of thermodynamics for a control volume is given by

$$\dot{m}_e\left(h_e + \frac{\mathbf{V}_e^2}{2} + gZ_e\right) - \dot{m}_i\left(h_i + \frac{\mathbf{V}_i^2}{2} + gZ_i\right) = \dot{Q} - \dot{W}_s - \frac{dE}{dt} \qquad (A)$$

While charging a cylinder the gas enters the cylinder only but no gas leaves the cylinder. Therefore $\dot{m}_e = 0$. No shaft work is associated with the control volume. That is $\dot{W}_s = 0$. It is reasonable to ingore the kinetic energy and potential energy changes. Then Eqn.(*A*) reduces to

$$-\dot{m}_i h_i = \dot{Q} - \frac{dE}{dt} = \dot{Q} - \frac{d(mu)}{dt} \qquad (B)$$

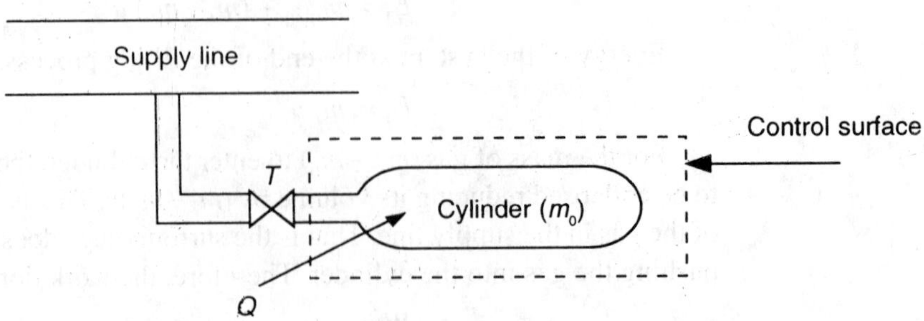


Fig.E 4.90. Control volume for charging of a cylinder.

where $m$ and $u$ denote the mass of gas and specific internal energy of gas inside the control volume at time *t*, respectively. Since the conditions, inside the supply line remain constant all the time $h_i$ = constant. Then Eqn.(*B*) can be rewritten as

$$-h_i \int_D^t \dot{m}_i \, dt = \int_0^t \dot{Q} \, dt - \int_0^t d\frac{(mu)}{dt} \, dt \qquad (C)$$

The principle of conservation of mass for the charging of a cylinder gives

$$\frac{dm}{dt} = \dot{m}_i \text{ or } \int_0^t \frac{dm}{dt} dt = \int_0^t \dot{m}_i \, dt = m_f - m_0 \qquad (D)$$

The first and second terms on the RHS of Eqn.(*C*) give

$$\int_0^t \dot{Q} \, dt = Q \qquad (E)$$

$$\int_0^t \frac{d[mu]}{dt} dt = m_f u_f - m_0 u_0 \qquad (F)$$

Substituting Eqns (*D*) – (*F*) in Eqn. (*C*), we get

$$(m_f - m_0)\, h_i = m_f u_f - m_0 u_0 - Q \qquad (G)$$

which is identical with Eqn.($A$) of Example 4.89,

### Example 4.91

An evacuated tank of 1 m$^3$ volume is connected to a line carrying steam at 10 bar and 400°C. If the valve in between the tank and the supply line is opened, steam flows into the tank till the steam pressure inside the tank is 10 bar. Calculate the final mass of steam in the tank and its temperature.

*Solution* :

Consider the tank shown in Fig.E 4.91 as the control volume. The first law of thermodynamics for the charging of a tank gives

$$(m_f - m_0)\, h_i = m_f\, u_f - m_0\, u_0 - Q \qquad (A)$$

Adiabatic tank ⇒ $Q = 0$

The tank is initially evaluated ⇒ $m_0 = 0$,

Then Eqn, ($A$) reduces to

$$m_f h_i = m_f u_f \text{ or } h_i = u_f \qquad (B)$$

Where $h_i$ = specific enthalpy of steam in supply line

$u_f$ = final specific internal energy of steam in the tank

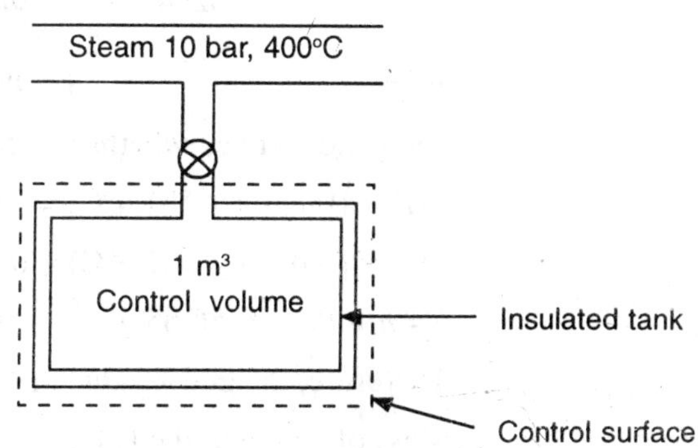


Fig.E 4.91. Sketch for Example 4.91.

From steam tables we find $h_i$ = 3264.4 kJ/kg at 10 bar and 400°C.

Therefore, $h_i = u_f = 3264.4$ kJ / kg

We also know that, the final pressure $P_f$ of steam in the tank is 10 bar. Knowing $P$ and $u$ one can determine the final state of steam in the tank. Since the properties of steam are available in tabular form, the determination of the final state involves a trial and error calculation. Assume the final temperature $T_f$ of steam is 550°C. Then

At $P = 10$ bar and $t = 550$°C; we have

$$h = 3587.85 \text{ kJ / kg}$$
$$v = 0.3775 \text{ m}^3\text{ / kg}$$

(The values of $h$ and $v$ are obtained by interpolation)

Then $u = h - Pv = 3587.85 \times 10^3 - 10 \times 10^5 \times 0.3775 = 3210.35$ kJ/kg

The calculated value $u = 3210.35$ kJ/kg $< u_f = 3264.4$ kJ/kg.

Now, assume $t_f = 575$°C. Then, from steam tables at $P = 10$ bar and $t = 575$°C, we get

$$h = 3642.63 \text{ kJ/kg} \text{ ; } v = 0.3893 \text{ m}^3\text{ / kg.}$$
$$u = h - Pv = 3642.63 \times 10^3 - 10 \times 10^5 \times 0.3893$$
$$= 3253.33 \text{ kJ/kg}$$

We find $u = 3210.35$ kJ / kg at 550°C and

$$u = 3253.33 \text{ kJ/kg at } 575°\text{C}$$

Let us extrapolate these results to find $t$ such that $u = 3264.4$ kJ/kg as shown below.

$$3264.4 = 3210.35 + \frac{(3253.33 - 3210.35)\Delta t}{575 - 550}$$

or $$\Delta t = 31.4°\text{C or } t = 581.4°\text{C}$$

Let us now check whether $t = 581.4$°C is acceptable or not.

$h$ ($P = 10$ bar, $t = 581.4$°C) $= 3656.65$ kJ/kg

$v$ ($P = 10$ bar, $t = 581.4$°C) $= 0.3923$ m³/kg

$u = h - Pv = 3656.65 \times 10^3 - 10 \times 10^5 \times 0.3923 = 3264.4$ kJ/kg

Therefore, the guess value of $t_f = 581.4$°C is correct.

Mass of steam in the tank,

$$m_f = \frac{V}{v} = \frac{1}{0.3923} = 2.549 \text{ kg}$$

**Example 4.92**

A rigid and insulated tank of volume 2 m³ contains an ideal gas ($\gamma = 1.4$) at 1 bar and 300 K. The tank is connected to a line carrying the same gas at 20 bar and 400 K. The valve in between the tank and the line is opened allowing the gas to enter the tank till the gas pressure in the tank rises to 20 bar and then closed. Determine the final temperature of the gas in the tank and the amount (moles) of gas that entered the tank.

*Solution* :

Choose the tank as the control volume and apply the first law of thermodynamics for the charging of a tank to obtain

$$(m_f - m_0)\, h_i = m_f\, u_f - m_0\, u_0$$

or
$$(n_f - n_0)\, h_i = n_f u_f - n_0 u_0 \qquad (A)$$

where $n_f = m_f/M$ = moles of gas in the tank at the end of filling operation

$n_0 = m_0/M$ = moles of gas in the tank at the beginning of filling operation

$M$ = molar mass of gas

We know that for an ideal gas

$$n = PV/RT;\; u = C_v\, T \text{ and } h = C_p T$$

Therefore, Eqn.($A$) can be rewritten as

$$\left(\frac{P_f V}{RT_f} - \frac{P_0 V}{RT_0}\right) C_p T_i = \frac{P_f V}{RT_f} C_v T_f - \frac{P_0 V}{RT_0} C_v T_0$$

or
$$\left(\frac{P_f}{T_f} - \frac{P_0}{T_0}\right) C_p T_i = P_f C_v - P_0 C_v = C_v\,(P_f - P_0)$$

or
$$\gamma\, T_i \left(\frac{P_f}{T_f} - \frac{P_0}{T_0}\right) = P_f - P_0 \qquad (B)$$

Where $P_0$ and $P_f$ are the initial and final pressure of gas in the tank, respectively, $T_0$ and $T_f$ are the initial and final temperature of the gas in the tank, respectively. $V$ is the volume of the tank and $T_i$ is the temperature of the gas in the supply line.

Substituting the values of $P_0$, $P_f$, $T_0$ and $T_i$ in Eqn.($B$) ,we get

$$1.4 \times 400 \left(\frac{20 \times 10^5}{T_f} - \frac{1 \times 10^5}{300}\right) = 20 \times 10^5 - 1 \times 10^5$$

or $T_f = 536.74$ K

Amount of gas that entered the tank

$$= (n_f - n_0) = \frac{P_f V}{RT_f} - \frac{P_0 V}{RT_0} = \frac{V}{R}\left(\frac{P_f}{T_f} - \frac{P_0}{T_0}\right)$$

$$= \frac{1}{8.314}\left(\frac{20\times10^5}{536.74} - \frac{1\times10^5}{300}\right) = 408.09 \text{ mol}$$

**Example 4.93**

A customer sent an empty hydrogen cylinder for refilling. Usually the hydrogen cylinders are filled with hydrogen at 200 bar and 300 K before they are delivered to the customers. A customer lodges a complaint with the supplier that the cylinder received by him is not filled upto the specified pressure. The supplier made an independent enquiry and found that the supply line is always maintained at 200 bar and 300 K during the filling of the cylinders and the cylinders are disconnected from the supply line only after pressure equalisation and wonders as to how a particular cylinder does not show the stated pressure. He has also ensured that the cylinder is leak proof. Analyze the situation and find whether there is a possibility of filling the cylinder with less amount of gas

*Solution* :

The first law of thermodynamics for charging of a cylinder gives

$$(m_f - m_0)\, h_i = m_f u_f - m_0 u_0 - Q$$

If the cylinder is initially empty, $m_0 = 0$. If the charging process is carried out rapidly such that there is not enough time to dissipate the energy by radiation or convection or if the charging process is carried out adiabatically $Q = 0$. Then, the above equation reduces to

$$m_f h_i = m_f u_f \text{ or } h_i = u_f \text{ or } C_p T_i = C_v T_f$$

or

$$T_f = \gamma T_i$$

Assume hydrogen (a diatomic gas) to be an ideal gas with $\gamma = 1.4$.Then

$$T_f = 1.4\, T_i = 1.4 \times 300 = 420 \text{ K}$$

That is the final temperature of gas immediately after filling is 420 K. When the cylinder is transported to the customer, its temperature reduces to 300 K at constant volume. Then

$$\frac{P_f V}{RT_f} = \frac{PV}{RT} \text{ or } P = P_f \frac{T}{T_f}$$

$$= \frac{200 \times 300}{420} = 142.86 \text{ bar}$$

where $P_f$ and $T_f$ denote the conditions immediately after filling and $P$ and $T$ denote the conditions as the customer receives.

Thus we find that the pressure of gas in the cylinder is 142.86 bar by the time the customer receives the cylinder. Hence, if the cylinder is filled adiabatically, there is every possibility that the customer's complaint is correct.

**Example 4.94**

An evacuated and insulated tank containing a piston and spring as shown in Fig.E. 4.94 is connected to a line carrying steam at 2 MPa and 300°C. Initially, the spring is just touching the piston exerting no force on it. When the valve is opened steam enters the tank till the pressure rises to 3 MPa. Determine the state of steam in the tank. Choose the tank, piston and spring as the control volume.

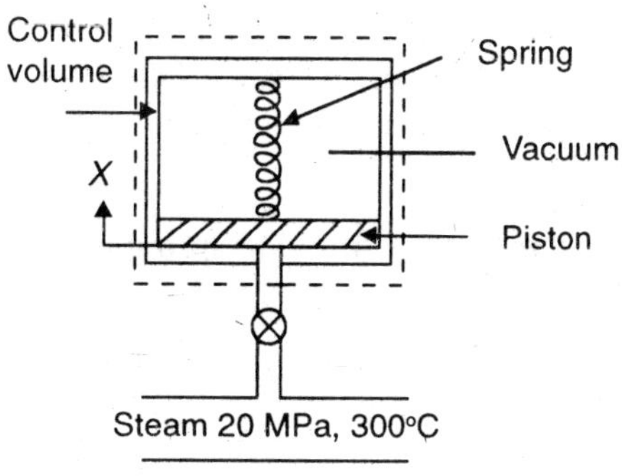


Fig.E 4.94. Sketch for Example 4.94.

*Solution* :

The first law of thermodynamics for a control volume is given by

$$\dot{m}_e \left(h_e + \frac{\mathbf{V}_e^2}{2} + gZ_e\right) - \dot{m}_i \left(h_i + \frac{\mathbf{V}_i^2}{2} + gZ_i\right) = \dot{Q} - \dot{W}_s - \frac{dE}{dt} \qquad (A)$$

Tank is insulated : $\dot{Q} = 0$. No shaft work is involved : $\dot{W}_s = 0$

No mass is escaping from the tank :

$$\dot{m}_e = 0$$

Ignore KE and PE terms: $\mathbf{V}_i = 0$ ; $Z_i = 0$

Then Eqn.($A$) reduces to

$$m_i \, h_i = dE/dt$$

or $h_i \, m_f = E_f =$ energy of steam + energy stored in spring ($B$)

Force due to spring, $F_s = KX$

where $K =$ spring constant

$X =$ compression in the spring

$A =$ cross sectional area of piston

$V_f =$ final volume of steam in the tank

$$\text{Energy stored in spring} = \frac{1}{2}KX^2 = \frac{1}{2}K\left(\frac{V_f}{A}\right)^2$$

A force balance on the piston in the final state gives

$$PA = KX = K\frac{V_f}{A} \text{ or } P = \frac{KV_f}{A^2}$$

$$\text{Hence, energy stored in spring} = \frac{1}{2}K\left(\frac{V_f}{A}\right)^2 = \frac{1}{2}PV_f = \frac{1}{2}Pm_f v_f \qquad (C)$$

where $m_f =$ final mass of steam in the tank

$v_f =$ final specific volume of steam in the tank

Energy of steam in the tank $= m_f u_f$ ($D$)

Substituting Eqns.($C$) and ($D$) in Eqn. ($A$) we get

$$m_f h_i = m_f u_f + P m_f v_f / 2$$

or $$h_i = 3025 \times 10^3 = u_f + P v_f / 2 \qquad (E)$$

$h_i$ **(2 MPa and 300°C) = 3025 kJ / kg.**

**Eqn.($E$) is to be solved by trial and error since we know $u_f$ and $P$.**

**Assume $t_f = 363$°C**

$h$ **(2 MPa and 363°C) = 3165.93 kJ/kg**

$v$ **(2 MPa and 363°C) = 0.1416 m³/kg.**

$u = h - Pv = 3165.93 \times 10^3 - 2 \times 10^6 \times 0.1416 = 2882.73$ **kJ/kg**

$$\text{RHS of Eqn }(E) = u_f + \frac{1}{2}Pv_f = 2882.73 \times 10^3 + (2 \times 10^6/2) \times 0.1416$$

$$= 3024.33 \text{ kJ/kg}$$

This value is approximately equal to $h_i$ and hence the assumed value $t_f$ = 363°C is correct. Hence, the final state of steam in the tank is 2 MPa and 363°C.

**Example 4.95**

Rework Example 4.94 by choosing the space below the piston only as the control volume. That is the piston and spring are not to be included in the control volume. (This is an example of control volume, the size of which does not remain constant).

*Solution* :

Choose the control–volume as the space below the piston as shown in Fig.E 4.95. As the steam enters the tank, the size of the control volume increases.

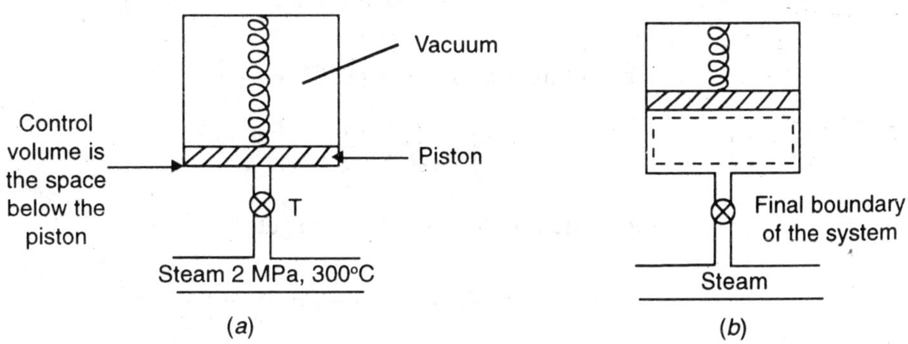


Fig.E 4.95. Sketch for Example 4.95.

The first law of thermodynamics for the control volume gives

$$\dot{m}_e\left(h_e + \frac{\mathbf{V}_e^2}{2} + gZ_e\right) - \dot{m}_i\left(h_i + \frac{\mathbf{V}_i^2}{2} + gZ_i\right) = \dot{Q} - \dot{W}_s - \frac{dE}{dt} \qquad (A)$$

Tank is insulated $\Rightarrow \dot{Q} = 0$

No mass leaves the control volume : $\dot{m}_e = 0$

Ignore KE and PE change. Then Eqn. (*A*) reduces to

$$-\dot{m}_e h_i = -\dot{W}_s - \frac{dE}{dT} \qquad (B)$$

The control volume does work in compressing the spring. The work done on the spring is given by

$$\int \dot{W}_s \, dt = \frac{1}{2} K X_f^{\,2} = \frac{1}{2} K \left( \frac{V_f}{A} \right)^2 \qquad (C)$$

Where $X_f$ = final compression in the spring

$V_f$ = final volume of the control volume = $m_f v_f$

$A$ = cross sectional area of the piston

$m_f$ = final mass of steam in the tank

$v_f$ = final specific volume of steam in the tank

A force balance on the piston, in the final state gives

$$(PA) = KX = K\frac{V_f}{A} \quad \text{or} \quad P = K\frac{V_f}{A^2} \qquad (D)$$

Substituting Eqn (*D*) in Eqn (*C*), we get

$$\int \dot{W}_s dt = \frac{PV_f}{2} = \frac{Pm_f v_f}{2} \qquad (E)$$

Substitute Eqn (*E*) in Eqn (*B*) to get

$$\int -\dot{m}_i h_i dt = -\frac{1}{2} P m_f v_f - \int \frac{dE}{dt} dt$$

or $$m_f h_i = m_f u_f + P m_f u_f / 2$$

or $$h_i = u_f + P v_f / 2 \qquad (F)$$

Eqn.(*F*) is identical with Eqn.(*E*) of Example 4.94. Now the rest of the solution will be identical to that given in Example 4.94.

**Example 4.96**

Rework Example 4.94 through control mass approach.

*Solution* :

Let us choose the control mass or system as shown in Fig.E 4.96.

The mass of steam $m_f$ which enters the tank is chosen as the system. Initially the system boundary is in the main line. As time progresses, the steam enters the tank and the system configuration keeps on changing and in the final state the system boundary is the space below the piston in the tank as shown in Fig.E 4.96. For the mass $m_f$ to enter the tank, work is done by the surroundings on the steam. If $v$ is the specific volume of steam in the supply line, then the reduction in the volume of the system in given by $m_f v$, then

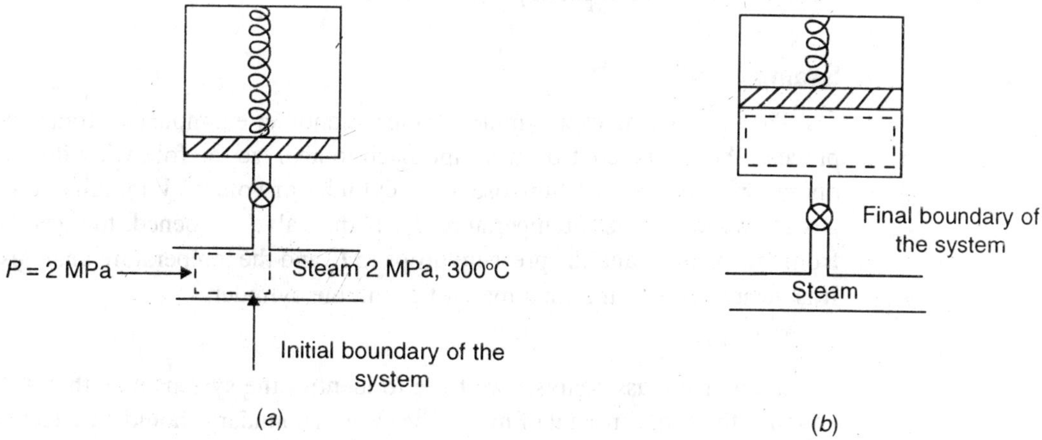


Fig.E 4.96. Sketch for Example 4.96

Work done by surroundings on steam in forcing it into the tank

$$= Pv\, m_f = -\, W_1$$

While steam enters the tank, the spring gets compressed. Thus the system does work on the spring. The work done on the spring is given by

$$W_2 = \frac{1}{2} KX^2 = \frac{1}{2} K\left(\frac{V_f}{A}\right)^2$$

A force balance on the piston in the final state gives

$$(PA) = KX = K\frac{V_f}{A} \text{ or } P = K\frac{V_f}{A^2}$$

$$\text{Therefore, } W_2 = \frac{PV_f}{2} = Pm_f v_f / 2$$

Thus net work done by system, $W = Pm_f v_f / 2 - Pm_f v$

Initial energy of system $E_1 = m_f u$

Where $u$ = specific internal energy of steam in the supply line.

Final energy of system, $E_2 = m_f u_f$

The first law of thermodynamics for a control mass gives

$$E_2 - E_1 = Q - W$$

or $$m_f u_f - m_f u = P m_f v - P m_f v_f / 2$$

or $$u_f + P v_f / 2 = u + P v = h \text{ or } h = u_f + P v_f / 2$$

which is identical with Eqn.(*F*) of Example 4.95. Hence, the final result will be identical to that of Example 4.94 or 4.95.

### Example 4.97

The discharging of a cylinder or tank is another example of a transient flow process. Apply the control mass approach to analyze the following discharging operation of a cylinder. Suppose a gas cylinder of volume $V$ initially contains a gas at pressure $P_0$ and temperature $T_0$. If the valve is opened, the gas escapes from the cylinder and the pressure drops to $P_f$ and the temperature changes to $T_f$. It is desired to find the final mass of gas in the cylinder.

*Solution* :

In control mass analysis we have to identify the system such that it always contains the same amount of mass. The system boundary should be selected such that it contains the same mass all the time. Therefore draw an imaginary envelope enclosing the cylinder and the gas that is escaping from the cylinder as shown in Fig.E 4.97 and this imaginary envelope is the system boundary. Initial mass of gas in the cylinder is $m_0$. At the end of the discharging operation, the mass of gas left in the cylinder is $m_f$ and the mass of gas that escaped from the cylinder is $(m_0 - m_f)$.

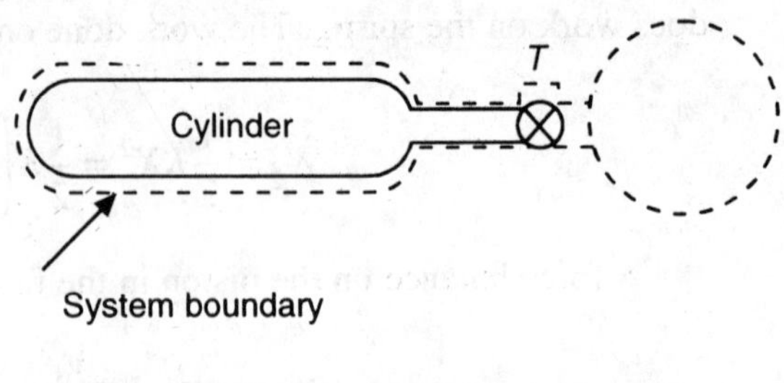


Fig.E 4.97. Control-mass for discharging of a cylinder.

Initial energy of system, $E_1 = m_0 u_0$

Final energy of system, $E_2 = m_f u_f + (m_0 - m_f)\, u$

where $u_0$ = specific internal energy of gas in the cylinder initially

$u$ = specific internal energy of gas in the surroundings.

When the gas escapes from the cylinder, it pushes the surrounding gas and does work on the surroundings. If $v$ denotes the specific volume of the gas in the surroundings at pressure $P$, then the change in the volume, external to the cylinder is $(m_0 - m_f)v$,

Then work done by gas, $W = P\,(m_0 - m_f)v$,

The first law of thermodynamics gives

$$E_2 - E_1 = Q - W$$

or $\quad m_f u_f + (m_0 - m_f)u - m_0 u_0 = Q - P\,(m_0 - m_f)\,v$

or $\quad (m_0 - m_f)\,(u + Pv) = Q + m_0 u_0 - m_f u_f$

or $\quad (m_0 - m_f)h = Q + m_0 u_0 - m_f u_f \qquad (A)$

where $h$ is the specific enthalpy of gas after it has escaped from the cylinder.

**Example 4.98**

Rework Example 4.97 through control volume analysis

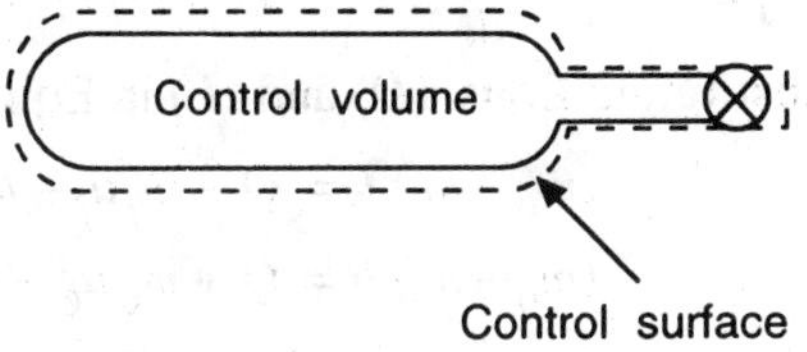


Fig.E 4.98. Control volume for discharging of a cylinder.

*Solution* :

Choose the control volume as shown in Fig.E 4.98. The principle of conservation of energy (or the first law of thermodynamics) for a control–volume is given by

$$\dot{m}_e\left(h_e + \frac{\mathbf{V}_e^2}{2} + gZ_e\right) - \dot{m}_i\left(h_i + \frac{\mathbf{V}_i^2}{2} + gZ_i\right) = \dot{Q} - \dot{W}_s - \frac{dE}{dt} \qquad (A)$$

No shaft work is involved; $\dot{W}_s = 0$

No mass is entering the cylinder : $\dot{m}_i = 0$

Ignore KE and PE terms: $\mathbf{V}_e = 0$ and $Z_e = 0$

Then Eqn. $(A)$ reduces to

$$\int \dot{m}_i h_e dt = -\int \frac{dE}{dt}dt = \int \dot{Q}\,dt - \int \frac{d(mu)}{dt}dt \qquad (B)$$

For the discharging of a tank , the principle of conservation of mass gives

$$\int_0^t \frac{dm}{dt}dt = \int_0^t \dot{m}_e dt = m_f - m_0 \quad \text{(C)}$$

While mass is escaping from the cylinder, the mass flow rate as well as the state of the gas leaving the cylinder change with time. However, we can assume that the gas leaves the cylinder at some constant value of enthalpy $h$, which may be taken as the average value of $h_0$ (initial enthalpy) and $h_f$ (final enthalpy) of the gas in the cylinder. Then the LHS of Eqn.($B$) gives

$$\int_0^t \dot{m}_e h_e dt = h_e \int_0^t \dot{m}_e dt = (m_0 - m_f)h \quad (D)$$

The RHS of Eqn ($B$) gives

$$\int \dot{Q}\,dt - \int \frac{d(mu)}{dt}dt = Q - (m_f u_f - m_0 u_0) \quad (E)$$

Substituting Eqns. ($D$) and ($E$) in Eqn ($B$), we get

$$(m_0 - m_f)\,h = Q - m_f u_f - m_0 u_0$$

or $$(m_0 - m_f)\,h = Q + m_0 u_0 - m_f u_f \quad (F)$$

This equation is identical with Eqn.($A$) of Example 4.97, which is obtained through control mass approach.

**Example 4.99**

A rigid and insulated tank of capacity $1 m^3$ contains dry saturated steam at 20 bar. A valve at the top of the tank is opened allowing only vapor to escape till the pressure in the tank is reduced to 10 bar. Estimate amount of steam withdrawn from the tank.

*Solution* :

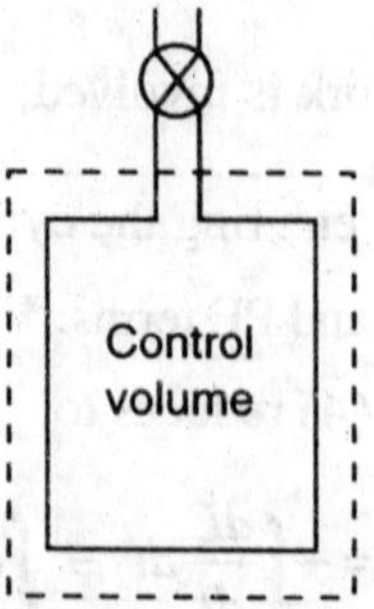


Fig.E. 4.99. Sketch for Example 4.99.

Choose the control–volume as shown in Fig.E. 4.99.

Insulated tank; $\dot{Q} = 0$

Then the first law of thermodynamics (for the control–volume) for the transient flow problem at hand reduces to

$$(m_0 - m_f)\, h = m_0 u_0 - m_f u_f \qquad (A)$$

For saturated steam at 20 bar

$$h_0 = 2797.2 \text{ kJ/kg}$$

$$v_0 = 0.099\,54 \text{ m}^3\text{/kg}$$

$$u_0 = h_0 - P_0\, v_0 = 2797.2 \times 10^3 - 20 \times 10^5 \times 0.099\,54 = 2598.12 \text{ kJ/kg}$$

$$m_0 = \frac{V}{v_0} = \frac{1}{0.099\,54} = 10.046 \text{ kg}$$

Assume $m_f = 5.537$ kg

Then $v_f = 1/5.537 = 0.1806 \text{ m}^3\text{/kg}$

At $P = 10$ bar ; $v_f = 0.001\,127\,4 \text{ m}^3\text{/kg}$ (saturated liquid) $h_f = 762.61$ kJ/kg

$$v_g = 0.1943 \text{ m}^3\text{/kg} \; ; \; h_g = 2776.2 \text{ kJ/kg}$$

Therefore, at $P = 10$ bar the final specific volume of steam lies in between the values for saturated liquid and saturated vapor. Hence, the quality of the mixture can be found as

$$0.1806 = 0.1943\, X_f + (1 - X_f)\, 0.001\,127\,4$$

or

$$X_f = 0.9291$$

$$h_f = 0.9291 \times 2776.2 + (1 - 0.9291)\, 762.61$$

$$= 2633.44 \text{ kJ/kg}$$

$$u_f = 2633.44 \times 10^3 - 10 \times 10^5 \times 0.1806 = 2452.84 \text{ kJ/kg}$$

Since $h_0$ and $h_f$ are close to each other, we may assume that the enthalpy of steam leaving the control volume is constant, and is given by the average value of saturated vapor enthalpies at 20 bar and 10 bar.

$$h = \frac{h_0 + h_f}{2} = \frac{2797.2 + 2776.2}{2} = 2786.7 \text{ kJ/kg}$$

Now let us check whether the assumed value of $m_f$ satisfies Eqn.($A$) or not.

LHS of Eqn ($A$) $= (m_0 - m_f)\, h = (10.046 - 5.537)\, 2786.7 \times 10^3$

$$= 12\,565.23 \text{ kJ}$$

RHS of Eqn.($A$) $= m_0 u_0 - m_f u_f = 10.046 \times 2598.12 \times 10^3$

$$- 5.537 \times 2452.84 \times 10^3 = 12\,519.34 \text{ kJ}$$

Since LHS ≅ RHS, we may assume that Eqn. (*A*) is satisfied.

Hence, the assumed value $m_f = 5.537$ kg is correct.

Amount of steam withdrawn from tank $= m_0 - m_f$

$= 10.046 - 5.537 = 4.509$ kg

**Example 4.100**

A rigid tank of $1 m^3$ volume is initially filled with wet steam of quality 0.8 at 30 bar. A pressure regulating valve located at the top of the tank is slowly opened allowing only dry saturated vapor to escape at 30 bar pressure, while energy is continously transferred to the tank as heat. The valve at the top of the tank is opened till the tank is completely filled with dry saturated steam at 30 bar. Neglecting the kinetic and potential energy changes of steam, calculate the amount of energy transferred as heat to the tank.

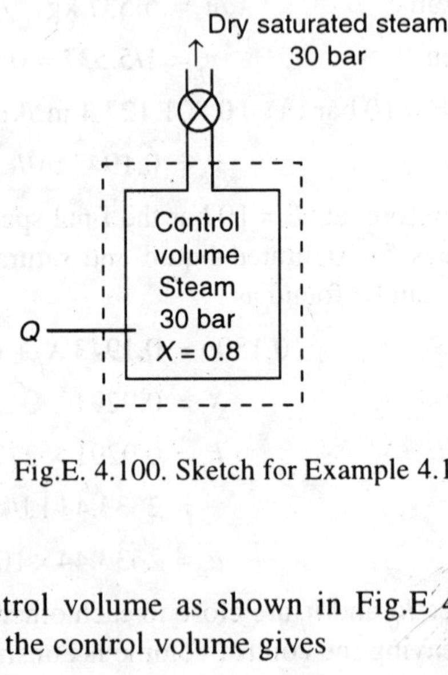


Fig.E. 4.100. Sketch for Example 4.100

*Solution* :

Choose the control volume as shown in Fig.E 4.100. The first law of thermodynamics for the control volume gives

$$\dot{m}_e\left(h_e + \frac{\mathbf{V}_e^2}{2} + gZ_e\right) - \dot{m}_i\left(h_i + \frac{\mathbf{V}_i^2}{2} + gZ_i\right) = \dot{Q} - \dot{W}_s - \frac{dE}{dt} \qquad (A)$$

No shaft work is involved : $\dot{W}_s = 0$

No mass enters the control volume: $\dot{m}_i = 0$

Ignore KE and PE terms : $\mathbf{V}_e = 0 : Z_e = 0$

Then Eqn. (*A*) reduces to

$$\int \dot{m}_e h_e dt = \int \dot{Q}\, dt - \int \frac{dE}{dt} dt = Q - \left(m_f u_f - m_0 u_0\right) \tag{B}$$

Since only dry saturated steam at 30 bar is withdrawn, the enthalpy of the escaping steam is constant at $h_e = h_g$ (at 30 bar ) = 2802.3 kJ/kg

Then Eqn. (*B*) reduces to

$$h_e \int \dot{m}_e dt = h_e (m_0 - m_f) = Q + m_0 u_0 - m_f u_f$$

or $$Q = h_e (m_0 - m_f) - m_0 u_0 + m_f u_f \tag{C}$$

At 30 bar: saturated liquid volume ($v_f$) = 0.001 216 3 m³/kg

Saturated vapor volume ($v_g$) = 0.066 63 m³/kg

Saturated liquid enthalpy ($h_f$) = 1008.4 kJ/kg

Saturated vapor enthalpy ($h_g$) = 2802.3 kJ/kg

$$v_0 = X_0 v_g + (1 - X_0) v_f = 8 \times 0.066\,63 + 0.2 \times 0.001\,216\,3$$
$$= 0.053\,55 \text{ m}^3/\text{kg}$$

$$h_0 = X_0 h_g + (1 - X_0) h_f = 0.8 \times 2802.3 + 0.2 \times 1008.4$$
$$= 2443.52 \text{ kJ/kg}$$

$$u_0 = h_0 - P_0 v_0 = 2443.52 \times 10^3 - 30 \times 10^5 \times 0.053\,55 = 2282.87 \text{ kJ/kg}$$

$$m_0 = \frac{1}{v_0} = \frac{1}{0.053\,55} = 18.674 \text{ kg}$$

$$m_f = \frac{1}{v_f} = \frac{1}{v_g} = \frac{1}{0.066\,63} = 15.008 \text{ kg}$$

$$u_f = 2802.3 \times 10^3 - 30 \times 10^5 \times 0.066\,63$$
$$= 2602.41 \text{ kJ/kg}$$

Substituting these values in Eqn (*C*), we get

$$Q = 2802.3 \times 10^3 (18.674 - 15.008)$$
$$-18.674 \times 2282.87 \times 10^3 + 15.008 \times 2602.41 \times 10^3$$
$$= 6699.89 \text{ kJ}$$

**Example 4.101**

A boiler produces steam at 1 MPa and 300°C. The steam from the boiler is used to operate a turbine. The turbine exhausts steam into an evacuated tank of volume 100 m³. The turbine operates till the pressure in the tank rises to 1 MPa at which point the temperature of steam in the tank is 250°C. Assuming that the turbine and tank are adiabatic, determine the work delivered by the turbine.

*Solution* :

Choose the combination of turbine and the tank as the control volume as shown in Fig.E 4.101. The first law of thermodynamics for the control volume gives

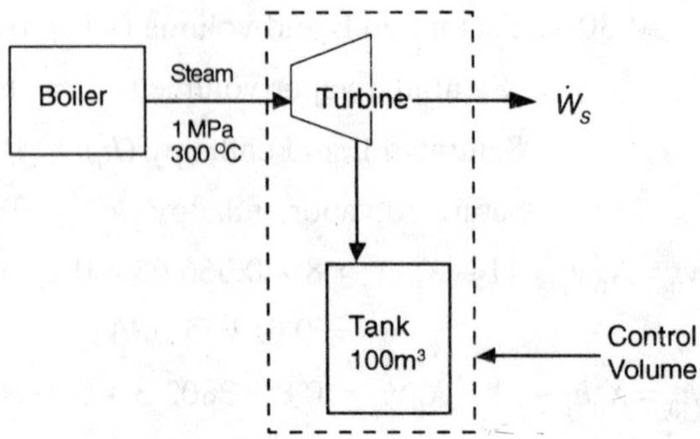


Fig.E 4.101. Sketch for Example 4.101.

$$\dot{m}_e(h_e + \frac{\mathbf{V}_e^2}{2} + gZ_e) - \dot{m}_i(h_i + \frac{\mathbf{V}_i^2}{2} + gZ_i) = \dot{Q} - \dot{W}_s - \frac{dE}{dt} \qquad (A)$$

No mass leaves the control volume : $\dot{m}_e = 0$

Control volume is adiabatic : $\dot{Q} = 0$

Ignore KE and PE changes. Then Eqn.(*A*) reduces to

$$-\int \dot{m}_i\, h_i\, dt = -\int \dot{W}_s\, dt - \int d\frac{(mu)}{dt} dt \qquad (B)$$

The principle of conservation of mass gives

$$\int \frac{dm}{dt} dt = \int \dot{m}_i\, dt = m_f - m_0 = m_f \qquad (C)$$

$$\int d\frac{(mu)}{dt} dt = m_f u_f - m_0 u_0 = m_f u_f \qquad (D)$$

$$\int \dot{W}_s\, dt = W$$

($m_0 = 0$ since the tank is initially evacuated).

Substituting Eqns. (*C*) and (*D*) in Eqn.(*B*), we get

$$m_f h_i = W + m_f u_f$$

or $W = m_f(h_i - u_f)$ (*E*)

For steam at 1MPa and 250°C:

$$v = 0.231\,95 \text{ m}^3/\text{kg} \;;\; h = 2939.45 \text{ kJ/kg}$$

$$u = 2939.45 \times 10^3 - 1 \times 10^6 \times 0.231\,95$$

$$= 2707.5 \text{ kJ/kg}$$

Therefore, $v_f = 0.231\,95$ m³/kg and $u_f = 2707.5$ kJ/kg

$$m_f = \frac{V}{v_f} = \frac{100}{0.231\,95} = 431.127 \text{ kg}$$

For steam at 1 MPa and 300°C,

$$h = h_i = 3052.1 \text{ kJ/kg}.$$

Substituting these values in Eqn. (*E*), we get

$$W = m_f\,(h_i - u_f) = 431.127\,(3052.1 - 2707.5) \times 10^3$$

$$= 148.566 \text{ MJ}$$

**Example 4.102**

It is a fact that during discharging of a gas from a tank, the conditions of the gas leaving the tank as well as that of the gas remaining in the tank change with time. Apply the control volume analysis and show that if an ideal gas discharges from an adiabatic tank, the conditions of the gas remaining in the tank are identical with those estimated as if the gas has undergone a reversible and adiabatic expansion.

*Solution* :

The first law of thermodynamics for a control volume is given by

$$\dot{m}_e\left(h_e + \frac{\mathbf{V}_e^2}{2} + gZ_e\right) - \dot{m}_i\left(h_i + \frac{\mathbf{V}_i^2}{2} + gZ_i\right) = \dot{Q} - \dot{W}_s - \frac{dE}{dt} \qquad (A)$$

Suppose a rigid and insulated cylinder is initially filled with an ideal gas at $P_0$ and $T_0$. If the valve of the cylinder is opened, the gas escapes from the cylinder. Let $P_f$ and $T_f$ denote the final pressure and temperature of the gas, left in the cylinder. Choose the control volume as shown in Fig.E 4.102.

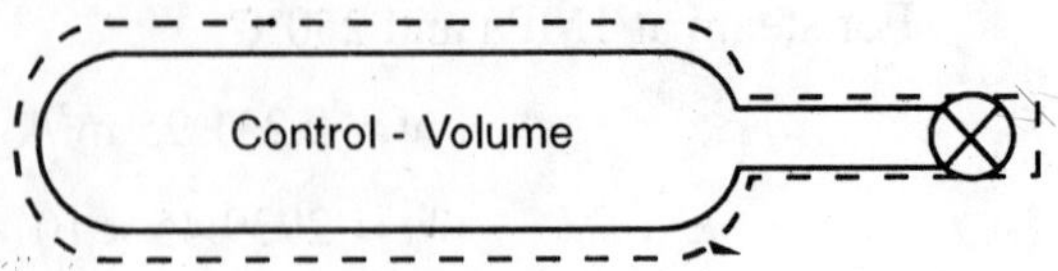


Fig.E 4.102. Sketch for Example 4.102.

Insulated cylinder : $\dot{Q} = 0$

No shaft work is done : $\dot{W}_s = 0$

No gas enters the control volume : $\dot{m}_i = 0$

Ignore KE and PE changes. Then Eqn. (*A*) reduces to

$$\dot{m}_e\, h_e = -\frac{dE}{dt} = -\frac{d(mu)}{dt} \tag{B}$$

For an ideal gas $h = h(T)$ only and $u = u(T)$ only. That is $h = C_pT$ and $u = C_vT$. At any instant of time, the gas left in the cylinder is at the same conditions as the gas leaving the cylinder. That is $T_e = T$ where $T$ is the temperature of the gas left in the cylinder and $m$ is the mass of gas in the control volume at time $t$. Then Eqn. (*B*) can be rewritten as

$$\dot{m}_e C_P T = -\frac{d}{dt}(mC_vT) \tag{C}$$

The principle of conservation of mass gives

$$dm/dt = -\dot{m}_e \tag{D}$$

Substituting Eqn. (*D*) in Eqn.(*C*), we get

$$C_P T\frac{dm}{dt} = \frac{d}{dt}(mC_vT)$$

$$\text{or} \quad \gamma\, T\frac{dm}{dt} = \frac{d}{dt}(mT) = \frac{mdT}{dt} + \frac{Tdm}{dt}$$

or $$(\gamma - 1)T\frac{dm}{dt} = m\frac{dT}{dt}$$

or $$\frac{dm}{m} = \frac{1}{\gamma - 1}\frac{dT}{T}$$

$$\ln\frac{m_f}{m_0} = \frac{1}{\gamma-1}\ln\frac{T_f}{T_0} \quad \text{or} \quad \ln\frac{n_f}{n_0} = \frac{1}{\gamma-1}\ln\frac{T_f}{T_0} \qquad (E)$$

where $n_0$ = moles of gas in the cylinder initially, and

$n_f$ = moles of gas in the cylinder at the end of discharging.

$V$ = volume of the cylinder

$$\frac{n_f}{n_0} = \left(\frac{P_f V}{RT_f}\right) \Big/ \left(\frac{P_0 V}{RT_0}\right) = \frac{P_f}{P_0}\cdot\frac{T_0}{T_f} \qquad (F)$$

Substituting Eqn.($F$) in Eqn. ($E$), we get

$$\frac{T_f}{T_0} = \left(\frac{P_f}{P_0}\right)^{\frac{\gamma-1}{\gamma}} \qquad (G)$$

Eqn.($G$) is identical with the relation for a reversible adiabatic expansion of an ideal gas. Thus, when an ideal gas escapes from a tank the temperature of the gas remaining in the tank can be estimated by using the relation which is true for a reversible adiabatic expansion.

### Example 4.103

An insulated tank of 100 litre capacity contains an ideal gas ($\gamma = 1.4$) at 10 bar and 400 K. A valve situated at the top of the tank is opened allowing the gas to escape till the pressure is reduced to 2 bar. Determine the final temperature of the gas remaining in the tank and the amount (in moles) of gas that left the tank.

*Solution* :

We know that the temperature of the gas remaining in the tank can be calculated as

$$\frac{T_f}{T_0} = \left(\frac{P_f}{P_0}\right)^{\frac{\gamma-1}{\gamma}} \quad \text{or} \quad T_f = T_0\left(\frac{P_f}{P_0}\right)^{\frac{\gamma-1}{\gamma}}$$

or $$T_f = 400\left(\frac{2}{10}\right)^{\frac{0.4}{1.4}} = 252.55\ \text{K}$$

$$\text{Amount of gas remaining in the tank} = \frac{P_f V}{RT_f} = \frac{2\times10^5\times100\times10^{-3}}{8.314\times252.55}$$

$$= 9.525\ \text{mol}$$

# 5

# Second Law Of Thermodynamics and its Applications

---

**Example 5.1**

Does the first law of thermodynamics deny the possibility of a spontaneous process reversing itself or does it specify the direction of a process ? Explain with the help of two examples.

*Solution* :

It is a matter of every day experience that spontaneous processes proceed in one direction only. Consider the spontaneous process of water flowing downhill. The potential energy of water at the top of the hill gets converted into kinetic energy of flowing water and finally the water falls on to the ground and the kinetic energy of the flowing water gets converted into its internal energy. The first law of thermodynamics tells that the increase in the kinetic energy of the flowing water is equal to the decrease in the potential energy of the water while it is falling. Finally when the water reaches the ground, the increase in the internal energy of the water is equal to the decrease in its kinetic energy. Now, imagine the reversal of the process. That is water flowing uphill on its own. Applying the first law of thermodynamics, we find that it would be well satisfied if the water at the ground level was to cool by transferring its internal energy into kinetic energy thus acquiring some velocity in the upward direction and then the water flowing uphill by transferring its kinetic energy into potential energy. However, in practice we never find water flowing uphill on its own. That is the first law of thermodynamics does not specify the direction of a process and does not deny the feasibility of a spontaneous process reversing on its own.

As a further example, if a running car is brought to rest by applying brakes, the kinetic energy of the car gets converted into the internal energy of the brakes and the wheels. According to the first law of thermodynamics, the increase in the internal energy of the brakes and wheels is equal to the decrease in the kinetic energy of the car. Imagine the reversal of the process, that is a car at rest starts running by cooling the brakes and wheels. The first law of thermodynamics would be still valid and says that the increase in the kinetic energy of the car is equal to the decrease in the internal energy of the brakes and wheels. However, we have never observed such a happening. Thus one can conclude that the first law of thermodynamics does not deny the feasibility of a spontaneous process reversing itself and does not specify the direction of a process.

**Example 5.2**

Suppose one is interested in producing industrial alcohol by vapor phase hydration of ethylene, at a specified temperature and pressure, according to the reaction

$$C_2H_4 + H_2O \rightarrow C_2H_5OH$$

Before initiating a commercial scale production it is necessary to have as much information as possible about the process. Does the first law of thermodynamics provide complete information?

*Solution* :

Usually the reaction $C_2H_4 + H_2O \rightarrow C_2H_5OH$ is carried out at elevated temperatures and pressures. The reactants $C_2H_4$ and $H_2O$ in the gas phase are to be raised to a high pressure and temperature before they are fed to a reactor. The reactor effluent is to be cooled and $C_2H_5OH$ is to be separated from the unreacted gases and the unreacted gases are to be recycled. Application of the first law of thermodynamics provides information regarding the work to be done in compressing the reactant gases, the energy required to raise the temperature of the reactant gases and to cool the reaction products. If the initial state of the reactants and the final state of the products are known, the first law of thermodynamics can be applied to determine the energy changes associated with the process. In other words, the first law of thermodynamics provides information regarding only the energy changes associated with the process.

In addition to the above information, it is essential to know, before initiating an expensive commercial production, whether the reaction is feasible at the specified temperature and pressure or not. If it is feasible, is complete conversion of $C_2H_4$ and $H_2O$ into $C_2H_5OH$ possible? Or does the reaction proceed to a certain extent only? If so, to what extent ? That is what fraction of the reactants can be converted into products at the specified conditions? How does the reaction conditions like pressure and temperature affect the extent of the reaction? These questions, cannot be answered by the application of the first law of thermodynamics.

**Example 5.3**

State the limitations of the first law of thermodynamics with the help of examples involving heat and work interactions.

*Solution* :

We know that kinetic energy and potential energy are interconvertible and the macroscopic modes of energy (KE and PE) can be readily converted into work. The conversion of microscopic modes of energy (that is energy associated with the random molecular motion of the matter) or internal energy into work requires a device called heat engine. Is it possible for the complete conversion of internal energy into work in a heat engine? Or is it possible to devise a heat engine, the efficiency (defined as the ratio of the net work done to the energy absorbed) of which is equal to one? Is it possible to transfer energy as heat spontaneously from a body at a lower temperature to a body at a higher temperature? All the above questions which deal with work and heat interactions cannot be answered by the application of the first law of thermodynamics and hence they reflect the limitations of the first law thermodynamics.

**Example 5.4**

Explain the terms – thermal reservoir, source and sink.

*Solution* :

A thermal reservoir is a large system to which a finite quantity of energy as heat can be added or from which a finite amount of energy as heat can be extracted without changing its temperature. The ambient atmosphere is an example of a thermal reservoir.

A source is a thermal reservoir at higher temperature from which energy in the form of heat is received by a heat engine.

A sink is a thermal reservoir at lower temperature to which energy as heat is rejected by a heat engine.

**Example 5.5**

What is meant by a heat engine and what are its characteristics ?

*Solution* :

A heat engine is an energy conversion device. It is a cyclically operating device and its primary objective is to convert the energy received as heat into work. It employs a working fluid which undergoes cyclic change. The working fluid absorbs energy as heat from a source and rejects energy as heat to a sink. The characteristics of a heat engine are :

1. It is a cyclically operating device.
2. Its primary purpose is to convert energy absorbed as heat into work.
3. It absorbs energy as heat from a high temperature source.
4. It rejects energy as heat to a low temperature sink.
5. It delivers some net work.

A heat engine can be represented as shown in Fig.E 5.5

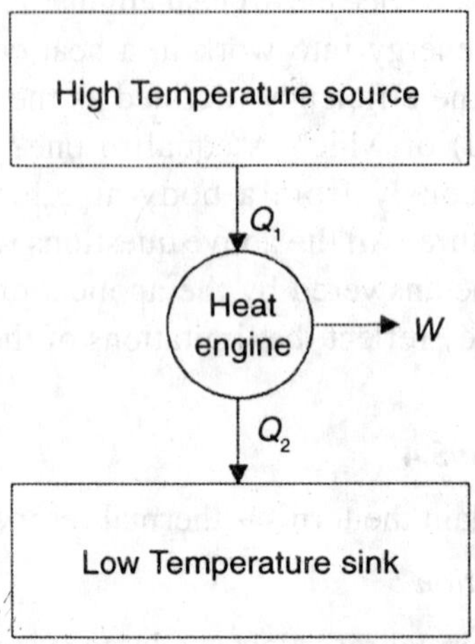


Fig.E 5.5. Schematic representation of a heat engine.

**Example 5.6**

Give two examples of a heat engine.

*Solution* :

A thermal power plant shown in Fig.E 5.6 (*a*) is a heat engine. The working fluid in a thermal power plant is water which undergoes the following cyclic change:

1. Water enters a boiler at high pressure where it receives energy as heat from the hot combustion products of coal or fuel oil and converts into superheated steam.
2. The superheated steam enters a turbine and expands to a low pressure. The turbine delivers work.
3. The low pressure steam enters a condenser where it rejects energy as heat to cooling water and emerges as liquid water at low pressure.

4. The low pressure water enters a pump which delivers water at high pressure to the boiler. Work is done on the pump.

Another example of a heat engine is an automobile engine which can be idealized as a piston–cylinder assembly shown in Fig.E 5.6 (*b*). The working fluid is a gas which undergoes the following cyclic change.

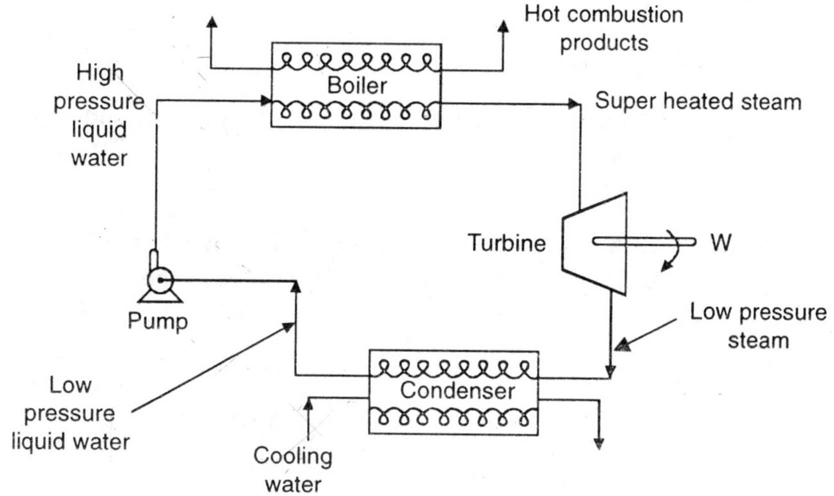


FigE. 5.6 (*a*) Schematic representation of a thermal power plant.

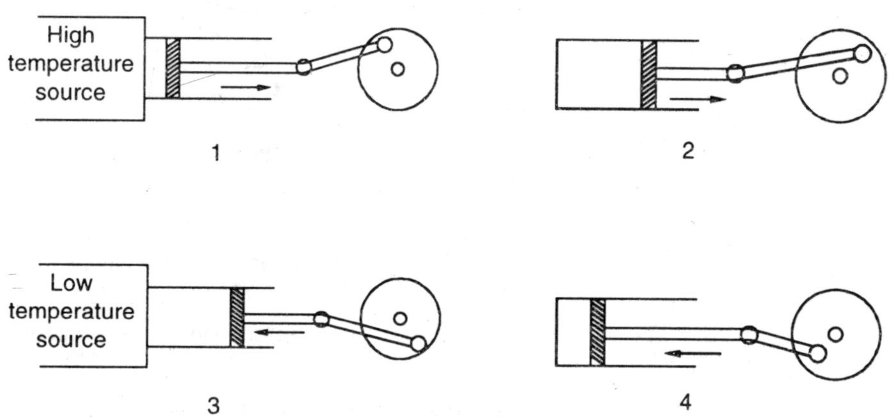


Fig.E 5.6 (*b*). Schematic representation of an automobile engine.

1. The cylinder is placed in thermal contact with a source. The working fluid absorbs energy as heat from the source and undergoes expansion.

2. The cylinder is isolated from the source and the working fluid undergoes further expansion adiabatically.
3. The cylinder is placed in thermal contact with a sink to which the working fluid rejects energy as heat when work is done on the gas.
4. Then the cylinder is isolated from the sink and the working fluid undergoes adiabatic compression till it reaches the original state.

**Example 5.7**

Define thermal efficiency of a heat engine and express it in terms of heat interactions.

*Solution* :

A schematic of a heat engine is shown is Fig.E 5.7. Consider the working fluid which undergoes the cyclic change as the system.

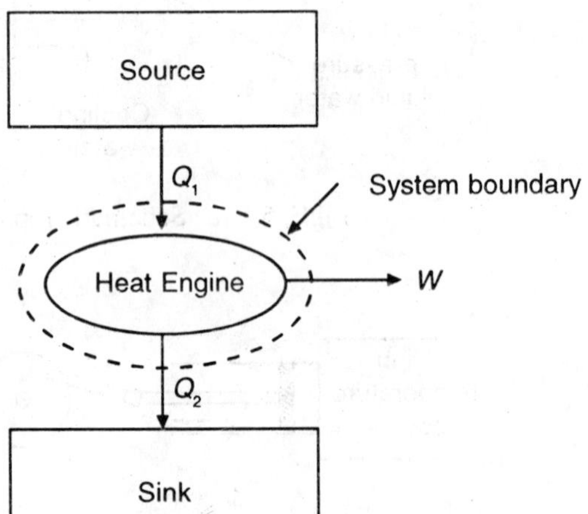


Fig.E 5.7. Schematic of a heat engine.

The following work and heat interactions are present across the system boundary.

$Q_1$ = energy absorbed as heat from the source

$Q_2$ = energy rejected as heat to the sink

$W$ = net work done

The system undergoes a cyclic change. The first law of thermodynamics for a cyclic process gives $\oint dQ = \oint dW$ or $Q_1 - Q_2 = W$.

The thermal efficiency or simply efficiency of a heat engine is defined as the ratio of the net work done to the energy absorbed as heat. That is the efficiency $\eta$ is given by

$$\eta = \frac{W}{Q_1} = \frac{Q_1 - Q_2}{Q_1} = 1 - \frac{Q_2}{Q_1}$$

**Example 5.8**

What is a heat pump or refrigerator and how the performance of these devices is evaluated ?

*Solution* :

A heat pump or a refrigerator is a device which works in a cycle, absorbing energy as heat from a low temperature reservoir and rejecting energy as heat to a high temperature reservoir, when work is done on the device. If the objective of the device is to reject energy in the form of heat to a high temperature reservoir, it is called a heat pump. Heat pumps are generally used to keep the rooms warm in winter. On the other hand, if the objective of the device is to extract energy in the form of heat from a low temperature body, it is called a refrigerator. Usually refrigerators are used to preserve food items and drugs at low temperature. In these devices Freon is usually used as a working fluid and it is called as refrigerant. The index of performance of these devices is called the coefficient of performance. The coefficient of performance (COP) is defined as the ratio of the energy effect sought to the work done on the device. A schematic of a heat pump or refrigerator is shown in Fig.E 5.8.

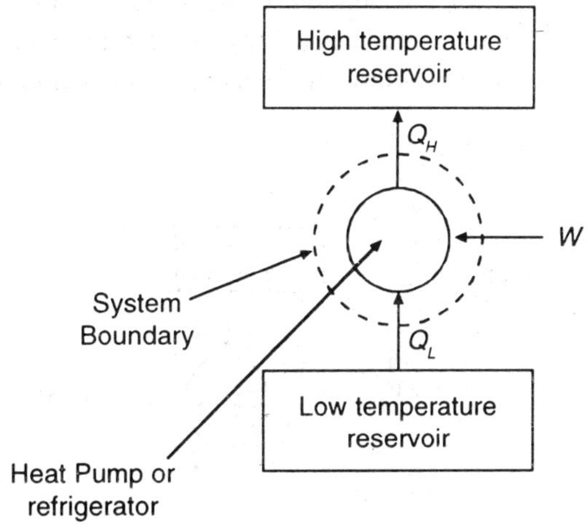


Fig.E 5.8. Schematic of heat pump or refrigerator.

Consider the working fluid in the heat pump or refrigerator as the system as shown in Fig.E 5.8 and apply the first law of thermodynamics to obtain

$$Q_H - Q_L = W$$

Where $Q_L$ = energy absorbed as heat from low temperature reservoir

$Q_H$ = energy rejected as heat to the high temperature reservoir

$W$ = work done on the device.

Then, the coefficient of performance in given by

$$(\text{COP})_{\text{Heat pump}} = \frac{Q_H}{W} = \frac{Q_H}{Q_H - Q_L}$$

$$\text{COP})_{\text{Refrigerator}} = \frac{Q_L}{W} = \frac{Q_L}{Q_H - Q_L}$$

**Example 5.9**

State the Kelvin– Planck statement of the second law of thermodynamics,

*Solution* :

The Kelvin–Planck statement of the second law of thermodynamics tells that it is impossible to devise a cyclically operating device which produces no other effect than the extraction of energy as heat from a single thermal–reservoir and delivers an equivalent amount of work. That is, it is impossible to devise a device shown in Fig.E 5.9. Considering the working fluid as a system as shown in Fig.E 5.9, application of the first law of thermodynamics gives $Q = W$. That is all the energy absorbed as heat by the device is completely converted into work. In other words, the thermal efficiency of such an engine is $\eta = W/Q = 1$. The Kelvin–Planck statement thus implies that no heat engine can have an efficiency equal to one.

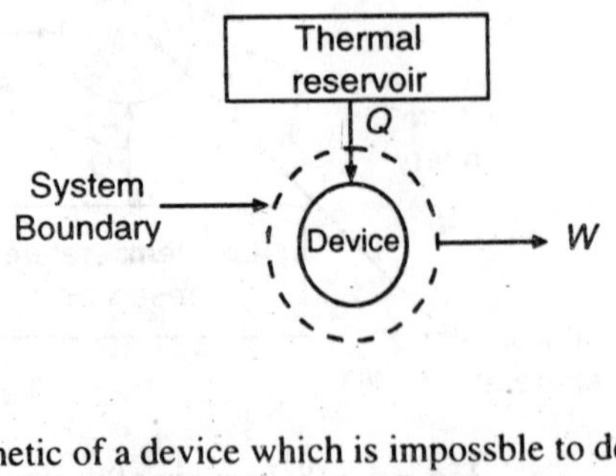


Fig.E 5.9. Schemetic of a device which is impossble to device according to Kelvin-Planck statement of the second law of thermodynamics.

**Example 5.10**

What is a perpetual motion machine of the second kind ?

*Solution* :

A Perpetual Motion Machine of the Second Kind (PMMSK) is a hypothetical device which operates in a cycle, absorbing energy as heat from a single thermal reservoir and delivering an equivalent amount of work. A schematic of a perpetual motion machine of the second kind in shown in Fig.E 5.10.

The Kelvin – Planck statement of the second law of thermodynamics denies the possibility of devising a PMMSK. The PMMSK does not violate the first law of thermodynamics, but it violates the second law of thermodynamics.

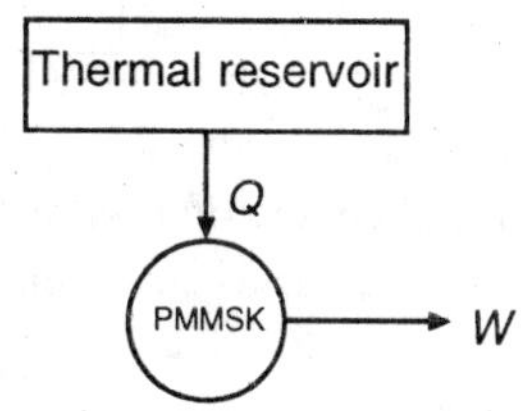


Fig.E 5.10. Schematic of a Perpetual Motion Machine of the Second Kind.

**Example 5.11**

State the Clausius statement of the second law of thermodynamics.

*Solution* :

The Clausius statement of the second law of thermodynamics tells that it is impossible to devise a selfacting (that is unaided by any external agency) device which working cyclically, will produce no other effect than the transfer of energy as heat from a low temperature body to a high temperature body. A schematic of the device which is impossible to devise according to the Clausius statement of the second law of thermodynamics is shown in Fig.E 5.11.

We know that the transfer of energy as heat from a high temperature body to a low temperature body occurs spontaneously. The Clausius statement of the second law of thermodynamics denies the possibility of self reversal of such a spontaneous process. In other words the second law of thermodynamics dictates the direction of a spontaneous process, The coefficient of performance of the device shown in Fig.E 5.11 is given by $\text{COP} = Q/W = Q/O = \infty$. Thus the second law of thermodynamics implies that the COP of a heat pump or refrigerator cannot be infinity.

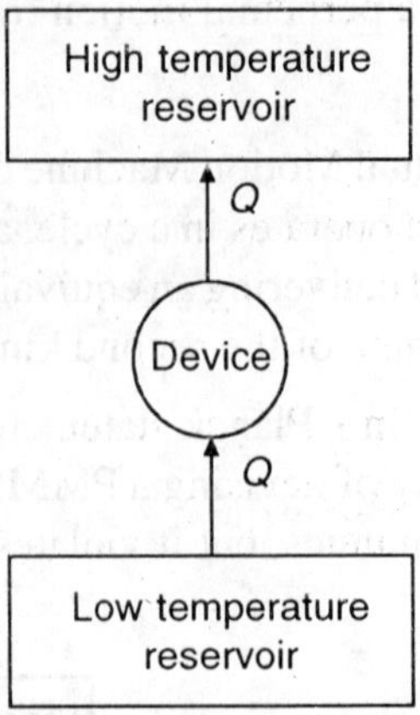


Fig.E 5.11. Schematic of a self acting device which is impossible to devise according to Clausius statement of the second law of thermodynamics.

**Example 5.12**

Does the Clausius statement of the second law of thermodynamics prohibit the transfer of energy as heat from a body at low temperature to a body at high temperature? A domestic refrigerator absorbs energy as heat from the cold space and rejects energy as heat to the ambient atmosphere at a higher temperature. Does the refrigerator violate the second law of thermodynamics ?

*Solution* :

The Clausius statement of the second law of thermodynamics does not prohibit the transfer of energy as heat from a body at low temperature to a body at high temperature if the transfer of energy is externally aided. However, spontaneous or externally unaided transfer of energy as heat from a low temperature body to a high temperature body is prohibited by the Clausius statement.

A domestic refrigerator does not violate the second law of thermodynamics, since work is done by an external agent on the refrigerator.

**Example 5.13**

Prove that violation of the Kelvin–Planck statement leads to violation of Clausius statement of the second law of thermodynamics.

*Solution* :

The Kelvin–Planck and Clausius statements of the second law of thermodynamics are equivalent though they appear to be altogether different.

Their equivalence can be proved by contradiction—that is violation of one statement implies the violation of the other statement and vice–versa. To prove that violation of Kelvin–Planck statement implies the violation of the Clausius statement, let us suppose that the Kelvin–Planck statement of the second law of thermodynamics is incorrect. That is, it is possible to devise a cyclically operating device A which absorbs energy $Q$ as heat from a thermal reservoir at high temperature $T_1$ and delivers an equivalent amount of work $W$ ($W = Q$) in a cycle as shown in Fig.E 5.13 (*a*). Now let us consider another cyclically operating device B which absorbs energy $Q_2$ as heat from a low temperature reservoir at $T_2$ ($T_2 < T_1$) and rejects energy $Q_1$ as heat to a high temperature reservoir at $T_1$ when work $W$ is done on the device in one cycle as shown in Fig.E 5.13 (*a*). The device B is aided and hence it is not in violation of the Clausius statement. The work done on the device B is given by $W = Q_1 - Q_2$. Let us consider the combination of the devices A and B as shown in Fig.E 5.13 (*b*). The work delivered by the device A is used to operate the device B and part of the energy $Q$ rejected by device B is absorbed by device A.

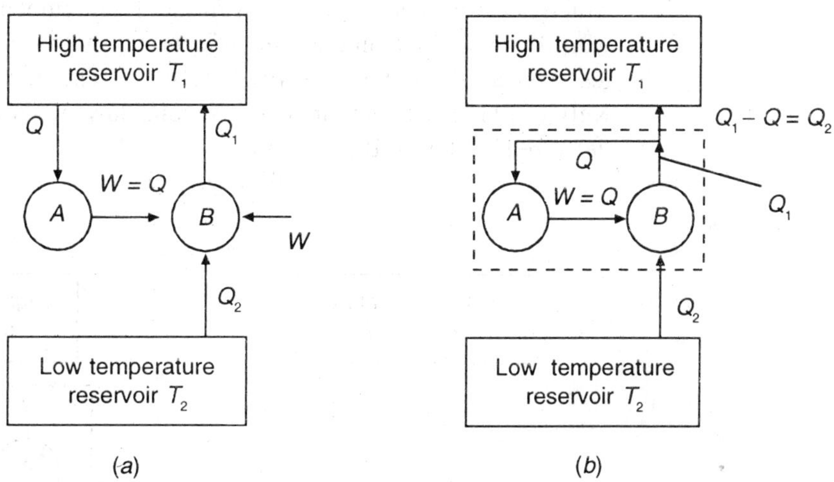


Fig.E 5.13. Sketch to illustrate that violation of Kelvin-Planck statement implies violation of Clausius statement.

$$W = Q \text{ and } W = Q_1 - Q_2$$

or $Q = Q_1 - Q_2$ or $Q_1 - Q = Q_2$

The combination of the devices A and B is shown in Fig.E 5.13(*b*). The combined device absorbs energy $Q_2$ as heat from a low temperature reservoir at $T_2$ and rejects energy $Q_2$ ($Q_2 = Q_1 - Q$) as heat to the high temperature reservoir at $T_1$, while it is unaided by any external agency. That is, it is possible to devise a selfacting cyclically operating device which transfers energy as heat from a low temperature body to a high temperature body. This is in violation of the Clausius statement of the second law of thermodynamics. Thus, violation of the Kelvin–Planck statement leads to violation of the Clausius statement.

### Example 5.14

Prove that violation of Clausius statement leads to violation of the Kelvin–Planck statement of the second law of thermodynamics.

*Solution* :

To prove that violation of Clausius statement leads to violation of Kelvin–Planck statement, let us assume that the Clausius statement of the second law of thermodynamics is incorrect. That is, it is possible to devise a selfacting (or externally unaided) cyclically operating device A which transfers energy $Q$ as heat from a low temperature reservior at $T_2$ to a high temperature reservoir at $T_1$ ($T_1 > T_2$) as shown in Fig.E 5.14 (*a*). Now, let us consider another device B which absorbs energy $Q_1$ as heat from a high temperature reservoir at $T_1$, does work $W$ on the surroundings and rejects energy $Q$ as heat to the low temperature reservoir at $T_2$ as shown in Fig.E 5.14 (*a*). The device B does not violate the Kelvin–Planck statement of the second law of thermodynamics. The work delivered by device B is given by

$$W = Q_1 - Q$$

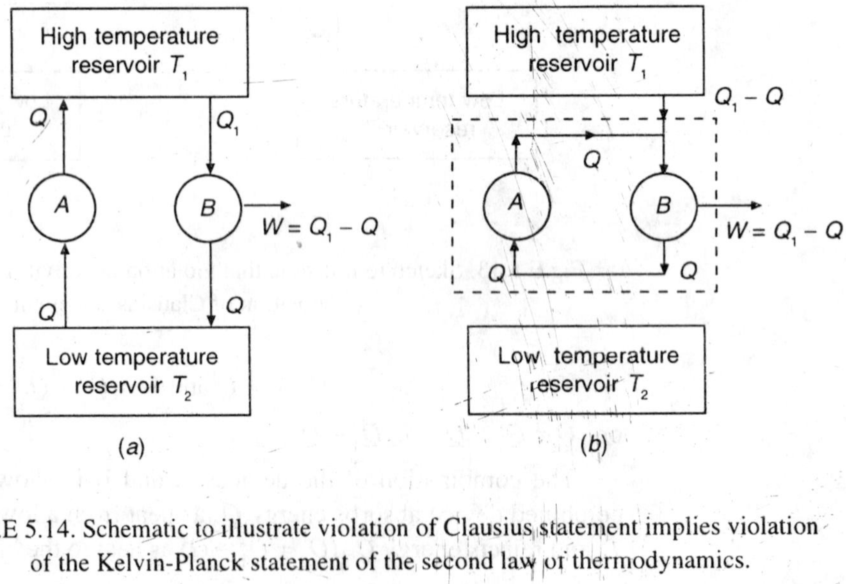


Fig.E 5.14. Schematic to illustrate violation of Clausius statement implies violation of the Kelvin-Planck statement of the second law of thermodynamics.

Now, let us combine the devices A and B such that the energy $Q$ rejected as heat at temperature $T_2$ by device B is directly fed to the device A and the energy $Q$ rejected as heat at temperature $T_1$ by device A is directly absorbed by device B. The difference in the energy $(Q_1 - Q)$ is absorbed by device B from the high temperature reservoir at $T_1$ and the work delivered by device B is $W$ $(W = Q_1 - Q)$ as shown in Fig.E 5.14 (*b*). The combined device effectively absorbs energy $(Q_1 - Q)$ as heat from a single thermal reservoir and delivers an equivalent amount of work $W$ $(W = Q_1 - Q)$ in violation of the Kelvin–Planck statement of the second law of thermodynamics. Thus violation of Clausius statement leads to violation of Kelvin–Planck statement. Since, the violation of Kelvin–Planck statement leads to violation of Clausius statement and violation of Clausius statement leads to violation of Kelvin–Planck statement, these two statements of the second law of thermodynamics are equivalent.

**Example 5.15**

Distinguish between reversible and irreversible processes and give some examples of irreversible processes.

*Solution* :

A process is said to be reversible if both the system and its surroundings can be restored to their respective initial states by reversing the direction of the process. If a process does not satisfy the above criterion, it is an irreversible process.

Examples of irreversible processes are :

1. Expansion or compression with finite pressure difference.
2. Energy transfer as heat with finite temperature difference.
3. Free expansion of a gas.
4. Mixing of nonidentical gases.
5. Mixing of matter at different states.
6. Motion with friction.
7. Viscous fluid flow.
8. Spontaneous chemical reactions.

**Example 5.16**

What is the importance of reversible processes in engineering thermodynamics ?

*Solution* :

A reversible process is an idealization. This is a concept which can be approximated closely at times by actual devices, but never followed.

The presence of friction, inelasticity and electrical resistance makes the processes irreversible. These elements can be reduced but cannot be completely eliminated. The presence of these elements of irreversibility makes a process irreversible. Since the real processes occur when these elements of irreversibility are present, the reversible process is a limiting process toward which all actual processes may approach in performance. The reversible processes deliver maximum work in engines and require minimum work in devices such as refrigerators, compressors etc. A thermodynamic analysis based on idealized reversible processes provides the limiting performance of the devices, against which the actual performance can be compared. This in turn provides an opportunity to improve the performance of the devices by reducing the sources of irreversibility.

**Example 5.17**

It is know that the second law of thermodynamics dictates the direction of a process. In other words, the second law distinguishes between reversible and irreversible processes. State the criterion which can be applied to identify whether a process is reversible or not.

*Solution* :

Since the second law of thermodynamics distinguishes between reversible and irreversible processes, one can use the following criterion to identify a reversible process. If a process can proceed in either direction without violating the second law of thermodynamics, then it is a reversible process. Suppose a process is proceeding in one direction and the assumption of reversibility of the process leads to a violation of the second law of thermodynamics, then the process can be called irreversible.

**Example 5.18**

Apply the criterion stated in Example 5.17 to identify irreversible process and show that free expansion of a gas is an irreversible process.

*Solution* :

Consider a rigid and insulated tank which is divided into two compartments by a removable partition. Let one compartment contain $n$ moles of an ideal gas at $P$ and $T$ and the second compartment is evacuated, If the partition is removed, the gas expands and occupies the entire volume of the tank. That is, the gas undergoes free expansion. Consider the gas as the system. Then $W = 0$ and $Q = 0$. Application of the first law of thermodynamics gives $\Delta U = nC_v \Delta T = Q - W = 0$ or $\Delta T = 0$.Hence the final temperature of the gas is the same as the initial temperature. To judge whether the free expansion process is reversible or not, let us apply the criterion for irreversibility. To start with let us assume that the process is reversible. That is the gas which is at some low pressure and occupying the entire volume of the tank flows on its own and occupies only one compartment at high pressure and leaves the second compartment under vacuum. Then the high pressure gas available in one compartment can be adiabatically expanded in a piston–cylinder assembly and the resulting low pressure gas can be fed to the second compartment. This process can be continued till the pressures in both the compartments are equalized. During adiabatic expansion, the gas does work on the surroundings at the expense of its own internal energy and the temperature of the gas reduces. Now, the gas at low temperature can be placed in thermal contact with a single thermal reservoir till its temperature raises to its original value. Then the gas moves on its own and occupies only one compartment and the cycle of operations can be repeated. The sequence of events are shown in Fig.E 5.18. The sequence of events enclosed by a dotted curve in Fig.E 5.18 constitutes a perpetual motion machine of the second kind which is impossible to devise according to the second law of thermodynamics, Hence our assumption of reversibility of the free expansion process leads to violation of the second law of thermodynamics. Therefore, we can conclude that free expansion of a gas is an irreversible process.

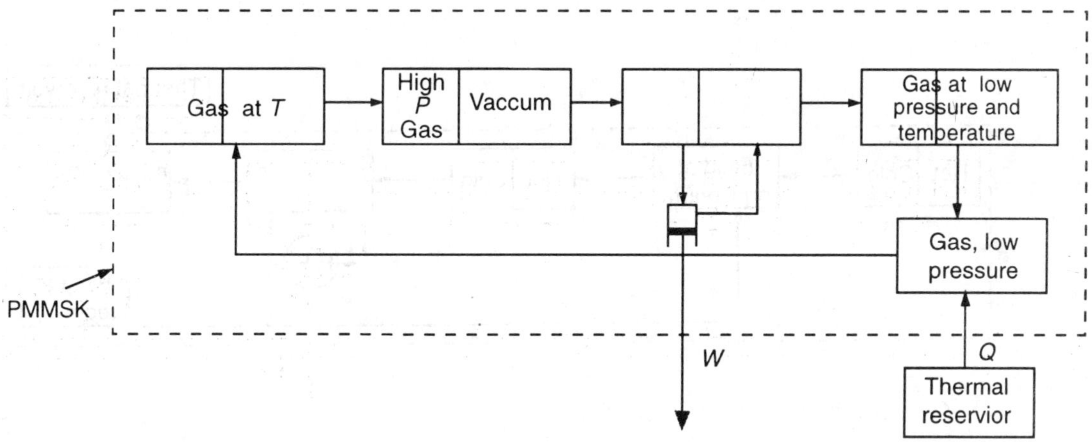


Fig.E 5.18. Sketch for Example 5.18.

### Example 5.19

Show that mixing of matter at different states is an irreversible process by applying the criterion for irreversibility.

*Solution :*

Consider an insulated tank divided into two compartments by a removable partition. Let one compartment be filled with a hot fluid and the other compartment be filled with cold fluid. If the partition between the two compartments is removed, the fluids mix with each other and attain an equilibrium temperature. Now let us apply the criterion to identify whether the process is reversible or not. Assume that the process is reversible. That is, the fluid at equilibrium temperature separates into two parts and one part of the fluid is available as hot while the other part is available as cold. Then, one can operate a heat engine using the hot fluid as source and the cold fluid as sink, thus obtaining some work. The engine can continue to deliver work till both the fluids reach a common temperature. Then the resulting fluid at low temperature can be placed in thermal contact with a thermal reservoir which transfers energy as heat to the fluid to raise it to the original temperature. The sequence of operations described above are shown in Fig.E 5.19 The combination of the processes enclosed by a dotted curve in Fig.E 5.19 constitutes a perpetual motion machine of the second kind which is impossible to devise according to the second law of thermodynamics. Hence our assumption of reversibility is incorrect. Therefore, mixing of fluids at different states is an irreversible process.

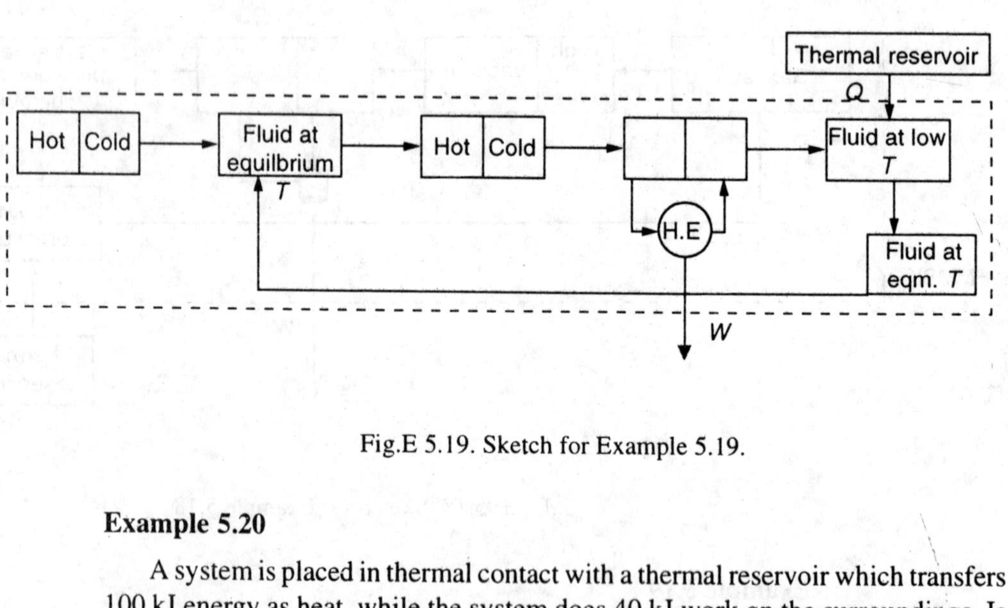


Fig.E 5.19. Sketch for Example 5.19.

**Example 5.20**

A system is placed in thermal contact with a thermal reservoir which transfers 100 kJ energy as heat, while the system does 40 kJ work on the surroundings. Is it possible to restore the system to the original state by following an adiabatic process ?

*Solution* :

During the first process, the system absorbs energy as heat from a single thermal reservoir and delivers some work. Application of the first law of thermodynamics gives $U_2 - U_1 = Q - W = 100 - 40 = 60$ kJ. In the second process, the system has to reach state 1 through an adiabatic process. Applying the first law of thermodynamics to the second process we get,

$$U_1 - U_2 = -60 = Q - W = -W \quad \text{or} \quad W = 60 \text{ kJ}.$$

Now, the system has undergone one cyclic change and the net work delivered is equal to 40 + 60 = 100 kJ. During this cycle, the system is absorbing 100 kJ energy as heat from a single thermal reservoir and delivering an equivalent amount of work. That is, it constitutes a PMMSK which is impossible to devise. Hence it is not possible to restore the system to the original state through an adiabatic path.

**Example 5.21**

Prove that two reversible adiabatic lines cannot intersect.

*Solution* :

To prove that two reversible adiabatic lines cannot intersect, let us assume the contrary. Suppose the two reversible adiabatic lines 1–*a* and *b*–1 intersect at point 1 as shown in Fig.E 5.21. Then draw an isotherm *a*–*b* connecting the two intersecting reversible adiabatic lines. Then 1–*a*–*b*–1 constitutes a cycle and the area enclosed by this cycle on a *P*–*V* diagram represents the work done. During this cycle, there are no heat interactions along the adiabatic paths 1–*a* and *b*–1.

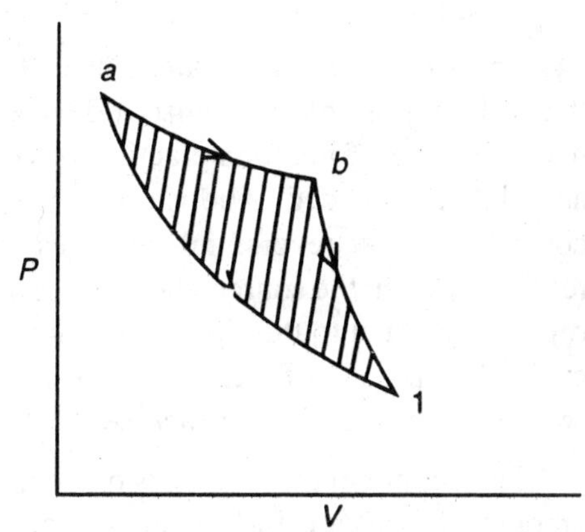


Fig.E 5.21. Two intersecting reversible adiabatic lines 1-*a* and *b*-1 on a pressure versus volume diagram.

The heat interaction is along the isothermal path *a–b* only. That is, the system which is undergoing the cyclic change absorbs energy as heat from a single thermal reservoir and delivers work, which is in violation of the Kelvin–Planck statement of the second law of thermodynamics. Hence our assumption is incorrect. Therefore, two reversible adiabatic lines cannot intersect.

**Example 5.22**

Does the Kelvin–Planck statement of the second law of thermodynamics prohibit the complete conversion of the energy absorbed as heat into work, by a device in a process? Support your answer with an example.

*Solution* :

The Kelvin–Planck statement of the second law of thermodynamics does not prohibit the complete conversion of energy absorbed as heat into work, by a device in a process, However, in a cyclic process it prohibits the complete conversion of energy absorbed as heat into work, by a device. Suppose an ideal gas undergoes isothermal expansion. Application of the first law of thermodynamics gives $\Delta U = Q - W = 0$ or $Q = W$. That is all the energy absorbed as heat by the gas is converted into work. However, the process cannot deliver work continuously because the gas pressure continuously decreases and expansion is not feasible after a certain pressure.

**Example 5.23**

Sketch the Carnot cycle on pressure versus volume diagram. Describe the practical difficulties associated with an engine which works on Carnot cycle.

*Solution* :

The Kelvin–Planck statement of the second law of thermodynamics tells that it is impossible to devise a device which interacts with only one thermal reservoir. Therefore, for a practical device at least two thermal reservoirs–source and sink–are needed. The Carnot cycle uses only two thermal reservoirs and consists of only reversible processes. If the energy is to be absorbed as heat reversibly from the source, the process has to be isothermal. Similarly, the energy rejection as heat to the sink is also through isothermal process. Thus the Carnot cycle shown in Fig.E 5.23 consists of two reversible isothermal processes and two reversible adiabatic processes.

The Carnot engine consists of a cylinder–piston assembly in which a certain amount of working fluid is enclosed. Initially, the working fluid is at state 1. Then the cylinder is placed in thermal contact with a source at $T_1$ and the working fluid absorbs energy $Q_1$ as heat through an isothermal process 1–2. Then the cylinder head is insulated and the working fluid undergoes reversible adiabatic expansion to reach state 3 at which point the temperature of the working fluid is $T_2$. Then the cylinder is placed in thermal contact with a sink at $T_2$ and the working fluid rejects energy $Q_2$ as heat to the sink in an isothermal process 3–4. Again the cylinder is insulted and the working fluid undergoes reversible adiabatic compression to reach state 1, thus completing a cycle.

The Carnot engine is a hypothetical device and it is not possible to devise it. Since it consists of reversible processes only, the energy absorption and rejection has to take place with infinitesimal temperature differences and hence the rate of energy transfer will be very low. Hence, the engine delivers only infinitesimal power. For any reasonable power output, the size of the engine has to be extremely large. Since the Carnot cycle consists of only reversible processes, it serves as a yardstick or standard of perfection against which the performance of any practical heat engine can be compared.

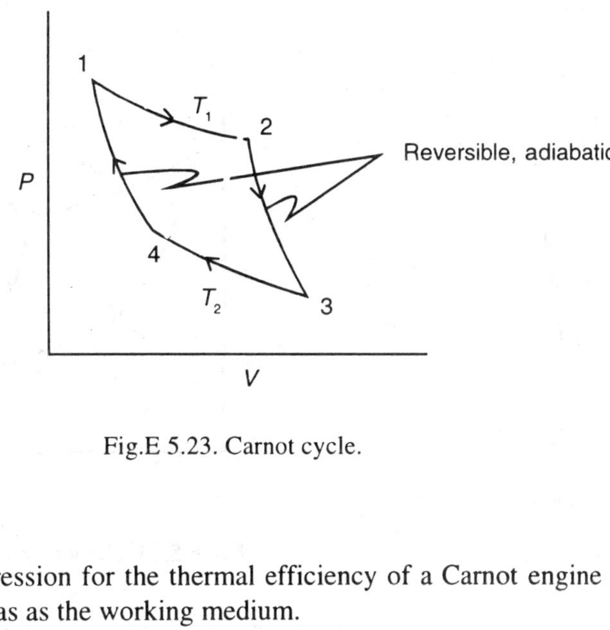


Fig.E 5.23. Carnot cycle.

**Example 5.24**

Derive an expression for the thermal efficiency of a Carnot engine which employs an ideal gas as the working medium.

*Solution :*

The Carnot cycle *abcda* is shown in Fig.E 5.24 on a *P–v* diagram. *a–b* and *c–d* are isothermal processes. *b–c* and *d–a* are reversible adiabatic processes.

Let $|Q_1|$ = amount of energy absorbed as heat from the source at $T_1$ during the reversible isothermal process $a - b$.

$|Q_2|$ = amount of energy rejected as heat to the sink at $T_2$ during the reversible isothermal process $c - d$.

There are no heat interactions along the reversible adiabatic processes $b–c$ and $d–a$. Application of the first law of thermodynamics for the complete cycle gives

$\oint dQ = \oint dW$ or $|Q_1| - |Q_2| = W$

The thermal efficiency, $\eta$ is given by

$$\eta = \frac{W}{Q_1} = \frac{|Q_1| - |Q_2|}{|Q_1|} = 1 - \frac{|Q_2|}{|Q_1|} \tag{A}$$

The first law of thermodynamics, when applied to the isothermal process $a–b$, gives

$$u_b - u_a = Q_{ab} - W_{ab} = |Q_1| - W_{ab} = 0 \quad \text{or} \quad |Q_1| = W_{ab}$$

or
$$|Q_1| = W_{ab} = \int_a^b Pdv = RT_1 \ln \frac{v_b}{v_a} \tag{B}$$

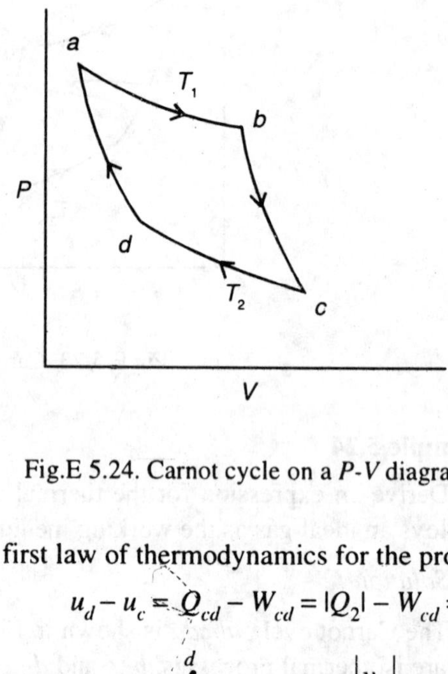


Fig.E 5.24. Carnot cycle on a *P-V* diagram.

Similarly, the first law of thermodynamics for the process $c–d$, gives

$$u_d - u_c = Q_{cd} - W_{cd} = |Q_2| - W_{cd} = 0 \quad \text{or} \quad |Q_2| = W_{cd}$$

or
$$|Q_2| = W_{cd} = \int_c^d Pdv = RT_2 \ln \left|\frac{v_d}{v_c}\right| = RT_2 \ln \frac{v_c}{v_d} \tag{C}$$

Substituting Eqns. (*B*) and (*C*) in Eqn. (*A*), we get

$$\eta = 1 - \frac{|Q_2|}{|Q_1|} = 1 - \frac{RT_2 \ln (v_c / v_d)}{RT_1 \ln (v_b / v_a)}$$

$$= 1 - \frac{T_2 \ln (v_c / v_d)}{T_1 \ln (v_b / v_a)} \tag{D}$$

We know that for an ideal gas undergoing reversible adiabatic process $Tv^{\gamma-1}$ = constant. Therefore, for the adiabatic process $b–c$

$$T_1 v_b^{\gamma-1} = T_2 v_c^{\gamma-1} \quad \text{or} \quad \left(\frac{v_c}{v_b}\right)^{\gamma-1} = \frac{T_1}{T_2} \tag{E}$$

Similarly for the adiabatic process $d–a$, we get

$$T_2 v_d^{\gamma-1} = T_1 v_a^{\gamma-1} \quad \text{or} \quad \left(\frac{v_d}{v_a}\right)^{\gamma-1} = \frac{T_1}{T_2} \tag{F}$$

From Eqns. (*E*) and (*F*), we get

$$\left(\frac{v_c}{v_b}\right)^{\gamma-1} = \frac{T_1}{T_2} = \left(\frac{v_d}{v_a}\right)^{\gamma-1} \quad \text{or} \quad \frac{v_c}{v_b} = \frac{v_d}{v_a}$$

or

$$\frac{v_c}{v_d} = \frac{v_b}{v_a} \tag{G}$$

Substitute Eqn. (*G*) in Eqn. (*D*), to obtain

$$\eta = 1 - \frac{|Q_2|}{|Q_1|} = 1 - \frac{T_2}{T_1}$$

**Example 5.25**

State the Carnot theorems.

*Solution* :

The second law of thermodynamics leads to several important consequences. The Carnot theorems are some of the consequences of the second law of thermodynamics. The Carnot theorems (proposed by Carnot) deal with the efficiency of a Carnot engine. The Carnot theorems are as follows :

*Carnot theorem 1:*

No heat engine operating between the two given thermal reservoirs, with fixed temperatures, can be more efficient than a reversible engine operating between the same two thermal reservoirs.

*Carnot theorem 2 :*

All reversible heat engines operating between the two given thermal reservoirs, with fixed temperatures, have the same efficiency. The efficiency of a reversible heat engine does not depend on the nature of the working fluid, but depends only on the temperatures of the reservoirs between which it operates.

**Example 5.26**

Prove the Carnot theorem–1 stated in Example 5.25.

*Solution* :

The Carnot theorem–1 states that no heat engine operating between two thermal reservoirs, with fixed temperatures, can be more efficient than a reversible engine operating between the same two thermal reservoirs. To prove this, let us assume the contrary. That is, it is possible to devise an engine $I$ (say an irreversible engine) such that its efficiency is greater than that of a reversible engine $R$, when both of them are operating between the same set of thermal reservoirs.

Let $Q_1$ = energy absorbed by engine $I$ and $R$, independently, from the source at temperature $T_1$.

$Q_2$ = energy rejected by engine $R$ to the sink at temperature $T_2$.

$Q$ = energy rejected by engine $I$ to the sink at temperature $T_2$.

$W_R$, $W_I$ = work delivered by engine $R$ and $I$, respectively

$\eta_R$ $\eta_I$ = efficiency of engines $R$ and $I$, respectively.

Then $W_I = Q_I - Q$ and $W_R = Q_1 - Q_2$

$\eta_I = 1 - Q/Q_1$ and $\eta_R = 1 - Q_2/Q_1$

We assumed that $\eta_I > \eta_R$, That is

$1 - Q/Q_1 > 1 - Q_2/Q_1$ or $Q < Q_2$

Therefore $W_I = Q_I - Q > W_R = Q_1 - Q_2$

That is, for the same amount of energy intake $Q_1$, the engine $I$ delivers more work than engine $R$. The extra work delivered by engine $I$ is given by $(W_I - W_R) = Q_2 - Q$.

A schematic of the engines $I$ and $R$, operating between the same two thermal reservoirs is shown in Fig.E 5.26 (*a*)

Since the engine $R$ is reversible, it can be operated to execute the cycle in the reverse. That is when work $W_R$ is done on the device, it absorbs energy $Q_2$ from the thermal reservoir at $T_2$ and rejects energy $Q_1$ as heat to the thermal reservoir at $T_1$. Now we can combine the devices $I$ and $R$ as shown in Fig.E 5.26 (*b*). The work needed $W_R$ to operate the device $R$ can be fed from the work delivered by engine $I$, and the energy $Q_1$ rejected by the device $R$ can be supplied to the engine $I$. The device $R$ receives energy $Q$, which is rejected by engine $I$ and the balance of energy $(Q_2 - Q)$ from the thermal reservoir at $T_2$ as shown in Fig.E 5.26 (*b*). The combination of the two devices $I$ and $R$ [enclosed by a dotted curve in

Fig.E 5.26 (*b*)] constitutes a perpetual motion machine of the second kind in violation of the second law of thermodynamics. Therefore, our assumption that $\eta_I > \eta_R$ is incorrect. Hence $\eta_R \geq \eta_I$. That is no heat engine can be more efficient than a reversible engine if both of them are operating between the same set of thermal reservoirs, with fixed temperatures.

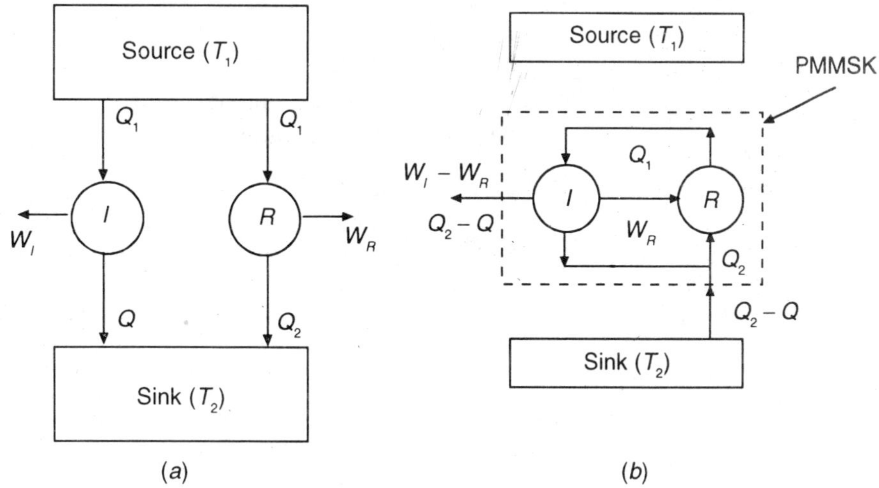


Fig.E 5.26. Schematic to prove the Carnot Theorem 1.

## Example 5.27

Prove the Carnot theorem–2 stated in Example 5.25

*Solution* :

Carnot theorem–2 states that all reversible heat engines operating between the same two thermal reservoirs, with fixed temperatures, have the same efficiency. The efficiency of a reversible heat engine is independent of the nature of the working fluid, but depends only on the temperatures of the reservoirs between which it operates. To prove the first part of the theorem, let us assume the contrary. That is, the efficiency $\eta_{R1}$ of the reversible engine $R_1$ is greater than the efficiency $\eta_{R2}$ of the reversible engine $R_2$.

Let $Q_1$ = amount of energy absorbed, independently, by the reversible engines $R1$ and $R2$ from the thermal reservoir at high temperature $T_1$.

$Q$, $Q_2$ = amount of energy rejected by the reversible engines $R1$ and $R2$ to the thermal reservior at low temperature $T_2$, respectively.

$W_{R1}$, $W_{R2}$ = work delivered by the reversible engine $R1$ and $R2$ respectively.

$$\eta_{R1} = 1 - \frac{Q}{Q_1} > \eta_{R2} = 1 - \frac{Q_2}{Q_1} \quad \text{or} \quad Q < Q_2$$

$$W_{R1} = Q_1 - Q \text{ and } W_{R2} = Q_1 - Q_2$$

Since $Q < Q_2$, we have $W_{R1} > W_{R2}$. That is, for the same amount of energy $Q_1$ absorbed, the reversible engine $R1$ delivers more work than the reversible engine $R2$. The extra work delivered by the reversible engine $R_1$ is given by

$$W_{R1} - W_{R2} = Q_1 - Q - (Q_1 - Q_2) = Q_2 - Q$$

A schematic of the reversible engines $R1$ and $R2$, operating between the same two thermal reservoirs is shown in Fig.E 5.27 (*a*)

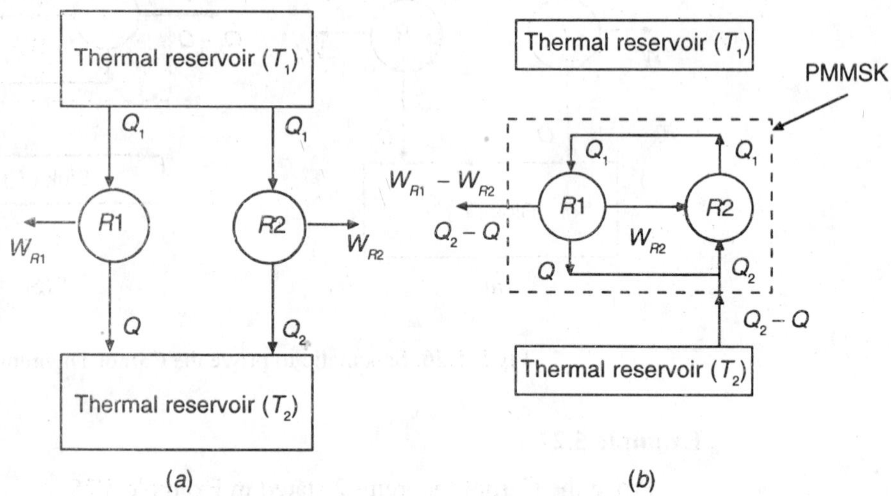


Fig.E 5.27. Schematic to prove the Carnot's second Theorem.

Since $R2$ is a reversible engine, it absorbs energy $Q_2$ as heat from the reservoir at $T_2$ and rejects energy $Q_1$ as heat to the reservoir at $T_1$ when work $W_{R2}$ is done on the device. The work $W_{R2}$ needed to operate the reversible engine $R2$, in the reverse can be supplied by the reversible engine $R1$ by combining the two devices $R1$ and $R2$ as shown Fig.E 5.27 (*b*). The net work delivered by the combination of the two devices is $W_{R1} - W_{R2} = Q_2 - Q$. This combined device absorbs energy $(Q_2 - Q)$ from the reservoir at $T_2$ and does not interact with the other reservoir at $T_1$. That is, the combined device is a PMMSK in violation of the second law of thermodynamics. Therefore, our assumption that $\eta_{R1} > \eta_{R2}$ is incorrect. Hence $\eta_{R2} \geq \eta_{R1}$.

Now, assume that the reversible engine $R2$ is more efficient than the reversible engine $R1$ and repeat the arguments as above to deduce that $\eta_{R1} \geq \eta_{R2}$. Therefore we can conclude that $\eta_{R1} = \eta_{R2}$.

That is, all reversible engines operating between the same two thermal reservoirs, will have the same efficiency.

To prove the second part of the theorem, that is the efficiency of a reversible heat engine is independent of the nature of the working fluid and depends only on the temperatures of the two reserviors between which it operates, assume that the reversible engine $R1$ and $R2$ are using two different working fluids. Also assume that the use of a different working fluid in engine $R1$ makes it more efficient than engine $R2$. Now repeat the arguments as above and reach the conclusion that the efficiency of a reversible heat engine does not depend on the nature of the working fluid. Since the engines $R1$ and $R2$ have the same efficiency while operating between the reservoirs at temperature $T_1$ and $T_2$, the efficiency of a reversible heat engine depends only on the temperatures of the two thermal reservoirs between which it works.

**Example 5.28**

What is a thermodynamic temperature scale or absolute temperature scale ?

*Solution* :

We know that a temperature scale can be established by assigning certain numerical values to two easily reproducible states and the temperature can be measured by measuring the change in the thermometric property of a substance such as the length of a mercury column in a capillary tube, the *e.m.f* generated at the junction of two dissimilar metals, or the electrical resistance of a wire. Since different substances have different relations between temperature and thermometric property, the temperature of a substance in a given state as measured by different thermometers will give different values. Therefore, it is necessary to establish a temperature scale which is independent of the nature of the thermometric fluid, Such a temperature scale was established based on the second law of thermodynamics and this temperature scale is called the thermodynamic temperature scale or absolute temperature scale.

**Example 5.29**

Which principle or fact is made use of in establishing the thermodynamic temperature scale ?

*Solution* :

The thermodynamic temperature scale is established based on the second law of thermodynamics. A consequence of the second law of thermodynamics is Carnot theorem. The Carnot theorem –2 tells that the efficiency of a reversible heat engine does not depend on the nature of the working medium but depends only on the temperatures of the two thermal reservoirs between which it operates.

This theorem is directly made use of in establishing the thermodynamic temperature scale.Hence, the establishment of a thermodynamic temperature scale is considered as a consequence of the second law of thermodynamics.

**Example 5.30**

Explain the establishment of a thermodynamic temperature scale.

*Solution :*

Let $\theta$ denote the temperature as measured on the thermodynamic temperature scale and consider three isotherms at $\theta_1$, $\theta_2$ and $\theta_3$ such that $\theta_1 > \theta_2 > \theta_3$. Draw two reversible adiabatics intersecting the three isotherms as shown in Fig.E 5.30 (*a*). Then ABCDA constitutes a Carnot cycle operating with thermal reservoirs at $\theta_1$ and $\theta_2$. Similarly DCEFD is another Carnot cycle employing the thermal reservoirs at $\theta_2$ and $\theta_3$. We can also have another Carnot cycle ABEFA which uses thermal reservoirs at $\theta_1$ and $\theta_3$ as shown in Fig.E 5.30 (*b*).

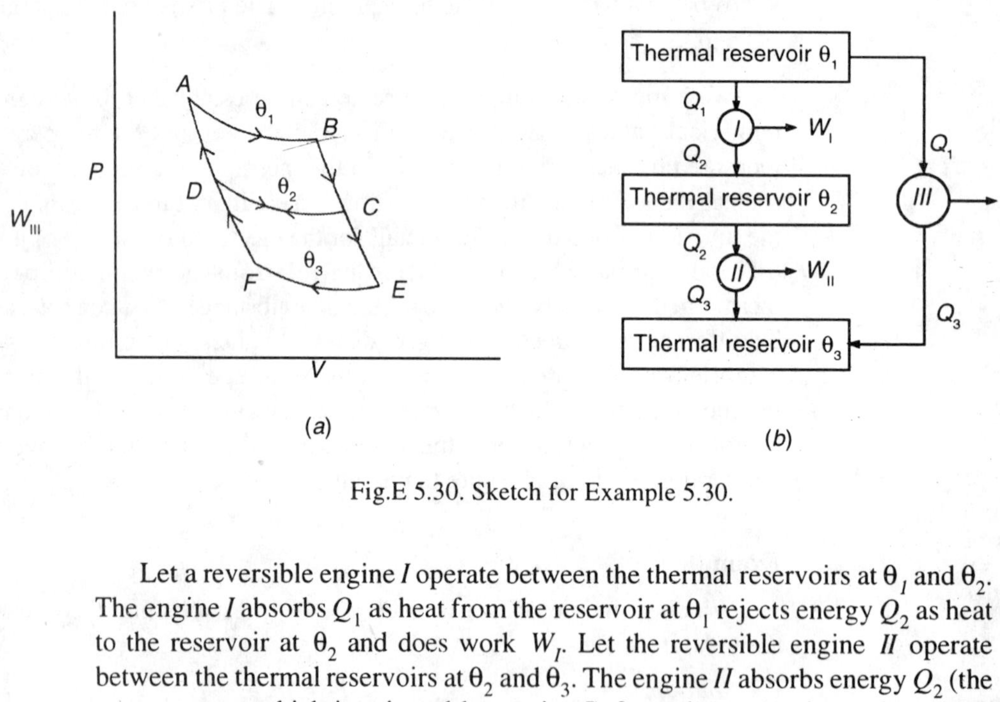


Fig.E 5.30. Sketch for Example 5.30.

Let a reversible engine *I* operate between the thermal reservoirs at $\theta_I$ and $\theta_2$. The engine *I* absorbs $Q_1$ as heat from the reservoir at $\theta_1$ rejects energy $Q_2$ as heat to the reservoir at $\theta_2$ and does work $W_I$. Let the reversible engine *II* operate between the thermal reservoirs at $\theta_2$ and $\theta_3$. The engine *II* absorbs energy $Q_2$ (the same amount which is rejected by engine I) from the reservoir at $\theta_2$, rejects energy $Q_3$ as heat to the reservoir at $\theta_3$ and does work $W_{II}$. The thermal reservoir at $\theta_2$ is unaffected, since it receives energy $Q_2$ from engine *I* and supplies the same amount to engine *II*. Let a third Carnot engine operate between the thermal reservoirs at $\theta_1$ and $\theta_3$. The engine *III* receives energy $Q_1$ from reservoir at $\theta_1$

and rejects energy $Q_3$ to the reservoir at $\theta_3$. Then the work done by these engines are given by

$$W_I = Q_1 - Q_2 \qquad W_{II} = Q_2 - Q_3$$

$$W_{III} = Q_1 - Q_3 = Q_1 - Q_2 + (Q_2 - Q_3) = W_I + W_{II}$$

All the three engines are reversible and hence their efficiencies do not depend on the nature of the working medium and depend only on the temperatures of the reservoirs between which they are operating. Hence the efficiencies of these engines are given by

$$\eta_I = 1 - \frac{Q_2}{Q_1} = f(\theta_1, \theta_2)$$

or $$\frac{Q_2}{Q_1} = 1 - f(\theta_1, \theta_2)$$

or $$\frac{Q_1}{Q_2} = F(\theta_1, \theta_2)$$

$$\eta_{II} = 1 - \frac{Q_3}{Q_2} = f(\theta_2, \theta_3)$$

or $$\frac{Q_2}{Q_3} = F(\theta_2, \theta_3)$$

$$\eta_{III} = 1 - \frac{Q_3}{Q_1} = f(\theta_1, \theta_3)$$

or $$\frac{Q_1}{Q_3} = F(\theta_1, \theta_3)$$

Where $f$ and $F$ denote two different functions.

Then $$\frac{Q_1}{Q_2} = \frac{Q_1 / Q_3}{Q_2 / Q_3}$$

or $$F(\theta_1, \theta_2) = \frac{F(\theta_1, \theta_3)}{F(\theta_2, \theta_3)} \qquad (A)$$

We find that the LHS of Eqn.($A$) is a function of $\theta_1$ and $\theta_2$. Therefore, the dependence on $\theta_3$ on the RHS of Eqn.($A$) must cancel out. Hence, the nature of the function $F$ should be of the form

$$F(\theta_1, \theta_2) = \frac{\psi(\theta_1)}{\psi(\theta_2)} \qquad (B)$$

The choice of the function $\psi$ defines the temperature scale. Kelvin choose the simplest choice for the function $\psi$ as $\psi(\theta) = \theta$. Then

$$\frac{Q_1}{Q_2} = F(\theta_1, \theta_2) = \frac{\psi(\theta_1)}{\psi(\theta_2)} = \frac{\theta_1}{\theta_2} \quad \text{or} \quad \frac{Q_1}{Q_2} = \frac{\theta_1}{\theta_2} \qquad (C)$$

That is, the ratio of the energy absorbed to the energy rejected as heat by a reversible engine is equal to the ratio of the temperatures of the source and sink. The Eqn.(*C*) can be used to determine the temperature of any reservoir by assigning a numerical value to one easily reproducible thermal reservoir (say the triple point of water 273.16 K) and by measuring the heat interactions $Q_l$ and $Q_2$.

Now the efficiency of a Carnot engine operating between two thermal reservoirs at $T_1$ and $T_2$ is given by

$$\eta = 1 - \frac{Q_2}{Q_1} = 1 - \frac{T_2}{T_1}$$

**Example 5.31**

Why is the thermodynamic temperature scale is also called absolute temperature scale?

*Solution* :

The efficiency of a Carnot engine is given by

$$\eta = 1 - \frac{T_2}{T_1}$$

Since the efficiency of the heat engine cannot be greater than one, the lowest possible temperature on this scale is zero ($\theta_2 = 0$). Hence, the thermodynamic temperature scale is called absolute temperature scale.

**Example 5.32**

How many fixed points or reference points are used in establishing the thermodynamic temperature scale and what are those fixed points ?

*Solution* :

Only one fixed point is used in establishing the thermodynamic temperature scale. The chosen fixed point is the triple point of water (0.01°C) which has been assigned a numerical value of 273.16 K.

**Example 5.33**

Show that the thermodynamic temperature scale and the ideal gas temperature scale are equivalent.

*Solution* :

Consider a Carnot heat engine, employing an ideal gas as the working medium, operating between two reservoirs at temperature $T_1$ and $T_2$ as measured by an ideal gas thermometer. Then, the efficiency of the Carnot engine is given by [See example 5.24]

$$\eta = 1-\frac{Q_2}{Q_1} = 1-\frac{T_2}{T_1} \text{ or } \frac{Q_1}{Q_2} = \frac{T_1}{T_2} \qquad (A)$$

The absolute or thermodynamic temperature scale is established through the relation

$$\frac{Q_1}{Q_2} = \frac{\theta_1}{\theta_2} \qquad (B)$$

Where $\theta_1$ and $\theta_2$ are the temperatures on the thermodynamic temperature scale. From Eqns. $(A)$ and $(B)$, we find

$$\frac{\theta_1}{\theta_2} = \frac{T_1}{T_2}$$

Since, the fixed point for the establishment of the ideal gas temperature scale is the same as the fixed point for the establishment of the thermodynamic temperature scale, these two temperature scales are equivalent.

**Example 5.34**

What is a Carnot heat pump or Carnot refrigerator ? Express the coefficient of performance of a Carnot heat pump and a Carnot refrigerator in terms of the temperatures of the thermal reservoirs between which they operate.

*Solution* :

A Carnot cycle consists of only reversible processes. Hence, one can hypothetically construct a device which executes the cycle in the reverse order. The reversed Carnot cycle is used as a heat pump or refrigerator cycle. A schematic of the Carnot heat pump or refrigerator is shown in Fig.E 5.34. The Carnot heat pump or refrigerator absorbs energy $Q_L$ as heat from a low temperature reservoir at $T_L$, rejects energy $Q_H$ as heat to a high temperature reservoir at $T_H$ when work $W$ is performed on it.

$$(COP)_{\text{Heat pump}} = \frac{Q_H}{W} = \frac{Q_H}{Q_H - Q_L} = \frac{T_H}{T_H - T_L}$$

$$(COP)_{\text{Refrigerator}} = \frac{Q_L}{W} = \frac{Q_L}{Q_H - Q_L} = \frac{T_L}{T_H - T_L}$$

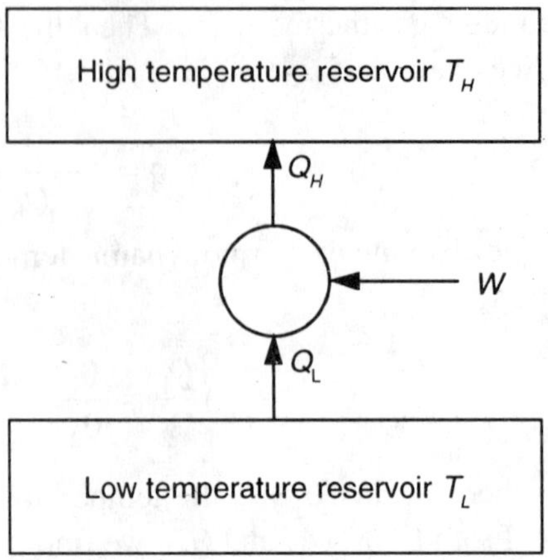


Fig.E 5.34. Schematic of Carnot heat pump or refrigerator.

**Example 5.35**

It is desired to set up a geothermal power plant at a site where the underground source of hot water is at 80°C and the ambient atmospheric air is at 30°C. Determine the maximum possible thermal efficiency of the power plant and the maximum power output of the plant if the rate at which energy can be supplied to the device is 1 kJ/s.

*Solution* :

We know that only a reversible heat engine can have the maximum efficiency and all reversible engines operating between the same set of thermal reservoirs will have the same efficiency. It is also known that the efficiency of a reversible engine does not depend on the nature of the working fluid, but depends only on the temperatures of the reservoirs between which it operates. The hot water at 80°C or 353.15 K can be used as source and the ambient atmospheric air at 30°C or 303.15 K can be used as sink.Then

$$\eta = 1 - \frac{T_2}{T_1} = 1 - \frac{303.15}{353.15} = 0.1416$$

$$\eta = \frac{\dot{W}}{\dot{Q}_1}$$

or $$0.1416 = \frac{\dot{W}}{1\times10^3} \quad \text{or} \quad \dot{W} = 0.1416 \text{ kW}$$

Therefore, the maximum power output of the plant = 0.1416 kW

Maximum thermal efficiency = 0.1416

**Example 5.36**

An inventor claims to have designed a heat engine which absorbs 1 kJ of energy as heat at 727°C and delivers 0.6 kJ of work when the ambient temperature is 27°C. Would you agree with this claim?

*Solution* :

The maximum possible efficiency of a heat engine operating between two thermal reservoirs is given by

$$\eta = \frac{W}{Q_1} = 1 - \frac{T_2}{T_1} = 1 - \frac{300}{1000} = 0.7$$

Efficiency claimed by the inventor, $\eta_{\text{(claim)}} = \frac{W}{Q_1} = \frac{0.6}{1} = 0.6$

The claimed efficiency (0.6) is less than the maximum possible efficiency (0.7) and hence it is feasible to devise such an engine.

**Example 5.37**

A reversible heat engine delivers 0.65 kW power and rejects energy at the rate of 0.4 kJ/s to a reservoir at 27°C. Determine the efficiency of the engine and the temperature at which energy is absorbed by the engine.

*Solution* :

We know that $$W = Q_1 - Q_2 \text{ or } Q_1 = W + Q_2$$

or $$Q_1 = 0.65 + 0.4 = 1.05 \text{ kJ/s}$$

$$\eta = \frac{W}{Q_1} = 1 - \frac{T_2}{T_1}$$

or $$\frac{0.65}{1.05} = 0.619 = 1 - \frac{300}{T_1} \quad \text{or} \quad T_1 = 787.4 \text{ K}$$

Therefore, efficiency of the engine = 0.619

Temperature of the source = 787.4 K

**Example 5.38**

A reversible heat engine has an efficiency of 0.6 when it absorbs 400 kJ of energy as heat from a reservoir at 537°C. Calculate the sink temperature and the energy rejected as heat to the sink.

*Solution* :

We know that $$\eta = 1 - \frac{T_2}{T_1} \quad \text{or} \quad 0.6 = 1 - \frac{T_2}{810}$$

or $$T_2 = 324 \text{ K or } 51^\circ\text{C} = \text{sink temperature}$$

We also know that $$\frac{Q_1}{Q_2} = \frac{T_1}{T_2} \quad \text{or} \quad Q_2 = \frac{Q_1 T_2}{T_1}$$

or $$Q_2 = 400 \times 324/810 = 160 \text{ kJ}$$

$$= \text{energy rejected to sink}$$

**Example 5.39**

It is proposed to design a cold storage for maintaining certain vegetables under frozen conditions at −20°C. The ambient temperature in summer is 37°C and the estimated energy transfer as heat into the cold storage through the doors, walls and roof is 3 kJ/s. Determine the minimum power required to operate a refrigeration plant for maintaining the cold storage.

*Solution* :

The coefficient of performance of the refrigerator working between two thermal reservoirs at temperatures $T_L$ (−20°C) and $T_H$ (37°C) is given by

$$(\text{COP})_{\text{Ref}} = \frac{Q_L}{W} = \frac{T_L}{T_H - T_L}$$

or $$(\text{COP})_{\text{Ref}} = \frac{T_L}{T_H - T_L} = \frac{253}{310 - 253} = 4.4386$$

$$(\text{COP})_{\text{Ref}} = 4.4386 = \frac{Q_L}{W} = \frac{3}{W} \quad \text{or} \quad W = 0.6759 \text{ kJ/s}$$

The maximum COP of a practical refrigerator is that of a Carnot refrigerator. Hence, the maximum possible (COP) of the refrigerator = 4.4386

A reversible refrigerator requires the minimum power to operate it. Therefore, the minimum power required $= 0.6759$ kW.

**Example 5.40**

The normal boiling point of liquid helium is 4.2 K and the enthalpy of vaporization at this temperature is 83.3 J/mol. It is required to produce liquid helium from saturated vapor at 4.2 K at the rate of one mole per second when the ambient atmospheric temperature is 37°C. Calculate the maximum possible coefficient of performance a refrigerator can have at these conditions and the minimum power consumption of the refrigerator.

*Solution :*

The maximum possible COP of a practicable refrigerating device is equal to the COP of a Carnot refrigerator.

$$(\text{COP})_{\text{Carnot refrigerator}} = \frac{T_L}{T_H - T_L} = \frac{4.2}{310.15 - 4.2}$$

$$= 1.3728 \times 10^{-2}$$

$$(\text{COP})_{\text{Refrigerator}} = \frac{Q_L}{W}$$

or

$$1.3728 \times 10^{-2} = \frac{83.3}{W} \quad \text{or} \quad W = 6.068 \text{ kJ/s}$$

The minimum power consumption = 6.068 kW

**Example 5.41**

It is proposed to produce 1000 kg of ice per hour from liquid water at 0°C in summer when the ambient atmospheric temperature is 37°C. It is planned to use a heat engine to operate the refrigeration plant. Hot water at 70°C which is produced by solar heating, may be used as a source to supply energy as heat to the heat engine and the engine uses the ambient atmospheric air as sink. Calculate (*a*) The power required by the refrigeration plant. (*b*) The ratio of the energy extracted as heat from the freezing water to the energy absorbed as heat by the heat engine and (*c*) The rate at which energy is rejected to the ambient atmosphere, by both the devices. The enthalpy of fusion of water at 0°C is 333.43 kJ / kg.

*Solution :*

A schematic of the heat engine and the refrigerator is shown in Fig.E 5.41.

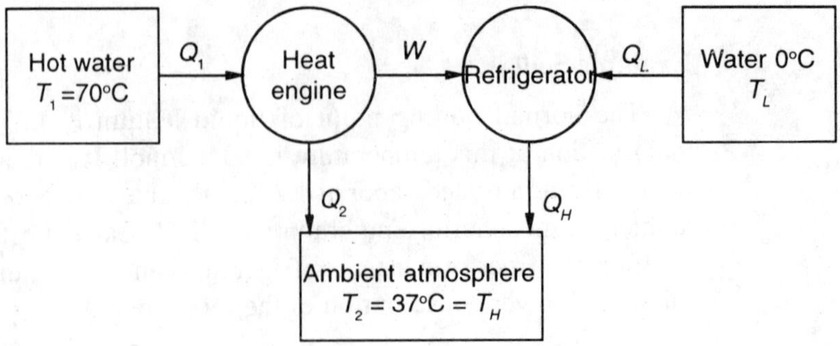


Fig.E 5.41. Sketch for Example 5.41.

$$(a)\ (\text{COP})_{\text{Ref}} = \frac{Q_L}{W} = \frac{Q_L}{Q_H - Q_L} = \frac{T_L}{T_H - T_L}$$

or $$(\text{COP})_{\text{Ref}} = \frac{273}{310-273} = 7.3784 = \frac{Q_L}{W}$$

$$Q_L = \frac{1000 \times 333.43}{3600} = 92.619 \text{ kJ/s}$$

Therefore, $$W = \frac{Q_L}{(\text{COP})_{\text{Ref}}} = \frac{92.619}{7.3784} = 12.553 \text{ kJ/s}$$

Power consumption of the refrigeration unit = 12.553 kW

$$(b)\ \eta = \frac{W}{Q_1} = \frac{Q_1 - Q_2}{Q_1} = 1 - \frac{T_2}{T_1}$$

$$= 1 - \frac{310}{343} = 0.0962$$

$$\frac{Q_L}{Q_1} = \frac{Q_L/W}{Q_1/W} = \frac{(\text{COP})_{\text{Ref}}}{1/\eta} = \frac{7.3784}{1/0.0962} = 0.7098$$

$$(c)\ \frac{Q_2}{Q_1} = \frac{T_2}{T_1} \quad \text{or} \quad Q_2 = Q_1 . \frac{T_2}{T_1} = \frac{Q_1 \times 310}{343} = 0.9038\, Q_1$$

$$\frac{Q_H}{Q_L} = \frac{T_H}{T_L} \quad \text{or} \quad Q_H = Q_L \frac{T_H}{T_L} = Q_L \frac{310}{273} = 1.1355\, Q_L$$

Rate of energy rejection to the ambient atmosphere = $Q_2 + Q_H$

$$Q_2 + Q_H = 0.9038 Q_1 + 1.1355 Q_L = \left(0.9038 + 1.1355 \frac{Q_L}{Q_1}\right) Q_1$$

$$= (0.9038 + 1.1355 \times 0.7098)\, Q_1$$

$$= 1.7098\, Q_1$$

$$Q_I = \frac{W}{\eta} = \frac{12.553}{0.0968} = 130.489 \text{ kJ/s}$$

Therefore, $\quad Q_2 + Q_H = 1.7098\, Q_1 = 1.7098 \times 130.489 = 223.11 \text{ kW}$

### Example 5.42

Two Carnot engines *I* and *II* operate in series between a high temperature reservoir at 1027°C and a low temperature reservoir at 27°C. The engine *I* absorbs energy from the high temperature reservoir and rejects energy to a reservoir at temperature *T*. The engine *II* receives energy from the reservoir at *T* and rejects energy to the low temperature reservoir. The amount of energy absorbed by engine *II* from the reservoir at *T* is the same as that rejected by engine *I* to the reservoir at *T*, If engines *I* and *II* are found to have the same efficiency, determine the temperature *T*. If engine *I* receives 100 kJ energy as heat from the high temperature reservoir, calculate the work delivered by engine *I* and engine *II*.

*Solution* :

A schematic of engines I and II is shown in Fig.E 5.42.

$$\eta_I = 1 - \frac{T}{T_1} \quad \text{and} \quad \eta_{II} = 1 - \frac{T_3}{T}$$

$$\eta_I = \eta_{II} \text{ or } 1 - \frac{T}{T_1} = 1 - \frac{T_3}{T} \text{ or } T = \sqrt{T_1 T_3} = \sqrt{1300 \times 300} = 624.5 \text{ K}$$

Therefore, the temperature of the intermediate reservoir = 624.5 K

$$\eta_I = 1 - \frac{T}{T_1} = 1 - \frac{624.5}{1300} = 0.5196$$

$$W_I = \eta_I Q_1 = 0.5196 \times 100 = 51.96 \text{ kJ}$$

$$\frac{Q_2}{Q_1} = \frac{T}{T_1} \text{ or } Q_2 = \frac{T}{T_1} Q_1 = \frac{624.5}{1300} \times 100 = 48.038 \text{ kJ}$$

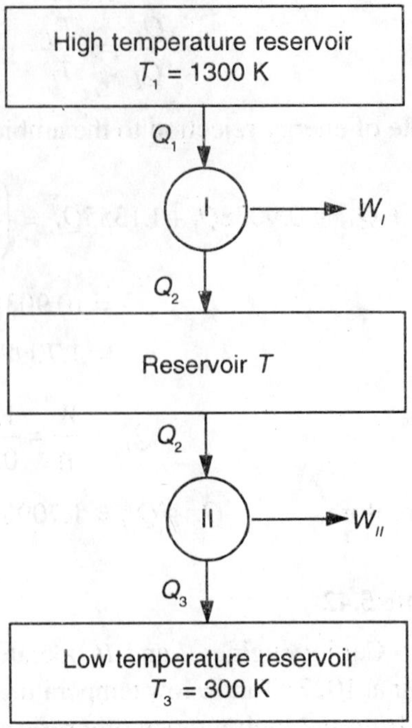


Fig.E 5.42. Sketch for Example 5.42

$$W_{II} = \eta_{II}\, Q_2 = 0.5196 \times 48.038 = 24.96 \text{ kJ}$$

**Example 5.43**

Two reversible engines are operating between three thermal reservoirs at temperature $T_1$,$T_2$ and $T_3$, respectively. The first engine operates with the reservoirs at $T_1$ and $T_2$ while the second engine operates between the reservoirs at $T_2$ and $T_3$. The second engine absorbs energy from the reservoir at $T_2$ such that the reservoir is unaffected. Determine a relation among $T_1$,$T_2$ and $T_3$ if: (*a*) the efficiencies of both the engines are same (*b*) the work output of both the engines are same.

*Solution* :

A schematic of the layout of the engines is shown in Fig.E 5.43.

(*a*) $\eta_I = 1 - \dfrac{T_2}{T_1};\ \eta_{II} = 1 - \dfrac{T_3}{T_2}$

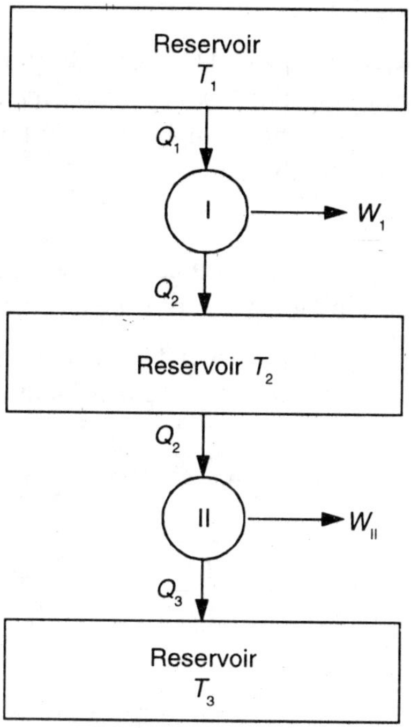


Fig.E 5.43. Sketch for Example 5.43.

$$\eta_I = \eta_{II} \text{ gives } \frac{T_2}{T_1} = \frac{T_3}{T_2} \quad \text{or} \quad T_2 = \sqrt{T_1 T_3}$$

$$(b)\ W_I = W_{II} \ \text{or}\ \eta_I Q_1 = \eta_{II} Q_2 \ \text{or}\ \eta_I \frac{Q_1}{Q_2} = \eta_{II} \qquad (A)$$

The second engine absorbs energy $Q_2$ from the reservoir at $T_2$ such that the reservoir is unaffected. That is the engine *II* absorbs the same amount of energy as the engine *I* rejects to the reservoir.

$$\frac{Q_1}{Q_2} = \frac{T_1}{T_2} \qquad (B)$$

Substitute Eqn. (*B*) in Eqn. (*A*), to obtain

$$\left(1 - \frac{T_2}{T_1}\right)\frac{T_1}{T_2} = 1 - \frac{T_3}{T_2} \quad \text{or} \quad \frac{T_1}{T_2} - 1 = 1 - \frac{T_3}{T_2}$$

$$\text{or} \quad \frac{T_1}{T_2} + \frac{T_3}{T_2} = 2 \quad \text{or} \quad T_2 = \frac{T_1 + T_3}{2}$$

**Example 5.44**

The efficiency of a Carnot heat engine can be increased either by increasing the source temperature while the sink temperature is held constant or by decreasing the sink temperature while the source temperature is held constant. Which of the above two possibilities is more effective ?

*Solution* :

Let $\eta_I$ = efficiency when the source temperature is increased by $\Delta T$ ($\Delta T > 0$) while the sink temperature is held constant.

$\eta_{II}$ = efficiency when the sink temperature is decreased by $\Delta T$ ($\Delta T > 0$) while the source temperature is held constant.

$$\eta_I = 1 - \frac{T_2}{T_1 + \Delta T} \quad \text{and} \quad \eta_{II} = 1 - \frac{T_2 - \Delta T}{T_1}$$

$$\eta_I - \eta_{II} = 1 - \frac{T_2}{T_1 + \Delta T} - 1 + \frac{T_2 - \Delta T}{T_1} = \frac{T_2 - \Delta T}{T_I} - \frac{T_2}{T_1 + \Delta T}$$

or
$$\eta_I - \eta_{II} = \frac{T_1 T_2 - T_1 \Delta T + T_2 \Delta T - (\Delta T)^2 - T_1 T_2}{T_1 (T_1 + \Delta T)}$$

$$= \frac{\Delta T (T_2 - T_1) - (\Delta T)^2}{T_1 (T_1 + \Delta T)} < 0$$

(The denominator is always positive. $T_2 - T_1 < 0$ and $\Delta T > 0$. Hence the numerator is negative.)

or $\eta_{II} > \eta_I$

Therefore, decreasing the sink temperature while the source temperature is held constant is more effective in increasing the efficiency of a Carnot engine.

**Example 5.45**

The COP of a Carnot refrigerator can be increased either by decreasing the temperature of the high temperature reservoir, while the low temperature reservoir is held at constant temperature or by increasing the temperature of the low temperature reservoir while the high temperature reservoir is held at constant temperature. Determine which of the above two possibilities is more effective.

*Solution* :

Let $(COP)_I$ = COP of refrigerator if the temperature of the high temperature reservoir is decreased by $\Delta T$ ($\Delta T > 0$) while the low temperature reservoir is held at constant temperature.

$(COP)_2$ = COP of refrigerator if the temperature of the low temperature reservoir is increased by $\Delta T$ ($\Delta T > 0$) while the high temperature reservoir is held at constant temperature.

Then $$(COP)_1 = \frac{T_L}{(T_H - \Delta T) - T_L} \quad \text{and}$$

$$(COP)_2 = \frac{T_L + \Delta T}{T_H - (T_L + \Delta T)}$$

$$(COP)_1 - (COP)_2 = \frac{T_L}{T_H - \Delta T - T_L} - \frac{T_L + \Delta T}{T_H - T_L - \Delta T} = \frac{-\Delta T}{T_H - T_L - \Delta T} < 0$$

Therefore, $(COP)_2 > (COP)_1$. That is, to increase the COP of a refrigerator it will be better to increase the temperature of the low temperature reservoir while the temperature of the high temperature reservoir is held constant.

### Example 5.46

It is desired to maintain an auditorium at 25°C throughout the year. Hence, it is planned to use a reversible device which can be used as a refrigerator in summer and as a heat pump in winter. The ambient temperature in summer is expected to reach 45°C and in winter the minimum temperature may touch 3°C.The energy loss through the walls, roof and doors is estimated at 50 kJ/s. Determine the minimum power required to operate the device in summer and in winter.

*Solution* :

A schematic of the reversible device which is used as a refrigerator in summer and as a heat pump in winter is shown in Fig.E 5.46.

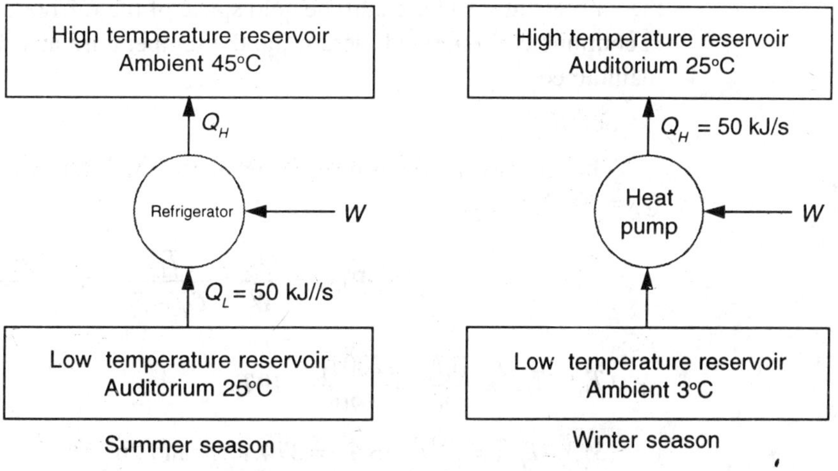


Fig.E 5.46. Sketch for Example 5.46.

**Summer season :**

Energy to be extracted from auditorium, $Q_L$= 50 kJ/s.

Ambient atmosphere acts as high temperature reservoir and the auditorium is the low temperature reservoir.

Therefore, $T_H$ = 318 K; $T_L$ = 298 K

$$(\text{COP})_{\text{Ref}} = \frac{Q_L}{W} = \frac{T_L}{T_H - T_L} \quad \text{or } W = \frac{Q_L(T_H - T_L)}{T_L}$$

$$\text{or} \quad W = \frac{50(318-298)}{298} = 3.356 \text{ kW}$$

**Winter season :**

The ambient atmosphere acts as low temperature reservoir, while the auditorium is high temperature reservoir. Therefore,

$$T_H = 298 \text{ K}; T_L = 276 \text{ K}; Q_H = 50 \text{ kJ/s}$$

$$(\text{COP})_{\text{Heat pump}} = \frac{Q_H}{W} = \frac{T_H}{T_H - T_L} \quad \text{or} \quad W = \frac{Q_H(T_H - T_L)}{T_H}$$

$$\text{or} \quad W = \frac{50(298-276)}{298} = 3.691 \text{ kW}$$

**Example 5.47**

A Carnot refrigerator consumes 200 W power in summer when the ambient atmosphere is at 40°C. The rate of energy leak into the refrigerator through the doors is estimated as 20 W per degree Celsius temperature difference between the ambient atmosphere and the cold space of the refrigerator. If the refrigerator is continuously operated, determine the temperature at which the cold space is maintained.

*Solution* :

The energy to be removed as heat from the cold space is given by $Q_L = 20\,(T_H - T_L)$

$$(\text{COP})_{\text{Ref}} = \frac{Q_L}{W} = \frac{T_L}{T_H - T_L} \quad \text{or} \quad \frac{20(T_H - T_L)}{W} = \frac{T_L}{(T_H - T_L)}$$

$$\text{or} \quad (T_H - T_L)^2 = \frac{WT_L}{20} = \frac{200T_L}{20} = 10T_L$$

$$\text{or} \quad (313 - T_L)^2 = 10\,T_L \text{ or } T_L = 374.17 \text{ K and } 261.83 \text{ K}$$

$T_L = 374.17$ K is in admissible, since it is greater then $T_H$.

Therefore, the cold space temperature $T_L = 261.83$ K or $-11.17$°C

**Example 5.48**

Show that $\Sigma\, Q/T \leq 0$ for any engine, where $Q$ is the heat interaction, (including the sign convention), with the reservoir at temperature $T$.

*Solution* :

We know that for a reversible engine

$$\eta = 1 - \frac{|Q_2|}{|Q_1|} = 1 - \frac{T_2}{T_1} \quad \text{or} \quad \frac{|Q_2|}{|Q_1|} = \frac{T_2}{T_1}$$

$$\text{or} \quad \frac{|Q_1|}{T_1} = \frac{|Q_2|}{T_2} \quad \text{or} \quad \frac{|Q_1|}{T_1} - \frac{|Q_2|}{T_2} = 0$$

Now, if we take into account the sign convention for heat interaction, the above expression can be written as

$$\frac{Q_1}{T_1} - \frac{(-Q_2)}{T_2} = 0 \quad \text{or} \quad \frac{Q_1}{T_1} + \frac{Q_2}{T_2} = 0$$

$$\text{or} \quad \Sigma \frac{Q}{T} = 0 \text{ for a reversible engine} \qquad (A)$$

Now let us consider an irreversible engine which receives energy $|Q_1|$ as heat from a reservoir at $T_1$ and rejects energy $|Q'_2|$ as heat to a reservoir at $T_2$. Since the efficiency $\eta$ of the irreversible engine is less then the efficiency $\eta$ of the reversible engine, we have

$$1 - \frac{|Q'_2|}{|Q_1|} < 1 - \frac{T_2}{T_1} \quad \text{or} \quad \frac{|Q'_2|}{|Q_1|} > \frac{T_2}{T_1}$$

$$\text{or} \quad \frac{|Q_1|}{T_1} < \frac{|Q'_2|}{T_2} \quad \text{or} \quad \frac{|Q_1|}{T_1} - \frac{|Q'_2|}{T_2} < 0$$

Taking the sign convention for heat interaction into account, the above expression can be written as

$$\frac{Q_1}{T_1} - \frac{(-Q'_2)}{T_2} < 0$$

$$\text{or} \quad \frac{Q_1}{T_1} + \frac{Q'_2}{T_2} < 0$$

or $\sum \frac{Q}{T} < 0$ for an irrevesible engine (*B*)

Eqns. (*A*) and (*B*) can be expressed in a general form as

$$\sum \frac{Q}{T} \leq 0 \text{ for any engine.}$$

**Example 5.49**

Show that any reversible process can be approximated by a series of reversible–isothermal and reversible–adiabatic processes.

*Solution* :

Consider a system undergoing a reversible process 1–*b*–2 as shown in Fig.E 5.49. Draw a reversible adiabatic 1–*a*, a reversible– isothermal *abc* and a reversible adiabatic 2–*c*. Suppose the combination of the processes 1–*a, a–b–c* and c–2 are such that the areas enclosed by 1–*a*–*b* and *b*–*c*–2 are equal on a *P–V* diagram. Then the actual reversible process 1–*b*–2 can be replaced by the combination of the above processes, resulting in the overall process 1–*a*–*b*– c–2. The processes 1*b*2 an 1*abc*2 are equivalent if and only if the heat and work interactions are identical for both the processes. To verify whether the processes are identical or not, let us apply the first law of thermodynamics for these two processes.

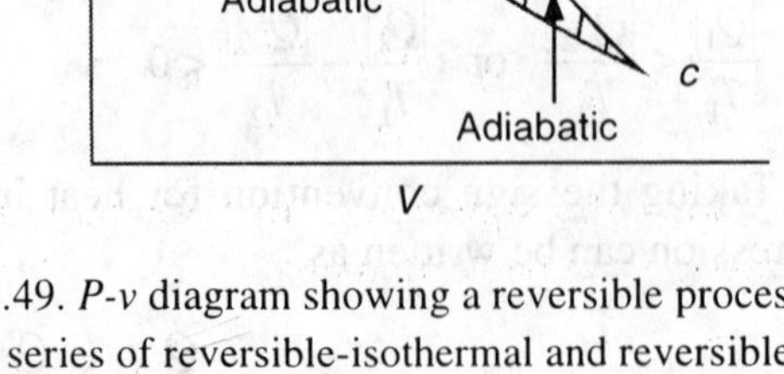


Fig.E 5.49. *P-v* diagram showing a reversible process 1-*b*-2 can be approximated as a series of reversible-isothermal and reversible-adiabatic processes.

For the process 1*abc*2, we get

$$U_2 - U_1 = Q_{1abc2} - W_{1abc2} = Q_{1a} + Q_{abc} + Q_{c2} - W_{1abc2}$$
$$= Q_{abc} - W_{1abc2} \quad (A)$$

(Since $1a$ and $c2$ are reversible– adiabatic processes)

For the process $1–b–2$ we get

$$U_2 - U_1 = Q_{1b2} - W_{1b2} \quad (B)$$

Now, consider the cycle $1–a–b–c–2–b–1$. The net work done in this cycle is zero since the area enclosed by $1–a–b$ is equal to the area enclosed by $b–c–2$ on a $P–V$ diagram. That is

$$\oint dW = W_{1abc2} + W_{2b1} = 0$$

or $W_{1abc2} = W_{1b2}$ $\quad (C)$

That is, the work done during the process $1–b–2$ is equal to the work done for the process $1–a–b–c–2$. From Eqns. $(A)$–$(C)$, we find

$$Q_{1abc2} = Q_{abc} = Q_{1b2}.$$

That is, the heat interactions along the processes $1b2$ and $1abc2$ are identical. Since work and heat interactions along the original reversible process $1–b–2$ are the same as those of the modified process $1–a–b–c–2$, the process $1–b–2$ can be replaced by the process $1–a–b–c–2$.

**Example 5.50**

State and prove the Clausius inequality.

*Solution* :

The Clausius inequality states that whenever a system undergoes a cyclic change the algebraic sum of the ratios of heat interactions to the temperature at which the heat interaction occurs, over the complete cycle is less than or equal to zero. Stated mathematically $\oint dQ/T \leq 0$. The equality sign holds good if the cycle consists of only reversible processes.

To prove the Clausius inequality, consider a system which undergoes the cyclic change as shown in Fig.E 5.50. Now, draw a family of closely spaced adiabatic lines similar to $a–3$ and $2–f$ covering the entire cycle. Then join the adjacent adiabatic lines $a–3$ and $2–f$ by the isotherms $a–b–c$ and $d–e–f$ as shown in Fig.E 5.50, such that the area of $1ab$ = area of $bc2$ and area of $3de$ = area of $ef4$.

Thus the segments $1b2$ and $3e4$ of the reversible cycle are replaced by $1abc2$ and $3def4$, respectively. Now the differential cycle $abc4feda$ is a Carnot cycle. Thus the original cycle can be transformed into a combination of several differential Carnot cycles. We know that for a Carnot cycles $\Sigma\, Q/T = 0$ or for a differential Carnot cycle $\oint dQ/T = 0$. Therefore, for the given reversible cycle, we get

$$\oint dQ/T = 0 \qquad \text{For a reversible cycle } (A)$$

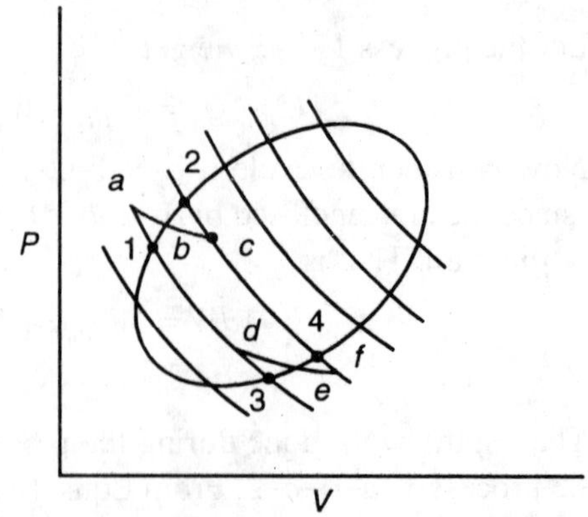


Fig.E 5.50. A reversible cycle is approximated as a combination of several Carnot cycles.

Now consider an irreversible cycle. That is, some processes of the cycle may be irreversible. We can follow the above argument and find that the given cycle splits into a large number of differential cycles of which some are irreversible. For the reversible cycles we have $\oint dQ/T = 0$ and for the irreversible cycles we have $\oint dQ/T < 0$. Then for the original irreversible cycle, we get

$$\oint dQ/T < 0 \qquad \text{For an irreversible cycle} \quad (B)$$

The Eqns. (*A*) and (*B*) can be generalized as

$$\oint dQ/T \leq 0$$

which is the Clausius inequality.

**Example 5.51**

A reversible heat engine interacts with three thermal reservoirs at 500 K, 400 K and 300 K, respectively. The engine does 300 kJ of net work and absorbs 900 kJ of energy as heat from the reservoir at 500 K. Determine the magnitude and direction of heat interactions of the engine with the other two reservoirs.

*Solution* :

A schematic of the heat engine which interacts with three thermal reservoirs is shown in Fig.E 5.51. Let $Q_1$, $Q_2$ and $Q_3$ denote the heat interactions with the reservoirs at 500 K, 400 K and 300 K, respectively with the direction of energy flow as shown in Fig.E 5.51. Application of the first law of thermodynamics to the heat engine gives

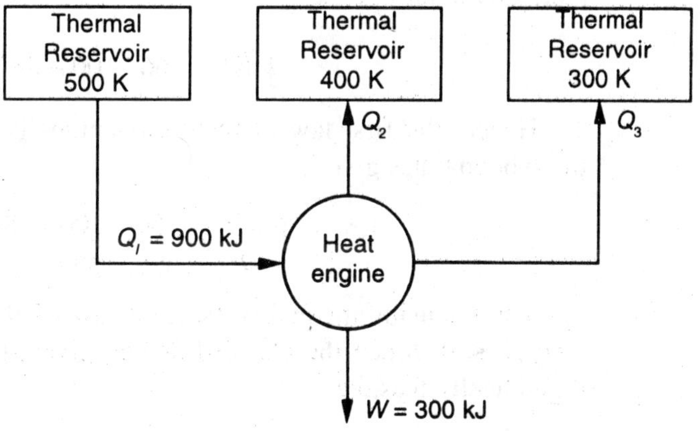


Fig.E 5.51. Sketch for Example 5.51.

$$Q_1 - Q_2 - Q_3 = W$$

or $$900 - Q_2 - Q_3 = 300 \text{ or } Q_2 + Q_3 = 600 \text{ kJ} \qquad (A)$$

Application of the second law of thermodynamics (in the form of Clausius inequality) to the reversible heat engine gives

$$\oint \frac{dQ}{T} = \frac{900}{500} - \frac{Q_2}{400} - \frac{Q_3}{300} = 0$$

or $$15\, Q_2 + 20\, Q_3 = 7200 \qquad (B)$$

Simultaneous solution of Eqns. (*A*) and (*B*) gives

$$Q_2 = 960 \text{ kJ and } Q_3 = -360 \text{ kJ}$$

That is the engine absorbs 360 kJ energy from the reservoir at 300 K and rejects 960 kJ to the reservoir at 400 K.

**Example 5.52**

An inventor claims to have designed a heat engine which absorbs 260 kJ of energy as heat from a reservoir at 52°C and delivers 72 kJ work. He also states that the engine rejects 100 kJ and 88 kJ of energy to the reservoirs at 27°C and 2°C, respectively. Judge whether the claim is acceptable or not.

*Solution* :

A schematic of the heat engine is shown in Fig.E 5.52. The claim can be verified from a thermodynamic point of view. If the proposed device satisfies the first and second laws of thermodynamics, then it is a theoretically feasible device.

If it violates any of the two laws, then it is an impossible device. The first law of thermodynamics gives

$$\oint dQ = 260 - 100 - 88 = 72 \text{ kJ} = \oint dW.$$

Hence, the first law of thermodynamics is satisfied. The second law of thermodynamics gives

$$\oint \frac{dQ}{T} = \frac{260}{325} - \frac{100}{300} - \frac{88}{275} = 0.1467 > 0$$

The Clausius inequality (second law of thermodynamics) requires that $\oint dQ/T \leq 0$. Since the claimed device gives $\oint dQ/T > 0$. The device is not theoretically feasible.

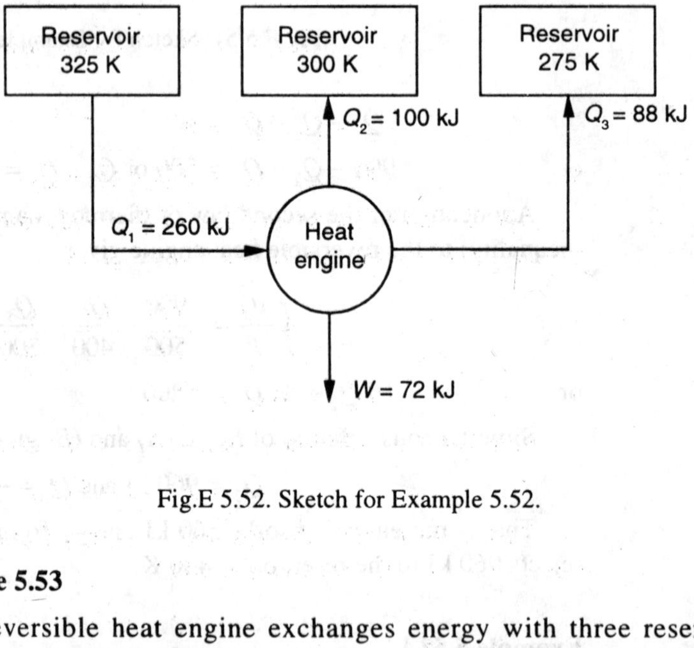


Fig.E 5.52. Sketch for Example 5.52.

**Example 5.53**

A reversible heat engine exchanges energy with three reservoirs at temperatures 1000 K, 800 K and 600 K, respectively. It receives 2000 kJ and 1200 kJ of energy from the reservoirs at 1000 K and 800 K, respectively. Calculate the energy exchanged with the third reservoir and the work output of the engine.

*Solution :*

A schematic of the reversible heat engine is shown in FigE. 5.53

The first law of thermodynamics gives

$$2000 + 1200 - Q_3 = W \quad \text{or} \quad 3200 - Q_3 = W \qquad (A)$$

The second law of thermodynamics gives

$$\frac{2000}{1000}+\frac{1200}{800}-\frac{Q_3}{600}=0$$

or $Q_3 = 2100$ kJ

From Eqn. ($A$), we get $W = 1100$ kJ

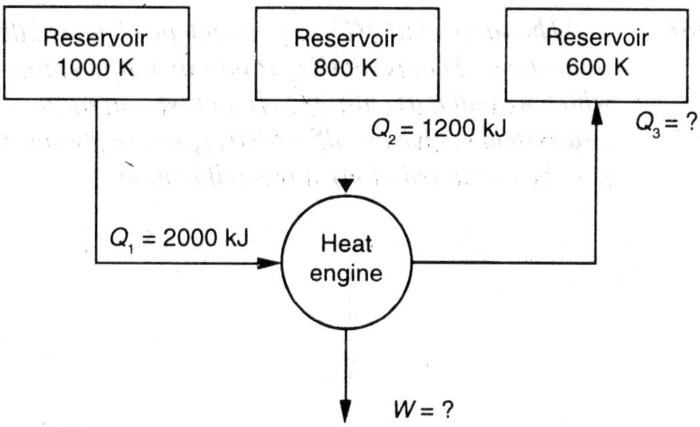


Fig.E 5.53. Sketch for Example 5.53.

**Example 5.54**

Suppose a system undergoes a change of state reversibly from the given initial state to a given final state. Show that $\int_1^2 dQ/T$ is the same for all reversible paths

*Solution :*

Suppose a system is taken from the initial state 1 to the final state 2 along the path A, reversibly and then it is restored to the initial state by following another reversible path B as shown in Fig.E 5.54. Then the system has undergone a cyclic change and it is a reversible cycle. Application of the Clausius inequality to the reversible cycle 1A2B1, we get

$$\oint \frac{dQ}{T} = \int_{1A2} \frac{dQ}{T} + \int_{2B1} \frac{dQ}{T} = 0$$

or

$$\int_{1A2} \frac{dQ}{T} = -\int_{2BI} \frac{dQ}{T} = \int_{1B2} \frac{dQ}{T}$$

That is, the value of $\int dQ/T$ is the same whether the system follows the reversible path 1A2 or the reversible path 1B2. In otherwords, the value of $\int dQ/T$ is independent of the path, as long as the path is reversible.

**Note** *The value of $(dQ/T)_R$ is path independent and depends only on the initial and final states. That is $(dQ/T)_R$ is an exact differential of some function or property which we call it as entropy S. Hence we can say that the change in the entropy dS, of a system is given by $dS = (dQ/T)_R$ where the subscript R denotes that the value is to be evaluated along a reversible path.*

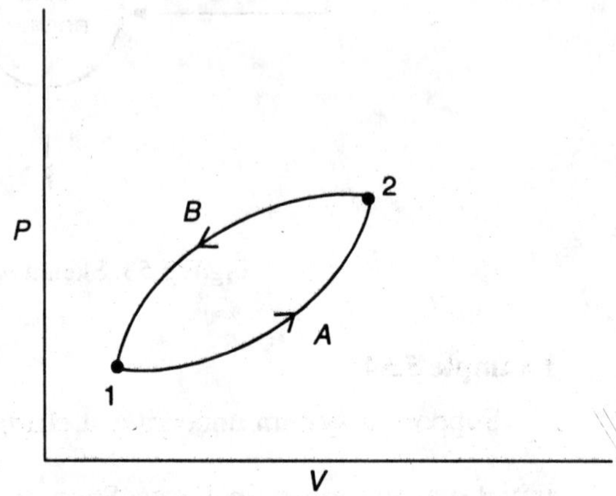


Fig.E 5.54. Two reversible paths by which a system can reach state 2 from state 1.

**Example 5.55**

Suppose a system undergoes a change of state. How do you estimate the entropy change of the system?

*Solution* :

We know that the change in entropy is given by

$$S_2 - S_1 = \int_1^2 dS = \int_1^{2} \left(\frac{dQ}{T}\right)_R$$

This relation can be used to calculate the change in entropy if the system changes from state 1 to state 2 through a reversible process only. We also know that entropy is a property or a state function. The change in property is determined by the initial and final states only and not on the path followed by the system. That is, if a system changes its state whether reversibly or irreversibly, the change in the property would be the same. Therefore, if a system undergoes an irreversible process, it is not necessary to evaluate the entropy change for the actual process. One can substitute the actual process by an imaginary reversible process connecting the initial and final states of the system and evaluate the entropy change for the imaginary reversible process by making use of the relation $\int dS = \int (dQ/T)_R$. The entropy change of the system for the actual irreversible path would be identical with that for the imaginary reversible process.

**Example 5.56**

The contents of a large constant temperature bath at 100°C is continuously stirred by a 200 W paddle wheel. Estimate the change in entropy of the bath in one second.

*Solution* :

Consider the contents of the constant temperature bath as the system. Paddle wheel work is done on the system which increases the energy of the system due to irreversible fluid friction. Clearly this process is irreversible. To estimate the entropy change of the system, one can imagine a reversible path in which an equivalent amount of energy is added to the system. Then the change in the entropy of the system is given by

$$\Delta S = \int \left(\frac{dQ}{T}\right)_R = \frac{Q}{T} = \frac{200}{373} = 0.5362 \text{ J/K}$$

**Example 5.57**

Calculate the change in entropy if :

(*a*) One kg of water at 30°C is heated to 90°C at constant pressure, by placing the container on a hot plate. The specific heat of water is 4.2 kJ/kg K.

(*b*) One kg saturated liquid at 30°C is converted into superheated steam at 1 bar and 200°C.

*Solution* :

(*a*) If water is heated from 30°C to 90°C by bringing it in contact with a hot plate, there is a large temperature difference between the water and the hot plate. Hence the process of heating is clearly irreversible. To evaluate the entropy change one can devise a reversible process by which the system is taken from the given initial state to the given final state. The water can be placed in contact with a series of thermal reservoirs, the temperatures of which change from 30°C to 90°C and add energy as heat reversibly. For this imaginary reversible process, the entropy change can be calculated through the relation $\Delta S = \int (dQ/T)_R$. We also notice that the heat interaction at constant pressure is equal to the change in the enthalpy. Hence, we can write

$$\Delta S = \int \frac{dH}{T} = \int \frac{C_p dT}{T} = C_p \ln \frac{T_2}{T_1}$$

$$= 4.2 \ln \frac{363}{303} = 0.7588 \text{ kJ / kg K}$$

(*b*) The initial state is saturated liquid at 30°C and the final state is superheated steam at 1 bar and 200°C. The process by which the change of state took place is not specified. It may be a reversible or irreversible process. Since the entropy is a state function, its change does not depend on the process followed by the system but depends only on the initial and final states of the system. Now one can read the values of entropy of water from the steam tables and obtain the following information.

Specific entropy of saturated liquid at 30°C = 0.4365 kJ/kg K

Specific entropy of superheated steam at 1 bar and 200°C = 7.8349 kJ/kg K

Therefore, $\Delta s = 7.8349 - 0.4365 = 7.3984$ kJ / kg K

**Example 5.58**

One mole of methane contained in a piston–cylinder assembly at 1 bar is heated at constant pressure from 300 K to 400 K. If the molar heat capacity of methane is given by

$$C_p = 17.449 + 60.449 \times 10^{-3}\, T + 1.117 \times 10^{-6}\, T^2 - 7.204 \times 10^{-9}\, T^3$$

Where $C_p$ is in J/mol K and $T$ is in K, calculate the change in entropy of methane.

*Solution* :

We know that for a constant pressure process $Q = \Delta H = \int C_p dT$

The entropy change, $$\Delta s = \int \frac{dq}{T} = \int \frac{dh}{T} = \int \frac{C_p dT}{T}$$

$$= \int \frac{(a + bT + cT^2 + dT^3)}{T} dT$$

or $\Delta s = a \ln \frac{T_2}{T_1} + b(T_2 - T_1) + \frac{c}{2}(T_2^2 - T_1^2) + \frac{d}{3}(T_2^3 - T_1^3)$ (A)

Substitute $a = 17.449$; $b = 60.449 \times 10^{-3}$; $c = 1.117 \times 10^{-6}$;

$d = -7.204 \times 10^{-9}$; $T_1 = 300$ and $T_2 = 400$ in Eqn (A) to obtain $\Delta s = 11.015$ J/mol K

**Example 5.59**

One mole of nitrogen contained in a rigid and insulated tank at 300 K is stirred by a paddle wheel, which is operated by a 60 W electric motor, for 2 minutes. If the molar heat capacity at constant volume for nitrogen is given by $C_v = 18.956 + 4.93 \times 10^{-3} T$ where $C_v$ is in J/mol K and $T$ is in Kelvin, calculate the change in the entropy of nitrogen.

*Solution* :

Consider the nitrogen gas as the system and apply the first law of thermodynamics to obtain

$$\Delta u = q - W = -W = 60 \times 2 \times 60 = 7200 \text{ J}$$

The process undergone by the system is irreversible. To estimate the entropy change we should know the final temperature of the gas.

$$\Delta u = \int du = \int C_v dT = \int_{T_1}^{T_2} (a + bT)\, dT$$

$$= a(T_2 - T_1) + \frac{b}{2}\left(T_2^2 - T_1^2\right)$$

or

$$7200 = 18.956\,(T_2 - 300) + \frac{4.93 \times 10^{-3}}{2}(T_2^{\,2} - 300)^2$$

or $T_2 = 638.5$ K (the second value $-$ 8328.5 K is inadmissible)

Now, the entropy change can be calculated as

$$\Delta s = \int \frac{dq}{T} = \int \frac{du}{T} = \int \frac{C_v dT}{T} = \int \frac{(a + bT)}{T} dT$$

$$= a \ln \frac{T_2}{T_1} + b(T_2 - T_1)$$

$$= 18.956 \ln \frac{638.5}{300} + 4.93 \times 10^{-3}\,(638.5 - 300)$$

or $\Delta s = 15.987$ J/mol K

### Example 5.60

The enthalpy of fusion (or the latent heat of fusion) of water at 0°C is 333.43 kJ/kg. Determine the change in entropy if one kg of ice melts to liquid water at 0°C when (*a*) ice is added to a mixture of ice and water at 0°C and energy is transferred as heat reversibly (*b*) ice is left in the surroundings which is at 30°C.

*Solution* :

(*a*) Pure substances undergo phase change at constant temperature and pressure. During the phase change energy is added at constant temperature and the process is reversible. For the reversible process the entropy change can be calculated through the relation

$$dS = \left(\frac{dQ}{T}\right)_R \quad \text{and for phase change } dQ = dH.$$

Therefore.

$$\Delta s = \frac{\Delta h}{T} = \frac{333.43 \times 10^3}{273} = 1.2214 \text{ kJ / kg K}$$

(*b*) If ice at 0°C is left in the surroundings at 30°C, the energy transfer as heat takes place with large temperature difference and the process becomes irreversible. Since entropy is a state function, its change is independent of the process and depends only on the initial and final states. The heat interaction for a constant pressure process is equal to the change in the enthalpy (the latent heat of fusion) and the enthalpy change can be determined form a knowledge of initial and final states. The entropy change in this case also is 1.2214 kJ/kg K, because it is the change in the property which does not depend on the path followed by the system.

### Example 5.61

The enthalpy of vaporization of water at 100°C and 200°C is 2256.94 kJ/kg and 1938.53 kJ/kg, respectively. Calculate the entropy change associated with the vaporization process at 100°C and 200°C and compare the results with the values given in steam tables.

*Solution* :

The vaporization of liquid occurs at constant temperature and pressure. The heat interaction for this process is equal to $h_{fg} = h_g - h_f$ = latent heat of vaporization. Then

$$\Delta s = \int \frac{dq}{T} = \frac{h_{fg}}{T}$$

$$\Delta s(\text{at } 100^\circ \text{C}) = \frac{2256.94}{373.15} = 6.0483 \text{ kJ / kg K}$$

$$\Delta s(\text{at } 200^\circ \text{C}) = \frac{1938.53}{473.15} = 4.0971 \text{ kJ / kg K}$$

The values given in the steam tables are

$\Delta s$ (at 100°C) = $s_g - s_f$ = 7.3554 – 1.3069 = 6.0485 kJ/kg K

$\Delta s$ (at 200°C) = $s_g - s_f$ = 6.4278 – 2.3307 = 4.0971 kJ/kg K

**Example 5.62**

A lump of steel of mass 10 kg at 627°C is dropped in 100 kg of oil at 30°C. The specific heats of steel and oil are 0.5 kJ/kg K and 3.5 kJ/kg K, respectively. Calculate the entropy change of steel, the oil and the universe.

*Solution* :

Let $T$ = final equilibrium temperature.

Then energy lost by steel = energy gained by oil

$$[mC_p \Delta T]_{\text{steel}} = [mC_p \Delta T]_{\text{oil}}$$

or $10 \times 0.5 \times (900 - T) = 100 \times 3.5 \times (T - 303)$

or $T = 320.43$ K

$$\Delta s_{\text{Steel}} = \int_{T_1}^{T} \frac{mC_p dT}{T} = mC_p \ln\frac{T}{T_1} = 10 \times 0.5 \ln\frac{320.43}{900}$$

$$= -5.1637 \text{ kJ/K}$$

$$\Delta s_{\text{oil}} = mC_p \ln\frac{T}{T_1} = 100 \times 3.5 \ln\frac{320.43}{303}$$

$$= 19.5759 \text{ kJ/K}$$

$$\Delta s_{\text{universe}} = \Delta s_{\text{steel}} + \Delta s_{\text{oil}} = -5.1637 + 19.5759.$$

$$= 14.4122 \text{ kJ/K}$$

**Example 5.63**

A lump of steel of mass 15 kg at 800°C is dropped in 10 kg of water at 30°C contained in an insulated container which is open to the atmosphere. If the specific heat of steel is 0.5 kJ/kg K, calculate the change in entropy of steel, water and the universe. Use steam tables for the properties of water.

*Solution* :

From steam tables we find the following data.

At 30°C : $h_f$ = 125.66 kJ/kg ; $s_f$ = 0.4365 kJ/kg K

At 100°C: $h_f$ = 419.06 kJ/kg ; $s_f$ = 1.3069 kJ/kg K

$h_g$ = 2676.0 kJ/kg ; $s_g$ = 7.3554 kJ/kg K

If the lump of steel reaches a final temperature of 100°C

The energy lost by steel = 15 × 0.5 (800 – 100) = 5250 kJ

If the water changes from 30°C to 100°C

The energy gained by water = 10 (419.06 – 125.66) = 2934 kJ

This clearly shows that some water vaporizes.

The energy available for vaporization = (5250 – 2934) = 2316 kJ

$$\text{Mass of water vaporized} = \frac{2316}{h_{fg}} = \frac{2316}{2676.0 - 419.06} = 1.026 \text{ kg}$$

Therefore, in the final state, the system is at 100°C and has 1.026 kg of saturated steam and 8.974 kg saturated liquid. The initial and final states are shown in Fig.E 5.63.

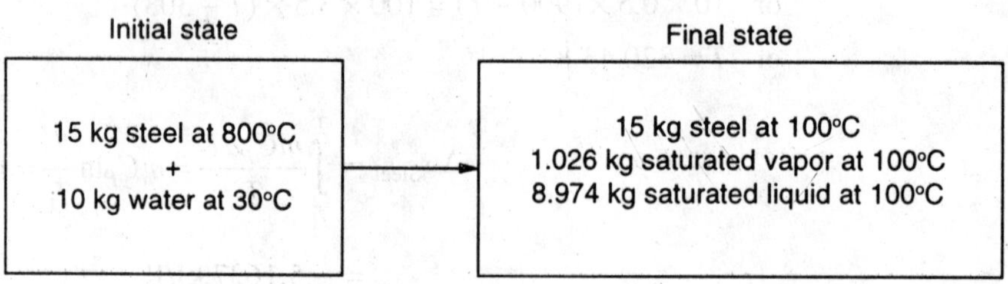


Fig.E 5.63. The initial and final states of system.

$$\Delta S_{\text{steel}} = mC_p \ln \frac{T_2}{T_1} = 15 \times 0.5 \ln \frac{373}{1073} = -7.9248 \text{ kJ/K}$$

$$\Delta S_{\text{water}} = 1.026\, s_g \text{ (at 100°C)} + 8.974\, s_f \text{(at 100°C)} - 10 s_f \text{ (at 30°C)}$$

$$= 1.026 \times 7.3554 + 8.974 \times 1.3069 - 10 \times 0.4365$$

$$= 14.9098 \text{ kJ/K}$$

$$\Delta S_{universe} = \Delta S_{steel} + \Delta S_{water} = -7.9248 + 14.9098 = 6.985 \text{ kJ/K}$$

**Example 5.64**

A mass $m$ of liquid at temperature $T_1$ is mixed with an equal amount of the same liquid at temperature $T_2$ in an insulated container. Show that the total entropy change due to mixing is given by

$$\Delta s = 2mC_P \ln \frac{(T_1 + T_2)/2}{\sqrt{T_1 T_2}}$$

Where $C_P$ is the specific heat of the liquid.

*Solution* :

Let $T$ = Final equilibrium temperature of the liquid

The principle of conservation of energy gives

$$mC_P (T - T_1) + mC_P (T - T_2) = 0 \text{ or } T = \frac{T_1 + T_2}{2}$$

Then, $$\Delta S = mC_P \ln \frac{T}{T_1} + mC_P \ln \frac{T}{T_2} = mC_P \ln \frac{T^2}{T_1 T_2}$$

$$= 2mC_P \ln \frac{T}{\sqrt{T_1 T_2}} = 2mC_P \ln \frac{(T_1 + T_2)/2}{\sqrt{T_1 T_2}}$$

**Example 5.65**

An insulated tank contains 5 kg supercooled liquid water at −10°C and 1 bar. If a crystal of ice of negligible mass is added to the water it acts as a seed for solidification and the water freezes. Given that the latent heat of fusion of water at 0°C is 333.43 kJ/kg and the specific heat of water is 4.2 kJ/kg K, calculate the mass of water frozen and the change in the entropy of water.

*Solution :*

Let $X$ = mass of water frozen into ice.

The principle of conservation of energy gives

[Energy gained by water in reaching 0°C from −10°C] = Energy released by water at 0°C during solidification of $X$ kg ice]

$$5 \times 4.2\,(273–263) = 333.43\,X$$

or $X = 0.63$ kg

The solidification process can be replaced by an imaginary path as given below for the purpose of calculating the entropy change.

(5kg water at –10°C) $\xrightarrow[\Delta S_1]{}$ (5kg water at 0°C) $\xrightarrow[\Delta S_2]{}$ (0.63kg ice a 0°C + 4.37 kg water at 0°C)

$$\Delta S_1 = 5 \times 4.2 \ln \frac{273}{263} = 0.7837 \text{ kJ / K}$$

$$\Delta S_2 = -\frac{0.63 \times 333.43}{273} = -0.7695 \text{ kJ / K}$$

$$\text{Total } \Delta S = \Delta S_1 + \Delta S_2 = 0.7837 - 0.7695 = 0.0142 \text{ kJ/K}$$

**Example 5.66**

Suppose an ideal gas initially at $(P_1, v_1, T_1)$ is changed to $(P_2, v_2, T_2)$. Devise two different convenient paths and derive relations for the estimation of entropy change.

*Solution* :

The initial state 1 at temperature $T_1$ and final state 2 at temperature $T_2$ of an ideal gas are shown in Fig.E 5.66. There are a large number of paths by which these two states can be connected. However, for the convenience of calculations, we use two paths,

First path : Constant volume heating followed by isothermal expansion.

path 1–*a*–2

Second path : constant pressure heating followed by isothermal compression

path 1–*b*–2.

For the constant volume heating process 1–*a*, the change in entropy is given by

$$\Delta s_{1a} = \int_1^a \frac{dq}{T} = \int_1^a \frac{du}{T} = \int_1^a \frac{C_v dT}{T} = C_v \ln \frac{T_2}{T_1}$$

For the isothermal expansion process *a* –2, the entropy change is given by

$$\Delta s_{a2} = \int_a^2 \frac{dq}{T} = \int_a^2 \frac{du + dW}{T} = \int_a^2 \frac{Pdv}{T} = \int_a^2 \frac{Rdv}{v} = R \ln\frac{v_2}{v_a} = R \ln\frac{v_2}{v_1}$$

($du = 0$ for an ideal gas when $T$ is constant)

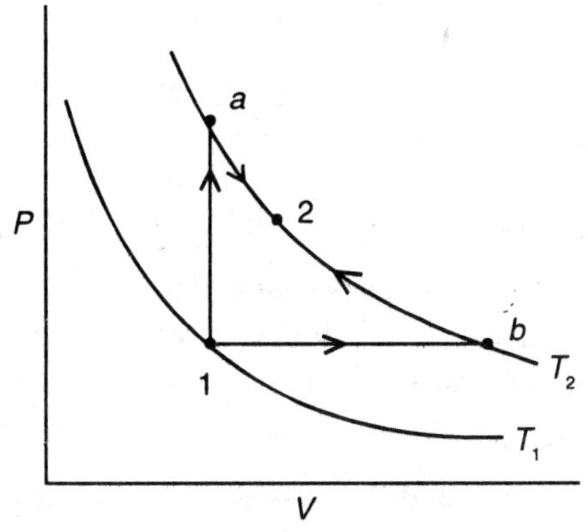


Fig.E 5.66. Two different reversible paths connecting the initial state 1 and the final state 2, for estimating entropy change.

Therefore, the entropy change for the path 1–$a$–2 is given by

$$\Delta s_{1a2} = \Delta s_{1a} + \Delta s_{a2} = C_v \ln\frac{T_2}{T_1} + R \ln\frac{v_2}{v_1}$$

For the constant pressure heating process $1b$, the change in entropy is given by

$$\Delta s_{1b} = \int_1^b \frac{dq}{T} = \int_1^b \frac{dh}{T} = \int_1^b \frac{C_p dT}{T} = C_p \ln\frac{T_2}{T_1}$$

For the isothermal compression process $b2$, the change in entropy is given by

$$\Delta s_{b2} = \int_b^2 \frac{dq}{T} = \int_b^2 \frac{du + dW}{T} = \int_b^2 \frac{Pdv}{T} = \int_b^2 \frac{Rdv}{v} = R \ln\frac{v_2}{v_b}$$

We know that for an ideal gas

$$\frac{P_2 v_2}{T_2} = \frac{P_b v_b}{T_b} \text{ or } \frac{P_2 v_2}{T_2} = \frac{P_1 v_b}{T_2} \text{ or } \frac{v_2}{v_b} = \frac{P_1}{P_2}$$

Therefore, $\Delta s_{b2} = R\ln\frac{v_2}{v_b} = R\ln\frac{P_1}{P_2} = -R\ln\frac{P_2}{P_1}$

Then the entropy change for the path $1b2$ is given by

$$\Delta s_{1b2} = \Delta s_{1b} + \Delta s_{b2} = C_P\ln\frac{T_2}{T_1} - R\ln\frac{P_2}{P_1}$$

**Example 5.67**

If an ideal gas is changed from $(P_1,v_1,T_1)$ to $(P_2,v_2,T_2)$ is it possible to obtain a relation to estimate the entropy change without devising a reversible path? If so, derive the relation.

*Solution* :

We know that the entropy change is given by $ds = (dq/T)_R$.

$$\Delta s = \int_1^2 \frac{dq}{T} = \int_1^2 \frac{du + Pdv}{T} = \int_1^2 \frac{du}{T} + \int_1^2 \frac{Pdv}{T} \qquad (A)$$

$$= \int_1^2 \frac{C_V dT}{T} + \int_1^2 \frac{Rdv}{v} = C_V\ln\frac{T_2}{T_1} + R\ln\frac{v_2}{v_1} \qquad (B)$$

We also know that $h = u + Pv$

or $dh = du + Pdv + vdP$

or $du + Pdv = dh - vdP$ $\qquad (C)$

Substituting Eqn $(C)$ in Eqn. $(A)$, we get

$$\Delta s = \int_1^2 \frac{du + Pdv}{T} = \int_1^2 \frac{dh - vdP}{T} = \int_1^2 \frac{dh}{T} - \int_1^2 \frac{vdP}{T}$$

$$= \int_1^2 \frac{C_P dT}{T} - \int_1^2 \frac{RdP}{P} = C_P\ln\frac{T_2}{T_1} - R\ln\frac{P_2}{P_1}$$

**Example 5.68**

Calculate the entropy change if one mole of an ideal gas with $\gamma = 1.4$ is changed from 1 bar and 300 K to 10 bar and 500 K.

*Solution* :

We know that the entropy change of an ideal gas is given by

$$\Delta s = C_P \ln\frac{T_2}{T_1} - R\ln\frac{P_2}{P_1} = \frac{R\gamma}{\gamma-1}\ln\frac{T_2}{T_1} - R\ln\frac{P_2}{P_1}$$

$$= \frac{8.314\times1.4}{1.4-1}\ln\frac{500}{300} - 8.314\ln\frac{10}{1} = -\,4.2792\ \text{J/mol K}$$

**Example 5.69**

Suppose one mole of an ideal gas (with $\gamma = 1.4$) at 1 bar and 300 K is compressed reversibly and adiabatically to 10 bar, calculate the final temperature and the change in entropy.

*Solution* :

If an ideal gas undergoes a reversible, adiabatic compression

$$\frac{T_2}{T_1} = \left(\frac{P_2}{P_1}\right)^{\frac{\gamma-1}{\gamma}} \text{ or } T_2 = T_1\left(\frac{P_2}{P_1}\right)^{\frac{\gamma-1}{\gamma}} = 300(10)^{\frac{0.4}{1.4}} = 579.21\ \text{K}$$

Then

$$\Delta s = C_P \ln\frac{T_2}{T_1} - R\ln\frac{P_2}{P_1} = \frac{R\gamma}{\gamma-1}\ln\frac{T_2}{T_1} - R\ln\frac{P_2}{P_1}$$

$$= \frac{8.314\times1.4}{1.4-1}\ln\frac{579.21}{300} - 8.314\ln(10) = 19.1437 - 19.1437 = 0$$

**Example 5.70**

If an ideal gas undergoes a reversible adiabatic process and changes from $P_1, T_1$ to $P_2$, $T_2$ show that the change in entropy is equal to zero.

*Solution* :

If an ideal gas undergoes a reversible adiabatic process, then

$$\frac{T_2}{T_1} = \left(\frac{P_2}{P_1}\right)^{\frac{\gamma-1}{\gamma}} \text{ or } \ln\frac{T_2}{T_1} = \frac{\gamma-1}{\gamma}\ln\frac{P_2}{P_1} \text{ or } \frac{\gamma}{\gamma-1}\ln\frac{T_2}{T_1} = \ln\frac{P_2}{P_1} \qquad (A)$$

The entropy change of an ideal gas is given by

$$\Delta s = C_P \ln\frac{T_2}{T_1} - R\ln\frac{P_2}{P_1} = \frac{R\gamma}{\gamma-1}\ln\frac{T_2}{T_1} - R\ln\frac{P_2}{P_1} \qquad (B)$$

Substituting Eqn. (*A*) in Eqn. (*B*), we get $\Delta s = 0$.

**Example 5.71**

Develop a relation to determine the entropy change associated with the mixing of non–identical gases at the same temperature and pressure.

*Solution* :

Consider a rigid and insulated tank which is divided into two compartments. One compartment holds $n_A$ moles of an ideal gas A at $T$ and $P$ while the second compartment holds $n_B$ moles of a different ideal gas $B$ at the same $T$ and $P$ as shown in Fig.E 5.71. If the partition between the two compartments is removed, the gases would mix. To calculate the entropy change associated with the mixing process let us devise the following method. Let the gases $A$ and $B$ expand reversibly and isothermally to the pressures $p_A$ and $p_B$, respectively where $p_A$ and $p_B$ denote the partial pressures of the gases $A$ and $B$, respectively in the final mixture. Then the gas $A$ enters the mixing chamber through a semipermeable membrane which allows the free passage of gas $A$ only, while the mixing chamber contains a mixture of gases $A$ and $B$ such that the partial pressures of the gases in the mixture are $p_A$ and $p_B$. The flow of the gases through the semipermeable membranes with infinitesimal difference in partial pressures is a reversible process and there is no change in the entropy of the gas. The entropy change of $n_A$ moles of gas $A$ is given by

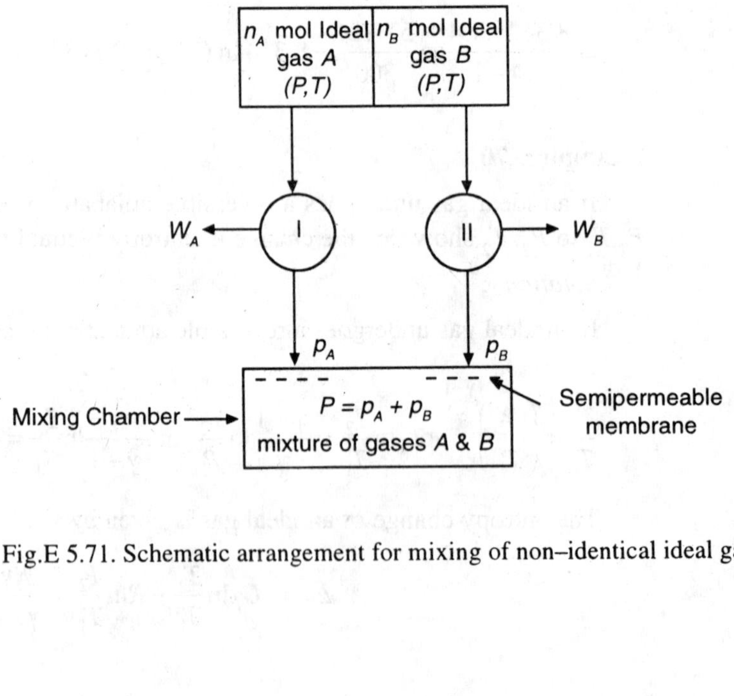


Fig.E 5.71. Schematic arrangement for mixing of non–identical ideal gases.

$$\Delta S_A = \int \frac{dQ}{T} = \int \frac{dH - VdP}{T} = -\int \frac{n_A v dP}{T} = -n_A \int \frac{RdP}{P} = -n_A R \ln \frac{p_A}{P}$$

Similarly, $\Delta S_B = -n_B R \ln (p_B/P)$

Then, the total change in entropy, $\Delta S$ is given by

$$\Delta S = \Delta S_A + \Delta S_B = -n_A R \ln \frac{p_A}{P} - n_B R \ln \frac{p_B}{P} \qquad (A)$$

Where $p_A$ and $p_B$ are partial pressures of gases $A$ and $B$, respectively in the final mixture. We know that

$$p_A = \frac{P n_A}{(n_A + n_B)} = P X_A \,;\, p_B = P X_B \qquad (B)$$

Where $X_A$ and $X_B$ denote the mole fraction of gases $A$ and $B$, respectively in the final mixture.

Substituting for $p_A$ and $p_B$ from Eqn.($B$) in Eqn.($A$) we get

$$\Delta S = -n_A R \ln X_A - n_B R \ln X_B$$

or $$\Delta s = \frac{\Delta S}{(n_A + n_B)} = -R X_A \ln X_A - R X_B \ln X_B = -R \Sigma X_i \ln X_i$$

Where $\Delta s$ is the entropy change due to mixing per mole of the mixture.

### Example 5.72

An insulated container has 21 moles of oxygen at 300 K and 1 bar. A second insulated container has 79 moles of nitrogen at the same temperature and pressure. These two containers are connected through a valve. If the valve is opened, the gases mix thoroughly. Estimate the entropy change associated with this process.

*Solution :*

The final mixture's temperature will be 300 K and the final pressure will be 1 bar. Since the gases are mixed at the same temperature and pressure, the entropy change due to mixing, $\Delta s$ is given by

$$\begin{aligned}\Delta s &= -R(X_1 \ln X_1 + X_2 \ln X_2)\\ &= -8.314\,(0.21 \ln 0.21 + 0.79 \ln 0.79)\\ &= 4.273 \text{ J/mol K}\end{aligned}$$

Where $X_1 = 0.21$ = mole fraction of oxygen in the mixture

$X_2 = 0.79$ = mole fraction of nitrogen in the mixture.

Therefore, $\Delta S = (n_1 + n_2)\, \Delta s = 427.3$ J/K

**Example 5.73**

The atomic mass of naturally available chlorine is $35.5 \times 10^{-3}$ kg/mol and is a mixture of the isotopes $^{35}Cl_2$ and $^{37}Cl_2$. Calculate the entropy of one mole of natural chlorine at 1 bar and 300 K with respect to its isotopes at the same temperature and pressure.

*Solution* :

Let $X_1$ = mole fraction of $^{35}Cl_2$ in natural chlorine. Then

$$35.5 \times 10^{-3} = 35 \times 10^{-3} X_1 + 37 \times 10^{-3}(1 - X_1)$$

or

$$X_1 = 0.75$$

The entropy of the mixture with respect to the entropy of its isotopes at the same $T$ and $P$ is equal to the entropy of mixing. Hence

$$s = \Delta s = -R \,\Sigma\, X_i \ln X_i$$

$$= -8.314\,(0.75 \ln 0.75 + 0.25 \ln 0.25)$$

$$= 4.675 \text{ J/mol K}$$

**Example 5.74**

A rigid insulated tank of volume 3 m$^3$ is divided into two compartments by a removable partition. One compartment of volume 1m$^3$ contains oxygen at 500 K and 10 bar while the second compartment of volume 2 m$^3$ is filled with nitrogen at 800 K and 20 bar. The partition is removed and the gases are allowed to mix. Assuming that both oxygen and nitrogen behave like ideal gases with $\gamma = 1.4$ determine the final temperature, pressure and the change in entropy.

*Solution* :

We know that the equation of state for an ideal gas is given by

$$PV = nRT \text{ or } n = PV/RT$$

$$n_{O_2} = \frac{10 \times 10^5 \times 1}{8.314 \times 500} = 240.558 \text{ mol}$$

$$n_{N_2} = \frac{20 \times 10^5 \times 2}{8.314 \times 800} = 601.395 \text{ mol}$$

Consider the contents (both oxygen and nitrogen) as the system. Since the system is held in a rigid and insulated tank $Q = 0$ and $W = 0$. Therefore

$$\Delta U = 0 = n_{O_2} C_{v_{O_2}} (T - 500) + n_{N_2} C_{v_{N_2}} (T - 800) \qquad (A)$$

For an ideal gas,

$$C_v = \frac{R}{\gamma - 1} = \frac{8.314}{1.4 - 1} = 20.785 \text{ J / mol K}$$

$$C_{v_{O_2}} = C_{v_{N_2}} = 20.785 \text{ J/mol K}; \; C_{P_{O_2}} = C_{P_{N_2}} = 29.099 \text{ J/mol K}$$

Substituting the values in Eqn.(*A*), we get

$$240.558 \times 20.785\,(T - 500) + 601.395 \times 20.785\,(T - 800) = 0$$

or
$$T = 714.3 \text{ K}$$

$$P = \frac{(n_{O_2} + n_{N_2})RT}{V} = \frac{(240.558 + 601.395)8.314 \times 714.3}{3} = 16.67 \text{ bar}$$

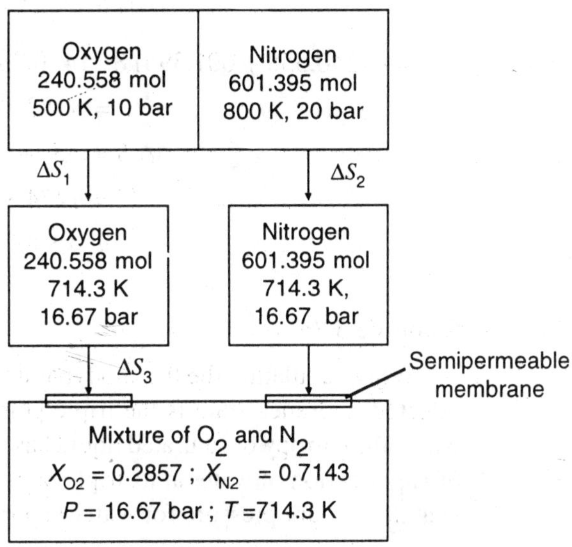


Fig.E 5.74. An imaginary convenient path to estimate the entropy change.

To calculate the entropy change let us devise a convenient path as shown in Fig.E 5.74. In the final mixture, we have

$$X_{O_2} = \frac{240.558}{240.558 + 601.395} = 0.2857;$$

$$X_{N_2} = 0.7143$$

$$\Delta S_1 = n_{O_2}\left[C_P \ln\frac{T_2}{T_1} - R\ln\frac{P_2}{P_1}\right]$$

Where subscripts 1 and 2 refer to the initial and final conditions, respectively of the species $O_2$.

$$\Delta S_1 = 240.558\left[29.099\ln\frac{714.3}{500} - 8.314\ln\frac{16.67}{10}\right] = 1474.813 \text{ J/K}$$

$$\Delta S_2 = 601.395\left[29.099\ln\frac{714.3}{800} - 8.314\ln\frac{16.67}{20}\right] = -1072.294 \text{ J/K}$$

$\Delta S_3$ = entropy change for mixing of non–identical gases at the same temperature and pressure.

$$\Delta S_3 = \left(n_{O_2} + n_{N_2}\right)\left[-R\left(X_{O_2}\ln X_{O_2} + X_{N_2}\ln X_{N_2}\right)\right]$$

$$= -(240.558 + 601.395)\, 8.314\, (0.2857 \ln 0.2857 + 0.7143 \ln 0.7143)$$

$$= 4187.794 \text{ J/K}$$

$$\Delta S = \Delta S_1 + \Delta S_2 + \Delta S_3$$

$$= 1474.813 - 1072.294 + 4187.794$$

$$= 4.5903 \text{ kJ/K}$$

**Example 5.75**

While tabulating the thermodynamic properties of water in steam tables, the selected reference state is the triple point of water ($t$ = 0.01°C, $P$ = 0.611kPa) where the entropy of saturated liquid has been assigned a zero value. The enthalpy of vaporization of water at its triple point is 2501.6 kJ/kg and the molar specific heat at constant pressure for water vapor is given by

$$C_P = 28.850 + 12.055\times10^{-3}T + \frac{1.006\times10^5}{T^2}$$

Where $C_P$ is in J/mol K and $T$ is in Kelvin. Assuming that water vapor behaves like an ideal gas estimate the entropy of dry saturated steam at 100°C and compare it with the value given is the steam tables.

*Solution* :

To calculate the entropy change let us choose the convenient path shown in Fig.E 5.75. Saturated liquid water at the triple point has the entropy $s_o = 0$. In the first step, the phase transition takes place at constant $T$ and $P$. The entropy change for the phase transition is given by

$$\Delta s_1 = \frac{h_{fg}}{T} = \frac{2501.6\times10^3}{273.16} = 9.158 \text{ kJ / kg K}$$

In the second step, the water vapor (assumed to be an ideal gas) changes from 273.16 K and 0.611 kPa to 373.15 K and 101.33 kPa and the entropy change for this process is given by

$$\Delta s_2 = \int_{T_1}^{T_2} C_P \frac{dT}{T} - R\ln\frac{P_2}{P_1}$$

$$= 28.850\ln\frac{373.15}{273.16} + 12.055\times10^{-3}\,(373.15 - 273.16)$$

$$-\frac{1.006\times10^{5}}{2}\left(\frac{1}{373.15^2} - \frac{1}{273.16^2}\right) - 8.314\ln\frac{101.33}{0.611}$$

$$= -\,31.976 \text{ J/mol K} = -1.7764 \text{ kJ/kg K}$$

Then

$s = s_0 + \Delta s_1 + \Delta s_2 = 0 + 9.158 - 1.7764 = 7.3816$ kJ/kg K

Therefore, the entropy of dry saturated steam at 100°C = 7.3816 kJ/kg K

The value reported in the steam tables = 7.3554 kJ / kg K

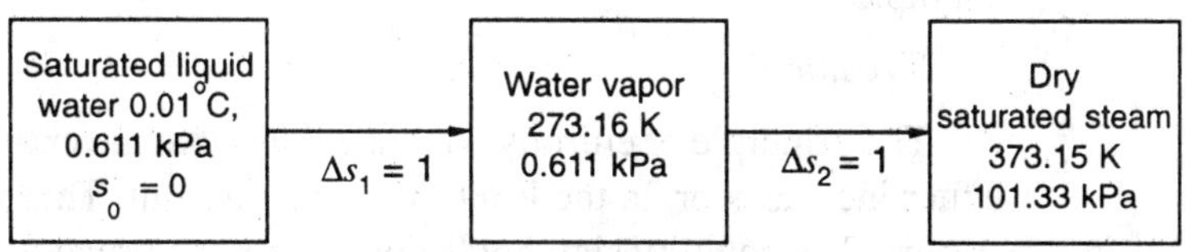


Fig.E 5.75. Sketch for Example 5.75.

## Example 5.76

Calculate the entropy change if one kg supercooled liquid water at −10°C and 1 bar is converted into ice at −10°C and 1 bar. The specific heats at constant pressure for water and the ice are 4.2 kJ/kg K and 2.1 kJ/kg K, respectively and the latent heat of fusion of water at 0°C is 333.43 kJ/kg

*Solution* :

To calculate the entropy change let us devise the convenient path shown in Fig.E 5.76 which indicates the following three steps:

1. Heating of water at constant pressure from −10°C to 0°C ($\Delta s_1$)
2. Freezing of water at 0°C ($\Delta s_2$)
3. Cooling of ice from 0°C to −10°C ($\Delta s_3$)

$$\Delta s_1 = C_P\ln\frac{T_2}{T_1} = 4.2\ln\frac{273.15}{263.15} = 0.1566 \text{ kJ / kg K}$$

$$\Delta s_2 = \frac{-333.43}{273.15} = -1.2207 \text{ kJ / kg K}$$

$$\Delta s_3 = 2.1 \ln \frac{263.15}{273.15} = -0.0783 \text{ kJ / kg K}$$

$$\Delta s = \Delta s_1 + \Delta s_2 + \Delta s_3 = 0.1566 - 1.2207 - 0.0783$$
$$= -1.1424 \text{ kJ/kg K}$$

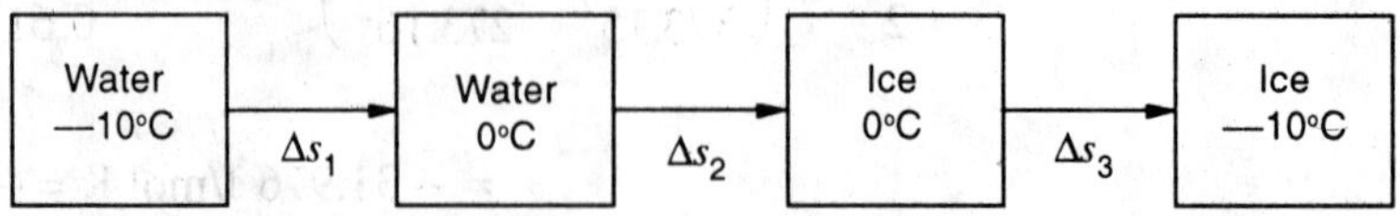


FigE. 5.76. Sketch for Example 5.76.

**Example 5.77**

State the principle of entropy increase and explain it with the help of an example.

*Solution* :

The principle of entropy increase states that the entropy of an isolated system either increases or, in the limit, remains constant. That is, $\Delta s \geq 0$ for an isolated system. The equality sign holds good when the process undergone by an isolated system is reversible and the inequality sign holds good if there is any irreversibility in the process.

Consider an isolated system, consisting of a system and its surroundings as shown in Fig.E 5.77.

Let $dQ$ = energy transferred as heat from system to surroundings

$T_{sys}$ = Temperature of the system.

$T_{sur}$ = Temperature of surroundings

If $T_{sys} > T_{sur}$, the energy transfer as heat from system to surroundings occurs irreversibly. Then the entropy changes of system and surroundings are given by

$\Delta S_{sys} = -dQ/T_{sys}$

$\Delta S_{sur} = dQ/T_{sur}$

$\Delta S$ (isolated system or universe) $= \Delta S_{sys} + \Delta S_{sur}$

$$= dQ\left(\frac{1}{T_{sur}} - \frac{1}{T_{sys}}\right) > 0$$

That is $\Delta S > 0$, if the isolated system undergoes an irreversible process, If $T_{sys} = T_{sur}$, then the process is reversible and we get $\Delta S$ (isolated system or universe) = 0. Therefore, we can generalize the statement as

$\Delta S \geq 0$ for an isolated system.

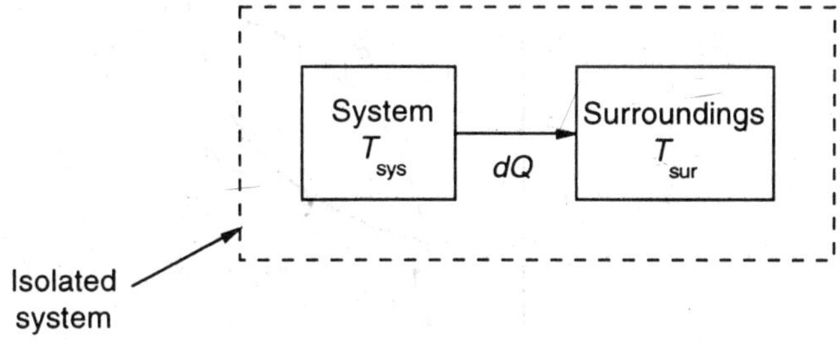


Fig.E 5.77. Sketch for Example 5.77.

**Example 5.78**

Apply the Clausius inequality for a system which undergoes an irreversible cyclic change and show that the entropy change of the system is given by $dS \geq dQ/T$.

*Solution* :

Let a system change from state 1 to state 2 by a reversible process A and return to the original state through an irreversible process B as shown in Fig.E 5.78. Then 1A2B1 constitutes an irreversible cycle. Apply the Clausius inequality to this cycle to obtain

$$\oint \frac{dQ}{T} = \int_{1A2} \frac{dQ}{T} + \int_{2B1} \frac{dQ}{T} < 0$$

We know that 1A2 is a reversible process and for a reversible process $\int (dQ/T)_R = dS$. Hence, the above equation can be rewritten as

$$S_2 - S_1 + \int_{2B1} \frac{dQ}{T} < 0$$

or

$$S_2 - S_1 > \int_{1B2} \frac{dQ}{T} \quad \text{or} \quad dS > \frac{dQ}{T}$$

That is, the entropy change of a system undergoing an irreversible process is given by $dS > dQ/T$ where $dQ$ is the heat interaction during the irreversible process. Now, one can generalize and state that

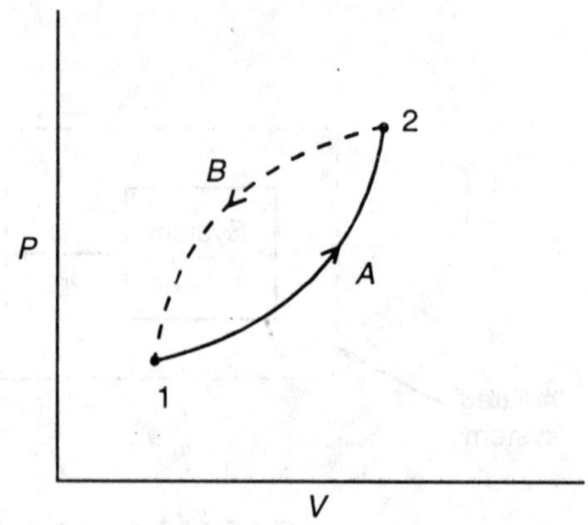


Fig.E 5.78. An irreversible cycle on a *P-v* diagram.

$$dS \geq dQ/T$$

Where the equality sign is true only when the system undergoes a reversible process.

**Example 5.79**

One kg superheated steam at 1 bar and 300°C contained in a piston–cylinder assembly is kept at ambient conditions of 27°C and the steam is allowed to condense to saturated liquid at 1 bar due to radiation to the surroundings. Calculate the change in the entropy of steam, the surroundings and the universe.

*Solution* :

For steam at 1 bar and 300°C;

$$h_1 = 3074.5 \text{ kJ/kg}; \; s = 8.2166 \text{ kJ/kg K}$$

Final state of steam is saturated liquid at 1 bar.

$$h_2 = 417.54 \text{ kJ/kg}; \; s_2 = 1.3027 \text{ kJ/kg K}$$

The pressure of the steam is held constant and hence the steam has undergone a constant pressure process. For a constant pressure process, the heat interaction is equal to the change in the enthalpy of the system. Hence

$$q = h_2 - h_1 = 417.54 - 3074.5$$
$$= -2656.96 \text{ kJ}$$

The change in the entropy of steam ($\Delta s$) is given by

$\Delta s_{steam} = s_2 - s_1 = 1.3027 - 8.2166 = -6.9139$ kJ/K

The surroundings are at 300 K. The change in the entropy of the surroundings is given by

$$\Delta s_{sur} = \frac{q}{T} = \frac{2656.96 \times 10^3}{300} = 8.8563 \text{ kJ/K}$$

$$\Delta s_{universe} = \Delta S_{steam} + \Delta S_{sur} = -6.9139 + 8.8563$$
$$= 1.9424 \text{ kJ/K}$$

**Example 5.80**

Sketch the temperature– entropy diagram, for a fluid, showing the essential features.

*Solution* :

While analyzing the processes from the second law of thermodynamics point of view, it is frequently helpful to use the temperature–entropy diagram on which one can locate the initial and final states of the system and follow the process. The main features of the $T$–$s$ diagram are shown in Fig.E 5.80. The constant quality lines, constant pressure lines and constant enthalpy lines are shown in Fig.E 5.80. We know that $dS = (dQ/T)_R$. Therefore, if a system undergoes a reversible process $dQ = TdS$. That is, the heat interaction during a reversible process is equal to the area under the process curve on a $T$–$s$ diagram.

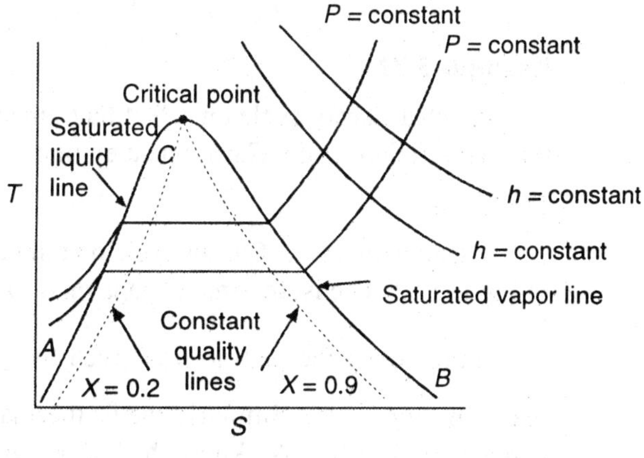


FigE. 5.80. Schematic of temperature-entropy diagram.

**Example 5.81**

Combine the two laws of thermodynamics and show that

$$dU = TdS - PdV$$

Is the above relation true for all processes?

*Solution* :

The first law of thermodynamics gives

$$dU = dQ - dW \tag{A}$$

Consider a system which undergoes a reversible process.

Then $$dW = PdV.$$

The second law of thermodynamics gives $dS = (dQ/T)_R$ or for a reversible process $dQ = TdS$. Substitute for $dW$ and $dQ$ in Eqn. (*A*) to get

$$dU = TdS - PdV \tag{B}$$

While deriving the Eqn. (*B*) we have assumed that the system is undergoing a reversible process. Hence Eqn. (*B*) is valid for reversible processes. Having stated this let us closely look at the Eqn (*B*). It expresses the change in the property ($dU$) in terms of the changes in other properties entropy and volume. Since $U$ is a property its differential is exact and the change in the value of the property does not depend on the path followed by the system. The change $\Delta U$ remains to be the same, irrespective of whether the path is reversible or irreversible, as long as the initial and final states of the system are fixed. Hence Eqn. (*B*) is true for all processes. However, in the case of irreversible processes $\int TdS$ does not represent the heat interaction and $\int PdV$ does not represent the work done.

**Example 5.82**

Sketch a Carnot cycle on a $T$–$s$ diagram and obtain a relation to calculate the thermal efficiency of a Carnot heat engine.

*Solution* :

A schematic of the Carnot cycle on temperature – entropy diagram is shown in Fig.E 5.82. The isothermal expansion $a$–$b$ occurs at temperature $T_1$. The heat interaction along the path $a$–$b$ is given by $Q_1 = \int_a^b T_1 dS$. That is $Q_1$ is the area under the curve $a$–$b$. Similarly the isothermal compression process is shown as $c$–$d$ in Fig.E 5.82. We know that $dS = (dQ/T)R$. For a reversible –adiabatic process $dS = 0$, since $dQ = 0$. That is a reversible–adiabatic process is an isentropic process. The isentropic processes are shown as $b$–$c$ and $d$–$a$ in Fig.E 5.82. Then

$$|Q_1| = T_1(S_b - S_a); |Q_2| = T_2(S_c - S_d) = T_2(S_b - S_a)$$

$$\eta = \frac{|Q_1| - |Q_2|}{|Q_1|}$$

$$= \frac{T_1(S_b - S_a) - T_2(S_b - S_a)}{T_1(S_b - S_a)} = \frac{T_1 - T_2}{T_1}$$

$$= 1 - \frac{T_2}{T_1}$$

***Note*** *The two laws of thermodynamics combined together gives*

$$dU = TdS - PdV$$

*Therefore, for a cyclic process the above equation gives*

$$\oint dU = 0 = \oint TdS - \oint PdV \text{ or } \oint TdS = \oint PdV$$

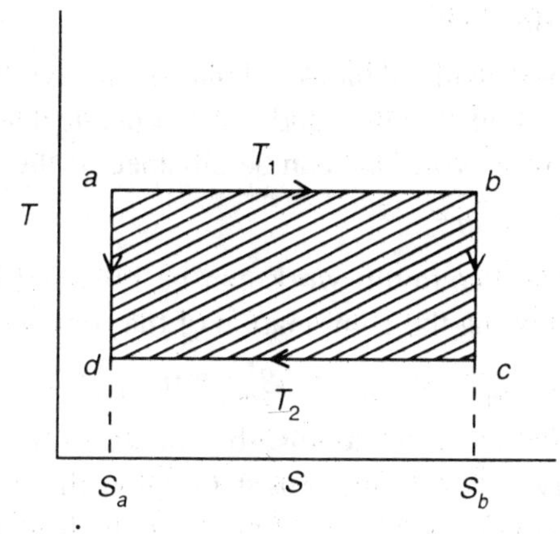


Fig.E 5.82. Carnot cycle on a temperature-entropy diagram.

**Example 5.83**

Atmospheric air is a mixture of 21 moles of oxygen for every 79 moles of nitrogen. It is desired to separate 100 moles of air at 300 K and 1 bar into its pure components at 300 K and 1 bar. Calculate the minimum amount of work to be done for this purpose. Treat $O_2$ and $N_2$ as ideal gases.

*Solution* :

We know that $dU = TdS - PdV$

The initial and final states of $O_2$ and $N_2$ are at the same temperature and $O_2$ and $N_2$ behave like ideal gases. Hence $\Delta U = 0$. Therefore, we have

$$TdS = PdV \text{ or } T\Delta S = \int PdV = W$$

We also know that the entropy change due to mixing is given by

$$\Delta s_{mixing} = s_{mixture} - \Sigma S_{pure\ gases} = -R\Sigma X_i \ln X_i$$

Where $X_i$ is the mole fraction of component $i$.

The separation of a mixture into its pure components is the reverse of mixing. Hence $\Delta s$ for separation is

$$\Delta s_{separation} = -\Delta s_{mixing} = R\,\Sigma X_i \ln X_i$$

Then work done, $W = T\,\Delta s = RT\,\Sigma\, X_i \ln X_i$ per mole

or $W = nRT\Sigma X_i \ln X_i$

$$= 100 \times 8.314 \times 300(0.21 \ln 0.21 + 0.79 \ln 0.79)$$

$$= -128.191 \text{ kJ}$$

Hence work to be done = 128.191 kJ

**Example 5.84**

Two identical blocks of mass $m$ are available at temperatures $T_1$ and $T_2$. They can be used as source and sink to operate a heat engine. Determine the maximum amount of work that can be obtained, if the specific heats of the blocks are $C$.

*Solution :*

The maximum work can be obtained by using a reversible heat engine. Considering the combination of the source and sink as the universe, we have

$$\Delta S_{uni} = \Delta S_{Source} + \Delta S_{sink} = 0 \qquad \text{(for reversible process)}$$

The heat engine absorbs energy as heat from the source at $T_1$ and rejects energy as heat to the sink at $T_2$. Since the source and sink are finite bodies, their temperatures keep changing when the heat engine works. The engine continues to deliver work as long as there is a difference in temperature between the source and sink. Let $T$ be the final temperature of the source and sink. Then

$$\Delta S_{source} = \int \frac{dQ}{T} = \int_{T_1}^{T} \frac{mCdT}{T} = mC \ln \frac{T}{T_1}$$

$$\Delta S_{sink} = \int \frac{dQ}{T} = \int_{T_2}^{T} \frac{mCdT}{T} = mC \ln \frac{T}{T_2}$$

$$\Delta S_{uni} = mC \ln \frac{T}{T_1} + mC \ln \frac{T}{T_2} = mC \ln \frac{T^2}{T_1 T_2} = 0$$

or $$T = \sqrt{T_1 T_2}$$

The work done, $$W = mC(T_1 - T) - mC(T - T_2) = mC(T_1 + T_2 - 2T)$$

$$= mC(T_1 + T_2 - 2\sqrt{T_1 T_2}) = (mC\sqrt{T_1} - \sqrt{T_2})^2$$

**Example 5.85**

Apply the second law of thermodynamics for a control volume and deduce a relation which can be used for the second law analysis of flow processes.

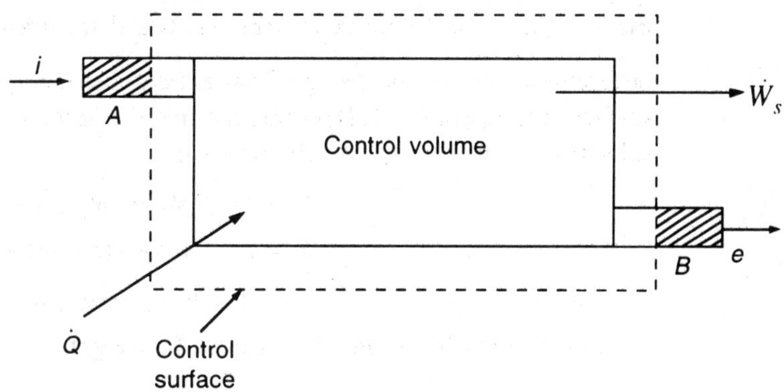


Fig.E 5.85. Schematic of a control-volume.

*Solution* :

Consider the control volume shown in Fig.E 5.85, through which mass enters at the rate $\dot{m}_i$ at the inlet $i$ and mass leaves the exit $e$ at the rate $\dot{m}_e$

Let $s_i$, $s_e$ = specific entropy of matter at the inlet and exit of the control volume, respectively

$\dot{Q}$ = rate of energy flow as heat into the control volume

$m(t)$ = mass inside the control volume at time $t$

$m(t + dt)$ = mass inside the control volume at time $(t + dt)$

$s(t)$ = specific entropy of matter at time $t$ in the control volume

$s(t + dt)$ = specific entropy of matter at time $(t + dt)$ in the control volume

$S(t)$ = entropy of matter in the control volume at time $t$

$S(t + dt)$ = entropy of matter in the control volume at time $(t + dt)$

We are familiar with the application of the second law of thermodynamics to a system and observed that whenever a system undergoes a change of state the entropy change of the system is given by

$$dS \geq \frac{dQ}{T} \quad \text{or} \quad S_2 - S_1 \geq \int \frac{dQ}{T} \tag{A}$$

In order to apply the above form of the second law of thermodynamics, we have to identify a system such that its mass remains constant all the time.

Therefore, let us choose the system at time $t$ as the mass inside the control volume and the mass inside the region marked as $A$ in Fig.E 5.85, which is about to enter the control volume in a differential time $dt$. During the differential time $dt$, the mass in the region $A$ enters the control volume and some mass leaves the control volume and occupies the region marked as $B$ in Fig.E 5.85. That is, the system configuration changes and the system at time $(t + dt)$ constitutes the mass in the control volume at time $(t+dt)$ and the mass in region $B$. During the differential time $dt$, the energy flow as heat into the system is $\dot{Q}dt$ through the surface at temperature $T$. Denoting the initial and final states of the system by the subscripts 1 and 2, respectively we have

$$S_1 = m(t)\,s(t) + \dot{m}_i s_i dt = S(t) + \dot{m}_i . s_i . dt$$

$$S_2 = m(t + dt)\, s(t + dt) + \dot{m}_e . s_e . dt$$

$$= S(t + dt) + \dot{m}_e s_e dt$$

Substituting for $S_1$ and $S_2$ in Eqn. ($A$), we get

$$[S(t + dt) + \dot{m}_e s_e dt] - [St + \dot{m}_i s_i\, dt] \geq - \dot{Q}\, dt/T$$

or $$\frac{S(t+dt) - S(t)}{dt} + \dot{m}_e s_e - \dot{m}_i s_i \geq \frac{\dot{Q}}{T}$$

or $$\frac{dS}{dT} + \dot{m}_e s_e - \dot{m}_i s_i \geq \frac{\dot{Q}}{T} \tag{B}$$

or $$\frac{dS}{dt} + \dot{m}_e s_e - \dot{m}_i s_i = \frac{\dot{Q}}{T} + \dot{S}_G$$

or $$\frac{dS}{dt} = (\dot{m}_i s_i + \frac{\dot{Q}}{T}) - \dot{m}_e s_e + \dot{S}_G \tag{C}$$

where $\dot{S}_G$ is the rate of entropy generation which cannot assume a negative value. $\dot{S}_G = 0$ if the process is reversible and $\dot{S}_G > 0$ if there is any irreversibility in the process.

Eqn. ($C$) which is commonly known as the entropy balance equation, can be stated in words, as

(Rate of accumulation of entropy) = (Rate of inflow of entropy) – (Rate of outflow of entropy) + (Rate of generation of entropy)

**Example 5.86**

Reduce the entropy balance equation for a steady–state, steady flow situation.

*Solution* :

The entropy balance equation is given by

$$\frac{dS}{dt} + \dot{m}_e s_e - \dot{m}_i s_i = \frac{\dot{Q}}{T} + \dot{S}_G$$

For steady–state, steady flow, $\dot{m}_e = \dot{m}_i = \dot{m}$ and there is no accumulation of entropy in the control volume. That is $dS/dt = 0$.

Then the above equation reduces to

$$\dot{m}(s_e - s_i) = \frac{\dot{Q}}{T} + \dot{S}_G \quad \text{or} \quad \dot{m}(s_e - s_i) \geq \frac{\dot{Q}}{T}$$

If the process is adiabatic, then $\dot{Q} = 0$ and we get

$$s_e \geq s_i$$

However, if the process is reversible( $\dot{S}_G = 0$) and adiabatic then $s_e = s_i$

**Example 5.87**

Superheated steam at 30 bar and 400°C enters a reversible and adiabatic turbine at a rate of 1kg/s and leaves at 0.3 bar. Neglecting the KE and PE terms, determine the power output of the turbine.

*Solution* :

Choose the turbine as the control volume. Application of the first law of thermodynamics to the control–volume gives

$$h_e - h_i = -\dot{W}_s / \dot{m}$$

Application of the second law of thermodynamics to the control volume gives

$$s_e = s_i$$

From steam tables, at 30 bar and 400°C, we get

$$h_i = 3232.5 \text{ kJ/kg} \ ; \ s_i = 6.9246 \text{ kJ/kg K}$$

For steam at 0.3 bar, we get

$$h_f = 289.30 \text{ kJ/kg} \ ; \ s_f = 0.9441 \text{ kJ/kg K}$$

$$h_g = 2625.4 \text{ kJ/kg} \ ; \ s_g = 7.7695 \text{ kJ/kg K}$$

The value of $s_e = s_i = 6.9246$ kJ/kg K lies in between the values of $s_f$ and $s_g$ at 0.3 bar. Hence it is wet steam of quality $X_e$

$$s_i = s_e = X_e s_g + (1 - X_e) S_f$$

or $6.9246 = 7.7695\, X_e + (1 - X_e)\, 0.9441$

or $X_e = 0.8762$

$$h_e = X_e h_g + (1 - X_e) h_f = 0.8762 \times 2625.4$$
$$+ (1 - 0.8762)\, 289.30$$
$$= 2336.19 \text{ kJ/kg}$$

Therefore, $-\dot{W}_s / \dot{m} = -\dot{W}_s = (2336.19 - 3232.5)10^3 = -\,896.31$ kW

Power out put of the turbine = 896.31 kW

**Example 5.88**

Rework Example 5.87 using a Mollier diagram.

*Solution* :

For a reversible and adiabatic turbine, we know that

$$-\dot{W}_s / \dot{m} = h_e - h_i \text{ and } s_e = s_i$$

Locate the state $i$ at the intersection of $P = 30$ bar and $t = 400$°C as shown in Fig.E. 5.88. Then read $h_i = 3232.5$ kJ/kg and $s_i = 6.9246$ kJ/kg K. Since $s_e = s_i$, draw a vertical line (constant entropy) from state $i$ and find the intersection point $e$, with $P = 0.3$ bar line as shown in Fig.E 5.88. Then read $h_e = 2336.19$ kJ/kg.

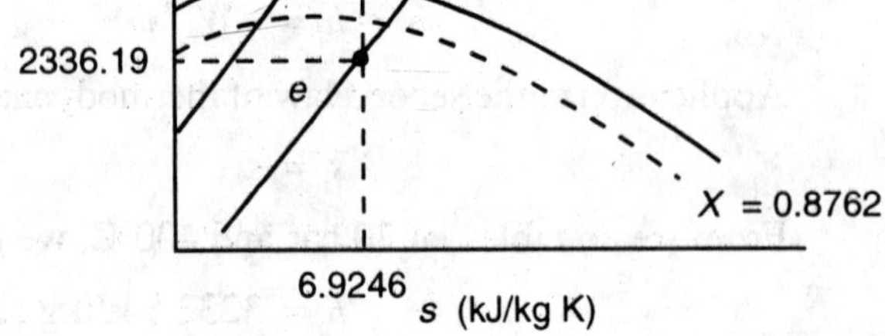


Fig.E 5.88. Sketch for Example 5.88.

Therefore $-\dot{W}_s = (2336.19 - 3232.5)10^3 = -896.31$ kW or $\dot{W}_s = 896.31$ kW.

**Example 5.89**

Calculate the rate of entropy generation and the power output if the steam leaves the turbine of Example 5.87 as dry saturated steam at 0.3 bar.

*Solution* :

We have $h_i = 3232.5$ kJ/kg and $s_i = 6.9246$ kJ/kg K

(see Example 5.87)

From steam tables, we find at 0.3 bar

$h_g = 2625.4$ kJ/kg $= h_e$ ; $s_g = 7.7695$ kJ/kg K $= s_e$

We know that $-\dot{W}_s / \dot{m} = h_e - h_i$

or $-\dot{W}_s = (2625.4 - 3232.5)\ 10^3 = -607.1$ kJ/s

or $-\dot{W}_s =$ Power output of turbine $= 607.1$ kW

We also know that

$$\dot{m}\ (s_e - s_i) = \dot{Q}\ / T + \dot{S}_G$$

or $s_e - s_i = \dot{S}_G$ or $\dot{S}_G = 7.7695 - 6.9246 = 0.8449$ kJ/K s.

**Example 5.90**

Define the isentropic efficiency of a turbine.

*Solution* :

The performance of a practical turbine is usually expressed in terms of isentropic efficiency, in which the performance of the actual turbine is compared with that of an ideal or a reversible– adiabatic turbine for the same inlet conditions and for the same exit pressure. The isentropic efficiency of a turbine $\eta_T$, is defined as

$$\eta_T = \frac{\text{Power output of the actual turbine per unit mass of fluid}}{\text{Power output of the turbine per unit mass, if it were isentropic}}$$

or

$$\eta_T = \frac{(\dot{W}_s / \dot{m})\text{of actual turbine}}{(\dot{W}_s / \dot{m})\text{of isentropic turbine}} = \frac{h_i - h_e}{h_i - h_{es}}$$

where $h_i$ = specific enthalpy of the inlet fluid

$h_e$ = specific enthalpy of the exit fluid in an actual turbine

$h_{es}$ = specific enthalpy of the exit fluid, when $s_e = s_i$

A schematic of the Mollier diagram showing the states *i*,.*e* and *es* is shown in Fig.E 5.90.

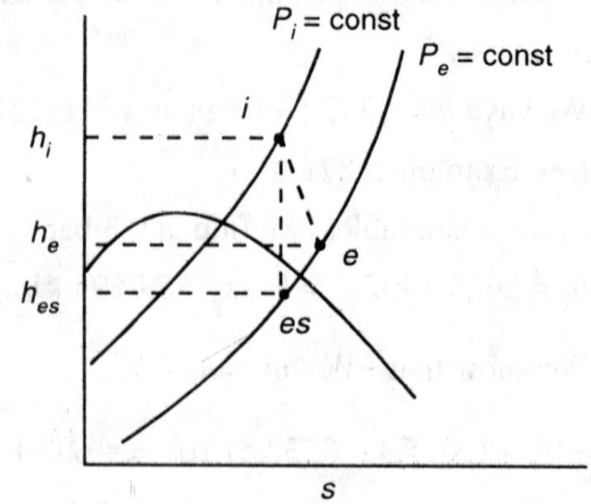


FigE. 5.90 Schematic of a Mollier diagram. $P_i$ and $P_e$ denote the inlet and outlet pressures of steam. *i-e* is the actual irreversible path followed by steam in the actual turbine. i-*es* is the path, the steam would have followed had the turbine been isentropic discharging steam of the same pressure as that of the actual turbine.

## Example 5.91

Superheated steam at 30 bar and 300°C enters a turbine and leaves as wet steam of quality 0.9 at 1 bar pressure. Calculate the isentropic efficiency of the turbine.

*Solution* :

From steam tables we read the following data :

Steam at 30 bar and 300°C : $h_i = 2995.1$ kJ/kg

$s_i = 6.5422$ kJ/kg K

Steam at 1 bar pressure:

$h_f = 417.54$ kJ/kg ; $h_g = 2675.4$ kJ/kg

$s_f = 1.3027$ kJ/kg K ; $s_g = 7.3598$ kJ/kg K

$$h_e = X_e h_g + (1 - X_e)\, h_f = 0.9 \times 2675.4 + 0.1 \times 417.54$$

$$= 2449.61 \text{ kJ/kg}$$

For an isentropic turbine $s_{es} = s_i = 6.5422$ kJ/kg K

$$s_{es} = X_{es}\, s_g + (1 - X_{es}) s_f$$

or $6.5422 = 7.3598\, X_{es} + (1 - X_{es})\, 1.3027$

or $X_{es} = 0.865$

$$h_{es} = 0.865 \times 2675.4 + (1 - 0.865)\, 417.54$$
$$= 2370.59 \text{ kJ/kg}$$

$$\eta_T = \frac{h_i - h_e}{h_i - h_{es}} = \frac{2995.1 - 2449.61}{2995.1 - 2370.59} = 0.8735$$

This problem can be conveniently solved by using a Mollier diagram as shown in Fig.E 5.91.

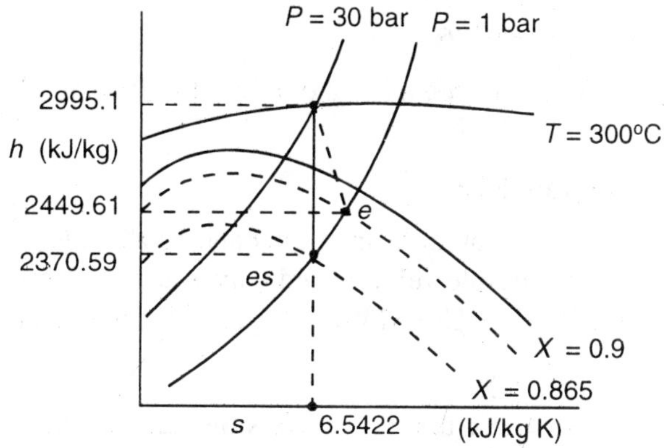


Fig. E 5.91.. Sketch for Example 5.91.

## Example 5.92

Superheated steam at 50 bar and 400°C enters a turbine and leaves at 1 bar. If the isentropic efficiency of the turbine is 0.8, determine the work output of the turbine and the state of the exit steam.

*Solution* :

Steam at 50 bar and 400°C; $h_i = 3198.3$ kJ/kg

$s_i = 6.6508$ kJ/kg K.

Steam at 1bar : $h_f = 417.54$ kJ/kg ; $h_g = 2675.4$ kJ/kg

$s_f = 1.3027$ kJ/kg K ; $s_g = 7.3598$ kJ/kg K

If the turbine were isentropic $s_{es} = s_i = 6.6508$ kJ/kg K

$$s_{es} = X_{es}\, s_g + (1 - X_{es})\, s_f$$

or $6.6508 = 7.3598X_{es} + (1 - X_{es})\, 1.3027$

or $X_{es} = 0.8829$

$$h_{es} = 0.8829 \times 2675.4 + (1 - 0.8829)\ 417.54$$
$$= 2411 \text{ kJ/kg}$$

$$\eta_T = \frac{h_i - h_e}{h_i - h_{es}} \quad \text{or} \quad 0.8 = \frac{3198.3 - h_e}{3198.3 - 2411}$$

or $h_e = 2568.46$ kJ/kg

Work output of the turbine $= h_i - h_e = 3198.3 - 2568.46$

$$= 629.84 \text{ kJ/kg}$$

$$h_e = X_e h_g + (1 - X_e) h_f$$

or $2568.46 = 2675.4\ X_e + (1 - X_e)\ 417.54$

or $X_e = 0.9526$

The exit steam has a quality of 0.9526 at 1 bar pressure.

**Example 5.93**

Superheated steam at 2 bar and 150°C enters a reversible and adiabatic nozzle with negligible velocity and leaves at 1 bar pressure. Calculate the exit velocity of the steam. Use Mollier diagram to solve the problem.

*Solution* :

We know that for a reversible and adiabatic process $s_e = s_i$ .A schematic of the Mollier diagram is shown in Fig.E 5.93. Locate the inlet state *i* of steam at 2 bar and 150°C. Then follow the vertical line (constant entropy) till it intersects with 1 bar line and locate state *e*. Then read $h_i = 2760$ kJ/kg and $h_e = 2640$ kJ/kg as shown in Fig.E 5.93.

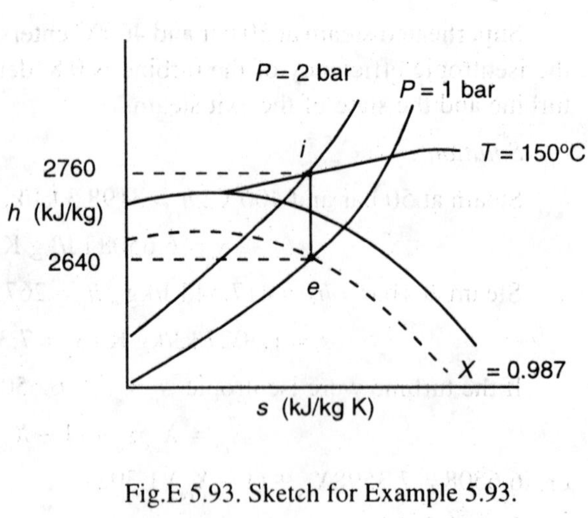


Fig.E 5.93. Sketch for Example 5.93.

Application of the first law of thermodynamics to the nozzle gives

$$\mathbf{V_e} = \sqrt{2(h_i - h_e)} = \sqrt{2(2760 - 2640)10^3} = 489.9 \text{ m/s}$$

The steam leaves the nozzle as wet steam of quality 0.987

**Example 5.94**

Define the isentropic efficiency of a nozzle.

*Solution* :

The isentropic efficiency of a nozzle is defined as the ratio of the actual specific kinetic energy of the gas leaving the nozzle $\mathbf{V}_e^2/2$, to the kinetic energy at the nozzle exit that would exist if the nozzle were isentropic and the expansion of the gas occurs to the same exit pressure. That is

$$\eta_N = \frac{\mathbf{V}_e^2/2}{\mathbf{V}_{es}^2/2}$$

Where $\mathbf{V}_{es}$ is the velocity of the gas leaving an isentropic nozzle.

**Example 5.95**

Superheated steam at 2 bar and 150°C enters an adiabatic nozzle with negligible velocity and leaves at 1 bar pressure. Calculate the exit velocity of the steam if the isentropic efficiency of the nozzle is 0.85.

*Solution* :

If the nozzle were isentropic, the exit velocity of the steam is 489.9 m/s (see Example 5.93)

$$\eta_N = \frac{\mathbf{V}_e^2/2}{\mathbf{V}_{es}^2/2} \quad \text{or} \quad 0.85 = \frac{\mathbf{V}_e^2}{(489.9)^2}$$

or $\mathbf{V}_e = 451.67$ m/s

**Example 5.96**

Define the isentropic efficiency of a compressor.

*Solution* :

The isentropic efficiency of a compressor is defined as the ratio of the power required per unit mass of the fluid by an isentropic compressor for the same inlet conditions of the fluid and for the same exit pressure of the fluid as in an actual compressor to the power required per unit mass of the fluid by the actual compressor. That is the isentropic efficiency $\eta_c$, of a compressor is given by

$$\eta_c = \frac{\text{power required per unit mass of the fluid by an isentropic compressor}}{\text{power required per unit mass of the fluid by an actual compressor}}$$

or $$\eta_c = \frac{(-\dot{W}_s / \dot{m})\text{for an isentropic compressor}}{(-\dot{W}_s / \dot{m})\text{for actual compressor}} = \frac{h_{es} - h_i}{h_e - h_i}$$

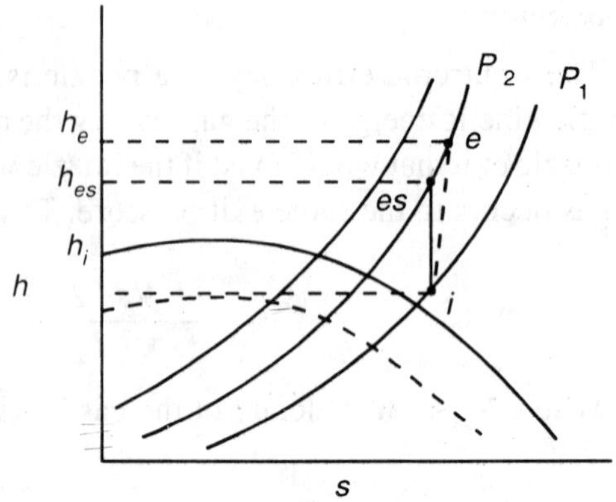


Fig.E 5.96. Schematic of Mollier diagram. The actual irreversible compression in an actual compressor is denoted by *i-e*, whereas *i-es* denotes the compression process if the compressor is isentropic.

A schematic of Mollier diagram is shown in Fig.E 5.96. Suppose a fluid at pressure $P_1$ enters an actual compressor at state $i$ and leaves at pressure $P_2$. The entropy of the exit fluid $s_e \geq s_i$ . If the compressor is irreversible, then $s_e > s_i$. The compression process in the actual compressor is represented by *i–e* in Fig.E 5.96. If the compressor is reversible and adiabatic and receives the fluid at the same inlet conditions represented by state $i$, and the fluid leaves at the same pressure $P_2$, then $s_{es} = s_i$ where *es* represents the state of the exit fluid in an isentropic compressor. The isentropic compression process is represented as *i–es* in Fig.E 5.96.

### Example 5.97

An adiabatic compressor with an isentropic efficiently of 0.8 is used to compress air (ideal gas with $\gamma = 1.4$) from 1 bar and 300 K to 6 bar pressure. Determine the exit temperature of the air and the power consumption, per mole of air, of the compressor.

*Solution* :

We know that for reversible, adiabatic compression of an ideal gas

$$T_{es} = T_i\left(\frac{P_e}{P_i}\right)^{\frac{\gamma-1}{\gamma}}$$

or $$T_{es} = 300\left(\frac{6}{1}\right)^{\frac{0.4}{1.4}} = 500.55\text{ K}$$

Application of the first law of thermodynamics to an adiabatic compressor (control volume) gives

$$\frac{\dot{W}_s}{\dot{m}} = h_e - h_i = C_P(T_{es} - T_i) = \frac{R\gamma}{\gamma - 1}(T_{es} - T_i)$$

$$= \frac{8.314\times 1.4}{1.4-1}(500.55 - 300) = 5.8358\text{ kJ / s}$$

$$\eta_c = \frac{h_{es} - h_i}{h_e - h_i} \quad \text{or} \quad 0.8 = \frac{5.8358}{h_e - h_i}$$

or $$\left(\frac{-\dot{W}_s}{\dot{m}}\right)\text{for actual compressor} = h_e - h_i = 7.2948\text{ kW}$$

$$(h_e - h_i) = 7.2948\times 10^3 = \frac{R\gamma}{\gamma-1}(T_e - T_i) = \frac{8.314\times 1.4}{1.4-1}(T_e - 300)$$

or $T_e = 550.69$ K

Therefore, exit temperature of air = 550.69 K

**Example 5.98**

Air at 1 bar and 25°C enters an adiabatic compressor and leaves at 10 bar and 350°C. Calculate the isentropic efficiency of the compressor. Treat air as an ideal gas with $\gamma$ = 1.4.

*Solution* :

We know that the isentropic efficiency of a compressor $\eta_c$ is given by

$$\eta_c = \frac{h_{es} - h_i}{h_e - h_i}$$

We also know that for an ideal gas $h = C_P T$.

Therefore, $$\eta_c = \frac{h_{es} - h_i}{h_e - h_i} = \frac{C_P(T_{es} - T_i)}{C_P(T_e - T_i)} = \frac{T_{es} - T_i}{T_e - T_i}$$

For isentropic compression of the ideal gas $T_{es}$ is given by

$$T_{es} = T_i\left(\frac{P_e}{P_i}\right)^{\frac{\gamma-1}{\gamma}} \quad \text{or} \quad T_{es} = 298(10)^{\frac{0.4}{1.4}} = 575.35\text{ K}$$

Therefore, $$\eta_c = \frac{575.35 - 298}{623 - 298} = 0.8534$$

**Example 5.99**

Define the isentropic efficiency of a pump.

*Solution* :

The isentropic efficiency of a pump $\eta_p$ is defined as the ratio of the power required per unit mass of the fluid by an isentropic pump, (which receives the fluid at the same inlet conditions as that of an actual pump and delivers the fluid at the same exit pressure as that of an actual pump) to the power required per unit mass of the fluid by the actual pump. That is

$$\eta_p = \frac{(-\dot{W}_s / \dot{m})\text{ of isentropic pump}}{(-\dot{W}_s / \dot{m})\text{ of actual pump}} = \frac{h_{es} - h_i}{h_e - h_i}$$

Where $h_i$ = specific enthalpy of fluid at the inlet pressure $P_i$

$h_e$ = specific enthalpy of fluid at the exit of the actual pump at pressure $P_e$

$h_{es}$ = specific enthalpy of fluid at the exit of the isentropic pump at pressure $P_e$

**Example 5.100**

In a steam power plant, saturated liquid water at 30°C enters a pump with an isentropic efficiency of 0.75 and leaves at 30 bar pressure. Calculate the power consumed by the pump per kg of liquid water.

*Solution* :

From steam tables we read the following data at 30°C.

$v_f = 0.001\ 004\ 3\ \text{m}^3/\text{kg}$ ; $P = 0.042\ 41$ bar

Assume the liquid to be incompressible. Then for an isentropic pump, we have

$$-\dot{W}_s / \dot{m}_s = \int_i^{es} v dP = v(P_e - P_i)$$

$$= 0.001\ 004\ 3(30 - 0.042\ 41)10^5$$

$$= 3.0086 \text{ kW}$$

$$\eta_P = \frac{(-\dot{W}_s / \dot{m}) \text{ for isentropic pump}}{(-\dot{W}_s / \dot{m}) \text{ for actual pump}} = 0.75$$

Therefore, power consumption of the actual pump per kg of liquid = 3.0086/0.75 = 4.0115 kW

**Example 5.101**

Air is steadily flowing through an insulated pipe and the flow direction is not known. An engineer measures the pressure and temperature of the air at two different stations *A* and *B* which are separated by sufficient distance and records the data as given below:

| | Station *A* | Station *B* |
|---|---|---|
| Pressure | 1.3 bar | 1.0 bar |
| Temperature | 50°C | 13°C |

Determine the flow direction of air, treating it as an ideal gas with $\gamma = 1.4$.

*Solution* :

We know that the second law of thermodynamics for a steady flow process gives

$$s_e \geq s_i \text{ or } s_e - s_i \geq 0$$

That is, the entropy of the exit fluid has to be greater than or equal to that of the inlet fluid. Thus, the second law of thermodynamics dictates the direction of a process as given below.

If $s_e - s_i > 0$ Process is feasible in the forward direction and the process is irreversible

$s_e - s_i = 0$ Process is feasible and reversible.

$s_e - s_i < 0$ Process is not feasible in the forward direction but it is spontaneous in the reverse direction.

This criterion can be used to determine the flow direction.

Let us suppose that air flows from station $A$ ($i$) to station $B$($e$)

Then
$$s_e - s_i = s_B - s_A = C_P \ln \frac{T_B}{T_A} - R \ln \frac{P_B}{P_A}$$

$$= \frac{R\gamma}{\gamma - 1} \ln \frac{T_B}{T_A} - R \ln \frac{P_B}{P_A}$$

$$= \frac{8.314 \times 1.4}{1.4 - 1} \ln \frac{286}{323} - 8.314 \ln \frac{1}{1.3}$$

$$= -1.3589 \text{ J/mol K}$$

Since $s_e - s_i < 0$, our assumption that air flows from station $A$ to Station $B$ is incorrect. Therefore, air flows from station $B$ to station $A$.

**Example 5.102**

An inventor claims to have designed a flow device which gives equal amounts of cold air at 250 K and hot air at 350 K at 1 bar pressure when the device is fed with air at 30 bar and 300 K. He further claims that the device does not require any energy input to operate it, Treat air as an ideal gas with $\gamma = 1.4$ and judge whether such a device is feasible or not.

*Solution* :

A schematic of the flow device is shown in Fig.E 5.102. The device is not receiving any energy either as heat or work and hence it is isolated. Any device cannot violate the laws of thermodynamics.Hence let us verify whether the device satisfies the laws of thermodynamics or not. Consider the device as the control volume and check whether the first law of thermodynamics (the principle of conservation of energy) is satisfied or not.

$$\dot{m}_i C_p T_i \; ? \; \dot{m}_{e1} C_p T_{e1} + \dot{m}_{e2} C_p T_{e2}$$

If $\dot{m}_i = 1$ mol/s, then $\dot{m}_{e1} = \dot{m}_{e2} = 0.5$ mol/s

Therefore $\quad C_P(300) \; ? \; 0.5 \; C_P (250) + 0.5 \; C_p(350) \qquad (A)$

The Eqn.($A$) is satisfied. Hence the device satisfies the first law of thermodynamics.

The second law of thermodynamics for the control volume gives

$$\dot{m}_{e1} s_{e1} + \dot{m}_{e2} s_{e2} - \dot{m}_1 s_i \geq 0$$

or $\quad 0.5(s_{e1} + s_{e2}) - s_i \geq 0$

or $\quad 0.5(s_{e1} - s_i) + 0.5(s_{e2} - s_i) \geq 0 \qquad (B)$

The LHS of Eqn.($B$) gives

$$0.5\left(C_P\ln\frac{T_{e1}}{T_i} - R\ln\frac{P_{e1}}{P_i}\right) + 0.5\left(C_P\ln\frac{T_{e2}}{T_i} - R\ln\frac{P_{e2}}{P_i}\right)$$

$$= 0.5\left[C_P\ln\frac{T_{e1}T_{e2}}{T_i^2} - R\ln\frac{P_{e1}P_{e2}}{P_i^2}\right]$$

$$= 0.5\ R\left[\frac{\gamma}{\gamma-1}\ln\frac{T_{e1}T_{e2}}{T_i^2} - \ln\frac{P_{e1}P_{e2}}{P_i^2}\right]$$

$$= 0.5\times 8.314\left[\frac{1.4}{0.4}\ln\frac{250\times 350}{300^2} - \ln\frac{1}{30^2}\right] = 27.8677 \text{ J / mol K}$$

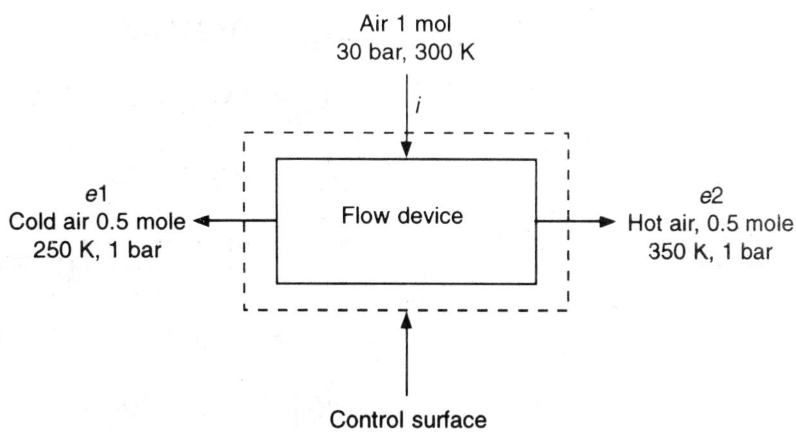


Fig.E. 5.102. Schematic of the flow device.

Therefore the LHS of Eqn($B$) is positive. Hence, the second law of thermodynamics is also satisfied. Therefore, the device is thoretically feasible.

**Example 5.103**

In a particular chemical plant air at 800°C and 1 bar leaves a unit at the rate of 10 mol/s whereas in another unit air is required at 200°C and 1 bar at a rate of 10 mol/s. An engineer plans of use the hot air as a source and the ambient atmosphere at 25°C as a sink to operate a heat engine and there by obtain some work and then supply the air at 200°C for use in the other unit. Assuming air to be an ideal gas with $\gamma = 1.4$ calculate the maximum power that can be delivered by the engine.

*Solution :*

The maximum power can be obtained by using a reversible engine .

We know that for a reversible engine

$$\Delta S_{sur} = \Delta S_{source} + \Delta S_{sink} = 0 \text{ or } \Delta S_{sink} = -\Delta S_{Source}$$

Consider the air flowing in one second.

$$\Delta S_{source} = nC_P \ln\frac{T_2}{T_1} - R\ln\frac{P_2}{P_1} = \frac{nR\gamma}{\gamma - 1}\ln\frac{T_2}{T_1}$$

$$= \frac{10 \times 8.314 \times 1.4}{1.4 - 1}\ln\frac{473}{10273}$$

$$= -238.355 \text{ J/K}$$

Therefore, $\Delta S_{sink} = 238.355$ J/K

$$\left|\dot{Q}_2\right| = T\Delta S_{sink} = 298 \times 238.355 = 71.03 \text{ kJ/s}$$

$$\left|\dot{Q}_1\right| = nC_P(T_1 - T_2)$$

$$= \frac{10 \times 8.314 \times 1.4}{1.4 - 1}(1073 - 473)$$

$$= 174.59 \text{ kJ/s}$$

Net power delivered $= \left|\dot{Q}_1\right| - \left|\dot{Q}_2\right| = 103.56$ kW

**Example 5.104**

Define $C_p$ and $C_v$ in terms of entropy change.

*Solution :*

The specific heat of a substance is defined as

$$C_P = \left(\frac{dq}{dT}\right)_P \text{ and } C_v = \left(\frac{dq}{dT}\right)_v$$

The second law of thermodynamics gives

$$ds = \left(\frac{dq}{T}\right)_R \text{ or } dq = Tds$$

Therefore, one can define the specific heat of a substance as

$$\frac{C_p}{T} = \left(\frac{\partial s}{\partial T}\right)_p \text{ and } \frac{C_v}{T} = \left(\frac{\partial s}{\partial T}\right)_v$$

**Example 5.105**

In a process industry hot gases are delivered by different units .One unit delivers a gas A at 1 bar and 1000 K at a rate of 1kmol/s while a second unit delivers a gas *B* at 1 bar and 800 K at a rate of 2 kmol/s. These hot gases are usually cooled to 300 K in heat exchangers. The ambient atmosphere is at 300 K. An engineer plans to use the hot gases as source and ambient atmosphere as sink to operate a heat engine and thus obtain some power. Calculate the maximum power that can be obtained if

(*a*) gases *A* and *B* are used as separate sources, and

(*b*) gases *A* and *B* are mixed and the mixture is used as a source.

Assume both the gases *A* and *B* are ideal with $\gamma = 1.4$

*Solution :*

(*a*) Maximum power can be obtained by employing a reversible engine. We know that for a reversible engine

$$\Delta S_{\text{sur}} = \Delta S_{\text{source}} + \Delta S_{\text{sink}} = 0$$

The ambient atmosphere is at constant temperature. However, the source temperature continuously decreases till 300 K,while energy is extracted as heat. If the gases *A* and *B* are used as separate sources, then for the heat engine which uses gas *A*

$$\Delta S_A + \Delta S_{\text{sink}} = 0$$

or $$nC_p \ln\frac{T_2}{T_1} + \frac{\dot{Q}_2}{T_2} = 0$$

or $$\frac{nR\gamma}{\gamma - 1}\ln\frac{T_2}{T_1} + \frac{\dot{Q}_2}{T_2} = 0$$

or $$\frac{1000 \times 8.314 \times 1.4}{1.4 - 1}\ln\frac{300}{1000} + \frac{\dot{Q}_2}{300} = 0$$

or $$\dot{Q}_2 = 10.5103 \times 10^6 \text{ J/s}$$

Where $\dot{Q}_2$ is the energy rejected to the ambient atmosphere.

Energy absorbed from source (gas $A$) = $\dot{Q}_1 = nC_P(T_1 - T_2)$

$$= \frac{1000 \times 8.314 \times 1.4}{1.4 - 1}(1000 - 300)$$

$$= 20.3693 \times 10^6 \text{ J/s}$$

Power delivered by the engine which uses gas $A$ = $\dot{Q}_1 - \dot{Q}_2$ = (20.3693 – 10.5103)$10^6$ = 9.859 MW

Similarly for the engine which uses gas $B$, we get

$$\Delta S_B + \Delta S_{\text{sink}} = 0$$

$$nC_P \ln\frac{T_2}{T_1} + \frac{\dot{Q}_2}{T_2} = 0$$

or
$$\frac{2000 \times 8.314 \times 1.4}{1.4 - 1}\ln\frac{300}{800} + \frac{\dot{Q}_2}{300} = 0$$

or
$$\dot{Q}_2 = 17.1247 \times 10^6 \text{ J/s}$$

Energy absorbed from source (gas $B$) = $\dot{Q}_1 = nC_p\ (T_1 - T_2)$

$$\dot{Q}_1 = \frac{2000 \times 8.314 \times 1.4}{1.4 - 1}(800 - 300)$$

$$= 29.099 \times 10^6 \text{ J/s}$$

Power delivered by the engine which uses gas $B$ = $\left|\dot{Q}_1\right| - \dot{Q}_2$ = (29.099 – 17.1247) × $10^6$ = 11.9743 MW

Therefore, the maximum power that can be obtained, when the gases $A$ and $B$ are separately used as sources

$$= 9.859 + 11.9743 = 21.8333 \text{ MW}$$

(*b*) If the gases $A$ and $B$ are mixed together, the equilibrium temperature ($T$) of the mixture is given by

$$[nC_P(T_1 - T)]_A = [nC_P(T - T_1)]_B$$

or $1000C_P(1000 - T) = 2000C_p(T - 800)$

or $T = 866.67$ K

If this mixture (3 kmol/s) at 866.67 K is used as a source to operate a reversible engine,

$$\Delta s_{gas} + \Delta s_{sink} = 3000\, C_P \ln \frac{300}{866.67} + \frac{\dot{Q}_2}{300}$$

or $$\frac{3000 \times 8.314 \times 1.4}{1.4 - 1} \ln \frac{300}{866.67} + \frac{\dot{Q}_2}{300} = 0$$

or $$Q_2 = 27.7834 \times 10^6 \text{ J/s}$$

Energy absorbed from source $\dot{Q}_1 = nC_p\,(866.67 - 300)$

$$\dot{Q}_1 = \frac{3000 \times 8.314 \times 1.4}{1.4 - 1}(866.67 - 300)$$

$$= 49.4686 \times 10^6 \text{ J/s}$$

Therefore, maximum power delivered by the engine $= \dot{Q}_1 - \dot{Q}_2$

$$= (49.4686 - 27.7834)10^6$$

$$= 21.6852 \text{ MW}$$

# 6

# Thermodynamic Potentials and Availability

**Example 6.1**

What does the principle of maximum entropy state ? Explain it with the help of an example.

*Solution* :

The principle of maximum entropy states that in a state of stable equilibrium, the entropy of an isolated system is maximum for a given value of internal energy.

Consider a rigid and insulated tank which is divided into two compartments by a partition as shown in Fig.E 6.1. The first compartment contains a hot fluid while the second compartment contains a cold fluid. If the partition is removed the fluids mix and attain a state of equilibrium.

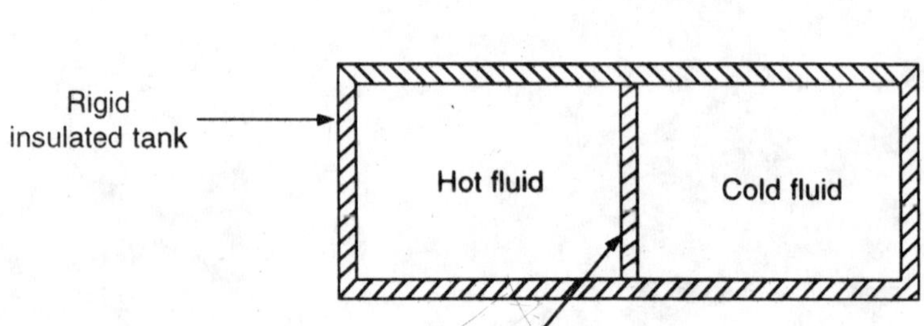


Fig.E 6.1. Sketch for Example 6.1.

Consider the fluid contained in both the compartments as the system. Since the tank is rigid and adiabatic, we have $W = 0$ and $Q = 0$. Application of the first law of thermodynamics gives $\Delta U = 0$ or $U_1 = U_2$.

That is, the final internal energy of the system is equal to the initial internal energy. In other words, the system occupies only that state for which the internal energy is equal to the initial value. There may be several states satisfying this condition. Which of the possible states satisfying the relation $U_1 = U_2$ is the actual final equilibrium state ? This can be answered by the application of the second law of thermodynamics. The second law of thermodynamics (the principle of entropy increase) tells that the entropy of an isolated system either increases or remains constant. If there is any irreversibility in the process, the entropy of the system must increase. Since mixing of fluids at different states is an irreversible process, the entropy of the system increases. The system changes from one state to another of higher entropy and finally settles in that state for which the entropy is maximum for the given value of internal energy. This state of maximum entropy is the stable equilibrium state of the system.

## Example 6.2

What can be inferred from the combined statement of the two laws of thermodynamics ?

*Solution* :

The first law of thermodynamics gives $dU = dQ - dW$. If a system undergoes a reversible process $dW = PdV$. The second law of thermodynamics gives $dS = (dQ/T)_R$. Combining these two statements, we obtain.

$$dU = TdS - PdV$$

This tells that for a system of constant composition, the internal energy $U$ is a function of $S$ and $V$. That is $U = U(S, V)$. If we consider a multicomponent system consisting of $c$ constituents, then

$$U = U(S, V, n_1, n_2, .., n_c)$$

Where $n_i$ = moles of constituent $i$.

## Example 6.3

Name the widely used thermodynamic potentials and define Helmholtz free energy and Gibbs free energy.

*Solution* :

The widely used thermodynamic potentials are internal energy $u$, enthalpy $h$, Helmholz free energy $a$ and Gibbs free energy $g$. We know that

$$du = Tds - Pdv \qquad (A)$$

or $u = u(s, v)$

which expresses the differential change in $u$ when the independent variables $s$ and $v$ are simultaneously changed. We also know that the enthalpy $h$, is defined as

$$h = u + Pv$$

or $dh = du + Pdv + vdP$

Substituting for $du$ from Eqn. ($A$). We get

$$dh = Tds + vdP \quad (B)$$

or $h = h\,(s, P)$

The Helmholtz free energy $a$ is defined as

$$a = u - Ts$$

or $da = du - Tds - sdT$

Substituting for $du$ from Eqn. ($A$), we get

$$da = -\,sdT - Pdv$$

or $a = a\,(T, v)$

The Gibbs free energy $g$ is defined as

$$g = h - Ts = u + Pv - Ts$$

or $dg = dh - Tds - sdT$

Subsituting for $dh$ from Eqn($B$) we get

$$dg = -\,sdT + vdP \quad (D)$$

or $g = g\,(T,P)$

**Note** *We have observed that $u = u\,(s, v)$; $h = h\,(s, P)$; $a = a\,(T, v)$ and $g = g\,(T, P)$. That is, the natural independent variables of u are s and v, the natural independent variables for h are s and P. Similarly the natural independent variables are T and v for a and T and P for g. The Eqns.(A) – (D) express the differential changes of the thermodynamic potentials in terms of their natural independent variables.*

**Example 6.4**

Why the energy functions, internal energy $u$ and Helmholtz free energy $a$, are called thermodynamic potentials ?

*Solution* :

Suppose a system undergoes a change of state by following an adiabatic process. Then $Q = 0$. Application of the first law of thermodynamics gives $\Delta U = -W$ or $W = -\,\Delta U$. That is, the work done by the system is equal to the decrease in its internal energy. Thus internal energy of a system represents the potential to do work in an adiabatic process. Hence internal energy $U$ is called a thermodynamic potential. If work is done on a system during an adiabatic process, the increase in the internal energy of the system is exactly equal to the work done on the system.

Now, let us consider a system which undergoes a process while it is interacting with the surroundings at $T_0$. Let $ds$ and $ds_0$ represent the change in the entropy of the system and surroundings, respectively. Then the first law of thermodynamics gives

$$du = dq - dW \quad \text{or} \quad dW = -du + dq \tag{A}$$

The second law of thermodynamics gives

$$ds + ds_0 \geq 0 \tag{B}$$

The change in the entropy of the surroundings $ds_0$, is given by

$$ds_0 = \frac{-dq}{T_0} \tag{C}$$

Substituting Eqn. ($C$) in Eqn. ($B$) gives

$$ds = \frac{-dq}{T_0} \geq 0 \quad \text{or} \quad dq \leq T_0 ds \tag{D}$$

Then from Eqns .($A$) and ($D$), we get

$$dW = -du + dq \leq -du + T_0\,ds$$

or $\quad W \leq (u_1 - u_2) - T_0\,(s_1 - s_2)$

or $\quad W \leq (u_1 - T_0\,s_1) - (u_2 - T_0\,s_2) \qquad (E)$

If the initial and final temperatures of the system are equal to the surrounding temperature, that is $T_1 = T_2 = T_0$, then $W \leq (u_1 - T_1\,s_1) - (u_2 - T_2\,s_2)$

or $\quad W \leq a_1 - a_2$

The equality sign holds good if the process is reversible. Thus we find that the maximum work that can be obtained from a system during a given process, in which the initial and final temperatures of the system are equal to the surrounding temperature, is equal to the decrease in the Helmholtz free energy of the system. The maximum work can be obtained only when the process is reversible. If there is any irreversibility in the process the work delivered by the system is less than the decrease in the Helmholtz free energy of the system. On the other hand if the work is done on the system, the minimum work to be done on the system is equal to the increase in the Helmholtz free energy of the system. The work to be done on the system is minimum only when the process is reversible. If there is any irreversibility in the process, the work to be done on the system is greater than the increase in the Helmholtz free energy of the system. Since the decrease in the Helmholtz function or Helmholtz free energy represents the potential to do work by a system during a given process, in which the initial and final temperatures of the system are equal to the surrounding temperature, it is called a thermodynamic potential.

**Example 6.5**

It is desired to compress air (an ideal gas with $\gamma = 1.4$) in a piston - cylinder from 1 bar and 300 K to a pressure of 10 bar and the final temperature of the air is 300 K. The air interacts with the surroundings which is also at 300 K. Determine the minimum amount of work to be done per mole of air.

*Solution* :

The initial and final temperatures of the air are same as the surroundings temperature 300 K. The work done on the system will be minimum only when the process is reversible. The minimum work required is equal to the increase in the Helmholtz free energy of the system. That is

$$W_{min} = \Delta a = a_2 - a_1 = (u_2 - T_2 s_2) - (u_1 - T_1 s_1)$$

We also know that $T_2 = T_1 = T_0 = 300$ K.

For an ideal gas $(u_2 - u_1) = C_v (T_2 - T_1)$.

Since $T_1 = T_2$, $(u_2 - u_1) = 0$.

Therefore $\quad W_{min} = a_2 - a_1 = - T_0 (s_2 - s_1)$

For an ideal gas $\quad s_2 - s_1 = \dfrac{-dq}{T_0} \geq 0 \quad \text{or} \quad dq \leq T_0 ds$

Therefore $W_{min} = -T_0\left(-R\ln\dfrac{P_2}{P_1}\right) = RT_0 \ln\dfrac{P_2}{P_1}$

$$= 8.314 \times 300 \times \ln\frac{10}{1} = 5.7431 \text{ kJ/mol}$$

**Example 6.6**

A piston - cylinder assembly contains 0.1 kg superheated steam at 30 bar and 300°C. This assembly is placed in thermal contact with a reservoir at 300°C and the steam is allowed to expand to a pressure of 1 bar. Calculate the maximum work that can be obtained from the steam.

*Solution* :

The initial and final temperatures of the steam are equal to the surrounding temperature. $T_1 = T_2 = T_0 = 300$°C. The maximum work can be obtained only if the process is reversible. The maximum work obtained is equal to the decrease in the Helmholtz free energy of the steam. That is

$$W_{min} = A_1 - A_2 = (U_1 - T_1S_1) - (U_2 - T_2S_2)$$
$$= m\,[(u_1 - T_1s_1) - (u_2 - T_2s_2)] = m(a_1 - a_2)$$

For steam at 30 bar and 300°C:

$$h_1 = 2995.1 \text{ kJ/kg}: v_1 = 0.081\ 16 \text{ m}^3\text{/kg} ;$$
$$s_1 = 6.5422 \text{ kJ/kg K}$$

Therefore $a_1 = u_1 - T_1 s_1 = h_1 - P_1 v_1 - T_1 s_1$

$$= 2995.1 \times 10^3 - 30 \times 10^5 \times 0.081\ 16 - 573 \times 6.5422 \times 10^3$$
$$= -\ 997.06 \text{ kJ/kg}$$

For steam at 1 bar and 300°C :

$$h_2 = 3074.5 \text{ kJ/kg}; v_2 = 2.6390 \text{ m}^3\text{/kg} ;$$
$$s_2 = 8.2166 \text{ kJ/kg K}$$

Therefore

$$a_2 = h_2 - P_2 v_2 - T_2 s_2$$
$$= 3074.5 \times 10^3 - 1 \times 10^5 \times 2.6390$$
$$- 573 \times 8.2166 \times 10^3$$
$$= -\ 1897.512 \text{ kJ/kg}$$

Then

$$W_{max} = m (a_1 - a_2) = 0.1 (-\ 997.06 + 1897.512) \times 10^3$$
$$= 90.045 \text{ kJ}$$

**Example 6.7**

Why the energy functions enthalpy and Gibbs free energy are called as thermodynamic potentials ?

*Solution* :

Consider a device like turbine which is operating in steady state, steady-flow conditions through which material continuously flows in and flows out as shown in Fig.E 6.7. Consider the device as the control - volume and apply the first law of thermodynamics to the control - volume to obtain

$$\dot{m}_e \left( h_e + \frac{\mathbf{V}_e^2}{2} + gZ_e \right) - \dot{m}_i \left( h_i + \frac{\mathbf{V}_i^2}{2} + gZ_i \right) = \dot{Q} - \dot{W}_s - \frac{dE}{dt} \qquad (A)$$

If the device is operating in steady - state, steady - flow conditions we have

$$\frac{dE}{dt} = 0 \text{ and } \dot{m}_i = \dot{m}_e = \dot{m}$$

Suppose the device is adiabatic. Then $\dot{Q} = 0$

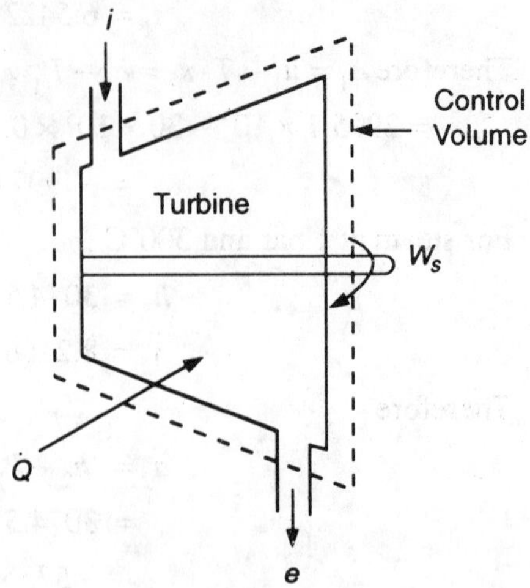


Fig.E 6.7. Control-Volume.

Ignore KE and PE changes. Then Eqn. (*A*) reduces to

$$(h_e - h_i) = -\dot{W}_s / \dot{m}$$

or $\dot{W}_s / \dot{m} = h_i - h_e$

That is the power delivered per unit mass of fluid flowing through the device is equal to the decrease in the specific enthalpy of the fluid flowing through an adiabatic device. Therefore, enthalpy represents the potential to do work and hence it is called a thermodynamic potential.

Now, consider a non-adiabatic device operating in steady-state, steady flow conditions. Ignore the KE and PE changes and apply the first law of thermodynamics to the device. Then Eqn. (*A*) gives

$$\dot{m}(h_e - h_i) = \dot{Q} - \dot{W}_s$$

or $\dot{W}_s = \dot{Q} - \dot{m}(h_e - h_i)$

Suppose energy flows at a rate of $\dot{Q}$ from the surroundings at $T_0$ into the control volume. The second law of thermodynamics gives

$$\Delta S + \Delta S_0 \geq 0$$

$$\Delta S - \frac{\dot{Q}}{T_0} \geq 0 \quad \text{or} \quad \dot{Q} \leq T_0 \Delta S \qquad (C)$$

Where $\Delta S_0$ is the change in the entropy of the surroundings. Substituting Eqn. (*C*) in Eqn. (*B*), we get

$$\dot{W}_s = \dot{Q} - \dot{m}\,(h_e - h_i) \leq T_0\,\Delta\,S - \dot{m}\,(h_e - h_i)$$

or $\dot{W}_s \leq \dot{m}\; T_0\,\Delta\, s - \dot{m}\;(h_e - h_i)$

or $\dot{W}_s \leq \dot{m}\;\{T_0(s_e - s_i) - (h_e - h_i)\}$

or $$\frac{\dot{W}_s}{\dot{m}} \leq (h_i - T_0 s_i) - (h_e - T_0 s_e) \qquad (D)$$

If the temperatures of the fluid at the inlet and outlet of the control - volume are equal to the surrounding temperature, that is

$$T_i = T_e = T_0$$

Eqn. (*D*) reduces to

$$\frac{\dot{W}_S}{\dot{m}} \leq (h_i - T_i s_i) - (h_e - T_e s_e)$$

or $$\frac{\dot{W}_S}{\dot{m}} \leq (g_i - g_e) \qquad (E)$$

The equality sign in Eqn. (*E*) is true for reversible process only. Thus we find that the maximum work per unit mass that can be obtained from a device which is operating in steady- state, steady flow conditions such that the inlet and outlet temperatures of the fluid are same as that of the surroundings, is equal to the decrease in the specific Gibbs free energy of the fluid flowing through the device. If there is any irreversibility in the device, the work that can be obtained per unit mass of the fluid is less than the decrease in the specific Gibbs free energy of the fluid.

If the device is an energy consuming device like compressor or pump, then the minimum work required per unit mass of fluid flowing through it is equal to the increase in the Gibbs free energy of the fluid, if the inlet and outlet temperatures of the fluid are equal to the surroundings temperature. From this we observe that the Gibbs free energy represents the potential to do work and hence it is called a thermodynamic potential.

### Example 6.8

The atmospheric air is a mixture of nitrogen and oxygen in the mole ratio 79:21. It is required to separate 1 mol/s of air at 1 bar 300 K into its pure components at 1 bar and 300 K. Calculate the minimum power required to separate the air. The surroundings are at 300 K. Treat air as an ideal gas.

*Solution* :

The air flows continuously at the rate of 1 mol/s through the device. The inlet and outlet temperatures are equal to the surroundings temperature. The minimum power is consumed by the device only when it is reversible. Then, the minimum power required is equal to the increase in the Gibbs free energy.

$$\dot{W}_{min} = g_e - g_i = (h_e - T_0 s_e) - (h_i - T_0 s_i)$$

For an ideal gas $h = h\,(T)$ only. Therefore $h_e - h_i = 0$.

Then

$$W_{min} = -T_0\,(s_e - s_i) = T_0\,(s_i - s_e)$$

The initial state $i$ is a mixture of nitrogen and oxygen and the final state is pure components at the same temperature and pressure. We know that the entropy change due to mixing is given by

$$\Delta s = -R\Sigma X_i \ln X_i = (s_i - s_e)$$

Therefore

$$\dot{W}_{min} = -RT_0 \Sigma X_i \ln X_i$$

$$= -8.314 \times 300 \times \{0.79 \ln 0.79 + 0.21 \ln 0.21\} = 1281.91 \text{ J/s}$$

Therefore minimum power required = 1.282 kW

**Example 6.9**

State the energy minimum principle.

*Solution* :

**The energy minimum principle states that a system in a state of stable equilibrium will have the minimum internal energy for given values of entropy and volume. That is $U$ is minimum for given values of $S$ and $V$ or $dU = 0$ when $S$ and $V$ are held constant.**

**Example 6.10**

**Define the intensive parameters in terms of thermodynamic potentials.**

*Solution* :

We know that $u = u\,(s, v)$

or $$du = Tds - Pdv = \left(\frac{\partial u}{\partial s}\right)_v ds + \left(\frac{\partial u}{\partial v}\right)_s dv$$

Therefore $T = \left(\frac{\partial u}{\partial s}\right)_v$ and $-P = \left(\frac{\partial u}{\partial v}\right)_s$

We know that $h = h\,(s, P)$

or $dh = Tds + vdP = \left(\frac{\partial h}{\partial s}\right)_P ds + \left(\frac{\partial h}{\partial P}\right)_s dP$

Therefore, $T = \left(\frac{\partial h}{\partial s}\right)_P$

We know that $a = (T, v)$

or $da = -\,sdT - Pdv = \left(\frac{\partial a}{\partial T}\right)_V dT + \left(\frac{\partial a}{\partial v}\right)_T dv$

Therefore, $-P = \left(\frac{\partial a}{\partial v}\right)_T$

## Example 6.11

We know that $u = u\,(s, v)$ and it is a thermodynamic property. We also know that its differential is exact. What additional information can be obtained from the fact that $du$ is an exact differential?

*Solution:*

If $z = z\,(x, y)$. Then

$$dz = \left(\frac{\partial z}{\partial x}\right)_y dx + \left(\frac{\partial z}{\partial y}\right)_x dy$$

$$= Mdx + Ndy \text{ is an exact differential}$$

if $\left(\frac{\partial M}{\partial y}\right)_x = \left(\frac{\partial N}{\partial x}\right)_y$

Then if $u = u(s, v)$

or $du = \left(\frac{\partial u}{\partial s}\right)_v ds + \left(\frac{\partial u}{\partial v}\right)_s dv = Tds - Pdv$

Since $du$ is an exact differential, it is true that

$$\left(\frac{\partial T}{\partial v}\right)_s = -\left(\frac{\partial P}{\partial s}\right)_v$$

which is known as a Maxwell relation. Thus every thermodynamic potential yields one commonly used Maxwell relation.

**Example 6.12**

What is the importance of Maxwell relations in thermodynamics ?

*Solution* :

The entropy plays an important role in the analysis of thermodynamic processes. Quite often one has to estimate the change in entropy of a system associated with changes in the independent, measurable variables $P$, $v$ and $T$. The entropy can be expressed in one of the following ways.

$$s = s\,(P, T) \quad \text{or} \quad ds = \left(\frac{\partial s}{\partial P}\right)_T dP + \left(\frac{\partial s}{\partial T}\right)_P dT$$

$$s = s\,(T, v) \quad \text{or} \quad ds = \left(\frac{\partial s}{\partial T}\right)_v dT + \left(\frac{\partial s}{\partial v}\right)_T dv$$

$$s = s\,(P, v) \quad \text{or} \quad ds = \left(\frac{\partial s}{\partial P}\right)_v dP + \left(\frac{\partial s}{\partial v}\right)_P dv$$

The entropy change can then be estimated by integrating the above relations. For this one should know the values of the six partial derivatives of entropy, namely $\left(\frac{\partial s}{\partial T}\right)_P, \left(\frac{\partial s}{\partial T}\right)_v, \left(\frac{\partial s}{\partial P}\right)_T, \left(\frac{\partial s}{\partial P}\right)_v, \left(\frac{\partial s}{\partial v}\right)_T$ and $\left(\frac{\partial s}{\partial v}\right)_P$

Of these, the partial derivatives of entropy with respect to temperature are measurable and are given by

$$\left(\frac{\partial s}{\partial T}\right)_P = \frac{C_P}{T} \quad \text{and} \quad \left(\frac{\partial s}{\partial T}\right)_v = \frac{C_v}{T}$$

The remaining four partial derivatives of entropy with respect to pressure and volume cannot be measured directly. The Maxwell relations express the partial derivatives of entropy with respect to pressure and volume in terms of measurable quantities, thus providing a means for estimating the entropy change. The Maxwell relation obtained from the thermodynamic potential internal energy, is given by [see Example 6.11]

$$\left(\frac{\partial T}{\partial v}\right)_s = -\left(\frac{\partial P}{\partial s}\right)_v \quad \text{or} \quad \left(\frac{\partial s}{\partial P}\right)_v = -\left(\frac{\partial v}{\partial T}\right)_s$$

That is $(\partial s/\partial P)_v$ is equal to the negative of the slope of $v$ versus $T$ curve for a reversible, adiabatic process which can be easily measured.

### Example 6.13

Derive the most commonly used Maxwell relations starting from, the thermodynamic potentials expressed in terms of their natural variables.

*Solution* :

The thermodynamic potentials are given by

$$u = u(s, v) \quad \text{or} \quad du = Tds - Pdv$$

Therefore, we get

$$\left(\frac{\partial T}{\partial v}\right)_s = -\left(\frac{\partial P}{\partial s}\right)_v \tag{A}$$

$$h = h(s, P) \quad \text{or} \quad dh = Tds + vdP$$

Therefore,
$$\left(\frac{\partial T}{\partial P}\right)_s = \left(\frac{\partial v}{\partial s}\right)_P \tag{B}$$

$$a = a(T, v) \quad \text{or} \quad da = -sdT - Pdv$$

Therefore,
$$\left(\frac{\partial s}{\partial v}\right)_T = \left(\frac{\partial P}{\partial T}\right)_v \tag{C}$$

$$g = g(T, P) \quad \text{or} \quad dg = -sdT + vdP$$

Therefore,
$$\left(\frac{\partial s}{\partial P}\right)_T = -\left(\frac{\partial v}{\partial T}\right)_P \tag{D}$$

Eqns. $(A) - (D)$ are the required Maxwell relations.

### Example 6.14

Present a thermodynamic Mnemonic diagram and explain the method of writing the differential expressions for the thermodynamic potentials and Maxwell relations based on the diagram.

*Solution* :

The Mnemonic diagram shown in Fig.E 6.14 ($a$) consists of a square with two diagonal arrows pointing upwards. The sides of the square are labelled with the thermodynamic potentials. Starting from the top, the potentials are placed in alphabetical order in clockwise direction. The natural variables of the potentials are then placed such that each potential is flanked by its own natural variables. The differential expressions for the potentials can be expressed in terms of the differentials of its natural variables. The coefficients associated with the natural variables are indicated by the diagonal arrows. The sign associated with the coefficient is positive if the arrow points away from the natural variable and the coefficient carries a negative sign if the arrow points towards the natural variable.

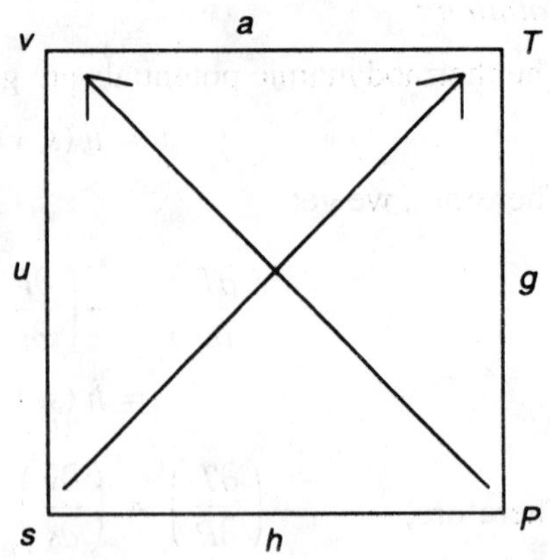


Fig.E 6.14. (*a*) Thermodynamic Mnemonic diagram.

Thus we can write as

$$du = \text{(sign) (coefficient) } ds + \text{(sign) (coefficient) } dv$$

$$du = \text{(sign) } Tds + \text{(sign) } Pdv$$

$$du = +\,Tds + (-)\,Pdv$$

or $du = Tds - Pdv$

Similarly for the potential $h$, we can write

$$dh = \text{(sign) (coefficient) } ds + \text{(sign) (coefficient) } dP$$

$$= \text{(sign) } Tds + \text{(sign) } vdP$$

$$= (+)\,Tds + (+)\,vdP$$

or $dh = Tds + vdP$

Similarly, one can write

$$da = -\,sdT - Pdv$$

$$dg = -\,sdT + vdP$$

To obtain the Maxwell relations we need to concentrate on the direction of arrows and the natural variables only as shown in Fig.E 6.14 (*b*)

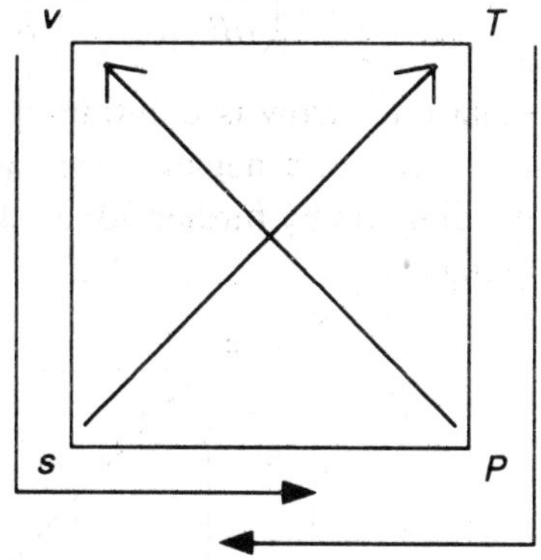


$$\left(\frac{\partial v}{\partial s}\right)_P = \left(\frac{\partial T}{\partial P}\right)_s$$

Fig.E 6.14. (*b*) Thermodynamic Mnemonic diagram.

Then move from the left hand top corner in the anticlockwise direction and write $\left(\frac{\partial v}{\partial s}\right)_P$ and equate it with $\left(\frac{\partial T}{\partial P}\right)_s$ obtained by moving in the clockwise direction starting from the top right hand corner.

If both the arrows are pointing in the same direction (either upwards or downwards), then there is no need to introduce an additional sign. However, if the arrows are pointing in opposite directions, a negative sign should be introduced in the equation. From Fig.E. 6.14 (*b*), we can write the Maxwell relation

$$\left(\frac{\partial v}{\partial s}\right)_P = \left(\frac{\partial T}{\partial P}\right)_s$$

To obtain the next Maxwell relation, rotate the diagram clockwise by 90° as shown in Fig.E 6.14 (*c*). Then we obtain the Maxwell relation.

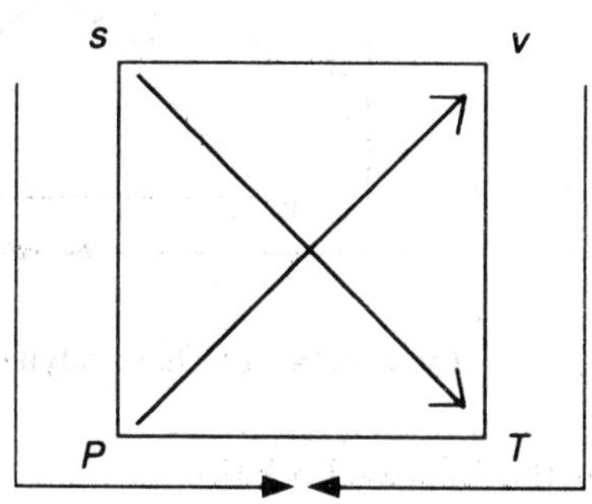


$$\left(\frac{\partial s}{\partial P}\right)_T = -\left(\frac{\partial v}{\partial T}\right)_P$$

Fig.E 6.14. (*c*) Thermodynamic Mnemonic diagram.

$$\left(\frac{\partial s}{\partial P}\right)_T = -\left(\frac{\partial v}{\partial T}\right)_P$$

Note that one arrow is pointing upwards while the other arrow is pointing downwards. Hence, a negative sign is introduced in the above equation. By rotating the diagram by further 90° as shown in FigE. 6.14 (*d*) we can obtain the Maxwell relation

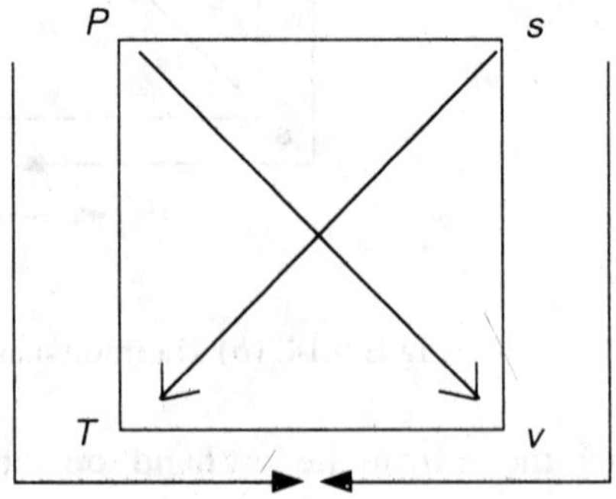


$$\left(\frac{\partial P}{\partial T}\right)_v = \left(\frac{\partial s}{\partial v}\right)_T$$

Fig.E 6.14. (*d*) Thermodynamic Mnemonic diagram.

$$\left(\frac{\partial P}{\partial T}\right)_v = \left(\frac{\partial s}{\partial v}\right)_T$$

Further rotation of the diagram by 90° as shown in Fig.E. 6.14 (*e*)

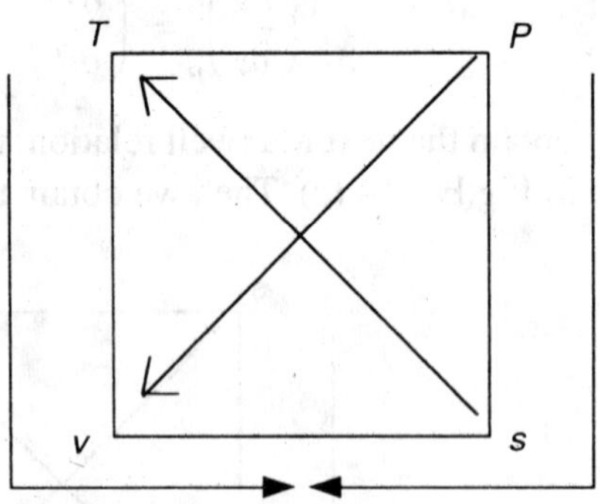


$$\left(\frac{\partial T}{\partial v}\right)_s = -\left(\frac{\partial P}{\partial s}\right)_v$$

Fig.E 6.14. (*e*) Thermodynamic Mnemonic diagram.

yields the Maxwell relation

$$(\partial T/\partial v)_s = -(\partial P/\partial s)_v$$

**Example 6.15**

How do we express the thermodynamic potentials for a multicomponent system ? Define the chemical potential of a component in terms of thermodynamic potentials.

*Solution* :

For a system of constant composition, the molar internal energy $u$ is given by $u = u(s, v)$. For a system consisting of $n$ moles, the internal energy $U (= nu)$ is given by

$$U = U(S, V, n)$$

If a system contains $n_1$ moles of species 1, $n_2$ moles of species 2, etc. then the internal energy of such a system is given by

$$U = U(S, V, n_1, n_2, \ldots.. n_c)$$

Then the differential change in internal energy can be expressed as

$$dU = \left(\frac{\partial U}{\partial S}\right)_{V, n_i} dS + \left(\frac{\partial U}{\partial V}\right)_{S, n_i} dV + \sum_{i=1}^{C} \left(\frac{\partial U}{\partial n_i}\right)_{S, V, n_J} dn_i$$

or $dU = TdS - Pdv + \Sigma\mu_i\, dn_i$

where $$\mu_i = \left(\frac{\partial U}{\partial n_i}\right)_{S, V, n_J} = \text{Chemical potential of species } i.$$

Similarly, the enthalpy $H$ is given by

$$H = H(S, P, n_1, n_2, \ldots, n_c)$$

or $$dH = \left(\frac{\partial H}{\partial S}\right)_{P, n_i} dS + \left(\frac{\partial H}{\partial P}\right)_{S, n_i} dP + \sum_{i=1}^{C} \left(\frac{\partial H}{\partial n_i}\right)_{S, P, n_J} dn_i$$

or $dH = Tds + VdP + \Sigma\mu_i dn_i$

where $$\mu_i = \left(\frac{\partial H}{\partial n_i}\right)_{S, P, n_J} \quad (J \neq i)$$

The Helmholtz free energy $A$ is given by

$$A = A(T, V, n_1, n_2, \ldots, n_c)$$

$$dA = \left(\frac{\partial A}{\partial T}\right)_{V, n_i} dT + \left(\frac{\partial A}{\partial V}\right)_{T, n_i} dV + \sum_{i=1}^{C} \left(\frac{\partial A}{\partial n_i}\right)_{T, V, n_J} dn_i$$

or $dA = -SdT - PdV + \sum_{i=1}^{C} \mu_i dn_i$

Where $$\mu_i = \left(\frac{\partial A}{\partial n_i}\right)_{T,V\,n_j}$$

The Gibbs free energy $G$ of the multicomponent system is given by

$$G = G(T, P, n_1, n_2, ..., n_c)$$

$$dG = \left(\frac{\partial G}{\partial T}\right)_{P,\,n_i} dT + \left(\frac{\partial G}{\partial P}\right)_{T,n_i} dP + \sum_{i=1}^{C}\left(\frac{\partial G}{\partial n_i}\right)_{T,\,P,\,n_j} dn_i$$

$$dG = -SdT + VdP + \sum_{i=1}^{C} \mu_i\, dn_i$$

Where $\mu_i = \left(\frac{\partial G}{\partial n_i}\right)_{T,\,P,\,n_j}$

**Example 6.16**

Explain the meaning of the terms available energy and unavailable energy.

*Solution* :

We know that no heat engine can have an efficiency equal to one. In other words all the energy absorbed as heat by an engine cannot be converted into work. The second law of thermodynamics dictates the limit to which the energy absorbed by an engine can be converted into work. Suppose a thermal reservoir at constant temperature $T$ is available. If a quantity of energy $Q$ is absorbed from the reservoir by a Carnot engine operating between the thermal reservoir at $T$ and the ambient atmosphere at $T_0$ then the work delivered by the engine is given by

$$W = \eta Q = \left(1 - \frac{T_0}{T}\right) Q = Q - T_0 \frac{Q}{T} = Q - T_0 |\Delta S|$$

where $\Delta S$ is the change in the entropy of the reservoir. The Carnot cycle is shown on a *T-s* diagram in Fig.E 6.16 (*a*). The work done by the engine is the area enclosed by 1 – 2 – 3 – 4. This represents the available energy or the portion of the energy which can be converted into work. The area enclosed by 4 – 3 – $B$ – $A$ represents the energy which is rejected to a sink. This denotes the unavailable energy or the portion of the energy which cannot be converted into work. The available energy is given by $Q - T_0 |\Delta S|$

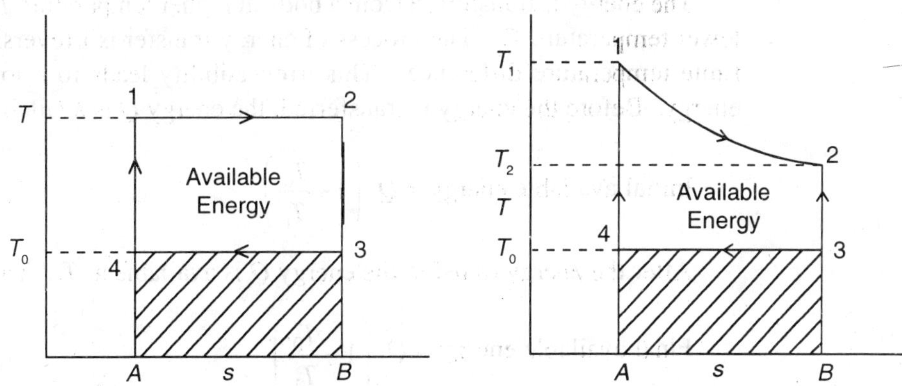


(*a*) (*b*)

Fig.E 6.15. Heat engine cycle on *T*-*s* diagram with

(*a*) a constant temperature thermal reservoir as a source and

(*b*) a finite body as a source.

Suppose a finite body which is initially at temperature $T_1$ is used as a source till its temperature reduces to $T_2$, then the heat engine cycle is shown in Fig.E 6.16 (*b*). This cycle can be treated as a sum of a large number of differential Carnot cycles. If $dQ$ is the energy absorbed as heat at temperature $T$ during a differential cycle, then the work $dW$ that can be obtained is given by

$$dW = \eta dQ = \left(1 - \frac{T_0}{T}\right) dQ$$

Then, the total work that can be obtained is given by

$$W \int_{T_1}^{T_2} \left(1 - \frac{T_0}{T}\right) dQ = \int dQ - T_0 \int \frac{dQ}{T} = Q - T_0 \left|\Delta S\right|$$

Therefore, the available energy is $Q - T_0 \left|\Delta S\right|$

### Example 6.17

Suppose a certain quantity of energy $Q$ is transferred from a body at constant temperature $T_1$ to another body at constant temperature $T_2$ ($T_1 > T_2$). Is there any loss in the available energy due to the process of energy transfer, if the ambient temperature is $T_0$ ? If so why and what is the loss in the available energy ?

*Solution :*

The energy is transferred from a body at higher temperature $T_1$ to a body at a lower temperature $T_2$. The process of energy transfer is irreversible because of finite temperature difference. This irreversibility leads to a loss of available energy. Before the energy is transferred, the energy $Q$ is available at $T_1$. Then

$$\text{Initial available energy} = Q\left(1-\frac{T_0}{T_1}\right)$$

After the energy transfer, the energy $Q$ is available at $T_2$. Therefore

$$\text{Final available energy} = Q\left(1-\frac{T_0}{T_2}\right)$$

$$\text{Loss in available energy} = Q\left(1-\frac{T_0}{T_1}\right) - Q\left(1-\frac{T_0}{T_2}\right)$$

$$= Q\left[1-\frac{T_0}{T_1}-1+\frac{T_0}{T_2}\right] = T_0\left(\frac{Q}{T_2}-\frac{Q}{T_1}\right)$$

$$= T_0\,(\Delta S_{\text{Source}} + \Delta S_{\text{Sink}}) = T_0 \Delta S_{\text{uni}}$$

where $\Delta S_{\text{uni}}$ is the total change in the entropy of the universe (the combination of the two interacting systems). If the process is irreversible, $\Delta S_{\text{uni}} > 0$ and hence there is a loss in available energy.

**Example 6.18**

Two identical steel blocks each of mass 1000 kg are available at two different temperatures. The first block is at 1200 K while the second block is at 600 K. The ambient temperature is 300 K. The specific heat of steel is 0.5 kJ/kg K. The blocks can be used as a source of energy to obtain work, till they are cooled to 300 K. Calculate (*a*) the available energy of each block (*b*) the available energy if the blocks are initially placed in thermal contact to reach a state of equilibrium and (*c*) the loss in available energy by allowing the blocks to reach thermal equilibrium.

*Solution :*

(*a*) The energy that can be extracted from each block is given by

$$Q = mC_P\,\Delta T$$

Energy extracted from First block, $Q_1 = 1000 \times 0.5 \times (1200 - 300) = 450$ MJ

Entropy change of First block,

$$\Delta S_1 = mC \ln \frac{T_0}{T_1} = 1000 \times 0.5 \ln \frac{300}{1200} = -693.15 \text{ kJ/K}$$

Available energy of First block

$= Q_1 - T_0 |\Delta S_1| = 450\,000 - 300 \times 693.15 = 242.055$ MJ

Energy extracted from Second block, $Q_2 = 1000 \times 0.5 \times (600 - 300)$

$= 150$ MJ

Entropy change of second block, $\Delta S_2 = 1000 \times 0.5 \ln \dfrac{300}{600} = -346.57$ kJ/K

Available energy of second block $= Q_2 - T_0 |\Delta S_2|$

$= 150\,000 - 300 \times 346.57 = 46.029$ MJ

Total available energy of both the blocks

$= 242.055 + 46.029 = 288.084$ MJ

(*b*) Let $T_f$ = Equilibrium temperature of the blocks.

Energy balance gives

$mC (T_f - 1200) + mC (T_f - 600) = 0$

or $T_f = 900$ K

Energy extracted from both the blocks $= 2\, mC (T_f - 300)$

$= 2 \times 1000 \times 0.5\, (900 - 300) = 600$ MJ

Entropy change of blocks

$$= 2mC \ln \frac{T_0}{T_f} = 2 \times 1000 \times 0.5 \ln \frac{300}{900} = -1098.61 \text{ kJ/K}$$

Available energy $= Q - T_o |\Delta S| = 600\,000 - 300 \times 1098.61$

$= 270.417$ MJ

(*c*) Loss in available energy due to energy transfer between the blocks

$= 288.084 - 270.417 = 17.667$ MJ

**Example 6.19**

In a heat exchanger 100 kg saturated liquid water at 200°C is converted into saturated vapor at the same temperature by passing air at 2 bar and 350°C. The air leaves the heat exchanger at 250°C and at the same pressure. If $C_P$ of air is 1.005 kJ/kg K and the ambient temperature is 300 K, determine the loss in the available energy due to the process of exchanging energy from air to water.

*Solution* :

At 200°C, enthalpy of saturated liquid water = 852.37 kJ/kg

Enthalpy of saturated vapor= 2790.9 kJ/kg

Energy gained by water = 100 × (2790.9 – 852.37) × $10^3$ = 193.853 MJ

Energy lost by air = 193.853 MJ = $(mC_P \Delta T)_{air}$

or
$$m = \frac{193.853 \times 10^6}{1.005 \times 10^3 \times (350 - 250)} = 1928.9 \text{ kg}$$

$$\Delta S_{air} = mC_P \ln \frac{T_2}{T_1} = 1928.9 \times 1.005 \ln \frac{523}{623} = -339.18 \text{ kJ/K}$$

$$\Delta S_{water} = m\,(s_g - s_f) = 100 \times (6.4278 - 2.3307) = 409.71 \text{ kJ/K}$$

Loss in available energy = $T_0 \Delta S_{uni} = T_0 (\Delta S_{air} + \Delta S_{water})$

$$= 300 \times (-339.18 + 409.71) \times 10^3 = 21.159 \text{ MJ}$$

**Example 6.20**

What is availability ?

*Solution* :

The availability of a given system is defined as the maximum useful work that can be obtained in a process in which the system comes to equilibrium with the surroundings or attains a dead state.

Suppose a system is initially available at temperature $T$ and pressure $P$ while the surroundings are $T_0$ and $P_0$. The system can undergo a process in which it reaches a state of thermal equilibrium and mechanical equilibrium with the surroundings and delivers some work. When the system is in equilibrium with the surroundings, there is no possibility of obtaining further work from such a system. In other words, the system attains a *dead state*. Then the total work delivered by the system will be maximum. A part of the total work delivered by the system is used in pushing the surrounding atmosphere, which will not be available to move a body. Therefore, the maximum useful work that can be obtained from a system is less than the maximum work done by the system during a given process. This maximum useful work delivered by the system is called the availability of the system.

**Example 6.21**

What is the availability function or availability for a non-flow process ?

*Solution* :

We know that the availability is the maximum useful work that can be obtained from a system during a given process. The availability is less than the maximum work delivered by the system by the amount of work that is spent in pushing back the atmosphere. Suppose a system does work while it is interacting with the surroundings at $T_0$ in such a way that the initial and final temperatures of the system are identical with the surroundings temperature. Then the maximum work that can be obtained is equal to the decrease in the Helmholtz free energy of the system [see Example 6.4]. Suppose the system is initially in the state $P_1$, $V_1$ and reaches the final state $P_0$, $V_0$ which is in equilibrium with the surroundings. Then

$$W_{max} = A_1 - A_0 = (U_1 - T_1 S_1) - (U_0 - T_0 S_0) \quad \text{where } T_1 = T_0$$

The change in the volume of the system = $V_0 - V_1$

The work done by the system in pushing the atmosphere = $P_0 (V_0 - V_1)$.

Then, the maximum useful work or availability is given by

$$\begin{aligned} \text{Availability} = W_{\text{Maximum useful}} &= W_{max} - P_0(V_0 - V_1) \\ &= (U_1 - T_0 S_1) - (U_0 - T_0 S_0) - P_0 (V_0 - V_1) \\ &= (U_1 + P_0 V_1 - T_0 S_1) - (U_0 + P_0 V_0 - T_0 S_0) \\ &= \phi_1 - \phi_0 \end{aligned}$$

where $\phi = U + P_0 V - T_0 S$ = Availability function for non-flow process

Suppose a system is in the initial state 1, where the availability is $(\phi_1 - \phi_0)$ and reaches the final state 2, where the availability is $(\phi_2 - \phi_0)$ during a given process, then the change in the availability or the change in the maximum useful work associated with the process is

$$(\phi_2 - \phi_0) - (\phi_1 - \phi_0) = \phi_2 - \phi_1$$

**Example 6.22**

If the ambient conditions are 1 bar and 25°C, what is the availability of steam at 5 bar and 300°C ?

*Solution* :

At 5 bar and 300°C : $v_1 = 0.5226$ m³/kg ; $h_1 = 3064.8$ kJ/kg

$$s_1 = 7.4614 \text{ kJ/kg K}$$

$$\begin{aligned} \phi_1 &= u_1 + P_0 v_1 - T_0 s_1 = h_1 - P_1 v_1 + P_0 v_1 - T_0 s_1 \\ &= h_1 - v_1 (P_1 - P_0) - T_0 s_1 \end{aligned}$$

At 25° and 1 bar, the water is compressed liquid. However, the properties of the compressed liquid can be taken as equal to that of the saturated liquid at 25°C.

At 25°C : $v_f = 0.001\ 002\ 9$ m³/kg ; $h_f = 104.77$ kJ/kg $= u_f$

$$s_f = 0.3670 \text{ kJ/kg K}$$

$$\phi_0 = u_0 + P_0 v_1 - T_0 s_0 = h_0 + P_0 v_0 - T_0 s_0$$

Then, availability $= \phi_1 - \phi_0 = \{h_1 - v_1 (P_1 - P_0) - T_0 s_1\} - \{h_0 + P_0 v_0 - T_0 s_0\}$

$$= (h_1 - h_0) + P_0(v_1 - v_0) - T_0 (s_1 - s_0) - P_1 v_1$$

or $\phi_1 - \phi_0 = (3064.8 - 104.77)10^3 + (0.5226 - 0.001\ 002\ 9) \times 10^5$

$- 298\ (7.4614 - 0.3670) \times 10^3 - 5 \times 10^5 \times 0.5226$

$$= 636.76 \text{ kJ/kg}$$

**Example 6.23**

In a process plant steam at 30 bar and 300°C is available. In some other section of the plant, it is necessary to use steam at 1 bar and 200°C. It is proposed to obtain some work from the steam at 30 bar and 300°C in a suitable process and reduce its conditions to 1 bar and 200°C, before it is supplied to the second section. Determine the maximum useful work that can be obtained per kg of steam in the process if the ambient conditions are 1 bar and 300 K.

*Solution* :

The maximum useful work that can be obtained $= \phi_1 - \phi_2$

$$\phi_1 - \phi_2 = (u_1 + P_0 v_1 - T_0 s_1) - (u_2 + P_0 v_2 - T_0 s_2)$$

$= (h_1 - P_1 v_1 + P_0 v_1 - T_0 s_1) - (h_2 - P_2 v_2 + P_0 v_2 - T_0 s_2)$

$= (h_1 - h_2) + P_0 (v_1 - v_2) - T_0 (s_1 - s_2) - P_1 v_1 + P_2 v_2$

For steam at 30 bar and 300°C :

$v_1 = 0.081\ 16$ m³/kg ; $h_1 = 2995.1$ kJ/kg ; $s_l = 6.5422$ kJ/kg K

For steam at 1 bar and 200°C :

$v_2 = 2.1720$ m³/kg ; $h_2 = 2875.4$ kJ/kg ; $s_2 = 7.8349$ kJ/kg K

Therefore,

$\phi_1 - \phi_2 = (2995.1 - 2875.4)10^3 + 1 \times 10^5\ (0.081\ 16 - 2.1720)$

$- 300\ (6.5422 - 7.8349) \times 10^3 - 30 \times 10^5 \times 0.081\ 16 + 1 \times 10^5 \times 2.1720$

$$= 270.83 \text{ kJ/kg}$$

**Example 6.24**

What is the availability function for a flow process ?

*Solution* :

We know that the maximum work per unit mass that can be obtained from a device operating in steady – state, steady flow conditions in which the inlet and outlet temperatures of the fluid flowing through it are equal to the surroundings temperature with which it is interacting, is equal to the decrease in the specific Gibbs free energy of the fluid flowing through the device [see Example 6.7]. That is

$$\frac{\dot{W}_{s\,max}}{\dot{m}} = (g_i - g_e) = (h_i - T_i\, s_i) - (h_e - T_e\, s_e)$$

where $T_i = T_e = T_0$ = surroundings temperature

Therefore

$$\frac{\dot{W}_{s\,max}}{\dot{m}} = (h_i - T_0 s_i) - (h_e - T_0 s_e)$$

or $\dot{W}_{S\,max} = \dot{m}\,(h_i - T_0 s_i) - \dot{m}\,(h_i - T_0 s_e) = (\dot{H}_i - T_0 \dot{S}_i) - (\dot{H}_e - T_0 \dot{S}_e)$

or $$W_{S\,max} = (H_i - T_0 S_i) - (H_e - T_0 S_e)$$

$\dot{W}_s$ represents the shaft power output which is useful power output only and no shaft power output is consumed in pushing back the atmosphere. The maximum work can be obtained from the device if the fluid leaving the device is in equilibrium with the surroundings or attains the dead state. Therefore, the maximum useful work done by a steady flow device is given by

$$W_{S\,useful} = (H_i - T_0 S_i) - (H_0 - T_0 S_0) = B_i - B_0$$

where $B = H - T_0 S$ = Availability for a flow process.

The function $B$ is sometimes called as Darrieus function.

Quite often an engineer encounters a situation in which the material flowing through a device undergoes a change from state 1 to state 2. Then the useful work that can be obtained is given by

$$W_{S\,useful} = (B_1 - B_0) - (B_2 - B_0) = B_1 - B_2$$

In several industrial processes the material flowing through a device interacts with some other fluids or bodies instead of with the surroundings. In analyzing such cases one should consider the combination of both the interacting fluids or bodies to estimate the change in the availability.

### Example 6.25

Air at 50 bar and 700 K enters a turbine operating in steady state and leaves at 2 bar and 350 K. Considering air to be an ideal gas with $\gamma$ = 1.4 (*a*) calculate the power delivered by the turbine if the flow rate of air is 100 mol/s. (*b*) If the ambient conditions are 1 bar and 300 K, determine the maximum power that can be delivered by the turbine.

*Solution* :

(*a*) We know that

$$W_S = \dot{m}\,(h_i - h_e) = \dot{m}\, C_P\,(T_i - T_e)$$

and $C_P = \dfrac{R\gamma}{\gamma - 1} = \dfrac{8.314 \times 1.4}{0.4} = 29.099$ J / mol K

Therefore

$$\dot{W}_S = 100 \times 29.099\,(700 - 350) = 1.018 \text{ MW}$$

(*b*) The maximum useful power delivered is given by

$$\dot{W}_{S\text{ useful}} = (B_1 - B_2) = \dot{m}\,[(h_1 - T_0 s_1) - (h_2 - T_0 s_2)]$$

$$= \dot{m}\,[(h_1 - h_2) - T_0\,(s_1 - s_2)] = \dot{m}\,[(C_P\,(T_1 - T_2) + T_0\,(s_2 - s_1)]$$

$$= \dot{m}\left[C_P\,(T_1 - T_2) + T_0\left\{C_P \ln\frac{T_2}{T_1} - R\ln\frac{P_2}{P_1}\right\}\right]$$

$$= 100 \times \left[29.099\,(700 - 350) + 300 \times \left\{29.099 \ln\frac{350}{700} - 8.314 \ln\frac{2}{50}\right\}\right]$$

$$= 1.216 \text{ MW}$$

**Example 6.26**

A heat exchanger is designed to cool 10 mol/s of chlorine gas from 1 bar and 700 K to 350 K and 1 bar by a stream of air entering at 1 bar and 300 K. The specific heats of chlorine gas and air are 34 J/mol K and 29 J/mol K, respectively. The chlorine gas and air enter the heat exchanger at opposite ends. The air leaves the heat exchanger at 400 K and 1 bar. Assume that the heat exchanger is perfectly insulated and the ambient temperature is 300 K. Calculate

(*a*) the entropy change of chlorine gas and air.

(*b*) the change in availability of chlorine gas.

(*c*) the change in availability of air, and

(*d*) the loss in available energy because of energy transfer from chlorine gas to air.

*Solution* :

A schematic of the heat exchanger is shown in Fig.E 6.26.

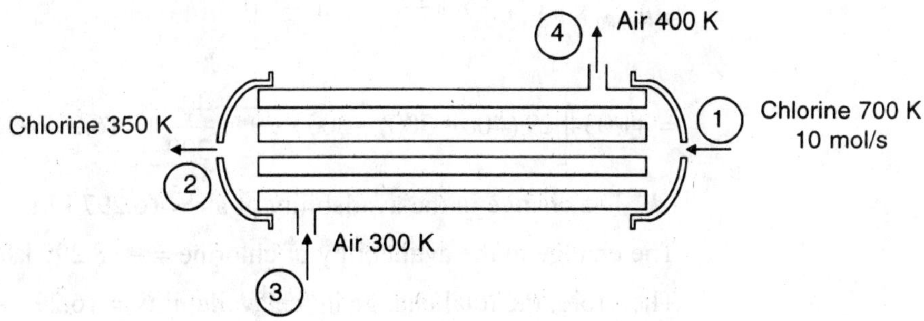


Fig.E 6.26. Schematic of a heat exchanger.

(*a*) Energy balance gives

Energy lost by chlorine gas = Energy gained by air

or $$[\dot{m}\, C_P\, (T_1 - T_2)]_{Cl_2} = [\dot{m} C_P\, (T_4 - T_3)]_{air}$$

$$10 \times 34 \times (700 - 350) = \dot{m}_{air} \times 29 \times (400 - 300)$$

or Flow rate of air, $\dot{m}_{air} = 41.034$ mol/s

$$\Delta S_{Cl_2} = \dot{m}\left[C_P \ln \frac{T_2}{T_1} - R \ln \frac{P_2}{P_1}\right] = \dot{m} C_P \ln \frac{T_2}{T_1} \quad (\text{Since } P_2 = P_1)$$

$$= 10 \times 34 \ln \frac{350}{700} = -\,235.67 \text{ J/K s}$$

$$\Delta \dot{S}_{air} = m_{air} C_P \ln \frac{T_4}{T_3} = 41.034 \times 29 \times \ln \frac{400}{300} = 342.338 \text{ J/ks}$$

(*b*) The change in availability of chlorine gas = $\dot{m}\, (B_2 - B_1)$

$$\dot{m}\, (B_2 - B_1) = \dot{m}\, [(h_2 - T_0 s_2) - (h_1 - T_0 s_1)]$$

$$= \dot{m}\left[C_P\, (T_2 - T_1) - T_0\, C_P \ln \frac{T_2}{T_1}\right]$$

$$= 10\left[34\,(350 - 700) - 300 \times 34 \ln \frac{350}{700}\right]$$

$$= -\,48.299 \text{ kJ/s}$$

(*c*) The change in availability of air = $\dot{m}_{air}(B_4 - B_3)$

$$\dot{m}_{air}(B_4 - B_3) = \dot{m}_{air}[(h_4 - T_0 s_4) - (h_3 - T_0 s_3)]$$

$$= \dot{m}_{air}\left[C_P(T_4 - T_3) - T_0 C_P \ln\frac{T_4}{T_3}\right]$$

$$= 41.034\left[29(400 - 300) - 300 \times 29 \ln\frac{400}{300}\right] = 16.297 \text{ kJ/s}$$

(*d*) The change in the availability of air = 16.297 kJ/s

The change in the availability of chlorine = – 48.299 kJ/s

Therefore, the total change in the availability = 16.297 – 48.299

= – 32.002 kJ/s

Hence, the loss in the availability = 32.002 kW

**Example 6.27**

What is meant by second law efficiency and what is the necessity of introducing the term second law efficiency ?

*Solution* :

We know that the thermal efficiency or simply the efficiency of a heat engine cycle is defined as the ratio of the net work done to the energy absorbed. This efficiency is based on the conversion of energy supplied as heat into work and it is usually called as the first law efficiency because the first law of thermodynamics deals with the conservation of energy. Suppose a certain quantity of energy $Q$ is available as heat at a higher temperature $T_1$ while the ambient temperature is $T_0$. Then the portion of $Q$ which can be converted into work, or the available energy is given by

$$Q\eta = Q\left(1 - \frac{T_0}{T_1}\right)$$

If the same amount of energy $Q$ is available as heat at temperature $T_2$ ($T_2 < T_1$), then the available energy is given by

$$Q\eta = Q\left(1 - \frac{T_0}{T_2}\right)$$

Since $T_2 < T_1$, we find that

$$Q\left(1 - \frac{T_0}{T_2}\right) < Q\left(1 - \frac{T_0}{T_1}\right)$$

That is, the available energy is high if the energy $Q$ as heat is available at a higher temperature. Even though the magnitude of $Q$ is identical, the potential or quality of $Q$ is higher if it is available at higher temperature.

To take into account the quality of energy, the term second law efficiency is introduced. The available energy depends on the entropy change of the source supplying the energy. Since the concept of entropy is based on the second law of thermodynamics, the performance of a device in maintaining the available energy is expressed in terms of second law efficiency. The second law efficiency ($\eta_2$) of a device is defined as

$$\eta_2 = \frac{\text{The change in the available energy of a system}}{\text{The change in the available energy of the source}}$$

**Example 6.28**

Define the second law efficiency of a compressor or pump.

*Solution* :

Suppose a fluid enters a compressor or pump at pressure $P_1$ and leaves at pressure $P_2$ when work $W$ is done on it. The compressor or pump receives work $W$ from an external agent such as an engine or electric motor, the work done on it represents the change in the available energy of the source. The path followed by the compression process is shown as 1–2 on a $T-s$ diagram in Fig.E 6.28.

The work done on the compressor or pump is greater than the minimum energy required because of the irreversibilities in the device. It is possible to change the state of the fluid from 1 to 2 by following the path $1-a-b-2$ which requires the minimum amount of energy. This minimum work is the change in the available energy of the system which is equal to $(B_2 - B_1)$. Then, the second law efficiency of a compressor or pump is given by

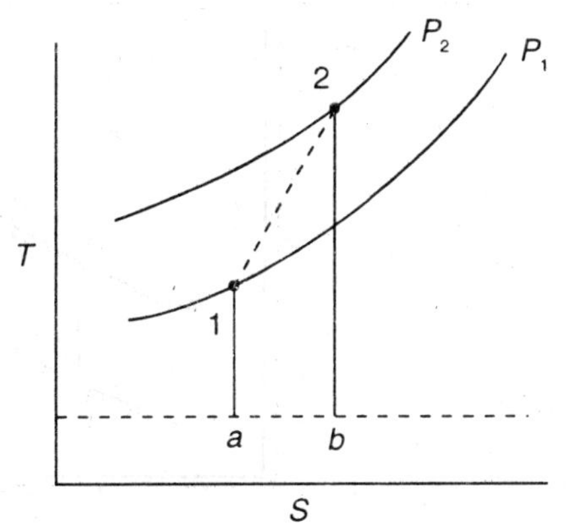


Fig.E 6.28. *T- s* diagram showing the compression process 1-2.

$$\eta_2 = \frac{\text{The change in the available energy of system}}{\text{The change in the available energy of source}} = \frac{B_2 - B_1}{W}$$

$$= \frac{(h_2 - T_0\, s_2) - (h_1 - T_0\, s_1)}{W} = \frac{(h_2 - h_1) - T_0\,(s_2 - s_1)}{W}$$

**Example 6.29**

Air at 1 bar and 300 K enters an adiabatic compressor and leaves at 10 bar and 650 K while the ambient temperature is 300 K. Calculate the isentropic efficiency and the second law efficiency of the compressor. For air $C_P = 29$ J/mol K.

*Solution* :

The $T - s$ diagram showing the actual compression process 1 – 2 and the isentropic compression process 1–2′ is shown in Fig.E 6.29.

For an isentropic compression $s_{2'} - s_1 = 0$.

$$s_{2'} - s_1 = C_P \ln \frac{T_{2'}}{T_1} - R \ln \frac{P_2}{P_1} = 0$$

or $29 \ln \dfrac{T_{2'}}{300} - 8.314 \ln \dfrac{10}{1} = 0$

or $T_{2'} = 580.5$ K

$$\text{Isentropic efficiency } \eta_C = \frac{\text{Work needed by an isentropic compressor}}{\text{Work done on actual compressor}}$$

$$\eta_C = \frac{h_{2'} - h_1}{h_2 - h_1} = \frac{C_P\,(T_{2'} - T_1)}{C_P\,(T_2 - T_1)} = \frac{580.5 - 300}{650 - 300} = 0.8014$$

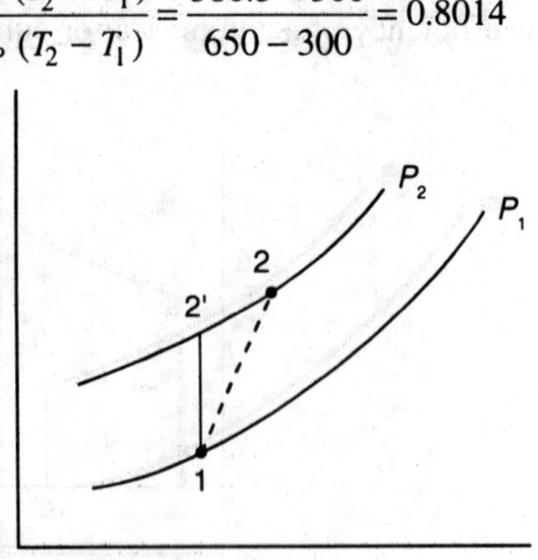


Fig.E 6.29. Sketch for Example 6.29. The path 1- 2 shows the actual compression process and the path 1- 2′ denotes the compression process if the compressor were isentropic.

The second law efficiency $\eta_2 = \dfrac{B_2 - B_1}{W} = \dfrac{(h_2 - h_1) - T_0 (s_2 - s_1)}{(h_2 - h_1)}$

or $$\eta_2 = 1 - \frac{T_0 \left\{ C_P \ln \dfrac{T_2}{T_1} - R \ln \dfrac{P_2}{P_1} \right\}}{C_P (T_2 - T_1)}$$

$$= 1 - \frac{300 \left\{ 29 \ln \dfrac{650}{300} - 8.314 \ln \dfrac{10}{1} \right\}}{29\,(650 - 300)} = 0.9031$$

**Example 6.30**

Define the second law efficiency of a turbine.

*Solution* :

Suppose a fluid at pressure $P_1$ enters a turbine operating in steady–state and leaves at pressure $P_2$ and the turbine delivers work $W$ per unit mass of the fluid flowing through it. The $T$– $s$ diagram showing the actual expansion process 1–2 and a hypothetical process 1–$a$–$b$–2 which delivers the maximum work is shown in Fig.E 6.30.

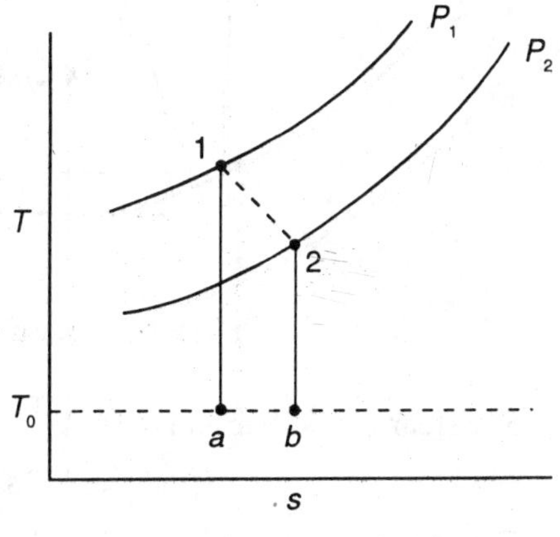


Fig.E 6.30. *T-s* diagram showing adiabatic expansion 1-2 in a turbine. The process 1-*a*-*b*-2 is a hypothetical process which delivers maximum work.

The actual work done, $W = h_1 - h_2$ is less than the maximum possible because of inherent irreversibility in the turbine. The actual work delivered by the turbine represents the change in the availability of the system. The maximum possible work represents the change in the availability ($B_1 - B_2$) of the source. Therefore, the second law efficiency of the turbine is given by.

$$\eta_2 = \frac{\text{The change in the availability of the system}}{\text{The change in the availability of the source}} = \frac{W_{actual}}{B_1 - B_2}$$

$$\eta_2 = \frac{h_1 - h_2}{(h_1 - T_0\, s_1) - (h_2 - T_0\, s_2)} = \frac{h_1 - h_2}{(h_1 - h_2) + T_0\,(s_2 - s_1)}$$

**Example 6.31**

In a thermal power plant superheated steam at 50 bar and 400°C enters an adiabatic turbine and leaves as wet steam of quality 0.90 at 0.1 bar while the ambient temperature is 30°C. Calculate the second law efficiency of the turbine.

*Solution* :

The $T - s$ diagram showing the adiabatic expansion process is shown in Fig.E 6.31.

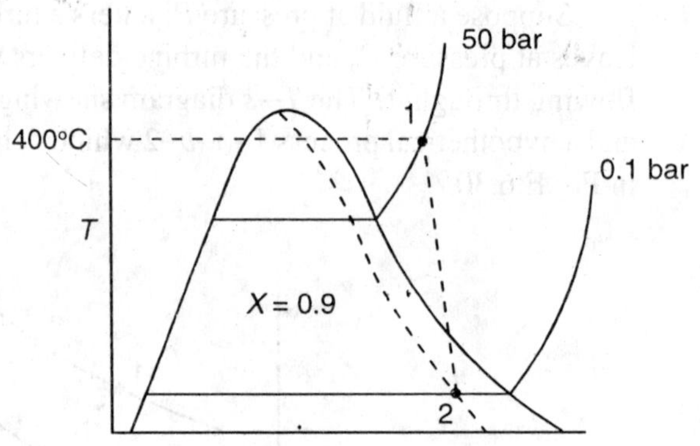


Fig.E 6.31. Sketch for Example 6.31.

For steam at 50 bar and 400°C.

$$h_1 = 3198.3 \text{ kJ/kg} \; ; \; s_1 = 6.6508 \text{ kJ/kg K}$$

For steam at 0.1 bar:

$$h_f = 191.83 \text{ kJ/kg} \; ; \; h_g = 2584.8 \text{ kJ/kg}$$

$$s_f = 0.6493 \text{ kJ/kg K} \; ; \; s_g = 8.1511 \text{ kJ/kg K}$$

$$h_2 = X_2 h_g + (1 - X_2)\, h_f = 0.9 \times 2584.8 + 0.1 \times 191.83 = 2345.50 \text{ kJ/kg}$$

$$s_2 = X_2 s_g + (1 - X_2)\, s_f = 0.9 \times 8.1511 + 0.1 \times 0.6493 = 7.4009 \text{ kJ/kg K}$$

The second efficiency $\eta_2$ of the turbine is given by

$$\eta_2 = \frac{h_1 - h_2}{(h_1 - h_2) + T_o (s_2 - s_1)}$$

$$= \frac{3198.3 - 2345.50}{(3198.3 - 2345.50) + 303\,(7.4009 - 6.6508)}$$

$$= 0.7896$$

**Example 6.32**

Heat exchangers are widely used in process industries for exchanging energy as heat from one fluid to another. It is neither a work delivering device like a heat-engine or turbine nor a work consuming device like a compressor. How does one define the second law efficiency of a heat exchanger ?

*Solution* :

Consider a heat exchanger shown in Fig.E 6.32 which is in steady – state operation.

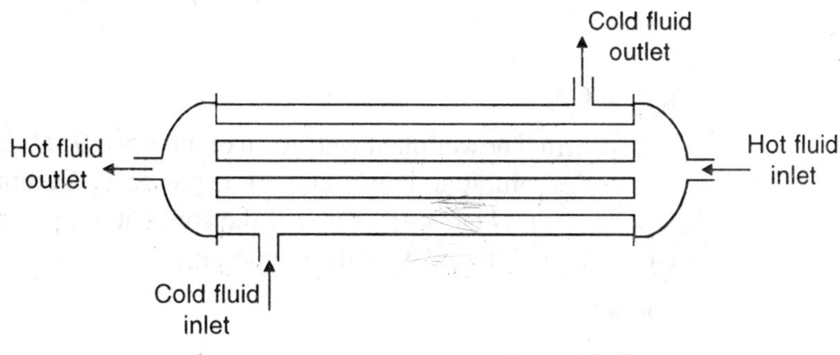


Fig.E 6.32. Schematic of heat exchanger.

Let

$\dot{m}_H$ = mass flow rate of hot fluid through the heat exchanger.

$\dot{m}_C$ = mass flow rate of cold fluid through the heat exchanger.

$h_{H1}, h_{H2}$ = specific enthalpy of the entering and leaving hot fluid, respectively.

$h_{C1}, h_{C2}$ = specific enthalpy of the entering and leaving cold fluid, respectively.

$s_{H1}, s_{H2}$ = specific entropy of the entering and leaving hot fluid, respectively.

$s_{C1}, s_{C2}$ = specific entropy of the entering and leaving cold fluid, respectively.

$T_0$ = ambient temperature

If the heat exchanger is adiabatic, an energy balance gives

$$\dot{m}_H (h_{H1} - h_{H2}) = \dot{m}_C (h_{C2} - h_{C1})$$

The decrease in the availability of the source is the change in the availability of the hot fluid and it is given by

$$(B_1 - B_2)_H = \dot{m}_H \left[(h_{H1} - T_0 s_{H1}) - (h_{H2} - T_0 s_{H2})\right]$$
$$= \dot{m}_H \left[(h_{H1} - h_{H2}) - T_0 (s_{H1} - s_{H2})\right]$$

The change in the availability of the system is equal to the change in the availability of the cold fluid and is given by

$$(B_2 - B_1)_C = \dot{m}_C \left[(h_{C2} - T_0 s_{C2}) - (h_{C1} - T_0 s_{C1})\right]$$
$$= \dot{m}_C \left[(h_{C2} - h_{C1}) - T_0 (s_{C2} - s_{C1})\right]$$

The second law efficiency $\eta_2$ of a heat exchanger is defined as

$$\eta_2 = \frac{\text{The change in the availability of the system}}{\text{The change in the availability of the source}} = \frac{(B_2 - B_1)_C}{(B_1 - B_2)_H}$$

$$= \frac{\dot{m}_C \left[(h_{C2} - h_{C1}) - T_0 (s_{C2} - s_{C1})\right]}{\dot{m}_H \left[(h_{H1} - h_{H2}) - T_0 (s_{H1} - s_{H2})\right]}$$

## Example 6.33

In a thermal power plant wet steam of quality 0.9 at 0.1 bar enters a condenser and leaves as saturated liquid at the same pressure. Cooling water at 25°C enters the condenser and leaves at 30°C. If the ambient temperature is 300 K, calculate the second law efficiency of the condenser.

*Solution* :

For steam at 0.1 bar:

$$h_f = 191.83 \text{ kJ/kg} \; ; \; h_g = 2584.8 \text{ kJ/kg}$$
$$s_f = 0.6493 \text{ kJ/kg K} \; ; \; s_g = 8.1511 \text{ kJ/kg K}$$
$$h_{H1} = 0.9 \times 2584.8 + 0.1 \times 191.83 = 2345.50 \text{ kJ/kg}$$
$$s_{H1} = 0.9 \times 8.1511 + 0.1 \times 0.6493 = 7.4009 \text{ kJ/kg K}$$
$$h_{H2} = 191.83 \text{ kJ/kg} \; ; \; s_{H2} = 0.6493 \text{ kJ/kg K}$$

At 25°C : $h_f = 104.77 \text{ kJ/kg} = h_{C1}$

$s_f = 0.3670 \text{ kJ/kg K} = s_{C1}$

At 30°C : $h_f = 125.66 \text{ kJ/kg} = h_{C2}$

$s_f = 0.4365 \text{ kJ/kg K} = s_{C2}$

Let $\dot{m}_C$ = mass flow rate of cooling water per kg of steam. An energy balance on the condenser gives

$$(h_{H1} - h_{H2}) = \dot{m}_C (h_{C2} - h_{C1})$$

or $2345.50 - 191.83 = \dot{m}_C (125.66 - 104.77)$

or $\dot{m}_C = 103.1$ kg cooling water per kg steam

$$\eta_2 = \frac{103.1[125.66 - 104.77 - 300(0.4365 - 0.3670)]}{2345.50 - 191.83 - 300(7.4009 - 0.6493)}$$

$$= \frac{103.1[125.66 - 104.77 - 300(0.4365 - 0.3670)]}{2345.50 - 191.83 - 300(7.4009 - 0.6493)}$$

or $\eta_2 = 0.0322$

**Example 6.34**

Define the second law efficiency of a heat engine.

*Solution :*

Consider the heat engine cycle 1–2–3–4 shown in FigE. 6.34. It recevies energy $Q_H$ as heat from the reservoir at $T$ and rejects energy to a low temperature reservoir at $T_0$ while it does work $W$. The available energy needed to deliver a certain amount of work by a reversible engine is less than the available energy needed by an irreversible engine to deliver the same amount of work because the loss in available energy in a reversible engine is zero. Therefore, the work delivered by an engine represents the minimum available energy required by a heat engine or the change in the available energy of the system. The available energy needed by the actual engine represents the change in the available energy of the source. The second law efficiency of a heat engine cycle can be defined as

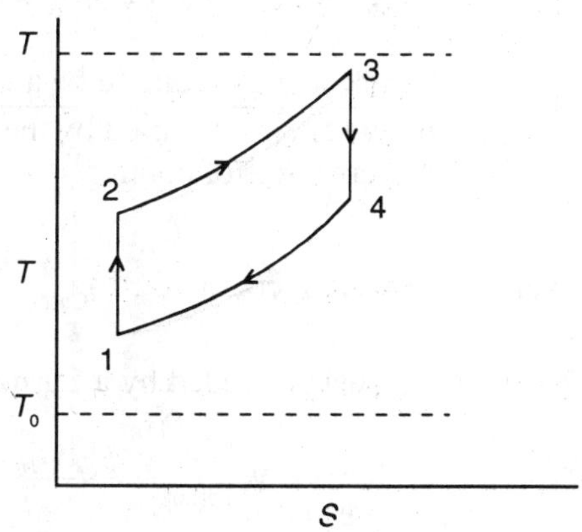


Fig.E 6.34. *T-s* diagram of a hypothetical heat engine cycle.

$$\eta_2 = \frac{\text{Work delivered (or the minimum available energy needed) by an engine}}{\text{Available energy required by the actual engine}}$$

The available energy absorbed by the heat engine $= Q_H\left(1 - \frac{T_0}{T}\right)$

Therefore, the second law efficiency $\eta_2$ of a heat engine is given by

$$\eta_2 = \frac{W}{Q_H\left(1 - \frac{T_0}{T}\right)} = \frac{W / Q_H}{1 - \frac{T_0}{T}} = \frac{\eta}{\eta_{\text{Carnot}}}$$

where $\eta = \frac{W}{Q_H}$ = thermal efficiency of the heat engine.

**Example 6.35**

Define the second law efficiency of a refrigerator and heat pump.

*Solution :*

The objective of a refrigerator is to absorb energy in the form of heat from a low temperature body and reject energy as heat to a high temperature body when work is done on the device. The work to be done on a refrigerator, for a given rate of cooling, will be minimum only when the refrigerator operates on a reversible cycle. A Carnot refrigerator, operating between two thermal reservoirs, requires the minimum available energy for a given rate of cooling. The amount of available energy (or work) required by any practical refrigerator will be larger compared to a Carnot refrigerator. Hence, a Carnot refrigerator is used as a standard of perfection against which the performance of practical refrigerators can be compared. The second law efficiency $\eta_2$ of a refrigerator is defined as

$$\eta_2 = \frac{\text{Available energy required by a Carnot refrigerator}}{\text{Available energy required by the actual refrigerator}}$$

for the same rate of cooling.

For a Carnot refrigerator $(\text{COP})_{\text{Carnot}} = \frac{Q_L}{W} = \frac{T_L}{T_H - T_L}$

or the available energy needed by a Carnot refrigerator,

$$W_{\text{Carnot}} = \frac{Q_L\,(T_H - T_L)}{T_L}$$

The coefficient of performance COP of the actual refrigerator is given by

$$(COP)_R = \frac{Q_L}{W_{actual}} \text{ or } W_{actual} = \frac{Q_L}{(COP)_R}$$

Hence, the second law efficiency $\eta_2$ of a refrigerator is given by

$$\eta_2 = \frac{W_{Carnot}}{W_{actual}} = \frac{Q_L\,(T_H - T_L)/T_L}{Q_L/(COP)_R} = \frac{(COP)_R}{(COP)_{Carnot}}$$

Similarly, the second law efficiency $\eta_2$ of a heat pump is defined as

$$\eta_2 = \frac{\text{Available energy needed by a Carnot heat pump}}{\text{Available energy needed by the actual heat pump}}$$

to perform the same task

$$= \frac{W_{Carnot\ H.P}}{W_{actual\ H.P}}$$

The COP of a heat pump is given by

$$(COP)_{H.P} = \frac{Q_H}{W_{actual\ H.P}} \quad \text{or} \quad W_{actual\ H.P} = \frac{Q_H}{(COP)_{H.P}}$$

For a Carnot heat pump, we have

$$(COP)_{Carnot H.P} = \frac{Q_H}{W_{Carnot\ H.P}} \text{ or } W_{Carnot\ H.P} = \frac{Q_H}{(COP)_{Carnot\ H.P}}$$

Hence the second law efficiency $\eta_2$ of a heat pump is given by

$$\eta_2 = \frac{Q_H/(COP)_{Carnot\ H.P}}{Q_H/(COP)_{H.P}} = \frac{(COP)_{H.P}}{(COP)_{Carnot\ H.P}}$$

**Example 6.36**

A vapor compression refrigerator employing Freon – 12 as the refrigerant operates between the temperature levels – 20°C and 40°C with a COP of 3.112. Calculate the second law efficiency of the refrigerator.

*Solution :*

Given $(COP)_R = 3.112$

$$(COP)_{Carnot} = \frac{T_L}{T_H - T_L} = \frac{253}{313 - 253} = 4.2167$$

Therefore
$$\eta_2 = \frac{(COP)_R}{(COP)_{Carnot}} = \frac{3.112}{4.2167} = 0.738$$

# 7

# Thermodynamic Relations

**Example 7.1**

Suppose there exists a relationship of the type $f(x, y, z) = 0$ among the three variables $x$, $y$ and $z$, show that the reciprocal relationship can be written as

$$\left(\frac{\partial x}{\partial y}\right)_Z = \frac{1}{(\partial y/\partial x)_Z}$$

*Solution* :

If there exists a relationship $f(x, y, z) = 0$ among the three variables $x$, $y$ and $z$, then one can write

$$x = x(y, z)$$

or
$$dx = \left(\frac{\partial x}{\partial y}\right)_z dy + \left(\frac{\partial x}{\partial z}\right)_y dz \qquad (A)$$

and
$$y = y(x, z) \quad \text{or} \quad dy = \left(\frac{\partial y}{\partial x}\right)_z dx + \left(\frac{\partial y}{\partial z}\right)_x dz \qquad (B)$$

Substitute for $dy$ from Eqn.($B$) in Eqn.($A$) to obtain

$$dx = \left(\frac{\partial x}{\partial y}\right)_z \left[\left(\frac{\partial y}{\partial x}\right)_z dx + \left(\frac{\partial y}{\partial z}\right)_x dz\right] + \left(\frac{\partial x}{\partial z}\right)_y dz$$

or
$$dx = \left(\frac{\partial x}{\partial y}\right)_z \left(\frac{\partial y}{\partial x}\right)_z dx + \left[\left(\frac{\partial x}{\partial y}\right)_z \left(\frac{\partial y}{\partial z}\right)_x + \left(\frac{\partial x}{\partial z}\right)_y\right] dz \qquad (C)$$

In Eqn. ($C$) the independent variables are $x$ and $z$ and hence

$$\left(\frac{\partial x}{\partial y}\right)_z\left(\frac{\partial y}{\partial x}\right)_z = 1 \quad \text{and} \tag{D}$$

$$\left(\frac{\partial x}{\partial y}\right)_z\left(\frac{\partial y}{\partial z}\right)_x + \left(\frac{\partial x}{\partial z}\right)_y = 0 \tag{E}$$

The Eqn. ($D$) can be rewritten as

$$\left(\frac{\partial x}{\partial y}\right)_z = \frac{1}{(\partial y/\partial x)_z}$$

Which is the reciprocal relation.

**Example 7.2**

Suppose the three variables $x$, $y$ and $z$ are related as $f(x, y, z) = 0$, show that the cyclic relation among the three variables can be written as

$$\left(\frac{\partial x}{\partial y}\right)_z\left(\frac{\partial y}{\partial z}\right)_x\left(\frac{\partial z}{\partial x}\right)_y = -1$$

*Solution* :

We have

$$\left(\frac{\partial x}{\partial y}\right)_z\left(\frac{\partial y}{\partial z}\right)_x + \left(\frac{\partial x}{\partial z}\right)_y = 0 \qquad \text{[see Eqn. (}E\text{) of Example 7.1]}$$

or
$$\left(\frac{\partial x}{\partial y}\right)_z\left(\frac{\partial y}{\partial z}\right)_x = -\left(\frac{\partial x}{\partial z}\right)_y \tag{A}$$

We know that the reciprocal relation gives

$$\left(\frac{\partial x}{\partial z}\right)_y = \frac{1}{(\partial z/\partial x)_y}$$

Substituting this in Eqn. ($A$), we get

$$\left(\frac{\partial x}{\partial y}\right)_z\left(\frac{\partial y}{\partial z}\right)_x = -\frac{1}{(\partial z/\partial x)_y} \quad \text{or} \quad \left(\frac{\partial x}{\partial y}\right)_z\left(\frac{\partial y}{\partial z}\right)_x\left(\frac{\partial z}{\partial x}\right)_y = -1$$

Which is the desired cyclic relation.

**Example 7.3**

Consider a single component simple thermodynamic system. For such a system we deal with the three measurable properties $P, V$ and $T$ and the four thermodynamic potentials $U, H, A$ and $G$ in addition to the important property $S$. Thus we normally deal with 8 variables and their partial derivatives. How many relations are possible among these derivatives ?

*Solution :*

Consider the three measurable variables $P, V$ and $T$. There exists a relationship $f(P, V, T) = 0$ among these three variables. Then one can express $V$ as a function of $P$ and $T$ as

$$V = V(P, T)$$

or
$$dV = \left(\frac{\partial V}{\partial P}\right)_T dP + \left(\frac{\partial V}{\partial T}\right)_P dT \qquad (A)$$

Dividing Eqn. *(A)* by $dP$ holding $S$ constant, we get

$$\left(\frac{\partial V}{\partial P}\right)_S = \left(\frac{\partial V}{\partial P}\right)_T + \left(\frac{\partial V}{\partial T}\right)_P \left(\frac{\partial T}{\partial P}\right)_S \qquad (B)$$

That is, the partial derivative $\left(\frac{\partial V}{\partial P}\right)_S$ is expressed in terms of three other partial derivatives $\left(\frac{\partial V}{\partial P}\right)_T$, $\left(\frac{\partial V}{\partial T}\right)_P$ and $\left(\frac{\partial T}{\partial P}\right)_S$. Thus it is always possible to express one partial derivative in terms of three other partial derivatives. If we are dealing with eight thermodynamic properties ($P, V, T, U, H, A, G$ and $S$), the number of partial derivatives of the type $\left(\frac{\partial V}{\partial P}\right)_S$ that can be obtained are $3! \times {}^8C_3 = 336$.

The number of relations of the type given in Eqn. ($B$) among the 336 possible partial derivatives are equal to the number of ways of choosing 4 out of the 336 partial derivatives. That is, the number of possible relation are ${}^{336}C_4 = 5.216 \times 10^8$.

**Example7.4**

Define the Jacobian of $x$ and $y$ with respect to $a$ and $b$.

*Solution :*

It $x = x(a, b)$ and $y = y(a, b)$ where $a$ and $b$ are independent variables, the Jacobian of $x$ and $y$ with respect to $a, b$ is defined in terms of the partial derivatives of $x$ and $y$ as

$$J\left[\frac{x,y}{a,b}\right] = \begin{vmatrix} \left(\frac{\partial x}{\partial a}\right)_b & \left(\frac{\partial x}{\partial b}\right)_a \\ \left(\frac{\partial y}{\partial a}\right)_b & \left(\frac{\partial y}{\partial b}\right)_a \end{vmatrix}$$

$$= \left(\frac{\partial x}{\partial a}\right)_b \left(\frac{\partial y}{\partial b}\right)_a - \left(\frac{\partial x}{\partial b}\right)_a \left(\frac{\partial y}{\partial a}\right)_b$$

For simplicity of notation, the Jacobian $J\left[\frac{x,y}{a,b}\right]$ can be written as

$$J\left[\frac{x,y}{a,b}\right] = \frac{[x,y]}{[a,b]}$$

**Example 7.5**

State some important properties of Jacobians.

*Solution* :

A few important properties of the Jacobians are

1. $\frac{[x,y]}{[a,b]} \times \frac{[a,b]}{[c,d]} = \frac{[x,y]}{[c,d]}$

2. $[x, y] = -[y, x]$

3. $[x,x] = 0$

4. $\frac{[x,z]}{[y,z]} = \left(\frac{\partial x}{\partial y}\right)_z$

Where $\frac{[x,y]}{[a,b]}$ is written as $[x, y]$ only by omitting the independent variables $a$ and $b$ for simplicity in writing. In the relation $[x, y] = -[y, x]$ one can introduce two independent variables like $a$ and $b$ on both sides of the equation, if needed.

**Example 7.6**

Suppose there exists a relation $f(x, y, z) = 0$ among the three variables $x$, $y$ and $z$. Then show that the partial differentials of these variables can be expressed in a cyclic relation form as

$$[x, y]\, dz + [y, z]\, dx + [z, x]\, dy = 0$$

*Solution* :

If $f(x, y, z) = 0$, then one can write

$$z = z(x, y)$$

or $$dz = \left(\frac{\partial z}{\partial x}\right)_y dx + \left(\frac{\partial z}{\partial y}\right)_x dy$$

Express the partial differentials of $z$ in terms of Jacobians to obtain

$$dz = \frac{[z, y]}{[x, y]} dx + \frac{[z, x]}{[y, x]} dy$$

or $[x, y]\, dz = [z, y]\, dx - [z, x]\, dy$

or $[x, y]\, dz + [y, z]\, dx + [z, x]\, dy = 0$

**Example 7.7**

Suppose one considers that four variables $x$, $y$, $z$ and $a$ are related as $f(x, y, z, a) = 0$ then show that a cyclic relation of the type

$$[x, y]\,[z, a] + [y, z]\,[x, a] + [z, x]\,[y, a] = 0$$

Can be written in terms of the Jacobians.

*Solution* :

Consider $z = z(x, y)$. Then

$$dz = \left(\frac{\partial z}{\partial x}\right)_y dx + \left(\frac{\partial z}{\partial y}\right)_x dy$$

$$= \frac{[z, y]}{[x, y]} dx + \frac{[z, x]}{[y, x]} dy$$

Dividing the above equation by $db$, holding $a$ constant, we get

$$\left(\frac{\partial z}{\partial b}\right)_a = \frac{[z, y]}{[x, y]}\left(\frac{\partial x}{\partial b}\right)_a + \frac{[z, x]}{[y, x]}\left(\frac{\partial y}{\partial b}\right)_a$$

or $$\frac{[z, a]}{[b, a]} = \frac{[z, y]}{[x, y]}\frac{[x, a]}{[b, a]} + \frac{[z, x]}{[y, x]}\frac{[y, a]}{[b, a]} \qquad (A)$$

or $[z, a]\,[x, y] = [z, y]\,[x, a] - [z, x]\,[y, a]$

or $[x, y]\,[z, a] + [y, z]\,[x, a] + [z, x]\,[y, a] = 0$

**Example 7.8**

We know that if $dz = Mdx + Ndy = \left(\frac{\partial z}{\partial x}\right)_y dx + \left(\frac{\partial z}{\partial y}\right)_x dy$ is an exact differential, $z$ is a property. We also know that the thermodynamic potentials $U, H, A$ and $G$ are thermodynamic properties. We can write

$$\left(\frac{\partial z}{\partial b}\right)_a = M\left(\frac{\partial x}{\partial b}\right)_a - N\left(\frac{\partial y}{\partial b}\right)_a$$

or $\frac{[z,a]}{[b,a]} = M\frac{[x,a]}{[b,a]} + N\frac{[y,a]}{[b,a]}$

or $[z\,,\,a] = M[x\,,\,a] + N[y\,,\,a]$

Where $a$ is any independent variable. Express similar relations for the thermodynamic potentials also.

*Solution* :

We know that the combined statement of the two laws of thermodynamics gives

$$dU = TdS - PdV.$$

Since $U$ is a property,its differential $dU$ is exact and hence we can write

$$[U\,,\,X] = T\,[S\,,\,X] - P\,[V\,,\,X]$$

We also know that $\qquad H = U + PV$

or $dH = dU + PdV + VdP$

or $dH = TdS + VdP$

The enthalpy is also a property and hence $dH$ is exact. Therefore, we can write

$$[H, X] = T\,[S, X] + V\,[P, X]$$

The Helmholtz free energy $A$ is defined as $A = U - TS$

or $dA = dU - TdS - SdT$

or $dA = -SdT - PdV$

Therefore, $\qquad [A, X] = -S\,[T, X] - P\,[V, X]$

Similarly, the Gibbs free energy $G$ is defined as $G = U + PV - TS$

or $dG = dU + PdV + VdP - TdS - SdT$

or $dG = -\,SdT + VdP$

Therefore, $\qquad [G, X] = -S\,[T, X] + V\,[P, X]$

In the above relations $X$ is any variable except $U$, $H$, $A$ and $G$.

**Example 7.9**

Show that the four important Maxwell relations are equivalent to $[T, S] = [P, V]$ in the Jacobian notation.

*Solution* :

We know that the four important Maxwell relations are :

$$\left(\frac{\partial V}{\partial S}\right)_P = \left(\frac{\partial T}{\partial P}\right)_S;$$

$$\left(\frac{\partial S}{\partial P}\right)_T = -\left(\frac{\partial V}{\partial T}\right)_P$$

$$\left(\frac{\partial S}{\partial V}\right)_T = \left(\frac{\partial P}{\partial T}\right)_V \quad \text{and} \quad \left(\frac{\partial P}{\partial S}\right)_V = -\left(\frac{\partial T}{\partial V}\right)_S$$

These Maxwell relations can be obtained from $[T, S] = [P, V]$ as given below.

Suppose we would like to find the value of $\left(\frac{\partial V}{\partial S}\right)_P$. Then express the partial differential in Jacobian notation as

$$\left(\frac{\partial V}{\partial S}\right)_P = \frac{[V,P]}{[S,P]}$$

Then substitute the relation $[T,S] = [P,V]$ to obtain

$$\left(\frac{\partial V}{\partial S}\right)_P = \frac{[V,P]}{[S,P]} = \frac{[S-T]}{[S,P]} = \left(\frac{\partial T}{\partial P}\right)_S$$

which is the first Maxwell relation. Similarly,

$$\left(\frac{\partial S}{\partial P}\right)_T = \frac{[S,T]}{[P,T]} = -\frac{[P,V]}{[P,T]} = -\left(\frac{\partial V}{\partial T}\right)_P$$

$$\left(\frac{\partial S}{\partial V}\right)_T = \frac{[S,T]}{[V,T]} = -\frac{[P,V]}{[V,T]} = \frac{[V,P]}{[V,T]} = \left(\frac{\partial P}{\partial T}\right)_V$$

$$\left(\frac{\partial P}{\partial S}\right)_V = \frac{[P,V]}{[S,V]} = \frac{[T,S]}{[S,V]} = -\frac{[T,S]}{[V,S]} = -\left(\frac{\partial T}{\partial V}\right)_S$$

**Example 7.10**

Express the definitions of $C_P$, $C_v$, $\beta$ and $\kappa$ in Jacobian notation.

*Solution* :

We know that $C_P = T\left(\frac{\partial S}{\partial T}\right)_P$. This can be expressed as

$$C_P = T\frac{[S,P]}{[T,P]}$$

Similarly, $$C_V = T\left(\frac{\partial S}{\partial T}\right)_V = T\frac{[S,V]}{[T,V]}$$

$$\beta = \frac{1}{V}\left(\frac{\partial V}{\partial T}\right)_P = \frac{[V,P]}{V[T,P]}$$

$$\kappa = \frac{-1}{V}\left(\frac{\partial V}{\partial P}\right)_T = -\frac{[V,T]}{V[P,T]} = \frac{[V,T]}{V[T,P]}$$

$$\frac{\beta}{\kappa} = \frac{[V,P]}{V[T,P]} \times \frac{V[T,P]}{[V,T]} = \frac{[V,P]}{[V,T]}$$

**Example 7.11**

Derive the following important $TdS$ relation following the method of partial differentials.

$$TdS = C_P\, dT - \beta\, VTdP$$

*Solution* :

Consider $S = S(T, P)$. Then

$$dS = \left(\frac{\partial S}{\partial T}\right)_P dT + \left(\frac{\partial S}{\partial P}\right)_T dP \qquad (A)$$

We know that the specific heat at constant pressure is defined as

$$C_P = T\left(\frac{\partial S}{\partial T}\right)_P \qquad (B)$$

and the Maxwell relation

$$\left(\frac{\partial S}{\partial P}\right)_T = -\left(\frac{\partial V}{\partial T}\right)_P \qquad (C)$$

Substituting Eqns. ($B$) and ($C$) in Eqn. ($A$), we get

$$dS = \frac{C_P}{T}dT - \left(\frac{\partial V}{\partial T}\right)_P dP \qquad (D)$$

We also know that the coefficient of volume expansion β is defined as

$$\beta = \frac{1}{V}\left(\frac{\partial V}{\partial T}\right)_P \qquad (E)$$

Substituting Eqn. ($E$) in Eqn. ($D$), we get

$$dS = \frac{C_P}{T}dT - \beta\, VdP$$

or $TdS = C_P dT - \beta\, VTdP$

**Example 7.12**

Following the method of partial differentials, derive the $TdS$ relation which is given by $TdS = C_v dT + [T\beta/\kappa]\, dV$

*Solution* :

Consider entropy $S$ as a function of temperature $T$ and volume $V$.

That is $S = S(T, V)$. Then

$$dS = \left(\frac{\partial S}{\partial T}\right)_V dT + \left(\frac{\partial S}{\partial V}\right)_T dV \qquad (A)$$

We know that the specific heat at constant volume is defined as

$$C_v = T\left(\frac{\partial S}{\partial T}\right)_V \qquad (B)$$

and the Maxwell relation

$$\left(\frac{\partial S}{\partial V}\right)_T = \left(\frac{\partial P}{\partial T}\right)_V \qquad (C)$$

Substitute Eqns, ($B$) and ($C$) in Eqn. ($A$) to obtain

$$dS = \frac{C_V}{T}dT + \left(\frac{\partial P}{\partial T}\right)_V dV \qquad (D)$$

We also know that β and κ are defined as

$$\beta = \frac{1}{V}\left(\frac{\partial V}{\partial T}\right)_P \text{ and } \kappa = -\frac{1}{V}\left(\frac{\partial V}{\partial P}\right)_T$$

Therefore, $$\frac{\beta}{\kappa} = \frac{1}{V}\left(\frac{\partial V}{\partial T}\right)_P \times \left[\frac{-V}{(\partial V/\partial P)_T}\right]$$

$$= -\left(\frac{\partial V}{\partial T}\right)_P \left(\frac{\partial P}{\partial V}\right)_T \qquad (E)$$

$$\left[\text{using the recipocal relation} \left(\frac{\partial P}{\partial V}\right)_T = \frac{1}{(\partial V/\partial P)_T}\right]$$

We know that the cyclic relation among the variables $P, V$ and $T$ is given by

$$\left(\frac{\partial P}{\partial V}\right)_T \left(\frac{\partial V}{\partial T}\right)_P \left(\frac{\partial T}{\partial P}\right)_V = -1$$

or $$-\left(\frac{\partial V}{\partial T}\right)_P \left(\frac{\partial P}{\partial V}\right)_T = \frac{1}{(\partial T/\partial P)_V} = \left(\frac{\partial P}{\partial T}\right)_V = \frac{\beta}{\kappa} \qquad (F)$$

Substituting Eqn. ($F$) in Eqn. ($D$), we get

$$dS = \frac{C_V}{T} dT + \frac{\beta}{\kappa} dV$$

or $$TdS = C_V dT + \frac{T\beta}{\kappa} dV$$

### Example 7.13

The entropy of a pure substance can be expressed as a function of the two independent variables $P$ and $V$. Following the method of partial differentials, derive the relation

$$TdS = C_V \frac{\kappa}{\beta} dP + \frac{C_P}{V\beta} dV$$

*Solution* :

Consider the entropy $S$ as a function of pressure $P$ and volume $V$. That is

$S = S(P, V)$. Then

$$dS = \left(\frac{\partial S}{\partial P}\right)_V dP + \left(\frac{\partial S}{\partial V}\right)_P dV \qquad (A)$$

We know the Maxwell relations

$$\left(\frac{\partial S}{\partial P}\right)_V = -\left(\frac{\partial V}{\partial T}\right)_S \text{ and } \left(\frac{\partial s}{\partial V}\right)_P = \left(\frac{\partial P}{\partial T}\right)_S \qquad (B)$$

Subsituting Eqn ($B$) in Eqn ($A$) we get.

$$dS = -\left(\frac{\partial V}{\partial T}\right)_S dP + \left(\frac{\partial P}{\partial T}\right)_S dV \qquad (C)$$

Now consider the relation $f(S, T, V) = 0$. Then we can write the cyclic relation

$$\left(\frac{\partial T}{\partial V}\right)_S \left(\frac{\partial S}{\partial T}\right)_V \left(\frac{\partial V}{\partial S}\right)_T = -1$$

or $$-\left(\frac{\partial V}{\partial T}\right)_S = \left(\frac{\partial S}{\partial T}\right)_V \left(\frac{\partial V}{\partial S}\right)_T \qquad (D)$$

We know the Maxwell relation

$$\left(\frac{\partial V}{\partial S}\right)_T = \left(\frac{\partial T}{\partial P}\right)_V \qquad (E)$$

and the definition of specific heat at constant volume as

$$C_v = T\left(\frac{\partial S}{\partial T}\right)_V \qquad (F)$$

Substituting Eqns. ($E$) and ($F$) in Eqn. ($D$), we get

$$-\left(\frac{\partial V}{\partial T}\right)_S = \frac{C_v}{T}\left(\frac{\partial T}{\partial P}\right)_V \qquad (G)$$

We know that $$\beta = \frac{1}{V}\left(\frac{\partial V}{\partial T}\right)_P \qquad (H)$$

and $$\kappa = -\frac{1}{V}\left(\frac{\partial V}{\partial P}\right)_T \qquad (I)$$

From Eqns. ($H$) and ($I$), we get

$$\frac{\kappa}{\beta} = -\left(\frac{\partial V}{\partial P}\right)_T \left(\frac{\partial T}{\partial V}\right)_P \qquad (J)$$

Now, consider the cyclic relation

$$\left(\frac{\partial P}{\partial V}\right)_T \left(\frac{\partial V}{\partial T}\right)_P \left(\frac{\partial T}{\partial P}\right)_V = -1 \quad \text{or} \quad -\left(\frac{\partial V}{\partial P}\right)_T \left(\frac{\partial T}{\partial V}\right)_P = \left(\frac{\partial T}{\partial P}\right)_V \qquad (K)$$

From Eqns.($J$) and ($K$), we get

$$\left(\frac{\partial T}{\partial P}\right)_V = \frac{\kappa}{\beta} \tag{L}$$

Substituting Eqn.($L$) in Eqn. ($G$) we get

$$-\left(\frac{\partial V}{\partial T}\right)_S = \frac{C_v}{T}\left(\frac{\partial T}{\partial P}\right)_V = \frac{C_v\kappa}{T\beta} \tag{M}$$

Now consider $f(S,T,P) = 0$. Then

$$\left(\frac{\partial S}{\partial T}\right)_P\left(\frac{\partial T}{\partial P}\right)_S\left(\frac{\partial P}{\partial S}\right)_T = -1 \quad \text{or} \quad \left(\frac{\partial P}{\partial T}\right)_S = -\left(\frac{\partial S}{\partial T}\right)_P\left(\frac{\partial P}{\partial S}\right)_T \tag{N}$$

We know the Maxwell relation

$$\left(\frac{\partial P}{\partial S}\right)_T = -\left(\frac{\partial T}{\partial V}\right)_P \tag{O}$$

and the definition for specific heat at constant pressure

$$C_P = T\left(\frac{\partial S}{\partial T}\right)_P \text{ or} \left(\frac{\partial S}{\partial T}\right)_P = \frac{C_P}{T} \tag{P}$$

Substituting Eqns. ($O$) and ($P$) in Eqn.($N$), we get

$$\left(\frac{\partial P}{\partial T}\right)_S = \frac{C_P}{T}\left(\frac{\partial T}{\partial V}\right)_P \tag{Q}$$

Substitute Eqn.($H$) in Eqn.($Q$) to obtain

$$\left(\frac{\partial P}{\partial T}\right)_S = \frac{C_P}{T\beta V} \tag{R}$$

Now substituting Eqns. ($M$) and ($R$) in Eqn. ($C$), we get

$$dS = \frac{C_v\kappa}{T\beta}dP + \frac{C_P}{T\beta V}dV \text{ or } TdS = \frac{C_v\kappa}{\beta}dP + \frac{C_P}{V\beta}dV$$

**Example 7.14**

It was observed in Example 7.13 that the method of partial differentials for deriving a thermodynamic relation at times is quite tedious and it is likely that one can make some mistakes also. Suggest a procedure for deriving the desired relation by the method of Jacobians.

*Solution* :

There are no definite rules or procedures for deriving the thermodynamic relations. However, the procedure suggested below can be adopted to derive the desired relation by following the method of Jacobians.

1. If the variation of a particular property is to be expressed in terms of the variations of two independent measurable quantities consider that particular property as a function of the two chosen independent variables and express the differential change of the property in terms of its partial derivatives.
2. Express the partial derivatives in Jacobian notation.
3. If any Jacobian contains only one of the thermodynamic potentials *U.H,A* and *G* they should be eliminated by substituting the following relations:

$$[U, X] = T[S, X] - P[V, X]$$

$$[H, X] = T[S, X] + V[P, X]$$

$$[A, X] = -S[T, X] - P[V, X]$$

$$[G, X] = -S[T, X] + V[P, X]$$

Where *X* is any variable like *P, V, T* or *S*.

4. If a Jacobian contains the combination of *S* and *T* it should be eliminated by using the Maxwell relations

$$[T, S] = [P, V].$$

5. If a Jacobian contains the combination of *S* either with *P* or *V*, it should be eliminated in terms of specific heats as

$$\frac{C_P}{T} = \frac{[S,P]}{[T,P]} \text{ and } \frac{C_V}{T} = \frac{[S,V]}{[T,V]}$$

6. If a Jacobian contains only *P, V* and *T*, then they can be expressed by using the relations

$$\beta = \frac{[V,P]}{V[T,P]}; \kappa = \frac{[V,T]}{V[T,P]} \text{ and } \frac{\beta}{\kappa} = \frac{[P,V]}{[T,V]}$$

   At the end of this one gets an expression containing only the measurable properties like *P, V, T,* $\beta$, $C_P$ and $C_V$.

7. Sometimes it is possible to find Jacobians containing two thermodynamic potentials. Then use the following type cyclic relation

$$[x, y][z, a] + [y, z][x, a] + [z, x][y, a] = 0$$

   to express the Jacobian having two thermodynamic potentials in terms of Jacobians containing only one thermodynamic potential. In such cases it is advisable to include two independent measurable variables (one from each of the potentials) in the cyclic relation. Then follow the steps 3 to 6. In case two thermodynamic potentials appear in a single partial derivative, the final result may contain some nonmeasurable property also.

### Example 7.15

Derive the relation $TdS = C_P dT - \beta VTdP$ following the method of Jacobians.

*Solution* :

The given relation expresses the entropy $S$ as a function of the independent variables $T$ and $P$. Therefore, consider $S = S(T, P)$

Then $$dS = \left(\frac{\partial S}{\partial T}\right)_P dT + \left(\frac{\partial S}{\partial P}\right)_T dP$$

Now express the partial derivatives in Jacobian notation. Then

$$dS = \frac{[S,P]}{[T,P]}dT + \frac{[S,T]}{[P,T]}dP \qquad (A)$$

In Eqn.($A$) the Jacobian containing $S$ and $P$ is related to $C_P$ as $\frac{C_P}{T} = \frac{[S,P]}{[T,P]}$

and the Jacobian containing $S$ and $T$ can be eliminated using Maxwell relation $[T, S] = [P, V]$. Then

$$dS = \frac{C_P}{T}dT + \frac{[V,P]}{[P,T]}dP \qquad (B)$$

In Eqn.($B$) the Jacobian has $P$, $V$ and $T$ only. This is related to $\beta$ as

$$\beta V = \frac{[V,P]}{[T,P]} \qquad (C)$$

Substitute Eqn. ($C$) in Eqn.($B$) to obtain

$$dS = \frac{C_P}{T}dT - \beta\, VdP$$

or $TdS = C_P dT - \beta\, VTdP$

### Example 7.16

Following the method of Jacobians show that

$$TdS = C_V dT + \frac{T\beta}{\kappa}dV$$

*Solution* :

Consider $S = S(T, V)$. Then

$$dS = \left(\frac{\partial S}{\partial T}\right)_V dT + \left(\frac{\partial S}{\partial V}\right)_T dV \qquad (A)$$

We know that $\dfrac{[S,V]}{[T,V]} = \dfrac{C_V}{T}$ and $[T, S] = [P, V]$ $\qquad (B)$

Substitute Eqns. (*B*) in Eqn. (*A*) to get

$$dS = \frac{C_V}{T} dT + \frac{[V,P]}{[V,T]} dV \qquad (C)$$

We also know that $\dfrac{[V,P]}{[V,T]} = \dfrac{\beta}{\kappa}$ $\qquad (D)$

Substitute Eqn. (*D*) in Eqn. (*C*) to obtain

$$dS = \frac{C_V}{T} dT + \frac{\beta}{\kappa} dV$$

or

$$TdS = C_V dT + \frac{T\beta}{\kappa} dV$$

## Example 7.17

Rework Example 7.13 using the method of Jacobians.

*Solution :*

Consider $S = S(P, V)$. Then

$$dS = \left(\frac{\partial S}{\partial P}\right)_V dP + \left(\frac{\partial S}{\partial V}\right)_P dV$$

$$= \frac{[S,V]}{[P,V]} dP + \frac{[S,P]}{[V,P]} dV \qquad (A)$$

Eliminate the Jacobians containing $S$ by substituting the definitions of $C_V$ and $C_P$.

$$\frac{[S,V]}{[T,V]} = \frac{C_V}{T} \text{ and } \frac{[S,P]}{[T,P]} = \frac{C_P}{T} \qquad (B)$$

Substituting Eqns.($B$) in Eqn.($A$), we get

$$dS = \frac{C_V}{T}\frac{[T,V]}{[P,V]}\,dP + \frac{C_P}{T}\frac{[T,P]}{[V,P]}dV \qquad (C)$$

We Know that $$\frac{\beta}{\kappa} = \frac{[P,V]}{[T,V]} \text{ and } \beta = \frac{[V,P]}{V[T,P]} \qquad (D)$$

Substitute Eqns. ($D$) in Eqn.($C$) to get

$$dS = \frac{C_V}{T}\frac{k}{\beta}dP + \frac{C_P}{T}\frac{1}{\beta V}dV$$

or $$TdS = \frac{\kappa}{\beta}C_V dP + \frac{C_P}{V\beta}dV$$

**Example 7.18**

The enthalpy of a substance is usually estimated by measuring the independent variables $T$ and $P$. Derive a relation to estimate the change in enthalpy from a knowledge of temperature and pressure of a substance.Use the method of partial differentials.

*Solution :*

Consider $h = h(T, P)$. Then

$$dh = \left(\frac{\partial h}{\partial T}\right)_P dT + \left(\frac{\partial h}{\partial P}\right)_T dP \qquad (A)$$

we know that $$dh = Tds + vdP.$$

Divide the above relation with $dP$ while $T$ is held constant to obtain

$$\left(\frac{\partial h}{\partial P}\right)_T = T\left(\frac{\partial s}{\partial P}\right)_T + v \qquad (B)$$

Substitute the Maxwell relation

$$\left(\frac{\partial s}{\partial P}\right)_T = -\left(\frac{\partial v}{\partial T}\right)_P$$

in Eqn ($B$) to get

$$\left(\frac{\partial h}{\partial P}\right)_T = -T\left(\frac{\partial v}{\partial T}\right)_p + v \qquad (C)$$

We know that $$\beta = \frac{1}{v}\left(\frac{\partial v}{\partial T}\right)_p \quad (D)$$

Substitute Eqn.($D$) in Eqn.($C$) to get

$$\left(\frac{\partial h}{\partial P}\right)_T = -\beta v T + v \quad (E)$$

We also know that $$C_P = \left(\frac{\partial h}{\partial T}\right)_P \quad (F)$$

Substituting Eqns.($E$) and ($F$) in Eqn ($A$), we get

$$dh = C_p dT + (v - v\,T\,\beta)\,dP = C_P dT + v\,(1 - T\,\beta)dP$$

**Example 7.19**

Rework Example 7.18 by the method of Jacobians.

*Solution* :

Consider $h = h\,(T, P)$. Then

$$dh = \left(\frac{\partial h}{\partial T}\right)_P dT + \left(\frac{\partial h}{\partial P}\right)_T dP$$

$$= \frac{[h,P]}{[T,P]}dT + \frac{[h,T]}{[P,T]}dP \quad (A)$$

We know that $$[h, P] = T\,[s, P] + v\,[P, P] = T\,[s, P] \quad (B)$$

$$[h, T] = T\,[s, P] + v\,[P, T] \quad (C)$$

Substitute Eqns. *(B)* and *(C)* in Eqn. *(A)* to get

$$dh = \frac{T[s,P]}{[T,P]}dT + \left[\frac{T[s,T]}{[P,T]} + v\right]dP \quad (D)$$

Eliminate the Jacobian Containing $s$ and $P$ by the relation

$$\frac{T[s,P]}{[T,P]} = C_P \quad (E)$$

and the Jacobian containing $s$ and $T$ by the Maxwell relation

$$[T, s] = [P, v] \tag{F}$$

Then we get

$$dh = C_P\,dT + \left[\frac{T[v,P]}{[P,T]} + v\right]dP \tag{G}$$

The Jacobian containing $P$, $v$ and $T$ only can be eliminated in terms of $\beta$ since

$$\beta = \frac{[v,P]}{v[T,P]} \tag{H}$$

Substituting Eqn. *(H)* in Eqn. *(G)*, we get

$$\begin{aligned} dh &= C_P\,dT + (v - \beta\,vT)\,dP \\ &= C_P\,dT + v\,(1 - T\,\beta)\,dP \end{aligned}$$

**Example 7.20**

The internal energy of a substance is usually expressed as a function of $T$ and $v$. Using the method of partial differentials derive a relation to estimate the change in the internal energy of a substance in terms of measurable quantities.

*Solution* :

Consider $u = u\,(T, v)$. Then

$$du = \left(\frac{\partial u}{\partial T}\right)_v dT + \left(\frac{\partial u}{\partial v}\right)_T dv$$

$$= C_v\,dT + \left(\frac{\partial u}{\partial v}\right)_T dv \tag{A}$$

We Know that $\quad du = Tds - Pdv$

Dividing the above relation by $dv$ holding $T$ constant, we get

$$\left(\frac{\partial u}{\partial v}\right)_T = T\left(\frac{\partial s}{\partial v}\right)_T - P \tag{B}$$

The Maxwell relation gives

$$\left(\frac{\partial s}{\partial v}\right)_T = \left(\frac{\partial P}{\partial T}\right)_v \tag{C}$$

Substituting Eqn. *(C)* in Eqn. *(B)*, we get

$$\left(\frac{\partial u}{\partial v}\right)_T = T\left(\frac{\partial P}{\partial T}\right)_v - P \tag{D}$$

Consider the cyclic relation

$$\left(\frac{\partial P}{\partial v}\right)_T\left(\frac{\partial v}{\partial T}\right)_P\left(\frac{\partial T}{\partial P}\right)_v = -1$$

or
$$\left(\frac{\partial P}{\partial T}\right)_v = -\left(\frac{\partial P}{\partial v}\right)_T\left(\frac{\partial v}{\partial T}\right)_P \tag{E}$$

We know that
$$\beta = \frac{1}{v}\left(\frac{\partial v}{\partial T}\right)_P \text{ and } \kappa = -\frac{1}{v}\left(\frac{\partial v}{\partial P}\right)_T \tag{F}$$

Substituting Eqn. *(F)* in Eqn *(E)*, we get

$$\left(\frac{\partial P}{\partial T}\right)_v = \beta / \kappa \tag{G}$$

Substituting Eqn. *(G)* in Eqn. *(D)* we get

$$\left(\frac{\partial u}{\partial v}\right)_T = \frac{T\beta}{\kappa} - P \tag{H}$$

Now subsitute Eqn *(H)* in Eqn *(A)* to obtain

$$du = C_v dT + \left(\frac{T\beta}{\kappa} - P\right)dv$$

**Example 7.21**

Rework Example 7.20 using the method of Jacobians.

*Solution* :

Consider $u = u\ (T, v)$.Then

$$du = \left(\frac{\partial u}{\partial T}\right)_v dT + \left(\frac{\partial u}{\partial v}\right)_T dv$$

$$= \frac{[u,v]}{[T,v]}dT + \frac{[u,T]}{[v,T]}dv \tag{A}$$

We know that
$$[u, v] = T\,[s, v] + P\,[v\,,\,v] = T\,[s, v] \tag{B}$$
$$[u, T] = T\,[s, T] - P\,[v, T] \tag{C}$$

Substitute Eqns.($B$) and ($C$) in Eqn. ($A$) to get

$$du = T\frac{[S,v]}{[T,v]}dT + \left[\frac{T[s,T]}{[v,T]} - P\right]dv$$

$$= C_v dT + \left[T\frac{[v,P]}{[v,T]} - P\right]dv = C_v dT + \left[T\frac{\beta}{\kappa} - P\right]dv$$

**Example 7.22**

Show that the internal energy of an ideal gas is a function of temperature only. That is $u = u\,(T)$ only.

*Solution* :

We know that $$du = C_v dT + \left(\frac{T\beta}{\kappa} - P\right)dv$$

For an ideal gas $$Pv = RT \text{ or } v = \frac{RT}{P}$$

$$\beta = \frac{1}{v}\left(\frac{\partial v}{\partial T}\right)_P = \frac{1}{v}\frac{R}{P} = \frac{1}{T}$$

and $$\kappa = -\frac{1}{v}\left(\frac{\partial v}{\partial P}\right)_T = -\frac{1}{v}\frac{(-RT)}{P^2} = \frac{1}{P}$$

Therefore, $$\frac{T\beta}{\kappa} - P = \left(\frac{TP}{T} - P\right) = 0$$

Hence $$du = C_v\, dT + \left(\frac{T\beta}{\kappa} - P\right)dv = C_v dT$$

Therefore, the internal energy of an ideal gas is independent of $P$ and $v$ and is a function of temperature only.

**Example 7.23**

Show that the enthalpy of an ideal gas is independent of $P$ and $v$ and depends only on temperature.

*Solution* :

We know that for any substance, the change in enthalpy is given by

$$dh = C_p dT + v(1 - \beta)\, dP$$

For an ideal gas $\quad Pv = RT$ or $v = RT/P$ and

$$\beta = \frac{1}{v}\left(\frac{\partial v}{\partial T}\right)_P = \frac{1}{v}\frac{R}{P} = \frac{1}{T}$$

Therefore, $dh = C_P\, dT + v\,(1 - T\beta)\, dP$

$$= C_P\, dT + v\,(1 - T/T)\, dP$$

or $\quad dh = C_P\, dT$

That is, for an ideal gas the enthalpy depends only on temperature.

**Example 7.24**

Show that the entropy change for an ideal gas can be calculated from any of the following two relations.

$$\Delta S = C_P \ln \frac{T_2}{T_1} - R \ln \frac{P_2}{P_1}$$

$$\Delta S = C_v \ln \frac{T_2}{T_1} + R \ln \frac{v_2}{v_1}$$

*Solution* :

We know the following two $TdS$ equations

$$TdS = C_p dT - \beta\, vT dP \qquad (A)$$

$$TdS = C_v dT + \frac{T\beta}{\kappa}\, dv \qquad (B)$$

For an ideal gas, we know that $\beta = \dfrac{1}{T}$ and $\kappa = \dfrac{1}{P}$. Substituting the value of $\beta$ in Eqn. *(A)*, we get

$$TdS = C_P\, dT - v dP$$

or $\quad \int dS = \int C_P \frac{dT}{T} - \int \frac{v}{T} dP = \int C_P \frac{dT}{T} - \int \frac{R}{P} dP$

If $C_P$ is assumed to be independent of temperature, the above equation can be integrated to obtain

$$\Delta S = C_P \ln \frac{T_2}{T_1} - R \ln \frac{P_2}{P_1}$$

Substituting the values of $\beta$ and $\kappa$ in Eqn *(B)*, we get

$$TdS = C_v dT + \frac{TP}{T} dv = C_v dT + P dv$$

or
$$\Delta S = \int dS = \int C_v \frac{dT}{T} + \int \frac{P}{T} dv = \int C_v \frac{dT}{T} + \int R \frac{dv}{v}$$

If $C_v$ is assumed to be independent of temperature, the equation can be integrated to obtain

$$\Delta S = C_v \ln \frac{T_2}{T_1} + R \ln \frac{v_2}{v_1}$$

**Example 7.25**

The internal energy of a pure substance can be expressed as a function of any two of the independent variables $P$, $v$ and $T$. Therefore, a knowledge of the partial derivatives of internal energy with respect to the independent variables enables one to estimate the change in the internal energy. Derive the following partial derivatives of internal energy using the method of partial differentials.

$$(a)\left(\frac{\partial u}{\partial v}\right)_T (b)\left(\frac{\partial u}{\partial v}\right)_P (c)\left(\frac{\partial u}{\partial T}\right)_v (d)\left(\frac{\partial u}{\partial T}\right)_P (e)\left(\frac{\partial u}{\partial P}\right)_v (f)\left(\frac{\partial u}{\partial P}\right)_T$$

*Solution :*

(*a*) We know that $du = Tds - Pdv$

Dividing the above equation by $dv$ holding $T$ constant,we get

$$\left(\frac{\partial u}{\partial v}\right)_T = T\left(\frac{\partial s}{\partial v}\right)_T - P$$

Substitute the Maxwell relation $\left(\frac{\partial s}{\partial v}\right)_T = \left(\frac{\partial P}{\partial T}\right)_v$ in the above relation to obtain

$$\left(\frac{\partial u}{\partial v}\right)_T = T\left(\frac{\partial P}{\partial T}\right)_v - P$$

We know that $\left(\frac{\partial P}{\partial v}\right)_T \left(\frac{\partial v}{\partial T}\right)_P \left(\frac{\partial T}{\partial P}\right)_v = -1$

or
$$\left(\frac{\partial P}{\partial T}\right)_v = -\left(\frac{\partial P}{\partial v}\right)_T \left(\frac{\partial v}{\partial T}\right)_P \qquad (A)$$

We know that $\beta = \frac{1}{v}\left(\frac{\partial v}{\partial T}\right)_p$ and $\kappa = -\frac{1}{v}\left(\frac{\partial v}{\partial P}\right)_T$

Substituting for $\left(\frac{\partial v}{\partial T}\right)_p$ and $\left(\frac{\partial v}{\partial P}\right)_T$ in terms of β and κ ,Eqn. (*A*) reduces to

$$(\partial P/\partial T)\ v = \beta/\kappa \tag{B}$$

Therefore $\left(\frac{\partial u}{\partial v}\right)_T = \frac{T\beta}{\kappa} - P$

(*b*) We know that $du = Tds - Pdv$

Divide it with $dv$ holding $P$ constant to obtain

$$\left(\frac{\partial u}{\partial v}\right)_P = T\left(\frac{\partial s}{\partial v}\right)_P - P$$

Substitute the Maxwell relation $\left(\frac{\partial s}{\partial v}\right)_P = \left(\frac{\partial P}{\partial T}\right)_s$ in the above equation to get

$$\left(\frac{\partial u}{\partial v}\right)_P = T\left(\frac{\partial P}{\partial T}\right)_s - P \tag{C}$$

Now consider $f(s, T, P) = 0,$

Then

$$\left(\frac{\partial s}{\partial T}\right)_P\left(\frac{\partial T}{\partial P}\right)_s\left(\frac{\partial P}{\partial s}\right)_T = -1$$

or $$\left(\frac{\partial P}{\partial T}\right)_s = \left(\frac{\partial s}{\partial T}\right)_P\left(\frac{\partial P}{\partial s}\right)_T = \frac{C_p}{T}\left(\frac{\partial T}{\partial v}\right)_P \tag{D}$$

[Obtained by substituting the definition of $C_P$ and the Maxwell relation $(\partial P/\partial s)_T = -(\partial T/\partial v)_P$]

or $$\left(\frac{\partial P}{\partial T}\right)_s = \frac{C_P}{Tv\beta} \tag{E}$$

[Obtained by using the relation $\beta = \frac{1}{v}\left(\frac{\partial v}{\partial T}\right)_P$ in Eqn. *(D)*]. Substituting Eqn. *(E)* in Eqn *(C)*, we get

$$(\partial u / \partial v)_P = \frac{C_P}{v\beta} - P$$

(*c*) We know that $du = Tds - Pdv$

Divide it with $dT$ holding $v$ constant to obtain

$$\left(\frac{\partial u}{\partial T}\right)_v = T\left(\frac{\partial s}{\partial T}\right)_v$$

We also know that $C_v = T\left(\frac{\partial s}{\partial T}\right)_v$

Hence

$$\left(\frac{\partial u}{\partial T}\right)_v = C_v$$

(*d*) We know that $du = Tds - Pdv$

Divide it with $dT$ while $P$ is held constant to obtain

$$\left(\frac{\partial u}{\partial T}\right)_P = T\left(\frac{\partial s}{\partial T}\right)_P - P\left(\frac{\partial v}{\partial T}\right)_P = C_P - Pv\beta$$

(*e*) We know that $du = Tds - Pdv$

or $\left(\frac{\partial u}{\partial P}\right)_v = T\left(\frac{\partial s}{\partial P}\right)_v$

Then substitute the Maxwell relation

$$\left(\frac{\partial s}{\partial P}\right)_v = -\left(\frac{\partial v}{\partial T}\right)_s \text{ to get } \left(\frac{\partial u}{\partial P}\right)_v = -T\left(\frac{\partial v}{\partial T}\right)_s \qquad (F)$$

Now consider $f(s, T, v) = 0$. Then

$$\left(\frac{\partial s}{\partial T}\right)_v\left(\frac{\partial T}{\partial v}\right)_s\left(\frac{\partial v}{\partial s}\right)_T = -1$$

or $$-\left(\frac{\partial v}{\partial T}\right)_s = \left(\frac{\partial s}{\partial T}\right)_v\left(\frac{\partial v}{\partial s}\right)_T = \frac{C_v}{T}\left(\frac{\partial T}{\partial P}\right)_v = \frac{C_v}{T}\frac{\kappa}{\beta} \qquad (G)$$

[Obtained by substituting the definition of $C_v$, the Maxwell relation

$$\left(\frac{\partial s}{\partial v}\right)_T = \left(\frac{\partial P}{\partial T}\right)_v \text{ and } \frac{\beta}{\kappa} = \left(\frac{\partial P}{\partial T}\right)_v$$
]

Substituting Eqn. (*G*) in Eqn. (*F*),we get

$$\left(\frac{\partial u}{\partial P}\right)_v = C_v \frac{\kappa}{\beta}$$

(*f*) We know that $du = Tds - Pdv$

or $$\left(\frac{\partial u}{\partial P}\right)_T = T\left(\frac{\partial s}{\partial P}\right)_T - P\left(\frac{\partial v}{\partial P}\right)_T$$

Substitute the Maxwell relation $\left(\frac{\partial s}{\partial P}\right)_T = -\left(\frac{\partial v}{\partial T}\right)_P$

Then

$$\left(\frac{\partial u}{\partial P}\right)_T = -T\left(\frac{\partial v}{\partial T}\right)_P - P\left(\frac{\partial v}{\partial P}\right)_T$$

$$= -Tv\beta + Pv\kappa = v\,(P\kappa - T\beta)$$

**Example 7.26**

Rework Example 7.25 using the Jacobian method.

*Soltion* :

(*a*) $$\left(\frac{\partial u}{\partial v}\right)_T = \frac{[u,T]}{[v,T]} = \frac{T[s,T] - P[v,T]}{[v,T]} = \frac{T[s,T]}{[v,T]} - P$$

$$= \frac{T[v,P]}{[v,T]} - P = \frac{T\beta}{\kappa} - P$$

(*b*) $$\left(\frac{\partial u}{\partial v}\right)_P = \frac{[u,P]}{[v,P]} = \frac{T[s,P] - P[v,P]}{[v,P]} = \frac{T[s,P]}{[v,P]} - P$$

or $$\left(\frac{\partial u}{\partial v}\right)_P = \frac{T[s,P]}{[T,P]}\frac{[T,P]}{[v,P]} - P = T\frac{C_P}{T}.\frac{1}{\beta v} - P = \frac{C_P}{v\beta} - P$$

(*c*) $$\left(\frac{\partial u}{\partial T}\right)_v = \frac{[u,v]}{[T,v]} = \frac{T[s,v] - P[v,v]}{[T,v]} = \frac{T[s.v]}{[T,v]} = C_v$$

(*d*) $$\left(\frac{\partial u}{\partial T}\right)_p = \frac{[u,P]}{[T,P]} = \frac{T[s,P] - P[v,P]}{[T,P]} = \frac{T[s,P]}{[T,P]} - P\frac{[v,P]}{[T,P]}$$

or $$\left(\frac{\partial u}{\partial T}\right)_P = C_P - Pv / \beta$$

(e) $$\left(\frac{\partial u}{\partial P}\right)_v = \frac{[u,v]}{[P,v]} = \frac{T[s,v] - P[v,v]}{[P,v]} = \frac{T[s,v]}{[P,v]}$$

$$= \frac{T[s,v]}{[T,v]}\frac{[T,v]}{[P,v]} = C_v \frac{\kappa}{\beta}$$

(f) $$\left(\frac{\partial u}{\partial P}\right)_T = \frac{[u,T]}{[P,T]} = \frac{T[s,T] - P[v,T]}{[P,T]} = \frac{T[s,T]}{P,T} - \frac{P[v,T]}{[P,T]}$$

$$= -\frac{T[P,v]}{[P,T]}\frac{-P[v,T]}{[P,T]} = -\,Tv\beta + P\kappa v$$

or $$\left(\frac{\partial u}{\partial P}\right)_T = v\,(P\kappa - T\beta)$$

## Example 7.27

The enthalpy of a pure substance can be expressed as a function of any two of the independents variables $P$, $v$ and $T$. Therefore, a knowledge of the partial derivatives of enthalpy with respect to the independent variables enables one to estimate the enthalpy change. Using the method of partial differentials derive expressions for determining the following partial derivatives in terms of measurable quantities.

$$(a)\left(\frac{\partial h}{\partial T}\right)_P (b)\left(\frac{\partial h}{\partial T}\right)_v (c)\left(\frac{\partial h}{\partial P}\right)_T (d)\left(\frac{\partial h}{\partial P}\right)_v (e)\left(\frac{\partial h}{\partial v}\right)_T \text{ and } (f)\left(\frac{\partial h}{\partial v}\right)_P$$

*Solution* :

(*a*) We know that $dh = Tds + vdP$ (*A*)

or $$\left(\frac{\partial h}{\partial T}\right)_P = T\left(\frac{\partial s}{\partial T}\right)_P = C_P$$

(*b*) We know that $dh = Tds + vdP$

or $$\left(\frac{\partial h}{\partial T}\right)_v = T\left(\frac{\partial s}{\partial T}\right)_v + v\left(\frac{\partial P}{\partial T}\right)_v = C_v + \frac{v\beta}{\kappa}$$

(*c*) From Eqn, (*A*) we get

$$\left(\frac{\partial h}{\partial p}\right)_T = T\left(\frac{\partial s}{\partial P}\right)_T + v = -T\left(\frac{\partial v}{\partial T}\right)_P + v$$

$$\left[\text{Obtained by subsituting the Maxwell relation } \left(\frac{\partial s}{\partial P}\right)_T = -\left(\frac{\partial v}{\partial T}\right)_P\right]$$

or $$\left(\frac{\partial h}{\partial P}\right)_T = -Tv\beta + v = v(1 - T\beta)$$

(*d*) From Eqn. (*A*) We get

$$\left(\frac{\partial h}{\partial P}\right)_T = T\left(\frac{\partial s}{\partial P}\right)_v + v$$

We know the Maxwell relation

$$\left(\frac{\partial s}{\partial P}\right)_v = -\left(\frac{\partial v}{\partial T}\right)_s$$

Therefore, $$\left(\frac{\partial h}{\partial P}\right)_v = -T\left(\frac{\partial v}{\partial T}\right)_s + v \qquad (B)$$

Consider $f(s, T, v) = 0$. Then

$$\left(\frac{\partial s}{\partial T}\right)_v \left(\frac{\partial T}{\partial v}\right)_s \left(\frac{\partial v}{\partial s}\right)_T = -1$$

or $$-\left(\frac{\partial v}{\partial T}\right)_s = \left(\frac{\partial s}{\partial T}\right)_v \left(\frac{\partial v}{\partial s}\right)_T = \frac{C_v}{T}.\left(\frac{\partial T}{\partial P}\right)_v = \frac{C_v}{T}.\frac{\kappa}{\beta} \qquad (C)$$

$$\left[\text{Obtained by using the definition of } C_v \text{ and the Maxwell relation } \left(\frac{\partial s}{\partial v}\right)_T = \left(\frac{\partial P}{\partial T}\right)_v\right]$$

Substitute Eqn. (*C*) in Eqn. (*B*) to obtain

$$\left(\frac{\partial h}{\partial P}\right)_v = C_v\frac{\kappa}{\beta} + v$$

(*e*) From Eqn. (*A*) we get

$$\left(\frac{\partial h}{\partial v}\right)_T = T\left(\frac{\partial s}{\partial v}\right)_T + v\left(\frac{\partial P}{\partial v}\right)_T$$

Substitute the Maxwell relation $\left(\frac{\partial s}{\partial v}\right)_T = \left(\frac{\partial P}{\partial T}\right)_v$ to obtain

$$\left(\frac{\partial h}{\partial v}\right)_T = T\left(\frac{\partial P}{\partial T}\right)_v + v\left(\frac{\partial P}{\partial v}\right)_T = \frac{T\beta}{\kappa} - v.\frac{1}{\kappa v}$$

$$= \frac{T\beta}{\kappa} - \frac{1}{\kappa} = \frac{1}{\kappa}(T\beta - 1)$$

(Obtained by subsituting the definitions of $\beta$ and $\kappa$)

(*f*) From Eqn. (*A*) we get

$$\left(\frac{\partial h}{\partial v}\right)_P = T\left(\frac{\partial s}{\partial v}\right)_P = T\left(\frac{\partial P}{\partial T}\right)_s$$

[Obtained by subsitituting the Maxwell relation $\left(\frac{\partial s}{\partial v}\right)_P = \left(\frac{\partial P}{\partial T}\right)_s$ ]

Now consider $f(s, T, P) = 0$

or $$\left(\frac{\partial s}{\partial T}\right)_P \left(\frac{\partial T}{\partial P}\right)_s \left(\frac{\partial P}{\partial s}\right)_T = -1$$

or $$\left(\frac{\partial P}{\partial T}\right)_s = -\left(\frac{\partial s}{\partial T}\right)_P \left(\frac{\partial P}{\partial s}\right)_T = \frac{C_P}{T}\left(\frac{\partial T}{\partial v}\right)_P$$

[Obtained by using the definition of $C_P$ and the Maxwell relation $\left(\frac{\partial s}{\partial P}\right)_T = -\left(\frac{\partial v}{\partial T}\right)_P$ ]

or $$\left(\frac{\partial P}{\partial T}\right)_s = \frac{C_P}{T}\left(\frac{\partial T}{\partial v}\right)_P = \frac{C_P}{T\beta v}$$

Therefore, $$\left(\frac{\partial h}{\partial v}\right)_P = \frac{C_P}{v\beta}$$

**Example 7.28**

Rework Example 7.27 using the Jacobian method.

*Solution* :

(*a*) $$\left(\frac{\partial h}{\partial T}\right)_P = \frac{[h,P]}{[T,P]} = \frac{T[s,P]+v[P,P]}{[T,P]} = \frac{T[s,P]}{[T,P]} = C_P$$

(*b*) $$\left(\frac{\partial h}{\partial T}\right)_v = \frac{[h,v]}{[T,v]} = \frac{T[s,v]+v[P,v]}{[T,v]} = \frac{T[s,v]}{[T,v]} + \frac{v[P,v]}{[T,v]}$$

or $$\left(\frac{\partial h}{\partial T}\right)_v = C_v + \frac{v\beta}{\kappa}$$

(*c*) $$\left(\frac{\partial h}{\partial P}\right)_T = \frac{[h,T]}{[P,T]} = \frac{T[s,T]+v[P,T]}{[P,T]}$$

$$= \frac{T[s,T]}{[P,T]} + v = -\frac{T[P,v]}{[P,T]} + v$$

$$= v - \beta vT = v(1 - T\beta)$$

(*d*) $$\left(\frac{\partial h}{\partial P}\right)_v = \frac{[h,v]}{[P,v]} = \frac{T[s,v]+v[P,v]}{[P,v]}$$

$$= \frac{T[s,v]}{[P,v]} + v = \frac{T[s,v]}{[T,v]}\frac{[T,v]}{[P,v]} + v = C_v\frac{\kappa}{\beta} + v$$

(*e*) $$\left(\frac{\partial h}{\partial v}\right)_T = \frac{[h,T]}{[v,T]} = \frac{T[s,T]+v[P,T]}{[v,T]}$$

$$= \frac{T[v,P]}{[v,T]} + v\frac{[P,T]}{[v,T]} = \frac{T\beta}{\kappa} - v\frac{1}{\kappa v} = \frac{T\beta}{\kappa} - \frac{1}{\kappa}$$

or $$\left(\frac{\partial h}{\partial v}\right)_T = \frac{T\beta - 1}{\kappa}$$

(*f*) $$\left(\frac{\partial h}{\partial v}\right)_P = \frac{[h,P]}{[v,P]} = \frac{T[s,P]+v[P,P]}{[v,P]} = \frac{T[s,P]}{[v,P]}$$

$$= \frac{T[s,P]}{[T,P]}\frac{[T,P]}{[v,P]} = \frac{C_P}{\beta v}$$

**Example 7.29**

Derive the relations $(a)\left(\frac{\partial h}{\partial s}\right)_v = T\left(1+\frac{v\beta}{C_v\kappa}\right)$ and $(b)\left(\frac{\partial h}{\partial s}\right)_T = T-\frac{1}{\beta}$ by the method of partial differentials.

*Solution* :

(*a*) We know that $dh = Tds + vdp$ (*A*)

Divide the above relation with *ds* holding *v* constant to obtain

$$\left(\frac{\partial h}{\partial s}\right)_v = T+v\left(\frac{\partial P}{\partial s}\right)_v = T+v\left(\frac{\partial P}{\partial T}\right)_v\left(\frac{\partial T}{\partial s}\right)_v$$

$$= T+v\frac{\beta}{\kappa}.\frac{T}{C_v} \quad \text{or} \quad \left(\frac{\partial h}{\partial s}\right)_v = T\left(1+\frac{v\beta}{C_v\kappa}\right)$$

(*b*) From Eqn. (*A*) we obtain

$$\left(\frac{\partial h}{\partial s}\right)_T = T+v\left(\frac{\partial P}{\partial s}\right)_T$$

Now substitute the Maxwell relation

$$\left(\frac{\partial s}{\partial P}\right)_T = -\left(\frac{\partial v}{\partial T}\right)_P \text{ to get } \left(\frac{\partial h}{\partial s}\right)_T$$

$$= T-v\left(\frac{\partial T}{\partial v}\right)_P = T-\frac{v}{\beta v} = T-\frac{1}{\beta}$$

**Example 7.30**

Rework Example 7.29 using the Jacobian method.

*Solution :*

(*a*) $\left(\frac{\partial h}{\partial s}\right)_v = \frac{[h,v]}{[s,v]}$

We know that $[h, v] = T[s, v] + v[P, v]$. Therefore,

$$\left(\frac{\partial h}{\partial s}\right)_v = \frac{[h,v]}{[s,v]} = \frac{T[s,v]+v[P,v]}{[s,v]}$$

$$= T+v\frac{[P,v]}{[s,v]} = T+v\frac{[P,v]}{[s,v]}\frac{[T,v]}{[T,v]}$$

$$= T + \frac{vT\,[P,v]}{C_v\,[T,v]} = T + \frac{vT\,\beta}{C_v\,\kappa}$$

or $$\left(\frac{\partial h}{\partial s}\right) = T\left(1 + \frac{v\beta}{C_v\kappa}\right)$$

(b) $$\left(\frac{\partial h}{\partial s}\right)_T = \frac{[h,T]}{[s,T]} = \frac{T[s,T] + v[P,T]}{[s,T]}$$

$$= T + v\frac{[P,T]}{[s,T]} = T + \frac{v[P,T]}{[v,P]}$$

$$= T - \frac{v}{\beta v} = T - \frac{1}{\beta}$$

**Example 7.31**

Using the method of partial differentials prove that

(a) $$\left(\frac{\partial u}{\partial T}\right)_s = \frac{P\,C_v\kappa}{T\beta} \text{ and}$$

(b) $$\left(\frac{\partial T}{\partial s}\right)_u = \frac{T}{C_v}\left(1 - \frac{T\beta}{P\kappa}\right)$$

*Solution* :

(a) We know that $du = Tds - Pdv$

or $$\left(\frac{\partial u}{\partial T}\right)_s = -P\left(\frac{\partial v}{\partial T}\right)_s = P\left(\frac{\partial v}{\partial s}\right)_T\left(\frac{\partial s}{\partial T}\right)_v$$

[Obtained by substituting the cyclic relation $\left(\frac{\partial s}{\partial T}\right)_v\left(\frac{\partial T}{\partial v}\right)_s\left(\frac{\partial v}{\partial s}\right)_T = -1$]

or $$\left(\frac{\partial u}{\partial T}\right)_s = P\left(\frac{\partial T}{\partial P}\right)_v\left(\frac{\partial s}{\partial T}\right)_v$$

[Obtained by substituting the Maxwell relation $\left(\frac{\partial v}{\partial s}\right)_T = \left(\frac{\partial T}{\partial P}\right)_v$]

or $$\left(\frac{\partial u}{\partial T}\right)_s = \frac{P\kappa}{\beta}\frac{C_v}{T} = \frac{PC_v\kappa}{T\beta}$$

[Obtained by substituting the definition of $C_v$. β and κ]

(*b*) Consider $f(s, u, T) = 0$. Then

$$\left(\frac{\partial s}{\partial u}\right)_T\left(\frac{\partial u}{\partial T}\right)_s\left(\frac{\partial T}{\partial s}\right)_u = -1$$

or $$\left(\frac{\partial T}{\partial s}\right)_u = -\left(\frac{\partial u}{\partial s}\right)_T\left(\frac{\partial T}{\partial u}\right)_s \qquad (A)$$

We know that $$du = Tds - Pdv \qquad (B)$$

or $$\left(\frac{\partial u}{\partial s}\right)_T = T - P\left(\frac{\partial v}{\partial s}\right)_T = T - P\left(\frac{\partial T}{\partial P}\right)_v = T - P\frac{\kappa}{\beta} \qquad (C)$$

[ Since the Maxwell relation gives $\left(\frac{\partial s}{\partial v}\right)_T = \left(\frac{\partial P}{\partial T}\right)_v$ and we know that

$(\partial P/\partial T)_v = \beta/\kappa$]

From the result of part (*a*) we get $\left(\frac{\partial u}{\partial T}\right)_s = \frac{PC_v\kappa}{T\beta}$

or $$\left(\frac{\partial T}{\partial u}\right)_s = \frac{T\beta}{PC_v\kappa}. \qquad (D)$$

Substituting Eqns. (*C*) and (*D*) in Eqn . (*A*), we get

$$\left(\frac{\partial T}{\partial s}\right)_u = \left(\frac{P\kappa}{\beta} - T\right)\frac{T\beta}{PC_v\kappa} = \frac{T}{C_v} - \frac{T^2\beta}{PC_v\kappa} = \frac{T}{C_v}\left(1 - \frac{T\beta}{P\kappa}\right)$$

**Example 7.32**

Rework Example 7.31 using the Jacobian method.

*Solution* :

(*a*) $$\left(\frac{\partial u}{\partial T}\right)_s = \frac{[u,s]}{[T,s]} = \frac{T[s,s] - P[v,s]}{[T,s]}$$

$$= -\frac{P[v,s]}{[T,s]} = \frac{P[s,v]}{[P,v]} = \frac{P[s,v][T,v]}{[T,v][P,v]} = \frac{PC_v\kappa}{T\beta}$$

(b) $$\left(\frac{\partial T}{\partial s}\right)_u = \frac{[T,u]}{[s,u]} = \frac{[u,T]}{[u,s]} = \frac{T[s,T]-P[v,T]}{u[s,s]-P[v,s]} = \frac{T[s,T]-P[v,T]}{-P[v,s]}$$

or $$\left(\frac{\partial T}{\partial s}\right)_u = \frac{T[s,T]}{P[s,v]} + \frac{[v,T]}{[v,s]} = \frac{T[v,P]}{P[s,v]} + \frac{[v,T]}{[v,s]}$$

$$= \frac{T[v,P][T,v]}{P[s,v][T,v]} + \frac{[v,T]}{[v,s]} = \frac{T}{P}\frac{[T,v][v,P]}{[s,v][T,v]} + \frac{[v,T]}{[v,s]}$$

$$= -\frac{T^2\beta}{PC_v\kappa} + \frac{T}{C_v} = \frac{T}{C_v}\left(1-\frac{T\beta}{P\kappa}\right)$$

**Example 7.33**

Prove that on a pressure versus volume diagram the slope of a reversible, adiabatic process is equal to $\gamma$ times the slope of a reversible isothermal process.

*Solution* :

We know that the ratio of specific heats is given by

$$\gamma = \frac{C_P}{C_v} = \frac{T(\partial s/\partial T)_P}{T(\partial s/\partial T)_v} = \frac{[s,P]}{[T,P]}\frac{[T,v]}{[s,v]}$$

$$= \frac{[s,P]}{[s,v]}\frac{[T,v]}{[T,P]} = \left(\frac{\partial P}{\partial v}\right)_s\left(\frac{\partial v}{\partial P}\right)_T$$

or $$\left(\frac{\partial P}{\partial v}\right)_s = \gamma\left(\frac{\partial P}{\partial v}\right)_T$$

**Example 7.34**

Derive an expression to determine the slope of a reversible adiabatic process on a temperature versus pressure diagram, in terms of measurable quantities. Suppose an ideal gas undergoes a reversible adiabatic process how does the temperature ratio depend on the pressure ratio ?

*Solution* :

The slope of a reversible adiabatic process on a temperature versus pressure diagram is given by $(\partial T/\partial P)_s$.

$$\left(\frac{\partial T}{\partial P}\right)_s = \frac{[T,s]}{[P,s]} = \frac{[T,s]}{[P,T]}\frac{[P,T]}{[P,s]} = \frac{[P,v][P,T]}{[P,T][P,s]}$$

$$= \frac{\beta v T}{C_P}$$

We know that for an ideal gas, $Pv = RT$ and $\beta = 1/T$. Then

$$\left(\frac{\partial T}{\partial P}\right)_s = \frac{\beta v T}{C_P} = \frac{v}{C_P} = \frac{RT}{P}\frac{(\gamma - 1)}{R\gamma} = \frac{(\gamma - 1)T}{\gamma P}$$

or $\dfrac{dT}{T} = \left(\dfrac{\gamma - 1}{\gamma}\right)\dfrac{dP}{P}$ or $\ln(T_2 / T_1) = \left(\dfrac{\gamma - 1}{\gamma}\right) \ln\left(\dfrac{P_2}{P_1}\right)$ or $\dfrac{T_2}{T_1} = \left(\dfrac{P_2}{P_1}\right)^{\frac{\gamma-1}{\gamma}}$

### Example 7.35

Suppose liquid water at 25°C is isentropically compressed such that its volume decreases by 10% . What will be the rise in the temperature of the water. Assume that $\beta = 2 \times 10^{-4}$ K$^{-1}$, $\kappa = 4.85 \times 10^{-4}$ MPa$^{-1}$ and. $C_v = 4.2$ kJ / kg K for liquid water.

*Solution* :

We would like to know the change in temperature with volume for an isentropic process. That is we would like to find $\left(\dfrac{\partial T}{\partial v}\right)_s$.

$$\left(\frac{\partial T}{\partial v}\right)_s = \frac{[T,s]}{[v,s]} = \frac{[P,v]}{[v,s]} = \frac{[P,v]\,[v,T]}{[v,s]\,[v,T]}$$

$$= \frac{[P,v]\,[v,T]}{[v,T]\,[v,s]} = -\frac{\beta T}{\kappa C_v}$$

Therefore, for liquid water

$$\left(\frac{\partial T}{\partial v}\right)_s = -\frac{\beta T}{\kappa C_v} = \frac{-2\times 10^{-4}\times 298}{(4.85\times 10^{-4}\times 10^{-6})\times 4.2\times 10^{3}} = -\,2.9259 \times 10^4 \text{ K/m}^3$$

$$\Delta v = -\,0.1 \times 0.001\,002\,9 = -1.0029 \times 10^{-4}$$

[$v_f$ at 25°C = 0.001 002 9 m$^3$/kg for water]

$$\left(\frac{\partial T}{\partial v}\right)_s = \left(\frac{\Delta T}{\Delta v}\right)_s = -2.9259 \times 10^4 \text{ K/m}^3$$

or $\Delta T = 2.9259 \times 10^4 \times 1.0029 \times 10^{-4} = 2.93$ K

Hence, the temperature rises by 2.93 K

**Example 7.36**

Derive a relation to estimate the difference between the two specific heats $C_P$ and $C_v$ of a substance and show that it is always positive. Use the method of Jacobians.

*Solution :*

We know that $C_P = T(\partial s / \partial T)_P$ and $C_v = T(\partial s / \partial T)_v$.

From these relations it is obvious that we have to deal with the four variables $s$, $T$, $P$ and $v$. The cyclic relation in these variables can be written as

$[s, T]\,[P, v] + [T, P]\,[s, v] + [P, S]\,[T, v] = 0$

Divide the above relation with $[T, P]\,[T, v]$ to obtain

$$\frac{[s\,T][P,v]}{[T,P][T,v]} + \frac{[s,v]}{[T,v]} + \frac{[P,s]}{[T,P]} = 0$$

or $$\frac{[s,P]}{[T,P]} - \frac{[s,v]}{[T,v]} = \frac{[s,T][P,v]}{[T,P][T,v]}$$

or $$\frac{C_P}{T} - \frac{C_v}{T} = \frac{[v,P][P,v]}{[T,P[T,v]]} = \beta v \frac{\beta}{\kappa}$$

or $C_P - C_v = Tv\,\beta^2/\kappa$

Since $\kappa$ is always positive $C_P - C_v > 0$.

**Example 7.37**

Rework Example 7.36 using the partial differentials method.

*Solution* :

We know that

$$C_P = \left(\frac{\partial s}{\partial T}\right)_P \quad \text{and} \quad C_v = T\left(\frac{\partial s}{\partial T}\right)_v$$

Then $$C_P - C_v = T\left[\left(\frac{\partial s}{\partial T}\right)_p - \left(\frac{\partial s}{\partial T}\right)_v\right] \qquad (A)$$

consider $s = s\,(T, v)$. Then

$$ds = \left(\frac{\partial s}{\partial T}\right)_v dT + \left(\frac{\partial s}{\partial v}\right)_T dv$$

Dividing the above equation with $dT$ holding $P$ constant, we get

$$\left(\frac{\partial s}{\partial T}\right)_P = \left(\frac{\partial s}{\partial T}\right)_v + \left(\frac{\partial s}{\partial v}\right)_T \left(\frac{\partial v}{\partial T}\right)_P$$

or $$\left(\frac{\partial s}{\partial T}\right)_P - \left(\frac{\partial s}{\partial T}\right)_v = \left(\frac{\partial s}{\partial v}\right)_T \left(\frac{\partial v}{\partial T}\right)_P = \left(\frac{\partial P}{\partial T}\right)_v \left(\frac{\partial v}{\partial T}\right)_P \qquad (B)$$

[ Obtained by substituting the Maxwell relation $\left(\frac{\partial s}{\partial v}\right)_T = \left(\frac{\partial P}{\partial T}\right)_v$ ]

Substituting Eqn. ($B$) in Eqn. ($A$), we get

$$C_P - C_v = T\left(\frac{\partial P}{\partial T}\right)_v \left(\frac{\partial v}{\partial T}\right)_P = \frac{T\beta}{\kappa}.\beta v = Tv\beta^2 / \kappa$$

**Example7.38**

($a$) Given that $\beta = 2 \times 10^{-4}$ K$^{-1}$, $\kappa = 4.85 \times 10^{-4}$ MPa$^{-1}$ and $v = 0.001\ 002\ 9$ m$^3$/ kg, for liquid water determine the difference in $C_P$ and $C_v$ for water.

($b$) Determine the value of $C_p - C_v$ for an ideal gas.

*Solution* :

($a$) We know that $C_P - C_v = Tv\beta^2 / \kappa$

Therefore, for liquid water,

$$C_P - C_v = \frac{298 \times 0.001\,002\,9 \times (2 \times 10^{-4})^2}{4.85 \times 10^{-4} \times 10^{-6}}$$

or $C_P - C_v = 0.0246$ kJ/kg K

($b$) We know that for an ideal gas $Pv = RT$; $\beta = 1/T$ and $\kappa = 1/P$.

Therefore, $$C_P - C_v = \frac{Tv\beta^2}{\kappa} = Tv\frac{P}{T^2} = \frac{Pv}{T} = R$$

**Example 7.39**

Derive a relation by the method of partial differentials to estimate the Joule – Thomson coefficient in terms of measurable quantities.

*Solution :*

The Joule–Thomson coefficient is defined as $\mu_{JT} = \left(\dfrac{\partial T}{\partial P}\right)_h$

From the definition of $\mu_{JT}$, it follows that the variables of interest are $h$, $T$ and $P$. Therefore, consider the relationship $f(h, T, P) = 0$.

Then the cyclic relation gives $\left(\dfrac{\partial h}{\partial T}\right)_P \left(\dfrac{\partial T}{\partial P}\right)_h \left(\dfrac{\partial P}{\partial h}\right)_T = -1$

or $$\left(\frac{\partial T}{\partial P}\right)_h = -\left(\frac{\partial T}{\partial h}\right)_P \left(\frac{\partial h}{\partial P}\right)_T \qquad (A)$$

We know that $$\left(\frac{\partial h}{\partial T}\right)_P = C_P \qquad (B)$$

The change in enthalpy is given $dh = Tds + vdp$

or $$\left(\frac{\partial h}{\partial P}\right)_T = T\left(\frac{\partial s}{\partial P}\right)_T + v = v - T\left(\frac{\partial v}{\partial T}\right)_P \qquad (C)$$

[Obtained by substituting the Maxwell relation $\left(\dfrac{\partial s}{\partial P}\right)_P = -\left(\dfrac{\partial v}{\partial T}\right)_P$ ]

Then $$\left(\frac{\partial h}{\partial P}\right)_T = v - T\left(\frac{\partial v}{\partial T}\right)_P = v - \beta v T = v(1 - \beta T) \qquad (D)$$

Substitute Eqns. (*B*) and (*D*) in Eqn. (*A*) to obtain

$$\left(\frac{\partial T}{\partial P}\right)_h = -\frac{v}{C_P}(1 - \beta T) = \frac{v}{C_P}(\beta T - 1)$$

or $$\mu_{JT} = \frac{v}{C_P}(\beta T - 1)$$

**Example 7.40**

Rework Example7.39 using the Jacobian method.

*Solution :*

The Joule–Thomson coefficient is defined as $\mu_{JT} = \left(\dfrac{\partial T}{\partial P}\right)_h$

$$\left(\frac{\partial T}{\partial P}\right)_h = \frac{[T,h]}{[P,h]} = \frac{[h,T]}{[h,P]} = \frac{T[s,T]+v[P,T]}{T[s,P]+v[P,P]}$$

$$= \frac{T[s,T]+v[P,T]}{T[s,P]} = \frac{[s,T]}{[s,P]} + \frac{v[P,T]}{T[s,P]}$$

$$= \frac{[v,P]}{[s,P]} + \frac{v[P,T]}{T[s,P]} = \frac{[v,P][T,P]}{[s,P]+v[T,P]} + \frac{v[P,T][T,P]}{T[s,p][T,P]}$$

$$= \frac{[T,P]}{[s,P]}\left\{\frac{[v,P]}{[T,P]} + \frac{v[P,T]}{T[T,P]}\right\} = \frac{T}{C_P}\left\{\frac{[v,P]}{[T,P]} - \frac{v}{T}\right\}$$

$$= \frac{T}{C_P}\left(\beta v - \frac{v}{T}\right) = \frac{v}{C_P}(T\beta - 1)$$

or
$$\mu_{JT} = \frac{v}{C_P}(\beta T - 1)$$

**Example 7.41**

Show that $\mu_{JT} = 0$ for an ideal gas at all temperatures and pressures,

*Solution :*

We know that
$$\mu_{JT} = \frac{v}{C_P}(T\beta - 1)$$

For an ideal gas: $Pv = RT$ and hence $\beta = 1/T$. Therefore,

$$\mu_{JT} = \frac{v}{C_P}\left(\frac{T}{T} - 1\right) = 0$$

Hence $\mu_{JT} = 0$ for an ideal gas.

**Example 7.42**

Prove that

(*a*) $\left(\frac{\partial C_v}{\partial v}\right)_T = T\left(\frac{\partial^2 P}{\partial T^2}\right)_v$ and

(*b*) $\left(\frac{\partial C_P}{\partial P}\right)_T = -T\left(\frac{\partial^2 v}{\partial T^2}\right)_P$

*Solution :*

(*a*) We know that $C_v = T\left(\frac{\partial s}{\partial T}\right)_v$

Then $\left(\frac{\partial C_v}{\partial v}\right)_T = \frac{\partial}{\partial v}\left[T\left(\frac{\partial s}{\partial T}\right)_v\right]_T = T\frac{\partial}{\partial v}\left[\left(\frac{\partial s}{\partial T}\right)_v\right]_T = T\frac{\partial}{\partial T}\left[\left(\frac{\partial s}{\partial v}\right)_T\right]_v$

Now substitute the Maxwell relation $\left(\frac{\partial s}{\partial v}\right)_T = \left(\frac{\partial P}{\partial T}\right)_v$ to obtain

$$\left(\frac{\partial C_v}{\partial v}\right)_T = T\frac{\partial}{\partial T}\left[\left(\frac{\partial P}{\partial T}\right)_v\right]_v = T\left(\frac{\partial^2 P}{\partial T^2}\right)_v$$

(*b*) We know that $C_P = T\left(\frac{\partial s}{\partial T}\right)_P$. Then

$$\left(\frac{\partial C_P}{\partial P}\right)_T = \frac{\partial}{\partial P}\left[T\left(\frac{\partial s}{\partial T}\right)_P\right]_T = T\frac{\partial}{\partial T}\left[\left(\frac{\partial s}{\partial P}\right)_T\right]_P$$

Now substitute the Maxwell relation $\left(\frac{\partial s}{\partial P}\right)_T = -\left(\frac{\partial v}{\partial T}\right)_P$ to get

$$\left(\frac{\partial C_P}{\partial P}\right)_T = T\frac{\partial}{\partial T}\left[-\left(\frac{\partial v}{\partial T}\right)_P\right]_P = -T\left(\frac{\partial^2 v}{\partial T^2}\right)_P$$

**Example 7.43**

For a van der Waals gas derive an expression for $C_P - C_v$

*Solution :*

The van der Waals equation of state gives

$$P = \frac{RT}{v-b} - \frac{a}{v^2} \text{ Then}$$

$$\left(\frac{\partial P}{\partial v}\right)_T = -\frac{RT}{(v-b)^2} + \frac{2a}{v^3} = \frac{2a(v-b)^2 - RTv^3}{v^3(v-b)^2} \qquad (A)$$

$$\kappa = -\frac{1}{v}\left(\frac{\partial v}{\partial P}\right)_T = -\frac{v^3(v-b)^2}{v\left[2a(v-b)^2 - RTv^3\right]} = \frac{v^2(v-b)^2}{RTv^3 - 2a(v-b)^2} \qquad (B)$$

$$\left(\frac{\partial P}{\partial T}\right)_v = \frac{R}{v-b} \qquad (C)$$

Consider the cyclic relation $\left(\frac{\partial P}{\partial v}\right)_T \left(\frac{\partial v}{\partial T}\right)_P \left(\frac{\partial T}{\partial P}\right) = -1$

or $$\left(\frac{\partial v}{\partial T}\right)_P = -\left(\frac{\partial v}{\partial P}\right)_T \left(\frac{\partial P}{\partial T}\right)_v = -\frac{(\partial P/\partial T)_v}{(\partial P/\partial v)_T} \qquad (D)$$

Substituting Eqns. ($A$) and ($C$) in Eqn . ($D$), we get

$$\left(\frac{\partial v}{\partial T}\right)_P = -\frac{R/(v-b)}{\{2a(v-b)^2 - RTv^3\}/\{v^3(v-b)^2\}} = \frac{Rv^3(v-b)}{RTv^3 - 2a(v-b)^2}$$

$$\beta = \frac{1}{v}\left(\frac{\partial v}{\partial T}\right)_P = \frac{Rv^2(v-b)}{RTv^3 - 2a(v-b)^2} \qquad (E)$$

We know that $C_P - C_v = Tv\beta^2/\kappa$. Substituting Eqns. ($B$) and ($E$), we get

$$C_P - C_v = \frac{Tv\left[Rv^2(v-b)\right]^2\left[RTv^3 - 2a(v-b)^2\right]}{\left[RTv^3 - 2a(v-b)^2\right]^2 v^2(v-b)^2}$$

or $$C_P - C_v = \frac{TR^2v^3}{RTv^3 - 2a(v-b)^2}$$

**Example 7.44**

For a van der Waals gas prove that $C_v$ is a function of temperature only.

*Solution* :

We know that $\left(\frac{\partial C_v}{\partial v}\right)_T = T\left(\frac{\partial^2 P}{\partial T^2}\right)_v$

For a van der Waals gas

$$P = \frac{RT}{v-b} - \frac{a}{v^2} \text{. Then}$$

$$\left(\frac{\partial P}{\partial T}\right)_v = \frac{R}{v-b} \text{ and } \left(\frac{\partial^2 P}{\partial T^2}\right)_v = 0$$

Therefore $$\left(\frac{\partial C_v}{\partial v}\right)_T = T\left(\frac{\partial^2 P}{\partial T^2}\right)_v = 0$$

$$\left(\frac{\partial C_v}{\partial P}\right)_T = \left(\frac{\partial C_v}{\partial v}\right)_T\left(\frac{\partial v}{\partial P}\right)_T = 0$$

Therefore, $C_v$ is independent of $P$ and $v$ and is a function of temperature only.

**Example 7.45**

For a van der Waals gas show that

(*a*) $du = C_v dT + \dfrac{a}{v^2}dv$

(*b*) $ds = C_v \dfrac{dT}{T} + \dfrac{Rdv}{v-b}$

*Solution* :

(*a*) Consider $u = u\,(T\,,\,v)$. Then

$$du = \left(\frac{\partial u}{\partial T}\right)_v dT + \left(\frac{\partial u}{\partial v}\right)_T dv = C_v dT + \left(\frac{\partial u}{\partial v}\right)_T dv \quad \text{(A)}$$

We know that $du = Tds - Pdv$

or $$\left(\frac{\partial u}{\partial v}\right)_T = T\left(\frac{\partial s}{\partial v}\right)_T - P = T\left(\frac{\partial P}{\partial T}\right)_v - P$$

[Obtained by using the Maxwell relation $\left(\frac{\partial s}{\partial v}\right)_T = \left(\frac{\partial P}{\partial T}\right)_v$]

For a van der Waals gas

$$P = \frac{RT}{v-b} - \frac{a}{v^2}$$

or $\left(\frac{\partial P}{\partial T}\right)_v = \frac{R}{v-b}$

Therefore, $\left(\frac{\partial u}{\partial v}\right)_T = \frac{RT}{v-b} - P = \frac{a}{v^2}$ (*B*)

Substitute Eqn.(*B*) in Eqn.(*A*) to obtain

$$du = C_v dT + \frac{a}{v^2} dv$$

(*b*) Consider s = *s* (*T*, *v*). Then

$$ds = \left(\frac{\partial s}{dT}\right)_v dT + \left(\frac{\partial s}{\partial v}\right)_T dv$$

$$= \frac{C_v dT}{T} + \left(\frac{\partial P}{\partial T}\right)_v dv$$

[ Obtained by substituting the definition of $C_v$ and the Maxwell relation

$$\left(\frac{\partial s}{\partial v}\right)_v = \left(\frac{\partial P}{\partial T}\right)_v$$ ]

We know that for a van der Waals gas

$$\left(\frac{\partial P}{\partial T}\right)_v = \frac{R}{v-b}$$

Therefore,

$$ds = \frac{C_v}{T} dT + \left(\frac{\partial P}{\partial T}\right)_v dv = \frac{C_v}{T} dT + \frac{Rdv}{v-b}$$

**Example 7.46**

Show that $C_P$ of a gas obeying the Clausius equation of state, given by $P(v - b) = RT$, is independent of pressure.

*Solution* :

We know that $\left(\frac{\partial C_P}{\partial P}\right)_T = -T\left(\frac{\partial^2 v}{\partial T^2}\right)_P$

The Clausius equation of state gives $P(v-b) = RT$ or $v = \dfrac{RT}{P} + b$

Then $$\left(\frac{\partial v}{\partial T}\right)_P = \frac{R}{P} \text{ and } \left(\frac{\partial^2 v}{\partial T^2}\right)_P = 0$$

Hence $$\left(\frac{\partial C_P}{\partial P}\right)_T = -T\left(\frac{\partial^2 v}{\partial T^2}\right)_P = 0$$

That is, $C_P$ is independent of pressure.

**Example 7.47**

The Euken coefficient $\xi$ is defined as $\xi = \left(\dfrac{\partial T}{\partial v}\right)_u$. Derive an expression to calculate $\xi$ in terms of measurable properties and find its value for an ideal gas.

*Solution :*

We know that $$\xi = \left(\frac{\partial T}{\partial v}\right)_u = \frac{[T,u]}{[v,u]} = \frac{[u,T]}{[u,v]}$$

or $$\xi = \frac{T[s,T] - P[v,T]}{T[s,v] - P[v,v]} = \frac{T[s,T] - P[v,T]}{T[s,v]}$$

$$= \frac{[s,T]}{[s,v]} - \frac{P[v,T]}{T[s,v]} = \frac{[v,P]}{[s,v]} - \frac{P[v,T]}{T[s,v]} = \frac{[T,v]}{[s,v]}\left\{\frac{[v,P]}{[T,v]} + \frac{P}{T}\right\} = \frac{T}{C_v}\left(\frac{-\beta}{\kappa} + \frac{P}{T}\right)$$

$$= \frac{1}{C_v}\left(P - \frac{T\beta}{\kappa}\right)$$

For an ideal gas $Pv = RT$; $\beta = 1/T$ and $\kappa = 1/P$.

Therefore, $$\xi = \frac{1}{C_v}\left(P - \frac{T\beta}{\kappa}\right) = \frac{1}{C_v}\left(P - \frac{TP}{T}\right) = 0$$

**Example 7.48**

The acoustic velocity $\mathbf{V}$ is defined as $\mathbf{V} = \sqrt{\left(\dfrac{\partial P}{\partial \rho}\right)_s}$ where $\rho$ is the mass density.

Develop a relation to determine the acoustic velocity. Calculate the value of the acoustic velocity in air (ideal gas with $\gamma = 1.4$ and molar mass is $28.97 \times 10^{-3}$ kg/mol) at 300 K.

*Solution* :

The acoustic velocity **V** is defined as

$$\mathbf{V}^2 = \left(\frac{\partial P}{\partial \rho}\right)_s = -v^2\left(\frac{\partial P}{\partial v}\right)_s$$

Where, $\rho = \frac{1}{v}$ and $v$ = specific volume.

$$\left(\frac{\partial P}{\partial v}\right)_s = \frac{[P,s]}{[v,s]} = \frac{[P,s]}{[P,T]}\frac{[P,T]}{[v,T]}\frac{[v,T]}{[v,s]}$$

$$= \left(\frac{C_P}{T}\right)\left(\frac{-1}{\kappa v}\right)\left(\frac{T}{C_v}\right)$$

or $$\left(\frac{\partial P}{\partial v}\right)_s = \frac{-\gamma}{\kappa v}$$

$$\mathbf{V}^2 = -v^2\left(\frac{\partial P}{\partial v}\right)_s = \frac{v^2\gamma}{\kappa v} = \frac{\gamma v}{\kappa} \text{ or } \mathbf{V} = \sqrt{\frac{\gamma v}{\kappa}}$$

For an ideal gas, $\kappa = \frac{1}{P}$ .Therefore,

$$\mathbf{V} = \sqrt{\gamma vP} = \sqrt{\frac{\gamma RT}{M}} \text{ where } M = \text{Molar mass}$$

or $$\mathbf{V} = \sqrt{\frac{1.4 \times 8.314 \times 300}{28.97 \times 10^{-3}}} = 347.18 \text{ m/s}$$

**Example 7.49**

Derive relations to determine the slopes of constant pressure process and constant volume process on temperature versus entropy diagram. Determine their values for an ideal gas and show that the constant volume line is steeper than the constant pressure line.

*Solution* :

Slope of constant pressure line on $T-s$ diagram $= \left(\frac{\partial T}{\partial s}\right)_P$

$$\left(\frac{\partial T}{\partial s}\right)_P = \frac{[T,P]}{[s,P]} = \frac{T}{C_P}$$

Slope of constant volume line on $T$–$s$ diagram = $\left(\frac{\partial T}{\partial s}\right)_P$

$$\left(\frac{\partial T}{\partial s}\right)_v = \frac{[T,v]}{[s,v]} = \frac{T}{C_v}$$

We know that $C_P - C_v = R > 0$ for an ideal gas. Hence

$\frac{T}{C_v} > \frac{T}{C_P}$. That is, the constant volume line is steeper than a constant pressure line on $T$–$s$ diagram.

**Example 7.50**

An air–standard Diesel cycle is shown in Fig.E 7.50 (*a*) on pressure versus volume diagram. Assume that an ideal gas with constant specific heats undergoes the cycle shown in the figure and redraw the cycle on temperature versus entropy diagram.

*Solution* :

We know that

$$\left(\frac{\partial T}{\partial s}\right)_P = \frac{T}{C_P} \text{ and } \left(\frac{\partial T}{\partial s}\right)_v = \frac{T}{C_v} \quad \text{(See Example 7.49)}$$

1–2 is an isentropic compression process. Hence $s_1 = s_2$ and $T_2 > T_1$. Therefore, on $T$ versus $s$ diagram the process 1–2 is a vertical line going upwards.

2–3 is a constant pressure expansion process with $T_3 > T_2$. The slope of this line on $T$– $s$ diagram is $T/C_P$ with increasing $T$. Hence the slope is positive and increasing from state 2 to state 3.

3–4 is an isentropic expansion process. Hence $s_3 = s_4$ and $T_4 < T_3$. Therefore, it is a vertical line going downwards.

4–1 is a constant volume process with $T_4 > T_1$. Its slope on $T$–$s$ diagram is $\frac{T}{C_v}$ > 0 and is decreasing in going from 4 to 1. It is steeper than the line 1–2. These processes are shown on a $T$–$s$ diagram in Fig.E 7.50 (*b*)

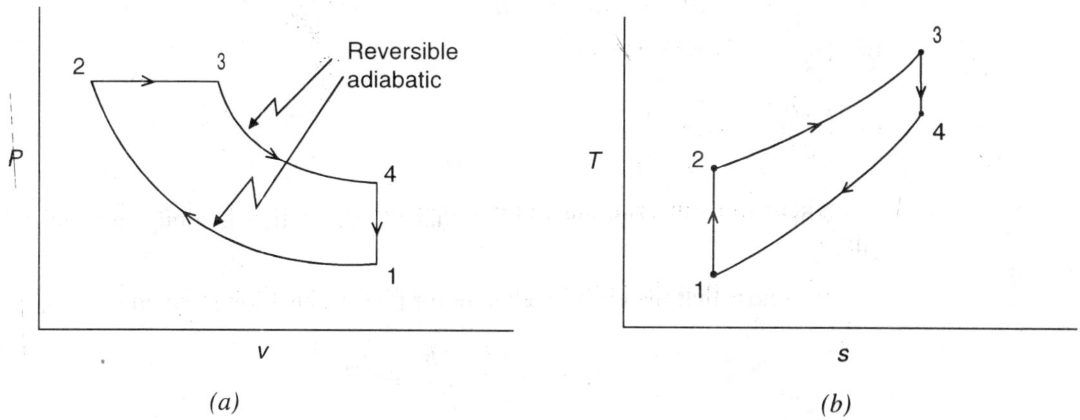


Fig.E 7.50 (*a*) Deisel cycle on *P*-*v* daigram (*b*) Deisel cycle on *T*-*s* diagram.

**Example 7.51**

Consider a single component simple system existing in two phases in a state of equilibrium –say liquid and vapor at a specified pressure. If the temperature of the system is changed its saturation pressure also changes. Apply the criterion for equilibrium and deduce the Clapeyron equation.

*Solution* :

Consider a single component simple system. The phase diagram (pressure versus temperature diagram) for such a system is shown in Fig.E 7.51. Along the vaporization curve both liquid and vapor phases coexist in equilibrium. Since both the phases are existing along the vaporization curve, the system has only one degree of freedom. If the pressure of the system is specified, its temperature is fixed at the saturation temperature. If the pressure of the system is changed, its saturation temperature also changes. If the system is initially in state 1 at pressure $P$ and temperature $T$, and if the pressure is changed to $(P + dP)$ its saturation temperature changes to $(T + dT)$. Consider a liquid in state $1f$ which is in equilibrium with vapor in state $1v$. If the temperature is changed from $T$ to $(T + dT)$, the system moves along the saturation curve and the liquid reaches the state $2f$ while the vapor reaches the state $2v$. If the system is in a state of equilibrium at constant temperature and pressure, then the criterion for equilibrium is given by

$$dg = -sdT + vdP = 0 \text{ (at constant } T \text{ and } P).$$

Where $g$ is the Gibbs free energy.

At state 1, $g_{1f} = g_{1v}$ and at state 2 also $g_{2f} = g_{2v}$,

Therefore $g_{2f} - g_{1f} = g_{2v} - g_{1v}$ or $dg_f = dg_v$

or $-s_f dT + v_f dP = -s_g dT + v_g dP$

or $(s_g - s_f)\, dT = (v_g - v_f)\, dP$

or $$\left(\frac{\partial P}{\partial T}\right)_{\text{sat}} = \frac{s_g - s_f}{v_g - v_f}$$

where the subscript sat implies that the derivative is along the saturation curve.

We know that the entropy change for phase transition is given by

$$s_g - s_f = \frac{h_g - h_f}{T} = \frac{h_{fg}}{T}$$

Therefore, $$\left(\frac{\partial P}{\partial T}\right)_{\text{sat}} = \frac{h_{fg}}{T(v_g - v_f)} = \frac{h_{fg}}{Tv_{fg}} = \frac{\Delta h}{T\Delta v}$$

Which is know as the Clapeyron equation,

Where, $h_{fg} = h_g - h_t$ = enthalpy (latent heat) of vaporization

$v_{fg} = v_g - v_f$ = change in volume due to vaporization.

$T$ = Phase transition temperature.

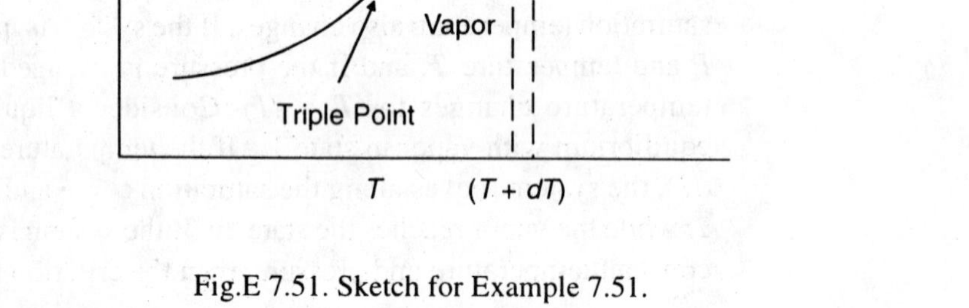


Fig.E 7.51. Sketch for Example 7.51.

**Example 7.52**

Is the Clapeyron equation applicable to solid–liquid phase transition (fusion) as well as to sublimation processes also? Is the slope of the coexistence curve always positive ?

*Solution :*

The Clapeyron equation is applicable to any change of state. It is applicable for fusion, vaporization and sublimation processes also. $\Delta h$ is always positive during phase transition (solid –liquid, liquid–vapor and solid –vapor) and hence $(\partial P / \partial T)_{sat}$ will be positive or negative depending on whether the phase transition is accompanied by expansion ($\Delta v > 0$) or contraction ($\Delta v < 0$). For solid–vapor and liquid–vapor phase transition ($\Delta v > 0$) and hence $(\partial P / \partial T)_{sat} > 0$. That is an increase in pressure increases the boiling point of the liquid and the sublimation temperature of a solid. For solid liquid phase transition for some substances (especially water, bismuth and antimony) $\Delta v < 0$. Therefore, if a substance contracts on melting, an increase in pressure decreases the melting point.

**Example 7.53**

Deduce the Clausius–Clapeyron equation from the Clapeyron equation.

*Solution :*

Consider the liquid–vapor phase transition at low pressures. Then the vapor phase may be approximated as an ideal gas. More over, the specific volume of the liquid is negligible compared to the specific volume of vapor ($v_f << v_g$) and hence

$$v_{fg} = v_g - v_f \approx v_g = RT/P.$$

The Clapeyron equation gives

$$\left(\frac{\partial P}{\partial T}\right)_{sat} = \frac{h_{fg}}{Tv_{fg}} = \frac{h_{fg}}{T(RT/P)} = \frac{h_{fg}P}{RT^2}$$

or $$\frac{d \ln P}{dT} = \frac{h_{fg}}{RT^2}$$

Which is known as the Clausius - Clapeyron equation.

**Example 7.54**

Is it possible to approximate the (saturation) vapor pressure of a substance as an exponential function of temperature? Does the Clausius– Clapeyron equation suggest such an approximation ?

*Solution :*

The Clausius - Clapeyron equation is given by

$$\frac{d \ln P}{dT} = \frac{h_{fg}}{RT^2} \text{ or } d \ln P = \frac{h_{fg}\, dT}{RT^2}$$

If $h_{fg}$ is constant over a small temperature range, the above equation can be integrated to obtain

$$\ln P = \frac{-h_{fg}}{RT} + \text{constant} \qquad (A)$$

or $P = A \exp\left(\frac{-h_{fg}}{RT}\right)$ where $A$ is a constant.

Therefore, it is possible to approximate the vapor pressure of a substance as an exponential function of temperature. Eqn.($A$) suggests that a plot of $\ln P$ versus $1/T$ yields a straight line with a slope of $(-h_{fg}/R)$. One can use the Clausius Clapeyron equation to estimate the enthalpy of vaporization from a knowledge of the vapor pressure of the substance.

**Example 7.55**

A domestic pressure cooker is a common kitchen appliance which is usually used for cooking meat or pulses which are hard to boil at atmospheric pressure. The boiling point of water in the cooker increases because of the pressure developed by the steam. The cooker is provided with a lid which can be tightly sealed with the help of a gasket or O–ring . In view of the safety in the kitchen, the pressure cooker is so designed that it maintains a constant pressure of 2 bar while it is in use. Assuming that no other data except that water boils at 100°C at 1 bar pressure with a latent heat of vaporization of 2256.94 kJ/kg, estimate the boiling point of water in the pressure cooker.

*Solution* :

In absence of any other information, one can assume that the latent heat of vaporization of water is constant in the pressure range 1 bar to 2 bar and at these pressures the water vapor may be assumed to behave as an ideal gas to determine the approximate boiling point of water. With these approximations, the Clausius - Clapeyron equation can be used to estimate the saturation temperature of water at 2 bar. The Clausius - Clapeyron equation gives

$$\frac{d\ln P}{dT} = \frac{h_{fg}}{RT^2}$$

Integrating the above equation,we get

$$\ln\frac{P_2}{P_1} = \frac{h_{fg}}{R}\left(\frac{1}{T_1} - \frac{1}{T_2}\right)$$

or $$\ln\frac{2}{1} = \frac{2256.94\times10^3\times18\times10^{-3}}{8.314}\left(\frac{1}{373} - \frac{1}{T_2}\right)$$

or $T_2 = 393.84$ K or $t_2 = 120.84$°C

(The molar mass of water is $18 \times 10^{-3}$ kg/mol)

Therefore, water boils at 120.84°C at 2 bar pressure.

**Example 7.56**

Suppose a person of mass 60 kg intends to skate on ice at −2°C and carries with him skates, which have 15 mm$^2$ as the area of contact with the ice surface when placed on ice. The specific volumes of ice and water at 0°C are $1.091 \times 10^{-3}$ m$^3$/kg and $1.0 \times 10^{-3}$ m$^3$/kg, respectively and the latent heat of fusion of water is 333.43 kJ/kg. Determine whether the person would be able to skate or not.

*Solution* :

The Clapeyron equation gives

$$\frac{dP}{dT} = \frac{\Delta h}{T\Delta v} \quad \text{or} \quad \frac{\Delta P}{\Delta T} = \frac{\Delta h}{T\Delta v}$$

$$\Delta P = \frac{mg}{A} = \frac{60 \times 9.81}{15 \times 10^{-6}} = 39.24 \text{ MPa}$$

$$\Delta v = v_{\text{liquid}} - v_{\text{solid}} = (1.0 - 1.091) \times 10^{-3} \times 18 \times 10^{-3} = -1.638 \times 10^{-6} \text{ m}^3\text{/mol}$$

(Molar mass of water = $18 \times 10^{-3}$ kg /mol)

Then
$$\frac{\Delta P}{\Delta T} = \frac{\Delta h}{T\Delta v} = \frac{333.43 \times 10^3 \times 18 \times 10^{-3}}{273(-1.638 \times 10^{-6})}$$

$$= -13.4215 \text{ MPa/K}$$

or
$$\Delta T = \frac{\Delta P}{-13.4215} = \frac{39.24}{-13.4215} = -2.92\text{°C}$$

Hence, ice melts at −2.92°C which is less than the temperature of the ice, Therefore, the ice melts due to the additional pressure applied by the person and provides a thin layer of water which helps him to easily slide or skate on ice. Thus the person would be able to skate.

**Example 7.57**

The saturation pressures of water at 75°C and 100°C are 0.3855 bar and 1.0133 bar, respectively. Estimate the approximate enthalpy of vaporization of water and compare with the values from steam tables.

*Solution* :

The Clausius - Clapeyron equation is given by

$$\ln\frac{P_2}{P_1}=\frac{h_{fg}}{R}\left(\frac{1}{T_1}-\frac{1}{T_2}\right)$$

or $\ln\dfrac{1.0133}{0.3855}=\dfrac{h_{fg}}{8.314}\left(\dfrac{1}{348}-\dfrac{1}{373}\right)$

or $h_{fg}$ = 41.7183 kJ / mol or 2317.68 kJ/kg

From steam tables we find

$$h_{fg}\text{ (at 75°C)} = 2321.46\text{ kJ/kg}$$

$$h_{fg}\text{ (at 100°C)} = 2256.94\text{ kJ/kg.}$$

**Example 7.58**

It is found that at a particular hill station water boils at 95°C. It is known that at mean sea level where the pressure is 1 bar, water boils at 100°C with $h_{fg}$ = 2256.94 kJ/kg. Assuming that the atmosphere is isothermal at 25°C, estimate the altitude of the hill station above the mean sea level.

*Solution* :

Consider a differential element of the atmosphere as shown in Fig.E 7.58. A force balance on the differential volume element gives

$$(P+dP)A+\frac{Agdh}{v} = PA$$

or $dP=-\dfrac{gdh}{v}$ (A)

Where $A$ = the cross–sectional area of the differential element

$v$ = specific volume of air

For an isothermal atmosphere $PV=P_0v_0$

Then Eqn.($A$) can be rewrittten as

$$dP = -\frac{gdhP}{P_0v_0}$$

or $\displaystyle\int_{P_0}^{P}\frac{dP}{P}=-\int_0^h\frac{gdh}{P_0v_0}$

or $$\ln\frac{P}{P_0}=-\frac{gh}{P_0v_0}=-\frac{ghM}{RT_0} \qquad (B)$$

Where $M = 28.97\times10^{-3}$ kg/mol, molar mass of air and the subscript 0 denotes conditions at the mean sea level.

We know that the Clausius – Clapeyron equation is given by

$$\ln\frac{P}{P_0}=\frac{h_{fg}}{R}\left(\frac{1}{T_0}-\frac{1}{T}\right)$$

$$=\frac{2256.94\times10^3\times18\times10^{-3}}{8.314}\left(\frac{1}{373}-\frac{1}{368}\right)$$

$$=-0.178$$

Then from Eqn.(*B*), we get

$$\ln\frac{P}{P_0}=-0.178=-\frac{ghM}{RT_0}$$

or $$h=\frac{0.178\times8.314\times298}{9.81\times28.97\times10^{-3}}=1551.78\text{ m}$$

Therefore, the hill station is 1551.78 m above the mean sea level.

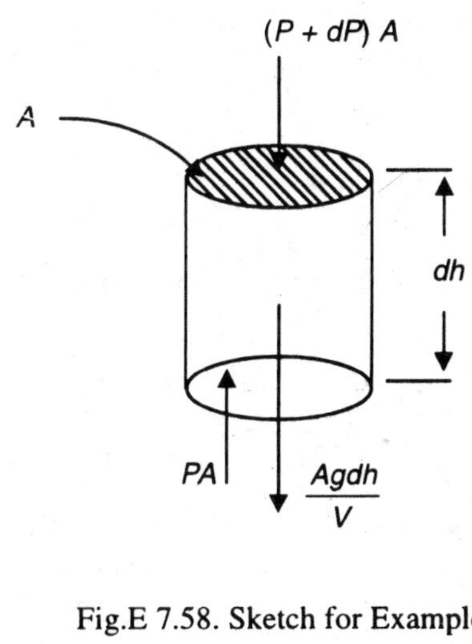


Fig.E 7.58. Sketch for Example 7.58.

### Example 7.59

Rework Example 7.58 assuming that the atmosphere is adiabatic, that is the relation between $P$ and $v$ of air is given by $Pv^\gamma = P_0 v_0^\gamma$ with $\gamma = 1.4$

*Solution* :

A force balance on a differential volume element of the atmosphere gives

$$dP = -\frac{gdh}{v} \qquad \text{[See Example 7.58]} \qquad (A)$$

For an adiabatic atmosphere $Pv^{\gamma} = P_0 v_0^{\gamma}$

or $dP = P_0 v_0^{\gamma}(-\gamma)v^{-\gamma-1}\,dv$ (B)

Substituting Eqn.(*B*) in Eqn.(*A*). we get

$$-\gamma P_0 v_0^{\gamma} v^{-\gamma-1}\,dv = -\frac{gdh}{v}$$

or $$v^{-\gamma}\,dv = \frac{gdh}{\gamma P_0 v_0^{\gamma}}$$

Integrating the above relation, we get

$$v^{1-\gamma} - v_0^{1-\gamma} = \left(\frac{1-\gamma}{\gamma}\right)\frac{gh}{P_0 v_0^{\gamma}}$$

or $$v^{1-\gamma} = v_0^{1-\gamma}\left[1+\left(\frac{1-\gamma}{\gamma}\right)\frac{gh}{P_0 v_0}\right]$$

or $$\left(\frac{v}{v_0}\right)^{1-\gamma} = 1+\left(\frac{1-\gamma}{\gamma}\right)\frac{gh}{P_0 v_0^{\gamma}}$$

or $$\left(\frac{P_0}{P}\right)^{\frac{1-\gamma}{\gamma}} = \left(\frac{P}{P_0}\right)^{\frac{\gamma-1}{\gamma}} = 1+\left(\frac{1-\gamma}{\gamma}\right)\frac{gh}{P_0 v_0}$$

or $$\frac{P}{P_0} = \left[1+\left(\frac{1-\gamma}{\gamma}\right)\frac{gh}{P_0 v_0}\right]^{\frac{\gamma}{\gamma-1}} = \left[1+\left(\frac{1-\gamma}{\gamma}\right)\frac{ghM}{RT_0}\right]^{\frac{\gamma}{\gamma-1}} \qquad (C)$$

By applying the Clausius–Clapeyron equation we get

$$\ln\frac{P}{P_0} = -0.178 \qquad \text{[see Example 7.58]}$$

or $$\frac{P}{P_0} = 0.8369$$

From Eqn.($C$), we have

$$\frac{P}{P_0} = 0.8369 = \left[1 + \left(\frac{1-1.4}{1.4}\right)\frac{9.81 \times 28.97 \times 10^{-3} h}{8.314 \times 298}\right]^{\frac{1.4}{1.4-1}}$$

or $0.8369 = (1 - 3.2774 \times 10^{-5} h)^{3.5}$

or $h = 1513.4$ m

Therefore, the hill station is 1513.4 m above the mean sea level.

**Example 7.60**

It is a known fact that the enthalpy of vaporization of a liquid decreases with increasing temperature and attains a value of zero at the critical point. Derive a relation,known as the Kirchoff relation, for estimating the enthalpy of vaporization at any temperature.

*Solution* :

We know that the change in enthalpy of a substance is given by

$$dh = C_P\, dT + v\,(1 - \beta\, T)dP$$

and the enthalpy of vaporization $h_{fg}$ is given by

$$h_{fg} = h_g - h_f$$

If a liquid is vaporizing at temperature $T$, its saturation pressure at the specified temperature is $P$. If the temperature is increased by $dT$, then the associated increase in saturation pressure is $dP$. The change in the enthalpy of vaporization, then can be written as

$$dh_{fg} = dh_g - df_g = (C_{Pg} - C_{Pf})\, dT + [(v_g - v_f) - v_g\,\beta_g T - v_f \beta_f T)]\, dP$$

or $$\frac{dh_{fg}}{dT} = (C_{Pg} - C_{Pf}) + \{(v_g - v_f) - (v_g\beta_g - v_f\beta_f)T\}\, \frac{dP}{dT} \qquad (A)$$

We know that the change in saturation pressure with a change in temperature is given by the Clapeyron equation

$$\frac{dP}{dT} = \frac{h_{fg}}{T(v_g - v_f)} \qquad (B)$$

Substituting Eqn.($B$) in Eqn. ($A$), we get

$$\frac{dh_{fg}}{dT} = (C_{Pg} - C_{Pf}) + \left[(v_g - v_f) - (v_g\beta_g - v_f\beta_f)T\right]\frac{h_{fg}}{T(v_g - v_f)}$$

or $$\frac{dh_{fg}}{dT} = (C_{pg} - C_{pf}) + \frac{h_{fg}}{T} - h_{fg}\frac{(v_g\beta_g - v_f\beta_f)}{(v_g - v_f)}$$

Which is known as Kirchoff relation.

**Example 7.61**

Consider liquid to vapor phase transition and simplify the Kirchoff relation which predicts the effect of temperature on enthalpy of phase transition.

*Solution* :

The Kirchoff relation is given by

$$\frac{dh_{fg}}{dT} = (C_{Pg} - C_{Pf}) + \frac{h_{fg}}{T} - h_{fg}\frac{(v_g\beta_g - v_f\beta_f)}{v_g - v_f} \qquad (A)$$

We know that the specific volume of vapor is much larger than that of the liquid phase. That is $v_g >> v_f$ and hence $v_g - v_f \approx v_g$. The coefficient of volume expansion of the vapor phase $\beta_g$ is much larger than the coefficient of volume expansion of the liquid phase. That is, $\beta_g >> \beta_f$. Hence $v_g\,\beta_g >> v_f\beta_f$.

Therefore,one can approximate

$$v_g\beta_g - v_f\,\beta_f \equiv v_g\beta_g \qquad (B)$$

The vapor phase may be assumed to behave like an ideal gas. Then

$$\beta_g = 1/T \qquad (C)$$

Substituting Eqns. (*B*) and (*C*) in Eqn. (*A*), we get

$$\frac{dh_{fg}}{dT} = (C_{Pg} - C_{Pf}) + \frac{h_{fg}}{T} - \frac{h_{fg}}{T} = C_{Pg} - C_{Pf}$$

or $(h_{fg})_2 - (h_{fg})_1 = (C_{Pg} - C_{Pf})\,(T_2 - T_1)$

**Example 7.62**

The enthalpy of vaporization of water at 100°C is 2256.94 kJ / kg. Determine the enthalpy of vaporization of water at 150°C if the specific heats of liquid and vapor at constant pressure are 4.26 kJ/kg K and 1.388 kJ/kg K, respectively.

*Solution :* We know that the Kirchoff relation for liquid – vapor phase transition is given by

$$\frac{dh_{fg}}{dT} = C_{pg} - C_{pf}$$

or $(h_{fg})_2 - (h_{fg})_1 = (C_{Pg} - C_{Pf})\,(T_2 - T_1)$

or $(h_{fg})_2 = (h_{fg})_1 + (C_{Pg} - C_{Pf})(T_2 - T_1)$

$$= 2256.94 + (1.388 - 4.26)50 = 2113.34 \text{ kJ/kg.}$$

**Example 7.63**

Consider the solid to liquid phase transition and reduce the Kirchoff relation so that it can be directly used to estimate the enthalpy of fusion of a pure substance.

*Solution* :

The Kirchoff equation for estimating the enthalpy of phase transition is given by

$$\frac{dh_{fg}}{dT} = (C_{Pg} - C_{Pf}) + \frac{h_{fg}}{T} - \frac{h_{fg}(v_g\beta_g - v_f\beta_f)}{v_g - v_f} \qquad (A)$$

Equation (*A*) is applicable for all phase transitions. For solid–liquid phase transition, it can be written as

$$\frac{dh_{sf}}{dT} = (C_{Pf} - C_{Ps}) + \frac{h_{sf}}{T} - \frac{h_{sf}(v_f\beta_f - v_s\beta_s)}{v_f - v_s} \qquad (B)$$

where $h_{sf}$ = enthalpy of fusion

$C_{Ps}$ = specific heat of solid phase

$v_s$ = specific volume of solid phase

$\beta_s$ = coefficient of volume expansion of solid phase

Eqn, (*B*) can be rewritten as

$$\frac{d(h_{sf}/T)}{dT} + \frac{(v_f\beta_f - v_s\beta_s)h_{sf}}{T(v_f - v_s)} = \frac{C_{Pf} - C_{Ps}}{T} \qquad (C)$$

Since solids and liquids are relatively incompressible, it is reasonable to approximate $v_f \cong v_s$ and $\beta_f \cong \beta_s$. Then Eqn.(C) reduces to

$$\frac{d(h_{sf}/T)}{dT} = \frac{C_{Pf} - C_{Ps}}{T}$$

which is the simplified form of the Kirchoff relation applicable for solid-liquid phase transition.

**Example 7.64.**

Given that the latent heat of fusion of water at 0°C is 333.43 kJ/kg and the specific heats of liquid and ice are 4.2 kJ/kg K and 2.1 kJ/kg K, estimate the latent heat of fusion of water at –10°C.

*Solution* :

We know that the enthalpy of fusion can be estimated using the Kirchoff relation given by

$$\frac{d(h_{sf}/T)}{dT} = \frac{C_{Pf} - C_{Ps}}{T}$$

or $$(h_{sf}/T)_2 - (h_{sf}/T)_1 = (C_{Pf} - C_{Ps}) \ln \frac{T_2}{T_1}$$

or $$(h_{sf}/T)_2 = (h_{sf}/T)_1 + (C_{Pf} - C_{Ps}) \ln \frac{T_2}{T_1}$$

$$= \frac{333.43 \times 10^3}{273} + (4.2 - 2.1) \times 10^3 \ln \frac{263}{273} = 1142.988$$

or $h_{sf}$ (at –10°C) = 1142.988 × 263 = 300.606 kJ/kg

**Example 7.65**

Calculate the entropy change if one kg liquid water at –10°C is converted into ice at –10°C. The latent heat of fusion of water at 0°C is 333.43 kJ/kg and the specific heats of liquid water and ice are 4.2 kJ/kg K and 2.1 kJ/kg K, respectively.

Solution :

In Example 7.64, the Kirchoff relation is used to estimate the latent heat of fusion of water at –10°C and we found that $h_{sf}$(–10°C) = 300.606 kJ/kg. Then

$$\Delta s = -\frac{h_{sf}}{T} = -\frac{300.606 \times 10^3}{263} = -1.143 \text{ kJ/kg K}$$

# 8

# Power and Refrigeration Cycles

**Example 8.1**

Describe a Carnot power cycle employing a condensable vapor as a working medium and mention the drawbacks associated with the design of an engine based on Carnot cycle.

*Solution* :

We know that a Carnot cycle consists of only reversible processes and hence is an ideal heat engine cycle. The $T-s$ diagram of a Carnot power cycle is shown in Fig.E 8.1. Assume that water is used as a working medium in the power cycle. Then wet steam at low temperature $T_2$ at state 1 is isentropically compressed to state 2 where it is available at saturated liquid. The saturated liquid at state 2 enters a boiler and leaves as saturated vapor at state 3, at a higher temperature $T_1$. Energy addition as heat takes place in the boiler at constant temperature $T_1$. The dry saturated steam at state 3 then undergoes reversible adiabatic expansion in a turbine and leaves as wet steam at a lower temperature $T_2$ at state 4. Then energy is rejected at constant temperature $T_2$ and the steam leaves at state 1, thus completing a cycle. The energy addition in the cycle takes place at constant temperature and pressure and the working medium undergoes a phase change from liquid to vapor. Similarly, the energy rejection also takes place at constant temperature and pressure.

The energy addition and rejection as heat at constant temperature is desirable. Since these processes are associated with vaporization and condensation, they can be practical. However, the process 1 – 2 is an isentropic compression of wet steam where both liquid and vapor phases are to be handled by the compressor. The design of compressors which handle a two phase mixture is very difficult and such a compressor is not practical. Therefore, the process 1–2 is highly undesirable from a practical point of view. Similarly, the turbine also has to handle a two phase mixture. The presence of water droplets in the vapor phase erodes the turbine blades and hence the process 3– 4 is also undesirable and impractical.

Moreover the rate at which work is delivered by such a reversible device is so low that it is not worthwhile to pursue it.

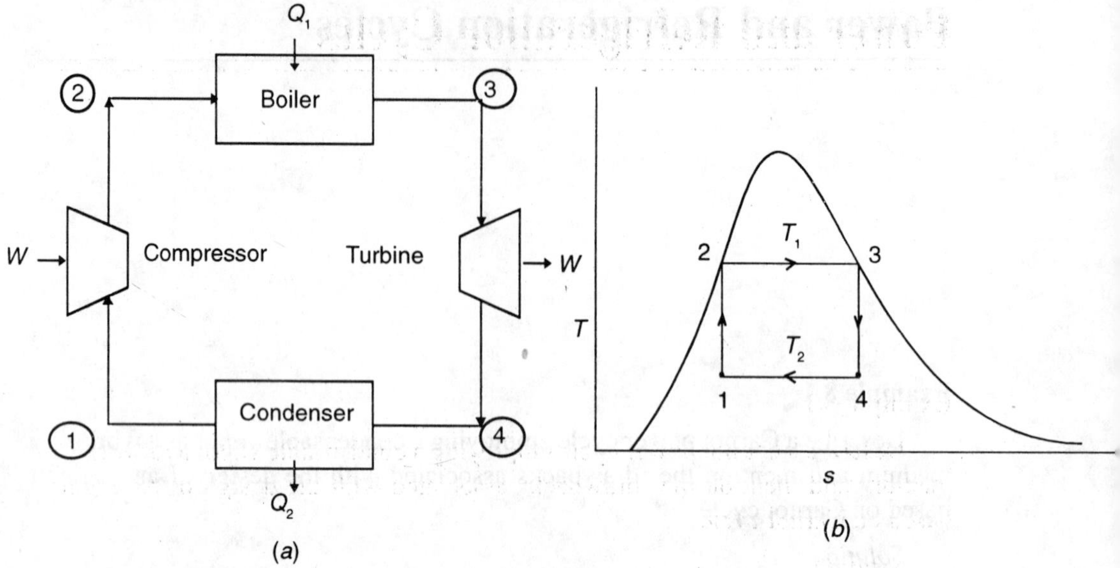


Fig.E 8.1. Sketch for Example 8.1.

(*a*) Schematic of a Carnot power cycle.

(*b*) *T- s* diagram of Carnot power cycle employing condensable vapor as working medium.

**Example 8.2**

It is possible to avoid the practical difficulties associated with the turbine operation by superheating the steam at constant temperature,as shown in Fig.E 8.2. Suppose such a modification is incorporated in the cycle what other practical difficulties can be foreseen?

*Solution* :

The energy addition during the process 2–3 takes place at constant temperature which is desirable. Moreover, superheated steam at state 3 enters the turbine and leaves as dry saturated steam at state 4. Such a modification is desirable because, the turbine now handles only vapor phase. However, this modification is undesirable and not practical in view of the following reason. It can be observed from Fig.E 8.2. that the pressure of the steam is continuously decreasing from $P_3$ to $P_1$ during superheating. That is, the steam is expanding doing work. The design of boilers which permit the expansion of steam at constant temperature is highly complicated and impractical. Hence this modification is not feasible.

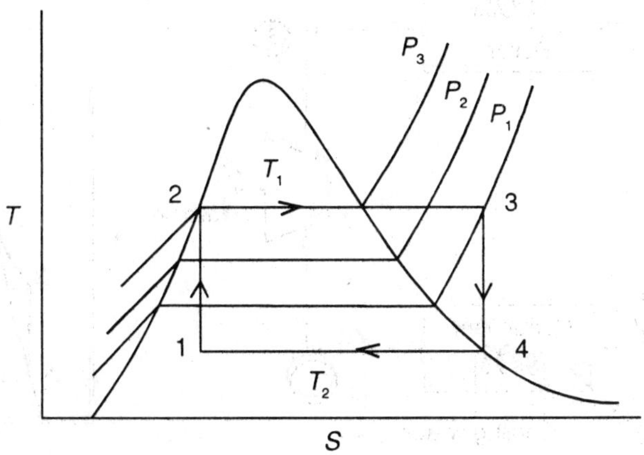


Fig.E 8.2. Sketch for Example 8.2.

*T*- *s* diagram showing modified Carnot cycle with superheating of steam at constant temperature. $P_3 > P_2 > P_1$

## Example 8.3

Professor Rankine who has made significant contributions to the engineering and scientific development modified the Carnot cycle and presented a technically feasible cycle. What are the modifications suggested by Rankine.? Sketch the ideal Rankine cycle on a $T-s$ diagram.

*Solution* :

The major difficulty associated with the Carnot cycle is the isentropic compression of the low pressure wet steam to the boiler pressure and the isentropic expansion in the turbine which has to handle a two phase mixture. Rankine suggested that instead of compressing the wet steam, the steam can be completely condensed into liquid water and the water can be easily compressed to the boiler pressure by employing a pump. In addition to this, the steam can be superheated at constant pressure instead of at constant temperature which eliminates the complexities in the design of the boiler. The Carnot cycle thus modified is presented in Fig.E 8.3 which is known as the ideal Rankine cycle. In the Rankine cycle, saturated liquid water leaving the condenser at low pressure (state 1) enters a pump. The pump compresses the liquid to the boiler pressure (state 2) and feeds water to the boiler. In the boiler, constant pressure energy addition takes place and superheated steam (state 3) leaves the boiler. The superheated steam, then undergoes reversible, adiabatic expansion to a low pressure (state 4). The low pressure steam is then completely condensed into liquid water at state 1, thus completing the cycle.

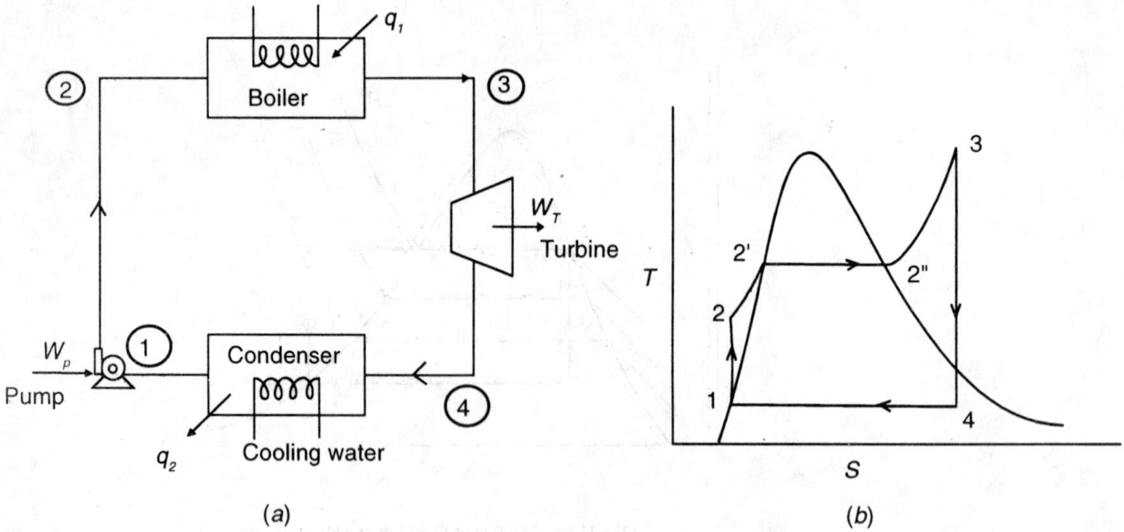


Fig.E 8.3. (*a*) Schematic of Rankine power cycle (*b*) $T - s$ diagram of ideal Rankine cycle

**Example 8.4**

Derive an expression to estimate the thermal efficiency of a Rankine cycle. Is the thermal efficiency of a Rankine cycle equal to that of a Carnot cycle? If it differs, explain the reasons for the same.

*Solution* :

The thermal efficiency $\eta$ of a power cycle is defined as

$$\eta = \frac{\text{Net work done}}{\text{Energy added as heat}}$$

Let 1 kg of water undergo the cyclic change in the Rankine cycle. The Rankine cycle is shown in Fig.E 8.4.

Let $q_1$ = energy added as heat in the boiler

$q_2$ = energy rejected as heat in the condenser

$W_p$ = work done on the pump

$W_T$ = work delivered by the turbine,

Application of the first law of thermodynamics to the control volume, pump, gives

$$W_p = h_2 - h_1$$

Treat the boiler as the control volume. Then the first law of thermodynamics gives

$$q_1 = h_3 - h_2$$

Similarly, application of the first law of thermodynamics to the control volumes, turbine and condenser give

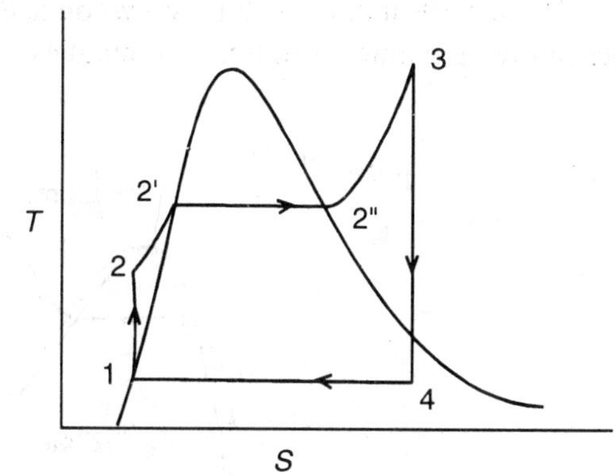


Fig.E 8.4. Rankine cycle on a $T-s$ diagram.

$$W_T = h_3 - h_4$$
$$q_2 = h_4 - h_1$$

The working medium water undergoes the cyclic change 12341.

The first law of thermodynamics gives

$$\oint dq = \oint dW \quad \text{or} \quad q_1 - q_2 = W_T - W_P$$

Net work delivered $= W = W_T - W_P$

Then $$\eta = \frac{W_T - W_P}{q_1} = \frac{(h_3 - h_4) - (h_2 - h_1)}{h_3 - h_2} = \frac{q_1 - q_2}{q_1}$$

The efficiency of a Rankine cycle is less than the efficiency of a Carnot cycle. It can be observed from Fig.E 8.4 that during $2-2'$ the temperature is not constant. Similarly, during $2''-3$ the temperature is continuously changing. Though most of the energy is added during $2-2''$ where the temperature is constant, all the energy added $q_1$ is not at constant temperature. The maximum temperature in the cycle is at state 3 and energy addition does not take place at the maximum temperature. Hence, the efficiency of a Rankine cycle is less than the efficiency of a Carnot cycle working between the same temperature levels.

**Example 8.5**

In a thermal power plant, employing ideal Rankine cycle, superheated steam at 20 bar and 400°C is produced in the boiler and the condenser is operated at 0.2 bar. Calculate the quality of steam at the turbine outlet and the thermal efficiency of the cycle.

*Solution* :

The ideal Rankine cycle is shown on a $T-S$ diagram in Fig.E 8.5. We read the following values from the steam tables.

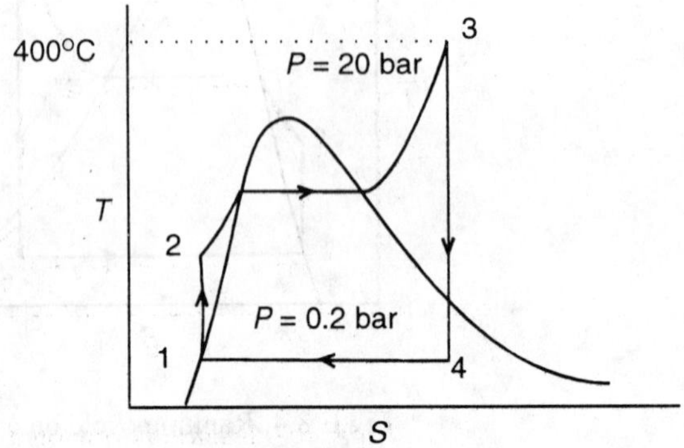


Fig.E 8.5. Sketch for Example 8.5.

At $P = 20$ bar and $T = 400$°C

$h_3 = 3248.7$ kJ/kg ; $s_3 = 7.1296$ kJ/kg K

At $P = 0.2$ bar

$v_f = 0.001\,017\,2$ m$^3$/ kg ; $h_f = 251.45$ kJ/kg

$h_g = 2609.9$ kJ/kg

$s_f = 0.8321$ kJ/kg K;

$s_g = 7.9094$ kJ/kg K;

The turbine is isentropic. Therefore,

$$s_3 = s_4 = 7.1296 \text{ kJ/kg K}$$

$$s_4 = X_4 s_g + (1 - X_4) s_f$$

or $7.1296 = 7.9094\, X_4 + (1 - X_4)\, 0.8321$

or $X_4 = 0.8898$

Hence, the quality of steam at the turbine outlet, $X_4 = 0.8898$

$$h_4 = X_4 h_g + (1 - X_4)\, h_f$$

$= 0.8898 \times 2609.9 + (1 - 0.8898)\ 251.45 = 2350$ kJ/kg

The pump is isentropic. Therefore the work done on the pump is given by.

$$W_p = h_2 - h_1 = \int v_f\, dP = v_f\, \Delta P$$

or $h_2 - h_1 = 0.001\ 017\ 2 \times (20 - 0.2)\ 10^5 = 2.01$ kJ/kg

or $h_2 = h_1 + 2.01 = 251.45 + 2.01 = 253.46$ kJ/kg

$$\text{Thermal efficiency, } \eta = \frac{(h_3 - h_4) - (h_2 - h_1)}{h_3 - h_2}$$

$$= \frac{(3248.7 - 2350) - 2.01}{3248.7 - 253.46} = 0.299$$

**Example 8.6**

An ideal Rankine cycle producing 20 MW net power is equipped with a boiler which generates steam at 50 bar and 500°C and a condenser which operates at 0.1 bar. Calculate the energy added in the boiler per kg of water, the thermal efficiency of the cycle, the mass flow rate of steam in kg /s and the mass flow rate of cooling water in the condenser if the cooling water enters the condenser at 30°C and leaves at 40°C.

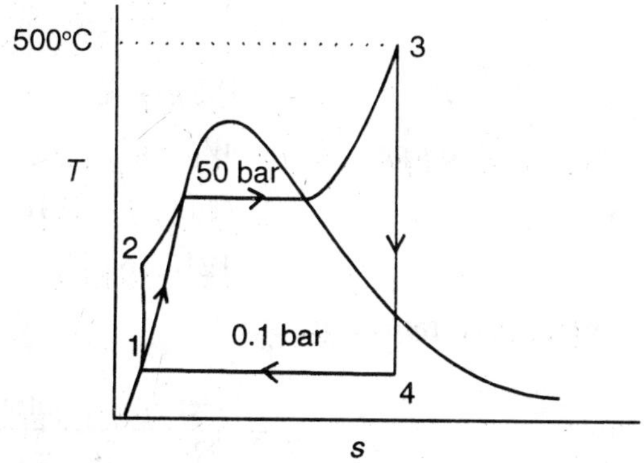


Fig.E 8.6. Sketch for Example 8.6.

*Solution* :

The ideal Rankine cycle is shown on a $T-s$ diagram in Fig.E 8.6. For steam at 50 bar and 500°C: $h_3 = 3433.7$ kJ/kg; $s_3 = 6.9770$ kJ/kg K

For steam at 0.1 bar:

$v_f = 0.001\ 010\ 2\ \text{m}^3/\text{kg}$ ; $h_f = 191.83$ kJ/kg $= h_1$ ; $h_g = 2584.8$ kJ/kg

$s_f = 0.6493$ kJ/kg K; $s_g = 8.1511$ kJ/kg K

At 30°C: $h_f = 125.66$ kJ/kg

At 40°C : $h_f = 167.45$ kJ/kg

We know that $s_3 = s_4 = s_g X_4 + (1 - X_4)\, s_f$

or $6.9770 = 8.1511\, X_4 + (1 - X_4)\, 0.6493$

or $X_4 = 0.8435$

$$h_4 = X_4 h_g + (1 - X_4)\, h_f$$
$$= 0.8435 \times 2584.8 + (1 - 0.8435)\, 191.83$$
$$= 2210.3 \text{ kJ/kg}$$

We know that $W_p = h_2 - h_1 = v_f\, \Delta P$

$$= 0.001\,010\,2\,(50 - 0.1)\, 10^5 = 5.04 \text{ kJ/kg}$$

$$h_2 = h_1 + W_p = 191.83 + 5.04 = 196.87 \text{ kJ/kg}$$

Energy added in boiler per kg of water

$$q = h_3 - h_2 = 3433.7 - 196.87 = 3236.83 \text{ kJ/kg}$$

Thermal efficiency, $\eta = \dfrac{(h_3 - h_4) - (h_2 - h_1)}{h_3 - h_2}$

$$= \frac{(3433.7 - 2210.3) - 5.04}{3236.86}$$

$$= 0.3764$$

Net work output, $= W_T - W_P = (h_3 - h_4) - (h_2 - h_1)$

$$= (3433.7 - 2210.3) - 5.04$$
$$= 1218.36 \text{ kJ/kg}$$

Mass flow rate of water,

$$\dot{m} = \frac{\text{Net Power output}}{\text{Net work per kg}} = \frac{20 \times 10^6}{1218.36 \times 10^3} = 16.42 \text{ kg/s}$$

Energy gained by cooling water $= h_f\,(40°\text{C}) - h_f\,(30°\text{C})$

$$= 167.45 - 125.66 = 41.79 \text{ kJ/kg}$$

Energy rejected in condenser

$$= \dot{m}\,(h_4 - h_1) = 16.42\,(2210.3 - 191.83)$$
$$= 33\,143.28 \text{ kJ/s}$$

Mass flow rate of cooling water

$$= \frac{33\,143.28}{41.79} = 793.09 \text{ kg/s}$$

**Example 8.7**

A thermal power plant operating on an ideal Rankine cycle has a mass flow rate of 10 kg/s through it. Water is fed to the boiler at 50 bar and leaves the condenser as saturated liquid at 0.1 bar. The mass flow rate of cooling water through the condenser is 500 kg/s. The cooling water enters the condenser at 30°C and leaves at 40°C. Calculate the temperature at which steam enters the turbine, the thermal efficiency of the power plant and the power output of the turbine.

*Solution* :

For Water at 30°C, $h_f = 125.66$ kJ/kg and

at 40°C, $h_f = 167.45$ kJ/kg

Energy removed in condenser = 500 (167.45 – 125.66) = 20 895 kJ / s

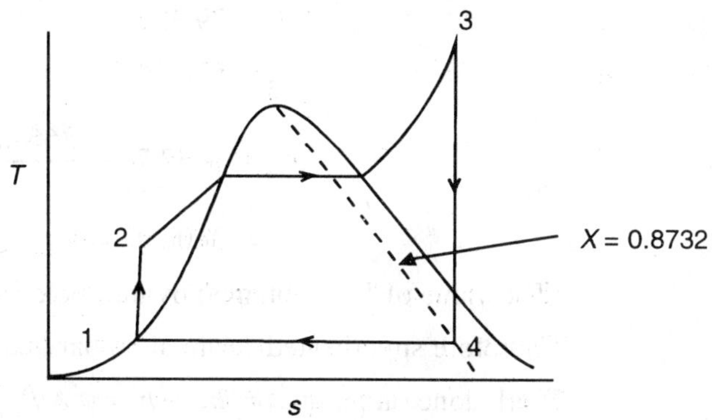


Fig.E 8.7. $T-s$ diagram showing an ideal Rankine cycle.

Mass flow rate of working fluid = 10 kg / s

Energy rejected by unit of mass of working fluid

$$= 20\,895 / 10 = 2089.5 \text{ kJ/kg} = h_4 - h_1$$

(Energy is rejected in the condenser)

Saturated liquid at 0.1 bar : $h_f = 191.83$ kJ/kg $= h_1$

$$h_g = 2584.8 \text{ kJ/kg} ; v_f = 0.001\,010\,2 \text{m}^3 / \text{kg}$$

$$s_f = 0.6493 \text{ kJ/kg K} ; s_g = 8.1511 \text{ kJ/kg K}$$

$$h_4 = h_1 + (h_4 - h_1) = 191.83 + 2089.5 = 2281.33 \text{ kJ/kg}$$

At state 4,we know $h_4 = 2281.33$ kJ/kg and $P = 0.1$ bar.

$$h_4 = X_4 h_g + (1 - X_4) h_f$$

or $2281.33 = 2584.8X_4 + (1 - X_4)\ 191.83$

or $X_4 = 0.8732$

Therefore, $s_4 = s_g X_4 + (1 - X_4)\ s_f$

$= 8.1511 \times 0.8732 + (1 - 0.8732) \times 0.6493 = 7.1999$ kJ/kg K

The turbine is isentropic. Therefore,

$$s_3 = s_4 = 7.1999 \text{ kJ/kg K}$$

State 3 is at $P = 50$ bar and $s = 7.1999$ kJ/kg K.

This corresponds to a temperature between 500°C and 600°C. By interpolation, we can find the temperature of steam at state 3.

$$7.1999 = 6.9770 + \left(\frac{7.2578 - 6.9770}{100}\right)\Delta T$$

or $\Delta T = 79.38$ K

or $t_3 = 579.38$°C

$$h_3 = 3433.71 + \frac{(3664.5 - 3443.7)}{100}\ (79.38)$$

$$= 3608.97 \text{ kJ/kg}$$

(The value of $h_3$ is obtained by interpolation in between 500°C and 600°C)

Therefore superheated steam at 50 bar and 579.38°C enters the turbine.

Work done on pump $= h_2 - h_1 = v\,\Delta P$

$= 0.001\,010\,2\ (50 - 0.1)10^5 = 5.04$ kJ/kg

$$h_2 = h_1 + (h_2 - h_1)$$

$$= 191.83 + 5.04 = 196.87 \text{ kJ/kg}$$

$$\text{Thermal efficiency, } \eta = \frac{(h_3 - h_4) - (h_2 - h_1)}{h_3 - h_2}$$

$$= \frac{(3608.97 - 2281.33) - 5.04}{3608.97 - 196.87}$$

Power output of the turbine= $\dot{m}\ (h_3 - h_4) = 10\ (3608.97 - 2281.33)10^3$

$= 13.2764$ MW

**Example 8.8**

In actual practice, it is impossible to find pumps and turbines operating under isentropic conditions because of inherent irreversibilites. Sketch a practical Rankine cycle on $T$– $s$ diagram.

*Solution* :

The second law of thermodynamics for a steady state, steady flow process gives $s_e \geq s_i$. If there is any irreversibility in the device $s_e > s_i$. That is, the entropy of the exit fluid is greater than the entropy of the inlet fluid. In a pump, the saturated liquid water enters at state 1. If the pump is reversible and adiabatic it leaves the pump at state 2' where $s_{2'} = s_1$. In the actual pump, the fluid leaves the pump at state 2 such that $s_2 > s_1$. The practical Rankine cycle is shown in Fig.E 8.8 on a $T$–$s$ diagram. Similarly the steam leaves an isentropic turbine in state 4′ but in a practical turbine the steam leaves at state 4, such that $s_4 > s_{4'} = s_3$.

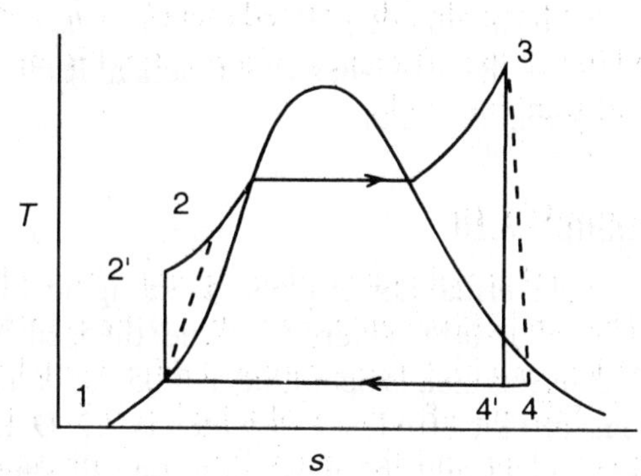


Fig.E 8.8. Practical Rankine cycle on $T$– $s$ daigram.

**Example 8.9**

Develop a relation to estimate the thermal efficiency of a practical Rankine cycle.

*Solution* :

The $T$–$s$ diagram of a practical Rankine cycle is shown in Fig.E 8.8. Let the isentropic efficiencies of the turbine and pump be denoted by $\eta_T$ and $\eta_P$, respectively. We know that $\eta_T$ and $\eta_P$ are defined as

$$\eta_T = \frac{h_3 - h_4}{h_3 - h_{4'}}$$

and
$$\eta_P = \frac{h_{2'} - h_1}{h_2 - h_1}$$

Since $\eta_T$ and $\eta_P$ are less than unity, we find

$$h_3 - h_4 < h_3 - h_{4'}$$

and $$h_2 - h_1 > h_{2'} - h_1$$

That is, the turbine delivers less work than an isentropic turbine while the pump consumes more work than an isentropic pump. The thermal efficiency $\eta$ of a practical Rankine cycle is given by

$$\eta = \frac{\text{net work delivered}}{\text{Energy absorbed}}$$

$$= \frac{(h_3 - h_4) - (h_2 - h_1)}{h_3 - h_2}$$

We have already noticed that $(h_3 - h_4) < (h_3 - h_{4'})$ and $(h_2 - h_1) > (h_{2'} - h_1)$ and hence the efficiency of a practical Rankine cycle will be less than that of an ideal Rankine cycle.

**Example 8.10**

In a thermal power plant operating on a Rankine cycle, superheated steam at 50 bar and 500°C enters a turbine, the isentropic efficiency of which is 0.8. The condenser which is operating at 0.05 bar delivers saturated liquid to a feed pump, the isentropic efficiency of which is 0.7. Determine the thermal efficiency of the power plant and the mass flow rate of steam required for 50 MW net power generation.

*Solution* :

Refer Fig.E 8.8 for a $T$–$s$ diagram.

Steam at 50 bar and 500°C : $h_3 = 3433.7$ kJ/kg; $s_3 = 6.9770$ kJ/kg K;

At 0.05 bar : $v_f = 0.001\,005\,2$ m$^3$ / kg

$h_f = 137.77$ kJ/kg $= h_1$;

$h_g = 2561.6$ kJ/kg

$s_f = 0.4763$ kJ/kg K;

$s_g = 8.3960$ kJ/kg K;

If the turbine is isentropic $s_3 = s_{4'} = 6.9770$ kJ/kg K;

$$s_{4'} = X_{4'} s_g + (1 - X_{4'}) s_f$$

or $6.9770 = 8.3960\, X_{4'} + (1 - X_{4'})\, 0.4763$

or $X_{4'} = 0.8208$ = Quality of steam leaving isentropic turbine.

$$h_{4'} = 0.8208 \times 2561.6 + (1 - 0.8208)\, 137.77$$
$$= 2127.25 \text{ kJ/kg}$$

$$\eta_T = \frac{h_3 - h_4}{h_3 - h_{4'}}$$

or $(h_3 - h_4) = \eta_T \ (h_3 - h_{4'})$

$0.8\ (3433.7 - 2127.25) = 1045.16$ kJ/kg

Work delivered by the actual turbine = 1045.16 kJ/kg

For an isentropic pump,

$$(h_{2'} - h_1) = v\,\Delta P = 0.001\ 005\ 2\ (50 - 0.05)\ 10^5 = 5.02 \text{ kJ/kg}$$

$$\eta_P = \frac{h_{2'} - h_1}{h_2 - h_1}$$

or $$(h_2 - h_1) = \frac{h_{2'} - h_1}{\eta_P} = \frac{5.02}{0.7} = 7.17 \text{ kJ/kg}$$

Work consumed by the actual pump $= h_2 - h_1 = 7.17$ kJ/kg

$$h_2 = h_1 + (h_2 - h_1) = 137.77 + 7.17 = 144.94 \text{ kJ/kg}$$

Net work output $= 1045.16 - 7.17$

$= 1037.99$ kJ/kg

$$\eta = \frac{(h_3 - h_4) - (h_2 - h_1)}{h_3 - h_2} = \frac{1045.16 - 7.17}{3433.7 - 144.94} = 0.3156$$

$$\text{Mass flow rate of steam} = \frac{50 \times 10^6}{1037.99 \times 10^3} = 48.17 \text{ kg/s}$$

**Example 8.11**

Suppose the turbine and pump of Example 8.10 are isentropic. Then the power plant will be operating on an ideal Rankine cycle. Compare the thermal efficiency of the ideal Rankine cycle with the thermal efficiency of the practical Rankine cycle.

*Solution* :

For isentropic turbine, $(h_3 - h_{4'}) = (3433.7 - 2127.25) = 1306.45$ kJ/kg

For isentropic pump, $(h_{2'} - h_1) = 5.02$ kJ/kg [see Example 8.10]

$$h_{2'} = h_1 + (h_{2'} - h_1) = 137.77 + 5.02 = 142.79 \text{ kJ/kg}$$

$$\eta \text{ (ideal Rankine cycle)} = \frac{(h_3 - h_{4'}) - (h_{2'} - h_1)}{h_3 - h_{2'}}$$

$$= \frac{1306.45 - 5.02}{3433.7 - 142.79} = 0.3955$$

or η (Practical Rankine cycle) = 0.3156 [See Example 8.10]

**Example 8.12**

In a thermal power plant operating on an ideal Rankine cycle, it is desired to study the effect of condenser pressure on the thermal efficiency of the power plant and on the quality of steam leaving the turbine,when superheated steam at 50 bar and 500°C is supplied to the turbine. Compare the results if the condenser is maintained at 1 bar, 0.5bar and 0.05bar.

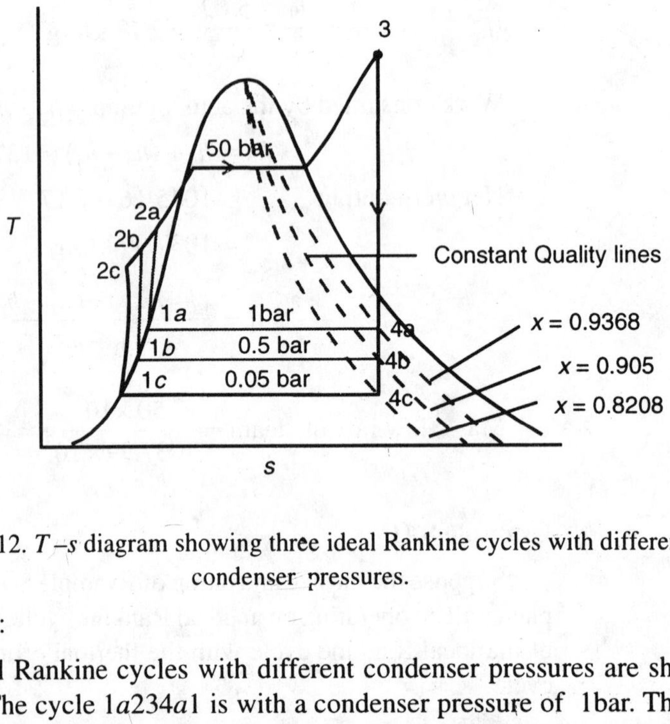


Fig.E 8.12. *T–s* diagram showing three ideal Rankine cycles with different condenser pressures.

*Solution* :

The ideal Rankine cycles with different condenser pressures are shown in Fig.E 8.12. The cycle $1a234a1$ is with a condenser pressure of 1bar. The cycle $1b234b1$ corresponds to a condenser pressure of 0.5 bar while the cycle $1c234c1$ corresponds to a condenser pressure of 0.05 bar. It can be immediately observed from Fig.E 8.12 that the quality of steam at the turbine outlet decreases with decreasing condenser pressure.

Steam at 50 bar and 500°C;

$$h_3 = 3433.7 \text{ kJ/kg}; \quad s_3 = 6.9770 \text{ kJ/kg K}$$

Condenser pressure = 1 bar

$$h_f = 417.54 \text{ kJ/kg} = h_{1a} \quad ; \quad h_g = 2675.4 \text{ kJ/kg}$$

$$s_f = 1.3027 \text{ kJ/kg K}; \quad s_g = 7.3598 \text{ kJ/kg K}$$

$$v_f = 0.001\,043\,4 \text{ m}^3/\text{kg}$$

$$s_3 = 6.9770 = s_{4a} = X_{4a}s_g + (1 - X_{4a})s_f$$

or $6.9770 = 7.3598\, X_{4a} + (1 - X_{4a})\, 1.3027$

or $X_{4a} = 0.9368$

$$h_{4a} = 2675.4 \times 0.9368 + (1 - 0.9368)\, 417.54$$

$$= 2532.7 \text{ kJ/kg}$$

$$h_{2a} - h_{1a} = v\, \Delta P = 0.001\,04\,3\,(50 - 1)\, 10^5 = 5.11 \text{ kJ/kg}$$

$$h_{2a} = h_{1a} + (h_{2a} - h_{1a}) = 417.54 + 5.11 = 422.65 \text{ kJ/kg}$$

$$\eta = \frac{(h_3 - h_{4a})(h_{2a} - h_{1a})}{h_3 - h_{2a}} = \frac{(3433.7 - 2532.7) - 5.11}{3433.7 - 422.65}$$

$$= 0.2975$$

Condenser pressure = 0.05 bar

$$h_f = 340.56 \text{ kJ/kg} = h_{1b} \quad ; \quad h_g = 2646 \text{ kJ/kg}$$

$$s_f = 1.0912 \text{ kJ/kg K} \quad ; \quad s_g = 7.5947 \text{ kJ/kg K};$$

$$v_f = 0.001\,030\,1 \text{ m}^3/\text{kg}$$

$$s_3 = s_{4b} = 6.9770 = 7.5947 X_{4b} + (1 - X_{4b})\, 1.0912$$

or $X_{4b} = 0.905$

$$h_{4b} = 0.905 \times 2646 + (1 - 0.905) 340.56 = 2426.98 \text{ kJ/kg}$$

$$h_{2b} - h_{1b} = 0.001\,030\,1\,(50 - 0.5) 10^5 = 5.1 \text{ kJ/kg}$$

$$h_{2b} = 345.66 \text{ kJ/kg}$$

$$\eta = \frac{(h_3 - h_{4b}) - (h_{2b} - h_{1b})}{h_3 - h_{2b}} = \frac{(3433.7 - 2426.98) - 5.1}{3433.7 - 345.66} = 0.3244$$

Condenser pressure = 0.05bar

$$X_{4c} = 0.8208 \qquad \text{[see Example 8.10]}$$

$$\eta = 0.3955 \qquad \text{[see Example 8.11]}$$

A summary of the final results is given below:

| | Condenser Pressure | | |
|---|---|---|---|
| | 1 bar | 0.5bar | 0.05bar |
| $X$ | 0.9368 | 0.905 | 0.8208 |
| $\eta$ | 0.2975 | 0.3244 | 0.3955 |

**Example 8.13**

Why does the efficiency of a Rankine cycle increase with decreasing condenser pressure ?

*Solution* :

The energy rejection as heat takes place in the condenser. The exhaust steam from the turbine undergoes a phase change at constant temperature and pressure in the condenser. As the condenser pressure is decreased, the saturation temperature of steam also decreases and hence the energy rejection takes place at lower temperature. We know that the efficiency of a power cycle increases by decreasing the sink temperature. Therefore, decreasing the condenser pressure increases the efficiency of a Rankine cycle.

**Example 8.14**

In a Rankine cycle, instead of condensing the steam, (leaving the turbine), in a condenser which requires additional expenditure, is it not possible to feed the steam to the boiler and thereby effect some saving in the energy and money ?

*Solution* :

In a Rankine cycle energy is supplied as heat in the boiler, and some energy is rejected in the condenser and some net work is obtained. If the steam is directed to the boiler instead of condensing and rejecting energy to the cooling water, the working fluid is undergoing a cyclic change while it is interacting with only one reservoir and absorbing energy and hence the device becomes a perpetual motion machine of the second kind which is impossible according to the second law of thermodynamics, Therefore,it is not possible to get a continuous supply of work unless the steam is condensed thereby rejecting energy to a sink.

**Example 8.15**

We know that the efficiency of a Rankine cycle can be increased by decreasing condenser pressure. Is there any limit on the lowest pressure at which the condenser can be operated ? What are the disadvantages associated with maintaining low pressure in a condenser ?

*Solution* :

The primary objective of the condenser is to reject energy as heat from the fluid undergoing cyclic change in the Rankine cycle. For energy to be transferred as heat, the temperature of the cooling water has to be lower than the saturation temperature of steam at the condenser pressure. Since large quantity of cooling water is needed, the easily available river water at ambient temperature which is around 20°C to 30°C (depending on the season) is employed in condensers. Moreover, the rate of energy transfer depends on the temperature difference between the condensing steam and cooling water. If the condenser is maintained at 0.05 bar, the saturation temperature of water at this pressure is 32.9°C which is slightly larger than the cooling water temperature. Therefore, the condenser pressure cannot be decreased at will to very low pressures, but it is governed by the availability of cooling water.

The specific volume of steam increases as the pressure is reduced. Therefore, the size of a condenser and hence the investment increases by decreasing the condenser pressure. Moreover, there is a large pressure difference between the atmosphere and the condenser, and hence there is every possibility of air leaking into the condenser which inturn acts as an insulating blanket on the condenser tubes, thus reducing the energy transfer rate. These are the major drawbacks associated with maintenance of condensers at low pressure.

### Example 8.16

In a thermal power plant operating on ideal Rankine cycle, the condenser is maintained at 0.05 bar and the steam generator produces steam at 500°C. It is desired to study the effect of boiler pressure on the performance of the power plant. Calculate the quality of steam at the turbine outlet and the thermal efficiency of the cycle if the boiler pressure is 50 bar and 20 bar.

*Solution* :

Two ideal Rankine cycles – one with a boiler pressure of 20 bar and another with a boiler pressure of 50 bar are shown in Fig.E 8.16. It is evident from the figure that the quality of steam at the turbine outlet–increases with decreasing pressure.

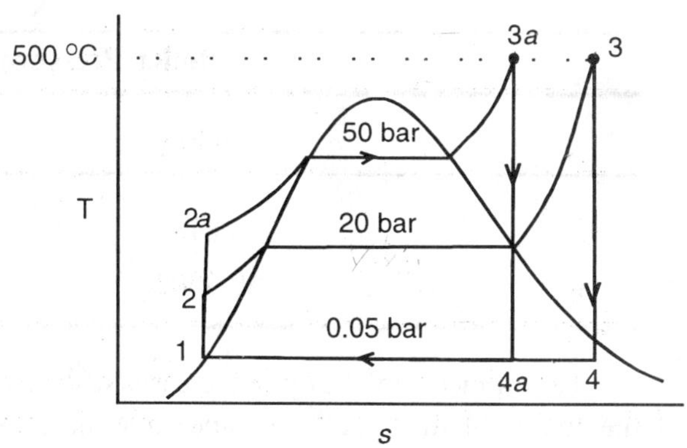


Fig.E 8.16. $T$–$s$ diagram showing ideal Rankine cycle 12341 with a boiler pressure of 20 bar and a Rankine cycle $12a3a4a1$ with boiler pressure of 50 bar.

Boiler Pressure = 50 bar

$X_{4a} = 0.8208$ [See Example 8.10]

$\eta = 0.3955$ [See Example 8.11]

Boiler Pressure = 20 bar

Steam at 30 bar and 500°C :

$$h_3 = 3467.3 \text{ kJ/kg}: s_3 = 7.4323 \text{ kJ/kg K}$$

Steam at 0.05 bar :

$$v_f = 0.001\,005\,2 \text{ m}^3/\text{kg};\ h_f = 137.77 \text{ kJ/kg} = h_1:$$

$$h_g = 2561.6 \text{ kJ/kg};$$

$$s_f = 0.4763 \text{ kJ/kg K};\ s_g = 8.3960 \text{ kJ/kg K}$$

Work done on pump

$$= h_2 - h_1 = v\,\Delta P = 0.001\,005\,2\,(20 - 0.05)\,10^5 = 2 \text{ kJ/kg}$$

$$h_2 = 137.77 + 2 = 139.77 \text{ kJ/kg}$$

$$s_3 = 7.4323 = s_4 = s_g X_4 + s_f (1 - X_4)$$

or $7.4323 = 8.3960 X_4 + (1 - X_4)\,0.4763$: or $X_4 = 0.8783$

$$h_4 = 0.8783 \times 2561.6 + (1 - 0.8783)\,137.77 = 2266.62 \text{ kJ/kg}$$

$$\eta = \frac{(h_3 - h_4) - (h_2 - h_1)}{h_3 - h_2} = \frac{(3467.3 - 2266.62) - 2}{3467.3 - 139.77}$$

$$= 0.3602$$

Summary of Results :

| | Boiler Pressure | |
|---|---|---|
| | **20 bar** | **50 bar** |
| $X$ | 0.8783 | 0.8208 |
| $\eta$ | 0.3602 | 0.3955 |

The efficiency of the cycle increases with increasing boiler pressure. However, the quality of steam at the turbine outlet decreases.

**Example 8.17**

Why does the efficiency of a Rankine cycle increase with increasing boiler pressure when the condenser is maintained at constant pressure ?

*Solution* :

In a Rankine cycle energy addition as heat takes place in a boiler which is held at constant pressure. We know that an increase in pressure is associated with a corresponding increase in the saturation temperature. A major fraction of the energy added in the boiler is during phase transition and hence an increase in boiler pressure increases the average temperature at which energy is added. We know that the efficiency of a heat engine increases with increasing source temperature when the sink is held at constant temperature. When the condenser is maintained at constant pressure, the temperature at which energy is rejected is held constant. Therefore, the efficiency of a Rankine cycle increases with increasing boiler pressure when the condenser is held at constant pressure.

### Example 8.18

In a thermal power plant operating on an ideal Rankine cycle steam at 50 bar enters a turbine. The condenser is maintained at 0.05 bar. It is desired to study the effect of superheat of steam on the quality of steam at the turbine outlet and the thermal efficiency of the cycle. Calculate the quality of steam leaving the turbine and the thermal efficiency of the power plant if the steam is superheated to 300°C and 500°C.

*Solution* :

The two ideal Rankine cycles – one using superheated steam at 300°C and the other using superheated steam at 500°C are show in Fig.E 8.18. It can be immediately observed from the figure that the quality of steam at the turbine exit increases with increasing superheat.

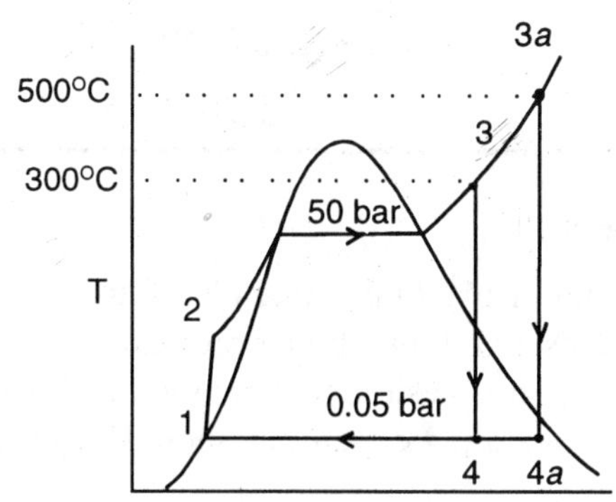


Fig.E 8.18. *T*– *S* diagram showing two ideal Rankine cycles operating between the same pressure levels. The cycle 12341 uses superheated steam at 300°C while the cycle 123a 4a1 employs superheated steam at 500°C.

Steam at 50 bar and 500°C. $X_{4a} = 0.8208$ [See Example 8.10]

$\eta = 0.3955$ [See Example 8.11]

Steam at 50 bar and 300°C; $h_3 = 2925.5$ kJ/kg; $s_3 = 6.2105$ kJ/kg K

Steam at 0.05 bar : $v_f = 0.001\ 005\ 2$ m$^3$/kg ; $h_f = 137.77$ kJ/kg $= h_1$;

$h_g = 2561.6$ kJ/kg

$s_f = 0.4763$ kJ/kg K; $s_g = 8.3960$ kJ/kg K;

For the isentropic turbine $s_3 = 6.2105 = s_4 = X_4 s_g + (1 - X_4)s_f$

or $6.2105 = 8.3960\, X_4 + (1 - X_4)\, 0.4763$ or $X_4 = 0.7240$

$$h_4 = 0.7240 \times 2561.6 + (1 - 0.7240)\ 137.77 = 1892.6 \text{ kJ/kg}$$

$$h_2 - h_1 = v\,\Delta P = 0.001\ 005\ 2\ (50 - 0.05)10^5 = 5.02 \text{ kJ/kg}$$

or $h_2 = h_1 + (h_2 - h_1) = 137.77 + 5.02 = 142.79$ kJ/kg

$$\eta = \frac{(h_3 - h_4) - (h_2 - h_1)}{h_3 - h_2} = \frac{(2925.5 - 1892.6) - 5.02}{2925.5 - 142.79}$$

$$= 0.3694$$

Summary of results :

| | Superheated Steam | |
|---|---|---|
| | 300°C | 500°C |
| $X$ | 0.7240 | 0.8208 |
| $\eta$ | 0.3694 | 0.3955 |

**Example 8.19**

In a power plant operating on a Rankine cycle several operating conditions like condenser pressure, boiler pressure and extent of superheating of steam can be altered. If it is desired to increase the efficiency of a Rankine cycle which of the above operating conditions should be changed and what are the associated drawbacks ?

*Solution* :

We know that the efficiency of a heat engine can be increased either by decreasing the sink temperature while the source temperature is held constant or by increasing the source temperature while the sink is held at constant temperature. By reducing the condenser pressure in a Rankine cycle, keeping the other operating conditions constant, the sink temperature reduces and hence the efficiency of the cycle increases. However, there is a limit on the lowest pressure that can be employed in a practical condenser and this limit is dictated by the availability of the cooling water.

The second alternative for increasing the efficiency of the cycle is to increase the source temperature while the sink temperature is held constant. This can be accomplished in two different ways—that is either by increasing the boiler pressure, which in turn raises the temperature at which energy addition takes place, while maintaining the same temperature of the steam leaving the steam generator or by maintaining the boiler at same pressure but by increasing the temperature at which steam leaves the boiler. Suppose the boiler pressure is increased to raise the temperature at which energy addition takes place (see Fig.E 8.16), then the steam leaves the turbine with a low quality. That is in the low pressure region of the turbine, the moisture content will be high. If the moisture content exceeds 10% (or quality is less than 0.9) it causes serious erosion of the turbine blades and hence it is undesirable. On the other hand, if the boiler is maintained at low pressure and the steam is superheated to high temperature, the steam in the low pressure region of the turbine will have high quality and the efficiency also increases. However, the metallurgical conditions of the boiler and turbine blades dictates that the temperature cannot be increased beyond 600°C, which is the maximum normally allowed.

### Example 8.20

Describe the reheat modification of the Rankine cycle and derive an expression to estimate the thermal efficiency of the reheat Rankine cycle.

*Solution* :

The reheat modification of the Rankine cycle is introduced to obtain the advantage of increased efficiency by increasing the temperature at which energy addition as heat takes place and at the same time to avoid the excessive moisture content in the low pressure region of the turbine. In this cycle, the steam is not allowed to expand completely to the condenser pressure in a single stage. After partial expansion to an intermediate pressure, the steam is withdrawn from the turbine and reheated to the original temperature at constant pressure. Then the steam is returned to the turbine for further expansion to the condenser pressure as shown in Fig.E 8.20. The $T-s$ diagram of the reheat cycle is shown in Fig.E 8.20 (*b*). The superheated steam leaving the steam generator at state 3 (high pressure) enters the high pressure turbine and leaves at some intermediate pressure at state 4. This steam is sent to the reheat zone of the steam generator and heated at constant pressure to state 5 (the temperatures of states 4 and 5 are the same). This steam then enters a low pressure turbine and expands to the condenser pressure (state 6). The low pressure steam enters a condenser and leaves as saturated liquids (state 1) which enters a pump. The pump raises the pressure (state2) of the liquid water and feeds the boiler. Usually the pressure ratio $P_4/P_3$ is kept in the rage 0.15 to 0.35.

Consider one kg of steam undergoing the cyclic change. Application of the first law of thermodynamics to the control volumes–boiler and reheat zone gives

$$q_1 = (h_3 - h_2) + (h_5 - h_4) \text{ (Energy is added in two stages)}$$

Work delivered by the turbine = $(h_3 - h_4) + (h_5 - h_6)$

Work done on pump = $(h_2 - h_1)$

Net work output = $(h_3 - h_4) + (h_5 - h_6) - (h_2 - h_1)$

$$\eta = \frac{\text{Net work done}}{\text{Energy absorbed}} = \frac{(h_3 - h_4) + (h_5 - h_6) - (h_2 - h_1)}{(h_3 - h_2) + (h_5 - h_4)}$$

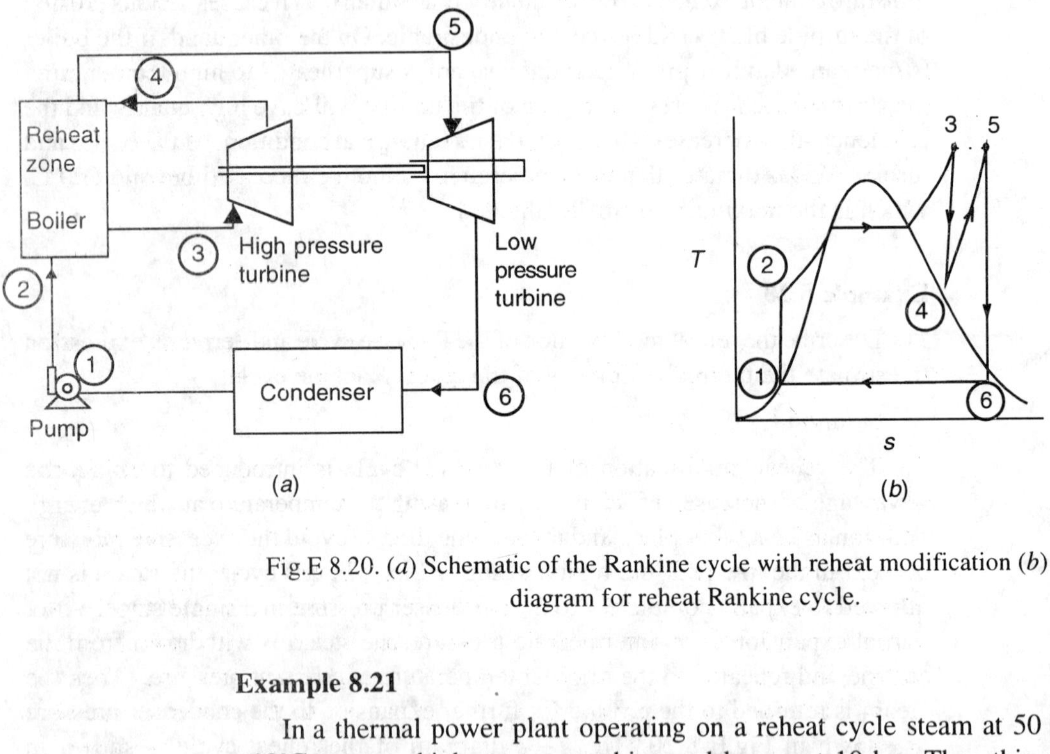


Fig.E 8.20. (*a*) Schematic of the Rankine cycle with reheat modification (*b*) $T-s$ diagram for reheat Rankine cycle.

**Example 8.21**

In a thermal power plant operating on a reheat cycle steam at 50 bar and 500°C enters a high pressure turbine and leaves at 10 bar. Then this steam is reheated to 500°C before it is fed to a low pressure turbine. The condenser is maintained at 0.05 bar. Calculate the thermal efficiency of the power plant, the mass flow rate of steam for a net power output of 20 MW and the quality of steam at the exit of the low pressure turbine.

*Solution* :

Refer Fig.E 8.21 for the $T-s$ diagram of reheat cycle.

Steam at 50 bar and 500°C: $h_3$= 3433.7 kJ/kg; $s_3$ = 6.9770 kJ/kg K

Steam at 10 bar: Saturation temperature = 179.88°C

$h_g$= 2776.2 kJ/kg; $s_g$ = 6.5862 kJ/kg K

Steam at 10 bar and 500°C; $h_5$= 3478.3 kJ/kg; $s_5$ = 7.7627 kJ/kg K

Steam at 0.05 bar : $v_f$ = 0.001 005 2 m$^3$/ kg; $h_f$ = 137.77 kJ/kg = $h_1$

$h_g$ = 2561.6 kJ/kg

$s_f$ = 0.4763 kJ/kg K; $s_g$ = 8.396 kJ/kg K;

$$h_2-h_1 = v\,\Delta P = 0.001\,005\,2\,(50-0.05)\,10^5 = 5.02 \text{ kJ/kg}$$

$$h_2 = h_1 + v\,\Delta P = 137.77 + 5.02 = 142.79 \text{ kJ/kg}$$

$s_3$ = 6.9770 = $s_4$ For the high pressure isentropic turbine.

The value of $s_4$ = 6.9770 > $s_g$ (10bar) = 6.5862

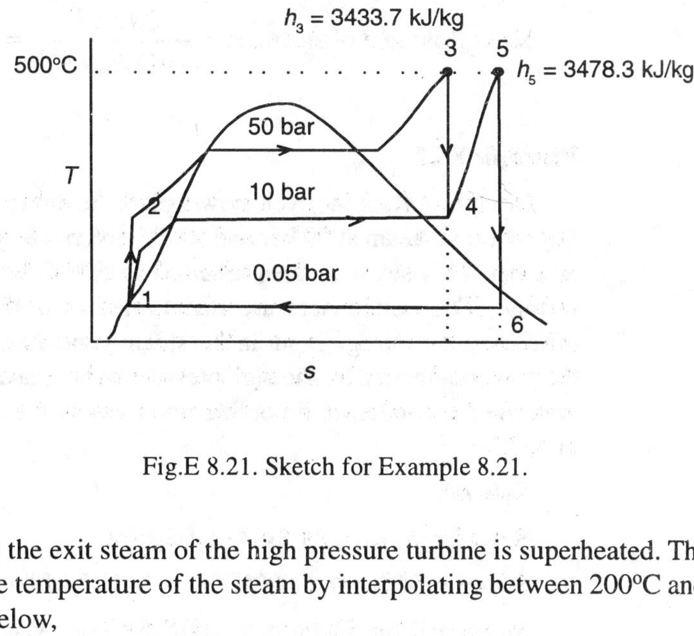


Fig.E 8.21. Sketch for Example 8.21.

Hence the exit steam of the high pressure turbine is superheated. Therefore, we find the temperature of the steam by interpolating between 200°C and 300°C as given below,

$$6.9770 = 6.6922 + \frac{(7.1251 - 6.6922)\,\Delta T}{100}$$

or $\Delta T$ = 65.79°C or $t_4$ = 265.79°C

$$h_4 = 2826.8 + \frac{(3052.1 - 2826.8)\,65.97}{100} = 2975.43 \text{ kJ/kg}$$

$s_5 = 7.7627 = s_6$ For the low pressure isentropic turbine.

$$s_6 = X_6\, s_g + (1 - X_6)s_f$$

or $7.7627 = 8.3960\, X_6 + (1 - X_6)\, 0.4763$

or $X_6 = 0.92$ = Quality of steam leaving the low pressure turbine

$$h_6 = 0.92 \times 2561.6 + 0.08 \times 137.77 = 2367.69 \text{ kJ/kg}$$

$$\begin{aligned} \text{Network done} &= (h_3 - h_4) + (h_5 - h_6) - (h_2 - h_1) \\ &= (3433.7 - 2975.43) + (3478.3 - 2367.69) - 5.02 \\ &= 1563.86 \text{ kJ/kg} \end{aligned}$$

$$\begin{aligned} q_1 &= (h_3 - h_2) + (h_5 - h_4) \\ &= (3433.7 - 142.79) + (3478.3 - 2975.43) \\ &= 3793.78 \text{ kJ/kg} \end{aligned}$$

$$\text{Thermal effciency, } \eta = \frac{\text{Net work done}}{\text{Energy absorbed}} = \frac{1563.86}{3793.78} = 0.4122$$

$$\text{Mass flow rate of steam, } \dot{m} = \frac{20 \times 10^6}{1563.86 \times 10^3} = 12.79 \text{ kg / s}$$

**Example 8.22**

In a reheat Rankine cycle power plant the turbines produce 100 MW of power. Superheated steam at 50 bar and 500°C enters a high pressure turbine and leaves at 8 bar. This steam is then reheated to 500°C before it enters a low pressure turbine. The condensers are maintained at 0.05 bar. Calculate the thermal efficiency, the energy input in the steam generator, the mass flow rate of steam, the power delivered by the high pressure turbine and the mass flow rate of cooling water in the condenser if cooling water enters the condenser at 20°C and leaves at 30°C

*Solution* :

Refer Fig.E 8.21 for the $T$–$s$ diagram.

Steam at 50 bar and 500°C: $h_3 = 3433.7$ kJ/kg ; $s_3 = 6.9770$ kJ/kg K

Steam at 8 bar 500°C: $h_5 = 3480.5$ kJ/kg ; $s_5 = 7.8678$ kJ/kg K

Steam at 0.05bar: $v_f = 0.001\,005\,2$ m³/kg ; $h_f = 137.77$ kJ/kg $= h_1$

$h_g = 2561.6$ kJ/kg

$s_f = 0.4763$ kJ/kg K ; $s_g = 8.3960$ kJ/kg K;

Work done on pump $= h_2 - h_1 = v\Delta P = 0.001\,005\,2(50 - 0.05)\, 10^5$

$$= 5.02 \text{ kJ/kg}$$

$$h_2 = h_1 + (h_2 - h_1) = 137.77 + 5.02 = 142.79 \text{ kJ/kg}$$

For the high pressure turbine, we have $s_3 = 6.9770 = s_4$

or $s_4 = 6.9770$ kJ/kg K

This value of $s_4$ at 8 bar lies in between 200°C and 300°C. Hence the steam leaving the high pressure turbine is superheated. We interpolate the data given in steam tables to find the temperature of steam at state 4 as follows.

$$6.9770 = 6.8148 + \frac{(7.2348 - 6.8148)\,\Delta T}{100}$$

or $\Delta T = 38.62$°C or $t_4 = 238.62$°C.

$$h_4 = 2838.6 + \frac{(3057.3 - 2838.6)\,38.62}{100} = 2923.06 \text{ kJ/kg}$$

For the low pressure turbine, we have $s_5 = 7.8678 = s_6$

$$s_6 = X_6 s_{6g} + (1 - X_6)\, s_f$$

or $7.8678 = 8.3960\, X_6 + (1 - X_6)\, 0.4763$ or $X_6 = 0.9333$

$$h_6 = 0.9333 \times 2561.6 + (1 - 0.9333)137.77$$

$$= 2399.93 \text{ kJ/kg}$$

$$\text{Thermal efficiency, } \eta = \frac{(h_3 - h_4) + (h_5 - h_6) - (h_2 - h_1)}{(h_3 - h_2) + (h_5 - h_4)}$$

$$= \frac{(3433.7 - 2923.06) + (3480.5 - 2399.93) - 5.02}{(3433.7 - 142.79) + (3480.5 - 2923.06)}$$

$$= 0.4122$$

Power output of the turbines = $(h_3 - h_4) + (h_5 - h_6)$

$$= (3433.7 - 2923.06) + (3480.5 - 2399.93)$$

$$= 1591.21 \text{ kJ/kg}$$

$$\text{Mass flow rate of steam, } m = \frac{100 \times 10^6}{1591.21 \times 10^3} = 62.85 \text{ kg / s}$$

Energy input in steam generator = $Q_1 = \dot{m}[(h_3 - h_2) + (h_5 - h_4)]$

$= 62.85\ [\ (3433.7 - 142.79) + (3480.5 - 2923.06)] \times 10^3 = 241.869$ MJ / s

Power delivered by high pressure turbine = $\dot{m}\ (h_3 - h_4)$

$$= 62.85\ (3433.7 - 2923.06) \times 10^3 = 32.093 \text{ MW}$$

$h_f(20°C) = 83.86$ kJ/kg; $h_f$ (30°C) = 125.66 kJ/kg

Energy rejected in the condenser $= Q_2 = \dot{m}(h_6 - h_1)$

$= 62.85(239.93 - 137.77)10^3 = 142.1768$ MJ / s

$$\text{Mass flow rate of cooling water in condenser} = \frac{142.1768 \times 10^6}{(125.66 - 83.86)10^3}$$

$$= 3401.36 \text{ kJ/s}$$

**Example 8.23**

Rework Example 8.21, if the isentropic efficiency of each turbine is 0.8 and the isentropic efficiency of the pump is 0.7.

*Solution* :

The reheat Rankine cycle (12*a*34*a*56*a*1) with irreversibilities is shown in Fig.E 8.23

We have $h_3 = 3433.7$ kJ/kg;

$s_3 = 6.9770$ kJ/kg K $\quad h_5 = 3478.3$ kJ/kg;

$s_5 = 7.7627$ kJ/kg K $\quad h_1 = 137.77$ kJ/kg;

$h_2 = 142.79$ kJ/kg $\quad h_4 = 2975.43$ kJ/kg;

$h_6 = 2367.69$ kJ/kg [See Example 8.21]

$$\text{Isentropic efficiency of pump, } \eta_P = 0.7 = \frac{h_2 - h_1}{h_{2a} - h_1} = \frac{142.79 - 137.77}{h_{2a} - h_1}$$

or $(h_{2a} - h_1) = 7.17$ kJ/kg,

Therefore, $h_{2a} = 144.94$ kJ/kg

$$\text{Isentropic efficiency of high pressure turbine, } \eta_T = \frac{h_3 - h_{4a}}{h_3 - h_4}$$

$$\text{or } 0.8 = \frac{h_3 - h_{4a}}{3433.7 - 2975.43}$$

or $h_3 - h_{4a} = 366.62$ kJ/kg.

Therefore, $h_{4a} = 3067.08$ kJ/kg

$$\text{Isentropic efficiency of low pressure turbine, } \eta_T = \frac{h_5 - h_{6a}}{h_5 - h_6}$$

or $0.8 = \dfrac{h_5 - h_{6a}}{3478.3 - 2367.69}$

or $(h_5 - h_{6a}) = 888.49$ kJ/kg.

Therefore $h_{6a} = 2589.81$ kJ/kg

$$h_{6a} = 2589.81 \text{ kJ/kg} > h_g \text{ at } 0.05 \text{ bar } = 2561.6 \text{ kJ/kg}.$$

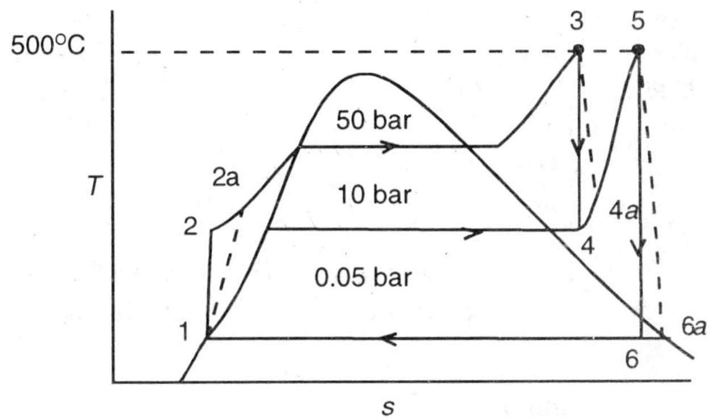


Fig.E 8.23 Reheat Rankine cycle with irriversibilities on a $T-s$ diagram.

Therefore, the steam leaving the low pressure turbine is superheated.

$$\eta = \frac{(h_3 - h_{4a}) + (h_5 - h_{6a}) - (h_{2a} - h_1)}{(h_3 - h_{2a}) + (h_5 - h_{4a})}$$

$$= \frac{366.62 + 888.49 - 7.17}{(3433.7 - 144.94) + (3478.3 - 3067.08)}$$

$$= \frac{1247.94}{3699.98} = 0.3373$$

Net work done $= (h_3 - h_{4a}) + (h_5 - h_{6a}) - (h_{2a} - h_1) = 1247.94$ kJ/kg

Mass flow rate of steam, $\dot{m} = \dfrac{20 \times 10^6}{1247.94 \times 10^3} = 16.03$ kg / s

**Example 8.24**

Describe a regenerative Rankine cycle, which employs an open feed water heater with the help of a $T$–$s$ diagram and derive a relation to estimate the thermal efficiency of the cycle.

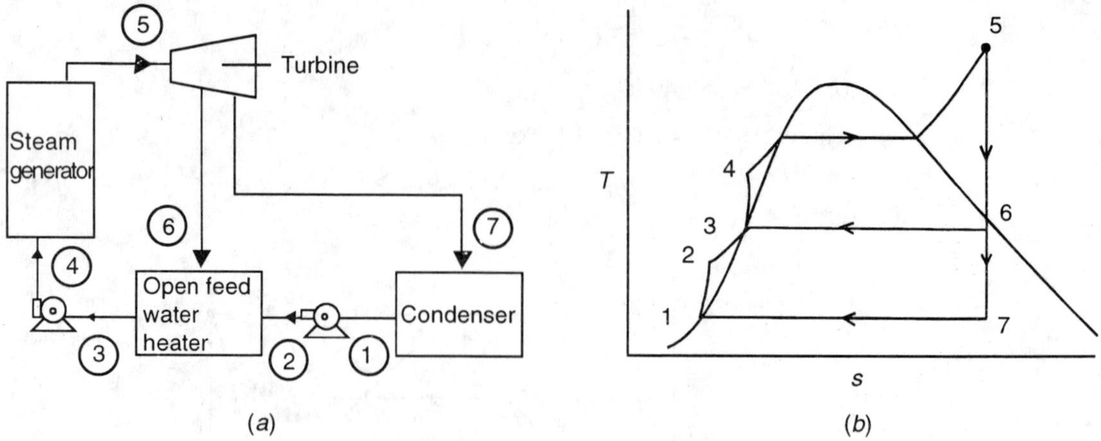


Fig.E 8.24. (*a*) Schematic of a regenerative Rankine cycle (*b*) $T-s$ diagram of a regenerative Rankine cycle with open feed water heater

*Solution* :

In a Rankine cycle, the average temperature at which energy is added in a steam generator can also be increased by preheating the feed water to the steam generator. A schematic of the regenerative Rankine cycle is shown in Fig.E 8.24. Superheated steam leaving the steam generator enters a turbine at state 5. Part of the steam in the turbine is extracted at an intermediate pressure (state 6) and the rest of the steam undergoes expansion to the condenser pressure (state 7). The steam extracted at state 6 is directly mixed, with the condensate (state1) which is raised to the same intermediate pressure as that of the steam at state 6, in an open feed water heater. In an ideal open feed water heater the resultant mixture is saturated liquid at state 3. This liquid is raised to the steam generator pressure by a pump which feeds it to the steam generator. In this arrangement the energy needed for preheating the feed water is not supplied from an external source but it is internally transferred from one stream to the other.

Now let us consider the open feed water heater shown in Fig.E 8.24 (*a*). The principle of conservation of mass gives

$$\dot{m}_2 + \dot{m}_6 = \dot{m}_3 \qquad (A)$$

and the principle of conservation of energy gives

$$\dot{m}_2 h_2 + \dot{m}_6 h_6 = \dot{m}_3 h_3 \qquad (B)$$

or $$(\dot{m}_3 - \dot{m}_6)\, h_2 + \dot{m}_6 h_6 = \dot{m}_3 h_3$$

or $$\left(1 - \frac{\dot{m}_6}{\dot{m}_3}\right) h_2 + \frac{\dot{m}_6}{\dot{m}_3} h_6 = h_3$$

or $(1-y)\,h_2 + yh_6 = h_3$ (C)

Where $y = \dfrac{\dot{m}_6}{\dot{m}_3}$ = fraction of the steam extracted from the turbine at an intermediate pressure.

Energy added in steam generator, $q_1 = h_5 - h_4$

Work done by the turbine = $(h_5 - h_6) + (1 - y)(h_6 - h_7)$

Energy spent on pumps = $(h_4 - h_3) + (1 - y)(h_2 - h_1)$

Net work delivered = $[(h_5 - h_6) + (1 - y)(h_6 - h_7)] - [(h_4 - h_3) + (1 - y)(h_2 - h_1)$

Energy rejected in the condenser, $q_2 = (1 - y)(h_7 - h_1)$

Thermal efficiency, $\eta = \dfrac{q_1 - q_2}{q_1} = 1 - \dfrac{q_2}{q_1} = 1 - \dfrac{(1-y)(h_7 - h_1)}{(h_5 - h_4)}$

**Example 8.25**

In a thermal power plant operating on a regenerative Rankine cycle superheated steam at 50 bar and 500°C enters a turbine. A suitable fraction of the steam is withdrawn from the turbine at 10 bar and the rest of the steam expands to the condenser pressure 0.05 bar. Calculate the thermal efficiency of the steam power plant and the mass flow rate of steam if the net power output of the plant is 100 MW.

*Solution* :

See Fig.E 8.24 for a $T$–$s$ diagram of regenerative cycle.

Steam at 50 bar and 500°C : $h_5 = 3433.7$ kJ/kg ; $s_5 = 6.9770$ kJ/kg K;

For the turbine $s_5 = s_6 = 6.9770$ kJ/kg K;

At 10 bar, this value of entropy corresponds to superheated steam at 265.79°C and $h_6 = 2975.43$ kJ/kg [see Example 8.21]

For the same turbine, we have $s_5 = s_7 = 6.9770$ kJ/kg K

Steam at 0.05 bar: $v_f = 0.001\ 005\ 2\ \text{m}^3/\text{kg}$;

$s_f = 0.4763$ kJ/kg K; $s_g = 8.3960$ kJ/kg K

$h_f = 137.77$ kJ/kg $= h_1$; $h_g = 2561.6$ kJ/kg

$$h_2 - h_1 = v\,\Delta P = 0.001\ 005\ 2\ (10 - 0.05)\ 10^5 = 1 \text{ kJ/kg.}$$

Therefore, $h_2 = 138.77$ kJ/kg

$$s_7 = s_g X_7 + (1 - X_7) s_f$$

or $6.9770 = 8.3960\, X_7 + (1 - X_7)0.4763$

or $X_7 = 0.8208$

$$h_7 = 0.8208 \times 2561.6 + (1 - 0.8208)137.77$$

$$= 2127.25 \text{ kJ/kg}$$

Steam at 10 bar. $v_f = 0.001\,127\,4 \text{ m}^3/\text{kg}; h_f = 762.61 \text{ kJ/kg} = h_3$

$$h_4 - h_3 = v\,\Delta P = 0.001\,127\,4\,(50 - 10)\,10^5 = 4.51 \text{ kJ/kg}$$

$$h_4 = h_3 + (h_4 - h_3) = 762.61 + 4.51 = 767.12 \text{ kJ/kg}$$

The fraction of steam extracted at the intermediate pressure can be calculated from the relation

$$(1 - y)\,h_2 + yh_6 = h_3$$

or $$y = \frac{h_3 - h_2}{h_6 - h_2} = \frac{762.61 - 138.77}{2975.43 - 138.77} = 0.2199$$

$$\eta = 1 - \frac{(1-y)(h_7 - h_1)}{(h_5 - h_4)} = 1 - \frac{(1 - 0.2199)(2127.25 - 137.77)}{3433.7 - 767.12} = 0.582$$

$$\eta = \frac{\text{Net work done}}{\text{Energy absorbed}} = 0.582$$

Net work done = $0.582\,(3433.7 - 767.12)10^3 = 1551.95$ kJ/kg

Mass flow rate of steam, $\dot{m} = \dfrac{100 \times 10^6}{1551.95 \times 10^3} = 64.44 \text{ kg/s}$

**Example 8.26**

Sketch a supercritical Rankine cycle on a *T–s* diagram and what are the advantages of this cycle over the conventional subcritical Rankine cycle ?

*Solution* :

If a Rankine cycle employs a boiler at 50 bar, a major fraction of energy is added at the saturation temperature (263.91°C), even though superheated steam can be produced at a temperature as high as 600°C which is the maximum permissible limit based on the material of construction of the steam generators. Usually the required energy is supplied by circulating the hot gases produced by burning coal, fuel oil etc. The hot gases so obtained will be at temperature of 900 – 1000°C and hence a large temperature difference exists between the steam and hot gases which introduce irreversibility in the cycle. This irreversibility can be reduced by adding energy at a higher temperature. Since the critical temperature for water is 374.15°C, the energy addition at a higher temperature level calls for a raise in the steam generator pressure above the critical pressure (221.2 bar) for water. The Rankine cycle in which the energy addition takes place at a pressure greater than the critical pressure is called a supercritical Rankine cycle. Modern thermal power plants operate on a cycle in which the turbine inlet pressure is usually in the range 250 to 325 bar. A supercritical Rankine cycle in which energy rejection is at low pressure is shown in Fig.E 8.26(*a*). It may be noted that there is no phase change during the energy addition process 2–3 and the temperature of the supercritical fluid continuously change in going from state 2 to state3. Since the average temperature at which energy is added is increased in a supercritical Rankine cycle, its efficiency will be higher than the subcritical Rankine cycles. The efficiency of supercritical Rankine cycle can be close to 40 percent. In fact in a supercritical Rankine cycle, it is not necessary to operate the condensers at low pressure. The energy rejection also can be done at a pressure greater than the critical pressure. Such a cycle is usually called a totally supercritical Rankine cycle which is shown in Fig.E 8.26(*b*).

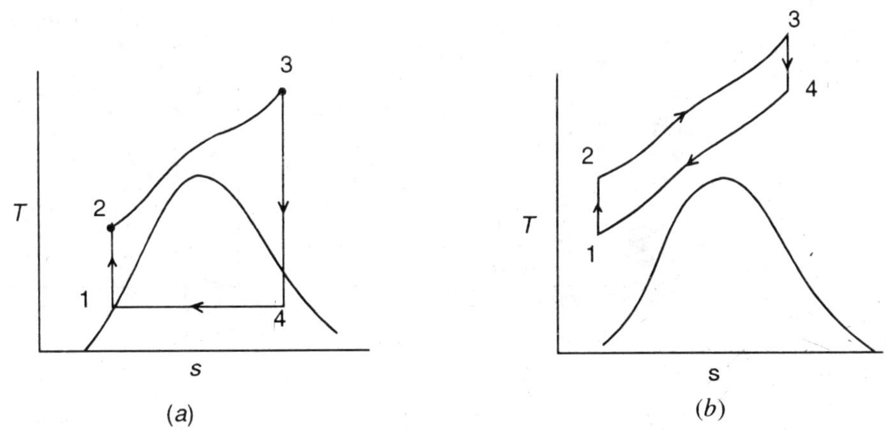


Fig.E 8.26. (*a*) Supercritical Rankine cycle in which energy rejection is at subcritical pressure (*b*) Supercritical Rankine cycle with energy rejection at supercritical pressure.

## Example 8.27

What are the drawbacks associated with water as a working fluid in a Rankine cycle which is to be designed to achieve higher efficiency?

*Solution* :

Water is used as a working fluid in almost all Rankine cycles because it is easily available in abundance, nontoxic, chemically stable and noncorrosive. However, if a higher efficiency is sought in a power cycle, the energy addition must take place at the highest possible temperature. The material of construction used in all the steam generators can easily withstand a temperature of approximately 550 - 600°C.However, the critical temperature for water is 374.15°C only which is about 200°C less than the maximum allowable temperature of the steam generators. In order to raise the average temperature of energy addition one has to employ supercritical ($P > 221.2$ bar) power cycle which require costly piping. Another disadvantage with water is that its saturation pressure at the available cooling water temperature is very low which requires the condensers to be operated at 0.05 bar. As a result air may leak into the condensers.

### Example 8.28

Explain the principle of binary vapor power cycle and sketch its *T–s* diagram.

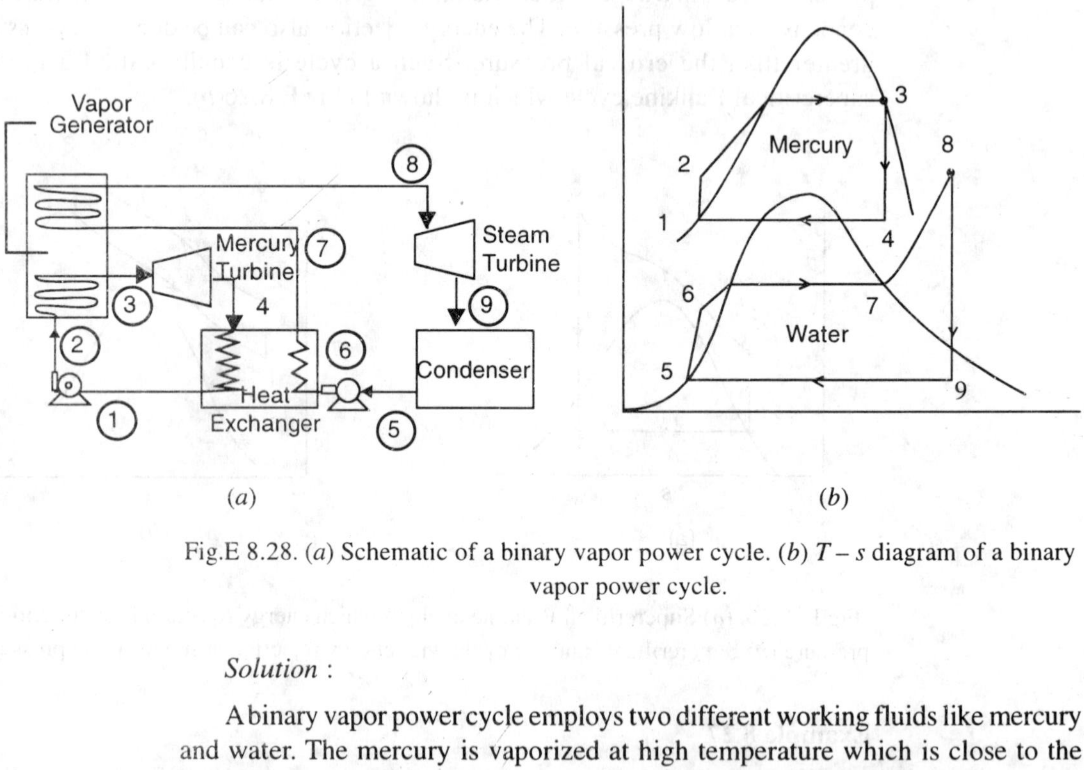


Fig.E 8.28. (*a*) Schematic of a binary vapor power cycle. (*b*) $T-s$ diagram of a binary vapor power cycle.

*Solution* :

A binary vapor power cycle employs two different working fluids like mercury and water. The mercury is vaporized at high temperature which is close to the maximum permissible limit of a boiler (say 550°C) and then the mercury vapor is allowed to expand in a turbine to a low pressure obtaining some work. Then the mercury is condensed in a heat exchanger in which it exchanges energy with water producing steam and this steam is used in another Rankine cycle. A schematic of the binary vapor power cycle is shown in Fig.E 8.28. In this saturated mercury at state 1 is pumped to high pressure (state 2 ) and then it is fed to a vapor generator. The saturated vapor at state 3 enters a mercury turbine and expands to low pressure (state 4). The low pressure mercury vapor passes through a heat exchanger in which the feed water to a steam power cycle is vaporized at approximately 250°C. The condensed mercury at state 1 enters the pump which raises it to the vapor generator pressure, thus completing a cycle. The saturated liquid water at state 5 enters a pump which raises its pressure (state 6) to that of the steam turbine inlet. The liquid water at state 6 passes through the heat exchanger and vaporizes (state 7). Then the saturated steam at state 7 passes through the vapor generator for superheating and comes out at state 8. The superheated steam at state 8 is fed to a steam turbine and it expands to the condenser pressure (state 9). The low pressure steam is then condensed to saturated liquid at state 5, thus completing a cycle. In these power cycles the two fluids do not mix with each other. Essentially a binary vapor power cycle is a combination of two Rankine cycles.

**Example 8.29**

What is meant by an air-standard cycle and where is it used ?

*Solution* :

Power cycles which employ a gas as a working medium are called gas power cycles. In these cycles, the gas does not undergo any phase change. The gas is obtained by burning a fuel inside the cylinder of an engine. The composition of the working fluid does not remain constant throughout the cycle. In the begining the working fluid contains a mixture of fuel vapor and air. But in the subsequent processes the working fluid constitutes the combustion products obtained by the burning of the fuel. The working fluid does not undergo a cyclic change and it is exhausted after some time and a fresh mixture of fuel and air is sucked in. For thermodynamic analysis of internal combustion engines, we make the following approximations.

1. The engine is assumed to operate on a cycle with a fixed mass of air which does not undergo any chemical change.
2. The actual combustion of the fuel is replaced by an equivalent energy addition process from an external source.
3. The exhaust process is replaced by an equivalent energy rejection process to surroundings by means of which the working fluid is restored to the initial state.
4. The air is assumed to behave like an ideal gas with constant specific heats.

The cycles on which these approximations are applicable are called air–standard cycles.

**Example 8.30**

Sketch an air–standard Otto cycle on $P-v$ and $T-s$ diagrams.

*Solution* :

In 1867 Nikolaus A.Otto developed a practical four stroke engine which is named after him. The Otto engines employ gasoline (petrol) as the fuel. The Otto cycle shown in Fig.E 8.30 consists of the following processes :

**0–1** : The piston executes an outward stroke and sucks a mixture of fuel vapor and air into the cylinder.

**1–2** : The compression stroke is initiated and the mixture of fuel and air is compressed reversibly and adiabatically and at the end of the compression stroke, the temperature of the gas will be sufficiently high.

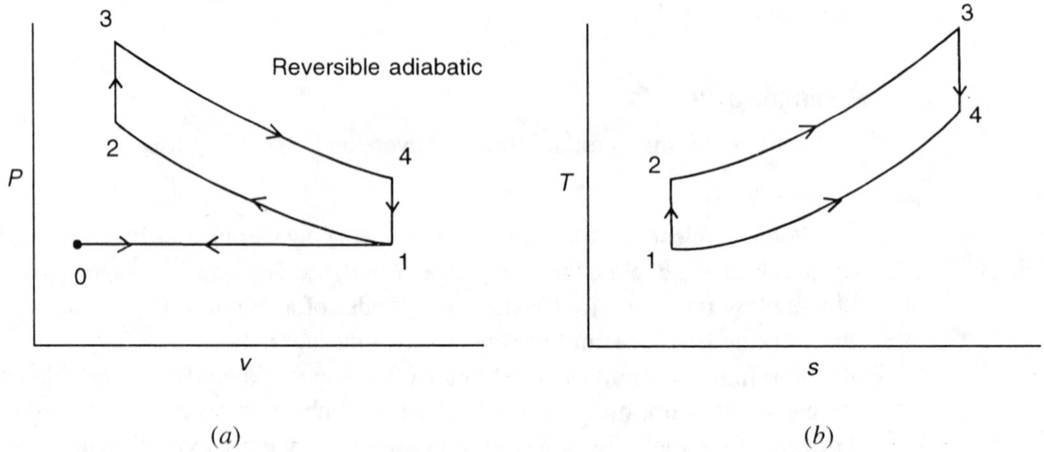


Fig.E 8.30. Air–Standard Otto cycle on (*a*) *P–v* diagram (*b*) *T – s* diagram

**2–3** : An electric spark is introduced to ignite the fuel–air mixture. Since gasoline at high temperature is highly inflammable, the combustion proceeds instantaneously and there is not enough time for the mechanical motion of the piston. Energy is liberated by the spontaneous exothermic chemical reaction and the temperature of the gas increases. This spontaneous combustion process is conveniently replaced by a constant volume energy addition process in which an equivalent amount of energy is added from external source.

**3–4** : The hot combustion products undergo reversible, adiabatic expansion and the piston executes an outward power stroke.

**4–1** : The exhaust valve of the engine cylinder opens and the spent gases are thrown out of the cylinder by reduction in pressure at constant volume.

**1–0** : The cylinder which is still full with the spent gases are thrown out by the piston which executes an inward exhaust stroke at constant pressure.

There are four strokes namely, suction, compression, expansion and exhaust strokes in an Otto cycle. It can be observed from the *P– v* diagram that the net work of the suction (0–1) and exhaust (1–0) strokes is zero. Hence, for thermodynamic analysis we consider the cycle 12341 only and we call it as the air–standard Otto cycle.

### Example 8.31

Develop an expression to estimate the thermal efficiency of an air–standard Otto cycle.

*Solution* :

The $P-v$ diagram of an air–standard Otto cycle is shown in Fig.E 8.31. Consider a unit mass of air undergoing the cyclic change. The processes 1–2 and 3–4 are adiabatic and hence there is no heat interaction in these processes. Energy is absorbed as heat during 2–3 and rejected during the process 4–1 at constant volume. Therefore,

$$q_1 = u_3 - u_2 = C_v (T_3 - T_2)$$
$$q_2 = u_4 - u_1 = C_v (T_4 - T_1)$$

Thermal efficiency, $$\eta = 1 - \frac{q_2}{q_1} = 1 - \frac{C_v(T_4 - T_1)}{C_v(T_3 - T_2)} = 1 - \frac{T_4 - T_1}{T_3 - T_2} \qquad (A)$$

For an ideal gas undergoing reversible adiabatic process $Tv^{\gamma-1}$ = Constant

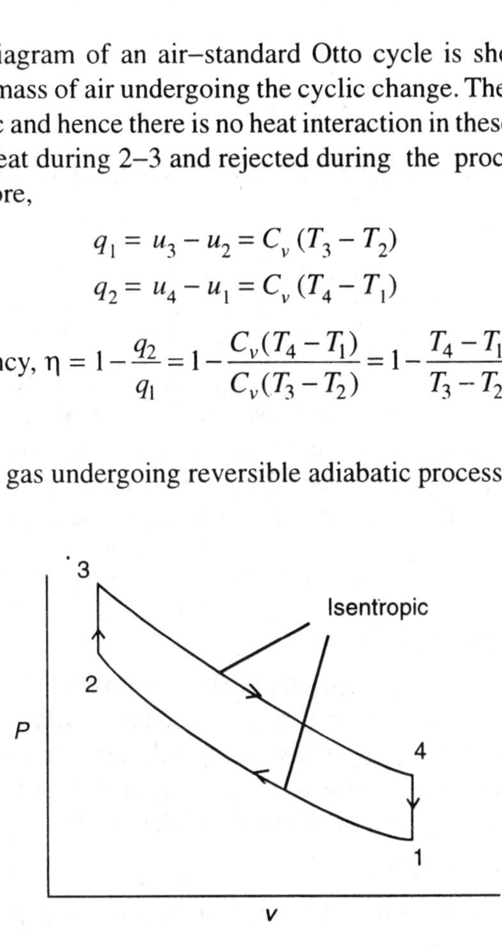


Fig.E 8.31. *P-v* diagram of air-standard Otto cycle.

Therefore, $$\frac{T_2}{T_1} = \left(\frac{v_1}{v_2}\right)^{\gamma-1} \text{ and } \frac{T_3}{T_4} = \left(\frac{v_4}{v_3}\right)^{\gamma-1}$$

But $v_2 = v_3$ and $v_4 = v_1$. Hence

$$\frac{T_2}{T_1} = \frac{T_3}{T_4} \quad \text{or} \quad \frac{T_4}{T_1} = \frac{T_3}{T_2} \quad \text{or} \quad \frac{T_4}{T_1} - 1 = \frac{T_3}{T_2} - 1$$

$$\text{or} \quad \frac{T_4 - T_1}{T_1} = \frac{T_3 - T_2}{T_2} \quad \text{or} \quad \frac{T_4 - T_1}{T_3 - T_2} = \frac{T_1}{T_2} \qquad (B)$$

Substituting Eqn. (*B*) in Eqn. (*A*), we get

$$\eta = 1 - \frac{T_1}{T_2}$$

The compression ratio $r_0$ is defined as $r_0 = \frac{v_1}{v_2}$

Then $\eta = 1 - \frac{T_1}{T_2} = 1 - \left(\frac{v_2}{v_1}\right)^{\gamma-1} = 1 - \left(\frac{1}{r_0}\right)^{\gamma-1}$

Since $\gamma > 1$, the efficiency of an Otto engine increases with increasing compression ratio. However, in an actual engine the compression ratio cannot be increased indefinitely since higher compression ratio results in higher temperature of the air after the compression stroke which may result in spontaneous and uncontrolled combustion of the fuel in the cylinder. This situation is usually termed as knocking.

**Example 8.32**

The compression ratio of an Otto cycle is 8. The pressure and temperature of the air at the beginning of the compression stroke are 1 bar and 300 K, respectively. The amount of energy added to the air as a result of combustion is 1500 kJ/kg air. Determine the pressure and temperature of the air at the end of each process of the cycle, and the thermal efficiency of the cycle.

*Solution* :

Refer Fig.E 8.31 for *P–v* diagram of Otto cycle.

For air (ideal gas), we know that $\gamma = 1.4$

and $$C_v = \frac{R}{\gamma - 1} = \frac{8.314}{1.4 - 1} = 20.785 \text{ J/mol K}$$

$$= 0.7175 \text{ kJ/kg K}$$

(Molar mass of air = $28.97 \times 10^{-3}$ kg / mol)

We know that $T_2 = T_1\left(\frac{v_1}{v_2}\right)^{\gamma-1} = T_1(r_0)^{\gamma-1}$

or $T_2 = 300\ (8)^{0.4} = 689.22$ K

$$P_2 = \frac{P_1 v_1^{\gamma}}{v_2^{\gamma}} = P_1\left(\frac{v_1}{v_2}\right)^{\gamma} = P_1\ (r_0)^{\gamma} = 1 \times (8)^{1.4} = 18.379 \text{ bar}$$

$$q_1 = u_3 - u_2 = C_v\ (T_3 - T_2)$$

or $1500 = 0.7175\ (T_3 - 689.22)$ or $T_3 = 2779.81$ K

2–3 is a constant volume process. Therefore, $\frac{P_2}{T_2} = \frac{P_3}{T_3}$

or $P_3 = P_2 . \frac{T_3}{T_2} = \frac{18.379 \times 2779.81}{689.22} = 74.1275$ bar

We know that $T_4 = T_3\left(\frac{v_3}{v_4}\right)^{\gamma-1} = T_3\left(\frac{v_2}{v_1}\right)^{\gamma-1} = T_3\left(\frac{1}{r_0}\right)^{\gamma-1}$

or $T_4 = 2779.81\left(\frac{1}{8}\right)^{0.4} = 1209.98$ K

$$P_4 = P_3\left(\frac{v_3}{v_4}\right)^{\gamma} = P_3\left(\frac{v_2}{v_1}\right)^{\gamma} = P_3\left(\frac{1}{r_0}\right)^{\gamma} = 74.1275\ \left(\frac{1}{8}\right)^{1.4} = 4.0332 \text{ bar}$$

Thermal effiency, $\eta = 1 - \left(\frac{1}{r_0}\right)^{\gamma-1} = 1 - \left(\frac{1}{8}\right)^{0.4} = 0.5647$

**Example 8.33**

Define the Mean Effective Pressure (MEP) of a reciprocating engine.

*Solution :*

The Mean Effective Pressure (MEP) is a useful parameter in the study of reciprocating engines. It is defined as the average pressure which, if acted on the piston during the power stroke, would produce the same work output as the net work delivered for the actual cycle. That is

$$W = \oint P dv = \text{MEP}\ (v_1 - v_2)$$

or $W$ = Mean effective pressure × displacement volume

**Example 8.34**

Calculate the mean effective pressure for the Otto engine of Example 8.32.

*Solution* :

In Example 8.32, we are given $q_1 = 1500$ kJ/kg and

we have calculated $\eta = 0.5647$

$$\eta = \frac{W}{q_1} \text{ or } W = \eta q_1 = 0.5647 \times 1500 = 847.05\text{kJ}$$

$$V_1 - V_2 = V_1\left(1-\frac{V_2}{V_1}\right) = V_1\left(1-\frac{1}{r_0}\right) = V_1\left(\frac{r_0-1}{r_0}\right) = \frac{7}{8}V_1$$

and $V_1 = \dfrac{nRT_1}{P_1} = \dfrac{m}{M}\dfrac{RT_1}{P_1} = \dfrac{1\times 8.314\times 300}{28.97\times 10^{-3}\times 1\times 10^5} = 0.861\,\text{m}^3$

Therefore, $V_1 - V_2 = \dfrac{7}{8}V_1 = 0.7534\ \text{m}^3$

Mean effective pressure, MEP $= \dfrac{W}{V_1 - V_2} = \dfrac{847.05\times 10^3}{0.7534} = 11.243$ bar

**Example 8.35**

In an air–standard Otto cycle, the conditions of the air at the start of the compression stroke are 1 bar and 300 K. The maximum pressure and temperature in the cycle are 70 bar and 2500 K, respectively. Calculate the compression ratio, the thermal efficiency of the cycle and the net work done per kg air. Take $C_v$ of air as 0.7175 kJ/kg K.

*Solution* :

Compression ratio $r_0 = \dfrac{v_1}{v_2} = \dfrac{P_2T_1}{P_1T_2}$

For the constant volume process 2–3 : $\dfrac{P_2}{T_2} = \dfrac{P_3}{P_3}$

Therefore, $r_0 = \dfrac{P_3T_1}{T_3P_1}$

The maximum pressure and temperature in an Otto cycle correspond to state 3.

Therefore, $P_3 = 70$ bar and $T_3 = 2500$ K.

Hence, $r_0 = \dfrac{70 \times 300}{2500 \times 1} = 8.4$

Thermal efficiency, $\eta = 1 - \left(\dfrac{1}{r_0}\right)^{\gamma-1} = 1 - \left(\dfrac{1}{8.4}\right)^{0.4} = 0.5731$

$$T_2 = T_1\left(\frac{v_1}{v_2}\right)^{\gamma-1} = T_1(r_0)^{\gamma-1} = 300\ (8.4)^{0.4} = 702.8\ \text{K}$$

Energy added as heat $q_1 = u_3 - u_2 = C_v\ (T_3 - T_2) = 0.7175\ (2500 - 702.8)$
$= 1289.49$ kJ/kg

Net work done, $W = \eta\ q_1 = 0.5731 \times 1289.49 = 739$ kJ/kg

**Example 8.36**

In an air–standard Otto cycle with a compression ratio of 8, the pressure and temperature at the end of the compression stroke are 15 bar and 750 K, respectively. After energy addition as heat the pressure rises to 40 bar. Determine the thermal efficiency of the cycle and the network done per kg air.

*Solution* :

Given data : $r_0 = 8$ ; $P_2 = 15$ bar ; $T_2 = 750$ K ; $P_3 = 40$ bar

Thermal efficiency, $\eta = 1 - \left(\dfrac{1}{r_0}\right)^{\gamma-1} = 1 - \left(\dfrac{1}{8}\right)^{0.4} = 0.5647$

For the constant volume process 2–3: $T_3 = T_2 \dfrac{P_3}{P_2} = 750 \times \dfrac{40}{15} = 2000$ K

Energy added as heat, $q_1 = C_v\ (T_3 - T_2) = 0.7175\ (2000 - 750)$
$= 896.88$ kJ/kg

Net work done, $W = \eta\ q_1 = 0.5647 \times 896.88 = 506.47$ kJ/kg

**Example 8.37**

An air–standard Otto cycle delivers a net work of 600 kJ/kg and the maximum temperature in the cycle is found to be 2200 K. The temperature of the gases at the end of the compression stroke is found to be 700 K. Calculate the compression ratio of the cycle.

*Solution* :

Given data : $W = 600$ kJ/kg; $T_3 = 2200$ K; $T_2 = 700$ K;

Energy added as heat, $q_1 = C_v (T_3 - T_2) = 0.7175\ (2200 - 700)$

$= 1076.25$ kJ/kg

$$\eta = \frac{W}{q_1} = \frac{600}{1076.25} = 0.5575$$

$$\eta = 1 - \left(\frac{1}{r_0}\right)^{\gamma - 1} \quad \text{or} \quad 0.5575 = 1 - \left(\frac{1}{r_0}\right)^{0.4}$$

or compression ratio, $r_0 = 7.677$

**Example 8.38**

An air–standard Otto cycle with a compression ratio of 8 and energy addition as heat of 1500 kJ/kg sucks air at 1 bar and 300 K before the compression stroke. Calculate the thermal efficiency of the cycle, the network done, the mean effective pressure and the maximum temperature and pressure in the cycle.

*Solution* :

Given data : $P_1 = 1$ bar ; $T_1 = 300$ K; $r_0 = 8$; $q_1 = 1500$ kJ/kg

$$\text{Thermal efficency, } \eta = 1 - \left(\frac{1}{r_0}\right)^{\gamma - 1} = 1 - \left(\frac{1}{8}\right)^{0.4} = 0.5647$$

Net work done, $W = \eta\, q_1 = 0.5647 \times 1500 = 847.05$ kJ/kg

$$T_2 = T_1 \left(\frac{v_1}{v_2}\right)^{\gamma - 1} = T_1 (r_0)^{\gamma - 1} = 300\ (8)^{0.4} = 689.22 \text{ K}$$

$$q_1 = C_v (T_3 - T_2)$$

or $1500 = 0.7175\ (T_3 - 689.22)$

or $T_3 = 2779.8$ K

Maximum temperature in the cycle $= T_3 = 2779.8$ K

$$P_2 = P_1 \left(\frac{v_1}{v_2}\right)^{\gamma} = P_1\, r_0^{\gamma}$$

$$\frac{P_2}{T_2} = \frac{P_3}{T_3} \text{ or } P_3 = \frac{P_2 T_3}{T_2} = \frac{P_1 r_0^{\gamma} T_3}{T_2}$$

$$\text{or } \quad P_3 = \frac{1 \times (8)^{1.4} \times 2779.8}{689.22} = 74.1279 \text{ bar}$$

Maximum Pressure in the cycle = 74.1279 bar

$$V_1 - V_2 = V_1\left(1 - \frac{V_2}{V_1}\right) = V_1\left(1 - \frac{1}{r_0}\right) = \frac{nRT_1}{P_1} = \frac{RT_1}{MP_1}$$

$$V_1 = \frac{nRT_1}{P_1} = \frac{mRT_1}{MP_1} = \frac{1 \times 8.314 \times 300}{28.97 \times 10^{-3} \times 1 \times 10^5} = 0.861 \text{ m}^3/\text{kg}$$

$$\text{Mean effective pressure MEP} = \frac{W}{V_1 - V_2} = \frac{8W}{7V_1} = \frac{8 \times 847.05 \times 10^3}{7 \times 0.861}$$

$$= 11.2434 \text{ bar}$$

**Example 8.39**

In an air–standard Otto engine, the minimum temperature $T_1$ is governed by the ambient atmosphere and the maximum temperature ($T_3$) is dictated by the material of construction of the piston and cylinder. For fixed values of $T_1$ and $T_3$ determine the compression ratio $r_0$ for obtaining maximum net work per unit mass of air undergoing the cyclic change.

*Solution* :

The net work done, $W = q_1 - q_2 = C_v (T_3 - T_2) - C_v (T_4 - T_1)$

$$= C_v (T_1 - T_2) + C_v (T_3 - T_4)$$

$$= C_v T_1\left(1 - \frac{T_2}{T_1}\right) + C_v T_3\left(1 - \frac{T_4}{T_3}\right)$$

The process 1–2 is isentropic. Therefore,

$$\frac{T_2}{T_1} = \left(\frac{v_1}{v_2}\right)^{\gamma - 1} = r_0^{\ \gamma - 1}$$

Similarly, the process 3–4 is isentropic. Hence

$$\frac{T_3}{T_4} = \left(\frac{v_4}{v_3}\right)^{\gamma - 1} = \left(\frac{v_1}{v_2}\right)^{\gamma - 1} = r_0^{\gamma - 1}$$

$$\text{Therefore, } W = C_v T_1 \left(1 - r_0^{\gamma-1}\right) + C_v\, T_3 \left[1 - \left(\frac{1}{r_0}\right)^{\gamma-1}\right]$$

For obtaining the compression ratio $r_0$ for maximum net work.

$$\frac{dW}{dr_0} = -C_V T_1 (\gamma - 1) r_0^{-\gamma-2} - C_v T_3 (1-\gamma) r_0^{-\gamma} = 0$$

$$\text{or } \quad T_1 (\gamma - 1)\, r_0^{\gamma-2} = T_3 (\gamma - 1) r_0^{-\gamma}$$

$$\text{or } \quad \frac{T_3}{T_1} = r_0^{2(\gamma-1)}$$

$$\text{or } \quad r_0 = \left(\frac{T_3}{T_1}\right)^{\frac{1}{2(\gamma-1)}}$$

**Example 8.40**

Sketch the air–standard Diesel cycle on $P-v$ and $T-s$ diagrams. What is the advantage of a Diesel cycle over the Otto cycle ?

*Solution* :

The Diesel cycle was developed by Rudolph Diesel in 1893. This cycle is extensively used and most of the present day transport vehicles like buses, cars, trucks and a good number of locomotives are operating on Diesel cycle. Since the efficiency of an engine increases with increasing compression ratio, the Diesel cycle is developed in which the compression ratio is much higher than in the Otto cycle. In this cycle only air is sucked into the cylinder at the start. Then it is compressed to a high pressure by using higher compression ratio. At the end of the compression, the fuel is injected which undergoes combustion. In this engine an electric spark is not used to initiate ignition, but it occurs because of the high temperature attained due to compression. The engines which use Diesel cycle are also known as compression – ignition engines. The $P-v$ and $T-s$ diagrams of Diesel cycle are shown in Fig.E 8.40. The Diesel cycle consists of the following processes.

**0–1** : Constant Pressure suction during which the piston executes an outward stroke and sucks fresh air into the cylinder.

**1–2** : Reversible–adiabatic compression with higher compression ratio. Because of the high compression ratio, the temperature of gas is sufficiently high so that when fuel is injected into the hot air it spontaneously undergoes combustion releasing energy as heat.

**2–3** : The fuel (Diesel oil) is injected into the hot air. The combustion of Diesel oil in hot air is not as spontaneous as the combustion of gasoline. Therefore, as the combustion proceeds releasing energy in the form of heat, the piston starts moving outwards. The combustion process is conveniently approximated as a constant–pressure energy addition.

**3–4** : The hot combustion products undergo reversible, adiabatic expansion and the piston executes an outward motion.

**4–1** : The combustion products are discharged into the surroundings while the volume is held constant and the pressure drops to the original value.

**1–0** : The remaining combustion products are exhausted at constant pressure when the piston executes an inward motion.

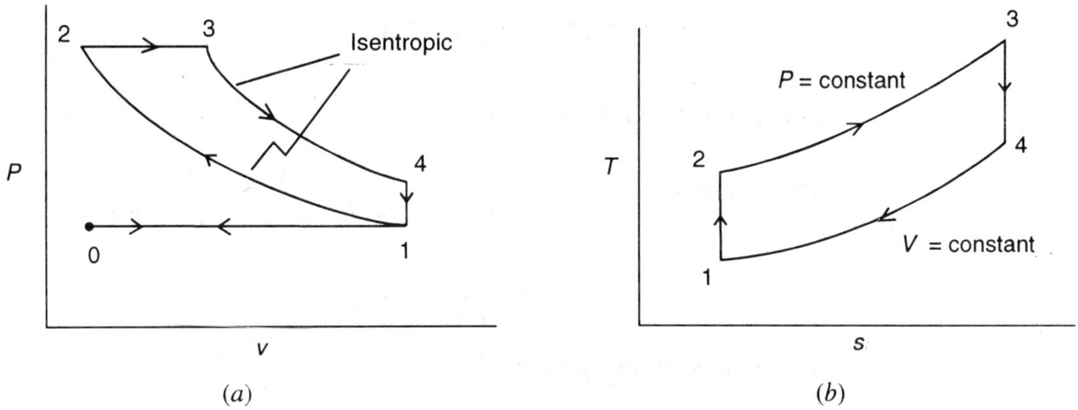


Fig.E 8.40. The $P - v$ and $T$ - $s$ diagrams for air–standard Diesel cycle.

It can be observed from Fig.E 8.40 ($a$) that the net work done in the processes 0–1 and 1–0 is equal to zero. Therefore, they can be ignored for the purpose of thermodynamic analysis of the cycle.

### Example 8.41

Drive an expression to estimate the thermal efficiency of an air–standard Diesel cycle.

*Solution* :

The $P-v$ diagram of air–standard Diesel cycle is shown in Fig.E 8.41. For the analysis of Diesel cycle, we define the compression ratio

$$r_0 = \left(\frac{v_1}{v_2}\right) \text{ and the Cut–off ratio, } r_c = \frac{v_3}{v_2}$$

Consider a unit mass of air (ideal gas) undergoing the cyclic change. Then energy added as heat, $q_1 = C_p (T_3 - T_2)$

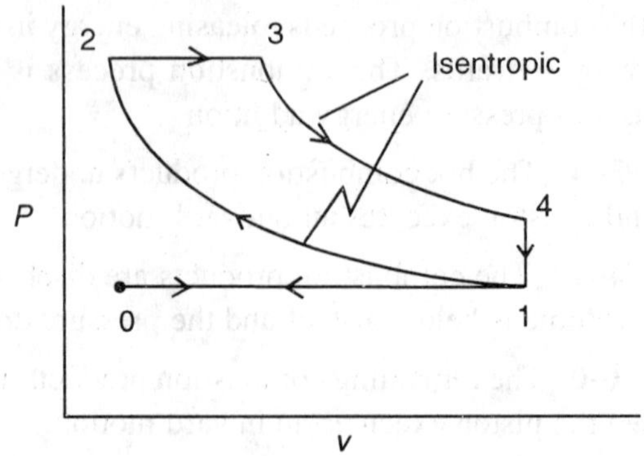


Fig.E 8.41. $P - v$ diagram of air–standard Diesel cycle.

Energy rejected as heat $q_2 = C_v (T_4 - T_1)$

$$\text{Thermal efficiency, } \eta = 1 - \frac{q_2}{q_1} = 1 - \frac{C_v (T_4 - T_1)}{C_P (T_3 - T_2)} = 1 - \frac{(T_4 - T_1)}{\gamma(T_3 - T_2)}$$

$$= 1 - \frac{T_1(T_4 / T_1 - 1)}{\gamma T_2 (T_3 / T_2 - 1)} \tag{A}$$

1–2 is an isentropic process. Therefore,

$$\frac{T_1}{T_2} = \left(\frac{v_2}{v_1}\right)^{\gamma - 1} = \left(\frac{1}{r_0}\right)^{\gamma - 1} \tag{B}$$

4–1 is a constant volume process. Therefore,

$$\frac{T_4}{T_1} = \frac{P_4}{P_1} = \left(\frac{P_4}{P_3}\right)\left(\frac{P_3}{P_1}\right)$$

2–3 is a constant pressure process. Therefore, $P_2 = P_3$ and

$$\frac{T_3}{T_2} = \frac{v_3}{v_2} = r_c \tag{C}$$

$$\text{Hence, } \frac{T_4}{T_1} = \left(\frac{P_4}{P_3}\right)\left(\frac{P_3}{P_1}\right) = \left(\frac{P_4}{P_3}\right)\left(\frac{P_2}{P_1}\right) \tag{D}$$

1–2 and 3–4 are reversible, adiabatic processes. Therefore,

$$\frac{P_2}{P_1} = \left(\frac{v_1}{v_2}\right)^{\gamma} \text{ and } \frac{P_4}{P_3} = \left(\frac{v_3}{v_4}\right)^{\gamma}$$

Then Eqn ($D$) can be written as

$$\frac{T_4}{T_1} = \left(\frac{v_3}{v_4}\right)^{\gamma}\left(\frac{v_1}{v_2}\right)^{\gamma} = \left(\frac{v_3}{v_2}\right)^{\gamma} = \rho_\chi^{\gamma} \text{ (Since } v_1 = v_4) \tag{E}$$

Substituting Eqns. ($B$), ($C$) and ($E$) in Eqn.($A$), we get

$$\eta = 1 - \frac{1}{\gamma\, r_0^{\gamma-1}}\left[\frac{r_c^{\gamma} - 1}{r_c - 1}\right]$$

**Example 8.42**

An air–standard Diesel cycle operates with a compression ratio of 16 and a cut–off ratio of 2. At the beginning of compression stroke, the air is at 1 bar and 300 K. Calculate the maximum temperature in the cycle, the energy added as heat, the net work done and the thermal efficiency.

*Solution* :

Thermal efficiency $\eta = 1 - \dfrac{1}{\gamma\, r_0^{\gamma-1}}\left(\dfrac{r_c^{\gamma} - 1}{r_c - 1}\right)$

$$= 1 - \frac{1}{1.4(16)^{0.4}}\left(\frac{2^{1.4} - 1}{2 - 1}\right) = 0.6138$$

$$T_2 = T_1\left(\frac{v_1}{v_2}\right)^{\gamma-1} = T_1\, r_0^{\gamma-1} = 300\,(16)^{\,0.4} = 909.43 \text{ K}$$

2–3 is a constant pressure process. Therefore,

$$\frac{T_3}{T_2} = \frac{v_3}{v_2} \text{ or } T_3 = T_2\left(\frac{v_3}{v_2}\right) = T_2\, r_c = 909.43 \times 2 = 1818.86 \text{ K}$$

Maximum temperature in the cycle = $T_3$ = 1818.86 K

Energy added as heat $q_1 = C_P(T_3 - T_2)$

$$C_P = \frac{R\gamma}{\gamma - 1} = \frac{8.314 \times 1.4}{0.4} = 20.099 \text{ J / mol K}$$

$$= 1.0045 \text{ kJ/kg K}$$

$$q_1 = 1.0045\,(1818.86 - 909.43) = 913.52 \text{ kJ/kg}$$

Net work done, $W = \eta\, q_1 = 0.6138 \times 913.52 = 560.72$ kJ/kg

**Example 8.43**

An engine operating on an air–standard Diesel cycle sucks in air at 1 bar and 300 K and compresses it to 40 bar before fuel injection. If the energy added (per cycle) as heat is 600 kJ/kg air, calculate the compression ratio, the cut off ratio, thermal efficiency and the work done per kg air. Take $C_P$ of air as 1.0045 kJ/kg K.

*Solution* :

We know that $P_1 v_1^{\gamma} = P_2 v_2^{\gamma}$ or $\dfrac{v_1}{v_2} = \left(\dfrac{P_2}{P_1}\right)^{\frac{1}{\gamma}} = \left(\dfrac{40}{1}\right)^{\frac{1}{1.4}} = 13.942$

Compression ratio, $r_0 = \dfrac{v_1}{v_2} = 13.942$

$$T_2 = T_1\left(\frac{v_1}{v_2}\right)^{\gamma-1} = T_1\, r_0^{\gamma-1} = 300\,(13.942)^{0.4} = 860.7 \text{ K}$$

Energy absorbed as heat, $q_1 = C_P\,(T_3 - T_2)$

or $600 = 1.0045\,(T_3 - 860.7)$

or $T_3 = 1458.01$ K

Cut–off ratio, $r_0 = \dfrac{v_3}{v_2} = \dfrac{T_3}{T_2} = \dfrac{1458.01}{860.7} = 1.694$

Thermal efficiency, $\eta = 1 - \dfrac{1}{\gamma\, r_0^{\gamma-1}}\left(\dfrac{r_c^{\gamma} - 1}{r_c - 1}\right)$

$$= 1 - \frac{1}{1.4\,(13.942)^{0.4}}\left\{\frac{1.694^{1.4} - 1}{0.694}\right\} = 0.60894$$

Net work done, $W = \eta\, q_1 = 0.6084 \times 600 = 365.04$ kJ/kg

**Example 8.44**

An air–standard Diesel cycle with a compression ratio of 16 takes in air at 1 bar and 300 K. If the maximum temperature in the cycle is 2100 K determine the cut off ratio, thermal efficiency, energy added as heat, net work done per kg air and the mean effective pressure.

*Solution* :

$$T_2 = T_1\left(\frac{v_1}{v_2}\right)^{\gamma-1} = T_1\, r_0^{\gamma-1} = 300\,(16)^{0.4} = 909.43 \text{ K}$$

$$\text{Cut–off ratio, } r_c = \frac{v_3}{v_2} = \frac{T_3}{T_2} = \frac{2100}{909.43} = 2.309$$

$$\text{Thermal efficiency, } \eta = 1 - \frac{1}{\gamma\, r_0^{\gamma-1}}\left(\frac{r_c^{\gamma}-1}{r_c-1}\right) = 1 - \frac{1}{1.4\,(16)^{0.4}}\left(\frac{2.309^{1.4}-1}{2.209}\right)$$

$$= 0.5991$$

Energy added as heat, $q_1 = C_P\,(T_3 - T_2) = 1.0045\,(2100 - 909.43)$

$$= 1195.93 \text{ kJ/kg}$$

Ner work done, $W = \eta\, q_1 = 0.5991 \times 1195.93 = 716.48$ kJ/kg

$$V_1 = \frac{nRT_1}{P_1} = \frac{mRT_1}{MP_1} = \frac{1\times 8.314\times 300}{28.97\times 10^{-3}\times 1\times 10^{5}}$$

$$= 0.861 \text{ m}^3/\text{kg}$$

$$V_1 - V_2 = V_1\left(1 - \frac{V_2}{V_1}\right) = V_1\left(1 - \frac{1}{r_0}\right) = V_1\left(1 - \frac{1}{16}\right) = \frac{15\,V_1}{16}$$

$$\text{Mean effective pressure, MEP} = \frac{W}{V_1 - V_2} = \frac{716.48\times 10^{3}\times 16}{15\times 0.861} = 8.8763 \text{ bar}$$

**Example 8.45**

An air-standard Diesel cycle has a compression ratio of 16. It is desired to study the effect of cut–off ratio on the thermal efficiency of the cycle. Calculate the thermal efficiency for cut–off ratio of 1.5, 2.5 and 4.

*Solution* :

The efficiency of a Diesel cycle is given by

$$\eta = 1 - \frac{1}{\gamma\, r_0^{\gamma-1}}\left(\frac{r_c^{\gamma}-1}{r_c-1}\right) = 1 - \frac{1}{1.4\,(16)^{0.4}}\left(\frac{r_c^{\gamma}-1}{r_c-1}\right)$$

$$= 1 - 0.2356\left(\frac{r_c^{\gamma} - 1}{r_c - 1}\right)$$

$$\text{If } r_c = 1.5; \eta = 1 - 0.2356\left(\frac{1.5^{1.4} - 1}{1.5 - 1}\right) = 0.6399$$

$$r_c = 2.5; \eta = 1 - 0.2356\left(\frac{2.5^{1.4} - 1}{2.5 - 1}\right) = 0.5906$$

$$r_c = 4; \eta = 1 - 0.2356\left(\frac{4^{1.4} - 1}{4 - 1}\right) = 0.5316$$

***Note*** *The efficiency of a Diesel cycle decreases with increasing cut–off ratio, for a given compression ratio.*

**Example 8.46**

An air–standard Diesel cycle has a compression ratio of 18 and a cut–off ratio of 2. At the start of the compression stroke, the pressure and temperature are 1 bar and 310 K, respectively. Determine the energy added as heat, the net work done and the maximum temperature of the cycle.

*Solution* :

$$T_2 = T_1\left(\frac{v_1}{v_2}\right)^{\gamma - 1} = T_1 r_0^{\gamma-1} = 310\,(18)^{0.4} = 985.08 \text{ K}$$

$$\frac{T_3}{T_2} = \frac{v_3}{v_2} = r_c$$

or $T_3 = r_c\, T_2 = 2 \times 985.08 = 1970.16$ K

Therefore, maximum temperature of the cycle = $T_3$ = 1970.16 K

Energy added as heat, $q_1 = C_P\,(T_3 - T_2)$

$$= 1.0045\,(1970.16 - 985.08) = 989.51 \text{ kJ/kg}$$

$$\text{Thermal efficiency, } \eta = 1 - \frac{1}{\gamma\, r_0^{\gamma-1}}\left(\frac{r_c^{\gamma} - 1}{r_c - 1}\right)$$

$$= 1 - \frac{1}{1.4(18)^{0.4}}\left(\frac{2^{1.4} - 1}{2 - 1}\right) = 0.6316$$

Net work done, $W = \eta q_1 = 0.6316 \times 989.51 = 624.97$ kJ/kg

### Example 8.47

Show that the mean effective pressure of an air–standard Diesel cycle is given by

$$\text{MEP} = \frac{P_1\left[\gamma\, r_0^{\gamma}\,(r_c - 1) - r_0\,(r_c^{\gamma} - 1\right]}{(\gamma - 1)(r_0 - 1)}$$

*Solution* :

The efficiency of an air–standard Diesel cycle is given by

$$\eta = 1 - \frac{1}{\gamma\, r_0^{\gamma-1}}\left(\frac{r_c^{\gamma} - 1}{r_c - 1}\right)$$

The net work done, $W = q_1\, \eta = C_P\,(T_3 - T_2)\left[1 - \frac{1}{\gamma r_0^{\gamma-1}}\left(\frac{r_c^{\gamma} - 1}{r_c - 1}\right)\right]$

$\left[\text{we know that } C_p = \frac{R\gamma}{\gamma - 1}\right]$

We know that $T_2 = T_1\left(\frac{v_1}{v_2}\right)^{\gamma-1} = T_1\, r_0^{\gamma-1}$ and $\frac{T_3}{T_2} = \frac{v_3}{v_2} = r_c$

Therefore, $$W = \frac{R\gamma}{\gamma - 1} T_1\, r_0^{\gamma-1}\,(r_c - 1)\left[1 - \frac{1}{\gamma\, r_0^{\gamma-1}}\left(\frac{r_c^{\gamma} - 1}{r_c - 1}\right)\right]$$

$$= \frac{RT_1}{\gamma - 1}\left[\gamma\, r_0^{\gamma-1}(r_c - 1) - (r_c^{\gamma} - 1\right]$$

We also know that $W = (\text{MEP})\,(v_1 - v_2) = (\text{MEP})\, v_1\left(1 - \frac{v_2}{v_1}\right)$

$$= (\text{MEP})\, v_1\left(1 - \frac{1}{r_0}\right) = \frac{(\text{MEP})\, RT_1\,(r_0 - 1)}{P_1\, r_0}$$

Therefore, $$\frac{(\text{MEP})\, RT_1\,(r_0-1)}{P_1 r_0} = \frac{RT_1}{\gamma-1}\left[\gamma\, r_0^{\gamma-1}\,(r_c-1)-(r_c^{\gamma}-1\right]$$

or $$\text{MEP} = \frac{P_1\ \gamma r_0^{\gamma}\ (r_c-1)-r_0\,(r_c^{\gamma}-1}{(\gamma-1)\,(r_0-1)}$$

**Example 8.48**

An air–standard Diesel cycle with a compression ratio of 16 and a cut–off ratio of 2 draws in air at 1 bar at the commencement of the suction stroke. Calculate the mean effective pressure of the engine.

*Solution* :

We know that the mean effective pressure is given by

$$\text{MEP} = \frac{P_1\ \gamma r_0^{\gamma}\ (r_c-1)-r_0\,(r_c^{\gamma}-1}{(\gamma-1)\,(r_0-1)}$$

$$= \frac{1\times 10^5\left\{1.4\times 16^{1.4}\,(2-1)-16\,(2^{1.4}-1)\right\}}{(1.4-1)\,(16-1)}$$

$$= 6.9466 \text{ bar}$$

**Example 8.49**

An air–standard Diesel engine with a compression ratio of 20 and a cut–off ratio of 2 is provided with a 3 litre cylinder. The engine draws air at 1 bar and 300 K and executes 1400 rpm. Calculate the power delivered by the engine.

*Solution* :

We know that $T_2 = T_1\, r_0^{\gamma-1} = 300\,(20)^{0.4} = 994.34$ K

$$\frac{T_3}{T_2} = \frac{v_3}{v_2} = r_c \text{ or } T_3 = r_C\, T_2$$

Energy absorbed as heat, $q_1 = C_P\,(T_3 - T_2) = C_P\,(r_c\, T_2 - T_2) = C_P\, T_2\,(r_c - 1)$

$= 1.0045 \times 994.34\ (2-1) = 998.81$ kJ/kg

$$\eta = 1-\frac{1}{\gamma\, r_0^{\gamma-1}}\left(\frac{r_c^{\gamma}-1}{r_c-1}\right) = 1-\frac{1}{1.4\,(20)^{0.4}}\left(\frac{2^{1.4}-1}{2-1}\right) = 0.6468$$

Net work done per cycle, $W = \eta\, q_1 = 0.6468 \times 998.81 = 646.03$ kJ/kg

$$\text{Mass of air in the cylinder} = \frac{P_1 V_1 M}{RT_1} = \frac{1\times10^5 \times 3\times10^{-3} \times 28.97\times10^{-3}}{8.314\times300}$$

$$= 3.4845 \times 10^{-3} \text{ kg}$$

Work done in one cycle = $3.4845 \times 10^{-3} \times 646.03 = 2.2511$ kJ

To complete one cycle, two complete revolutions are needed in a four stroke engine. Therefore, the number of cycles completed in one second is given by

$$\frac{1400}{60\times2} = 11.67 \text{ cycles/s}$$

Therefore, power output of the engine = $11.67 \times 2.2511 = 26.27$ kW

**Example 8.50**

For a given compression ratio and the initial state of the gas before compression, does a Diesel cycle has a higher efficiency than Otto cycle ?

*Solution* :

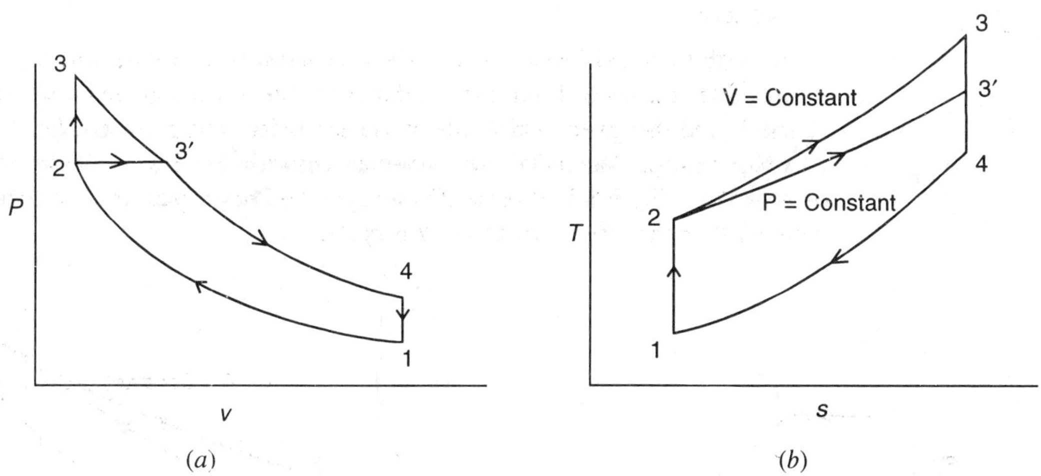


Fig.E 8.50. (*a*) *P–v* diagram showing the Otto cycle (1–2–3–4–1–) and a Deisel cycle (1–2–3′–4–1) (*b*) *T–s* diagram showing the Otto cycle (1–2–3–4–1) and a Diesel cycle (1–2–3'–4–1)

The initial state of the gas before compression (state 1) and compression ratio $r_0$ (= $v_1/v_2$) are same for both the Otto cycle and Diesel cycle implies that states 1 and 2 for these cycles are identical. The $P-v$ and $T-s$ diagrams for the Otto cycle and Diesel cycle having the same compression ratio and initial state of gas before compression are shown in Fig.E 8.50. The Otto cycle is represented by 1–2–3–4–1 and the Diesel cycle is shown by 1–2–3'–4–1. It can be easily observed from Fig.E 8.50 that the efficiency of the Otto cycle is more than that of the Diesel cycle.

We know that the efficiency of a Diesel cycle is given by

$$\eta \text{ (Diesel)} = 1 - \frac{1}{\gamma\, r_0^{\gamma-1}} \left( \frac{r_c^{\gamma} - 1}{r_c - 1} \right)$$

As $r_c \to 1$, we find $\eta$ (Diesel) $\to 1 - \dfrac{1}{r_0^{\gamma-1}}$ which is the efficiency of an Otto cycle. That is, for the same compression ratio, the limiting value of the efficiency of a Diesel cycle is the efficiency of an Otto cycle.

**Example 8.51**

Which of the two cycles – Otto cycle and Diesel cycle, will have higher efficiency for a given maximum pressure and temperature in the cycle ?

*Solution* :

In both Otto and Diesel cycles, the maximum temperature and pressure is attained after energy addition as heat, that is at state 3. For a given fixed values of $T_3$ and $P_3$ and the given initial state of the gas before compression, the $P-v$ and $T-s$ diagrams for Otto and Diesel cycles are shown in Fig.E 8.51. It can be easily observed from Fig.E 8.51 that the efficiency of the Diesel cycle at these conditions is higher than the efficiency of an Otto cycle.

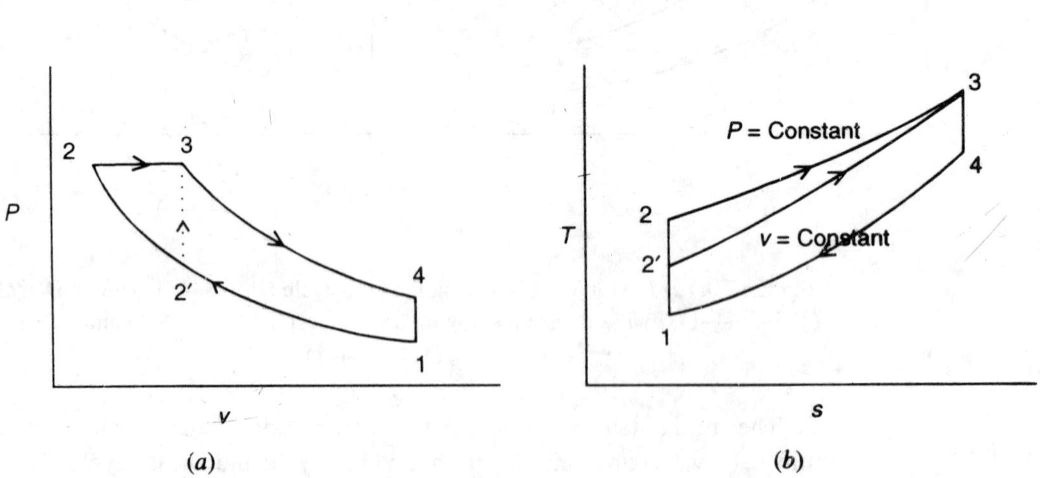


Fig.E 8.51. Otto and Diesel cycles on the same (*a*) $P-v$ diagram and (*b*) $T-s$ diagram 1–2–3–4–1 represents the Diesel cycle and (1–2'–3–4–1) represents the Otto cycle.

**Example 8.52**

In a heat engine cycle the minimum pressure and temperature are 1 bar and 300 K, respectively. The maximum pressure and temperature are 70 bar and 2000 K, respectively. Calculate the thermal efficiency if (*a*) an Otto cycle is operated and (*b*) a Diesel cycle is operated in between the minimum and maximum pressures and temperatures.

*Solution* :

Given $P_1 = 1$ bar; $T_1 = 300$ K; $P_3 = 70$ bar; $T_3 = 2000$ K

(*a*) For an Otto cycle $r_0 = \dfrac{v_1}{v_2} = \dfrac{P_2 T_1}{T_2 P_1}$

We know that 2–3 is a constant volume process. Therefore, $\dfrac{P_2}{T_2} = \dfrac{P_3}{T_3}$

$$\text{Therefore, } r_0 = \frac{P_2 T_1}{T_2 P_1} = \frac{P_3 T_1}{T_3 P_1} = \frac{70 \times 300}{2000 \times 1} = 10.5$$

$$\text{Thermal efficiency, } \eta = 1 - \left(\frac{1}{r_0}\right)^{\gamma-1} = 1 - \left(\frac{1}{10.5}\right)^{\gamma-1} = 0.6096$$

(*b*) For a Diesel cycle, $P_2 = P_3 = 70$ bar

$$\frac{P_2}{P_1}\left(\frac{v_1}{v_2}\right)^{\gamma} = r_0^{\gamma} \text{ or } \frac{70}{1} = r_0^{1.4} \text{ or } r_0 = 20.793$$

$$T_2 = T_1\left(\frac{v_1}{v_2}\right)^{\gamma-1} = T_1\, r_0^{\gamma-1} = 300 \times (20.793)^{0.4} = 1009.9 \text{ K}$$

$$\text{Cut-off ratio, } r_c = \frac{v_3}{v_2} = \frac{T_3}{T_2} = \frac{2000}{1009.9} = 1.9804$$

$$\text{Thermal efficiency, } \eta = 1 - \frac{1}{\gamma\, r_0^{\gamma-1}}\left(\frac{r_c^{\gamma} - 1}{r_c - 1}\right) = 1 - \frac{1}{1.4(20.793)^{0.4}}\left[\frac{(1.9804)^{1.4} - 1}{1.9804 - 1}\right]$$

$$= 1 - \frac{1}{1.4(20.793)^{0.4}}\left[\frac{(1.9804)^{1.4} - 1}{1.9804 - 1}\right]$$

$$= 0.6531$$

**Example 8.53**

Sketch the indicator diagram (or the pressure–displacement diagram) for a reciprocating four stroke internal combustion engine and explain the various zones in it.

*Solution* :

The pressure-displacement (or indicator) diagram for a four stroke internal combustion engine is shown in Fig.E 8.53. In this engine the piston executes four strokes for every two revolutions of the crank shaft. At state 1 the intake valve opens and the piston executes an outward motion (suction stroke) sucking in a fresh charge of air and the air reaches state 2. At state 2, the intake valve closes and the piston executes the compression stroke 2–3, raising the temperature and pressure of the gas. At state 3 combustion of the fuel starts and raises the temperature and pressure of the gas and the gas follows the path 3–4–5. At state 5, the power stroke 5–6 starts and the piston executes an outward stroke. At state 6, the exhaust valve opens and the combustion products leave the cylinder and the gases in the cylinder reach state 7. At state 7, the exhaust stroke 7 – 1 commences while the exhaust valve is still open and the piston executes an inward motion discharging the combustion products from the cylinder. The gases in the cylinder reach state 1, thus completing one cycle.

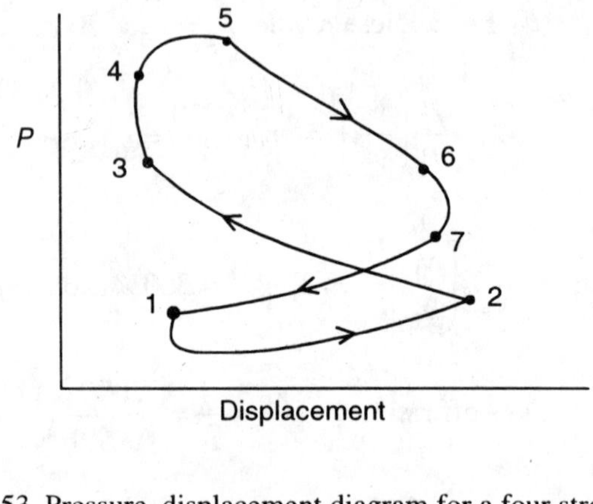


Fig.E 8.53. Pressure–displacement diagram for a four stroke internal combustion engine.

**Example 8.54**

Sketch the air–standard dual cycle on $P-v$ diagram and $T-s$ diagram.

*Solution* :

The indicator diagram shown in Fig.E 8.53 shows that energy addition as heat takes place during 3–4–5. The pressure–volume variation in this region clearly indicates that in a heat engine all the energy is not added either at constant volume or at constant pressure. In the analysis of Otto engine it was assumed that the combustion of gasoline is instantaneous and hence the combustion of the fuel is approximated as constant volume energy addition. In the analysis of Diesel cycle, it was assumed that the combustion takes place slowly and hence it was approximated as constant pressure energy addition. In reality the combustion process cannot be so fast that it can be approximated as constant–volume process, or it cannot be so slow that it can be approximated as constant– pressure process. A more realistic process should closely approximate the actual combustion process. This modification leads to the air–standard dual cycle shown in Fig.E 8.54. In the dual cycle, the energy addition occurs in two stages. Part of the energy is added at constant–volume and rest of the energy is added at constant pressure. The remaining processes are similar to Otto or Diesel cycle. Thus a dual cycle is a combination of Otto and Diesel cycle.

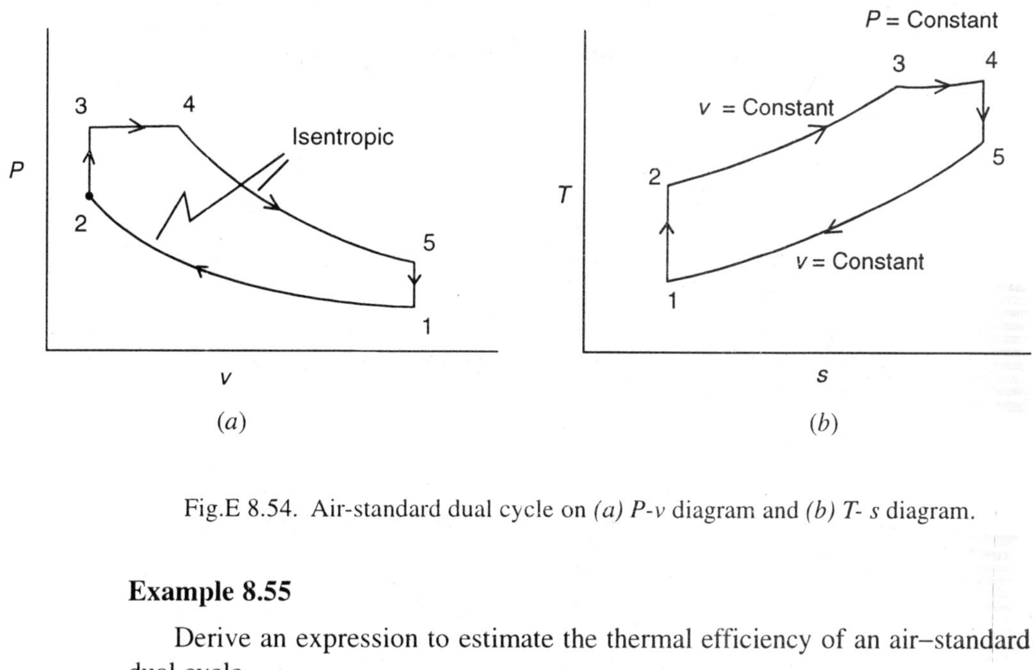


Fig.E 8.54. Air-standard dual cycle on *(a) P-v* diagram and *(b) T- s* diagram.

## Example 8.55

Derive an expression to estimate the thermal efficiency of an air–standard dual cycle.

*Solution* :

Consider the dual cycle shown in Fig.E 8.54. Assuming that 1 kg of air is undergoing the cyclic change, we can estimate the thermal efficiency by calculating the energy absorbed and rejected as heat. Then

$$q_1 = C_v (T_3 - T_2) + C_P (T_4 - T_3)$$

$$q_2 = C_v (T_5 - T_1)$$

$$\text{Thermal efficiency, } \eta = 1 - \frac{q_2}{q_1} = 1 - \frac{C_v(T_5 - T_1)}{C_v(T_3 - T_2) + C_P(T_4 - T_3)}$$

$$\text{or} \quad \eta = 1 - \frac{T_1(T_5/T_1 - 1)}{T_2\left[\left(\frac{T_3}{T_2} - 1\right) + \frac{\gamma . T_3}{T_2}\left(\frac{T_4}{T_3} - 1\right)\right]} \qquad (A)$$

Define, compression ratio, $r_0 = v_1/v_2$

Cut–off ratio, $r_c = \dfrac{v_4}{v_3}$ and

Pressure ratio for constant volume combustion, $r_{P,v} = \dfrac{P_3}{P_2}$

Then, for the isentropic process 1–2, we have

$$\frac{T_2}{T_1} = \left(\frac{v_1}{v_2}\right)^{\gamma-1} = r_0^{\gamma-1}$$

$$\text{or} \quad \frac{T_1}{T_2} = \left(\frac{1}{r_0}\right)^{\gamma-1} \qquad (B)$$

For the constant–pressure process 3–4, we have

$$\frac{T_4}{T_3} = \frac{v_4}{v_3} = r_c \qquad (C)$$

For the constant–volume process 2–3, we have

$$\frac{T_3}{T_2} = \frac{P_3}{P_2} = r_{Pv} \qquad (D)$$

For the constant–volume process 5–1, we have

$$\frac{T_5}{T_1} = \frac{P_5}{P_1} = \frac{P_5 P_4}{P_4 P_1} = \frac{P_5}{P_4}\left(\frac{P_3}{P_1}\right)$$

$$= \left(\frac{P_5}{P_4}\right)\left(\frac{P_3}{P_2}\right)\left(\frac{P_2}{P_1}\right) \quad (\text{Since } P_3 = P_4)$$

or $$\frac{T_5}{T_1} = \left(\frac{v_4}{v_5}\right)^{\gamma} r_{P,v} \left(\frac{v_1}{v_2}\right)^{\gamma} = r_{P,v}\left(\frac{v_4}{v_3}\right)^{\gamma} = r_{P,V} r_c^{\gamma} \qquad (E)$$

[ Since $v_1 = v_5 \quad v_2 = v_3$ ]

Substituting Eqns. (*B*) – (*E*) in Eqn. (*A*), we get

$$\eta = 1 - \frac{1}{r_0^{\gamma-1}}\left[\frac{r_{P_V} r_c^{\gamma} - 1}{\gamma\, r_{P,v}(r_c - 1) + (r_{P,v} - 1)}\right]$$

**Example 8.56**

In an air–standard dual cycle with a compression ratio of 15, the minimum pressure and temperature are 1 bar and 300 K, respectively. The maximum temperature and pressure are 3000 K and 70 bar, respectively. Determine the work done per kg of air, the energy added as heat per kg of air and the thermal efficiency of the cycle.

*Solution* :

The *P*– *v* diagram of the air–standard dual cycle is shown in Fig.E 8.56.

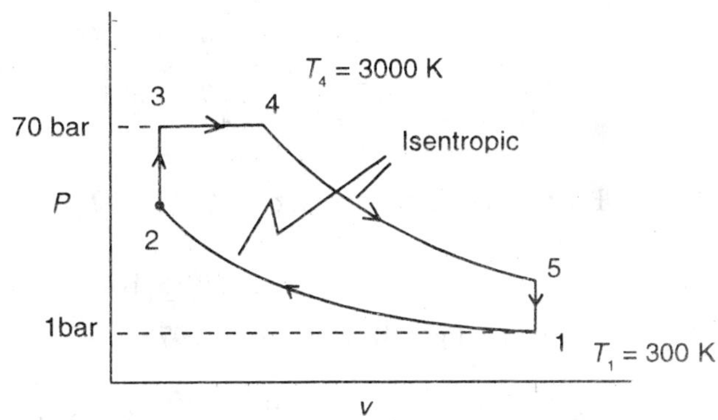


Fig.E 8.56. *P-v* diagram of air-standard dual cycle.

Given data : $P_1 = 1$ bar; $T_1 = 300$ K; $r_0 = 15$; $P_3 = P_4 = 70$ bar; $T_4 = 3000$ K

$$T_2 = T_1\, r_0^{\gamma-1} = 300\,(15)^{0.4} = 886.25 \text{ K}$$

$$P_2 = P_1\, r_0^{\gamma} = 1 \times (15)^{1.4} = 44.313 \text{ bar}$$

$$T_3 = \frac{T_2 P_3}{P_2} = \frac{886.25 \times 70}{44.313} = 1400 \text{ K}$$

Energy added as heat, $q_1 = C_v (T_3 - T_2) + C_P (T_4 - T_3)$

$$= \frac{R}{\gamma - 1}(T_3 - T_2) + \frac{R\gamma}{\gamma - 1}(T_4 - T_3)$$

$$= \frac{R}{\gamma - 1}[(T_3 - T_2) + \gamma(T_4 - T_3)]$$

$$= \frac{8.314}{1.4 - 1} [\ 1400 - 886.25 + 1.4\ (3000 - 1400)\ ]$$

$$= 5.7237 \times 10^4 \text{ J / mol} = 1975.7 \text{ kJ/kg}$$

$$\frac{T_4}{T_3} = \frac{v_4}{v_3} = \frac{v_4}{v_2}$$

$$T_5 = T_4\left(\frac{v_4}{v_5}\right)^{\gamma - 1} = T_4\left(\frac{v_4}{v_1}\right)^{\gamma - 1} = T_4\left[\frac{v_4}{v_2} \times \frac{v_2}{v_1}\right]^{\gamma - 1}$$

$$= T_4\left[\frac{v_4}{v_3} \times \frac{v_2}{v_1}\right]^{\gamma - 1}$$

$$= T_4\left(\frac{T_4}{T_3} \times \frac{1}{r_0}\right)^{\gamma - 1} = 3000\left(\frac{3000}{1400} \times \frac{1}{15}\right)^{0.4} = 1377.5 \text{ K}$$

Energy rejected as heat, $q_2 = C_v (T_5 - T_1) = \dfrac{R(1377.5 - 300)}{\gamma - 1}$

$$= 2.2396 \times 10^4 \text{ J / mol} = 773.07 \text{ kJ/kg}$$

Net work done, $W = q_1 - q_2 = 1975.7 - 773.07 = 1202.6$ kJ/kg

Thermal efficiency, $\eta = \dfrac{W}{q_1} = \dfrac{1202.6}{1975.7} = 0.6087$

**Example 8.57**

Calculate the mean effective pressure of the air–standard dual cycle of Example 8.56.

*Solution* :

In Example 8.56, we obtained

$$W = 1202.6 \text{ kJ/kg}$$

Mean effective pressure, $\text{MEP} = \dfrac{W}{v_1 - v_2}$

$$v_1 - v_2 = v_1\,(1 - v_2 / v_1) = \frac{nRT}{P_1}\left(1 - \frac{1}{r_0}\right) = \frac{mRT_1(r_0 - 1)}{MP_1\, r_0}$$

$$= \frac{1 \times 8.314 \times 300(15-1)}{28.97 \times 10^{-3} \times 1 \times 10^5 \times 15} = 0.8036 \text{ m}^3$$

$$\text{MEP} = \frac{W}{v_1 - v_2} = \frac{1202.6 \times 10^3}{0.8036} = 14.965 \text{ bar}$$

**Example 8.58**

An air–standard dual cycle with a compression ratio of 15, has the minimum temperature and pressure of 300 K and 0.95 bar, respectively. The cylinder volume is 3 litres and 6 kJ of energy is added as heat of which one–third is added at constant volume and the rest at constant pressure. Calculate the maximum pressure in the cycle, the temperature of the gas before and after the constant pressure energy addition, the energy rejected as heat and the thermal efficiency of the cycle. For air $C_v = 0.7175$ kJ / kg K.

*Solution* :

Given data : $r_0 = 15$; $T_1 = 300$ K; $P_1 = 0.95$ bar; $V_1 = 3 \times 10^{-3}$ m$^3$; $q_1 = 6$ kJ

$$T_2 = T_1\, r_0^{\gamma-1} = 300\,(15)^{0.4} = 886.25 \text{ K}$$

$$P_2 = P_1\, r_0^{\gamma} = 0.95\,(15)^{1.4} = 42.097 \text{ bar}$$

$$m = \frac{P_1 V_1 M}{RT_1} = \frac{0.95 \times 10^5 \times 3 \times 10^{-3} \times 28.97 \times 10^{-3}}{8.314 \times 300} = 3.31 \times 10^{-3} \text{ kg}$$

$$mC_v\,(T_3 - T_2) = \frac{6}{3}$$

or $(T_3 - T_2) = \dfrac{2}{mC_v} = \dfrac{2}{3.31 \times 10^{-3} \times 0.7175} = 842.13 \text{ K}$

or $T_3 = 842.13 + 886.25 = 1728.4$ K

Therefore, temperature of the gas before the constant pressure energy addition

$$= 1728.4 \text{ K}$$

$$mC_P\,(T_4 - T_3) = \frac{6 \times 2}{3} = 4 \text{ kJ}$$

or $mC_v \gamma\ (T_4 - T_3) = 4$ kJ

or $3.31 \times 10^{-3} \times 0.7175 \times 1.4\ (T_4 - 1728.4) = 4$ or $T_4 = 2931.4$ K

Therefore, the temperature of the gas after the constant pressure energy addition = 2931.4 K

$$\frac{P_3}{P_2} = \frac{T_3}{T_2} \text{ or } P_3 = P_2 \frac{T_3}{T_2} = \frac{42.097 \times 1728.4}{886.25} = 82.099 \text{ bar}$$

Maximum pressure = $P_4 = P_3 = 82.099$ bar

$$T_5 = T_4 \left( \frac{T_4}{T_3} \times \frac{1}{r_0} \right)^{\gamma - 1}$$ [see Example 8.56]

$$= 2931.4 \left( \frac{2931.4}{1728.4} \times \frac{1}{15} \right)^{0.4} = 1225.8 \text{ K}$$

Energy rejected as heat, $Q_2 = mC_v\ (T_5 - T_1)$

$$= 3.31 \times 10^{-3} \times 0.7175\ (1225.8 - 300) = 2.1987 \text{ kJ}$$

$$\text{Thermal efficiency, } \eta = \frac{W}{Q_1} = 1 - \frac{Q_2}{Q_1} = 1 - \frac{2.1987}{6} = 0.6336$$

**Example 8.59**

An air–standard dual cycle has a compression ratio of 16 and a cut–off ratio of 1.5. The minimum pressure and temperature of the cycle are 1 bar and 300 K. During the constant–volume energy addition as heat the pressure doubles. Calculate the thermal efficiency of the cycle, the energy added as heat per kg of air, the network done per kg of air and the mean effective pressure.

*Solution* :

Given data : $r_0 = 16$; $r_c = 1.5$; $r_{Pv} = 2$; $P_1 = 1$ bar; $T_1 = 300$ K

We know that the thermal efficiency $\eta$ of a dual cycle is given by

$$\eta = 1 - \frac{1}{r_0^{\gamma - 1}} \left[ \frac{r_{P,v}\, r_c^{\gamma} - 1}{\gamma_{r_{P,v}} (r_c - 1) + (r_{Pv} - 1)} \right]$$

[see Example 8.55]

$$= 1 - \frac{1}{(16)^{0.4}} \left[ \frac{2 \times (1.5)^{1.4} - 1}{1.4 \times 2 (1.5 - 1) + (2 - 1)} \right] = 0.6525$$

Energy added as heat, $q_1 = C_v (T_3 - T_2) + C_P (T_4 - T_3)$

$$= C_v [T_3 - T_2 + \gamma (T_4 - T_3)]$$

We know that $\frac{T_3}{T_2} = \frac{P_3}{P_2} = r_{pv} = 2$

or $T_3 - T_2 = T_2$.

$$\frac{T_4}{T_3} = \frac{v_4}{v_3} = r_c = 1.5$$

or $T_4 = 1.5\, T_3$

or $T_4 - T_3 = \frac{T_3}{2}$

or $(T_4 - T_3) = \frac{T_3}{2} = \frac{2T_2}{2} = T_2$

$$T_2 = T_1 \left(\frac{v_1}{v_2}\right)^{\gamma-1} = T_1 (r_0)^{\gamma-1} = 300\,(16)^{0.4} = 909.43 \text{ K}$$

Therefore, $q_1 = C_v [T_3 - T_2 + \gamma (T_4 - T_3)]$

$$= C_v (T_2 + \gamma T_2) = C_v T_2 (\gamma + 1)$$

$$= 0.7175 \times 909.43 \times 2.4 = 1566 \text{ kJ/kg}$$

Net work done, $W = \eta\, q_1 = 0.6525 \times 1566 \times 10^3 = 1021.8$ kJ/kg

$$v_1 - v_2 = v_1 \left(1 - \frac{1}{r_0}\right) = \frac{mRT_1}{P_1 M} \left(\frac{r_0 - 1}{r_0}\right)$$

$$= \frac{1 \times 8.314 \times 300 \times 15}{1 \times 10^5 \times 28.97 \times 10^{-3} \times 16} = 0.8072 \text{ m}^3/\text{kg}$$

Mean effective pressure, $\text{MEP} = \frac{W}{v_1 - v_2} = \frac{1021.8 \times 10^3}{0.8072} = 12.659$ bar

**Example 8.60**

Show that the efficiency of an air–standard dual cycle tends to the efficiency of (*a*) an air–standard Otto cycle as $r_c \to 1$ and (*b*) an air–standard Diesel cycle as $r_{pv} \to 1$.

*Solution* :

We know that the efficiency of an air–standard dual cycle is given by

$$\eta = 1 - \frac{1}{(r_0)^{\gamma-1}}\left[\frac{r_{P,v}\, r_c^{\gamma} - 1}{\gamma\, r_{P,v}\,(r_c - 1) + (r_{P,v} - 1)}\right] \qquad (A)$$

(*a*) Subsitute $r_c = 1$ in Eqn. (*A*) to obtain

$$\eta = 1 - \frac{1}{(r_0)^{\gamma-1}}\left[\frac{r_{P,v} - 1}{\gamma\, r_{P,v}(1-1) + r_{Pv} - 1}\right] = 1 - \left(\frac{1}{r_0}\right)^{\gamma-1}$$

which is the same as the efficiency of an air–standard Otto cycle.

(*b*) Substitute $r_{p,v} = 1$ in Eqn. (*A*) to obtain

$$\eta = 1 - \frac{1}{(r_0)^{\gamma-1}}\left[\frac{r_c^{\gamma} - 1}{\gamma(r_c - 1) + (1-1)}\right]$$

$$= 1 - \frac{1}{\gamma\,(r_0)^{\gamma-1}}\left(\frac{r_c^{\gamma} - 1}{r_c - 1}\right)$$

which is the same as the efficiency of an air–standard Diesel cycle.

**Example 8.61**

Sketch the air–standard Brayton cycle on $P-v$ and $T-s$ diagrams. What are the advantages of the Brayton cycle over the conventional heat engine cycles ?

*Solution* :

The Brayton cycle was developed by George Brayton in the 1870s. Compared with the Otto cycle and Diesel cycle, the Brayton cycle operates over small ranges of pressures and temperatures. The Brayton cycle essentially uses a compressor, a turbine and a combustion chamber. Because of the light weight of the equipment it is ideally suited for gas turbines. A schematic arrangement of the units for a Brayton cycle is shown in Fig.E 8.61 (*a*). Fig.E 8.61 (*b*) shows the $P-v$ diagram and Fig.E 8.61 (*c*) shows the $T-s$ diagram for an air–standard Brayton cycle. The Brayton cycle consists of the following processes :

**1–2** : Isentropic compression of the air.

**2–3** : Constant pressure energy addition as heat. Usually the hot gases leaving the compressor enter a combustion chamber where the fuel oil is injected and burned raising the temperature of the gas to a higher value at constant pressure.

**3–4** : Isentropic expansion of the hot gases in a turbine. Part of the work delivered by the turbine is used to operate the compressor.

**4–1** : The exit gas of the turbine is cooled at constant pressure, returning the gas to the original temperature. Usually the gas leaving the turbine is exhausted into the surroundings. However, for thermodynamic analysis, the cycle is completed by idealizing this process as a constant pressure energy rejection process.

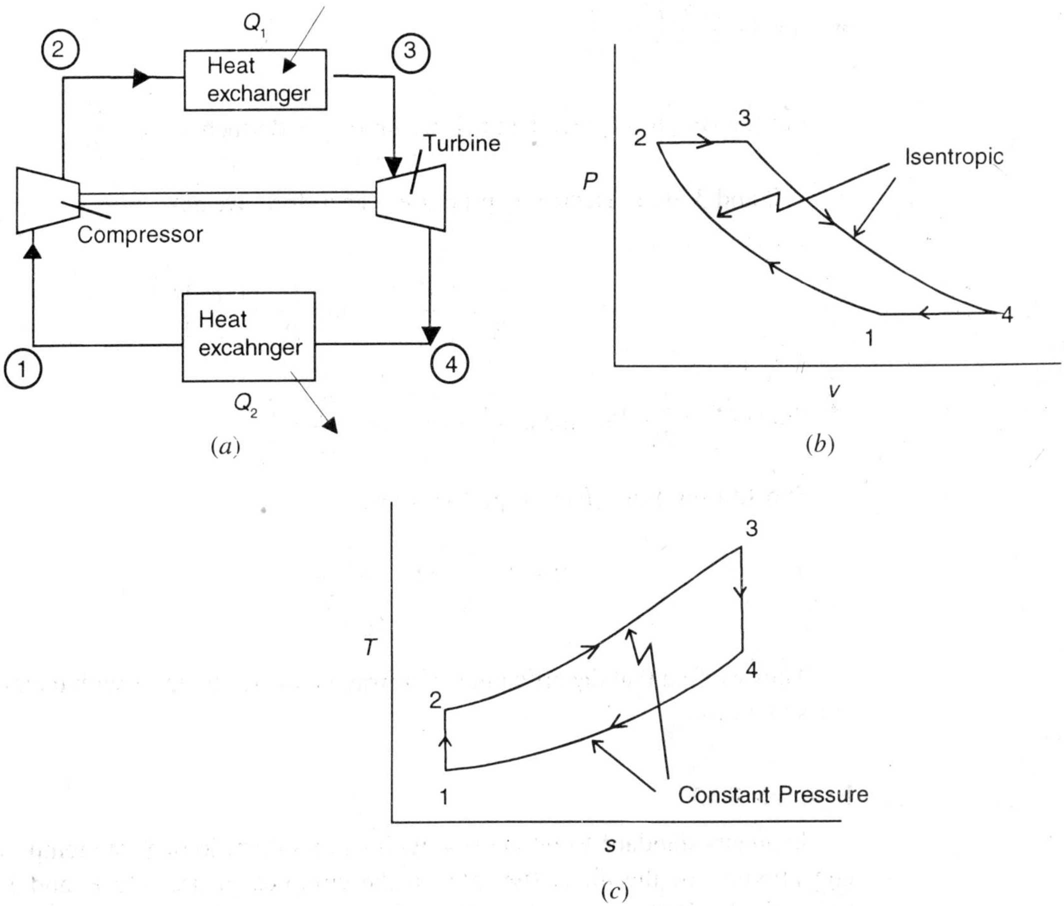


Fig.E 8.61. Air–standard Brayton cycle (*a*) schematic arrangement of equipment (*b*) on $P - v$ daigram (*c*) on $T - s$ diagram.

**Example 8.62**

Derive an expression to estimate the thermal efficiency of an air–standard Brayton cycle.

*Solution*:

Assume that a unit mass of gas is undergoing the cyclic change.

Then, energy added as heat, $q_1 = C_P (T_3 - T_2)$

Energy rejected as heat, $q_2 = C_P (T_4 - T_1)$

Thermal efficiency, $\eta = 1 - \dfrac{q_2}{q_1} = 1 - \dfrac{C_p(T_4 - T_1)}{C_p(T_3 - T_2)} = 1 - \dfrac{T_4 - T_1}{T_3 - T_2}$

or $$\eta = 1 - \frac{T_1(T_4/T_1 - 1)}{T_2(T_3/T_2 - 1)} \qquad (A)$$

For the Brayton cycle, the pressure ratio $r_P$ is defined as $r_P = \dfrac{P_2}{P_1}$

1–2 and 3–4 are isentropic processes. Therefore, we get

$$\frac{P_2}{P_1} = \left(\frac{T_2}{T_1}\right)^{\frac{\gamma}{\gamma-1}} \text{ and } \frac{P_3}{P_4} = \left(\frac{T_3}{T_4}\right)^{\frac{\gamma}{\gamma-1}}$$

Since $\dfrac{P_2}{P_1} = \dfrac{P_3}{P_4}$ We have, $\dfrac{T_2}{T_1} = \dfrac{T_3}{T_4}$ or $\dfrac{T_4}{T_1} = \dfrac{T_3}{T_2}$ $\qquad (B)$

Substituting Eqn. (*B*) in Eqn. (*A*) we get

$$\eta = 1 - \frac{T_1}{T_2} = 1 - \frac{1}{(r_P)^{\frac{\gamma-1}{\gamma}}}$$

Thus we find that the efficiency of a Brayton cycle increases with increasing pressure ratio $r_p$.

**Example 8.63**

In an air–standard Brayton cycle with a pressure ratio of 8, the temperature and pressure of the air at the inlet to the compressor are 300 K and 1 bar, respectively. If the maximum allowed temperature in the cycle is 1000 K, determine the energy added as heat, the net work done per kg of air and the thermal efficiency of the cycle. $C_P$ of air is 1.0045 kJ/kg K.

*Solution* :

Given data : $r_P = 8$; $P_1 = 1$ bar ; $T_1 = 300$ K ; $T_3 = 1000$ K

$$T_2 = T_1\left(\frac{P_2}{P_1}\right)^{\frac{\gamma-1}{\gamma}} = T_1(r_P)^{\frac{\gamma-1}{\gamma}} = 300 \times (8)^{\frac{0.4}{1.4}} = 543.4 \text{ K}$$

$q_1 = C_P (T_3 - T_2) = 1.0045\,(1000 - 543.4) = 458.65$ kJ/kg

Therefore energy added as heat, $q_1 = 458.65$ kJ/kg

$$\eta = 1 - \left(\frac{1}{r_P}\right)^{\frac{\gamma-1}{\gamma}} = 1 - \left(\frac{1}{8}\right)^{\frac{0.4}{1.4}} = 0.448$$

Net work done, $W = \eta\, q_1 = 0.448 \times 458.65 = 205.48$ kJ/kg

**Example 8.64**

In a Brayton cycle the ambient conditions limit the minimum temperature $T_1$ and the metallurgical limit for the turbine blades dictates the maximum temperature $T_3$. For fixed values of $T_1$ and $T_3$ determine the optimum pressure ratio $r_P$ which gives the maximum work in a Brayton cycle.

*Solution* :

Energy absorbed as heat, $q_1 = C_P (T_3 - T_2)$

Energy rejected as heat, $q_2 = C_P (T_4 - T_1)$

Net work delivered, $W = q_1 - q_2 = C_p (T_3 - T_2) - C_p (T_4 - T_1)$

$$= C_P (T_3 - T_4) - C_P (T_2 - T_1)$$

$$= C_p [(T_3 (1 - T_4/T_3)) - T_1 (T_2/T_1 - 1)] \qquad (A)$$

1–2 and 3–4 are isentropic processes. Therefore,

$$\frac{T_2}{T_1} = \left(\frac{P_2}{P_1}\right)^{\frac{\gamma-1}{\gamma}} \text{ and } \frac{T_3}{T_4} = \left(\frac{P_3}{P_4}\right)^{\frac{\gamma-1}{\gamma}}$$

But $\dfrac{P_2}{P_1} = \dfrac{P_3}{P_4} = r_P$

Therefore, Eqn. (*A*) can be written as

$$W = C_P\left[T_3\left\{1 - \left(\frac{1}{r_P}\right)^{\frac{\gamma-1}{\gamma}}\right\} - T_1\left\{(r_P)^{\frac{\gamma-1}{\gamma}} - 1\right\}\right]$$

The optimum pressure ratio is given by

$$\frac{dW}{dr_P} = 0 = -C_P T_3\left(\frac{1-\gamma}{\gamma}\right)(r_P)^{\frac{1-2\gamma}{\gamma}} - C_P T_1\left(\frac{\gamma-1}{\gamma}\right)(r_P)^{\frac{-1}{\gamma}}$$

$$\text{or} \quad T_3\left(\frac{\gamma-1}{\gamma}\right)(r_P)^{\frac{1-2\gamma}{\gamma}} = T_1\left(\frac{\gamma-1}{\gamma}\right)(r_P)^{-\frac{1}{\gamma}}$$

$$\text{or} \quad \frac{T_3}{T_1} = (r_P)^{\frac{2(\gamma-1)}{\gamma}} \quad \text{or} \quad r_P = \left(\frac{T_3}{T_1}\right)^{\frac{\gamma}{2(\gamma-1)}}$$

## Example 8.65

Sketch the air–standard Ericsson cycle on $P-v$ and $T-s$ diagrams and develop an expression to estimate the thermal efficiency of the cycle.

*Solution* :

The air–standard Ericsson cycle is shown in Fig.E 8.65. Air at temperature $T_1$ enters a compressor and undergoes isothermal compression to state 2, while it is rejecting energy as heat to a thermal reservoir at $T_L$. Then the air at state 2 enters a regenerator and leaves at state 3 while receiving energy as heat. The energy received by the gas is $C_P(T_3 - T_2)$. Then the gas at state 3 undergoes isothermal expansion in a turbine while receiving energy as heat from a thermal reservoir at $T_H$ and leaves at state 4, which then enters the regenerator. In the regenerator it rejects energy as heat to the incoming gas from the compressor. The energy rejected as heat is $C_p(T_4 - T_1)$. Since $T_4 = T_3$ and $T_1 = T_2$, the energy rejected is $C_P(T_4 - T_1) = C_P(T_3 - T_2)$ which is the same as the energy received by the gas leaving the compressor. There is no additional energy exchange in the regenerator. The regenerator receives energy as heat in one part of the cycle and the same amount of energy is given to the gas in the other part of the cycle.

$$\text{Energy absorbed as heat, } q_1 = W_{3-4} = RT_H \ln\frac{v_4}{v_3}$$

$$\text{Energy rejected as heat, } q_2 = W_{21} = RT_L \ln\frac{v_1}{v_2}$$

$$\text{Thermal efficiency, } \eta = 1 - \frac{q_2}{q_1} = 1 - \frac{RT_L \ln v_1/v_2}{RT_H \ln v_4/v_3} \qquad (A)$$

2–3 and 4–1 are constant–pressure processes. Therefore

$$\frac{v_2}{T_2} = \frac{v_3}{T_3} \quad \text{or} \quad \frac{v_2}{T_L} = \frac{v_3}{T_H} \quad \text{or} \quad \frac{T_H}{T_L} = \frac{v_3}{v_2} \qquad (B)$$

and

$$\frac{v_1}{T_1} = \frac{v_4}{T_4} \quad \text{or} \quad \frac{v_1}{T_L} = \frac{v_4}{T_H} \quad \text{or} \quad \frac{T_H}{T_L} = \frac{v_4}{v_1} \qquad (C)$$

From Eqns.(*B*) and (*C*), we get

$$\frac{v_3}{v_2} = \frac{v_4}{v_1} \quad \text{or} \quad \frac{v_1}{v_2} = \frac{v_4}{v_3} \qquad (D)$$

Substituting Eqn.(*D*) in Eqn.(*A*), we get

$$\eta = 1 - \frac{T_L}{T_H}$$

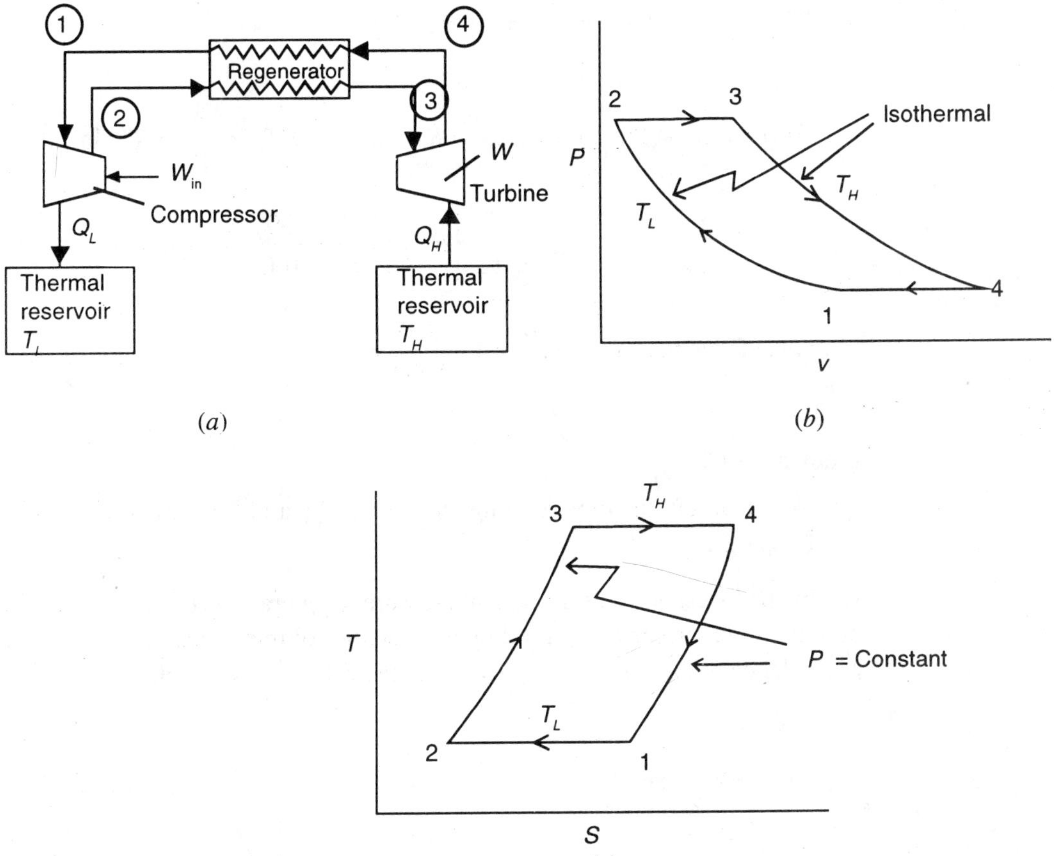


Fig.E 8.65. Air standard Ericssion cycle (*a*) schematic arrangement (*b*) on *P*-*v* diagram (*c*) on *T*-*s* diagram.

**Example 8.66**

In an air–standard Ericsson cycle, the air is at 1 bar at the inlet to the compressor. 1000 kJ of Energy is added as heat at 1000 K per kg of air and energy is rejected to the atmosphere at 300 K. Calculate the work done on the compressor, work done by the turbine and the thermal efficiency of the cycle.

*Solution* :

Given that $T_H = 1000$ K; $T_L = 300$ K; $P_1 = 1$ bar

$$T_1 = T_2 = T_L \text{ and } T_3 = T_4 = T_H$$

The expansion of gas in turbine is isothermal at $T_H$. Therefore,

Work done by turbine = energy absorbed as heat.

or $RT_H \ln \dfrac{v_4}{v_3} = RT_H \ln \dfrac{P_3}{P_4} = 1000$ kJ/kg

$$\text{Work done on the compressor} = RT_L \ln \frac{v_1}{v_2} = RT_L \ln \frac{P_2}{P_1} = RT_L \ln \frac{P_3}{P_4}$$

$$= \left(\frac{T_L}{T_H}\right) RT_H \ln \frac{P_3}{P_4} = \frac{300}{1000} \times 1000 = 300 \text{ kJ/kg}$$

$$\text{Thermal efficiency, } \eta = 1 - \frac{T_L}{T_H} = 1 - \frac{300}{1000} = 0.7$$

**Example 8.67**

Sketch an air-standard Stirling cycle on a *P-v* and *T-s* diagrams.

*Solution* :

In the Ericsson cycle a constant pressure regenerator is used. If the constant–pressure regenerator is replaced by a constant–volume regenerator, the Ericsson cycle degenerates to a Stirling cycle. In the Stirling cycle, the gas at state 1 is

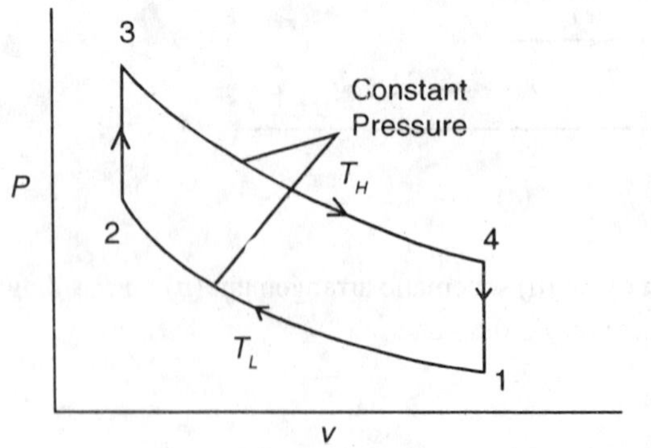


(*a*)

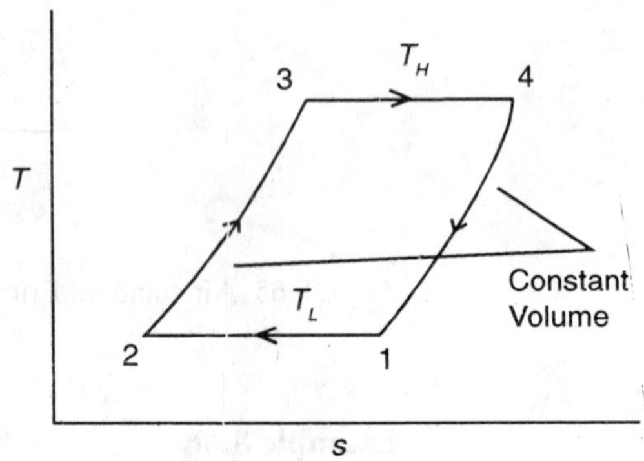


(*b*)

Fig.E 8.67. Air-standard Stirling cycle on (*a*) *P- v* diagram and (*b*) *T- s* diagram.

isothermally compressed to state 2 while rejecting energy to a thermal reservoir at $T_L$. Then the air passes through a constant volume regenerator receiving energy $C_v(T_3 - T_2)$ and reaches state 3. Then the gas undergoes isothermal expansion in a turbine which receives energy from a thermal reservoir at $T_H$. The gas leaving the turbine passes through the regenerator and rejects energy $C_v(T_4 - T_1)$. The thermal efficiency of Stirling cycle is given by $\eta = 1 - T_L/T_H$.

**Example 8.68**

A Stirling cycle uses air and at the beginning of the isothermal expansion the air is at 1000 K and 10 bar. The minimum pressure in the cycle is 2 bar and at the end of the isothermal compression the volume is 60% of the maximum volume. Calculate the thermal efficiency of the cycle and the mean effective pressure.

*Solution* :

The $P-v$ diagram of the Stirling cycle is shown in Fig.E 8.68. From the Fig.E 8.68, we can find that $P_1 = 2$ bar, $P_3 = 10$ bar

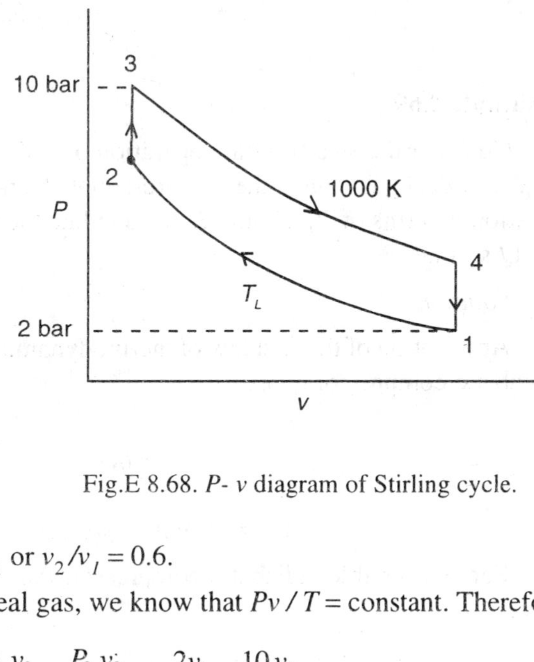


Fig.E 8.68. *P- v* diagram of Stirling cycle.

and $v_2 = 0.6v_1$ or $v_2/v_1 = 0.6$.

For an ideal gas, we know that $Pv/T$ = constant. Therefore

$$\frac{P_1 v_1}{T_L} = \frac{P_3 v_3}{T_H} = \frac{P_3 v_2}{T_H} \text{ or } \frac{2v_1}{T_L} = \frac{10 v_2}{1000}$$

$$\text{or} \quad T_L = \frac{1000 \times 2v_1}{10 v_2} = 200\left(\frac{v_1}{v_2}\right) = \frac{200}{0.6} = 333.33 \text{ K}$$

$$\text{Thermal efficiency, } \eta = 1 - \frac{T_L}{T_H} = 1 - \frac{333.33}{1000} = 0.6667$$

$$\text{Energy absorbed as heat, } q_1 = W_{3-4} = RT_H \ln \frac{v_4}{v_3} = RT_H \ln \frac{v_1}{v_2}$$

$$= 8.314 \times 1000 \times \ln\left(\frac{1}{0.6}\right) = 4.247 \text{ kJ / mol}$$

$$\text{Net work done} \quad W = \eta\, q_1 = 0.6667 \times 4.247 = 2.8315 \text{ kJ / mol}$$

$$v_1 - v_2 = v_1 (1 - v_2 / v_2) = v_1 (1 - 0.6) = 0.4\, v_1 = \frac{0.4\, RT_L}{P_1}$$

$$= \frac{0.4 \times 8.314 \times 333.33}{2 \times 10^5} = 5.5426 \times 10^{-3} \text{ m}^3 \text{ / mol}$$

$$\text{Mean effective pressure, MEP} = \frac{W}{v_1 - v_2} = \frac{2.8315 \times 10^3}{5.5426 \times 10^{-3}} = 5.1086 \text{ bar}$$

**Example 8.69**

Consider the steady– state operation of a steady flow adiabatic compressor. The air (ideal gas) enters the compressor at $P_1$ and $T_1$ and leaves at $P_2$. Derive a relation in terms of $P_1$, $P_2$ and $T_1$ to calculate the work done in compressing one mol / s of air.

*Solution* :

Application of the first law of thermodynamics for steady–state, steady flow, adiabatic compressor gives

$$-\frac{\dot{W}_s}{\dot{m}} = h_2 - h_1$$

If $\dot{m} = 1$ mol /s, we get $-W_s = h_2 - h_1 = C_p(T_2 - T_1)$

For a reversible–adiabatic compressor, the second law of thermodynamics gives

$$s_2 = s_1 \text{ or } s_2 - s_1 = 0$$

For an ideal gas, the change in entropy is given by

$$s_2 - s_1 = C_P \ln \frac{T_2}{T_1} - R \ln \frac{P_2}{P_1} = 0$$

or $\dfrac{R\gamma}{\gamma-1}\ln\dfrac{T_2}{T_1} = R\ln\dfrac{P_2}{P_1}$ or $\dfrac{T_2}{T_1} = \left(\dfrac{P_2}{P_1}\right)^{\frac{\gamma-1}{\gamma}}$

$$\text{Therefore, } -\dot{W}_S = C_P(T_2 - T_1) = \frac{R\gamma}{\gamma-1}(T_2 - T_1) = \frac{RT_1\gamma}{\gamma-1}(T_2/T_1 - 1)$$

or $$-\dot{W}_S = \frac{RT_1\gamma}{\gamma-1}\left[\left(\frac{P_2}{P_1}\right)^{\frac{\gamma-1}{\gamma}} - 1\right]$$

### Example 8.70

Consider a two-stage compression with intercooling between pressure limits $P_1$ and $P_2$ as shown Fig.E 8.70. The air enters the first stage at $P_1$, $T_1$ and leaves at an intermediate pressure $P$. Then the air is cooled at constant pressure to the original temperature $T_1$. The air is then compressed in the second stage to $P_2$. For fixed values of $P_1$, $P_2$ and $T_1$, determine the optimum value of the intermediate pressure $P$, which will minimize the work input.

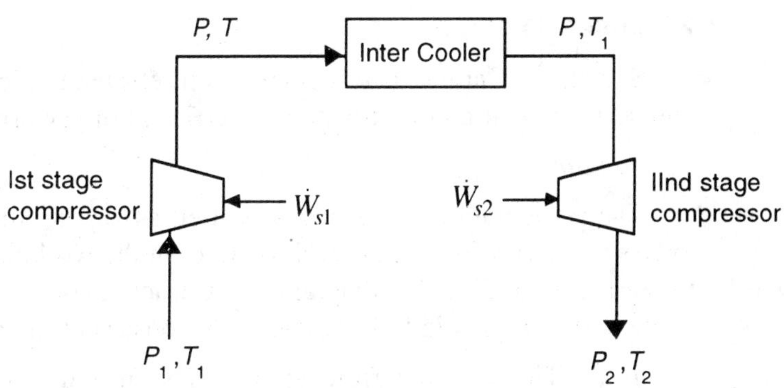


Fig.E 8.70. Schematic of a two-stage compressor with intercooling.

*Solution* :

The work done on the I stage compressor, $\dot{W}_{s1}$ is given by

$$-\dot{W}_{s1} = \frac{RT_1\gamma}{\gamma-1}\left[\left(\frac{P}{P_1}\right)^{\frac{\gamma-1}{\gamma}} - 1\right]$$

Similarly, the work done on the II stage compressor is given by

$$-W_{s2} = \frac{RT_1\gamma}{\gamma-1}\left[\left(\frac{P_2}{P}\right)^{\frac{\gamma-1}{\gamma}} - 1\right]$$

Then, the total work done ($W_s$) is given by

$$-W_s = \frac{RT_1\gamma}{\gamma-1}\left[\left(\frac{P}{P_1}\right)^{\frac{\gamma-1}{\gamma}} + \left(\frac{P_2}{P}\right)^{\frac{\gamma-1}{\gamma}} - 2\right]$$

The optimum pressure $P$ which minimizes the work done is given by

$$= -\frac{dW_s}{dP} = 0 = \frac{RT_1\gamma}{\gamma-1}\left[\left(\frac{\gamma-1}{\gamma}\right)\left(\frac{P}{P_1}\right)^{-1/\gamma}\frac{1}{P_1} + \left(\frac{\gamma-1}{\gamma}\right)\left(\frac{P_2}{P}\right)^{-1/\gamma}\left(\frac{-P_2}{P^2}\right)\right]$$

$$\text{or} \quad \frac{1}{P_1}\left(\frac{P}{P_1}\right)^{-1/\gamma} - \frac{P_2}{P^2}\left(\frac{P_2}{P}\right)^{-1/\gamma} = 0$$

$$\text{or} \quad P^{\frac{2(\gamma-1)}{\gamma}} = (P_1P_2)^{\frac{\gamma-1}{\gamma}} \quad \text{or} \quad P = \sqrt{P_1P_2}$$

**Example 8.71**

Sketch the Carnot vapor compression refrigeration cycle on a $T$–$s$ diagram and derive a relation to estimate its coefficient of performance.

*Solution* :

The refrigeration cycles are classified as vapor–compression refrigeration cycles and gas refrigeration cycles based on the working medium employed in the refrigerator. The $T$–$s$ diagram of a Carnot vapor–compression refrigeration cycle is shown the Fig.E 8.71. The cycle consists of the following processes.

**1–2** : The working fluid at state 1 (a mixture of liquid and vapor) is isentropically compressed to state 2 (saturated vapor).

**2–3** : Isothermal energy rejection as heat ($Q_H$) to the surroundings at $T_H$ in a condenser. At the end of the process, the working fluid is available at state 3 (saturated liquid).

**3–4** : Isentropic expansion of the working fluid to a low pressure. The temperature of the working fluid is $T_L$ at the end of the expansion process.

**4–1** : Isothermal energy transfer as heat ($Q_L$) from the low temperature body at $T_L$ to the working fluid.

The coefficient of performance (COP) of the Carnot refrigerator is given by

$$(\text{COP})_{\text{Carnot Ref.}} = \frac{Q_L}{W} = \frac{Q_L}{Q_H - Q_L} = \frac{T_L}{T_H - T_L}$$

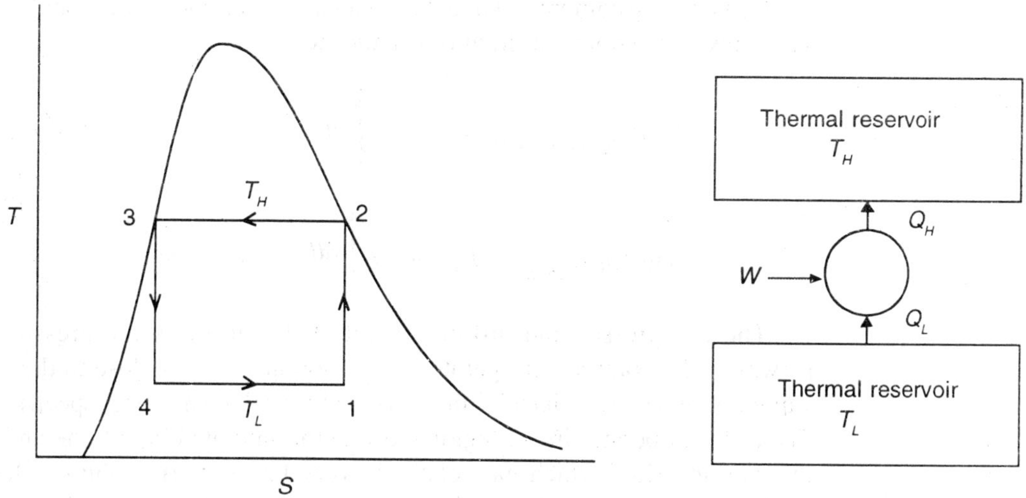


(*a*) (*b*)

Fig.E 8.71. (*a*) *T-s* diagram (*b*) schematic representation of Carnot vapor-compression refrigeration cycle.

**Example 8.72**

What is the significance of the Carnot refrigerator? What are the practical difficulties in devising a Carnot refrigerator?

*Solution* :

We know that no refrigerator can have a COP greater than that of a Carnot refrigerator, which consists of all reversible processes, when both of them are operating between the same two thermal reservoirs. Hence a Carnot refrigerator is considered as a standard of perfection against which the performance of practical refrigerators can be compared. The Carnot refrigeration cycle is not a practical cycle though the constant temperature energy absorption and rejection processes are desirable. The practical difficulty associated with a Carnot refrigerator is the isentropic expansion of saturated liquid to a low pressure. As the pressure reduces the liquid vaporizes and the expander or turbine has to handle a two phase mixture of relatively low quality.

**Example 8.73**

What modifications are made in the Carnot refrigerator in devising a technically feasible ideal vapor–compression refrigerator?

*Solution* :

Consider the compressor and the expander (or turbine) as the control volumes and apply the first law of thermodynamics to get

$$-(\dot{W}/\dot{m})_{\text{compressor}} = h_2 - h_1 = \int_1^2 v dP$$

$$-(\dot{W}/\dot{m})_{\text{Turbine}} = h_4 - h_3 = \int_3^4 v dP$$

The compressor and turbine operate between the same pressure limits. However, the compressor operates in the region which is close to the saturated vapor line and hence it handles the working fluid which has large specific volume. The turbine operates in the region close to the saturated liquid line and handles the working fluid which has relatively very low specific volume. Hence, the work delivered by the turbine is negligibly smaller than the work spent on the compressor. In view of the negligible work output and high cost of the turbine, it is replaced by an inexpensive throttle valve in a practical ideal vapor compression

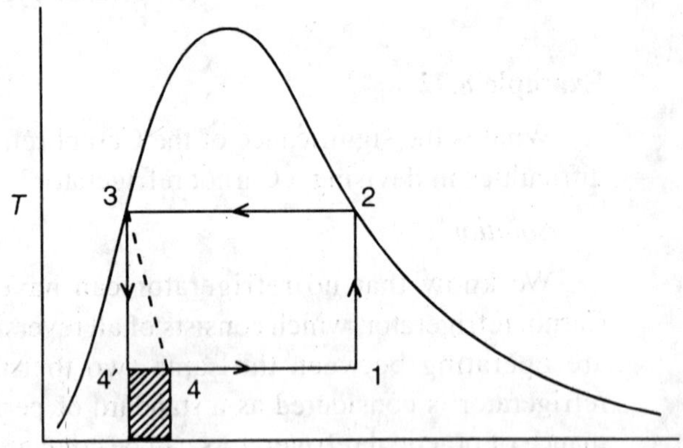


Fig.E 8.73. $T$–$s$ diagram of an ideal vapor compression refrigeration cycle (1–2–3–4–1) and Cannot refrigerator (1–2–3–4'–1).

refrigeration system. Since throttling is an irreversible process, the entropy of the exit fluid is larger than the entropy of the fluid entering the throttle valve. That is $h_4 = h_3$ and $s_4 > s_3$. The practical vapor–compression refrigeration cycle is shown as 1–2–3–4–1 in Fig.E 8.73 whereas the Carnot refrigeration cycle is represented as 1–2–3–4'. In a Carnot refrigerator, the energy extracted as heat $Q_L$ from the low temperature body is given by $T_L (s_1 - s_{4'})$ whereas in the practical vapor–compression refrigerator, the energy extracted as heat from the low temperature body is $T_L (s_1 - s_4)$. Since $s_4 > s_{4'}$, there is a loss in the refrigeration capacity. The loss in the refrigeration capacity is $T_L (s_4 - s_{4'})$ which is shown by the shaded area in Fig.E 8.73.

### Example 8.74

Present a schematic of the equipment for vapor compression refrigerator and derive a relation to estimate its COP.

*Solution :*

A schematic of the equipment for vapor–compression refrigerator is shown in Fig.E 8.74. The ambient atmosphere is usually used as the high temperature reservoir to which energy $Q_H$ is rejected by passing the working fluid through the condenser. The evaporator is the unit in which the working fluid absorbs energy $Q_L$ and vaporizes partially. If the flow rate of the refrigerant or working fluid through the device is $\dot{m}$, then

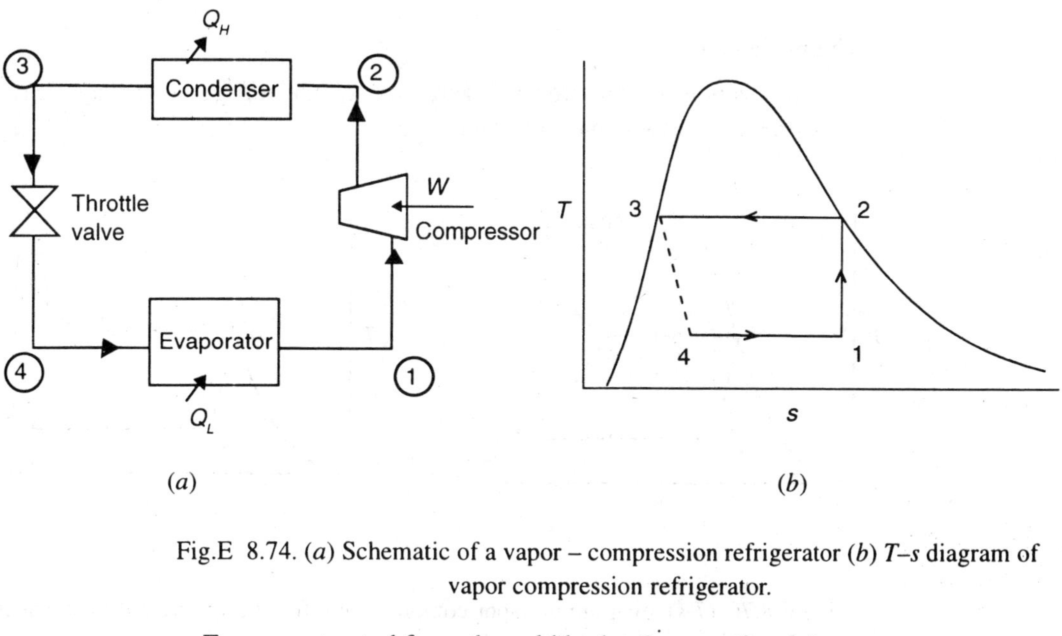


(*a*) (*b*)

Fig.E 8.74. (*a*) Schematic of a vapor – compression refrigerator (*b*) *T–s* diagram of vapor compression refrigerator.

Energy extracted from the cold body, $\dot{Q}_L = \dot{m}\,(h_1 - h_4)$

Work done on the compressor, $\dot{W} = \dot{m}\,(h_2 - h_1)$

$$\text{COP} = \frac{\dot{Q}_L}{\dot{W}} = \frac{\dot{m}\,(h_1 - h_4)}{\dot{m}\,(h_2 - h_1)} = \frac{h_1 - h_4}{h_2 - h_1}$$

Since 3–4 is a throttling process, we have $h_4 = h_3$.

**Example 8.75**

What is meant by a ton of refrigeration?

*Solution* :

The refrigeration capacity of a unit is commonly expressed in tons of refrigeration. If a refrigeration unit can convert one ton (2000 pound mass) of water at 0°C into ice at 0°C in one day, then its capacity is called one ton of refrigeration.

$$2000 \text{ pound mass} = \frac{2000}{2.205} = 907 \text{ kg}$$

Enthalpy of fusion of water at 0°C = 333.43 kJ/kg

$$\text{Therefore, one ton of refrigeration} = \frac{907 \times 333.43}{24 \times 60} = 210 \text{ kJ / min}$$

In practice energy removal rate of 211 kJ / min is called one ton of refrigeration.

**Example 8.76**

In a vapor–compression refrigerator does the refrigerant always leave the evaporator as a two phase mixture ?

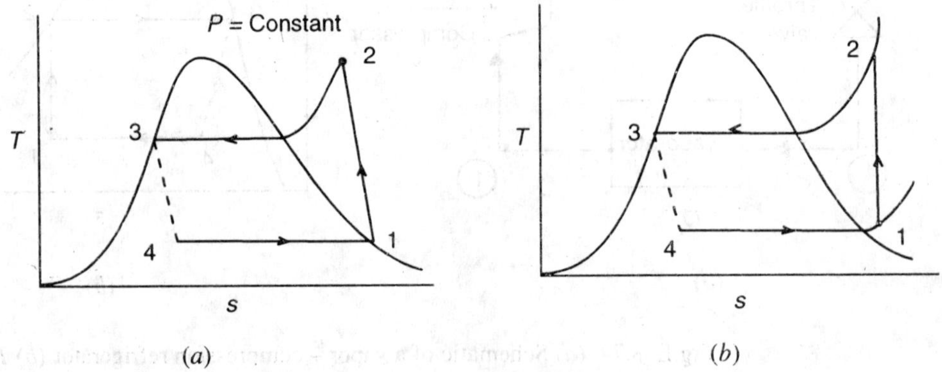


Fig.E 8.76. (*T-s*) diagram of vapor-compression refrigeration cycle if the refrigerant leaves the evaporator (*a*) as saturated vapor (*b*) as superheated vapor.

*Solution* :

In the ideal vapor–compression refrigeration cycle, we have shown that the refrigerant leaves the evaporator as a two–phase mixture such that upon compression it leaves the compressor as saturated vapor. This is purely an idealization. In an actual vapor–compression refrigerator, the refrigerant may leave the evaporator as saturated vapor or as superheated vapor, In such cases, the vapor–compression refrigeration cycles appear as shown in Fig.E 8.76.

### Example 8.77

What are the important factors which govern the choice of a refrigerant ?

*Solution* :

The temperatures of the refrigerant in the evaporator and condenser are determined by the temperatures of the cold body and surroundings, respectively. This in turn fixes the operating pressures in the evaporator and condenser. Therefore, the selection of a substance to be used as a refrigerant is based on the pressure–temperature relationship. The pressure at which the refrigerant leaves the evaporator should be higher than the atmospheric pressure so that no air leaks into the refrigerant. Then, it is also necessary that the boiling point of the refrigerant at the evaporator pressure has to be lower than the evaporator temperature. Moreover, the saturation temperature of the refrigerant at the condenser pressure has to be greater than the surrounding temperature at which energy can be rejected as heat to the surroundings. This necessitates the use of high pressure in the condenser and at this pressure the phase transition from vapor to liquid should be feasible. This in turn demands that the critical pressure of the refrigerant should be high. In addition to this the refrigerant has to be chemically stable as it undergoes several cyclic changes. It should also be non–toxic, non–flammable and should not be very expensive. The commonly used refrigerants are halogenated hydrocarbons like $CCl_2 F_2$ (dichloro difluoromethane which is known by the trade name Freon–12 or Genatron–12), and ammonia.

### Example 8.78

An ideal vapor–compression refrigerator with Freon–12 as the refrigerant is operating between the temperature levels –20°C and 40°C. If the flow rate of the refrigerant is 1.5 kg/min, calculate the COP of the refrigerator, the refrigeration capacity and the power consumption of the refrigerator.

*Solution* :

The *T–s* diagram of the ideal vapor compression refrigeration cycle is shown in Fig.E 8.78. The thermodynamic properties of Freon–12 are presented in Appendix tables 6–8.

At 40°C: $h_f = 75.1134$ kJ/kg $= h_3$ ;

$$h_g = 203.1063 \text{ kJ/kg} = h_2$$

$$s_g = 0.6825 \text{ kJ/kg K} = s_2$$

At $-20°C$: $h_f = 17.9517$ kJ/kg ; $h_g = 178.9017$ kJ/kg

$s_f = 0.0736$ kJ/kg K ; $s_g = 0.7094$ kJ/kg K

3–4 is a throttling process. Therefore, $h_4 = h_3 = 75.1134$ kJ/kg

1–2 is an isentropic process. Therefore, $s_2 = s_1 = X_1 s_g + (1 - X_1) s_f$

or $0.6825 = 0.7094\, X_1 + (1 - X_1)\, 0.0736$ or $X_1 = 0.9577$

$$h_1 = X_1 h_g + (1 - X_1) h_f$$
$$= 0.9577 \times 178.9017 + (1 - 0.9577)\, 17.9517$$
$$= 172.09 \text{ kJ/kg}$$

$$\text{COP} = \frac{\dot{Q}_L}{\dot{W}} = \frac{\dot{m}\,(h_1 - h_4)}{\dot{m}\,(h_2 - h_1)} = \frac{h_1 - h_4}{h_2 - h_1} = \frac{172.09 - 75.1134}{203.1063 - 172.09} = 3.1266$$

Refrigeration capacity $= \dot{Q}_L = \dot{m}\,(h_1 - h_4)$

$$= 1.5\,(172.09 - 75.1134) = 145.46 \text{ kJ / min}$$

or $\dfrac{145.46}{211} = 0.6891$ Tons of refrigeration

$$\text{Power Consumption,} = \frac{\dot{W}_L}{\text{COP}} = \frac{145.46 \times 10^3}{60 \times 3.1266} = 0.7754 \text{ kW}$$

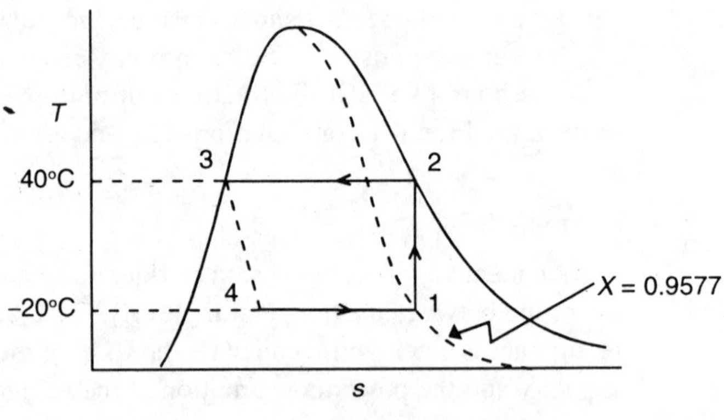


Fig.E 8.78. $T$–$s$ diagram of ideal Vapor Compresssion refrigerator.

**Example 8.79**

An ideal vapor–compression refrigerator with Freon –12 as the refrigerant is operating with a condenser pressure of 10 bar and the evaporator pressure is 1.5 bar. The condenser is rejecting 100 kJ / min of energy as heat while the evaporator is absorbing 70 kJ / min of energy as heat. Calculate the flow rate of refrigerant, the power consumption and the COP of the refrigerator. Also estimate the COP of Carnot refrigerator operating between the same temperature levels.

*Solution* :

At 10 bar: $t = 41{,}7°C$; $h_f = 76.7967$ kJ/kg $= h_3$

$h_g = 203.6486$ kJ/kg $= h_2$ ; $s_g = 0.6820$ kJ/kg K $= s_2$

At 1.5 bar: $t = -20.1°C$; $h_f = 17.8234$ kJ/kg; $h_g = 178.8372$ kJ/kg

$s_f = 0.0731$ kJ/kg K; $s_g = 0.7095$ kJ/kg K

We know that $h_4 = h_3 = 76.7967$ kJ/kg

$$s_2 = 0.6820 = s_1 = X_1 s_g + (1 - X_1) s_f$$

or $0.6820 = 0.7095\, X_1 + (1 - X_1)\, 0.0731$

or $X_1 = 0.9568$

$h_1 = 0.9568 \times 178.8372 + (1 - 0.9568)\, 17.8234 = 171.88$ kJ/kg

$$\text{Refrigerant flow rate, } \dot{m} = \frac{\dot{Q}_L}{h_1 - h_4} = \frac{70}{171.88 - 76.7967} = 0.736 \text{ kg / min}$$

$$\text{Power consumption, } \dot{W} = \dot{m}\,(h_2 - h_1) = \frac{0.736}{60}\,(203.6486 - 171.88)\,10^3$$

$$= 0.3897 \text{ kW}$$

$$\text{COP} = \frac{h_1 - h_4}{h_2 - h_1} = \frac{171.88 - 76.7967}{203.6486 - 171.88} = 2.993$$

$$(\text{COP})_{\text{Carnot}} = \frac{T_L}{T_H - T_L} = \frac{273 + (-20.1)}{(273 + 41.7) - (273 - 20.1)} = 4.0922$$

**Example 8.80**

It is proposed to use a reversed Carnot engine as a refrigerator and heat pump. The unit consumes 10 kW power. If the COP is 3.5 as a refrigerator determine the refrigeration capacity, COP if it is used as a heat pump and the heating capacity of the unit.

*Solution* :

A schematic of the reversed Carnot cycle is shown in Fig.E 8.80.

$$(\text{COP})_{\text{Ref}} = \frac{\dot{Q}_L}{\dot{W}} = \frac{T_L}{T_H - T_L}$$

or $3.5 = \frac{\dot{Q}_L}{\dot{W}}$ or $\dot{Q}_L = 3.5\,\dot{W} = 3.5 \times 10 \times 10^3 = 35$ kW

or $$\dot{Q}_L = \frac{35}{(211/60)} = 9.9526 \text{ Tons of refrigeration}$$

$$(\text{COP})_{\text{Heat Pump}} = \frac{\dot{Q}_H}{\dot{W}} = \frac{T_H}{T_H - T_L} = \frac{T_H / T_L}{\frac{T_H}{T_L} - 1}$$

$$(\text{COP})_{\text{Ref}} = \frac{T_L}{T_H - T_L} = \frac{1}{\frac{T_H}{T_L} - 1} = 3.5$$

or $$T_H / T_L = 1.2857$$

Therefore, $(\text{COP})_{\text{Heat Pump}} = \frac{1.2857}{1.2857 - 1} = 4.5002$

Heating capacity, $\dot{Q}_H = (\text{COP})_{\text{Heat Pump}} \times \dot{W}$

$= 4.5002 \times 10 \times 10^3 = 45.002$ kW

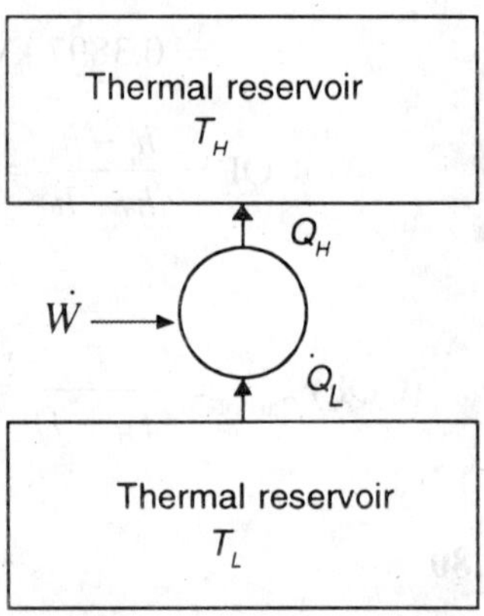


Fig.E 8.80. Schematic of the reversed Carnot cycle.

**Example 8.81**

Two identical bodies of mass $m$ having specific heat $C$ are available at temperature $T_1$. It is desired to operate a refrigerator between the two bodies to cool one of them to $T_2$. Determine the minimum amount of work to be done on the device and the final temperature of the second body.

*Solution* :

Minimum amount of work is required only when a reversible refrigerator is employed. Then

$$(\Delta S)_{universe} = (\Delta S)_1 + (\Delta S)_2 = 0$$

where $(\Delta S)_1$ and $(\Delta S)_2$ denote the change in the entropy of bodies 1 and 2, respectively. Let $T_f$ denote the final temperature of the second body. Then

$$mC \ln \frac{T_2}{T_1} + mC \ln \frac{T_f}{T_1} = 0$$

or $T_2 T_f = T_1^2$ or $T_f = T_1^2/T_2$

Work done, $W = Q_H - Q_L = mC\,(T_f - T_1) - mC\,(T_1 - T_2)$

$$= mC\,(T_f - 2T_l + T_2)$$

$$= mC\left(\frac{T_1^2}{T_2} - 2T_1 + T_2\right) = \frac{mC}{T_2}(T_1^2 - 2T_1T_2 + T_2^2)$$

$$= \frac{mC}{T_2}(T_1 - T_2)^2$$

**Example 8.82**

Describe an absorption–refrigeration system. How does it differ from vapor–compression refrigeration system ?

*Solution* :

In the vapor – compression refrigeration system, the energy needed for compressing the refrigerant from evaporator pressure to condenser pressure is supplied in the form of work. This work is converted into enthalpy and then it is rejected along with the energy absorbed from the evaporator, as heat to the surroundings in a condenser. There will be some saving in the quality of energy if the energy needed to operate the refrigerator is supplied in the form of heat rather than in the form of work. In the absorption refrigeration system, refrigeration is produced mainly by using energy in the form of heat.

In the absorption refrigeration system, the refrigerant (say ammonia) is usually dissolved in water. A schematic diagram of ammonia absorption refrigeration system is shown in Fig.E 8.82. A strong liquor of ammonia is boiled in a vapor generator producing high pressure $NH_3$ vapor, which passes through a condenser and liquefies. This liquid ammonia is then throttled to a low pressure producing low temperature and this cold ammonia passes through the evaporator absorbing energy as heat from the body to be refrigerated and leaves as low pressure ammonia vapor. This low pressure ammonia vapor is absorbed in the weak ammonia liquor, which comes from the vapor generator, in an absorber producing strong ammonia liquor. Then this strong ammonia liquor is pumped to the vapor generator.

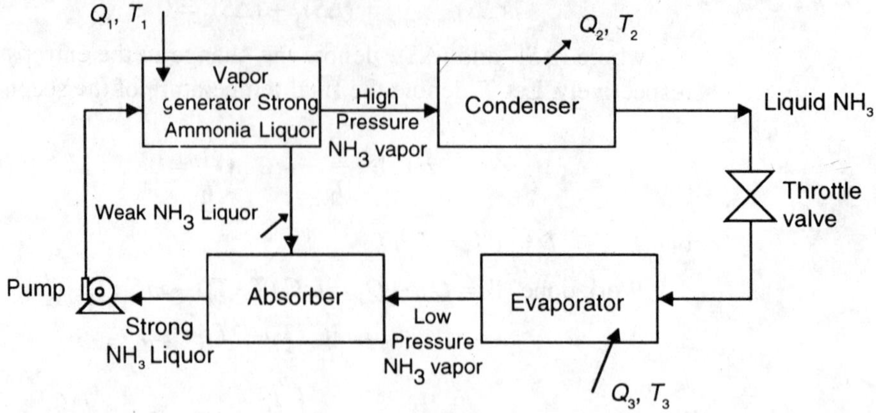


Fig.E 8.82 Schematic diagram of ammonia absorption refrigeration system.

## Example 8.83

Derive an expression to estimate the approximate COP of an absorption–refrigeration system.

*Solution* :

A detailed analysis of absorption–refrigeration system requires a knowledge of the thermodynamic properties of a binary mixture as a function of its composition. However, the approximate COP of an absorption–refrigeration system can be evaluated by considering the unit as a combination of a heat engine and heat pump as shown in Fig.E 8.83.

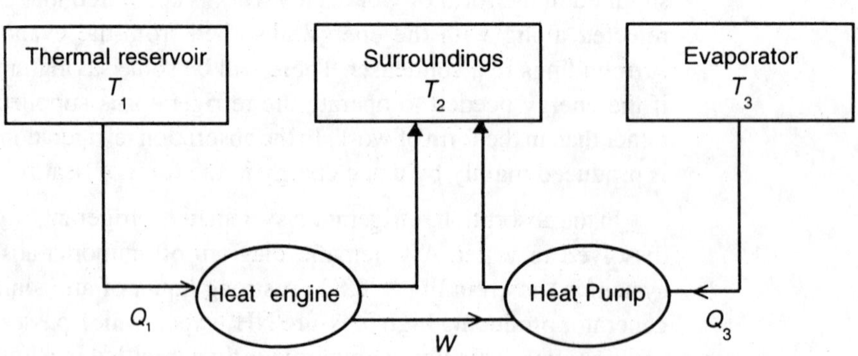


Fig.E 8.83. Schematic of an absorption refrigeration system viewed as a combination of a heat engine and heat pump.

Let $Q_1$ = energy supplied as heat in the vapor generator at $T_1$.

$W$ = Work done by the heat engine, which is utilized in operating the heat pump.

$T_2$ = Surrounding temperature.

$Q_3$ = energy absorbed as heat from the evaporator.

$T_3$ = evaporator temperature.

If the energy $Q_1$ which is supplied to the vapor generator at $T_1$, is supplied to a reversible heat engine which uses the surroundings at $T_2$ would deliver the work $W$ given by

$$W = \frac{Q_1(T_1 - T_2)}{T_1} \tag{A}$$

Let this work be made use of to operate the heat pump which works between the evaporator at $T_3$ and surroundings at $T_2$. Then for the heat pump, we have

$$\frac{Q_3}{W} = \frac{T_3}{T_2 - T_3}$$

or

$$W = \frac{Q_3(T_2 - T_3)}{T_3} \tag{B}$$

From Eqns. (*A*) and (*B*), we have

$$\frac{Q_1(T_1 - T_2)}{T_1} = \frac{Q_3(T_2 - T_3)}{T_3}$$

or

$$\frac{Q_3}{Q_1} = \frac{T_3(T_1 - T_2)}{T_1(T_2 - T_3)}$$

Ignoring the small amount of work spent on the pump, the COP of absorption – refrigeration system is given by

$$\text{COP} = \frac{Q_3}{Q_1} = \frac{T_3(T_1 - T_2)}{T_1(T_2 - T_3)}$$

**Example 8.84**

Describe a simple gas refrigeration cycle and sketch it on a *T–s* diagram.

*Solution* :

If a gas undergoes adiabatic expansion, its temperature decreases. Therefore, it is possible to produce cooling by expanding a gas from high pressure to low pressure. The simplest gas refrigeration cycle is based on this principle.

The Brayton power cycle can be operated in the reverse to yield the reverse Brayton cycle for refrigeration as shown in Fig.E 8.84. The atmospheric air at state 1 is compressed isentropically to state 2 in a compressor. Then the hot air leaving the compressor passes through a heat exchanger where it rejects energy as heat to the surrondings at constant pressure and leaves at state 3. Then the air at state 3 is allowed to expand isentropically to state 4 in a turbine. The low temperature air at state 4 then passes through the cooler (or evaporator) where it absorbs energy as heat, at constant pressure, from the body to be maintained at low temperature and leaves at state 1, thus completing a cycle.

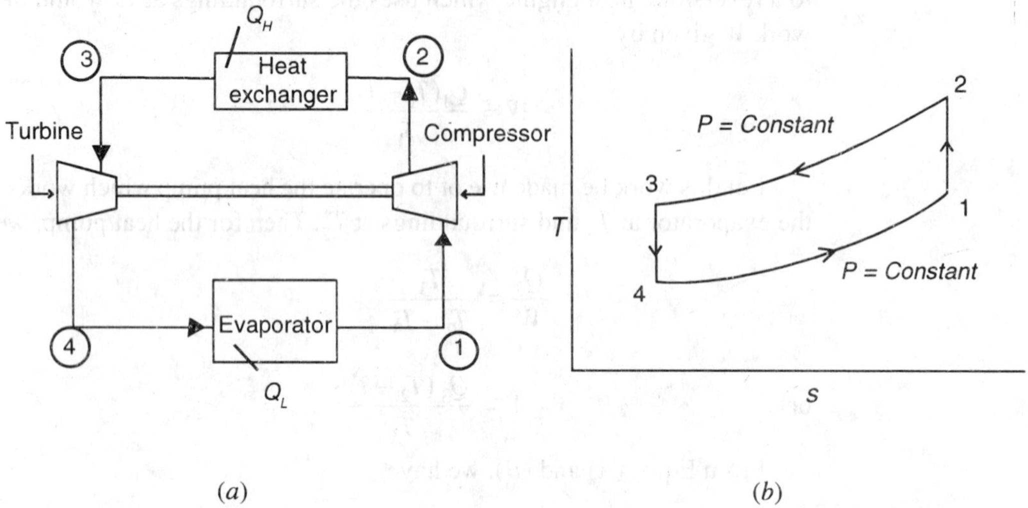


Fig.E 8.84. (*a*) Schematic of a gas refrigeration cycle (*b*) *T-s* diagram of a gas refrigeration cycle.

Work done on the compressor = $C_P(T_2 - T_1)$

Work delivered by the turbine = $C_P(T_3 - T_4)$

Net work done on the unit = $W = C_P(T_2 - T_1) - C_P(T_3 - T_4)$

The net work is supplied to the compressor from an external source.

Energy absorbed at low temperature, $Q_L = C_P\,(T_1 - T_4)$

Then the coefficient of performance COP is given by

$$\text{COP} = \frac{Q_L}{W} = \frac{T_1 - T_4}{(T_2 - T_1) - (T_3 - T_4)}$$

# 9

# Nonreacting Gas Mixtures and Psychrometry

**Example 9.1**

How does one specify the state of a nonreacting gas mixture and what are the ways of specifying its composition?

*Solution* :

The state of a nonreacting gas mixture is usually specified by two independent intensive parameters like pressure and temperature in addition to the composition of the mixture. The composition of a nonreacting gas mixture can be specified by giving the mass $m_i$ or the moles $n_i$ of each constituent gas in the mixture. Alternatively, one can describe the composition in terms of mass fractions $\phi_i$ or mole fractions $y_i$. The mass fraction $\phi_i$ of a component $i$ in the mixture is given by

$$\text{Mass fraction of component } i,\ \phi_i = \frac{m_i}{\Sigma m_i} = \frac{m_i}{m}$$

where $m = \Sigma\, m_i$ = the total mass of the mixture.

$$\text{The mole fraction } y_i \text{ of component } i = \frac{n_i}{\Sigma n_i} = \frac{n_i}{n}$$

where $n = \Sigma\, n_i$ = total moles of the mixture.

The number of moles $n_i$ of a component $i$ is given by

$$n_i = \frac{m_i}{M_i}$$

where $M_i$ = molar mass of component $i$

**Example 9.2**

What is meant by molar mass of a mixture? Air is a mixture of $O_2$, $N_2$, A, $CO_2$ and other gases. The composition of air on mole basis is $y_{O_2} = 0.2095$, $y_{N_2} = 0.7808$; $y_A = 0.0093$; $y_{CO_2} = 0.0003$ and the mole fraction of all other gases is 0.0001. Neglecting the presence of other gases calculate the molar mass of air.

*Solution* :

The molar mass of a mixture is defined as the mass of one mole of a mixture. The molar mass $M$ of a mixture is given by

$$M = \frac{\text{Mass of mixture}}{\text{Moles of mixture}} = \frac{\Sigma m_i}{\Sigma n_i} = \frac{\Sigma M_i n_i}{\Sigma n_i} = \frac{\Sigma M_i n_i}{n} = \Sigma y_i M_i$$

Ignoring the presence of other gases (which account for a mole fraction of 0.0001), the molar mass $M$ of air is given by

$M = [0.7808\ (28.02) + 0.2095(32.00) + 0.0093\ (39.94) + 0.0003\ (44.01)] \times 10^{-3} = 28.97 \times 10^{-3}$ kg / mol

**Example 9.3**

The combustion products of a certain hydrocarbon fuel on analysis is found to have the molar composition $CO_2 = 8.32\%$; $H_2O = 9.98\%$; $N_2 = 75.05\%$ and $O_2 = 6.65\%$. Determine the molar mass of the gas mixture and the composition in terms of mass fraction.

*Solution* :

The molar mass ($M$) of the mixture is given by

$M = \Sigma y_i M_i = [0.0832(44.01) + 0.0998\ (18) + 0.7505\ (28.02) + 0.0665(32)] \times 10^{-3} = 28.62 \times 10^{-3}$ kg / mol

Consider 100 mol of mixture. Then the mixture contains

| **Component** | **Moles ($n_i$)** | **Molar mass $M_i \times 10^3$ kg / mol** | **Mass $m_i = n_i M_i$ kg × $10^3$** | **Mass fraction $\phi_i = m / \Sigma m_i$** |
|---|---|---|---|---|
| $CO_2$ | 8.32 | 44.01 | 366.16 | 0.1279 |
| $H_2O$ | 9.98 | 18 | 179.64 | 0.0628 |
| $N_2$ | 75.05 | 28.02 | 2102.9 | 0.7348 |
| $O_2$ | 6.65 | 32 | 212.9 | 0.0735 |

$$m = \Sigma\, m_i = 2862 \times 10^{-3} \text{ kg}$$

$$\phi_{CO_2} = \frac{366.16 \times 10^{-3}}{2862 \times 10^{-3}} = 0.1279$$

Similarly for other components also $\phi_i$ is calculated as shown in the above table.

**Example 9.4**

A mixture of gases has the following analysis on mass basis.

| Component | $CO_2$ | $H_2O$ | $O_2$ | $N_2$ | $H_2$ |
|---|---|---|---|---|---|
| Mass % | 20 | 10 | 15 | 35 | 20 |

Calculate the composition in terms of mole fractions and the molar mass of the mixture.

*Solution* :

Consider 100 kg of mixture as the basis for calculation. Then the mixture contains

| Component | Mass ($m_i$) (kg) | Molar mass $M_i \times 10^3$ | Mole $n_i = m_i / M_i$ (kmol) | Mole fraction $y_i = n_i / \Sigma\, n_i$ |
|---|---|---|---|---|
| $CO_2$ | 20 | 44 | 0.4546 | 0.0357 |
| $H_2O$ | 10 | 18 | 0.5556 | 0.0436 |
| $O_2$ | 15 | 32 | 0.4688 | 0.0368 |
| $N_2$ | 35 | 28 | 1.2500 | 0.0983 |
| $H_2$ | 20 | 2 | 10.0000 | 0.7856 |

$$\Sigma\, n_i = n = 12.729 \text{ kmol}$$

$$\text{Molar mass of mixture, } M = \frac{\Sigma m_i}{\Sigma n_i} = \frac{100}{12.729 \times 10^{-3}} = 7.856 \times 10^{-3} \text{ kg/mol}$$

**Example 9.5**

Define partial pressure and partial volume of a gas.

*Solution* :

The partial pressure of a component $i$ in a gas mixture is defined as the pressure exerted by the component gas $i$ if it were alone occupying the entire volume of the mixture at the mixture temperature.

The partial volume of component $i$ in a gas mixture is defined as the volume occupied by the component gas $i$ if it were held at the mixture pressure and temperature.

**Example 9.6**

State the Dalton's law of partial pressures.

*Solution* :

The Dalton's law of partial pressures states that in a mixture of ideal gases, the pressure exerted by the gas mixture is equal to the sum of the partial pressures of the constituent gases. That is

$P = \Sigma p_i$ where $p_i$ = partial pressure of component $i$

$P$ = pressure of the gas mixture.

If the gases are ideal, then we have

$$p_i = \frac{n_i RT}{V}$$

$$\text{Then } \Sigma p_i = \frac{\Sigma n_i RT}{V} = \frac{RT\Sigma n_i}{V} = \frac{nRT}{V} = P$$

**Example 9.7**

State the Amagat's law of additive volumes.

*Solution* :

The Amagat's law of additive volumes states that the volume occupied by a mixture of ideal gases is equal to the sum of the partial volumes.

That is $V = \Sigma V_i$ where $V_i$ = partial volume of component $i$

$V$ = volume of the ideal gas mixture.

$$\text{We know that } V_i = \frac{n_i RT}{P}$$

$$\text{Therefore, } \Sigma V_i = \frac{\Sigma n_i RT}{P} = \frac{RT\Sigma n_i}{P} = \frac{nRT}{P} = V$$

**Example 9.8**

Define pressure fraction and volume fraction and find a relation between mole fraction, volume fraction and pressure fraction.

*Solution* :

The pressure fraction of component $i$ = $\dfrac{\text{Partial Pressure of component } i}{\text{Total Pressure}}$

$$= \frac{p_i}{P} = \frac{(n_i RT / V)}{(nRT / V)} = \frac{n_i}{n} = y_i$$

Volume fraction of component $i$ = $\dfrac{\text{Partial volume of component } i}{\text{Total Volume}}$

$$= \frac{V_i}{V} = \frac{(n_i RT / P)}{(nRT / P)} = \frac{n_i}{n} = y_i$$

For an ideal gas mixture, we find that the pressure fraction is equal to the mole fraction and the volume fraction is also equal to the mole fraction.

**Example 9.9**

How does one evaluates the internal energy, enthalpy and entropy of a mixture of ideal gases from a knowledge of the same properties of the constituent gases ?

*Solution*:

The internal energy, enthalpy and entropy are extensive properties. That is they depend on the quantity of the gas under consideration and they are additive. Then the internal energy $U$ of $n$ mol of a mixture is given by

$$U = n_1 u_1 + n_2 u_2 + \ldots = \Sigma n_i u_i$$

or $nu = \Sigma n_i u_i$ or $u = \dfrac{\Sigma n_i u_i}{n} = \Sigma y_i u_i$

where $u_i$ = molar internal energy of component $i$ and

$u$ = molar internal energy of mixture

(Similarly one can express the specific internal energy $u$ in terms of mass fraction $\phi_i$ also

$$u = \frac{U}{m} = \frac{\Sigma m_i u_i}{m} = \Sigma \phi_i u_i$$

Where $u_i$, now is the specific internal energy of component $i$

The enthalpy $H$ of $n$ mol of a mixture is given by

$$H = nh = \Sigma\, n_i\, h_i \text{ or } h = \frac{\Sigma n_i\, h_i}{n} = \Sigma\, y_i\, h_i$$

(Similarly, the specific enthalpy is given by $h = \Sigma\, \phi_i\, h_i$)

where $h$ = molar (specific) enthalpy of the mixture

$h_i$ = molar (specific) enthalpy of component $i$

The entropy $S$ of the mixture can also be found by adding the contribution of each constituent gas existing at the mixture temperature and at its partial pressure. That is

$$S = ns = \Sigma\, n_i\, s_i \text{ or } s = \frac{\Sigma n_i s_i}{n} = \Sigma y_i\, s_i$$

(Similarly, the specific entropy $s = \Sigma\, \phi_i\, s_i$.)

where $s$ = molar (specific) entropy of the mixture

$s_i$ = molar (specific) entropy of the constituent $i$ evaluated at the partial pressure of component of $i$ ($p_i$) and the mixture temperature.

**Example 9.10**

How does one evaluates the specific heats $C_P$ and $C_v$ of a mixture of ideal gases in terms of $C_P$ and $C_v$ of constituent gases ?

*Solution* :

We know that $u = \Sigma\, y_i\, u_i$ or $u = \Sigma\, \phi_i\, u_i$ and $h = \Sigma\, y_i\, h_i$ or $h = \Sigma\, \phi_i\, h_i$

Differentiating the above relations with respect to temperature, one gets

$$C_v = \Sigma\, y_i\, C_{vi} \text{ or } C_v = \Sigma \phi_i\, C_{vi}$$

$$C_P = \Sigma\, y_i\, C_{pi} \text{ or } C_P = \Sigma \phi_i\, C_{pi}$$

where $C_v$ = molar (specific) heat capacity at constant volume of the mixture

$C_{vi}$ = molar (specific) heat capacity at constant volume of the constituent $i$

$C_P$ = molar(specific) heat capacity at constant pressure of the mixture

$C_{Pi}$ = molar (specific) heat capacity at constant pressure of the constituent $i$

**Example 9.11**

A gaseous fuel with a volumetric analysis of 40% ethane, 40% butane and 20% pentane is flowing through a pipe line at the rate of 100 $m^3$/s at 1 bar and 300 K. Calculate the partial pressure of each component and the mass flow rate.

*Solution* :

We know that

$$\text{volume fraction}\left(\frac{V_i}{V}\right) = \text{pressure fraction}\left(\frac{p_i}{P}\right) = \text{mole fraction } y_i$$

The volumetric analysis gives the volume fraction in percentages. Hence

$$y_{\text{ethane}} = 0.4 = \left(\frac{p_i}{P}\right)$$

or $p_{\text{ethane}} = 0.4\,P = 0.4 \times 1 = 0.4$ bar

Similarly, we get $p_{\text{butane}} = 0.4 \times 1 = 0.4$ bar

$$p_{\text{pentane}} = 0.2 \times 1 = 0.2 \text{ bar}$$

Molar mass of mixture $M = \Sigma y_i M_i$

or $M = (0.4 \times 30 + 0.4 \times 58 + 0.2 \times 72)10^{-3} = 49.6 \times 10^{-3}$ kg/mol

(Molar masses of ethane, butane and pentane are $30 \times 10^{-3}$ kg/mol, $58 \times 10^{-3}$ kg/mol and $72 \times 10^{-3}$ kg/mol, respectively)

Volumetric flow rate = 100 m$^3$/s = $V$

$$\text{Therefore } \dot{n} = \frac{PV}{RT} = \frac{1\times10^5\times100}{8.314\times300} = 4.0094 \times 10^3 \text{ mol/s}$$

Mass flow rate, $\dot{m} = M\dot{n} = 49.6 \times 10^{-3} \times 4.0094 \times 10^3 = 198.87$ kg / s

**Example 9.12**

A cylinder of volume 0.25 m$^3$ initially contains a mixture of 20% propane ($C_3H_8$) and 80% butane ($C_4H_{10}$) on molar basis, at 1 bar and 300 K. The cylinder is connected to a line carrying nitrogen at 300 K and the cylinder is charged isothermally to a pressure of 1.5 bar. Determine the mass of nitrogen that entered the cylinder.

*Solution* :

Initially the cylinder contains $n_1$ moles of gas. Then

$$n_1 = \frac{P_1V}{RT} = \frac{1\times10^5\times0.25}{8.314\times300} = 10.023 \text{ mol}$$

At the end, the cylinder contains $n_2$ moles of gas. where

$$n_2 = \frac{P_2V}{RT} = \frac{1.5\times10^5\times0.25}{8.314\times300} = 15.035 \text{ mol}$$

Moles of nitrogen that entered the tank = $n_2 - n_1$ = 5.012 mol = 0.14 kg

**Example 9.13**

A mixture of gases having molar composition 25% $CO_2$; 20% $N_2$, 20% $O_2$; 20% $H_2O$ and 15% $H_2$ is at 300 K and 2 bar. This mixture is placed in contact with a silica gel, which absorbs the water vapor, for sufficient time and the final temperature of the mixture is again 300 K. Determine the final pressure and composition of the gas mixture.

*Solution* :

The partial pressures of the components in the initial mixture are

| Component | Mole fraction $y_i$ | $p_i = y_i\ P$ (bar) |
|---|---|---|
| $CO_2$ | 0.25 | 0.50 |
| $N_2$ | 0.20 | 0.40 |
| $O_2$ | 0.20 | 0.40 |
| $H_2O$ | 0.20 | 0.40 |
| $H_2$ | 0.15 | 0.30 |

When the gas is in contact with silica gel, it absorbs all the $H_2O$. Hence, in the final gas mixture, we have the following partial pressures in bars

$p_{CO_2} = 0.50;\ p_{N_2} = 0.40;\ p_{O_2} = 0.40;\ p_{H_2} = 0.30$

Therefore, the total pressure is given by

$P = \Sigma\, p_i = 0.50 + 0.40 + 0.40 + 0.30 = 1.6$ bar

The composition expressed in mole fractions is given by

$$y_i = p_i/P.$$

Therefore, $$y_{CO_2} = \frac{0.50}{1.60} = 0.3125\ ;\ y_{N_2} = \frac{0.40}{1.60} = 0.25\ ;$$

$$y_{O_2} = 0.25;\ y_{H_2} = 0.1875$$

**Example 9.14**

A mixture of gases having the molar composition 30% helium, 30% nitrogen and 40% carbon dioxide is used as a coolant to cool 1000 kg/hr of a hydrocarbon oil from 150°C to 50°C. If the coolant gas mixture enters the heat exchanger at 30°C and leaves at 40°C, calculate the mass flow rate of the gas mixture through the heat exchanger. The molar heat capacities, of helium, nitrogen and carbon dioxide, at constant pressure are (5/2) $R$ J/mol K; (7/2) $R$ J/mol K and (9/2) $R$ J/mol K, respectively, where $R = 8.314$. The specific heat of hydrocarbon oil is 1.2 kJ/kg K.

*Solution* :

Let us consider 1 mol of the gas mixture as the basis for calculation. Then, the mixture contains

| **Components** | **Moles** $n_i$ | **Molar mass** $M_i$ **kg/mol** | **Mass** $m_i = n_i M_i$ **(kg)** |
|---|---|---|---|
| Helium | 0.3 | $4 \times 10^{-3}$ | $1.2 \times 10^{-3}$ |
| Nitrogen | 0.3 | $28 \times 10^{-3}$ | $8.4 \times 10^{-3}$ |
| Carbon dioxide | 0.4 | $44 \times 10^{-3}$ | $17.6 \times 10^{-3}$ |

The molar mass of the mixture is given by $\Sigma\, m_i = 27.2 \times 10^{-3}$ kg/mol

The molar heat capacity at constant pressure of the mixture is given by

$$C_P = \Sigma\, Y_i\, C_{Pi} = 0.3\left(\frac{5}{2}\right)R + 0.3\left(\frac{7}{2}\right)R + 0.4\left(\frac{9}{2}\right)R = 3.6\, R$$

$$= 3.6 \times 8.314 = 29.93 \text{ kJ/mol K}$$

The principle of conservation of energy gives

$$1000 \times 1.2 \times (150 - 50) \times 10^3 = \dot{n} \times 29.93 \times (40 - 30)$$

or $\dot{n} = 4.0094 \times 10^5$ mol/hr

Mass flow rate of gas mixture, $\dot{m} = \dot{n}\, M = 4.0094 \times 10^5 \times 27.2 \times 10^{-3}$

$$= 10\,905.57 \text{ kg / hr}$$

**Example 9.15**

Calculate the ratio of heat capacities for a gas mixture having the molar composition 75% helium and 25% nitrogen. The ratio of heat capacities ($\gamma$) of helium and nitrogen are 5/3 and 7/5, respectively.

*Solution* :

We know that $C_v = \dfrac{R}{\gamma - 1}$ and $C_P = \dfrac{R\gamma}{\gamma - 1}$

$$\text{Therefore, } C_V \text{ (He)} = \frac{R}{5/3-1} = \frac{3}{2}R \text{ ; } C_P \text{ (He)} = \frac{5}{2}R$$

$$C_V \text{ (N}_2\text{)} = \frac{R}{7/5-1} = \frac{5}{2}R; \; C_P \text{ (N}_2\text{)} = \frac{7}{2}R$$

The molar heat capacities of the mixture are given by

$$C_v = \Sigma y_i \; C_{vi} = 0.75 \times \frac{3}{2}R + 0.25 \times \frac{5}{2}R = 1.75\,R$$

$$C_P = \Sigma y_i \; C_{Pi} = 0.75 \times \frac{5}{2}R + 0.25 \times \frac{7}{2}R = 2.75\text{R}$$

$$\text{Therefore, } \gamma \text{ (Mixture)} = C_P / C_v = \frac{2.75R}{1.75R} = 1.5714$$

**Example 9.16**

It is known that sonic velocity in a gas medium is given by $\sqrt{\gamma RT/M}$. A mixture of methane ($CH_4$) and ethane ($C_2H_6$) flowing through a pipe line is found to have a sonic velocity of 400 m / s at 300 K. The molar heat capacities of methane and ethane are given by

$$C_P \text{ (CH}_4\text{)} = 36.15 \text{ J / mol K; } C_v \text{ (CH}_4\text{)} = 27.84 \text{ J / mol K;}$$
$$C_P \text{ (C}_2\text{H}_6\text{)} = 53.11 \text{ J / mol K; } C_v \text{ (C}_2\text{H}_6\text{)} = 44.80 \text{ J/mol K}$$

Determine the molar composition of the gas.

*Solution* :

We know that

$$\text{V} = \sqrt{\frac{\gamma RT}{M}}$$

$$\text{or} \quad 400 = \sqrt{\frac{\gamma \times 8.314 \times 300}{M}}$$

$$\text{or} \quad \frac{\gamma}{M} = 64.149 \qquad (A)$$

where $M$ = Molar mass of the gas mixture.

$$M = \Sigma y_i \; M_i = [16y + 30\,(1-y)\,] \times 10^{-3}$$
$$= (30 - 14y) \times 10^{-3} \text{ kg/mol} \qquad (B)$$

where $y$ = mole fraction of $CH_4$ in the mixture.

$C_P$ (mixture) $= \Sigma\, y_i\, C_{Pi} = 36.15y + 53.11(1 - y)$

$C_v$ (mixture) $= \Sigma\, y_i\, C_{vi} = 27.84y + 44.8(1 - y)$

$$\text{Therefore, } \gamma = \frac{36.15y + 53.11(1-y)}{27.84y + 44.8(1-y)} = \frac{53.11 - 16.96y}{44.8 - 16.96y} \qquad (C)$$

From Eqns. (*A*) and (*B*), we get

$$\gamma = 64.149\,(30 - 14y) \times 10^{-3} = 1.9245 - 0.898\,09\,y \qquad (D)$$

From Eqns. (*C*) and (*D*), we get

$$1.9245 - 0.898\,09\,y = \frac{53.11 - 16.96y}{44.8 - 16.96y}$$

or $y^2 - 3.6708y + 2.1736 = 0$

or $y = 2.9286$ and $0.7422$

The value of $y = 2.9286 > 1$ is inadmissible, since the mole fraction cannot exceed unity, Therefore $y = 0.7422$.

Mole fraction of $CH_4 = 0.7422$ ; Mole fraction of $C_2H_6 = 0.2578$

Let us check the results as follows:

Substitute the value of $y$ in Eqn. (*B*) to obtain

$M = (30 - 14 \times 0.7422) \times 10^{-3} = 19.609 \times 10^{-3}$ kg/mol

Substitute the value of $y$ in Eqn. (*C*) to obtain $\gamma = 1.258$. Then

$$V = \sqrt{\frac{\gamma RT}{M}} = \sqrt{\frac{1.258 \times 8.314 \times 300}{19.609 \times 10^{-3}}} = 400 \text{ m/s}$$

### Example 9.17

An internal combustion engine is to be operated with methane as a fuel. During the suction stroke 10 mol of air enters with 1 mol of methane into the cylinder at 1 bar and 300 K. If the compression ratio of the engine is 10, estimate the pressure and temperature of the gas mixture at the end of the compression stroke. The molar heat capacities of methane and air are as follows :

$$C_P\,(CH_4) = 36.15 \text{ J / mol K}; \; C_v\,(CH_4) = 27.84 \text{ J / mol K};$$
$$C_P\,(\text{air}) = 29.1 \text{ J / mol K}; \; C_v\,(\text{air}) = 20.786 \text{ J / mol K}$$

*Solution* :

The mixture has the composition $y_{CH_4} = \dfrac{1}{11} = 0.0909$ and $y_{\text{air}} = 0.9091$

$C_P = \Sigma\, y_i\; C_{Pi}$ and $C_v = \Sigma\, y_i\; C_{vi}$

$$\gamma = \frac{C_P}{C_V} = \frac{\Sigma y_i\, C_{Pi}}{\Sigma y_i\, C_{vi}} = \frac{0.0909 \times 36.15 + 0.9091 \times 29.1}{0.0909 \times 27.84 + 0.9091 \times 20.786} = \frac{29.741}{21.427} = 1.388$$

For an isentropic compression

$P_2 = P_1\,(r_0)^{\gamma} = 1 \times (10)^{1.388} = 24.434$ bar

$T_2 = T_1\,(r_0)^{\gamma-1} = 300 \times (10)^{0.388} = 733.03$ K

**Example 9.18**

A rigid tank of $1 m^3$ volume contains a mixture of molar composition 30% $CH_4$, 25% $C_2H_6$ and 45% air at 1 bar and 300 K. It is necessary to change the composition such that the final mixture contains 60 mole percent methane and the tank is always held at 300 K. Determine the mass of methane to be added into the tank and the final pressure of the gas mixture.

*Solution* :

The initial composition of the gas mixture is

$y_{CH_4} = 0.3$ ; $y_{C_2H_6} = 0.25$ and $y_{air} = 0.45$

$$\text{Initial moles of gas, } n = \frac{PV}{RT} = \frac{1 \times 10^5 \times 1}{8.314 \times 300} = 40.093 \text{ mol}$$

Therefore, initially the tank contains

$$n_{CH_4} = y_{CH_4} n = 0.3 \times 40.093 = 12.028 \text{ mol}$$

$$n_{CH_4} = y_{C_2H_6} n = 0.25 \times 40.093 = 10.023 \text{ mol}$$

$$n_{air} = 18.042 \text{ mol}$$

Let $n_f$ = moles of $CH_4$ present in the tank after adding additional mass of $CH_4$. Then

$$y_{CH_4} = 0.6 = \frac{n_f}{(n_f + 10.023 + 18.042)}$$

or $n_f = 42.098$ mol

Therefore $CH_4$ added = 42.098 – 12.028 = 30.07 mol

Mass of $CH_4$ added = $30.07 \times 16 \times 10^{-3} = 0.481$ kg

$$P_f = \frac{nRT}{V} = \frac{(42.098 + 10.023 + 18.042) \times 8.314 \times 300}{1} = 1.75 \text{ bar}$$

**Example 9.19**

Reconsider Example 9.17 and calculate the change in the entropy of the gas mixture, methane and air.

*Solution* :

See Example 9.17. we have $P_2 = 24.434$ bar; $T_2 = 733.03$ K ; $P_1 = 1$ bar and $T_1 = 300$ K.

For the mixture $C_P = 29.741$ J / mol K; $\gamma = 1.388$

The change in entropy of the mixture is given by

$$s_2 - s_1 = C_P \ln (T_2/T_1) - R \ln (P_2/P_1) = 29.741 \ln \frac{733.03}{300} - 8.314 \ln \frac{24.434}{1}$$

$$= 0 \text{ J/mol K}$$

The change in entropy of methane is given by

$$(s_2 - s_1)_{CH_4} = 36.15 \ln \frac{733.03}{300} - 8.314 \ln \frac{24.434}{1} = 5.7256 \text{ J / mol K}$$

The change in entropy of air is given by

$$(s_2 - s_1)_{air} = 29.1 \ln \frac{733.03}{300} - 8.314 \ln \frac{24.434}{1} = -5.7294 \text{ J/mol K}$$

**Example 9.20**

A mixture of $CH_4$ and $C_2H_6$ with molar composition of 25% $CH_4$ and 75%. $C_2H_6$ is compressed in a reversible and adiabatic compressor from 1 bar and 300 K to 10 bar. Determine the final temperature of the gas, the change in entropy of the mixture, the change in entropy of $CH_4$ and the change in entropy of $C_2H_6$. See Example 9.16 for the molar heat capacities of $CH_4$ and $C_2H_6$.

*Solution* :

$$y_{CH_4} = 0.25; \; y_{C_2H_6} = 0.75$$

$$C_P = \Sigma y_i \, C_{Pi} = 0.25 \times 36.15 + 0.75 \times 53.11 = 48.87 \text{ J / mol K}$$

$$C_v = \Sigma y_i \, C_{vi} = 0.25 \times 27.84 + 0.75 \times 44.80 = 40.56 \text{ J / mol K}$$

$$\gamma = \frac{C_P}{C_v} = \frac{48.87}{40.56} = 1.2049$$

For reversible, adiabatic compression we know that

$$T_2 = T_1 \left(\frac{P_2}{P_1}\right)^{\frac{\gamma - 1}{\gamma}} \text{ or } T_2 = 300\,(10)^{\frac{0.2049}{1.2049}} = 443.79 \text{ K}$$

Final temperature of the gas = 443.79 K.

$$(s_2 - s_1)_{\text{mixture}} = C_P \ln \frac{T_2}{T_1} - R \ln \frac{P_2}{P_1} = 48.87 \ln \frac{433.79}{300}$$

$$- 8.314 \ln (10) = 0 \text{ J / mol K}$$

$$(s_2 - s_1)_{CH_4} = 36.15 \ln \frac{443.79}{300} - 8.314 \ln (10) = - 4.9888 \text{ J / mol K}$$

$$(s_2 - s_1)_{C_2H_6} = 53.11 \ln \frac{443.79}{300} - 8.314 \ln (10) = 1.6522 \text{ J / mol K}$$

***Note*** *The entropy change of the individual gases is not equal to zero. However, the net entropy change, which is given by*

[0.25 × (– 4.9888) + 0.75 (1.6522)] *is equal to zero.*

**Example 9.21**

A rigid and insulated tank is divided into two compartments. One compartment of volume 1 m³ contains air ($C_P$ = 29.1 J / mol K; $C_v$ = 20.786 J/mol K) at 300 K and 1 bar while the second compartment of volume 2 m³ contains helium ($C_P$ = 20.786 J/mol K; $C_v$ = 12.4717 J / mol K) at 100 K and 5 bar. The gases are allowed to mix by removing the partition. Determine (*a*) the molar composition of the mixture (*b*) the final temperature and pressure of the mixture (*c*) the change in entropy of helium and air and (*d*) the net entropy change.

*Solution* :

$$(a)\ \ n_{\text{air}} = \frac{PV}{RT} = \frac{1 \times 10^5 \times 1}{8.314 \times 300} = 40.093 \text{ mol}$$

$$n_{\text{He}} = \frac{5 \times 10^5 \times 2}{8.314 \times 1000} = 120.28 \text{ mol}$$

Therefore, $y_{\text{air}} = \dfrac{40.093}{40.093 + 120.28} = 0.25;\ y_{\text{He}} = 0.75$

(*b*) Since the tank is rigid and insulated, we have $Q = 0$ and $W = 0$. Therefore, the first law of thermodynamics gives $\Delta U = 0$. Let $T_f$ = final temperature of the gas mixture. Then

$$\Delta U = n_{\text{air}}\, C_{v\,\text{air}}\, (T_f - 300) + n_{\text{He}}\, C_{v_{\text{He}}}\, (T_f - 1000) = 0$$

or $40.093 \times 20.786\, (T - 300) + 120.28 \times 12.4717\, (T - 1000) = 0$ or $T_f = 750$ K

$$\text{The final pressure } P_f = \frac{(n_{\text{air}} + n_{\text{He}})RT_f}{V} = \frac{(40.093+120.28)\times 8.314\times 750}{3}$$

$$= 3.3334 \text{ bar}$$

(*c*) Partial pressure of air, $p_{\text{air}} = y_{\text{air}} P = 0.25 \times 3.3334 = 0.83335$ bar

Partial pressure of helium, $p_{\text{He}} = y_{\text{He}} P = 2.5001$ bar

$$\Delta s_{\text{air}} = C_P \ln \frac{T_f}{T_1} - R \ln \frac{p}{P_1} = 29.1 \ln \frac{750}{300} - 8.314 \ln \frac{0.83335}{1}$$

$$= 28.180 \text{ J / mol K}$$

$$\Delta s_{\text{He}} = 20.786 \ln \frac{750}{1000} - 8.314 \ln \frac{2.5001}{5} = -0.217\,27 \text{ J / mol K}$$

(*d*) Net entropy change $\Delta S = n_{\text{air}}\ \Delta s_{\text{air}} + n_{\text{He}}\ \Delta s_{\text{He}}$

$= 40.093 \times 28.180 + 120.28 \times (-0.217\,27) = 1.1037$ kJ / K

**Example 9.22**

A rigid and insulated tank of volume 1m$^3$ contains helium at 1 bar and 300 K. This tank is connected to a line carrying nitrogen at 10 bar and 100 K and the nitrogen is allowed to flow into the tank till the tank reaches a pressure of 10 bar. Calculate

(*a*) the final temperature of the gas in the tank

(*b*) the mass of nitrogen that enters the tank, and

(*c*) the final molar composition of the gas in the tank

$C_P$ (He) = 20.786 J / mol K; $C_v$ (He) = 12.4717 J / mol K;

$C_P$ ($N_2$) = 29.178 J / mol K; $C_v$ ($N_2$) = 20.864 J / mol K;

*Solution* :

(*a*) A schematic of the control-volume for filling a tank is shown in Fig.E 9.22. The first law of thermodynamics for a transient flow process is given by

$$n_e\left(h_e + \frac{\mathbf{V}_e^2}{2} + gZ_e\right) - \dot{n}_i\left(h_i + \frac{\mathbf{V}_i^2}{2} + gZ_i\right) = \dot{Q} - \dot{W}_s - \frac{dE}{dt}$$

Ignore KE and PE changes. $\dot{Q} = 0$ (since the tank is insulated) and $\dot{W}_s = 0$ (as no shaft work is associated). Since no gas is escaping from the tank $n_e = 0$. Then the above equation reduces to

$$-n_i h_i = -\frac{dE}{dt}$$

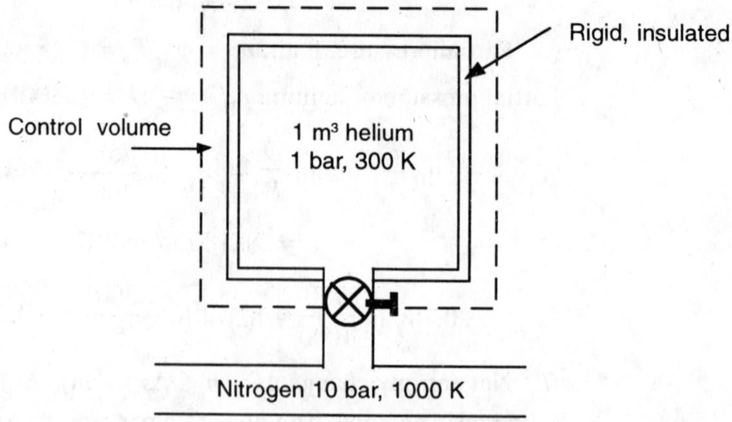


Fig.E 9.22. Schematic of the control volume for filling a tank.

or $(nC_P T)_{N_2} = U_f - U_0$ (A)

where $U_0 =$ initial internal energy of control-volume

$U_f =$ final internal energy of control volume

$n =$ moles of nitrogen that entered the tank

$$U_0 = n_{He} C_{v_{He}} T_{He} = \left(\frac{P_0 V}{RT_0}\right) C_{v_{He}} T_0 = \frac{1 \times 10^5 \times 12.4717}{8.314} = 150.01 \text{ kJ}$$

$$n_{He} = \frac{P_0 V}{RT_0} = \frac{1 \times 10^5 \times 1}{8.314 \times 300} = 40.093 \text{ mol}$$

$$U_f = \left(n_{H_e} C_{v_{He}} + n_{N_2} C_{v_{N_2}}\right) T_f \quad (B)$$

$$T_f = \frac{P_f V}{(n_{H_e} + n_{N_2})R} = \frac{10 \times 10^5 \times 1}{(40.093 + n_{N_2})8.314} = \frac{1.2028 \times 10^5}{40.093 + n_{N_2}} \quad (C)$$

Therefore, Eqns, (*B*) and (*C*) give

$$U_f = \frac{\left(40.093 \times 12.4717 + n_{N_2} \times 20.864\right) 1.2028 \times 10^5}{(40.093 + n_{N_2})}$$

$$\text{or } U_f = \frac{6.0143\times10^7 + 2.5095\times10^6 n_{N_2}}{40.093 + n_{N_2}} \qquad (D)$$

From Eqns. (*A*) and (*D*), we get

$$n_{N_2} \times 29.178 \times 1000 = \frac{6.0143\times10^7 + 2.5095\times10^6 n_{N_2}}{40.093 + n_{N_2}} - 150.01 \times 10^3$$

or $n_{N_2} = 68.039$ and $-27.265$,

Of the above two values $n_{N_2} = -27.265$ is inadmissible. Hence

$$n_{N_2} = 68.039 \text{ mol.}$$

Substituting this value of $n_{N_2}$ in Eqn. (*C*) , we get

$$T_f = \frac{1.2028\times10^5}{40.093 + 68.039} = 1112.3 \text{ K}$$

Therefore, final temperature of gas in the tank = 1112.3 K

(*b*) $n_{N_2} = 68.039$ mol

Mass of nitrogen that entered the tank = $68.039 \times 28 \times 10^{-3} = 1.905$ kg

(c) $y_{He} = \dfrac{n_{He}}{n_{He} + n_{N_2}} = \dfrac{40.093}{40.093 + 68.039} = 0.3708$ ; $y_{N_2} = 0.6292$

### Example 9.23

It is desired to compress a gas mixture having a molar composition of 25% helium and 75% nitrogen from 1 bar and 300 K to a pressure of 50 bar in a steady flow, steady-state adiabatic compressor. Calculate the work to be done on the compressor if the flow rate of the gas mixture is 10 mol/s. The molar heat capacities of helium and nitrogen are given in Example 9.22.

*Solution* :

We know that $C_P$ and $C_v$ of the mixture are given by

$C_p = \Sigma\, y_i C_{Pi} = 0.25 \times 20.786 + 0.75 \times 29.178 = 27.08$ J / mol K

$C_v = \Sigma\, y_i C_{vi} = 0.25 \times 12.4717 + 0.75 \times 20.864 = 18.766$ J / mol K

$$\text{Therefore, } \gamma = \frac{C_P}{C_v} = \frac{27.08}{18.766} = 1.443$$

Application of the first law of thermodynamics for steady flow, steady state adiabatic compressor gives

$$-\dot{W}_s = \dot{m}\,(h_2 - h_1)$$

or $-\dot{W}_s = \dot{n}\,(h_2 - h_1) = \dot{n}\; C_P\,(T_2 - T_1)$

For reversible and adiabatic compression, we know that

$$T_2 = T_1\left(\frac{P_2}{P_1}\right)^{\frac{\gamma-1}{\gamma}} \quad \text{or} \quad T_2 = 300\,(50)^{\frac{0.443}{1.443}} = 996.99 \text{ K}$$

Therefore, work to be done on the compressor, $(-\dot{W}_s)$ is given by

$$-\dot{W}_s = 10 \times 27.08\,(996.99 - 300) = 188.74 \text{ kW}$$

**Example 9.24**

What is the importance of gas and condensable vapor mixtures in engineering thermodynamics ?

*Solution* :

The combustion products in the cylinder of an internal combustion engine contain $CO_2$, CO and $H_2O$ of which water vapor is a condensable vapor at the temperatures of interest to an engineer. In the field of air-conditioning an engineer deals with a mixture of dry air and water vapor. In chemical process industries many of the gas streams contain condensable vapors. Therefore, a study of gas and condensable vapor mixtures is essential in the thermodynamic analysis of processes. At low pressures, the behavior of condensable vapor may be approximated as that of an ideal gas. Hence, the theory of ideal gas mixtures can be extended to study the behavior of gas and condensable vapor mixtures.

**Example 9.25**

What is moist air and saturated air ?

*Solution* :

Moist air refers to a mixture of dry air and water vapor in which dry air (which is a mixture of constant composition) is treated as if it were a pure component. If a mixture of dry air and water vapor behaves like an ideal gas, at the temperature $T$, total pressure $P$ and has $n_a$ moles of dry air and $n_w$ moles of water vapor occupying a volume $V$, then

$$P = \frac{(n_a + n_w)RT}{V} = \frac{n_a RT}{V} + \frac{n_w RT}{V} = p_a + p_w$$

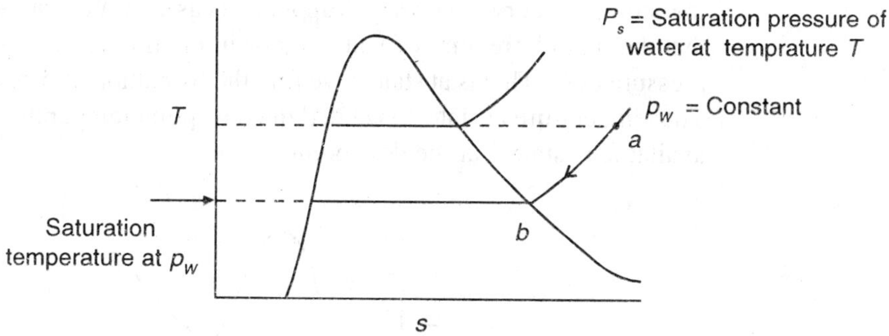


Fig.E 9.25. *T-s* diagram for water-vapor in air–water–vapor mixture.

where $p_a$ and $p_w$ denote the partial pressures of dry air and water vapor, respectively.

The state of water vapor in moist air is shown as *a* in Fig.E 9.25. At this state the partial pressure of water-vapor in the mixture is $p_w$. If the moist air is cooled at constant pressure, initially the composition of the mixture does not undergo any change and the water vapor follows the path *a-b*. At state *a* the temperature of the gas is *T* and at this temperature the saturation pressure of water is $P_S$. If the system reaches the state *b* which is lying on the saturation curve, the water-vapor exists as saturated water vapor. But at state *a*, the water-vapor is at a pressure $p_w$ and temperature *T* which is greater than the saturation temperature and hence it is in superheated state. Further cooling of the mixture results in condensation of water vapor. At state *b*, the air is saturated with water vapor. The air which is saturated with water-vapor or a mixture of dry air and water-vapor such that the partial pressure of water-vapor in the mixture is equal to the saturation pressure of water at the mixture temperature is called saturated air.

**Example 9.26**

What is dew point ?

*Solution* :

Suppose we have a mixture of air and water vapor which is initially not saturated (that is unsaturated air) as shown by 1 on a *T-s* diagram in Fig.E 9.26. Let the partial pressure of water vapor in the mixture at this temperature be $p_w$. If the mixture is cooled at constant pressure, the water vapor follows the path 1-2 along the constant pressure ($p_w$) line till it reaches state 2. That is, the partial pressure of water vapor remains constant during the path 1-2. At state 1, (temperature = $T_1$ ) the saturation pressure of water is $P_s$ which is greater than $p_w$. By the time the system reaches state 2 ($T = T_2$ and $P = p_w$ ) which is located on the saturation curve, the water vapor exists as saturated vapor. At state 2, we find that the partial pressure of water vapor in the mixture is equal to the saturation pressure at $T_2$. That is at state 2 we find the formation of droplets of water or dew. The temperature at state 2 is called the dew point temperature. The condensate is available at state 3 at the dew point.

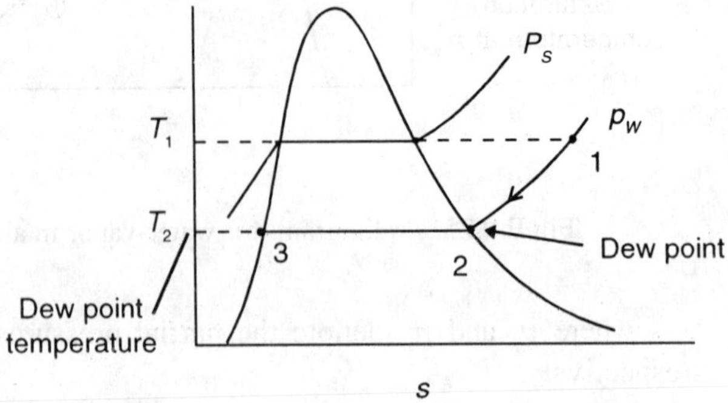


Fig.E 9.26. *T-s* diagram showing states of water vapor in air water-vapor mixture.

**Example 9.27**

Define specific humidity or humidity ratio and relative humidity.

*Solution* :

The composition of a given moist air can be expressed in terms of

(*a*) moles of dry air and water vapor

(*b*) mole fractions of dry air and water vapor

(*c*) masses of dry air and water vapor

(*d*) mass fractions of dry air and water vapor

(*e*) specific humidity of humidity ratio (ω)

(*f*) Relative humidity (ϕ)

The specific humidity or humidity ratio, ω is defined as the ratio of the mass of water vapor to the mass of dry air. That is,

$$\text{Humidity ratio (or specific humidity) } \omega = \frac{m_w}{m_a}$$

This can also be expressed in terms of partial pressure of water vapor $p_w$ as follows :

$$\omega = \frac{m_w}{m_a} = \frac{(M_w p_w V / RT)}{(M_a p_a V / RT)} = \frac{M_w p_w}{M_a P_a} = \frac{18 \times 10^{-3} p_w}{28.97 \times 10^{-3}(P - p_w)}$$

or $$\omega = \frac{0.622 p_w}{P - p_w}$$

where $m_w$, $m_a$ = mass of water vapor and dry air, respectively

$p_w$, $p_a$ = partial pressure of water vapor and dry air, respectively

$M_w$, $M_a$ = Molar mass of water and air, respectively

$P$ = Total pressure

The relative humidity $\phi$ is defined as the ratio of the partial pressure of water vapor in the mixture to the saturation pressure of water at the mixture temperature. That is,

Relative humidity, $\phi = p_w / P_S$,

where $P_S$ = Saturation pressure of water at the mixture temperature,

**Example 9.28**

A moist air has a relative humidity of 75% at 1bar and 40°C. Calculate the humidity ratio, the dew point, the masses of water vapor and dry air in 100 m$^3$ of the mixture.

*Solution* :

At 40°C, the saturation pressure of water, $P_S$ = 0.073 75 bar

Relative humidity, $\phi = \frac{p_w}{P_s} = \frac{P_w}{0.073\,75}$ or $p_w = 0.055\,313$ bar

Partial pressure of air, $p_a = 1 - 0.055\,313 = 0.944\,687$ bar

Humidity ratio, $\omega = \frac{0.622 p_w}{P - p_w} = \frac{0.622 p_w}{p_a} = \frac{0.622 \times 0.055\,313}{0.944\,687}$

or $\omega = 0.036\,42$ kg $H_2O$ / kg dry air

The dew point is the temperature at which the saturation pressure is equal to the partial pressure of water-vapor in the mixture. That is

$$P_S = p_w = 0.055\,313 \text{ bar}$$

From steam tables one can find the temperature corresponding to this pressure as 34.67°C. Therefore, dew point = 34.67°C

Mass of water vapor, $m_W = \frac{18 \times 10^{-3} \times 0.055\,313 \times 10^5 \times 100}{8.314 \times 313} = 3.826$ kg

$$\text{Mass of dry air, } m_a = \frac{m_w}{\omega} = \frac{3.826}{0.036\,42} = 105.05 \text{ kg}$$

**Example 9.29**

A 100 kg sample of moist air with a relative humidity of 0.75 initially at 40°C and 1 bar is cooled to 25°C while the pressure is held constant. Calculate the amount of water that condenses.

*Solution* :

For the air with $\phi = 0.75$ at 40°C and 1 bar, we have $\omega = 0.036\,42$ kg $H_2O$ / kg dry air and dew point = 34.67°C, (see Example 9.28)

Mass of moist air = Mass of dry air + Mass of water vapor

or $100 = m_{a1} + m_{w1}$

$$\text{We know that } \omega = \frac{m_w}{m_a} \text{ or } m_a = \frac{m_w}{\omega}$$

$$\text{Therefore,} \qquad 100 = m_{a_1} + m_{w_1} = \frac{m_{w_1}}{\omega} + m_{w_1} = m_{w_1}\left(1 + \frac{1}{\omega}\right)$$

$$\text{or} \quad 100 = m_{w_1}\left(1 + \frac{1}{0.036\,42}\right)$$

or $m_{w1} = 3.514$ kg

∴ Initial mass of water-vapor, $m_{w1} = 3.514$ kg

The final temperature of the air = 25°C < Dew point = 34.67°C. Therefore, some water vapor condenses and at the end, the water-vapor in the mixture exists in the saturated state. In other words, the air at 25°C is saturated. For saturated air, the partial pressure of water-vapor in the mixture is equal to the saturation pressure.

Saturation pressure at 25°C = $P_{S2}$ = 0.031 66 bar (from steam tables)

Therefore, $p_{w2} = P_{S2} = 0.031\,66$ bar.

$$\text{Then } \omega_2 = \frac{0.622\, p_{w2}}{P - p_{w2}} = \frac{0.622 \times 0.031\,66}{100 - 0.031\,66} = 0.000\,197 \text{ kg } H_2O / \text{kg dry air.}$$

Mass of dry air = $100 - m_{w1} = 100 - 3.514 = 96.486$ kg

The mass of dry air remains constant at 96.486 kg.

$$\text{We know that } \omega_2 = \frac{m_{w2}}{m_{a_2}} \text{ or } m_{w2} = m_{a2}\,\omega_2$$

or $m_{w2} = 96.486 \times 0.000\,197 = 0.019$ kg.

Mass of water condensed $= m_{w1} - m_{w2} = 3.514 - 0.019 = 3.495$ kg

**Example 9.30**

A gas mixture has the following molar composition; 30% $CO_2$, 40% $N_2$, 10% $O_2$ and 20% $H_2O$ at 1 bar and 150°C. If the mixture is cooled at constant pressure, determine the temperature at which the water begins to condense. If the temperature is reduced by 20°C below the condensation temperature, estimate the mass of water which condenses from one kmol of the original mixture.

*Solution* :

Consider 1 kmol of mixture. Then

| Component | Mole fraction ($y_i$) | Moles ($n_i$) | Partial pressure $p_i$ (bar) |
|---|---|---|---|
| $CO_2$ | 0.30 | 300 | 0.30 |
| $N_2$ | 0.40 | 400 | 0.40 |
| $O_2$ | 0.10 | 100 | 0.10 |
| $H_2O$ | 0.20 | 200 | 0.20 |

The water starts condensing at the dew point. At the dew point $P_S = p_w =$ 0.20 bar.

From steam tables at 0.20 bar, the saturation temperature = 60.086°C.

Therefore water begins to condense at 60.086°C.

Final temperature = 60.086 – 20 = 40.086°C

At 40.086°C; $P_S = 0.074\,13$ bar (from steam tables)

The final mixture will be saturated. Hence $p_{w2} = P_S = 0.074\,13$ bar

Let $n$ = final moles of water vapor in the mixture.

Then $p_{w2} = Py_{H_2O}$

Since $P$ is held constant at 1 bar, $y_{H_2O} = p_{w2} = 0.074\,13$

$$y_{H_2O} = \frac{n}{(300+400+100+n)} = 0.074\,13$$

or $$\frac{n}{800+n} = 0.074\,13 \text{ or } n = 64.052 \text{ mol}$$

Mass of water condensed = (initial moles of $H_2O$ – Final moles of $H_2O$) × Molar mass of $H_2O$ = (200 – 64.052) × 18 × $10^{-3}$ = 2.447 kg.

**Example 9.31**

An auditorium of dimensions 30 m × 50 m × 6 m has moist air with a relative humidity of 80% at 1 bar and 40°C. If the air is cooled at constant pressure to 25°C, calculate the mass of water that condenses.

*Solution* :

Saturation pressure at 40°C, $P_{S1} = 0.073\ 75$ bar

$$\phi = \frac{p_{w1}}{P_{Si}} \text{ or } p_{wi} = \phi_1\ P_{S1} = 0.8 \times 0.073\ 75 = 0.059 \text{ bar}$$

$$\omega_i = 0.622\ \frac{p_{w1}}{P - P_{w1}} = \frac{0.622 \times 0.059}{1 - 0.059} = 0.038\ 999 \text{ kg } H_2O/\text{ kg dry air}$$

Mole fraction of water vapor, $y_{H_2O} = \dfrac{P_{w1}}{P}$ or $y_{H_2O} = 0.038\ 999$

Volume of air = 30 × 50 × 6 = 9000 $m^3$

$$\text{Moles of mixture, } n = \frac{PV}{RT} = \frac{1 \times 10^5 \times 9000}{8.314 \times 313} = 3.4585 \times 10^5$$

Initial moles of water, $= n\ y_{H_2O} = 3.4585 \times 10^5 \times 0.038\ 999 = 13\ 487.8$ mol

Moles of air = 3.4585 × $10^5$ – 13 487.8 = 3.3236 × $10^5$

Dew point is given by the temperature at which $P_S = p_{w1} = 0.059$ bar.

From steam tables we find $t = 35.23$°C where $P_S = 0.059$ bar.

Hence dew point = 35.23°C.

Since the air is cooled below its dew point, it will exist as saturated air.

That is, $p_{w2} = P_S$ (25°C) = 0.031 66 bar

Total pressure is held constant at 1 bar. Hence $(y_{H_2O})_2 = p_{w2} / P = p_{w2}$

$$\text{or } (y_{H_2O}) = 0.03\ 166 = \frac{n_{H_2O}}{n_{air} + n_{H_2O}} = \frac{n_{H_2O}}{3.3236 \times 10^5 + n_{H_2O}}$$

or $n_{H_2O}$ = 10 866.6 = moles of water in the final air

Mass of water condensed = (13 487.8 – 10 866.6) × 18 × $10^{-3}$ = 47.182 kg

**Example 9.32**

What is an adiabatic saturator and how is it used to measure the specific humidity or relative humidity of a moist air ?

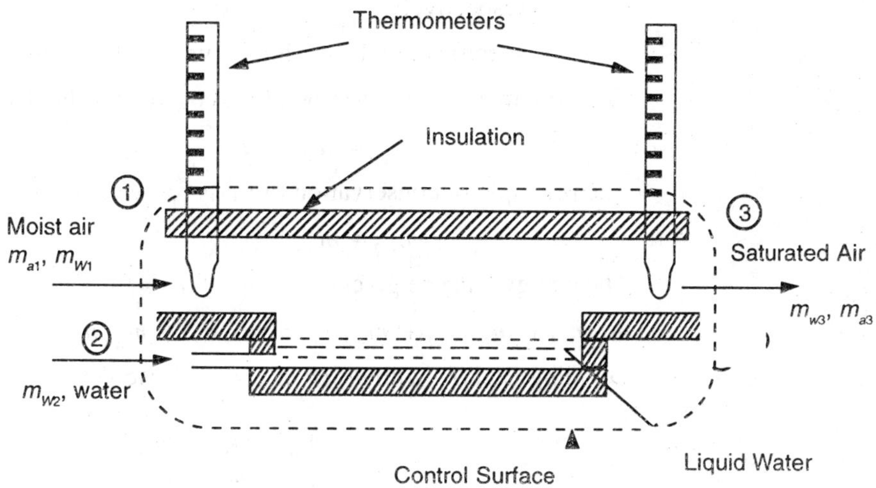


Fig.E 9.32. Schematic of adiabatic saturator.

*Solution* :

The specific humidity (or humidity ratio) or relative humidity of a moist air can be measured with the help of an adiabatic saturator. An adiabatic saturator,shown in Fig.E 9.32, is a device which is insulated and sufficiently long so that it provides an intimate contact between the moist air and liquid water such that the air leaving the device is saturated. The moist air, the specific humidity of which is to be determined, continuously flows into the device at a steady state. Liquid water also enters into the device. Inside the device, the moist air and liquid water interact for enough time and they reach a state of equilibrium. Then the moist air becomes saturated and leaves the device. By measuring the temperatures of the entering and leaving air, one can calculate the specific humidity or relative humidity of the entering moist air.

Consider the device shown in Fig.E 9.32 as the control volume.

Let

$\dot{m}_{a1}$ $\dot{m}_{a3}$ = mass flow rate of dry air into and out of the device, respectively.

$\dot{m}_{w1,}$ $\dot{m}_{w3}$ = mass flow rate of water vapor into and out of the device, respectively

$\dot{m}_{w2}$ = mass flow rate of liquid water into the device

$T_1, T_3$ = temperature of the air at the inlet and outlet of the device, respectively.

$h_{a1}, h_{a3}$ = enthalpy of dry air at the inlet and outlet of the device, respectively.

$h_{w1}, h_{w3}$ = specific enthalpy of water vapor at the inlet and outlet of the device, respectively.

$h_{w2}$ = specific enthalpy of liquid water entering the device.

The principle of conservation of mass gives for the dry air

$$\dot{m}_{a1} = \dot{m}_{a3} = \dot{m}_a \tag{A}$$

The principle of conservation of mass for water gives

$$\dot{m}_{w1} + \dot{m}_{w2} = \dot{m}_{w3} \tag{B}$$

The energy balance gives

$$\dot{m}_{a1} h_{a1} + \dot{m}_{w1} h_{w1} + \dot{m}_{w2} h_{w2} = \dot{m}_{a3} h_{a3} + \dot{m}_{w3} h_{w3} \tag{C}$$

Dividing Eqn ($B$) with $\dot{m}_a$ $(= \dot{m}_{a1} = \dot{m}_{a3})$, we get

$$\frac{\dot{m}_{w_1}}{\dot{m}a_1} + \frac{\dot{m}_{w_2}}{\dot{m}a} = \frac{\dot{m}_{w_3}}{\dot{m}a_3}$$

or $\omega_1 + \dfrac{\dot{m}_{w2}}{\dot{m}_a} = \omega_3$

or

$$\dot{m}_{w2} = \dot{m}_a (\omega_3 - \omega_1) \tag{D}$$

Dividing Eqn. ($C$) with $\dot{m}_a$ $(= \dot{m}_{a1} = \dot{m}_{a3})$, we get

$$h_{a1} + \omega_1 h_{w1} + \frac{\dot{m}_{w2} h_{w2}}{\dot{m}a} = h_{a3} + \omega_3 h_{w3} \tag{E}$$

Substituting Eqn.($D$) in Eqn. ($E$), we get

$$h_{a1} + \omega_1 h_{w1} + (\omega_3 - \omega_1) h_{w2} = h_{a3} + \omega_3 h_{w3}$$

or

$$\omega_1 = \frac{(h_{a3} - h_{a1}) + \omega_3 (h_{w3} - h_{w2})}{(h_{w1} - h_{w2})} \tag{F}$$

The Eqn. ($F$) can be used to determine $\omega_1$ from a knowledge of $T_1$ and $T_3$ which provide information regarding $h_{a1}, h_{a3}, h_{w1}, h_{w2}$ and $h_{w3}$.

If the air is considered to behave like an ideal gas, then $(h_{a3} - h_{a1}) = C_{pair} (T_3 - T_1)$

Suppose the liquid water at 2 enters the device at the same temperature as that of the air leaving the device, then $T_2 = T_3$. In that case $h_{w2} = h_{f2} = h_{f3}$

$h_{w1}$ and $h_{w3}$ refer to the specific enthalpies of water vapor at states 1 and 3 and hence, we can use the standard notation $h_g$ at these states. Now Eqn.(*F*) can be rewritten as

$$\omega_1 = \frac{C_P(T_3 - T_1) + \omega_3(h_{g_3} - h_{f3})}{h_{g_1} - h_{f3}} = \frac{C_P(T_3 - T_1) + \omega_3\, h_{fg3}}{h_{g_1} - h_{f3}} \qquad (G)$$

Knowing the values of $T_1$ and $T_3$ one can calculate $\omega_1$ from Eqn. (*G*).

**Example 9.33**

A sample of moist air enters an adiabatic saturator at 40°C and 1 bar and leaves at 30°C and 1 bar. Determine the specific humidity and relative humidity of the sample of moist air. $C_P$ of air is 1.005 kJ/kg K.

*Solution* :

Since the air leaves the adiabatic saturator as saturated air, $\phi_3 = 1$.

$$\phi_3 = \frac{p_{w3}}{P_{S3}} \text{ or } p_{w3} = P_{S3}$$

Saturation pressure of water at 30°C = $P_{S3}$ = 0.042 41 bar. Therefore,

$$p_{w3} = 0.042\,41 \text{ bar}$$

$$\omega_3 = \frac{0.622\, p_{w3}}{P - p_{w3}} = \frac{0.622 \times 0.042\,41}{1 - 0.042\,41} = 0.0275 \text{ kg } H_2O \text{ kg dry air}$$

From steam tables we read the following data

$h_{f3}$ = 125.66 kJ / kg; $h_{g3}$ = 2556.4 kJ / kg, $h_{g1}$ = 2574.4 kJ / kg.

$h_{fg3} = h_{g3} - h_{f3}$ = 2556.4 – 125.66 = 2430.74 kJ / kg

The specific humidity $\omega_1$ is given by

$$\omega_1 = \frac{C_P(T_2 - T_1) + \omega_3 h_{fg3}}{h_{g1} - h_{f3}} = \frac{1.005(-10) + 0.0275 \times 2430.74}{2574.4 - 125.66}$$

$$= 0.0232 \text{ kg } H_2O \text{ / kg dry air}$$

We can calculate $p_{w1}$ from a knowledge of $\omega_1$ as

$$\omega_1 = \frac{0.622\, p_{w1}}{P - p_{w1}}$$

$$\text{or} \quad 0.0232 = \frac{0.622\, p_{w1}}{1 - p_{w1}} \text{ or } p_{w1} = 0.035\,958 \text{ bar}$$

Saturation pressure at 40°C, $P_{S1} = 0.073\ 75$ bar.

Therefore, relative humidity, $\phi_1 = \dfrac{p_{w1}}{P_{S1}} = \dfrac{0.035\ 958}{0.073\ 75} = 0.4876$

**Example 9.34**

How is the specific enthalpy of moist air defined ?

*Solution* :

The specific enthalpy of moist air is expresed on the basis of dry air. That is, the enthalpy of 1 kg dry air and the enthalpy of the accompany water vapor is called the specific enthalpy of the moist air. Stated in words

$$h^* = h_a + \omega h_w$$

Where the enthalpy of dry air and water vapor is measured with reference to 0°C. This arbitrary choice of the reference temperature does not affect the results since we are interested in the enthalpy changes.

**Example 9.35**

Calculate the specific enthalpy of moist air of Example 9.33.

*Solution* :

The specific humidity $\omega_1$ of the moist air is found to be 0.0232 kg $H_2O$ / kg dry air in Example 9.33. We know that

$$\begin{aligned} h_1^* &= h_a + \omega_l\ h_w = C_P (t - 0) + \omega_1\ h_{g1} \\ &= 1.005 \times 40 + 0.0232 \times 2574.4 \\ &= 99.93 \text{ kJ /kg dry air} \end{aligned}$$

(Note that $h_w$ is taken as $h_g$ since the water is present as vapor in the air)

**Example 9.36**

In deriving the adiabatic saturation relation in Example 9.32, it was assumed that liquid water enters the device at the same temperature as the saturated air leaving the device. If the make up water enters the device at a temperature other than the saturated air leaving the device, how much error does it introduce in the value of specific humidity of air ?

*Solution* :

Based on mass and energy balances on the adiabatic saturator, we have obtained

$$\omega_l = \frac{(h_{a3} - h_{a1}) + \omega_3 (h_{w3} - h_{w2})}{(h_{w1} - h_{w2})} \qquad \text{[See Eqn } (F) \text{ of Example 9.32]}$$

$$\text{or} \quad \omega_1 = \frac{C_P(T_3 - T_1) + \omega_3(h_{g3} - h_{f2})}{(h_{g1} - h_{f2})} \qquad (A)$$

If the liquid water enters the adiabatic saturator at $T_2$ which is different from $T_3$ then $(h_{g3} - h_{f2}) \neq h_{fg3}$. The difference in $h_{fg3}$ and $(h_{g3} - h_{f2})$ is usually small and it is multiplied by $\omega_3$ which is also a small quality, resulting in a very little variation in the value of the numerator of Eqn. (*A*).

Similarly in the denominator $h_{f2}$ will be different from $h_{f3}$ resulting in a small variation of the denominator. The calculated value of $\omega_1$ will at the best differ by 1% only.

**Example 9.37**

Rework Example. 9.33, assuming that liquid water enters the adiabatic saturator at 35°C instead of at 30°C and compare the value of $\omega_1$ with that calculated in Example 9.33

*Solution* :

We found, $\omega_3 = 0.0275$ kg $H_2O$ / kg dry air [See Example 9.33]

If $T_2$ is different from $T_3$, the specific humidity $\omega_1$ is given by

$$\omega_1 = \frac{C_P(T_3 - T_1) + \omega_3(h_{g3} - h_{f2})}{(h_{g1} - h_{f2})}$$

At 40°C: $h_{g1}$ =2574.4 kJ / kg

At 30°C: $h_{g3}$= 2556.4 kJ / kg

At 35°C: $h_{f2}$= 146.56 kJ / kg.

Then,

$$\omega_1 = \frac{1.005\,(-10) + 0.0275\,(2556.4 - 146.56)}{(2574.4 - 146.56)} = 0.0232 \text{ kg } H_2O \text{ / kg dry air}$$

Which is same as $\omega_1$ calculated in Example 9.33.

**Note** *For all practical purposes, the adiabatic saturation temperature of air water-vapor mixture is independent of the make–up water temperature and depends on the humidity ratio and temperature of the moist air.*

**Example 9.38**

What is meant by dry bulb temperature and wet bulb temperature ? Describe a sling psychrometer.

*Solution* :

The principle of adiabatic saturation is used in measuring the humidity ratio ω of moist air in a laboratory by using a simple instrument called psychrometer. A schematic of a sling psychrometer is presented in Fig.E 9.38. It consists of two thermometers, one of which is surrounded by a wet - wick or a wet fabric. The thermometer having the wet wick is called the wet-bulb thermometer and the other is called dry-bulb thermometer. The instrument can be whirled around in the atmosphere of moist air, the humidity ratio of which is to be determined. This induces the flow of moist air around the thermometers. The dry - bulb thermometer reads the temperature of the moist air, which is called the dry - bulb temperature. That is, the dry-bulb temperature refers to the temperature that would be measured by a thermometer when it is placed in the air. If the air is not saturated, the water in the wick evaporates and its temperature falls below the temperature of the air. Eventually a steady-state condition will prevail and the wet - bulb thermometer reads this steady-state temperature of the wick which is called the wet - bulb temperature. The wet - bulb temperature depends on the rates of energy transfer and mass transfer from water to air, which in turn depends on the velocity of air over the thermometers and other factors. Therefore, the wet - bulb temperature is not a property of the moist air, where as the adiabatic saturation temperature is a property of the moist air. It has been experimentally found that for air water-vapor mixtures, the wet-bulb temperature and adiabatic saturation temperature are very close to each other. From a knowledge of the dry-bulb temperature and wet - bulb temperature, one can calculate the humidity ratio or the relative humidity of the moist air.

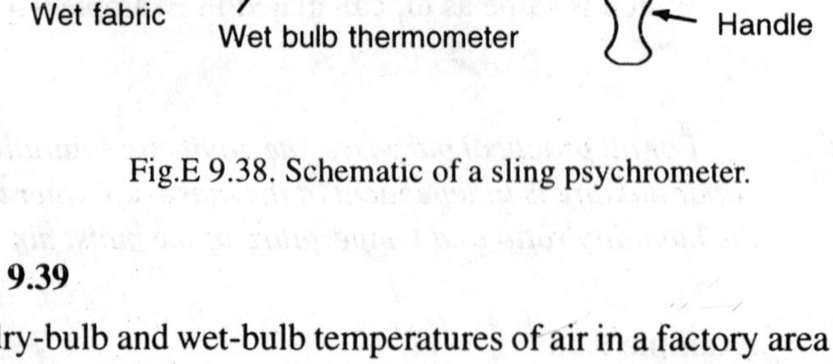


Fig.E 9.38. Schematic of a sling psychrometer.

**Example 9.39**

The dry-bulb and wet-bulb temperatures of air in a factory area are 45°C and 30°C, respectively and the atmospheric pressure is 1 bar. Calculate the humidity ratio and relative humidity of the air.

*Solution :*

The wet-bulb temperature is equal to the adiabatic saturation temperature. Therefore, we have $t_1 = 45°C$; $t_3 = 30°C$.

At 45°C : $h_{g1} = 2583.3$ kJ / kg ; $P_{S1} = 0.095\ 82$ bar

At 30°C: $h_{f3} = 125.66$ kJ / kg ; $h_{g3} = 2556.4$ kJ / kg ; $P_{S3} = 0.042\ 41$ bar

At wet-bulb temperature, air is saturated.

$$\text{Therefore, } \phi_3 = 1 = \frac{p_{w3}}{P_{S3}} \text{ or } p_{w3} = P_{S3} = 0.042\ 41 \text{ bar}$$

$$\text{Then, } \omega_3 = \frac{0.622 p_{w3}}{P - p_{w3}} = \frac{0.622 \times 0.042\ 41}{1 - 0.042\ 41} = 0.0275 \text{ kg } H_2O \text{ / kg dry air}$$

$$\omega_1 = \frac{C_P(T_3 - T_1) + \omega_3 h_{fg3}}{h_{g1} - h_{f3}} = \frac{1.005(-15) + 0.0275(2556.4 - 125.66)}{2583.3 - 125.66}$$

or $\omega_1 = 0.0211$ kg $H_2O$ / kg dry air

$$\omega_1 = \frac{0.622\, p_{w1}}{P - p_{w1}} \quad \text{or} \quad 0.0211 = \frac{0.622\, P_{w1}}{1 - p_{w1}}$$

or $p_{w1} = 0.032\ 81$ bar

$$\phi_1 = \frac{p_{w1}}{p_{S1}} = \frac{0.032\ 81}{0.095\ 82} = 0.3424$$

Therefore, humidity ratio, $\omega_1 = 0.0211$ kg $H_2O$ / kg dry air

Relative humidity, $\phi_1 = 0.3424$

**Example 9.40**

What is a psychrometric chart and what information does is provide ?

*Solution :*

The properties of moist air can be evaluated by making use of the adiabatic saturation relation. In the field of air-conditioning an engineer has to solve the adiabatic saturation relation quite often. A graphical representation of the solution of adiabatic saturation relation is called a psychrometric chart. The properties of interest like dry-bulb temperature, wet - bulb temperature, humidity ratio, relative humidity and specific enthalpy of moist air are shown on a psychrometric chart. These charts are usually prepared for particular values of pressure. A schematic of the psychrometric chart showing the salient features is presented in Fig.E 9.40. An enlarged version of the psychrometric chart at 1 bar pressure for use in solving the problems is presented in Appendix-9.

The mass and energy balance equations for an adiabatic saturator gives

$$h_{a1} + \omega_1 h_{w1} + \frac{\dot{m}_{w_2}}{\dot{m}_a} h_{w2} = h_{a3} + \omega_3 h_{w3} \text{ [ See Eqn (}E\text{) of Example 9.32]}$$

$$\text{or} \quad h_1^* + \frac{\dot{m}_{w2}}{\dot{m}_a} h_{w2} = h_3^*$$

$$\text{or} \quad h_1^* + (\omega_3 - \omega_1)\, hw_2 = h_3^* \qquad (A)$$

The value of the second terms in LHS of Eqn. (*A*) is very small compared to the other quantities as it is related to the difference in the humidity ratio. For many engineering applications it is reasonable to ignore this term and rewrite Eqn.(*A*) as

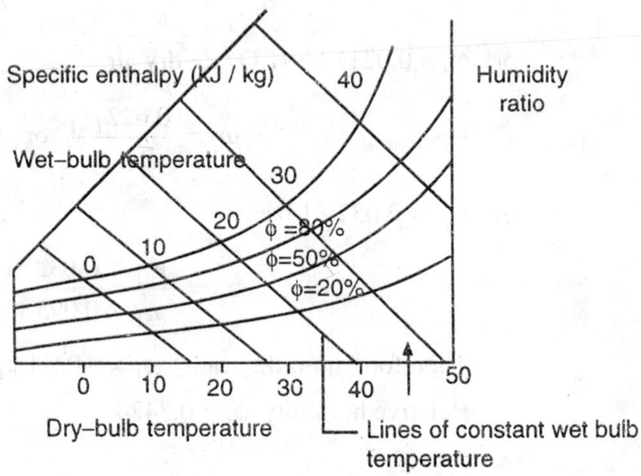


Fig.E 9.40. Schematic of psychrometric chart.

$$h_1^* = h_3^*$$

That is, in an adiabatic saturator, the enthalpy of moist air remains unaltered. The value of $h_3^*$ depends on $T_3$ and $\omega_3$, and $\omega_3$ depends on $p_{w3}$ which in turn depends on $T_3$. Therefore, $h_3^*$ is a function of $T_3$ (adiabatic saturation temperature) only. Hence the constant lines of wet-bulb temperature coincide with the constant specific enthalpy lines.

**Example 9.41**

A sample of moist air at 1 bar pressure has dry-bulb and wet-bulb temperatures of 35°C and 25°C, respectively. Using the psychrometric chart determine the humidity ratio, relative humidity and the specific enthalpy of the moist air.

*Solution :*

A schematic of the psychrometric chart is shown in Fig.E 9.41. Locate the point A at the intersection of the dry bulb temperature line of 35°C and the constant wet bulb temperature line of 25°C as shown in Fig.E 9.41. The state of the moist air is given by point A. Corresponding to state A one can read

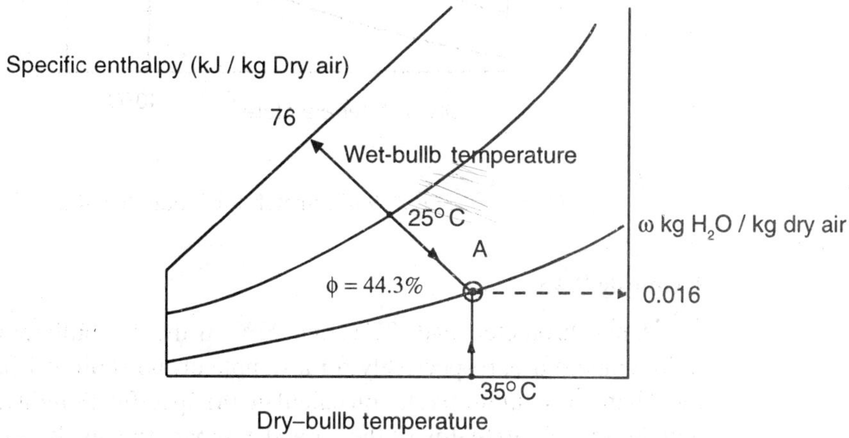


Fig.E 9.41. Sketch for Example 9.41.

$\omega = 0.016$ kg $H_2O$ / kg dry air ; $\phi = 44.3\%$ and $h^* = 76$ kJ / kg dry air as shown in Fig.E 9.41.

**Example 9.42**

On a particular day the atmospheric air is at 1 bar and 40°C and has a relative humidity of 50%. Using the psychrometric chart determine the humidity ratio, wet-bulb temperature and specific enthalpy of the air.

*Solution :*

Locate the point A at the intersection of the dry-bulb temperature line of 40°C and the constant relative humidity curve of 50% as shown in Fig.E 9.42. Then, from the chart corresponding to point A read the following values (as shown in Fig.E 9.42)

$\omega$ = 0.0235 kg $H_2O$ / kg dry air; Wet-bulb temperature = 31°C and $h^*$ = 101 kJ / kg dry air.

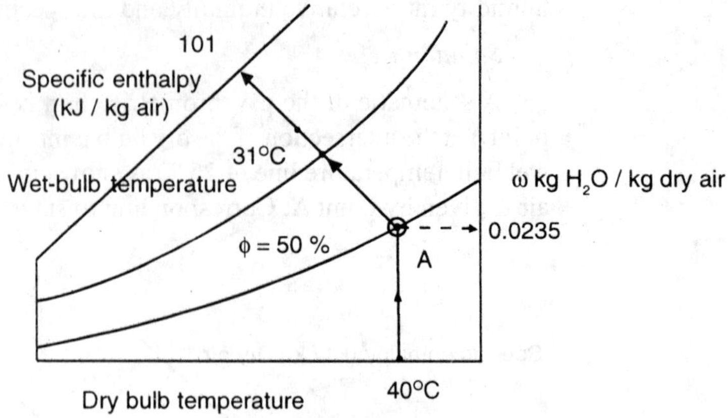


Fig.E 9.42. Sketch for Example 9.42.

**Example 9.43**

A psychrometer reads 37°C and 22°C as the dry-bulb temperature and wet-bulb temperature, respectively for a sample of moist air at 1 bar pressure. Using the adiabatic saturation relation calculate the specific humidity, relative humidity and the specific enthalpy of the air and compare the results with those read from the psychrometric chart.

*Solution* :

Given: Dry-bulb temperature, $T_1$ = 37°C

Wet-bulb temperature = Adiabatic saturation temperature or $T_3$ = 22°C

Pressure, $P$ = 1 bar

From steam tables, we find

at 37°C: $P_{S1}$ = 0.063 232 bar ; $h_{g1}$ = 2569 kJ / kg

at 22°C: $P_{S3}$ = 0.026 686 bar ; $h_{f3}$ = 92.22 kJ / kg

$h_{g3}$ = 2541.84 kJ / kg or $h_{fg3}$ = 2449.62 kJ / kg

At adiabatic saturation temperature $\phi_3 = 1 = \dfrac{p_{w3}}{P_{S3}}$

or $p_{w3} = P_{S3}$ = 0.026 686 bar

$$\omega_3 = \frac{0.622\, p_{W3}}{1 - p_{W3}} = \frac{0.622 \times 0.026\,686}{1 - 0.026\,686} = 0.0171 \text{ kg } H_2O / \text{kg dry air}$$

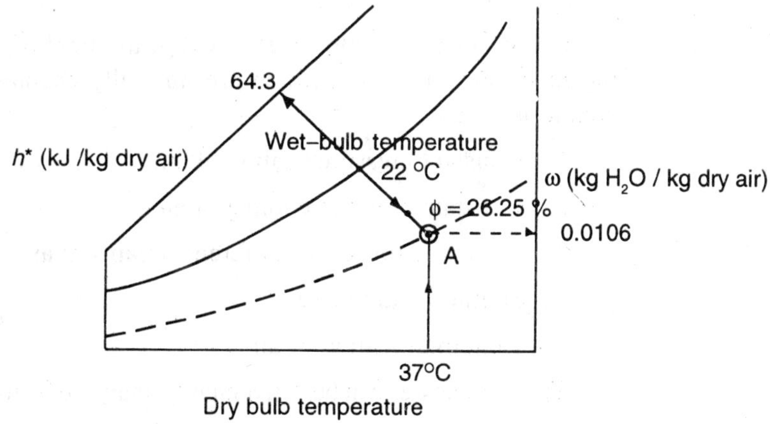


Fig.E 9.43. Sketch for Example 9.43.

The adiabatic saturation relation gives,

$$\omega_1 = \frac{C_P(T_3 - T_1) + \omega_3 h_{fg3}}{h_{g1} - h_{f3}} = \frac{1.005(-15) + 0.0171 \times 2449.62}{2569 - 92.22}$$

$$= 0.0108 \text{ kg } H_2O / \text{kg dry air}$$

We know that $\omega_1 = \dfrac{0.622\, p_{w1}}{P - p_{w1}}$

or $0.0108 = \dfrac{0.622\, p_{w1}}{P - p_{w1}}$ or $p_{w1} = 0.017\,067$ bar

$$\phi_1 = \frac{p_{w1}}{p_{s1}} = \frac{0.017\,067}{0.063\,232} = 0.2699$$

$h_1^* = C_p t_1 + \omega_1 h_{g1} = 1.005 \times 37 + 0.0108 \times 2569 = 64.93$ kJ / kg dry air

Now locate point A at the intersection of 37°C dry-bulb temperature line and 22°C wet-bulb temperature line as shown in Fig.E 9.43.

Then read, $\omega_1$ =0.0106 kg $H_2O$/kg dry air ; $\phi_1$= 26.25% and $h_1^*$ = 64.3 kJ/kg dry air

These values are in good agreement with the calculated values.

**Example 9.44**

Name a few important applications of psychrometry in the field of air-conditioning.

*Solution* :

Psychrometry is extensively used in the field of air-conditioning. Some of the important processes which are normally encountered in the field of air-conditioning are

1. Constant pressure heating of air.
2. Constant pressure cooling of air.
3. Adiabatic mixing of different streams of air.
4. Humidification of air
5. Dehumidification of air

These processes can be conveniently analyzed by using a psychrometric chart.

**Example 9.45**

An important process in the field of air-conditioning is the adiabatic mixing of different streams with different humidity ratios such that the resulting mixture has the desired humidity ratio. Develop an expression to determine the ratio in which two streams of air with different humidity ratios are to be mixed to obtain the desired humidity ratio.

*Solution* :

Let us consider an adiabatic mixer as shown in Fig.E 9.45. Suppose the two streams 1 and 2 of air enter the mixer at the conditions shown in Fig.E 9.45 and the mixed air leaves as stream 3. Choose the control-volume as shown in FigE 9.45. Then the material and energy balances give

Mass balance for dry air ; $\dot{m}_{a1} + \dot{m}_{a2} = \dot{m}_{a3}$ (*A*)

Mass balance for $H_2O$ ; $\dot{m}_{a1}\,\omega_1 + \dot{m}_{a2}\,\omega_2 = \dot{m}_{a3}\,\omega_3$ (*B*)

Energy balance ; $\dot{m}_{a1}\,h^*_1 + \dot{m}_{a2}\,h^*_2 = \dot{m}_{a3}\,h^*_3$ (*C*)

From Eqns. (*A*) and (*B*), we get

$$\dot{m}_{a1}\omega_1 + \dot{m}_{a2}\omega_2 = (\dot{m}_{ai} + \dot{m}_{a2})\omega_3$$

or $$\omega_3 = \frac{\dot{m}_{a1}\,\omega_1 + \dot{m}_{a2}\omega_2}{(\dot{m}_{ai} + \dot{m}_{a2})}$$

or $$\dot{m}_{a2}\,(\omega_2 - \omega_3) = \dot{m}_{a1}\,(\omega_3 - \omega_1)$$

or $$\frac{m_{a2}}{\dot{m}_{aL}} = \frac{\omega_3 - \omega_1}{\omega_2 - \omega_3} \qquad (D)$$

From Eqns.(*A*) and (*C*), we get,

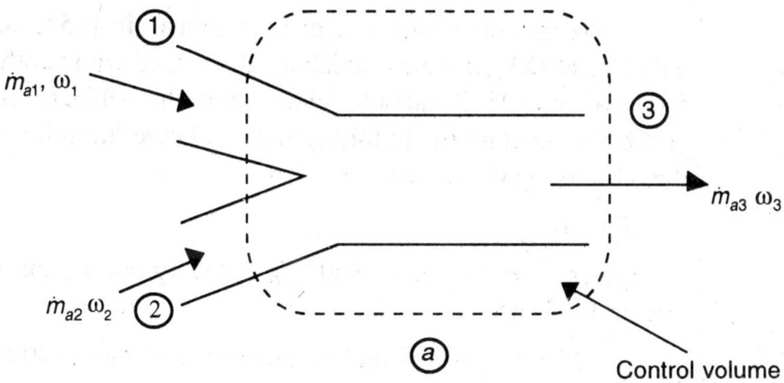


Fig.E 9.45. (*a*) Schematic of adiabatic mixing chamber.

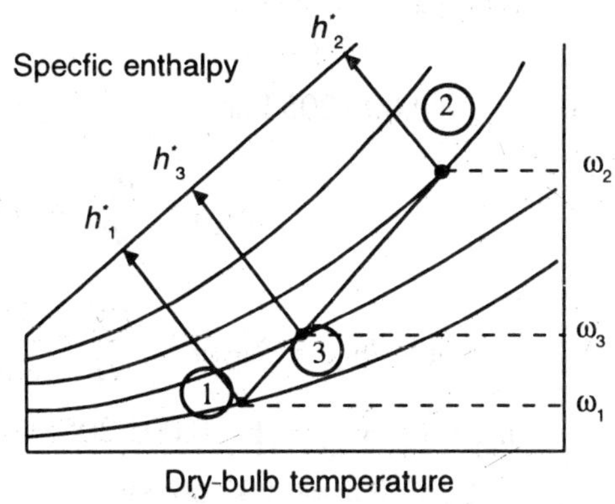


Fig.E 9.45. (*b*) Representation of adiabatic mixing of two air streams on a psychrometric chart

$$\dot{m}_{a1}\, h_1^* + \dot{m}_{a2}\, h_2^* = (\dot{m}_{a1} + \dot{m}_{a2}) h_3^*$$

or $$h_3^* = \frac{\dot{m}_{a1}\, h_1^* + \dot{m}_{a2}\, h_2^*}{(\dot{m}_{a1} + \dot{m}_{a2})}$$

or $$\dot{m}_{a2}\,(h_2^* - h_3^*) = \dot{m}_{a1}\,(h_3^* - h_1^*)$$

or $$\frac{\dot{m}_{a2}}{\dot{m}_{a1}} = \frac{h_3^* - h_1^*}{h_2^* - h_3^*} \qquad (E)$$

**Example 9.46**

A stream consisting of 2 $m^3/s$ of moist air at 5°C and a humidity ratio of 0.002 kg $H_2O$/kg dry air is adiabatically mixed with another stream consisting of 5 $m^3/s$ of air at 25°C and 60% relative humidity while the pressure is held constant at 1 bar. Calculate the humidity ratio, relative humidity and the temperature of the air leaving the mixer.

*Solution* :

Given $t_1 = 5°C$; $\omega_1 = 0.002$ kg $H_2O$ / kg dry air and volumetric flow rate of stream 1 = 2 $m^3/s$.

$t_2 = 25°C$; $\phi_2 = 60\%$ and volumetric flow rate of stream 2 = 5 $m^3/s$

We know that $$\omega_1 = \frac{0.622\,p_{w1}}{P - p_{w1}}$$

or $$0.002 = \frac{0.622\,p_{w1}}{1 - p_{w1}}$$

or $p_{w1} = 0.003\ 205$ bar

$$y_{H_2O} = \frac{p_{w1}}{P} = 0.003\ 205$$

$$\dot{n}_1 = \dot{n}_{H_2O} + \dot{n}_{dryair} = \frac{PV}{RT} = \frac{1 \times 10^5 \times 2}{8.314 \times 278} = 86.5317 \text{ mol}$$

$\dot{m}_{a1} = y_{air} M_{air} = (1 - 0.003\ 205)\ 86.5317 \times 28.97 \times 10^{-3} = 2.499$ kg/s

At 25°C $P_{S2} = 0.031\ 66$ bar

$$\phi_2 = \frac{p_{w2}}{P_{S2}} \text{ or } p_{w2} = \phi_2 P_{S2} = 0.60 \times 0.031\,66 = 0.018\ 996 \text{ bar}$$

$$y_{H_2O} = \frac{p_{w2}}{P} = 0.018\ 996$$

$$\omega_2 = \frac{0.622\,p_{w2}}{P - p_{w2}} = \frac{0.622 \times 0.018\,996}{1 - 0.018\,996} = 0.0120 \text{ kg } H_2O \text{ / kg dry air}$$

$$\dot{n}_2 = \frac{PV}{RT} = \frac{1 \times 10^5 \times 5}{8.314 \times 298} = 201.81 \text{ mol}$$

$\dot{m}_{a2} = (1 - 0.018\ 996) \times 201.81 \times 28.97 \times 10^{-3} = 5.735$ kg / s

Then, $$\omega_3 = \frac{\dot{m}_{a1}\,\omega_1 + \dot{m}_{a2}\,\omega_2}{\dot{m}_{a1} + \dot{m}_{a2}'} = \frac{2.499 \times 0.002 + 5.735 \times 0.012}{2.499 + 5.735}$$

or $\omega_3 = 0.009$ kg $H_2O$ kg dry air

Therefore, the humidity ratio of the resulting air mixture = 0.009 kg $H_2O$ / kg dry air

$$h_1^* = C_P t_1 + \omega_1 h_{g1} = 1.005 \times 5 + 0.002 \times 2510.75 = 10.047 \text{ kJ / kg dry air}$$

$$(h_{g1} \text{ at } 5°\text{C} = 2510.75 \text{ kJ /kg })$$

$$h_2^* = C_P t_2 + \omega_2 h_{g2} = 1.005 \times 25 + 0.012 \times 2547.3 = 55.693 \text{ kJ / kg dry air}$$

$$(h_{g2} \text{ at } 25°\text{C} = 2547.3 \text{ kJ / kg})$$

We know that, $h_3^* = \dfrac{\dot{m}_{a1} h_1^* + \dot{m}_{a2} h_2^*}{\dot{m}_{a1} + \dot{m}_{a2}}$

or $h_3^* = \dfrac{2.499 \times 10.047 + 5.735 \times 55.693}{2.499 + 5.735} = 41.84$ kJ / kg dry air

$$h_3^* = C_P t_3 + \omega_3 h_{g3}$$

or $41.84 = 1.005 t_3 + 0.009 h_{g3}$ (*A*)

Eqn,(*A*) has to be solved to determine the temperature of the resulting air stream. Since we have no mathematical relation which expresses $h_g$ as a function of temperature and this information is available in a tabular form (steam tables), we shall solve this by trial and error procedure. Let us assume $t$ = 19°C and check whether Eqn.(*A*) is satisfied or not.

At 19°C, $h_g$ =2536.38 kJ / kg

RHS of Eqn (*A*) = $1.005\, t_3 + 0.009\, h_{g3} = 1.005\,(19) + 0.009\,(2536.38)$

= 41.92 kJ/kg dry air

LHS of Eqn. (*A*) = 41.84 kJ / kg dry air

Let us say that LHS = RHS. Then Eqn.(*A*) is satisfied, Therefore, temperature of air leaving the mixer = 19°C,

At 19°C; $P_{S3} = 0.022\,104$ bar

$$\omega_3 = \frac{0.622\, p_{w3}}{P - p_{w3}}$$

or $0.009 = \dfrac{0.622\, p_{w3}}{1 - p_{w3}}$ or $p_{w3} = 0.01\,4263$ bar

$$\phi_3 = \frac{p_{w3}}{P_{S3}} = \frac{0.014\,263}{0.022\,104} = 0.6453$$

Relative humidity of the air mixture = 64.53%

### Example 9.47

Rework Example 9.46 by making use of the psychrometric chart.

*Solution* :

We have, $\dot{m}_{a1} = 2.499$ kg / s; $\dot{m}_{a2} = 5.735$ kg / s

$\omega_1 = 0.002$ kg $H_2O$ / kg dry air; $\omega_2 = 0.012$ kg $H_2O$ / kg dry air

[See example 9.46]

A schematic of the psychrometric chart is shown in Fig.E 9.47

We know that

$$\frac{\dot{m}_{a2}}{\dot{m}_{a1}} = \frac{\omega_3 - \omega_1}{\omega_2 - \omega_3} = \frac{h_3^* - h_1^*}{h_2^* - h_3^*}$$

That is, the state 3 lies on the line joining states 1 and 2.

Locate state 1 at the intersection of 5°C dry-bulb temperature line and $\omega = 0.002$ and state 2 at the intersection of 25°C dry-bulb temperature line and 60% relative humidity curve as shown in FigE. 9.47. Join states 1 and 2 by a straight line on the psychrometric chart.

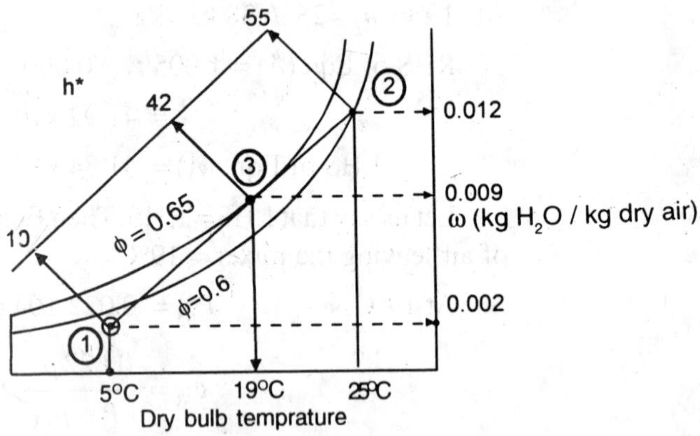


Fig.E 9.47. Sketch for Example 9.47.

Then we calculate, $\omega_3 = \dfrac{\dot{m}_{a1}\,\omega_1 + \dot{m}_{a2}\,\omega_2}{\dot{m}_{a1} + \dot{m}_{a2}} = 0.009$ kg $H_2O$ / kg dry air

Locate state 3 at the intersection of the straight line connecting states 1 and 2 with $\omega = 0.009$ as shown in Fig.E 9.47. At state 3, we read the following values from the figure,

$$t_3 = 19°C;\ \phi_3 = 65\%;\ h_3 = 42 \text{ kJ / kg dry air}$$

**Example 9.48**

A stream of air at 10°C with a relative humidity of 90% is adiabatically mixed with another stream of air at 40°C with 30% relative humidity at 1 bar pressure. If the flow rates of these streams are in the ratio of 1:2 on dry air basis, determine the temperature,specific enthalpy, humidity ratio and relative humidity of the resulting mixture.

*Solution* :

Locate state 1 at 10°C and $\phi_1 = 90\%$ and state 2 at 40°C and $\phi_2 = 30\%$ as shown in Fig.E 9.48. We know that

$$\frac{\dot{m}_{a2}}{\dot{m}_{a1}} = \frac{2}{1} = \frac{\omega_3 - \omega_1}{\omega_2 - \omega_3}$$

or $\dot{m}_{a2}(\omega_2 - \omega_3) = \dot{m}_{a1}(\omega_3 - \omega_1)$

Therefore, draw a line joining states 1 and 2. Then locate the point 3 such that the lengths of 1-3 and 2-3 are in the ratio of 2:1 as shown in FigE. 9.48. Then from point 3 read the following values from the figure.

$$t_3 = 29.5°C;\phi_3 = 45\%;\ h_3^* = 55 \text{ kJ / kg dry air}$$

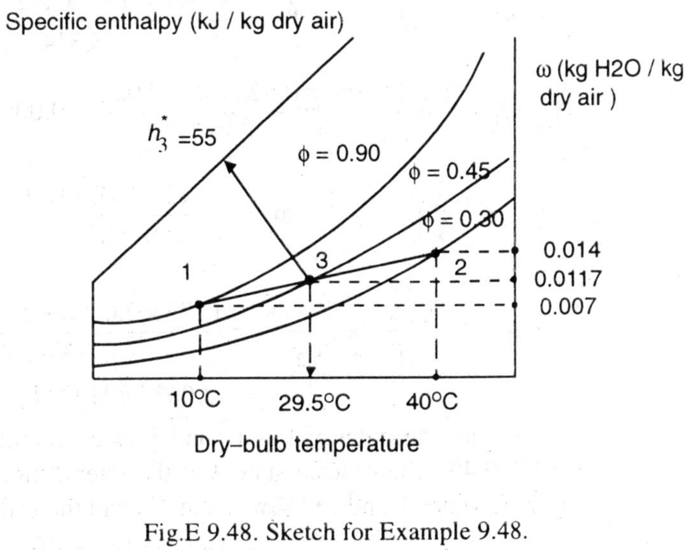


Fig.E 9.48. Sketch for Example 9.48.

**Example 9.49**

Two air-conditioners blow cold air into a common duct. The air conditioner A delivers $1 m^3$/min of air at 20°C and 1 bar with 60% relative humidity while the air-conditioner B delivers $2 m^3$/min of saturated air at 4°C and 1 bar. Determine the temperature and the relative humidity of the mixture.

*Solution* :

Given: $t_1 = 20°C$; $\phi_1 = 60\%$; volumetric flow rate = $1 m^3 / min$.

$t_2 = 4°C$ ; $\phi_2 = 100\%$ ; volumetric flow rate = $2 m^3 / min$

At 20°C ; $P_{S1} = 0.023\,37$ bar

$$\phi_1 = \frac{P_{w1}}{P_{S1}}$$

or $0.60 = \dfrac{p_{w1}}{0.02337}$ or $p_{w1} = 0.140\,22$ bar

$$p_{a1} = 1 - 0.014\,022 = 0.985\,978 \text{ bar}$$

$$\dot{m}_{a1} = \frac{p_{a1}\,VM_a}{RT} = \frac{0.985\,978 \times 10^5 \times 28.97 \times 10^{-3}}{8.314 \times 293} = 1.173 \text{ kg / min}$$

At 4°C : $P_{S2} = 0.008\,129$ bar

$$\phi_2 = 1 = \frac{p_{w2}}{P_{S2}} \text{ or } p_{w2} = 0.008\,129 \text{ bar}$$

$$p_{a2} = 1 - 0.008\,129 = 0.991\,871 \text{ bar}$$

$$\dot{m}_{a2} = \frac{0.991\,871 \times 10^5 \times 2 \times 28.97 \times 10^{-3}}{8.314 \times 277} = 2.495 \text{ kg/min}$$

$$\omega_1 = \frac{0.622 \times P_{w1}}{P_{a1}} = \frac{0.622 \times 0.014\,022}{0.985\,978} = 0.0088 \text{ kg } H_2O \text{ / kg air}$$

$$\omega_2 = \frac{0.622 \times 0.008\,129}{0.991\,871} = 0.0051 \text{ kg } H_2O \text{ / kg dry air}$$

$$\omega_3 = \frac{\dot{m}_{a1}\,\omega_1 + \dot{m}_{a2}\,\omega_2}{\dot{m}_{a1} + \dot{m}_{a2}} = \frac{1.173 \times 0.0088 + 2.495 \times 0.0051}{1.173 + 2.495}$$

$$= 0.0063 \text{ kg } H_2O \text{ / kg dry air}$$

Locate the states of 1 and 2 and draw a straight line joining them as shown in FigE. 9.49. Then locate state 3 at the intersection of $\omega_3 = 0.0063$ and the line joining states 1 and 2. Then at state 3 read the following values

$$t = 11°C \text{ and } \phi_3 = 81\%$$

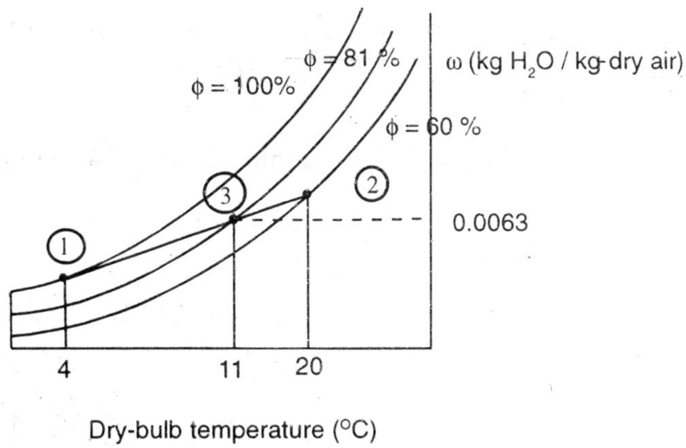


Fig.E 9.49. Sketch for Example 9.49.

## Example 9.50

It is required to blow air at 25°C and 1 bar into a room at the rate of 10 kg/min on dry air basis. For this purpose saturated air at 10°C is produced by blowing air over chilled water pipes. It is proposed to mix this air with atmospheric air which is available at 45°C and 30% relative humidity. Determine the proportions on dry air basis in which these air streams are to be mixed and the relative humidity of the resulting mixture.

*Solution* :

Stream 1: $t_1$ =10°C; $\phi_1 = 1$

Read $\omega_1 =$ 0.0076 kg $H_2O$ / kg dry air from chart

Stream 2: $t_2 =$ 45°C; $\phi_2 = 0.3$

Read $\omega_2 = 0.0182$ kg $H_2O$ / kg dry air from chart

Stream 3: $t_3$ = 25°C.

Locate states 1, and 2 on the psychrometric chart as shown in Fig.E. 9.50. Then find the intersection of the line joining 1 and 2 with $t$ = 25°C and read $\omega_3 = 0.0124$ kg $H_2O$ / kg dry air and $\phi_3 = 61\%$.

$$\frac{\dot{m}_{a2}}{\dot{m}_{a1}} = \frac{\omega_3 - \omega_1}{\omega_2 - \omega_3}$$

$$= \frac{0.0124 - 0.0076}{0.0182 - 0.0124} = 0.8276$$

$$\dot{m}_{a3} = 10 \text{ kg / min} = \dot{m}_{a1} + \dot{m}_{a2} = \dot{m}_{a1}\left(1+\frac{\dot{m}_{a2}}{\dot{m}_{a1}}\right)$$

or $10 = \dot{m}_{a1}\,(1 + 0.8276)$

or $\dot{m}_{a1} = 5.4717$ kg/min.and $\dot{m}_{a2} = 4.5283$ kg/mm.

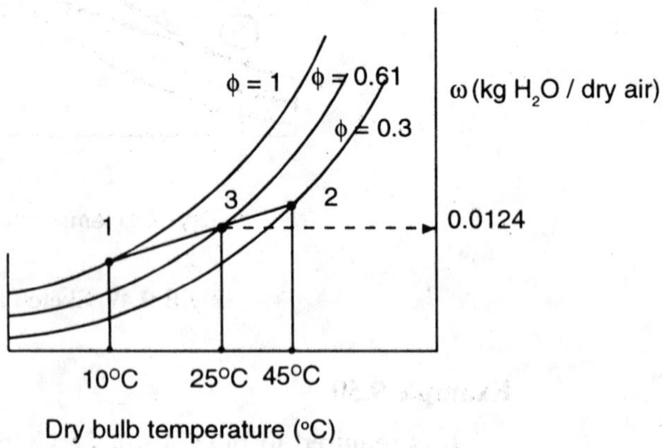


Fig.E 9.50. Sketch for Example 9.50.

Therefore, 5.4717 kg/min of (dry air) chilled air at 10°C is to be mixed with 4.5283 kg/min (dry air) of atmospheric air at 45°C.

**Example 9.51**

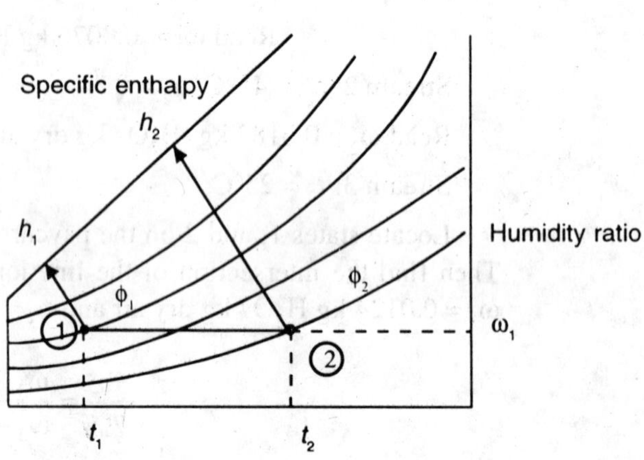


Fig. E 9.51. Constant pressure heating of moist air.

Suppose moist air at a given temperature and humidity ratio is heated at constant pressure. Sketch the process of constant pressure heating on a psychrometric chart.

*Solution* :

Let the moist air is at temperature $t_1$°C with a humidity ratio $\omega_1$. If the air is heated at constant pressure, the total pressure as well as the partial pressure of water do not change and hence the humidity ratio also does not change. However, as temperature increase the saturation pressure of water increases and hence the relative humidity $\phi$ which is given by the ratio of partial pressure of water to the saturation pressure of water ($\phi = p_w / P_S$) decreases. If the moist air is initially at state 1, it follows the path 1-2 shown in Fig.E 9.51, when it is heated at constant pressure.

## Example 9.52

If saturated air at 10°C and 1 bar is heated to 49°C at constant pressure determine the humidity ratio and relative humidity of the air. Also calculate the amount of energy to be transferred to 10 m$^3$/min of the saturated air.

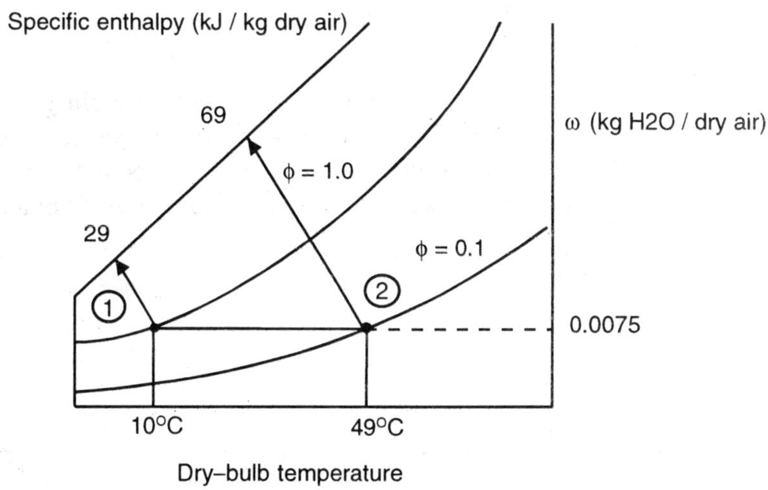


Fig.E 9.52. Sketch for Example 9.52.

*Solution* :

For saturated air at 10°C, $\omega_1 = 0.0075$ kg $H_2O$/ kg dry air and $h^*_1 = 29$ kJ / kg dry air (from chart)

$$\omega_1 = \frac{0.622\ p_{w1}}{P - p_{w1}}$$

$$\text{or} \quad 0.0075 = \frac{0.622\, p_{w1}}{1 - p_{w1}}$$

or $p_{w1} = 0.011\,914$ bar and $p_{a1} = 0.988\,086$ bar

$$\dot{m}_{a1} = \frac{p_{a_1} V M_a}{R T_1} = \frac{0.988\,086 \times 10^5 \times 10 \times 28.97 \times 10^{-3}}{8.314 \times 283} = 12.166 \text{ kg / min}$$

During heating at constant pressure, the humidity ratio does not change. Hence $\omega_2 = \omega_1 = 0.0075$ kg $H_2O$ / kg dry air.

Therefore at state 2, $t = 49$°C and $\omega = 0.0075$. This corresponds to $\phi_2 = 0.10$ and $h^*_2 = 69$ kJ / kg dry air (from chart).

The initial and final states of the air are shown in Fig.E 9.52.

Then energy added as heat $Q = \dot{m}_a (h_2 - h_1) = 12.166\,(69 - 29)$

$$= 486.64 \text{ kJ / min}$$

**Example 9.53**

Sketch the process of constant pressure cooling of a moist air on a psychrometric chart.

*Solution* :

Suppose moist air at state 1 enters a cooling unit and the air cools at constant pressure. While the air is cooling to a temperature above its dew point temperature, no condensation of water vapor occurs and hence the humidity ratio remains constant. However, the relative humidity of the air increases because at lower

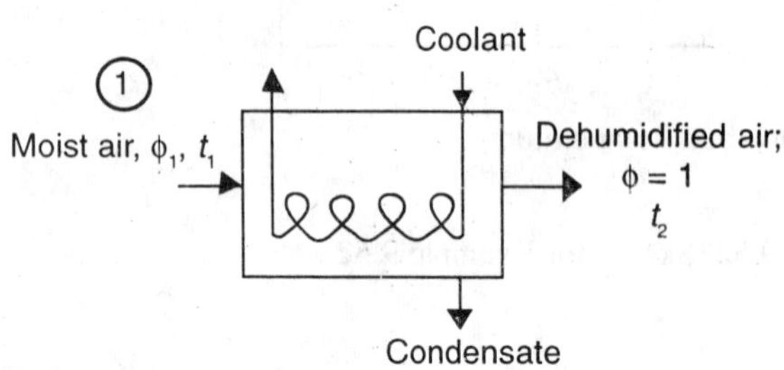


(*a*)

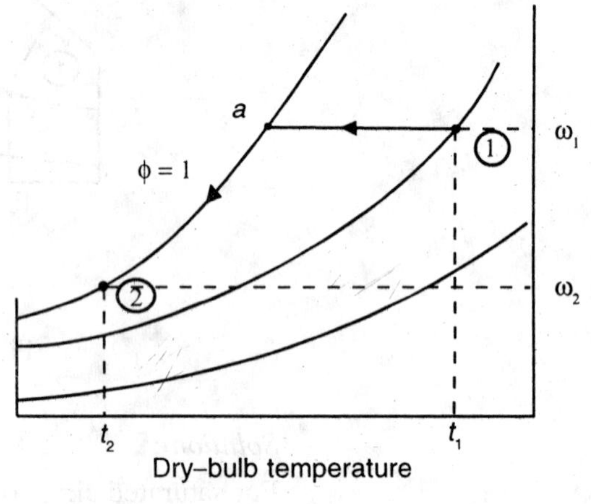


(*b*)

Fig.E 9.53. (*a*) Schematic of a cooling unit (*b*) Representation of constant pressure cooling on a psychrometric chart.

temperatures the saturation pressure of water will be low. If the moist air is cooled to a temperature below its dew point, some water vapor condenses and the air leaves as saturated. Although water condenses at various temperatures the condensed water leaves the cooling unit along with the saturated air. Therefore, the saturated air and the condensate leave the cooling unit at the same temperature. Since the water vapor condenses and the mass of air remains constant, the final humidity ratio of the air will be less than the initial humidity ratio. Therefore, one can dehumidify the air by cooling it to a temperature below its dew point. The constant pressure cooling of moist air is represented on a psychrometric chart in Fig.E 9.53. along with a schematic of the cooling unit. During 1-*a*, the temperature of the air decreases, the relative humidity increases but the humidity ratio remains constant. At state *a* the air gets saturated. Further cooling takes the air along the path *a*-2 during which the water vapor condenses and the humidity ratio decreases, but the air remains as saturated. The saturated air as well as the condensate leaving the cooling unit have the temperature $t_2$.

**Example 9.54**

Moist air at 1 bar and 40°C with a relative humidity of 50% enters a cooling device at the rate 10 $m^3/s$ and exits as saturated air at 10°C. Calculate the rate at which the condensate leaves the cooling unit and the cooling capacity of the unit. Also determine the temperature at which condensation of water vapor begins.

*Solution* :

The constant pressure cooling process is shown in Fig.E 9.54.

For the air entering the device: $t_1 = 40°C$; $\phi_1 = 0.5$

From chart we get; $\omega_1 = 0.0235$ kg $H_2O$ / kg dry air

$$\omega_1 = \frac{0.622\, p_{w1}}{P - p_{w1}}$$

or $0.0235 = \dfrac{0.622\, p_{w1}}{1 - p_{w1}}$ or $p_{w1} = 0.036\,406$ bar

and $p_{a1} = 0.963\,594$ bar

$$\dot{m}_{a1} = \frac{p_{a1}\,\dot{V}\,M_a}{RT} = \frac{0.963\,594 \times 10^5 \times 10 \times 28.97 \times 10^{-3}}{8.314 \times 313} = 10.727 \text{ kg / s}$$

Choose the cooling device as the control volume. Then mass and energy balances give

Mass balance for dry air: $\dot{m}_{a1} = \dot{m}_{a3}$ (*A*)

Mass balance for water : $\dot{m}_{w1} = \dot{m}_{w2} + \dot{m}_{w3}$ (*B*)

Energy balance : $\dot{m}_{a1} h_1^* = \dot{m}_{a3} h_3^* + \dot{m}_{w2} h_{w2} + \dot{Q}_C$ (C)

where $\dot{Q}_C$ = rate of energy removal as heat in the cooling device

$h_{w2}$ = specific enthalpy of liquid water (condensate)

Dividing Eqn. (B) with Eqn.(A), we get

$$\frac{\dot{m}_{w1}}{\dot{m}_{a1}} = \frac{\dot{m}_{w2}}{\dot{m}_{a1}} + \frac{\dot{m}_{w3}}{\dot{m}_{a1}}$$

or $\omega_1 = \dfrac{\dot{m}_{w2}}{\dot{m}_{w1}} + \omega_3$

or $\dot{m}_{w2} = \dot{m}_{a1}(\omega_1 - \omega_3)$ (D)

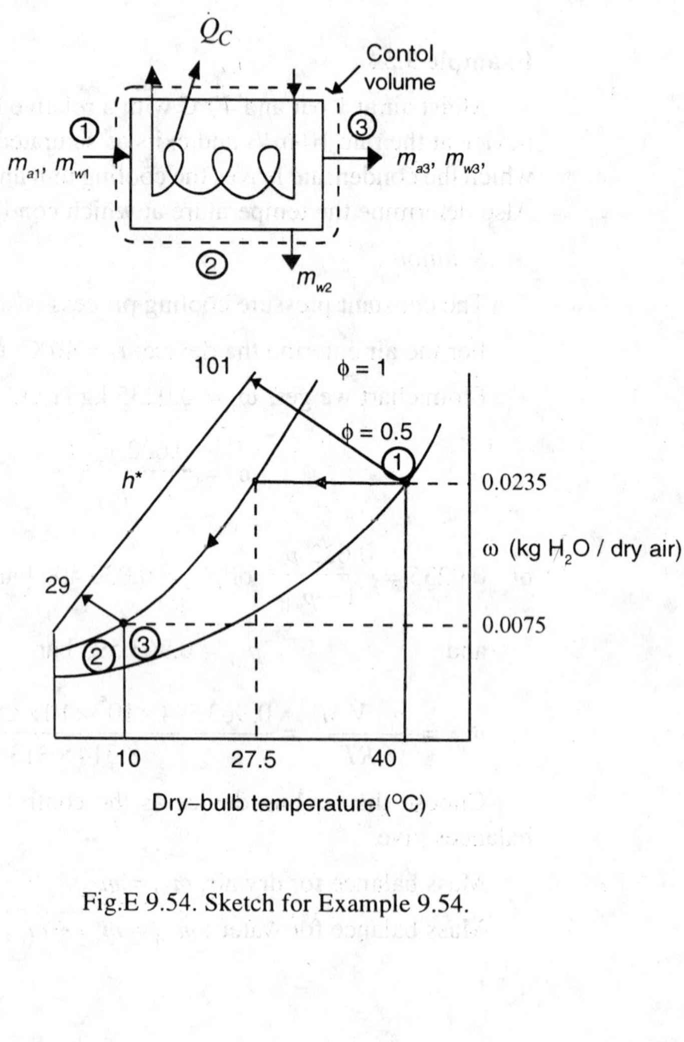


Fig.E 9.54. Sketch for Example 9.54.

Substituting Eqn.(*D*) in Eqn.(*C*) we get

$$\dot{Q}_c = \dot{m}_{a1}(h^*_1 - h^*_3) - \dot{m}_{a1}\, h_{w2}\,(\omega_1 - \omega_3)$$

or $$Q_c = m_{a1}\left[\left(h_1^* - h_3^*\right) - \left(\omega_1 - \omega_3\right)h_{w2}\right]$$

From Fig.E 9.45, we find $h_1^* = 101$ kJ / kg dry air; $h^*_3 = 29$ kJ / dry air;

$\omega_3 = 0.0075$kg $H_2O$/ kgdryair

Temperature at which condensation begin is 27.5°C

$h_{W2} = h_f(10°C) = 41.99$ kJ / kg.

From Eqn. (*D*), we get

$$\dot{m}_{w2} = \dot{m}_{a1}\left(\omega_1 - \omega_3\right) = 10.727(0.0235 - 0.0075) = 0.1716 \text{ kg/s}$$

Therefore, the rate at which condensate leaves the cooling unit = 0.1716 kg/s

From Eqn. (*E*), we get

$$Q_c = m_{a1}\left[\left(h_1^* - h_3^*\right) - \left(\omega_1 - \omega_3\right)h_{w2}\right]$$

= 10.727 [(101 - 29) - (0.0235 - 0.0075) 41.99] = 765.137 kW

Therefore, the coling capacity of the unit = 765.137 kW

**Example 9.55**

During the months of July and August the atmospheric air is humid and hot and it is desired to condition the air to a comfortable temperature and humidity level at a lower cost. Suggest a method and indicate the process to be adopted, on a psychrometirc chart,

*Solution* :

For comfortable living the desirable temperature is 25°C and a relative humidity of 40%. If the humidity of the atmospheric air is very high, it can be dehumidified to the desired level by cooling the air below its dew point. Then the air will be at a temperature lower than the desired value. Therefore, the air can then be reheated to the desired temperature at constant pressure. The process of cooling at constant pressure below the dew point of the initial moist air and subsequent heating at constant pressure is shown in Fig.E 9.55 on a psychrometric chart. The air which is initially at state 1 with humidity ratio $\omega_1$ can be cooled at constant pressure by following the path 1-*a*-3. During path 1-*a*, there is no reduction in the humidity ratio. At state *a*, the condensation of water vapor begins and continues till state 3 is reached. At state 3, the air has the desired humidity ratio $\omega_4$, but it is available at a temperature $t_2 < t_4$. Then the saturated air at state 3 can be heated at constant pressure to reach state 4. During the path 3-4 there is no change in the humidity ratio of air. By applying the mass and energy balances to the control volumes– cooling unit and heating unit–one can determine the cooling capacity and heating capacity of the unit.

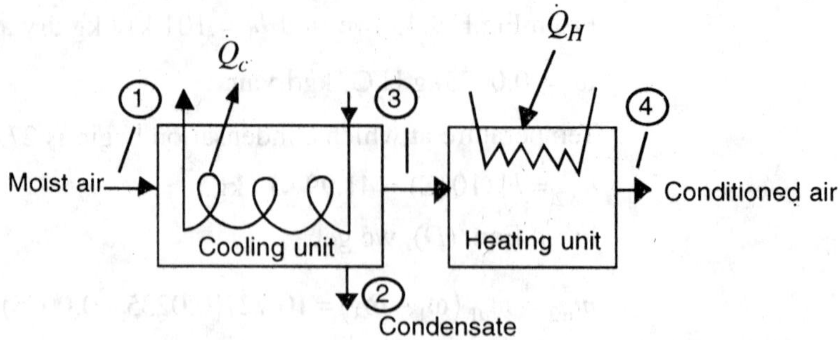


(*a*)

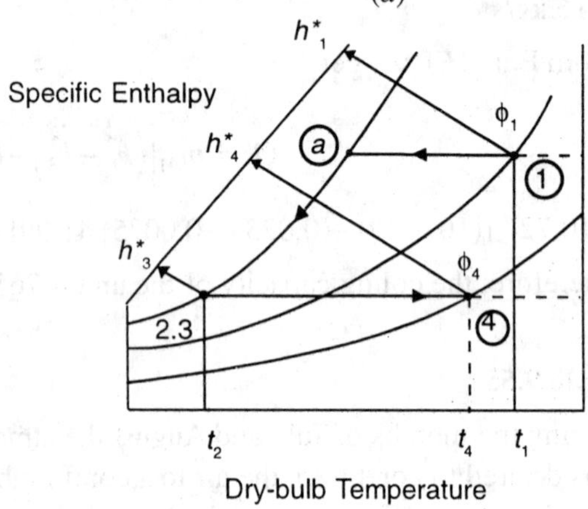


(*a*)

Fig.E 9.55. (*a*) Schematic of a dehumidifier (*b*) Psychrometric chart showing dehumidification of moist air.

### Example 9.56

Moist air at 40°C, 1 bar and 60% relative humidity is to be conditioned to 25°C with 40% relative humidity at 1 bar. If the flow rate of the moist air at 40°C is 5 m³/s, calculate the rate of cooling and the rate of heating of the dehumidifier and the rate at which water is condensed in the unit.

*Solution* :

A schematic of the dehumidifier and the process of dehumidification on a psychrometric chart are shown in Fig.E 9.56. We have $t_1 = 40$°C and $\phi_1 = 0.60$. At this state, we read from the chart

$$\omega_1 = 0.0285 \text{ kg H}_2\text{O / kg dry air},$$

and $$h_1^* = 114 \text{ kJ / kg dry air}.$$

$t_4 = 25°C$ and $\phi_4 = 0.40$. At this state, we read from the chart

$$\omega_4 = 0.008 \text{ kg H}_2\text{O / kg dry air},$$

and $$h_4^* = 46 \text{ kJ / kg dry air}.$$

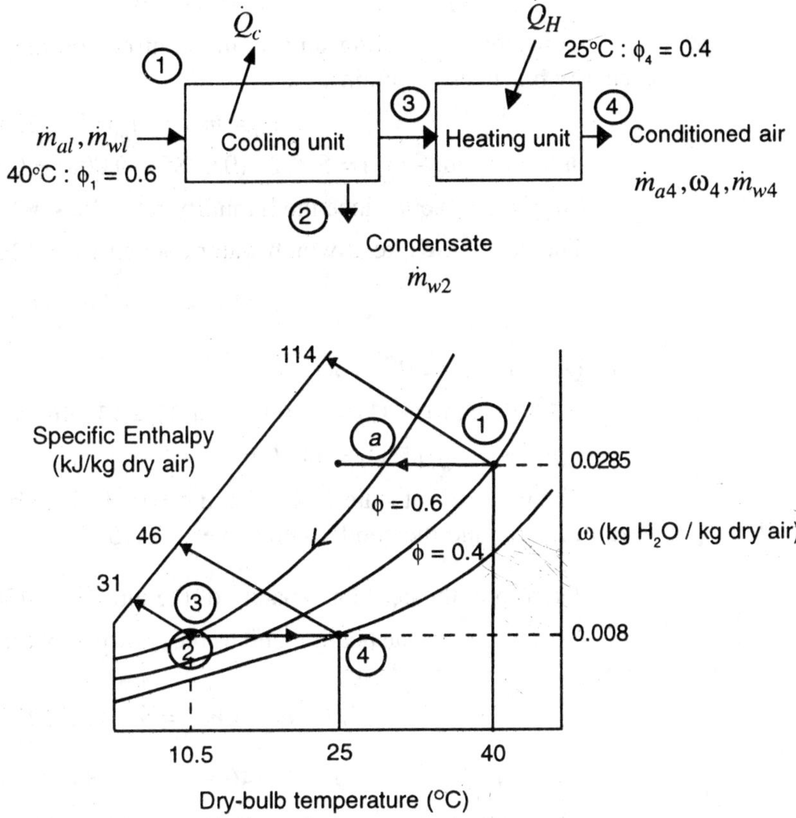


Fig.E 9.56. (*a*) Schematic of dehumidifier (*b*) Dehumidification process on a psychrometric chart

The moist air at state 1 is to be changed to state 4. For this purpose the air can be cooled at constant pressure following the path 1-*a*-3-4. The air leaves the cooling unit as saturated at state 3 and the condensed water at state 2 (at the same temperature as that of air at state 3) leaves the cooling unit. The air is then heated at constant pressure following the path 3-4.

We know that, $\omega_1 = \dfrac{0.622\, p_{w1}}{P - p_{w1}}$

or $0.0285 = \dfrac{0.622\ p_{w1}}{1 - p_{w1}}$ or $p_{w1} = 0.0438$ bar; $p_{a1} = 0.9562$ bar

$$\dot{m}_{a1} = \frac{P_{a1}\,\dot{V} M_a}{RT} = \frac{0.9562 \times 10^5 \times 5 \times 28.97 \times 10^{-3}}{8.314 \times 313} = 5.3225 \text{ kg/s}$$

Consider the cooling unit as the control volume and apply the mass and energy balances to obtain

$$\dot{m}_{a1} = \dot{m}_{a3} = \dot{m}_a = 5.3225 \text{ kg/s}$$

$\dot{m}_{w2} = \dot{m}_a(\omega_1 - \omega_3) = 5.3225\ (0.0285 - 0.008) = 0.1091$ kg / s

($\omega_3 = \omega_4 = 0.008$ since the humidity ratio does not change in the path 3-4).

Therefore, the rate at which water condenses = 0.1091 kg/s

$$\dot{m}_a h^*_1 = \dot{m}_a h^*_3 + \dot{m}_{w2}\ h_{w2} + \dot{Q}_c$$

or $\dot{Q}_c = \dot{m}_a (h_1^* - h_3^*) - \dot{m}_{w2}\ h_{w2}$

$= 5.3225\ (114 - 31) - 0.1091 \times 44.09 = 436.96$ kW

$[(h_{w2} = h_f\ (10.5°\text{C}) = 44.09 \text{ kJ / kg})]$

The moist air at state 1 (40°C and $\phi = 0.60$) is to be cooled to 10.5°C in the cooling unit and the condensate leaves at 10.5°C,

Therefore, the cooling capacity of the unit $\dot{Q}_c = 436.96$ kW.

Now, consider the heating unit as the control-volume and apply the mass and energy balances to obtain

$$\dot{m}_{a3} = \dot{m}_{a4} = \dot{m}_a = 5.3225 \text{ kg/s}$$

$\dot{Q}_H = \dot{m}_a (h_4^* - h_3^*) = 5.3225\ (46 - 31) = 79.84$ kW

Therefore, the heating capacity of the unit = 79.84 kW.

**Example 9.57**

It is required to supply 200 m$^3$ /min of air at 20°C and 1 bar with 40% relative humidity into an auditorium. For this purpose atmospheric air at 37°C with 60% relative humidity is to be conditioned in a dehumidifier unit. Calculate the temperature to which the air is to be initially cooled, and the capacities of the cooling and heating units. Also determine the rate at which water condenses in the cooling unit.

*Solution* :

For moist air at state 1, we have $t = 37°$C and $\phi = 0.60$. From the psychrometric chart we read $\omega_1 = 0.0244$ kg $H_2O$ / kg dry air and $h_1^* = 97$ kJ/kg dry air.

For the conditioned air at 20°C and $\phi = 0.40$, we read from the chart $\omega_4 = 0.006$ kg $H_2O$ / kg dry air and $h_4^* = 35$ kJ / kg dry air.

The dehumidification path 1-*a*-3-4 to be followed by the moist air is shown on a psychrometric chart in Fig.E 9.57.

From the figure, we find that the air is to be initially cooled to 6.5°C in the cooling unit, before it is fed to the heater. At 6.5°C; we find $h_3^* = 21$ kJ / kg dry air.

The condensate also leaves at 6.5°C.

At 6.5°C, $h_{w2} = 27.21$ kJ/kg

Air is supplied into the auditorium at $\omega_4 = 0.006$ kg $H_2O$ / kg dry air

$$\omega_4 = \frac{0.622\, p_{w4}}{P - p_{w4}}$$

or $$0.006 = \frac{0.622\, p_{w4}}{1 - p_{w4}}$$

or $p_{w4} = 0.0096$ bar and $p_{a4} = 0.9904$ bar

Therefore, $$\dot{m}_{a4} = \frac{p_{a4}\, \dot{V}_4 M_a}{RT} = \frac{0.9904 \times 10^5 \times 200 \times 28.97 \times 10^{-3}}{8.314 \times 293}$$

$$= 235.57 \text{ kg / min}$$

or $$\dot{m}_{a4} = \frac{235.57}{60} = 3.926 \text{ kg / s}$$

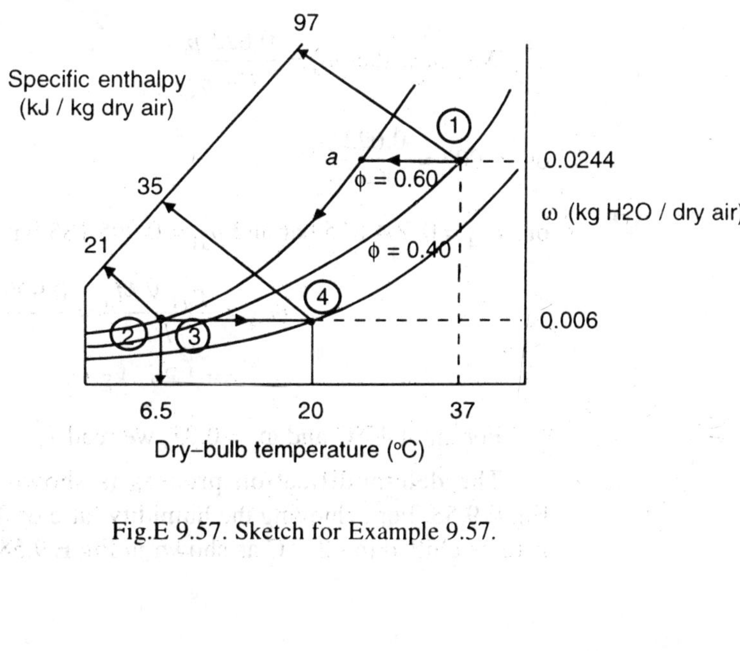


Fig.E 9.57. Sketch for Example 9.57.

From the figure, we find $\omega_3 = \omega_4 = 0.006$ kg $H_2O$ / kg dry air

Rate at which water condenses in the cooling unit, $\dot{m}_{w2}$ is given by

$$\dot{m}_{w2} = \dot{m}_a(\omega_1 - \omega_3) = 3.926\,(0.0244 - 0.006)$$
$$= 0.072 \text{ kg/s}$$

The cooling capacity of the cooling unit $\dot{Q}_c$ is given by

$$\dot{Q}_c = \dot{m}_a(h_1^* - h_3^*) - \dot{m}_{w2}\, h_{w2}$$
$$= 3.926\,(97 - 21) - 0.072 \times 27.21 = 296.42 \text{ kW}$$

The capacity of the heating unit $\dot{Q}_H$ is given by

$$\dot{Q}_H = \dot{m}_a(h_4^* - h_3^*) = 3.926\,(35 - 21) = 54.96 \text{ kW}$$

### Example 9.58

It is required to maintain a cold storage at 10°C and 1 bar with a relative humidity of 40%. On a particular summer day the atmospheric air is at 45°C with a humidity ratio of 0.03 kg $H_2O$/kg dry air. The atmospheric air is to be conditioned and supplied to the cold storage at 10°C at a rate of 50 $m^3$/s. Determine the temperature to which the atmospheric air is to be chilled before reheating and the capacities of the refrigeration and heating units. Also determine the rate of condensate removal from the refrigeration unit

*Solution* :

For air at 10°C and $\phi = 0.40$, we read $\omega_4 = 0.003$ kg $H_2O$ / kg dry air

$$h_4^* = 17 \text{ kJ / kg dry air (Read from chart)}$$

We know that $\omega_4 = \dfrac{0.622\, p_{w4}}{P - p_{w4}}$

or $0.003 = \dfrac{0.622\, p_{w4}}{1 - p_{w4}}$

or $p_{w4} = 0.004\,815$ bar and $p_{a4} = 0.995\,185$ bar

$$\dot{m}_{a4} = \frac{p_{a4}\,\dot{V}\,M_a}{RT} = \frac{0.995\,185 \times 10^5 \times 50 \times 28.97 \times 10^{-3}}{8.314 \times 283}$$
$$= 61.267 \text{ kg / s}$$

For air at 45°C and $\omega_1 = 0.03$, we read $h_1^* = 123$ kJ / kg dry air

The dehumidification process is shown on a psychrometric chart in Fig.E 9.58. For achieving the humidity ratio of 0.003 kg $H_2O$ / kg dry air the air is to be chilled to –2.5°C as shown in Fig.E 9.58

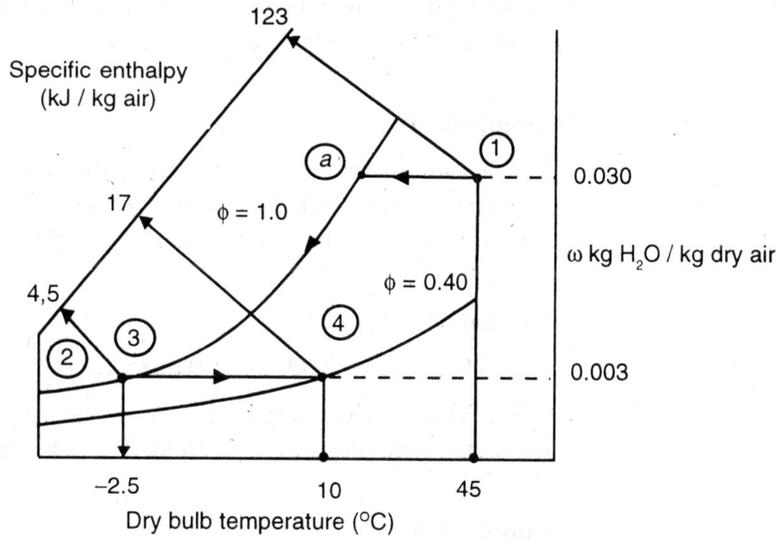


Fig.E 9.58 Sketch for Example 9.58.

Rate of condensate formation, $\dot{m}_{w2} = \dot{m}_a(\omega_1 - \omega_3) = \dot{m}_a(\omega_1 - \omega_4)$

or $\dot{m}_{w2} = 61.27\,(0.03 - 0.003) = 1.654$ kg/s

At –2.5°C, $h_{w2} = -\ 10.58$ kJ / kg (Extrapolated from steam tables)

Capacity of refrigeration unit, $\dot{Q}_c = \dot{m}_a(h_1^* - h_3^*) - \dot{m}_{w2}\,h_{w2}$

or $\dot{Q}_c = 61.27\,(123 - 4.5) - 1.654\,(-10.58) = 7.278$ MW

Capacity of heating unit, $\dot{Q}_H = \dot{m}_a(h_4^* - h_3^*)$

or $\dot{Q}_H = 61.27(17 - 4.5) = 765.88$ kW.

## Example 9.59

In addition to the use of refrigeration units, are there any other methods for dehumidification of moist air ?

*Solution* :

It is possible to dehumidify the moist air by spraying chilled water into the moist air. The fine spray provides adequate heat transfer area and the air leaves with reduced humidity ratio. It is also possible to pass the moist air through silica gel or anhydrous calcium carbonate which absorbs the water vapor, for reducing the humidity ratio. The moist air may also be brought into intimate contact with oleum or a strong liquor of lithium chloride to reduce the humidity ratio. However, the only method which is least expensive is to spray chilled water into the moist air and all other methods are quite expensive.

### Example 9.60

Moist air at 40°C and 1 bar with a relative humidity of 60% enters a chamber into which enough chilled water at 10°C is sprayed, reducing the temperature of the air to 10°C. Calculate the humidity ratio of the entering air and the air leaving the unit.

*Solution* :

At 40°C and $\phi = 0.60$; $\omega_1 = 0.0285$ kg $H_2O$ / kg dry air.

The air leaves as saturated at 10°C. Hence at 10°C and $\phi = 1.0$, the humidity ratio $\omega$ of the leaving air is 0.0075 kg $H_2O$ / kg dry air.

### Example 9.61

Is it possible to achieve cooling by humidifying the air ? If so sketch the path followed by the process of humidification with cooling, on a psychrometric chart.

*Solution* :

If moist air which is initially not saturated is made to flow through porous pads which are maintained wet and the unit is well insulated, then the air comes into thorough contact with liquid water and reaches a state of equilibrium. Since

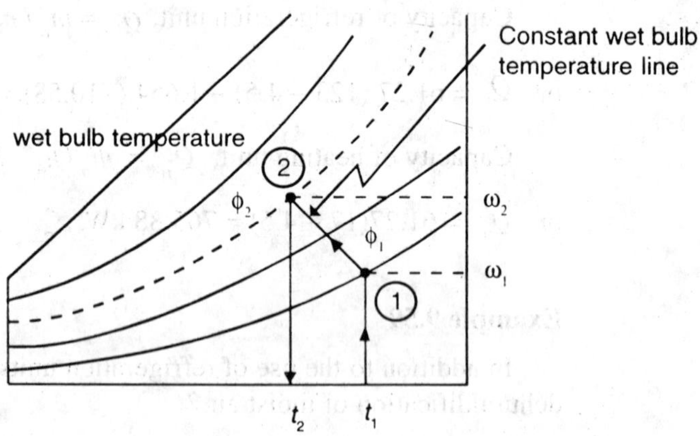


Fig.E 9.61. Psychrometric chart showing the process of humidification with cooling.

the air is not saturated initially, some water vaporizes and the air gets saturated. The energy required for the evaporation of water does not come from external sources (since the unit is well insulated), it receives the same from the air itself and hence the air gets cooled. The process is essentially equivalent to adiabatic saturation except that the air may not leave the unit in a saturated state. Suppose air at temperature $t_1$ with relative humidity $\phi_1$ enters the device it follows the path 1-2 as shown in Fig.E 9.61 and leaves at temperature $t_2 < t_1$ with $\omega_2 > \omega_1$.

### Example 9.62

What is the principle on which the desert coolers or evaporative coolers function ?

*Solution* :

The desert coolers or evaporative coolers are extensively used in hot and relatively dry climatic conditions. The desert coolers employ the principle of humidification with cooling. A schematic of a desert cooler is shown in Fig.E 9.62. It consists of an exhaust fan and soaked pads. The pads are kept wet all the time by spraying water on them. The exhaust fan draws the hot and dry atmospheric air through the pores of the soaked pads. While the air passes through the pads, the humidity of the air increases and its temperature decreases. Then the exhaust fan throws the cool air into the room. In principle it is possible to cool the air to its dew point temperature by employing a desert cooler.

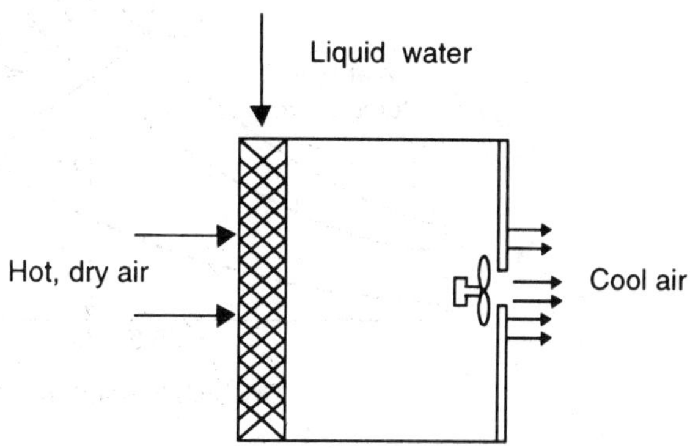


Fig.E 9.62. Schematic of a desert cooler.

### Example 9.63

On a hot summer day the atmospheric air is at 45°C and 1 bar with a relative humidity of 20%. (*a*) Estimate the minimum temperature of the air that can be obtained with a perfectly designed desert cooler. (*b*) Suppose a desert cooler, which is provided with an exhaust fan which can suck 1 m$^3$/s of air 45°C increases the relative humidity of the air to 80%, what is the temperature of the air delivered by the cooler and the amount of water required to operate the cooler continuously for 3 hours ?

*Solution* :

(*a*) The atmospheric air is at 45°C with $\phi = 0.20$. Locate the initial state of the air on the psychrometric chart as shown in Fig.E 9.63. The initial and final states of the air are represented as 1 and 2, respectively in the figure. If the desert cooler is perfectly designed it is possible to achieve adiabatic saturation and the air can leave the cooler at adiabatic saturation temperature. Follow the constant wet-bulb temperature line from state 1 till it intersects the saturation curve at state 2. At state 2, read the temperature as 25.2°C. Therefore, the minimum temperature that ~~can be~~ obtained is 25.2°C.

(*b*) Starting from state 1, follow the constant wet - bulb temperature line till it intersects the $\phi = 0.80$ curve (state 2'), At state 2' the temperature of the air is 27.8°C and $\omega_{2'} = 0.019$ kg $H_2O$ / kg dry air

Therefore, the cooler delivers air at 27.2°C.

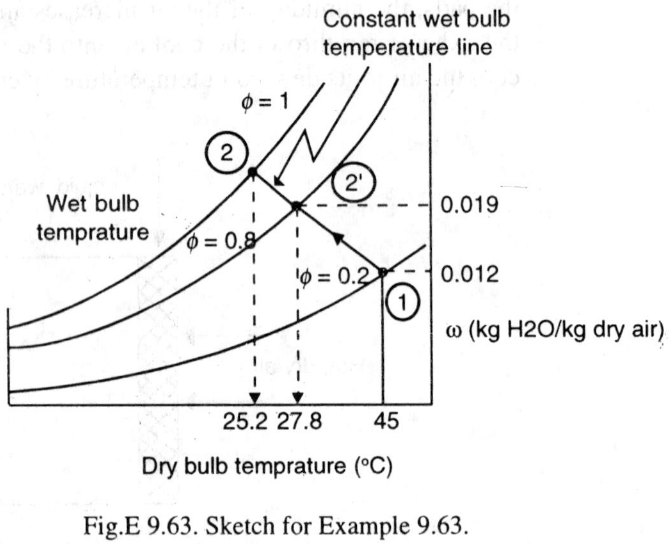


Fig.E 9.63. Sketch for Example 9.63.

The amount of water required can be obtained from $\dot{m}_w = \dot{m}_a\,(\omega_{2'} - \omega_1)$

At State 1, $\omega_1 = 0.012$ kg $H_2O$ / kg dry air

$$\text{We know that } \omega_1 = \frac{0.622\, p_{w1}}{P - p_{w1}}$$

$$\text{or } \quad 0.012 = \frac{0.622\, p_{w1}}{1 - p_{w1}}$$

or $p_{w1} = 0.018\,927$ bar and $p_{a1} = 0.981\,073$ bar

$$\dot{m}_a = \frac{p_{a1}\dot{V}M_a}{RT} = \frac{0.981\,073\times10^5\times1\times28.97\times10^{-3}}{8.314\times318}$$

$$= 1.075 \text{ kg / s}$$

Amount of water required for 3 hours of operation $= 3 \times 3600 \times 1.075$ $(0.019 - 0.012) = 81.27$ kg

## Example 9.64

Show that the process followed in a desert cooler is the same as the adiabatic saturation process.

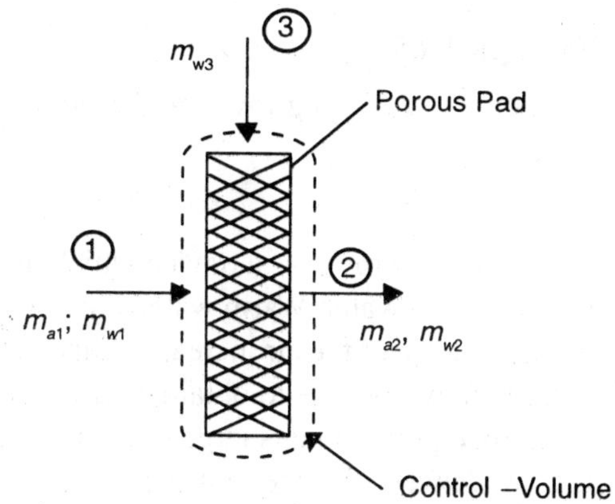


Fig.E 9.64. Schematic of a porous pad through which unsaturated air passes increasing its humidity

*Solution* :

A Schematic of the porous pad through which unsaturated moist air passes increasing its specific humidity is shown in Fig.E 9.64. Let

$\dot{m}_{a1}, \dot{m}_{a2}$ = mass flow rate of dry air into and out of the pad, respectively.

$\dot{m}_{w1}, \dot{m}_{w2}$ = mass flow rate of water along with the dry air into and out of the pad, respectively.

$h_1^*, h_2^*$ = specific enthalpy air at the inlet and outlet of the pad, respectively.

$\dot{m}_{w3}$ = flow rate of make up water.

Consider the control volume as shown in Fig.E 9.64 and apply the conservation of mass and energy to get:

Mass balance for dry air : $\dot{m}_{a1} = \dot{m}_{a2} = \dot{m}_a$

mass balance for water : $\dot{m}_{w1} + \dot{m}_{w3} = \dot{m}_{w2}$

or $\dot{m}_a \omega_1 + m_{w3} = \dot{m}_a \omega_2$

or $\dot{m}_{w3} = \dot{m}_a (\omega_2 - \omega_1)$ (*A*)

Energy balance : $\dot{m}_a h_{a1} + \dot{m}_{w1} h_{g1} + \dot{m}_{w3} h_{w3} = \dot{m}_a h_{a2} + \dot{m}_{w2} h_{g2}$

or $\dot{m}_a (h_{a1} + \omega_1 h_{g1}) + \dot{m}_a (\omega_2 - \omega_1) h_{w3} = \dot{m}_a (h_{a2} + \omega_2 h_{g2})$

or $h_1^* + (\omega_2 - \omega_1) h_{w3} = h_2^*$ (*B*)

In Eqn (*B*) the second term of the LHS is very small compared to the other terms and hence it can be neglected. Then Eqn. (*B*) reduces to

$$h_1^* = h_2^*$$

Which is the same as the adiabatic saturation relation.

**Example 9.65**

What is a cooling tower and how do we analyse the operation of a cooling tower ?

*Solution* :

A cooling tower is a device which uses the principle of evaporative cooling to cool warm water. Warm water is cooled by evaporation and in the process the humidity ratio of the air increases. Since cooling is produced by evaporation, it is theoretically possible to obtain cold water at the wet-bulb temperature of the air. In thermal power plants large quantities of cooling water is necessary to operate the condensers. Hence cooling towers are extensively used to provide large quantities of cold water. There are two types of cooling towers in common use. They are natural draft cooling towers and forced draft cooling towers. A schematic of a cooling tower is shown in Fig.E 9.65. The warm water to be cooled enters at 3 and it is sprayed from the top of the tower. The falling water passes through a series of baffles which break the stream into fine drops. Atmospheric air is drawn at 1 by the fan which is located at the top of the tower. The fine spray of water provides large surface area for energy transfer and evaporation. The air gets humidified and leaves the tower at 2. The outgoing air has a higher humidity ratio than the incoming air. The cool water leaves the tower at 4. Since some water evaporates to increase the humidity of air, some make up water is usually added in the cooling tower. In some region of the tower there is a temperature difference between the air and water and energy transfer as heat occurs from water to air. However, in some other region of the tower, the temperature of the water may be lower than that of the air and energy flows as heat from air to water. To analyze the complex processes that occur in a cooling tower a knowledge of energy transfer through various modes like conduction, convection and radiation, and mass transfer as well as fluid mechanics is needed in addition to a knowledge of thermodynamics. However, some information regarding the performance of cooling towers can be obtained by applying mass and energy balances to the cooling tower.

Let

$\dot{m}_{a1}, \dot{m}_{a2}$ = mass flow rate of dry air into and out of the cooling tower, respectively

$\omega_1, \omega_2$ = humidity ratio of air at the inlet and exit of the cooling tower, respectively.

$\dot{m}_{w3}$ = mass flow rate of warm water into the cooling tower.

$\dot{m}_{w4}$ = mass flow rate of cool water out of the cooling tower.

Consider the cooling tower as the control volume and apply the mass balance and energy balance to obtain

Mass balance for air: $\dot{m}_{a1} = \dot{m}_{a2} = \dot{m}_a$

Mass balance for water; $\dot{m}_{a1}\,\omega_1 + \dot{m}_{w3} = \dot{m}_{a2}\,\omega_2 + \dot{m}_{w4}$

or $$(\dot{m}_{w3} - \dot{m}_{w4}) = \dot{m}_a(\omega_2 - \omega_1) \qquad (A)$$

Energy balance : $\dot{m}_{a1}\,h_1^* + \dot{m}_{w3}\,h_{w3} = \dot{m}_{a2}\,h_2^* + \dot{m}_{w4}\,h_{w4}$

or $$\dot{m}_a(h_2^* - h_1^*) = \dot{m}_{w3}\,h_{w3} - \dot{m}_{w4}\,h_{w4} \qquad (B)$$

The Eqns. (*A*) and (*B*) can be simultaneously solved to obtain the required air flow rate and make up water $(\dot{m}_{w3} - \dot{m}_{w4})$.

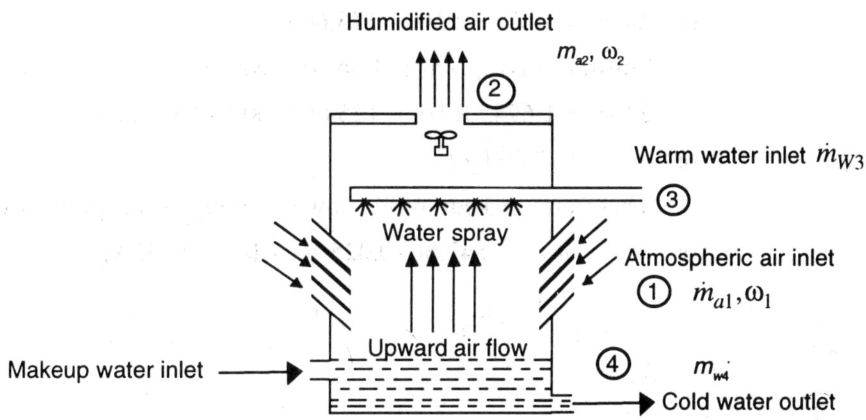


Fig.E 9.65. Schematic of a cooling tower.

## Example 9.66

A large thermal power plant uses a cooling tower to provide cold water at 30°C to the condensers. The warm water leaves the condensers at 40°C at a rate of 1000 kg/s. On a particular day in summer, the atmospheric air is at 40°C and 1 bar with 30% relative humidity. The air leaves the cooling tower at 35°C with 70% relative humidity. Calculate the required air flow rate and make up water flow rate.

*Solution* :

Given data: $\dot{m}_{w3} = 1000$ kg / s; $t_3 = 40$°C.

Therefore, $h_{w3} = 167.45$ kJ / kg

Cold water: $t_4 = 30$°C.

Therefore, $h_{w4} = 125.66$ kJ / kg

Entering air: $t_1 = 40$°C and $\phi_1 = 0.3$

Therefore, $\omega_1 = 0.014$ kg $H_2O$ / kg dry air; $h_1^* = 76$ kJ / kg dry air

Leaving air: $t_2 = 35$°C and $\phi_2 = 0.7$

Therefore, $\omega_2 = 0.025$ kg $H_2O$ / kg dry air; $h_2^* = 100$ kJ / kg dry air

We know that, $\dot{m}_{w3} - \dot{m}_{w4} = \dot{m}_a (\omega_2 - \omega_1)$

or $1000 - \dot{m}_{w4} = \dot{m}_a (0.025 - 0.014)$

or $1000 - \dot{m}_{w4} = 0.011\, \dot{m}_a$ (*A*)

We also know that $\dot{m}_a (h_2^* - h_1^*) = \dot{m}_{w3} h_{w3} - \dot{m}_{w4} h_{w4}$

or $\dot{m}_a (100 - 76) = 1000 \times 167.45 - \dot{m}_{w4} (125.66)$

or $24\, \dot{m}_a = 1.6745 \times 10^5 - 125.66\, \dot{m}_{w4}$ (*B*)

Substituting Eqn (*A*) in Eqn (*B*), we get

$24\, \dot{m}_a = 1.6745 \times 10^5 - 125.66 (1000 - 0.011 \dot{m}_a)$

or $\dot{m}_a = 1847.66$ kg / s

Therefore, makeup water flow rate $= \dot{m}_{w3} - \dot{m}_{w4} = \dot{m}_a (\omega_2 - \omega_1)$

or $\dot{m}_{w3} - \dot{m}_{w4} = 1847.66 (0.025 - 0.014) = 20.32$ kg / s

# 10

# Combustion and Chemical Thermodynamics

### Example 10.1

The energy needed to operate the heat engines, boilers etc. is obtained by the combustion of fuels. Name a few commonly used fuels and how does one specify the composition of a fuel.

*Solution* :

Most of the fuels contain carbon and hydrogen and they are available as solids, liquids or gases. The commonly used fuels are coal, gasoline, diesel oil, kerosene, natural gas, coke oven gas, producer gas and water gas. Usually the analysis of coal is given in terms of proximate analysis which gives the mass percent of volatile matter, water vapor, carbon, ash etc. or in terms of ultimate analysis or chemical analysis which specifies the mass percent of each constituent element. The liquid fuels are mixtures of several hydrocarbons. For combustion analysis usually a liquid fuel is treated as a single hydrocarbon with an empirical formula $C_xH_y$, even though it is a mixture of several hydrocarbons. For example gasoline is treated as octane ($C_8H_{18}$) and diesel oil is treated as dodecane, ($C_{12}H_{26}$). The natural gas is mostly methane, coal gas or coke oven gas is primarily a mixture of methane and hydrogen, water gas is essentially a mixture of CO and $H_2$, whereas producer gas is mainly a mixture of CO, $H_2$ and $N_2$. Usually the analysis of gaseous fuels is given in terms of volume percent or mole percent of the constituent chemical species.

### Example 10.2

How does one represent a combustion reaction ?

*Solution* :

The combustion of a fuel involves the oxidation of the constituents of a fuel and the process can be represented by a chemical reaction. In combustion process the reactants disappear and products appear as the reaction proceeds. Even though the masses of the fuel and oxidant do not remain constant in a combustion reaction, the masses of the constituent elements remain constant. Suppose methane $CH_4$, undergoes complete combustion in presence of oxygen forming $CO_2$ and $H_2O$, the combustion reaction can be represented as

$$CH_4 + 2O_2 \rightarrow CO_2 + 2H_2O$$

That is one mole of $CH_4$ reacts with 2 moles of $O_2$ yielding one mole of $CO_2$ and 2 moles of $H_2O$.

**Example 10.3**

What is stoichiometric or theoretical air ?

*Solution* :

In most of the combustion processes, the oxygen required for combustion is not supplied as pure oxygen but supplied in the form of air, which contains oxygen. Ignoring the presence of $CO_2$, argon and other gases which are present in negligible quantities in air, the air can be treated as a mixture of 21 moles of oxygen for every 79 moles of nitrogen. Then the mole ratio of nitrogen to oxygen is 79/21 = 3.76 in air. The minimum amount of air which supplies the required amount of oxygen for complete combustion of a fuel is called the stoichiometric or theoretical air. The complete combustion means all the carbon in the fuel is converted to $CO_2$ and all the hydrogen in the fuel is completely converted to $H_2O$ and nitrogen does not undergo any oxidation. That is, there is no free oxygen in the combustion products and nitrogen is treated as an inert gas.

**Example 10.4**

What is the meaning of air – fuel ratio?

*Solution* :

In the combustion of fuels, usually if the fuel is supplied with stoichiometric or theoretical air, the reaction does not go to completion. Therefore, the air is supplied in excess of the stoichiometric requirement. The air–fuel ratio (A/F) is the ratio of the mass of air used for combustion to the mass of fuel. That is

$$\text{Air–Fuel ratio } A/F = \frac{\text{mass of air}}{\text{mass of fuel}}$$

The air–fuel ratio can also be expressed on molar basis as

$$A/F = \frac{\text{moles of air}}{\text{moles of fuel}}$$

Usually the amount of air actually supplied is expressed in terms of percent theoretical air or percent excess air. The excess air denotes the amount of air supplied over and above the theoretical requirement. Thus 120 percent theoretical air means that the fuel is supplied with 1.2 times the stoichiometric air or the fuel is supplied with 20 percent excess air.

**Example 10.5**

Calculate the stoichiometric air–fuel ratio on mass basis and on molar basis for the combustion of butane $C_4H_{10}$.

*Solution* :

The complete combustion reaction for butane in presence of air can be written as

$$C_4H_{10} + 6.5O_2 + 6.5\,(3.76)\,N_2 \longrightarrow 4CO_2 + 5H_2O + 24.44\,N_2$$

Therefore, for the complete combustion of one mole of butane, 6.5 moles of $O_2$ and 24.44 moles of $N_2$ (that is 6.5 + 24.44 = 30.94 moles of air) is to be supplied. Then

$$\text{Stoichiometric A/F} = \frac{30.94}{1} = 30.94 \text{ mol air / mol fuel}$$

Molar mass of air $= 28.97 \times 10^{-3}$ kg / mol and

Molar mass of butane $C_4H_{10} = 58 \times 10^{-3}$ kg / mol.

Therefore, stoichiometric air fuel ratio on mass basis is given by

$$\text{Stoichiometric A/F} = \frac{30.94 \times 28.97 \times 10^{-3}}{58 \times 10^{-3}} = 15.45\ \frac{\text{kg air}}{\text{kg fuel}}$$

**Example 10.6**

Calculate the molar composition of the combustion products if butane is burned with theoretical air. Assume complete combustion.

*Solution* :

The combustion reaction is given by

$$C_4\,H_{10} + 6.5O_2 + 6.5\,(3.76)\,N_2 \longrightarrow 4\,CO_2 + 5\,H_2O + 24.44\,N_2$$

Total moles of combustion products $n = \Sigma\, n_i$

or $n = 4 + 5 + 24.44 = 33.44$

$$\text{Therefore, } CO_2 = \frac{4}{33.44} \times 100 = 11.96\%$$

$$H_2O = \frac{5 \times 100}{33.44} = 14.95\%;\ N_2 = \frac{22.44 \times 100}{33.44} = 73.09\%$$

Hence, the molar composition of combustion products is 11.96% $CO_2$ ; 14.95% $H_2O$ and 73.09% $N_2$.

**Example 10.7**

A sample of coke–oven gas has the following volumetric analysis: 6.5% CO; 2.5%$CO_2$; 54% $H_2$; 4% $N_2$ ; 32.5% $CH_4$ and 0.5% $O_2$.

Calculate (*a*) the theoretical air-fuel ratio on molar basis and (*b*) the composition of the combustion products if 150 per cent theoretical air is used for combustion. Assume complete combustion.

*Solution* :

(*a*) Consider 100 moles of fuel gas as the basis for calculation. Then the fuel contains 6.5 moles of CO. The combustion reaction for CO is

$$CO + \frac{1}{2} O_2 \rightarrow CO_2$$

That is, for the combustion of 1 mol CO, we need 1/2 mol $O_2$. Therefore, for the complete combustion of 6.5 mol CO, we need 6.5/2 mol $O_2$. Similarly, we can calculate for the other species also. The calculations are shown in the following table.

| Component | Moles | $O_2$ required moles | Products formed (moles) |
|---|---|---|---|
| CO | 6.5 | 6.5/2 | 6.5/2 $CO_2$ |
| $CO_2$ | 2.5 | 0 | 2.5 $CO_2$ |
| $H_2$ | 54 | 27 | 54 $H_2O$ |
| $N_2$ | 4 | 0 | 4 $N_2$ |
| $CH_4$ | 32.5 | 65 | 32.5 $CO_2$ + 65 $H_2O$ |
| $O_2$ | 0.5 | −0.5 | – |
| | | 94.75 | 38.25 $CO_2$ + 119 $H_2O$ + 4 $N_2$ |

For the combustion of 100 mol fuel, we need 94.75 mol $O_2$. When air is supplied to provide the oxygen for combustion it carries 3.76 × 94.75 = 356.26 mol $N_2$ with it. Then the combustion reaction can be written as

$6.5\ CO + 2.5\ CO_2 + 54\ H_2 + 4\ N_2 + 32.5\ CH_4 + 94.75\ O_2 + 356.26\ N_2$

$\rightarrow 38.25\ CO_2 + 119 H_2O + (4 + 356.26)\ N_2$

Total moles of air supplied = 94.75 + 356.26 = 451

Theoretical air-fuel ratio A/F = 451/100 = 4.51 mol air/mol fuel

(*b*) If excess air is used for combustion, the excess $O_2$ will appear in the combustion products. Then the combustion reaction with 150 percent theoretical air is given by

$6.5\ CO + 2.5\ CO_2 + 54\ H_2 + 4\ N_2 + 32.5\ CH_4 + 1.5\ (94.75)\ O_2 + 1.5\ (356.26)\ N_2$
$\rightarrow 38.25\ CO_2 + 119\ H_2O + 47.375\ O_2 + 538.39\ N_2$

$\Sigma n_i$ (Combustion products) = 38.25 + 119 + 43.375 + 538.39

= 739.015 mol

Then the molar composition of the combustion products is given by

$$CO_2 = \frac{38.25 \times 100}{739.015} = 5.18\%$$

$$H_2O = \frac{119 \times 100}{739.015} = 16.10\%$$

$$O_2 = \frac{43.375 \times 100}{739.015} = 5.87\%$$

$$N_2 = \frac{538.39 \times 100}{739.015} = 72.85\%$$

**Example 10.8**

What is an Orsat apparatus and how is it used ?

*Solution* :

An Orsat apparatus is a device which is used to analyze the products of combustion. A schematic of the Orsat apparatus is shown in FigE. 10.8. It consists of a measuring burette and three reagent pipettes which are used to successively absorb $CO_2$, $O_2$ and CO. The first pipette contains 20% solution of potassium hydroxide which absorbs $CO_2$, the second pipette contains an alkaline solution of pyrogallic acid which absorbs $O_2$ and the third pipette contains an acidic solution of cuprous chloride which absorbs CO. First a known volume of the sample of the combustion products to be analyzed is drawn into the measuring burette while the stop–cocks of all the reagent pipettes are closed. Then the apparatus is isolated by closing the sampling stop–cock. Next, the stop–cock connecting the reagent pipette containing 20% KOH solution is opened and the sample is forced through the solution absorbing all the $CO_2$. Then the sample is brought back into the measuring burette and the reduction in volume is measured. The procedure is repeated such that the sample is forced through the other two reagent pipettes also individually. The reduction in volume represents the partial volume of the component which has been absorbed. The composition of the sample is then estimated with a knowledge of the partial volumes of the constituents. The Orsat analysis is carried out at temperature which is lower than the dew point temperature of the combustion products and hence the water vapor is not present in the combustion products analyzed by the Orsat apparatus. Thus the Orsat analysis gives the analysis on dry basis. The combustion products contain $CO_2$, CO, $O_2$ and $N_2$ only and hence the unabsorbed gas is treated as $N_2$.

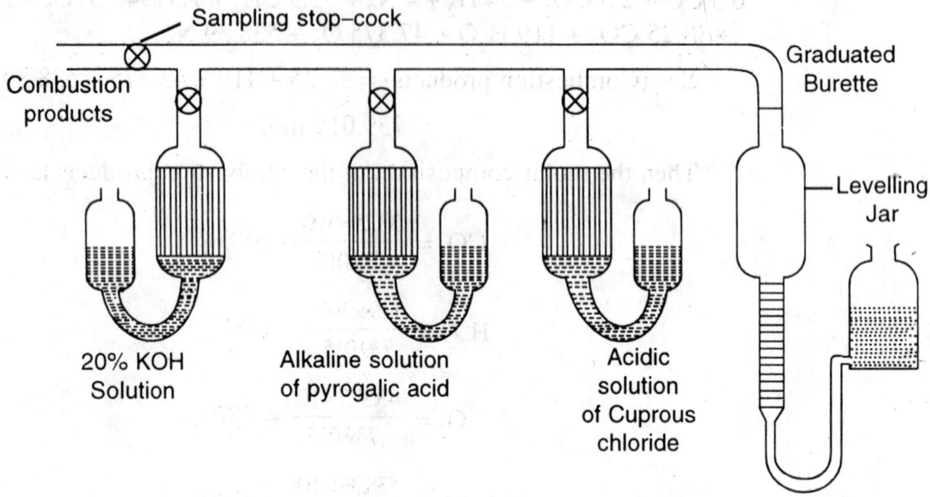


Fig.E 10.8. Schematic of an Orsat apparatus.

**Example 10.9**

In an internal combustion engine, gasoline is burned with air. An Orsat analysis of the combustion products gives 8.83% $CO_2$, 6.94% $O_2$; 1.26% CO and 82.97% $N_2$. Determine the air-fuel ratio on molar basis and the percent theoretical air used in the engine.

*Solution* :

For combustion analysis purpose, gasoline is treated as pure octane ($C_8H_{18}$). The combustion reaction of $C_8H_{18}$ can be written as

$$C_8H_{18} + 12.5O_2 + 12.5(3.76)N_2 \longrightarrow 8CO_2 + 9H_2O + 47N_2$$

Moles of air used $= 12.5 + 47 = 59.5$

Theoretical air-fuel ratio A/F = 59.5 mol air/mol fuel.

Now, let us consider 100 mol of combustion products. Then the combustion products contain 8.83 mol $CO_2$; 6.94 mol $O_2$; 1.26 mol CO; 82.97 mol $N_2$ and z mol $H_2O$ (Since the Orsat analysis is on dry basis, we do not know the amount of water vapor in the combustion products). Let us assume that $x$ mol of $C_8H_{18}$ is burned with $y$ mol of $O_2$ which carries with it 3.76$y$ mol $N_2$ with it (since air is used for combustion) to produce the above combustion products. Then the actual combustion reaction can be written as:

$$x\,C_8H_{18} + y\,O_2 + 3.76\,y\,N_2 \longrightarrow 8.83\,CO_2 + 6.94\,O_2 + 1.26\,CO + 82.97\,N_2 + z\,H_2O.$$

In a chemical reaction, the masses of the chemical elements are conserved even though the reactants disappear and new products are formed. Therefore, we can write mass balance for the constituent elements. Then

$N_2$ balance : $3.76\,y = 82.97$ or $y = 22.07$

Carbon atom balance: $8\,x = 8.83 + 1.26$ or $x = 1.26$

Hydrogen atom balance: $1: x = 2z$

or $18 \times 1.26 = 2z$ or $z = 11.34$

Having determined the values of $x$, $y$ and $z$, the actual combustion reaction can be rewritten as

$$1.26\,C_8H_{18} + 22.07\,O_2 + 22.07(3.76)\,N_2 \longrightarrow 8.83\,CO_2 + 6.94\,O_2 + 1.26\,CO + 82.97\,N_2 + 11.34\,H_2O$$

Moles of air used for combustion = 22.07 + 82.98 = 105.05

Moles of fuel used = 1.26

$$\text{Therefore, air-fuel ratio} = \frac{105.05}{1.26} = 83.37$$

$$\text{Percent theoretical air} = \frac{83.37}{59.5} \times 100 = 140$$

**Example 10.10**

The combustion of methane ($CH_4$) gas with dry air gave the following Orsat analysis: 8% $CO_2$; 4.32% $O_2$ and 1.7% CO. Determine the actual air-fuel ratio and the percent theoretical air used for combustion.

*Solution* :

The Orsat analysis gives the composition of $CO_2$, $O_2$, CO and $N_2$ only and the analysis is on dry–basis,Therefore,

$$N_2 = 100 - (8 + 4.32 + 1.7) = 85.98\%$$

Consider 100 mol combustion products as the basis for calculation. It contains $z$ mol $H_2O$ also. Let $x$ mol $CH_4$ be burned with ($y$ mol $O_2$ + 3.76 $y$ mol $N_2$) to yield the above combustion products. Then the combustion reaction can be written as :

$$x\,CH_4 + yO_2 + 3.76y\,N_2 \longrightarrow 8\,CO_2 + 4.32\,O_2 + 1.7\,CO + 85.98\,N_2 + z\,H_2O$$

The mass balance of the constituent elements give:

$N_2$ balance : $3.76\,y = 85.98$ or $y = 22.867$

Carbon atom balance: $x = 8 + 1.7$ or $x = 9.7$

Hydrogen atom balance : $4\,x = 2z$

or $4 \times 9.7 = 2z$ or $z = 19.4$

Then moles of air used for combustion $= y + 3.76\,y = 4.76\,y = 108.847$

Moles of fuel used $= x = 9.7$

$$\text{Actual air-fuel ratio} = \frac{108.94}{9.7} = 11.22 \frac{\text{mol air}}{\text{mol fuel}}$$

The stoichiometric or theoretical air-fuel ratio can be obtained from the following combustion reaction :

$$CH_4 + 2\,O_2 + 2(3.76)\,N_2 \longrightarrow CO_2 + 2\,H_2O + 7.52\,N_2$$

Theoretical air fuel ratio = 2 + 7.52 = 9.52 mol air/mol fuel.

$$\text{Percent theoretical air} = \frac{11.22}{9.52} \times 100 = 117.86$$

**Example 10.11**

Ethylene, $C_2H_4$ is burned with 20 percent excess air. An analysis of the products of combustion on a dry basis reveals 4.5% $O_2$ by volume and the remaining data of analysis is missing. Determine the percentage of carbon in the fuel which is converted into CO.

*Solution* :

The theoretical $O_2$ requirement for the complete combustion of $C_2H_4$ can be obtained from the reaction.

$$C_2H_4 + 3\,O_2 \longrightarrow 2\,CO_2 + 2\,H_2O$$

That is, 3 mol $O_2$ is required per mole of $C_2H_4$.

Let us consider 100 mol of burned gases as analyzed (on dry basis). This gas contains $b$ mol $CO_2$; $c$ mol CO; 4.5 mol $O_2$; $d$ mol $N_2$. Therefore, we have

$$b + c + 4.5 + d = 100 \text{ or } b + c + d = 95.5 \qquad (A)$$

It also carries with it $e$ mol $H_2O$.

Let us assume that $a$ mol $C_2H_4$ is burned with 20 percent excess air to yield the above combustion products. Then the combustion reaction is

$$a\,C_2H_4 + 1.2(3)\,a\,O_2 + (1.2)\,(3)\,(3.76)\,a\,N_2 \longrightarrow b\,CO_2 + c\,CO + 4.5O_2 + d\,N_2 + e\,H_2O$$

The mass balance of the constituent elements give:

Carbon atom balance : $2\,a = b + c$ $\quad (B)$

Hydrogen atom balance: $4\,a = 2e$ or $e = 2a$ $\quad (C)$

Nitrogen balance : $13.536\,a = d$ $\quad (D)$

Oxygen atom balance: $7.2\,a = 2b + c + 9 + e$ or $7.2\,a = b + 2a + 9 + 2a$

or $3.2a = 9 + b$ $\quad (E)$

We have 5 unknowns $a, b, c, d$ and $e$ and five Eqns. $(A) - (E)$. They can be simultaneously solved to determine the values of $a, b, c, d$ and $e$.

From Eqns. $(A)$ and $(B)$, we get

$$b + c + d = 100 \text{ or } 2\,a + d = 100 \qquad (F)$$

Substitute Eqn. $(D)$ in Eqn. $(E)$ to obtain

$$2\,a + d = 2\,a + 13.536\,a = 100 \text{ or } a = 6.4367 \qquad (G)$$

Substitute Eqn. $(G)$ in Eqn. $(E)$ to obtain

$$3.2\,a = 9 + b \text{ or } 3 \times 6.4367 = 9 + b$$

or $b = 10.3101$ $\quad (H)$

From Eqns.$(B)$ and $(H)$, we get

$$2\,a = b + c$$

or $2 \times 6.4367 = 10.3101 + c$ or $c = 2.5633$

Therefore, the combustion products contain 10.3101 mol $CO_2$ and 2.5633 mol CO when 6.4367 mol $C_2H_4$ is burned.

Carbon atoms present in fuel = $2a = 2 \times 6.4367 = 12.8734$

$$\text{Percentage of carbon converted into CO} = \frac{2.5633 \times 100}{12.8734} = 19.91$$

**Example 10.12**

The exhaust gas of a Diesel engine has the following volumetric analysis on a dry basis. 9.98% $CO_2$ ; 2% CO; 4.69% $O_2$ and 83.33% $N_2$. Determine the actual air-fuel ratio on mass basis and percent excess air used for combustion.

*Solution* :

The Diesel oil is treated as dodecane ($C_{12}H_{26}$) for combustion analysis purpose. Let us consider 100 moles of combustion products as the basis for calculations. Since the analysis is on dry basis, we should add an unknown amount of $H_2O$ also. Then the combustion reaction can be written as:

$$a\,C_{12}H_{26} + b\,O_2 + 3.76\,b\,N_2 \longrightarrow 9.98\,CO_2 + 2\,CO + 4.69\,O_2 + 83.33\,N_2 + cH_2O$$

Then the mass balance equations give:

Nitrogen balance: $3.76\,b = 83.33$ or $b = 22.16$

Carbon atom balance: $12\,a = 9.98 + 2$ or $a = 1$

Hydrogen atom balance: $26\,a = 2\,c$ or $26 = 2c$ or $c = 13$

$$\text{Actual air-fuel ratio} = \frac{\text{Moles of air}}{\text{Moles of fuel}} = \frac{b + 3.76b}{a} = 105.48\ \frac{\text{mol air}}{\text{mol fuel}}$$

Molar mass of air $= 28.97 \times 10^{-3}$ kg/mol

Molar mass of $C_{12}H_{26} = (12 \times 12 + 26 \times 1)10^{-3} = 170 \times 10^{-3}$ kg/mol

$$\text{Therefore, actual air-fuel ratio} = \frac{105.48 \times 28.97 \times 10^{-3}}{170 \times 10^{-3}} = 17.975\ \frac{\text{kg air}}{\text{kg fuel}}$$

The theoretical air-fuel ratio can be obtained from the combustion reaction.

$$C_{12}H_{26} + 18.5\,O_2 + 18.5\,(3.76)\,N_2 \longrightarrow 12CO_2 + 13H_2O + 69.56\,N_2$$

$$\text{Therefore, theoretical air-fuel ratio} = \frac{18.5 + 18.5 \times 3.76}{1} = 88.06\ \frac{\text{mol air}}{\text{mol fuel}}$$

$$\text{Percent excess air} = \frac{(105.48 - 88.06)}{88.06} \times 100 = 19.78$$

**Example 10.13**

A hydrocarbon fuel of unknown composition is burned with air and an Orsat analysis of the combustion products gave 7.26% $CO_2$; 2.42% CO; 7.5% $O_2$ and 82.82% $N_2$. Determine the composition of the fuel on mass basis and percent theoretical air used for combustion.

*Solution* :

Let the hydrocarbon fuel be represented by the empirical formula $C_xH_y$. Consider 100 moles of combustion products as the basis for calculations. Then, the combustion reaction can be expressed as

$$C_x\,H_y + a\,O_2 + 3.76\,a\,N_2 \longrightarrow 7.26\,CO_2 + 2.42\,CO + 7.5\,O_2 + 82.82\,N_2 + b\,H_2O$$

The mass balances of the constituent elements give:

Nitrogen balance : $3.76\,a = 82.82$ or $a = 22.066$

Carbon atom balance : $x = 7.26 + 2.42$ or $x = 9.68$

Oxygen atom balance : $2\,a = 2 \times 7.26 + 2.42 + 2 \times 7.5 + b$

or $2 \times 22.066 = 14.52 + 2.42 + 15 + b$ or $b = 12.192$

Hydrogen atom balance: $y = 2b$

or $y = 2 \times 12.192 = 24.384$

Therefore, the empirical formula of the fuel is

$$C_x H_y = C_{9.68} H_{24.384}$$

Composition of the fuel, on mass basis:

$$\text{Carbon} = \frac{9.68 \times 12 \times 100}{(9.68 \times 12 + 24.384 \times 1)} = 82.65\%$$

$$\text{Hydrogen} = 17.35\%$$

$$\text{Actual air-fuel ratio} = \frac{\text{Moles of air}}{\text{Moles of fuel}} = \frac{a + 3.76a}{1} = 105.03 \ \frac{\text{mol air}}{\text{mol fuel}}$$

The theoretical oxygen requirement can be obtained from the following combustion reaction.

$$C_{9.68} H_{24.384} + 15.776\, O_2 \longrightarrow 9.68\, CO_2 + 12.192\, H_2O$$

That is 15.776 mol $O_2$ is needed per mole of fuel.

Therefore,theoretical air-fuel ratio

$$= \frac{15.776 + 15.776(3.76)}{1} = 75.094 \ \frac{\text{mol air}}{\text{mol fuel}}$$

$$\text{Percent theoretical air} = \frac{105.03 \times 100}{75.094} = 139.86$$

**Example 10.14**

The Orsat analysis of the products of combustion of a hydrocarbon fuel of unknown composition is 12.37% $CO_2$; 2.47% $O_2$; 0.87% CO; and 84.29% $N_2$. It is also found that one mole of fuel on burning under identical conditions yields 7.11 mol $CO_2$. Determine the air–fuel ratio used for combustion and the empirical formula of the fuel.

*Solution* :

Let the hydrocarbon fuel be represented as $C_xH_y$. Now, consider 100 mol of the combustion products. One mole of fuel yields 7.11 mol $CO_2$.

Therefore 12.37 mol $CO_2$ corresponds to $\dfrac{12.37}{7.11} = 1.74$ mol fuel

Then the combustion reaction can be written as

$$1.74\, C_xH_y + a\, O_2 + 3.76a\, N_2 \longrightarrow 12.37\, CO_2 + 2.47\, O_2 + 0.87\, CO + 84.29\, N_2 + b\, H_2O$$

Mass balances of the constituent elements give :

Nitrogen balance: $3.76\,a = 84.29$ or $a = 22.42$

Carbon atom balance: $1.74\,x = 12.37 + 0.87$ or $x = 7.61$

Oxygen atom balance : $2\,a = 2 \times 12.37 + 2 \times 2.47 + 0.87 + b$

or $2 \times 22.42 = 24.74 + 4.94 + 0.87 + b$

or $b = 14.29$

Hydrogen atom balance: $1.74\,y = 2\,b = 2 \times 14.29$

or $y = 16.43$

$$\text{Air-fuel ratio} = \frac{a + 3.76a}{1.74} = 61.333 \ \frac{\text{mol air}}{\text{mol fuel}}$$

The fuel is $C_xH_y = C_{7.61}H_{16.43}$

**Example 10.15**

It is a known fact that chemical reactions are usually accompanied by absorption or evolution of energy as heat. Where from does the energy come in chemical reactions ?

*Solution* :

During a chemical reaction the reactants disappear and new products are formed. Consider the reaction between $H_2$ and $Cl_2$ to form HCl according to the reaction

$$H_2 + Cl_2 \rightarrow 2HCl$$

This reaction involves the breaking of H–H and Cl–Cl bonds and the formation of H–Cl bonds. The structure of the product molecules is different from the structure of reactant molecules. A molecule is associated with certain energy. Because of the difference in the structures of reactant and product molecules, the difference in the energies of the reactant and product molecules appears as energy liberated or absorbed during a chemical reaction.

**Example 10.16**

Sketch the potential energy diagram of a diatomic molecule. With the help of the potential energy diagram, explain the energy change associated with the reaction $H_2+Cl_2 \rightarrow 2HCl$.

*Solution* :

Consider an atom which is at rest and is at infinite distance from other atoms. Then the atom does not possess any translational kinetic energy. As we are considering a single atom, it does not possess any rotational kinetic energy and vibrational energy. Then the atom possesses zero energy. If two atoms of hydrogen are separated from each other and they are at rest, then the energy of the atoms is zero. If we bring the atoms closer to each other there will be attraction and repulsion forces between them. The potential energy possessed by the diatomic molecule then appears as shown the FigE. 10.16. The potential energy of the molecule is the sum of the energies due to repulsion and attraction which depend on the interatomic distance.The potential energy of a molecule reaches a minimum value at $r_e$, where $r_e$ is the equilibrium interatomic distance. This state corresponds to the equilibrium state of the molecule. The depth of the potential well ($D_e$) represents the bond energy. A diatomic molecule has one vibrational degree of freedom and the energy (ε) of the vibrational mode is given by

$$\varepsilon_n = \left(n+\frac{1}{2}\right)h\upsilon \,;\, n = 0,1,2..........$$

Where $h$ is Planck's constant and $\upsilon$ is the vibrational frequency of the molecule. If a diatomic molecule is held at a temperature of zero kelvin, it still possesses some amount of energy $h\upsilon/2$ which is called the zero point vibrational energy. Therefore, if we supply the energy $D_0$, $\left(D_0 = D_e - \frac{1}{2}h\upsilon\right)$ which is called the dissociation energy, the bond between the two atoms breaks.

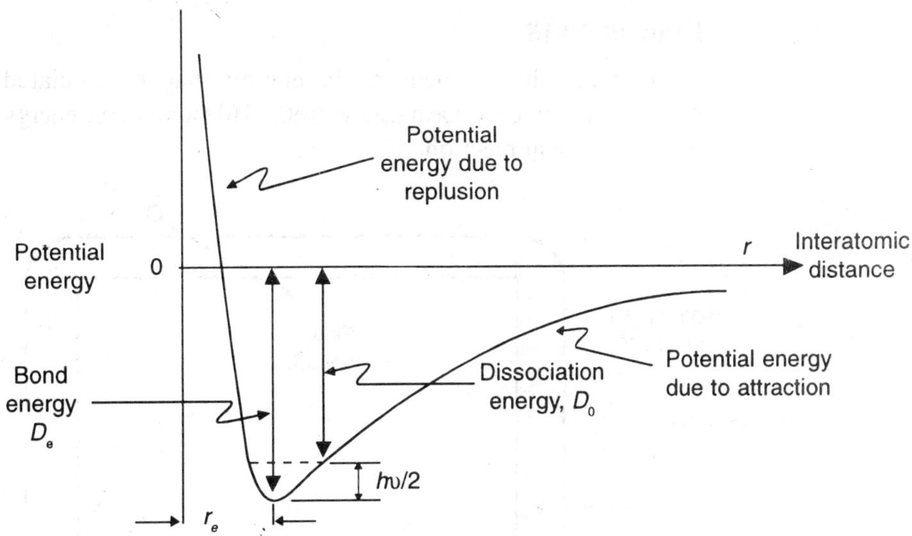


Fig.E 10.16. Potential energy diagram of a diatomic molecule.

We know that the dissociation energies of $H_2$, $Cl_2$ and HCl and they are 429.78 kJ/mol; 239.3 kJ/ mol and 427.47 kJ / mol, respectively. If the reaction $H_2$

$+ Cl_2 \rightarrow 2HCl$ is carried out at zero kelvin, the energy needed to break the bonds of one mole each of $H_2$ and $Cl_2$ is 429.78 + 239.3 = 669.08 kJ. While the bonds in 2 moles of HCl are formed, they liberate the energy $2 \times 4 \times 27.47 = 854.94$ kJ. Therefore, the net energy liberated is 854.94 – 669.08 =185.86 kJ or 92.93 kJ / mol of HCl.

**Example 10.17**

Does the energy liberated or consumed during a chemical reaction depend on the temperature? Is it practical to calculate the energy changes associated with a chemical reaction at a given temperature from a knowledge of dissociation energies of the reactant and product molecules ?

*Solution* :

The molecules are distributed among all the possible vibrational and rotational states according to the Maxwell–Boltzmann distribution. This distribution is different at different temperatures. The dissociation energy of a molecule depends on the energy state of the molecule. Moreover, when the products are formed, the product molecules also are distributed among all the possible vibrational and rotational states. Though it is possible, in principle, to calculate the energy change associated with a chemical reaction, it is next to impossible from a practical point of view.

**Example 10.18**

Is it possible to measure the energy change associated with a chemical reaction ? If yes, describe a simple method to measure the energy change associated with a chemical reaction.

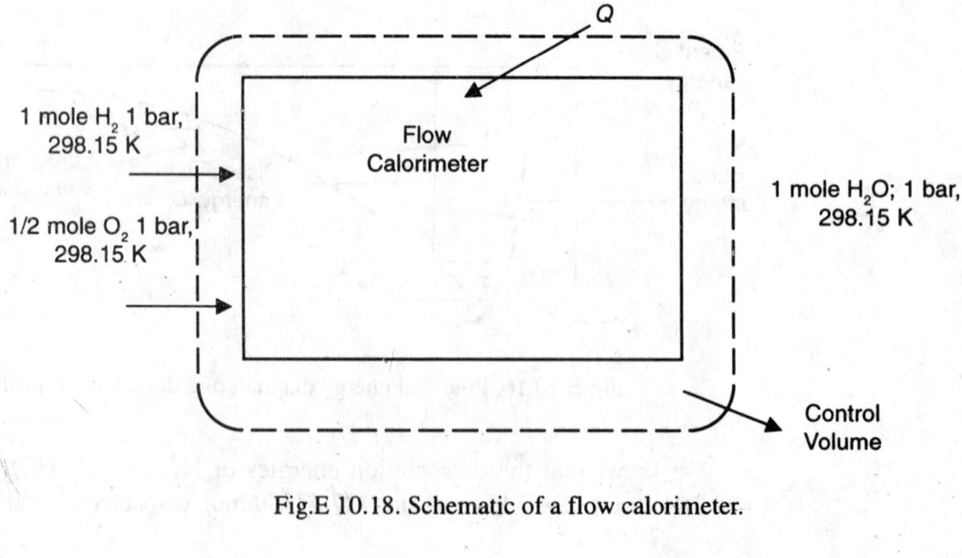


Fig.E 10.18. Schematic of a flow calorimeter.

*Solution* :

The energy change associated with a chemical reaction can be easily measured in a laboratory with the help of a flow calorimeter shown in FigE. 10.18. Suppose one mole of $H_2$ and 1/2 mole of $O_2$ at 1 bar and 298.15 K enter the reactor in which the reaction occurs between $H_2$ and $O_2$ to form $H_2O$ and $H_2O$ leaves the calorimeter at 1 bar and 298.15 K. Let $Q$ denote the energy added as heat to the calorimeter. Then applying the first law of thermodynamics to the control–volume (flow calorimeter), which is operating at steady – state, steady flow, we get

$$h_{H_2O} - \left( h_{H_2} + \frac{1}{2} h_{O_2} \right) = Q$$

Thus, by measuring $Q$, one can find the energy change (enthalpy change) associated with a chemical reaction.

## Example 10.19

In the tables of thermodynamic properties of substances, the values are reported with reference to some arbitrarily selected reference state. Is this approach (or the data reported in tables) satisfactory to analyze chemical reactions ?

*Solution* :

The reference state selected for water, in steam tables, is the triple point of water where the specific internal energy and entropy of saturated liquid has been arbitrarily assigned a zero value. Similarly the properties of Freon-12 have been reported with reference to saturated liquid at – 40°C. This approach is satisfactory while one evaluates the changes in the properties between two states of the same substance because while calculating the changes in the properties, the reference state value cancels. However, when a chemical reaction occurs, reactants disappear and products are formed, so differences cannot be calculated for all substances involved in the reaction. For chemically reacting systems it is necessary to evaluate the changes in the properties in such a way that there are no ambiguities or inconsistencies.

## Example 10.20

Why is it necessary to define standard states for analyzing chemically reacting systems?

*Solution* :

The energy liberated or consumed during a chemical reaction depends on the temperature of the reactants and products. There are a large number of chemical reactions and these reactions can be carried out under varying conditions of temperature and in different ways. Therefore, it is impossible to compile the data regarding the energy changes associated with chemical reactions carried out under different conditions. Hence, it is necessary to devise a method by which the energy changes associated with chemical reactions can be estimated from a limited amount of data which is based on a few reactions carried out in a standard way. In order to have a consistent basis it is convenient to measure the energy changes associated with chemical reactions when the reactants and products are in their standard states and at the same temperature. Therefore it is necessary to define standard states.

**Example 10.21**

What are the standard states in use for analyzing chemically reacting systems ?

*Solution* :

For analyzing chemically reacting systems, the following standard states are used.

Gases : Pure gas in the ideal gas state at 1 bar.

The gas may not really behave as an ideal gas at 1 bar. But the hypothetical state, if it behaves as an ideal gas at 1 bar, is taken as the standard state.

Liquids : Pure liquid at 1 bar.

Solids : Pure solid at 1 bar. If the solid can exist in different allotropic forms, then the most stable allotropic form is taken as the standard state. For example, carbon can be found as diamond and graphite. Since graphite is more stable than diamond, the standard state for carbon is graphite.

The choice of the standard state temperature is arbitrary. Usually the temperature at which reaction takes place is taken as the standard state temperature. For convenience of tabulating the data, the standard state temperature is taken as 298.15 K.

**Example 10.22**

What is meant by standard enthalpy change (or standard heat) of a reaction ?

*Solution* :

If a chemical reaction is carried out in which the reactants as well as the products are in their respective standard states and at the same temperature $T$, the enthalpy change associated with the reaction is called the standard enthalpy change (or standard heat) of the reaction. It is denoted by $\Delta H_T^0$, where the superscript '0' denotes that all the reactants and products are in their respective standard states and the subscript $T$ denotes the temperature at which the reaction is carried out. For chemically reacting systems it is essential to denote the state of the species as gas ($g$), liquid ($l$) or solid ($s$).

### Example 10.23

What is the sign convention adopted for enthalpy changes for chemically reacting systems? Is it possible to identity whether energy is liberated or consumed during a chemical reaction based on the sign of $\Delta H_T^0$ ?

*Solution* :

The sign convention adopted for enthalpy change of chemical reaction is identical with the sign convention for heat interaction. It the reactants enter a steady flow, steady state calorimeter in their standard states at temperature $T$, and the products leave the flow calorimeter in their standard states at the same temperature $T$, while energy $Q$ in the form of heat flows into the device then we get

$$(\Sigma n_i h_i)_{\text{Products}} - (\Sigma n_i h_i)_{\text{Reactants}} = \Delta H_T^0 = Q$$

If $Q > 0$ then energy flows as heat into the device. That is, if $\Delta H_T^0$ (or $Q$) > 0, then it is necessary to supply energy for the chemical reaction to take place. On the other hand if $\Delta H_T^0$ (or $Q$) < 0, then energy is liberated due to chemical reaction. Thus a knowledge of the sign of $\Delta H_T^0$ reveals whether energy is liberated or consumed during a chemical reaction.

$\Delta H_T^0 < 0$ — reaction is called exothermic.

$\Delta H_T^0 > 0$ — reaction is called endothermic.

$\Delta H_T^0 = 0$ — reaction is thermoneutral

### Example 10.24

Given the standard enthalpy change at 298.15 K for the reaction $N_2 (g) + 3H_2 (g) \rightarrow 2\,NH_3 (g)$ is – 96.11 kJ. Calculate the standard enthalpy change at 298.15 K for the following reactions.

($a$) $\frac{1}{2}N_2(g) + \frac{3}{2}H_2(g) \rightarrow NH_3(g)$

($b$) $NH_3(g) \rightarrow \frac{1}{2}N_2(g) + \frac{3}{2}H_2(g)$

($c$) $2NH_3(g) \rightarrow N_2(g) + 3H_2(g)$

*Solution* :

(*a*) The given data is

$$N_2(g) + 3H_2(g) \rightarrow 2NH_3(g) : \Delta H^0_{298} = -96.11 \text{ kJ} \quad (A)$$

In this reaction 2 mol $NH_3$ (*g*) is formed starting with 1 mol $N_2$ (*g*) and 3 mol $H_2$ (*g*)

Now the given reaction is

$$\frac{1}{2}N_2(g) + \frac{3}{2}H_2(g) \rightarrow NH_3(g) \quad (B)$$

Which yields only 1 mol $NH_3(g)$ consuming 1/2 mol $N_2$ (*g*) and 3/2 mol $H_2$ (*g*). Here, we find the given reaction (*B*) is $1/2 \times (A)$. Therefore, $\Delta H^0_{298}$ for reaction (*B*) is $1/2 \times (-96.11) = -48.055$ kJ

(*b*) The reaction $NH_3(g) \rightarrow \frac{1}{2} N_2(g) + \frac{3}{2} H_2(g)$ (*C*)

We find that the reaction (*C*) = – (*B*). Hence $\Delta H^0_{298}$ for reaction (*C*) is $-(-48.055) = 48.055$ kJ

(*c*) The reaction $2NH_3(g) \rightarrow N_2(g) + 3H_2(g)$ (*D*)

We find that the reaction (*D*) = – (*A*). Hence $\Delta H^0_{298}$ for reaction (*D*)

$$= -(-96.11) = 96.11 \text{ kJ}$$

**Example 10.25**

What is the meaning of stoichiometric coefficient ? Is there any sign convention for writing the stoichiometric coefficient ?

*Solution* :

Consider the reaction

$$CH_4 + 2H_2O \rightarrow CO + 3H_2$$

Which tells that 1 mol of $CH_4$ reacts with 2 mol $H_2O$ to produce 1 mol CO and 3 mol $H_2$. That is

$$(1)\, CH_4 + (2)\, H_2O \rightarrow (1)\, CO + (3)\, H_2$$

In this the coefficients 1, 2, 1 and 3 associated with the chemical species $CH_4$, $H_2O$, CO and $H_2$ are called stoichiometric coefficients. As the reaction proceeds, the reactants are consumed and products are produced. Therefore to indicate the species produced (or products) and consumed (or reactants) we adopt the following sign convention. The stoichiometric coefficients associated with products are assigned a positive value and the stoichiometric coefficients associated with reactants are assigned negative values. Thus, in the above example, the stoichiometric coefficients associated with $CH_4$ and $H_2O$ are –1 and –2, respectively whereas the stoichiometric coefficients associated with CO and $H_2$ are + 1 and +3, respectively. With the above sign convention, the above chemical reaction can be rewritten as

$$CO + 3H_2 + (-1)\,CH_4 + (-2)\,H_2O = 0$$

or $\Sigma \upsilon_i A_i = 0$

where $\upsilon_i$ is the stoichiometric coefficient associated with species $A_i$

**Example 10.26**

Express the following chemical reactions as $\Sigma\, \upsilon_i A_i = 0$.

(*a*) $4\,NH_3 + 5\,O_2 \rightarrow 4\,NO + 6\,H_2O$ (*b*) $2\,H_2S + 3\,O_2 \rightarrow 2\,H_2O + 2\,SO_2$

*Solution* :

(*a*) The given reaction can be written as

$$4\,NO + 6\,H_2O + (-4)\,NH_3 + (-5)\,O_2 = 0$$

or $\sum_{1}^{4} \upsilon_i A_i = 0$

where $\upsilon_1 = 4;\ \upsilon_2 = 6;\ \upsilon_3 = -4$ and $\upsilon_4 = -5$

$$A_1 = NO\ ;\ A_2 = H_2O;\ A_3 = NH_3\ ;\ A_4 = O_2$$

(*b*) The given reaction can be expressed as

$$2\,H_2O + 2\,SO_2 + (-2)\,H_2S + (-3)\,O_2 = 0 \quad \text{or} \quad \sum_{1}^{4} \upsilon_i A_i = 0$$

where $\upsilon_1 = 2;\ \upsilon_2 = 2;\ \upsilon_3 = -2$ and $\upsilon_4 = -3$

$$A_1 = H_2O\ ;\ A_2 = SO_2;\ A_3 = H_2S\ ;\ A_4 = O_2$$

**Example 10.27**

What is a formation reaction and what is meant by standard enthalpy (or heat) of formation? What convention is adopted in reporting the standard enthalpies of formation?

*Solution* :

A chemical reaction in which only one mole of a single compound is formed from its constituent elements is called a formation reaction. For example

$$C\,(s) + 2\,H_2\,(g) + \frac{1}{2}\,O_2\,(g) \rightarrow CH_3OH\,(g)$$

is a reaction in which only one mole of a single compound $CH_3OH$ is formed starting from the elements C, $H_2$ and $O_2$. Hence, it is a formation reaction. It is possible to produce $CH_3OH$ by the following reaction also.

$$CO(g) + 2H_2\,(g) \rightarrow CH_3OH\,(g)$$

However, the above reaction is not a formation reaction, because $CH_3OH$ is produced starting from CO and $H_2$, in which CO is not an element. Similarly the reaction

$$2\,C\,(s) + 4\,H_2\,(g) + O_2\,(g) \rightarrow 2\,CH_3OH\,(g)$$

is also not a formation reaction, because 2 mol $CH_3OH$ is obtained, even though $CH_3OH$ is produced starting from the elements C, $H_2$ and $O_2$.

The enthalpy change associated with a formation reaction, if the elements and product are in their respective standard states, is called standard enthalpy of formation. It is denoted by $\Delta H^0_{fT}$, where the subscript '*f*' denotes that it represents the formation reaction and the second subscript '*T*' denotes the temperature at which the reaction takes place. For convenience, a temperature of 298.15 K has been selected for reporting the standard enthalpies of formation and they are denoted as $\Delta H^0_{f298}$.

By convention, the standard enthalpy of formation of the element in its standard state has been assigned a zero value. Thus $\Delta H^0_{f298}$ for $H_2(g)$, $O_2(g)$ and $C(s)$ where *s* stands for solid (in this case graphite) is zero.

**Example 10.28**

Which of the following reactions can be called as formation reactions?

(*a*) $C_2H_4\,(g) + H_2O\,(g) \rightarrow C_2H_5OH\,(g)$

(*b*) $CO\,(g) + H_2O\,(g) \rightarrow CO_2\,(g) + H_2(g)$.

(*c*) $N_2\,(g) + 3\,H_2\,(g) \rightarrow 2\,NH_3\,(g)$.

(*d*) $\frac{1}{2}\,N_2\,(g) + \frac{3}{2}\,H_2\,(g) \rightarrow NH_3(g)$.

(*e*) $CH_3CHO\,(g) + H_2\,(g) \rightarrow C_2H_5OH\,(g)$.

*Solution* :

(*a*) Not a formation reaction, since $C_2H_5OH$ is formed from compounds $C_2H_4$ and $H_2O$..

(*b*) Not a formation reaction.

(*c*) Not a formation reaction, since 2 mol $NH_3$ is produced, even though the reactants are elements.

(*d*) Formation reaction.

(*e*) Not a formation reaction.

### Example 10.29

What is the advantage of compiling the standard enthalpies of formation of compounds ?

*Solution* :

We know that enthalpy is a state function and its change does not depend on the path followed by the system but depends only on the initial and final states of the system. Therefore, if one is interested in calculating the enthalpy change when a system changes from a given initial state to a given final state, then a most convenient path connecting the initial and final states can be devised and evaluate the enthalpy change for the imaginary path instead of evaluating the enthalpy change for the actual path. This observation suggests that a given chemical reaction also can be replaced by a set of reactions such that the overall or net change of the set of reactions is equal to the desired chemical reaction. Then the enthalpy change of the desired chemical reaction is given by the net enthalpy change of the set of reactions which are used to replace the given chemical reaction. Since all compounds are formed from elements only, it is possible to express a given chemical reaction as an algebraic sum of several formation reactions. Therefore, it is advantageous to compile the standard enthalpies of formation of compounds from which the standard enthalpy change of a chemical reaction can be estimated.

### Example 10.30

It is desired to calculate the standard enthalpy change at 298.15 K ($\Delta H^0_{298}$), for the reaction

$$4\,NH_3\,(g) + 5\,O_2\,(g) \longrightarrow 4\,NO\,(g) + 6\,H_2O\,(g)$$

from a knowledge of the standard enthalpies of formation of the species involved in the reaction. Devise a method for the estimation of $\Delta H^0_{298}$ and obtain its value.

*Solution*:

The given reaction is

$$4\,NH_3\,(g) + 5\,O_2\,(g) \longrightarrow 4\,NO\,(g) + 6\,H_2O\,(g)$$

The chemical species involved in this reaction are $NH_3$ (*g*), $O_2$ (*g*), NO (*g*) and $H_2O$(*g*) of which $O_2$ (*g*) is an element for which $\Delta H^0_{298} = 0$ by convention. Now, the formation reactions for the other compounds can be written as

$$\frac{1}{2}\,N_2\,(g) + \frac{1}{2}\,O_2\,(g) \rightarrow NO\,(g) \qquad (A)$$

$$H_2\,(g) + \frac{1}{2}\,O_2\,(g) \rightarrow H_2O\,(g) \qquad (B)$$

$$\frac{1}{2} N_2 (g) + \frac{3}{2} H_2 (g) \rightarrow NH_3 (g) \quad (C)$$

$$O_2 (g) \rightarrow O_2 (g) \quad (D)$$

Now the given chemical reaction is a result of

$$= [4 \times (A) + 6 \times (B)] - [4 \times (C) + 5 \times (D)] \quad (E)$$

Hence, the standard enthalpy change of the given reaction is the net enthalpy change obtained from (*E*). That is

$$\Delta H^0_{298} = \{4\Delta H^0_{f298} [NO (g)] + 6 \ \Delta H^0_{f298} [H_2O (g)] \}$$
$$- \{4 \ \Delta H^0_{f298} [NH_3 (g)] + 5 \ \Delta H^0_{f298} [N_2 (g)]\} \quad (F)$$

The standard enthalpies of formation, $\Delta H^0_{f298}$ of a few selected compounds are given in Appendix – 10. We read the following values of $\Delta H^0_{f298}$ from Appendix – 10.

NO (*g*) = 90.435 kJ/mol ; $H_2O$ (*g*) = –241.997 kJ/mol

$NH_3$ (*g*) = – 46.055 kJ/mol

Substituting these values in Eqn. (*E*), we get

$$\Delta H^0_{298} = [4 \times (90.435) + 6 \times (-241.997)] - [4 \times (-46.055) + 5 \times (0)]$$

or $\Delta H^0_{298} = -\ 906.022$ kJ

**Example 10.31**

Obtain a general expression to calculate the standard enthalpy change at 298 K, ($\Delta H^0_{298}$) for a chemical reaction given by

$$a\,A + b\,B \rightarrow r\,R + s\,S$$

*Solution* :

The given chemical reaction is

$$a\,A + b\,B \rightarrow r\,R + s\,S \quad (A)$$

This can be rewritten as

$$r\,R + s\,S - a\,A - b\,B = 0 \ \text{ or } \ \Sigma\, \upsilon_i A_i = 0 \quad (B)$$

where $\upsilon_1 = r$ ; $\upsilon_2 = s$; $\upsilon_3 = -a$; $\upsilon_4 = -b$ ; $A_1 = R$ ; $A_2 = S$; $A_3 = A$; $A_4 = B$

where $\upsilon_i$ is the stoichiometric coefficient associated with $A_i$. $A_i$ can be a compound or element.

The standard enthalpy change of the reaction (*A*) is given by

$$\Delta H^0_{298} = [r \times \Delta H^0_{f298} (R) + s \times \Delta H^0_{f298} (S)] - [a \times \Delta H^0_{f298} (A) + b \times \Delta H^0_{f298} (B)]$$

or $$\Delta H^0_{f298} = [\Sigma \,|\, \upsilon \,|\, \Delta H^0_{f298} ]_{\text{Products}} - [\Sigma \,|\, \upsilon \,|\, \Delta H^0_{f298} ]_{\text{Reactants}}$$

or $$\Delta H^0_{f298} = \Sigma \upsilon_i \ \Delta H^0_{f298} [A_i]$$

**Example 10.32**

Calculate the standard enthalpy change at 298.15 K for the following reactions, from the standard enthalpies of formation data given in Appendix – 10.

(*a*) $2H_2S\,(g) + 3\,O_2\,(g) \rightarrow 2\,H_2O\,(g) + 2\,SO_2\,(g)$

(*b*) $CH_4\,(g) + 2\,O_2\,(g) \rightarrow CO_2\,(g) + 2\,H_2O\,(g)$

*Solution*:

(*a*) The given reaction is

$2H_2S\,(g) + 3\,O_2\,(g) \rightarrow 2H_2O\,(g) + 2SO_2\,(g)$

This reaction can be written as $\Sigma\, \upsilon_i\, A_i = 0$, where

$$\upsilon_1 = 2;\ \upsilon_2 = 2;\ \upsilon_3 = -2 \text{ and } \upsilon_4 = -3$$
$$A_1 = H_2O\,(g);\ A_2 = SO_2\,(g)\ ;\ A_3 = H_2S\,(g)\ ;\ A_4 = O_2\,(g)$$

Then $\Delta H^0_{298}$ for the given reaction is given by

$$\Delta H^0_{298} = \Sigma\, \upsilon_i\, \Delta H^0_{298}\,(A_i)$$

From Appendix –10, we obtain the $\Delta H^0_{f298}$ of the compounds as :

$H_2O\,(g) = -241.997$ kJ / mol ; $SO_2\,(g) = -297.263$ kJ / mol;
$H_2S\,(g) = -20.097$ kJ / mol and we know $O_2\,(g) = 0$. Then

$$\Delta H^0_{298} = 2 \times (-241.997) + 2 \times (-297.263) + (-2) \times (-20.097) + (-3) \times (0)$$
$$= -1038.326 \text{ kJ}$$

(*b*) The given reaction is

$$CH_4\,(g) + 2\,O_2\,(g) \rightarrow CO_2\,(g) + 2\,H_2O\,(g)$$

This can be expressed as $\Sigma \upsilon_i A_i = 0$, where

$$\upsilon_1 = 1\ ;\ \upsilon_2 = 2;\ \upsilon_3 = -1;\ \upsilon_4 = -2\ ;\ A_1 = CO_2\,(g)$$
$$A_2 = H_2O\,(g)\ ;\ A_3 = CH_4\,(g) \text{ and } A_4 = O_2\,(g).$$

Then, the standard enthalpy change of the reaction at 298.15 K is given by

$$\Delta H^0_{298} = \Sigma\, \upsilon_i\, \Delta H^0_{f298}\ (A_i)$$

From Appendix – 10, we obtain the values of $\Delta H^0_{f298}$ of the compounds as

$CO_2\,(g) = -393.978$ kJ / mol ; $H_2O\,(g) = -241.997$ kJ / mol
$CH_4\,(g) = -74.943$ kJ / mol

Then $$\Delta H^0_{298} = 1 \times (-393.978) + 2 \times (-241.997) + (-1) \times (-74.943)$$

or $$\Delta H^0_{298} = -803.029 \text{ kJ}$$

**Example 10.33**

The standard enthalpies of formation at 298.15 K ($\Delta H^0_{f298}$) for the compounds like methanol, benzene, etc. which exist as liquid at 298.15 K are tabulated for the liquid state only and not for the gas state. Suppose these species appear in the gas phase in a chemical reaction, how does one estimate the value of $\Delta H^0_{f298}$ in the gas phase from a knowledge of $\Delta H^0_{f298}$ in the liquid phase.

*Solution* :

Suppose for a compound $A$, we are given $\Delta H^0_{f298}$ [$A$ ($l$)] and it is desired to calculate $\Delta H^0_{f298}$ [$A$ ($g$)]. The standard state for the liquid is pure liquid at 1 bar and the standard state for the gas is the pure gas in the ideal gas state at 1 bar and both of them are at the same temperature 298.15 K. Then the change can be expressed as

$$A\,(l;\ 1\text{ bar},\ 298.15\text{ K}) \rightarrow A\,(g;\ 1\text{ bar},\ 298.15\text{ K}) \qquad (A)$$

If the enthalpy change for this process is $\Delta H$ then $\Delta H^0_{298}$ [$A$ ($g$)] is given by

$$\Delta H^0_{f298}\ [A\,(g)] = \Delta H^0_{f298}\ [A\,(l)] + \Delta H \qquad (B)$$

At 1 bar and 298.15 K, the liquid is a compressed liquid. The required phase change represented by Eqn. ($A$) can be written as a sum of the following physical changes.

$$A\,(l,1\text{ bar},\ 298.15\text{K}) \rightarrow A(l,P_s,\ 298.15\text{ K}):\ \ \Delta H_1 \qquad (C)$$
$$A\,(l,P_s,298.15\text{K}) \rightarrow A\,(g,P_s,\ 298.15\text{ K}):\ \ \Delta H_2 \qquad (D)$$
$$A\,(g,\ P_s,\ 298.15\text{K}) \rightarrow A(g,1\text{bar},\ 298.15\text{ K}):\ \ \Delta H_3 \qquad (E)$$

---

$$A\,(l,1\text{ bar},\ 298.15\text{K}) \rightarrow A(g,1\text{bar},\ 298.15\text{ K});\ \ [\Delta H = \Delta H_1 + \Delta H_2 + \Delta H_3]$$

Eqn.($C$) denotes the change of state of a liquid from 1 bar to $P_s$ (saturation pressure) at 298.15 K. We know that

$$dH_l = C_p dT + v\,(1 - \beta\,T)dP$$

When $T$ is held constant, we get

$$dH_l = v\,(1 - \beta\,T)dP$$

For liquids $\beta \cong 10^{-4}\text{ K}^{-1}$ and $\beta T << 1$. Therefore $\beta T$ can be neglected compared to 1 and hence $dH_l = vdP$. It is a good approximation to consider $v$ as constant since liquids are incompressible. Therefore, $\Delta H_1 = \int dH_1 = v\int dP = v\Delta P$. Eqn.($D$) denotes the vaporization process at constant temperature and pressure and hence $\Delta H_2 = h_{fg}$ = molar enthalpy of vaporization at 298.15 K. The Eqn($E$). represents the change of pressure for an ideal gas at constant pressure. We know that enthalpy is a function of temperature only for an ideal gas and hence $\Delta H_3 = 0$. Therefore, we get

$$\Delta H = v\Delta P + h_{fg}\ (\text{at } 298.15\text{ K})$$

Hence Eqn.(*B*) gives

$$\Delta H^0_{f298}[A(g)] = \Delta H^0_{f298}[A(l)] + v\Delta P + h_{fg} \text{ (at 298.15 K)}$$

**Example 10.34**

The standard state enthalpy of formation of $H_2O$ (*l*) at 298.15 K is – 285.958 kJ/mol. Making use of the information given in steam tables calculate $\Delta H^0_{f298}$ for $H_2O(g)$ and compare it with the tabulated data.

*Solution* :

For water at 25° (298.15 K), we get the following information from steam tables.

$P_S$ = 0.031 66 bar; $v_f$ = 0.001 002 9 m³/kg;

$h_f$ = 104.77 kJ/kg and $h_g$ = 2547.3 kJ/kg

$$v_f = 0.001\,002\,9 \text{ m}^3/\text{kg} = 1.8052 \times 10^{-5} \text{ m}^3/\text{mol}$$

$$h_{fg} = 2547.3 - 104.77 = 2442.53 \text{ kJ/kg}$$

$$= 43.965\,54 \text{ kJ/mol}$$

Therefore,

$$\Delta H^0_{f298}[H_2O(g)] = \Delta H^0_{f298}[H_2O(l)] + v\,\Delta P + h_{fg}$$

$$= -285.958 \times 10^3 + 1.8052 \times 10^{-5} \times (0.03166 - 1) \times 10^5 + 43.965\,54 \times 10^3$$

$$= -241.994\,21 \text{ kJ/mol}$$

The value of $\Delta H^0_{f298}$ for $H_2O(g)$ is reported as – 241.997 kJ/mol in Appendix –10.

**Note** *It is reasonable to ignore the term $v\Delta P$ (that is the enthalpy change associated with the change of pressure from 1 bar to the saturation pressure at 298.15 K,) since the molar volume of liquid is of the order of $10^{-5}$ m³/mol, while calculating $\Delta H^0_{f298}$ for the liquid state of a substance.*

**Example 10.35**

Calculate the standard enthalpy change at 298.15 K for the reaction $CH_4(g) + 2O_2(g) \rightarrow CO_2(g) + 2H_2O(l)$ from the standard enthalpy of formation of the compounds in the gaseous state.

*Solution* :

The given reaction is

$$CH_4(g) + 2O_2(g) \rightarrow CO_2(g) + 2H_2O(l) \qquad \Delta H^0_{298} \qquad (A)$$

This reaction can be obtained from

$CH_4\,(g) + 2O_2\,(g) \longrightarrow CO_2\,(g) + 2H_2O\,(l) \qquad \Delta H^0_{298}$ (B)

$H_2O\,(g) \rightarrow H_2O\,(l)$

$$(B) + 2(C) = (A).$$

Therefore,

$$\Delta H^0_{298} = \Delta H^0_{298}\,(B) + 2\Delta H^0_{298}(C)$$

The values of $\Delta H^0_{298}$ of the compounds are obtained from Appendix −10. They are

$$CH_4(g) = -74.943 \text{ kJ/mol}; \; CO_2\,(g) = -393.978 \text{ kJ/mol};$$
$$H_2O(g) = -241.997 \text{ kJ/mol}.$$

Then $\Delta H^0_{298}\,(B) = -393.978 + 2 \times (-241.997) - (-74.943) = -803.029$ kJ

The value of $\Delta H^0_{298}\,(C)$ can be obtained from

$H_2O\,(g,\, 1 \text{ bar},\, 298.15 \text{ K}) \rightarrow H_2O\,(g, P_s,\, 298.15 \text{ K}): \qquad \Delta H_1$

$H_2O\,(g,\, P_s,\, 298.15 \text{ K}) \rightarrow H_2O\,(l; P_s,\, 298.15 \text{ K}): \qquad \Delta H_2$

$H_2O\,(l; P_s,\, 298.15 \text{ K}) \rightarrow H_2O\,(l; 1 \text{ bar } 298.15 \text{ K}): \qquad \Delta H_3$

Adding $H_2O(g) \rightarrow H_2O(l); \quad \Delta H^0_{298}\,(C) = \Delta H_1 + \Delta H_2 + \Delta H_3$

or $\Delta H^0_{298}\,(C) = 0 + (-43.966) + 0 = -43.966$ kJ/mol

[Enthalpy of vaporization of water at 298.15 K is 43.966 kJ/ mol and the enthalpy change of liquid for the change of pressure from saturation pressure to 1 bar is neglected].

Therefore,

$$\Delta H^0_{298} = \Delta H^0_{298}\,(B) + 2\Delta H^0_{298}\,(C) = -803.029 + 2 \times (-43.966) = -890.961 \text{ kJ}$$

**Example 10.36**

Define a combustion reaction. What is standard enthalpy of combustion of a compound or an element ?

*Solution* :

A combustion reaction is defined as a reaction between an element or a compound and oxygen to form specified products. In case of hydrocarbons containing carbon, hydrogen and oxygen only, the combustion products are $CO_2$ and $H_2O$. The $H_2O$ may be present either in the gas phase or liquid phase. The standard enthalpy change associated with a combustion reaction is called standard enthalpy (or standard heat) of combustion which is denoted by $\Delta H^0_C$,

**Example 10.37**

How are the standard enthalpies of combustion are useful in estimating the standard enthalpy change of a chemical reaction?

*Solution* :

The standard enthalpy change of a chemical reaction can be easily estimated from a knowledge of the standard enthalpies of formation of the compounds involved in the reaction. Consider the cracking reaction

$$C_3H_8\,(g) \rightarrow C_2H_4\,(g) + CH_4(g)$$

If one is interested in calculating the standard enthalpy change of the above reaction, a knowledge of the standard enthalpies of formation of propane $C_3H_8$ $(g)$, ethylene $C_2H_4(g)$ and methane $CH_4(g)$ are needed. The formation reactions given below cannot be carried out in practice to determine the standard enthalpies of formation of the compounds.

$$3C\,(s) + 4H_2(g) \rightarrow C_3H_8(g)$$

$$2C\,(s) + 2H_2(g) \rightarrow C_2H_4\,(g)$$

$$C\,(s) + 2H_2\,(g) \rightarrow CH_4(g)$$

Then the question that naturally arises is how to determine the standard enthalpy of formation of a compound, for which the formation reaction cannot be carried out in practice.

Now consider the formation reaction for propane given by

$$3C\,(s) + 4H_2(g) \rightarrow C_3H_8(g) \qquad (A)$$

It is possible to replace the reaction $(A)$ by the set of following reactions

$$C\,(s) + O_2(g) \rightarrow CO_2(g) \qquad (B)$$

$$H_2\,(g) + \frac{1}{2}\,O_2\,(g) \rightarrow H_2O\,(g) \qquad (C)$$

$$C_3H_8\,(g) + 5O_2\,(g) \rightarrow 3CO_2\,(g) + 4H_2O\,(g) \qquad (D)$$

Then, we can write

$$(A) = 3 \times (B) + 4 \times (C) - (D) \qquad (E)$$

Eqns $(B)$ – $(D)$ are combustion reactions. Thus we find that the required formation reaction $(A)$ is expressed in terms of combustion reactions $(B)$ – $(D)$. Hence, it is possible to determine the standard enthalpy of formation of a compound, for which the formation reaction cannot be carried out in practice, in terms of the standard enthalpies of combustion of some other elements or compounds. Similarly, the standard enthalpy change of a chemical reaction can be estimated from a knowledge of the standard enthalpies of combustion of some other elements or compounds.

### Example 10.38

Calculate the standard enthalpy of formation of n–butane($g$) at 298.15 K from a knowledge of the standard enthalpies of combustion.

*Solution* :

The formation reaction for n–butane, $C_4H_{10}(g)$ is given by

$$4C(s) + 5\,H_2(g) \rightarrow C_4H_{10}(g) \qquad (A)$$

In the above reaction, the participating species are $C(s)$, $H_2(g)$ and $C_4H_{10}(g)$. The combustion reactions of these species are

$$C(s) + O_2(g) \rightarrow CO_2(g) \qquad (B)$$

$$H_2(s) + \frac{1}{2}O_2(g) \rightarrow H_2O(l) \qquad (C)$$

$$C_4H_{10}(g) + \frac{13}{2}O_2(g) \rightarrow 4CO_2(g) + 5H_2O\,(l) \qquad (D)$$

Then the required formation reaction ($A$) is given by

$$(A) = 4 \times (B) + 5 \times (C) - (D)$$

That is

$$4C(s) + 4O_2(g) \rightarrow 4CO_2(g): \qquad 4 \times (B)$$

$$5H_2(g) + \frac{5}{2}O_2(g) \rightarrow 5H_2O(l): \qquad 5 \times (C)$$

$$4CO_2(g) + 5H_2O(l) \rightarrow C_4H_{10}(g) + \frac{13}{2}O_2(g): \qquad -1 \times (D)$$

$$4C(s) + 5H_2(g) \rightarrow C_4H_{10}(g): \qquad (A) = 4 \times (B) + 5 \times (C) - 1 \times (D)$$

The standard enthalpies of combustion $\Delta H^0_{C298}$ are presented in Appendix–11. We read the values of $\Delta H^0_{C298}$ from Appendix–11 as

$$C(s) = -\,393.78 \text{ kJ/mol};\; H_2(g) = -\,286.03 \text{ kJ/mol}$$

$$C_4H_{10}\,(g) = -\,2880.44 \text{ kJ/mol.}$$

Then

$$\Delta H^0_{f298}\,[C_4H_{10}\,(g)] = 4 \times (-393.78) + 5 \times (-286.03) - 1 \times (-2880.44)$$

$$= -\,124.83 \text{ kJ/mol}$$

### Example 10.39

Calculate the standard enthalpy change at 298.15 K for the reaction

$$C_3H_8(g) \rightarrow C_2H_4(g) + CH_4(g)$$

from the standard enthalpies of combustion.

*Solution* :

The given reaction is $C_3H_8(g) \rightarrow C_2H_4(g) + CH_4(g)$ $\qquad (A)$

The combustion reactions of the compounds are

$$C_3H_8(g) + 5O_2(g) \rightarrow 3CO_2\,(g) + 4H_2O(l) \qquad (B)$$

$$C_2H_4(g) + 3O_2(g) \rightarrow 2CO_2\,(g) + 2H_2O(l) \qquad (C)$$

$$CH_4(g) + 2O_2(g) \rightarrow CO_2\,(g) + 2H_2O(l) \qquad (D)$$

$$(B) - \{(C) + (D)\} = C_3H_8(g) \rightarrow C_2H_4(g) + CH_4(g) = (A)$$

That is the desired Eqn.($A$) is given by

$$(A) = (B) - \{(C) + (D)\}$$

Therefore, the enthalpy change for reaction ($A$) is given by

$$\Delta H^0_{298}\,(A) = \Delta H^0_{298}\,(B) - \{\Delta H^0_{298}\,(C) + \Delta H^0_{298}\,(D)\}$$

We find that reactions ($B$) – ($D$) are combustion reactions and $\Delta H^0_{298}$ for these reactions refer to $\Delta H^0_{C298}$.

From Appendix –11, we find the standard enthalpies of combustion as

$$\Delta H^0_{C298}\,(C_3H_8) = -2221.54 \text{ kJ/mol}$$

$$\Delta H^0_{C298}\,(C_3H_4) = -411.93 \text{ kJ/mol}$$

$$\Delta H^0_{C298}\,(CH_4) = -890.94 \text{ kJ/mol}$$

Therefore,

$$\Delta H^0_{C298} = -2221.54 - \{(-1411.93) + (-890.94)\} = 81.33 \text{ kJ}$$

### Example 10.40

What is higher heating value and what is lower heating value? For a particular substance why these two heating values differ ?

*Solution* :

The term heating value is commonly used. The heating value represents the amount of energy that can be transferred as heat from the reactor to the surroundings and it is a positive quantity which is equal in magnitude to the enthalpy of combustion. That is heating value is equal to $(-\Delta H^0_C)$. If the water in the products is in the liquid state, the standard enthalpy of combustion has a higher value and it is called as Higher Heating Value (HHV) or gross heating value. On the other hand, if the water in the products is in the gaseous state, the standard enthalpy of combustion has a lower value and it is called Lower Heating Value (LHV). The higher heating value exceeds the lower heating value by the amount of energy that would be liberated by condensing the water vapor into liquid. The heating value also depends on the state of the fuel whether it is in liquid state or gaseous state.

**Example 10.41**

Calculate the higher heating value and lower heating value of n–pentane $C_5H_{12}$ (*g*) at 25°C.

*Solution* :

The combustion reaction for n–pentane is given by

$$C_5H_{12}(g) + 8\,O_2(g) \longrightarrow 5\,CO_2(g) + 6\,H_2O(l)$$

If the water in the combustion products is in the liquid state, the standard enthalpy of combustion of $C_5H_{12}$ (*g*) is given by

$$\Delta H^0_{C298} = -\,3538.52 \text{ kJ/mol (see Appendix–11]}$$

$$\text{Higher Heating Value, HHV} = |\,\Delta H^0_{C298}\,| = 3538.52 \text{ kJ/mol}$$

If one mole of $C_5H_{12}$ (*g*) is burned, it produces 6 mol $H_2O$. The enthalpy of vaporization of water at 25°C is given by

$$h_g - h_f = 2547.3 - 104.77 = 2442.53 \text{ kJ/kg}$$

$$= 43.9655 \text{ kJ/mol [See steam tables]}$$

Therefore, the amount of energy carried away by 6mol $H_2O$ is

$$6 \times 43.9655 = 263.793 \text{ kJ}$$

Hence, the Lower Heating Value, LLV = 3538.52 – 263.793

$$= 3274.727 \text{ kJ/mol}$$

**Example 10.42**

The liquified petroleum gas (LPG) which is normally supplied for domestic use as cooking gas is a mixture of 80% propane and 20% butane by volume. In a kitchen stove only 50% of the heating value of the fuel is utilized and the rest is dissipated away to the surroundings. If it is desired to transfer 5000 kJ of energy as heat for cooking a particular item, estimate the volume of the fuel gas required at 1 bar and 25°C. Assume that the combustion products leave the stove at 25°C.

*Solution* :

When the fuel is burned in a kitchen stove, the water in the combustion products is in gaseous state. Therefore, the lower heating value only is given by the fuel. The composition of the fuel gas is given on volume basis which is equivalent to molar basis for ideal gas mixture. The combustion reaction for propane is given by

$$C_3H_8(g) + 5\,O_2(g) \rightarrow 3\,CO_2(g) + 4\,H_2O(l)$$

$$\Delta H^0_{C298} = -\,2221.54 \text{ kJ/mol [ see Appendix–11]}$$

$$\text{LHV} = 2221.54 - 4 \times 43.9655 = 2045.678 \text{ kJ/mol}$$

(Enthalpy of vaporization of water at 25°C = 43.9655 kJ/mol)

The combustion reaction for butane is given by

$$C_4H_{10}(g) + \frac{13}{2}O_2(g) \rightarrow 4\ CO_2(g) + 5\ H_2O\ (l)$$

$$\Delta H^0_{C298} = -\ 2880.44\ \text{kJ/mol [See Appendix} -11]$$

$$\text{LHV} = 2880.44 - 5 \times 43.9655 = 2660.613\ \text{kJ/mol}$$

One mole of fuel gas contains 0.8 mol propane and 0.2 mol butane.

Hence LHV of 1 mole fuel gas = 0.8 × 2045.678 + 0.2 × 2660.613

= 2168.665 kJ/mol

Energy required for cooking = 5000 kJ

Therefore, energy to be given by fuel = 5000 × 2 = 10,000 kJ

$$\text{Fuel gas required} = \frac{10{,}000}{2168.665} = 4.611\ \text{mol} = n$$

$$\text{Hence volume of fuel gas required} = V = \frac{nRT}{P}$$

$$= \frac{4.611 \times 8.314 \times 298}{1 \times 10^5} = 0.1142\ \text{m}^3$$

**Example 10.43**

Does the standard enthalpy of combustion at 25°C depend on whether the fuel is burned with pure oxygen or with air?

*Solution* :

The combustion reaction for a fuel (say $CH_4$) with pure oxygen is given by

$$CH_4\ (g) + 2O_2\ (g) \rightarrow CO_2(g) + 2H_2O\ (g)$$

The value of $\Delta H^0_{C298}$ for this reaction is given by

$$\Delta H^0_{C298} = \Delta H^0_{298} = \Delta H^0_{f298}\ [CO_2(g)] + 2\ \Delta H^0_{f298}\ [H_2O(g)] - \Delta H^0_{f298}\ [CH_4(g)] \quad (A)$$

If the fuel ($CH_4$) is burned in air, the combustion reaction is given by

$$CH_4\ (g) + 2O_2(g) + 2\ (3.76)\ N_2\ (g) \rightarrow CO_2(g) + 2H_2O\ (g) + 2(3.76)\ N_2\ (g)$$

The value of $\Delta H^0_{C298}$ for this reaction is given by

$$\Delta H^0_{C298} = \Delta H^0_{298} = \Delta H^0_{f298}\ [CO_2\ (g)] + 2\Delta H^0_{f298}\ [H_2O(g)] - \Delta H^0_{f298}[CH_4(g)]$$

which is identical with Eqn. (*A*). Therefore, the presence of $N_2$ or any inert does not change the standard enthalpy of combustion of a fuel.

**Example 10.44**

Does the standard enthalpy change ($\Delta H^0_{298}$) of a reaction depend on whether the reactants are in the stoichiometric ratio or not ?

*Solution* :

Consider the reaction

$$CH_4\,(g) + H_2O\,(g) \rightarrow CO\,(g) + 3\,H_2\,(g) \qquad (A)$$

in which the reactants are in stoichiometric ratio. The standard enthalpy change of the reaction (*A*) is given by

$$\Delta H^0_{298} = \Delta H^0_{f298}\,[CO(g)] - \Delta H^0_{f298}\,[CH_4(g)] + \Delta H^0_{f298}\,[H_2O\,(g)] \qquad (B)$$

Suppose the same reaction is carried out with excess $H_2O(g)$. Then the reaction can be written as

$$CH_4\,(g) + nH_2O\,(g) \rightarrow CO\,(g) + 3H_2\,(g) + (n-1)\,H_2O\,(g) \qquad (C)$$

Then $\Delta H^0_{298} = \{\Delta H^0_{f298}\,[CO\,(g)] + (n-1)\,\Delta H^0_{f298}\,[H_2O\,(g)]\} - \{\Delta H^0_{f298}\,[CH_4(g)] + n\Delta H^0_{f298}\,[H_2O\,(g)]\}$

or $\Delta H^0_{298} = \Delta H^0_{f298}\,[CO(g)] - \{\Delta H^0_{f298}\,[CH_4(g)] + \Delta H^0_{f298}\,[H_2O(g)]\} \qquad (D)$

We find that Eqns. (*B*) and (*D*) are identical. Hence the standard enthalpy change of a reaction does not depend on whether the reactants are in stoichiometric ratio or not.

**Example 10.45**

An internal combustion engine uses gasoline as a fuel. The engine is supplied with 150 percent theorietical air at 1 bar and 25°C. An analysis of the exhaust gases reveals that 75% of the fuel's carbon is converted to $CO_2$ and the rest to CO. The combustion products leave the engine at 400 K. Calculate the amount of energy transferred as heat to the engine per kg of gasoline. The standard enthalpy of formation $\Delta H^0_{f298}$ for n–octane is –208.75 kJ/mol.

*Solution* :

For combustion analysis gasoline may be treated as pure octane ($C_8H_{18}$). To estimate the theoretical air requirement, the combustion equation is written as

$$C_8H_{18}\,(g) + \frac{25}{2}\,O_2\,(g) + \frac{25}{2}\,(3.76)\,N_2\,(g) \longrightarrow 8\,CO_2\,(g) + 9\,H_2O\,(g) + \frac{25}{2}(3.76)N_2(g)$$

$$\text{Theorietical air-fuel ratio} = \frac{\frac{25}{2}\times 4.76}{1} = 59.5 \text{mol air / mol fuel}$$

The combustion equation with 150 percent theoretical air and 75% of carbon converted to $CO_2$ and the rest to CO is written as

$$C_8H_{18}(g) + 1.5\times\frac{25}{2}O_2(g) + 1.5\times\frac{25}{2}(3.76)N_2(g) \longrightarrow 6CO_2(g) + 2CO(g) + 9H_2O(g) + 7.25O_2(g) + 70.5N_2(g)$$

The standard enthalpies of formation $\Delta H^0_{f298}$ of the compounds are

$$C_8H_{18}(g) = -208.75 \text{ kJ / mol; } CO_2(g) = -393.978 \text{kJ / mol;}$$
$$CO(g) = -110.532 \text{ kJ / mol; } H_2O(g) = -241.997 \text{kJ / mol}$$

Then the standard enthalpy change of the reaction $\Delta H^0_{f298}$ in given by

$$\Delta H^0_{298} = 6\times(-393.978) + 2\times(-110.532) + 9\times(-241.997) - (-208.75)$$
$$= -4554.155 \text{ kJ}$$

The air-fuel mixture is entering the engine at 1 bar and 25°C and the combustion products are leaving at 400 K. Let us consider the engine cylinder as the control–volume as shown in Fig.E 10.45 (*a*) and apply the energy balance (ignoring KE and PE changes, steady state, steady flow) to obtain

$$h_e - h_i = Q \quad \text{or} \quad \Delta H = Q$$

To evaluate $\Delta H$, now let us choose a convenient path connecting the initial and final states as shown in FigE. 10.45 (*b*).

The actual process path is shown by a dotted line while the imaginary path is shown by solid line in FigE. 10.45 (*b*). Then

$$\Delta H = \Delta H^0_{298} + \Delta H_1.$$

The imaginary path indicates that chemical reaction takes place at 298.15 K and then the combustion products are raised to 400 K.

$$\Delta H_1 = \int_{298}^{400} (\Sigma n_i\, C_{P_i} dT)$$

The isobaric molar heat capacities of some selected gases in the ideal gas state are presented in Appendix –12. We obtain the following values from Appendix–12.

| Species | *a* | *b* | *e* |
|---|---|---|---|
| $CO_2$ | 45.369 | $8.688 \times 10^{-3}$ | $-9.619 \times 10^5$ |
| CO | 28.068 | $4.631 \times 10^{-3}$ | $-0.258 \times 10^5$ |
| $H_2O$ | 28.850 | $12.055 \times 10^{-3}$ | $1.066 \times 10^5$ |
| $O_2$ | 30.255 | $4.207 \times 10^{-3}$ | $-1.887 \times 10^5$ |
| $N_2$ | 27.270 | $4.930 \times 10^{-3}$ | $0.333 \times 10^5$ |

The molar heat capacities are given as

$$C_P^0 = a + bT + eT^{-2}$$

where $C_P^0$ is in J/mol K and $T$ is in Kelvin.

$$\text{Then } \Delta H_1 = \int_{298}^{400} (\Sigma n_i C_{P_i}^0 dT) = \int_{298}^{400} (\Sigma n_i (a_i + b_i T + e_i . T^{-2})\, dT$$

$$= \Delta a\,(400 - 298) + \frac{\Delta b}{2}(400^2 - 298^2) - \Delta e\left(\frac{1}{400} - \frac{1}{298}\right)$$

Where $\Delta a = \Sigma\, n_i a_i$; $\Delta b = \Sigma\, n_i b_i$ and $\Delta e = \Sigma\, n_i e_i$

$$\Delta a = 6 \times 45.396 + 2 \times 28.068 + 9 \times 28.850 + 7.25 \times 30.255 + 70.5 \times 27.27 = 2729.884$$

$$\Delta b = (6 \times 8.688 + 2 \times 4.631 + 9 \times 12.055 + 7.25 \times 4.207 + 70.5 \times 4.93) \times 10^{-3} = 547.951 \times 10^{-3}$$

$$\Delta e = \{6 \times (-9.615) + 2 \times (-0.258) + 9 \times 1.066 + 7.25 \times (-1.887) + 70.5 \times 0.333\} \times 10^5 = -38.816 \times 10^5$$

Therefore,

$$\Delta H_1 = 2729.884\,(400 - 298) + \frac{547.951 \times 10^{-3}}{2}(400^2 - 298^2) + 38.816 \times 10^5 \left(\frac{1}{400} - \frac{-1}{298}\right) = 297.954 \text{ kJ}$$

Then $\Delta H = \Delta H^0_{298} + \Delta H_1 = -4554.155 + 297.954$

or $\Delta H = -4256.2 \text{ kJ} = Q$

Molar mass of $C_8H_{18} = 114 \times 10^{-3}$ kg

$$1 \text{kg } C_8H_{18} = \frac{1}{114 \times 10^{-3}} = 8.772 \text{ mol}$$

Therefore, energy transferred to the engine = 8.772 × 4256.2 = 37.335 MJ for 8.772 mol of gasoline

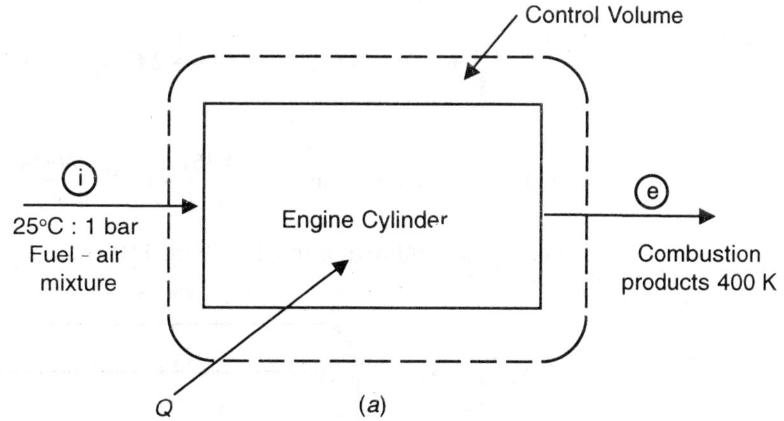


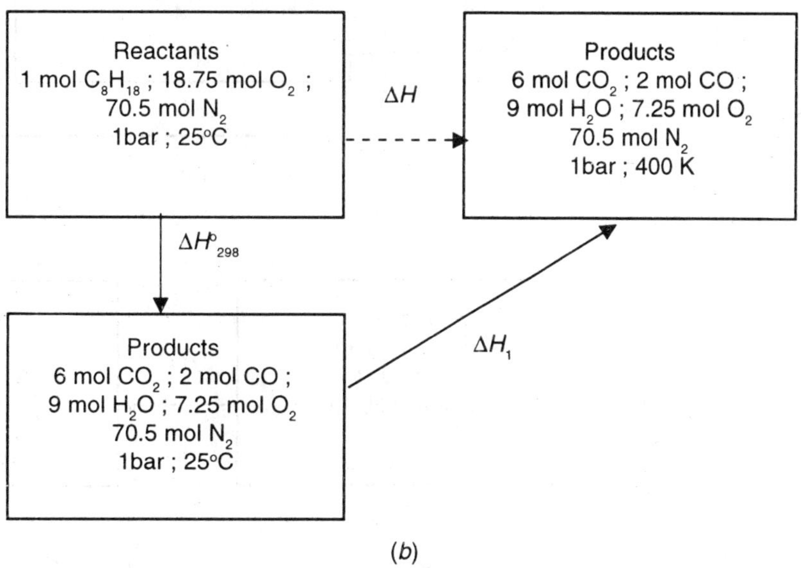


Fig.E 10.45 (*a*) Control volume for energy balance (*b*) Convenient path to calculate $\Delta H$.

### Example 10.46

Liquid ethanol ($C_2H_5OH$) at 1 bar and 25°C is continuously sprayed into a reactor operating at steady – state and burns completely with 150 percent theoretical air which is entering at 1 bar and 25°C. The products of combustion leave the reactor at 150°C and 1 bar. If the rate of energy transfer from the reactor is 1000 kW, calculate the mass flow rate of ethanol into the reactor.

*Solution* :

To estimate the theoretical air requirement, the combustion reaction is given by

$$C_2H_5OH + 3O_2 + 3\,(3.76)\,N_2 \longrightarrow 2\,CO_2 + 3\,H_2O + 3\,(3.76)\,N_2$$

$$\text{Theoretical air fuel ratio } = \frac{3(4.76)}{1} = 14.28\ \frac{\text{mol air}}{\text{mol fuel}}$$

The actual combustion reaction with 150% theoretical air is

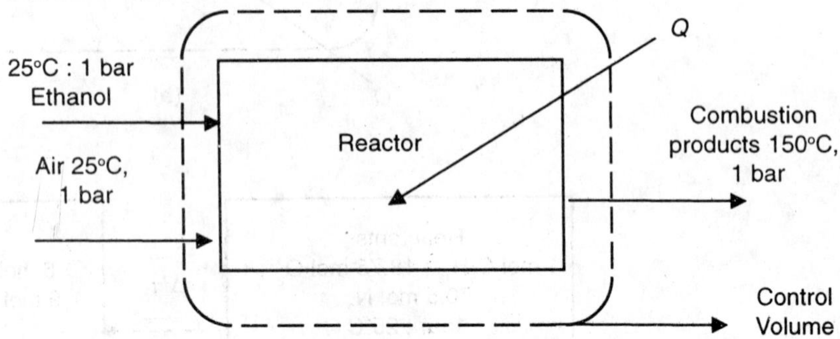


(*a*)

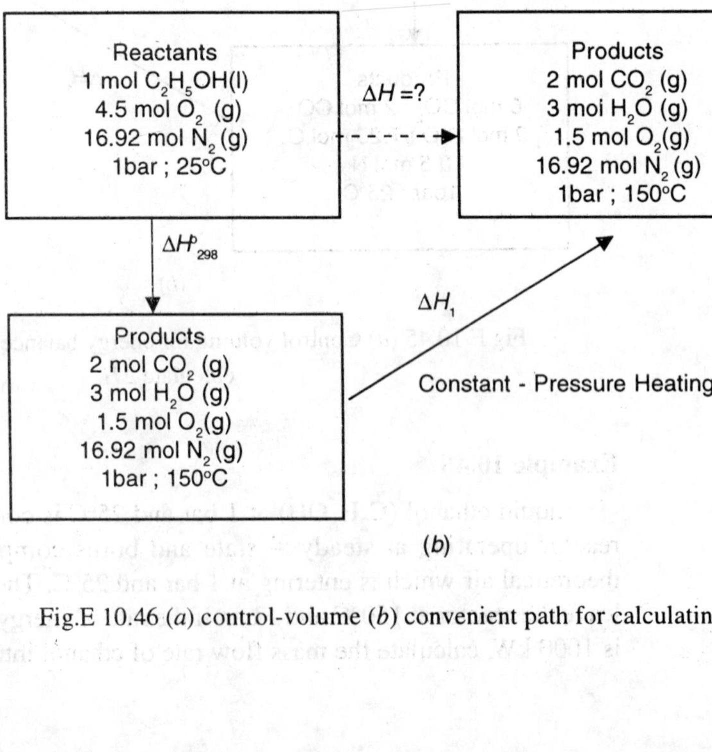


(*b*)

Fig.E 10.46 (*a*) control-volume (*b*) convenient path for calculating $\Delta H$.

$$C_2H_5OH(l) + 1.5(3)\ O_2\ (g) + 1.5 \times 3 \times 3.76\ N_2\ (g) \rightarrow 2\ CO_2(g) + 3\ H_2O\ (g) + 1.5\ O_2\ (g) + 16.92\ N_2\ (g)$$

The standard enthalpies of formation $\Delta H^0_{f298}$ of the compounds are

$C_2H_5OH\ (l) = -\ 277.819$ kJ/mol; $CO_2(g) = -\ 393.978$ kJ/mol;

$H_2O\ (l) = -\ 241.997$ kJ/mol;

$$\Delta H^0_{298} = 2 \times (-\ 393.978) + 3 \times (-\ 241.997) - (-\ 277.819)$$

$$= -\ 1236.128 \text{ kJ}$$

Now, let us choose the control–volume as shown in FigE. 10.46(*a*).

Applying the first law of thermodynamics for a steady–flow, steady state control volume and ignoring KE and PE changes, we get

$$\Sigma n_e h_e - \Sigma n_i h_i = Q = \Delta H$$

To evaluate $\Delta H$, we can choose the convenient process path as shown in FigE. 10.46(*b*). Then

$$\Delta H = \Delta H^0_{298} + \Delta H_1$$

$$\text{Where } \Delta H_1 = \int_{298}^{423} (\Sigma n_i\, C_i^0)\, dT = \int (\Delta a + \Delta bT + \Delta eT^{-2}) dT$$

where $\Delta a = \sum n_i a_i$ ; $\Delta b = \Sigma n_i b_i$; $\Delta e = \Sigma n_i e_i$

The constants of the molar heat capacities as read from Appendix–12 are

| Constituent | *a* | $b \times 10^3$ | $e \times 10^{-5}$ |
|---|---|---|---|
| $CO_2$ | 45.369 | 8.688 | –9.619 |
| $H_2O$ | 28.850 | 12.055 | 1.006 |
| $O_2$ | 30.255 | 4.207 | – 1.887 |
| $N_2$ | 27.270 | 4.930 | 0.333 |

$$\Delta a = 2 \times 45.369 + 3 \times 28.85 + 1.5 \times 30.255 + 16.92 \times 27.27 = 684.079$$

$$\Delta b = (2 \times 8.688 + 3 \times 12.055 + 1.5 \times 4.207 + 16.92 \times 4.93)\ 10^{-3}$$

$$= 143.267 \times 10^{-3}$$

$$\Delta e = \{2 \times (-9.619) + 3 \times 1.006 + 1.5(-1.887) + 16.92 \times 0.333\} \times 10^5 = 13.416 \times 10^5$$

Therefore,

$$\Delta H_1 = \int_{298}^{423} (684.079 + 143.267 \times 10^{-3} T - 13.416 \times 10^5 T^{-2})\, dT$$

$$= 684.079\ (423 - 298) + \frac{143.276 \times 10^{-3}}{2}(423^2 - 298^2) + 13.416 \times 10^5\left(\frac{1}{423} - \frac{1}{298}\right)$$

or $\Delta H_1 = 90.635$ kJ

Then $\Delta H = \Delta H^0_{298} + \Delta H_1 = -1236.128 + 90.635 = -1145.493$ kJ

Therefore, rate of energy transfer from the reactor = 1145.493 kW per mol of $C_2H_5OH(l)$

The rate of energy transfer required = 1000 kW

Hence flow rate of ethanol $= \dfrac{1000}{1145.493} = 0.873$ mol / s

Molar mass of ethanol ($C_2H_5OH$) = $46 \times 10^{-3}$ kg / mol

Mass flow rate of ethanol = $0.873 \times 46 \times 10^{-3} = 0.04$ kg/s

**Example 10.47**

Does the standard enthalpy change of a reaction depend on the temperature at which the chemical reaction occurs? If so devise a method to estimate the standard enthalpy change of a reaction at temperature $T$.

*Solution* :

The standard enthalpy change of a reaction is the change in the enthalpy of the products and reactants in their standard states. The standard state temperature is the temperature at which the reaction occurs. Therefore, if a reaction occurs at a temperature other than 298.15 K, then the standard enthalpy change of the reaction will be different from $\Delta H^0{}_{298}$. Since enthalpy is a property, its change does not depend on the path followed by the chemically reacting system but depends only on the initial and final states of the system. Therefore, one can devise an imiginary path, which is convenient from the calculation point of view, connecting the given initial and final states of the system and evaluate the enthalpy change for the imaginary path. The actual enthalpy change will be identical with that of the imaginary path. Suppose a chemical reaction occurs at temperature $T$ > 298 K. Then the standard enthalpy change of the reaction at temperature $T$ can be estimated by the three step procedure shown in Fig.E 10.47.

**Step 1** : The reactants in their standard states at temperature $T$ are cooled at constant pressure from $T$ to 298.15 K. The change in the enthalpy associated with this process is given by

$$\Delta H_R^0 = \int_T^{298} (\Sigma n_i C_{P_i}^0 \, dT) = \int_T^{298} (\Sigma \,|\, \upsilon_i \,|\, C_{P_i}^0 \, dT) = \int_{298}^{T} (\Sigma \upsilon_i C_{P_i}^0 \, dT)_R$$

where the summation is over all the reactants. Note that $\upsilon_i$ is negative for reactants.

**Step 2**: The reaction occurs at 298.15 K. The change in the enthalpy associated with this process is $\Delta H_{298}^0$.

**Step 3**: The products in their standard states are heated at constant pressure from 298.15 K to $T$. The change in the enthalpy associated with this process is given by

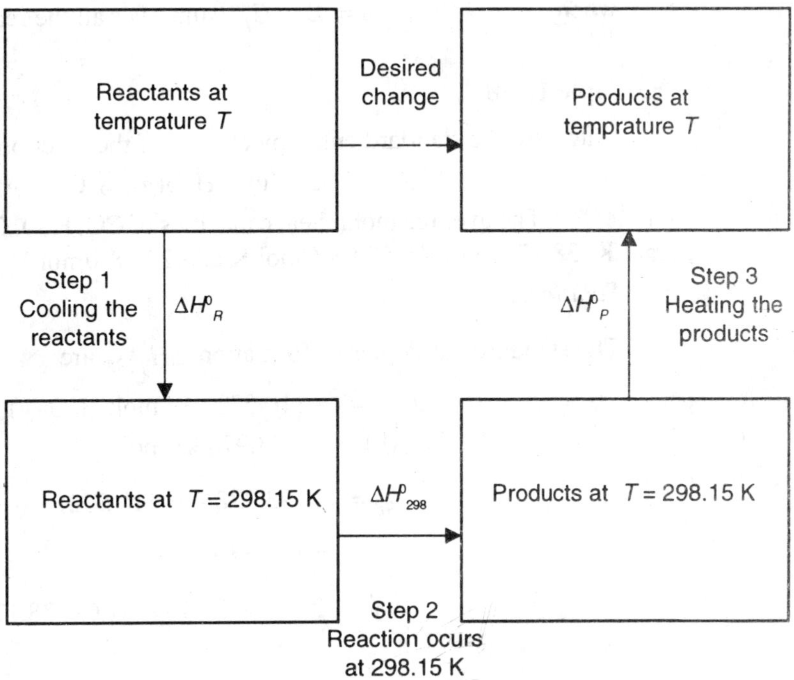


Fig.E 10.47. Three step path replacing the actual path for the evaluation of $\Delta H^0{}_T$

$$\Delta H_P^0 = \int_{298}^{T} (\Sigma \upsilon_i C_{P_i}^0 dT)_P$$

where the summation is over all the products. Then the desired $\Delta H_T^0$, the enthalpy change of the reaction at temperature $T$ is given by

$$\Delta H_T^0 = \Delta H_R^0 + \Delta H_{298}^0 + \Delta H_P^0$$

$$= \int_{298}^{T} (\Sigma \upsilon_i C_{P_i}^0 dT)_R + \Delta H_{298}^0 + \int_{298}^{T} (\Sigma \upsilon_i C_{P_i}^0 dT)_P$$

or
$$\Delta H_T^0 = \Delta H_{298}^0 + \int_{298}^{T} (\Sigma \upsilon_i \, C_{P_i}^0 dT) \quad (A)$$

where, the summation is over all the species participating in the reaction. The Eqn.($A$) can be rewritten as

$$\Delta H_T^0 = \Delta H_{298}^0 + \int_{298}^{T} (\Delta C_{P_i}^0 dT)$$

where $\Delta C_P^0 = \Sigma \, \upsilon_i \, C_{Pi}$ (sum over all the reactants and products).

**Example 10.48**

Calculate the standard enthalpy change of the reaction

$$CO\,(g) + H_2O(g) \rightarrow CO_2\,(g) + H_2\,(g)$$

at 1000°C. The average molar heat capacities of CO, $H_2O$, $CO_2$ and $H_2$ are 31.6 J/mol K; 38.17 J/mol K; 52.08 J/mol K and 27.28 J/mol K, respectively.

*Solution* :

The standard enthalpies of formation $\Delta H_{f298}^0$ are

$$CO\,(g) = -110.532 \text{ kJ / mol}; \; H_2O\,(g) = -241.997 \text{ kJ/mol};$$
$$CO_2\,(g) = -393.978 \text{ kJ/mol}$$

$$\Delta H_{298}^0 = -393.978 - (-110.532) - (-241.997)$$
$$= -41.449 \text{ kJ}$$

$$\Delta C_P^0 = 27.28 + 52.08 - 31.6 - 38.17 = 9.59 \text{ J/mol K}$$

We know that

$$\Delta H_T^0 = \Delta H_{298}^0 + \int_{298}^{T} (\Delta C_{P_i}^0 dT)$$

or $\Delta H_{1273}^0 = -41449 + 9.59\,(1273 - 298) = -40.474$ kJ

**Example 10.49**

The molar heat capacities of gases depend on temperature. If the temperature dependency of the molar heat capacity is given by

$$C_P^0 = a + bT + cT^2 + dT^3 + eT^{-2}$$

develop a general expression to estimate the standard enthalpy change of a reaction at temperature $T$.

*Solution* :

We know that

$$\Delta H_T^0 = \Delta H_{298}^0 \int_{298}^{T} (\Sigma \upsilon_i C_{P_i}^0 dT)$$

If $C_{P_i}^0 = a_i + b_i T + c_i T^2 + d_i T^3 + e_i T^{-2}$

Then $\Sigma\, \upsilon_i\, C_P^0 = \Sigma\, \upsilon_i a_i + \Sigma\, \upsilon_i b_i T + \Sigma\, \upsilon_i c_i T^2 + \Sigma\, \upsilon_i d_i T^3 + \Sigma\, \upsilon_i e_i T^{-2}$

$$= \Delta a + \Delta bT + \Delta cT^2 + \Delta dT^3 + \Delta eT^{-2}$$

Where $\Delta a = \Sigma\, \upsilon_i a_i$; $\Delta b = \Sigma\, \upsilon_i b_i$; $\Delta c = \Sigma\, \upsilon_i c_i$ ; $\Delta d = \Sigma\, \upsilon_i d_i$ and $\Delta e = \Sigma\, \upsilon_i e_i$

Then $\Delta H_T^0 = \Delta H_{298}^0 + \int_{298}^{T} (\Delta a + \Delta bT + \Delta cT^2 + \Delta dT^3 + \Delta eT^{-2})\, dT$

or $\Delta H_T^0 = \Delta H_{298}^0 + \Delta a\,(T - 298) + \frac{\Delta b}{2}(T^2 - 298^2) + \frac{\Delta c}{3}$

$$(T^3 - 298^3) + \frac{\Delta d}{4}(T^4 - 298^4) - \Delta e\left(\frac{1}{T} - \frac{1}{298}\right)$$

or $\Delta H_T^0 = \Delta H_0 + \Delta aT + \frac{\Delta b}{2}T^2 + \frac{\Delta c}{3}T^3 + \frac{\Delta d}{4}T^4 - \frac{\Delta e}{T}$

where $\Delta H_0$ is a constant, the value of which can be determined from a knowledge of $\Delta H_T^0$ at any temperature.

***Note*** *The molar heat capacities of substances are usually expressed as functions of temperature by one of the following relations.*

$$C_P^0 = a + bT + cT^2$$

$$C_P^0 = a + bT + cT^2 + dT^3$$

$$C_P^0 = a + bT + eT^{-2}$$

*These relations can be combined and expressed in a general form as*

$$C_P^0 = a + bT + cT^2 + dT^3 + eT^{-2}$$

*which reduces to one of the above relations depending on the availability of the constants.*

**Example 10.50**

Calculate the standard enthalpy change of the reaction

$$CO\,(g) + H_2O\,(g) \rightarrow CO_2\,(g) + H_2\,(g)$$

at 1000°C, taking into account the variation of the molar heat capacity of the gases as given in Appendix –12.

*Solution* :

We have $\Delta H^0_{298} = -41.449$ kJ [see Example 10.48]

From Appendix –12, the constants in the expression for the molar heat capacities of the gases are

| Species | *a* | *b*×10 ³ | *e*×10 ⁻⁵ |
|---|---|---|---|
| $CO\,(g)$ | 28.068 | 4.631 | – 0.258 |
| $H_2O\,(g)$ | 28.850 | 12.055 | 1.006 |
| $CO_2\,(g)$ | 45.369 | 8.688 | – 9.619 |
| $H_2\,(g)$ | 27.012 | 3.509 | 0.690 |

Then $\Delta a = 45.369 + 27.012 - 28.068 - 28.850 = 15.463$

$$\Delta b = (8.688 + 3.509 - 4.631 - 12.055) \times 10^{-3} = -4.489 \times 10^{-3}$$

$$\Delta e = -9.619 + 0.690 - (-0.258) - 1.006) \times 10^{5} = -9.677 \times 10^{5}$$

We know that $\Delta H^0_T = \Delta H_0 + \Delta a T + \dfrac{\Delta b}{2} T^2 - \dfrac{\Delta e}{T}$

or $$\Delta H^0_T = \Delta H_0 + 15.463T - \frac{4.489 \times 10^{-3}}{2} T^2 + \frac{9.677 \times 10^5}{T} \qquad (A)$$

We know the value of $\Delta H^0_{298}$. Hence substitute the value of $T = 298$ in Eqn.($A$) to determine the value of the constant $\Delta H_0$.

$$-41449 = \Delta H_0 + 15.463 \times 298 - \frac{4.489 \times 10^{-3}}{2} \times (298)^2 + \frac{9.677 \times 10^5}{298}$$

or $$\Delta H_0 = -49.104 \text{ kJ}$$

Now substitute the value of $\Delta H_0$ and $T = 1273$ in Eqn.($A$) to obtain the value of $\Delta H^0_{1273}$.

$$\Delta H^0_{1273} = -49104 + 15.463 \times 1273 - \frac{4.489 \times 10^{-3}}{2} (1273)^2 + \frac{9.677 \times 10^5}{1273}$$

or $\Delta H^0_{1273} = -32.297$ kJ

**Example 10.51**

Calculate the standard enthalpy change at 500°C for the reaction

$N_2(g) + 3H_2(g) \rightarrow 2NH_3(g)$

taking into account the variation of molar heat capacities of gases with temperature.

*Solution* :

The given reaction is

$$N_2(g) + 3H_2(g) \rightarrow 2NH_3(g)$$

The standard enthalpy change for this reaction at 298.15 K is

$$\Delta H^0_{298} = 2\Delta H^0_f\{NH_3(g)\} = 2 \times (-46.055) = -92.11 \text{ kJ}$$

The constants of the molar heat capacity expression are

| **Species** | ***a*** | ***b*×10 ³** | ***e*×10 ⁻⁵** |
|---|---|---|---|
| $NH_3(g)$ | 29.747 | 25.108 | −1.546 |
| $N_2(g)$ | 27.270 | 4.930 | 0.333 |
| $H_2(g)$ | 27.012 | 3.509 | 0.690 |

Then $\Delta a = 2 \times 29.747 - 27.270 - 3 \times 27.012 = -48.812$

$\Delta b = (2 \times 25.108 - 4.930 - 3 \times 3.509) \times 10^{-3} = 34.759 \times 10^{-3}$

$\Delta e = \{2 \times (-1.546) - 0.333 - 3 \times 0.690\} \times 10^5 = -5.495 \times 10^5$

We know that $\Delta H^0_T = \Delta H_0 + \Delta aT + \dfrac{\Delta b}{2}T^2 - \dfrac{\Delta e}{T}$

$$\text{or} \quad \Delta H^0_T = \Delta H_0 - 48.812T + \frac{34.759 \times 10^{-3}}{2}T^2 + \frac{5.495 \times 10^5}{T} \qquad (A)$$

Substitute for $\Delta H^0_T = \Delta H^0_{298}$ and $T = 298$ in Eqn. (*A*) to obtain the value of $\Delta H_0$.

$$-92110 = \Delta H_0 - 48.812 \times 298 + \frac{34.759 \times 10^{-3}}{2}(298)^2 + \frac{5.495 \times 10^5}{298}$$

or $\Delta H_0 = -80.951$ kJ

Now, substitute the value of $\Delta H_0$ and $T = 773$ to obtain

$$\Delta H^0_{773} = -80951 - 48.812 \times 773 + \frac{34.759 \times 10^{-3}}{2}(773)^2 + \frac{5.495 \times 10^5}{773}$$

or $\Delta H^0_{773} = -107.587$ kJ

### Example 10.52

What is meant by adiabatic flame temperature and on which factors does it depend ?

*Solution* :

Suppose a mixture of fuel and oxidant enter an adiabatic burner which is operating under steady–flow, steady–state conditions. As the burner does not exchange energy as heat with the surroundings, the combustion products leave the burner at high temperature. The maximum flame temperature that can be attained in an adiabatic burner when the fuel and oxidant enter the burner at a specified temperature, is called the adiabatic flame temperature.

A knowledge of the adiabatic flame temperature is essential because the permissible temperature of the combustion chamber is controlled by metallurgical considerations .

The adiabatic temperature reaches a maximum value when the fuel is burned with stoichiometric oxygen. The use of air or excess air reduces the flame temperature. In addition to this incomplete combustion and dissociation of the combustion products also reduce the flame temperature. The adiabatic flame temperature can be controlled by adjusting the air–fuel ratio.

### Example 10.53

Suppose a mixture of fuel and air enters an adiabatic combustion chamber at known temperature $T_0$ and it is desired to estimate the maximum flame temperature that can be attained with the given air-fuel ratio. Indicate a method of estimating the adiabatic flame temperature.

*Solution*:

The flame temperature attains a maximum value when the combustion is complete and no dissociation of the products occur. Therefore, the combustion reaction of the fuel is written for the given air-fuel ratio, assuming complete combustion occurs. Then the composition of the combustion products expressed in moles of each constituent, is estimated. Considering the combustion chamber as the control –volume,application of the first law of thermodynamics gives

$$H_e - H_i = 0 \text{ or } H_P - H_R = 0 \text{ or } \Delta H = 0$$

Where $H_P$ and $H_R$ denote the enthalpies of products and reactants, respectively. In other words, the change in the enthalpy is equal to zero. To determine the temperature of the combustion products (or the flame temperature) the following two step procedure shown in FigE. 10.53 can be adopted. Then

$$\Delta H = 0 = \Delta H_{298}^{0} + \Delta H_P$$

$$\text{or} \quad \Delta H_{(298)}^{0} + \int_{(298)}^{T} (\Sigma \upsilon_i C_{P_i}^{0} dT)_P = 0 \qquad (A)$$

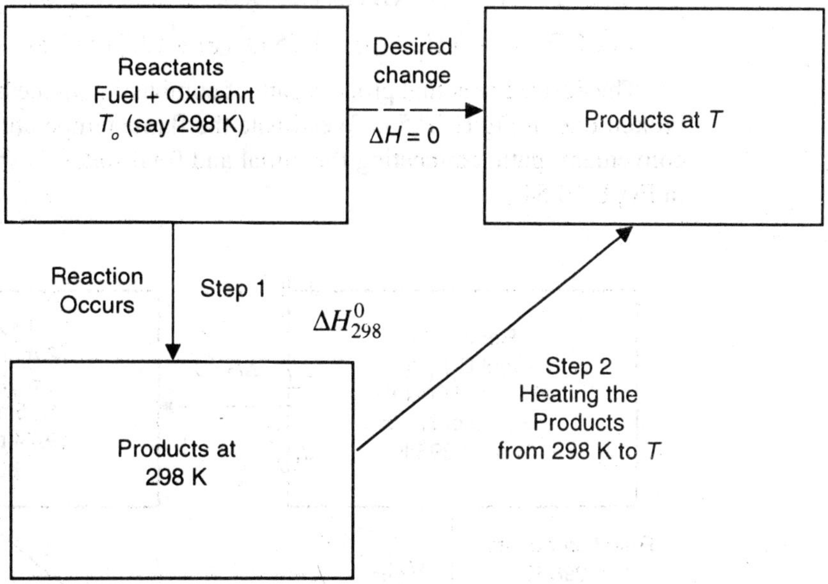


Fig.E 10.53. Two-step process to determine the adiabatic flame temperature.

where the summation is over all the product species. By solving Eqn.(*A*) one can determine the flame temperature.

## Example 10.54

Calculate the adiabatic flame temperature when ethane $C_2H_6(g)$ is burned with 150 percent theoretical air. The fuel and air enter the burner at 1 bar and 298 K and the combustion products leave the burner at 1 bar. The molar heat capacities of $CO_2$, $H_2O$, $O_2$ and $N_2$ are constant and are given by 53.1 J/mol K; 41 J/mol K; 34.3 J/mol K and 32.2 J/mol K, respectively.

*Solution* :

To determine the theoretical air-fuel ratio, the combustion reaction is written as

$$C_2H_6\,(g) + 3.5\,O_2\,(g) + 3.5\,(3.76)\,N_2\,(g) \rightarrow 2\,CO_2\,(g) + 3\,H_2O\,(g) + 13.16\,N_2\,(g)$$

Therefore, the theoretical air-fuel ratio $= \dfrac{3.5 + 3.5\,(3.76)}{1} = 16.66\,\dfrac{\text{mol air}}{\text{mol fuel}}$

Air-fuel ratio used (150%) = 16.66 × 1.5 = 24.99 mol air /mol fuel

The actual combustion reaction is given by

$$C_2H_6\,(g) + 1.5 \times 3.5\,O_2\,(g) + 1.5 \times 3.5 \times 3.76\,N_2\,(g)$$
$$\rightarrow 2\,CO_2\,(g) + 3\,H_2O\,(g) + 1.75\,O_2\,(g) + 19.74\,N_2\,(g)$$

The desired or actual process path of combustion in the burner is shown by a broken line in Fig.E 10.54. To estimate the flame temperature an imaginary but convenient path, connecting the initial and final states, is shown by a solid line in Fig.E 10.54.

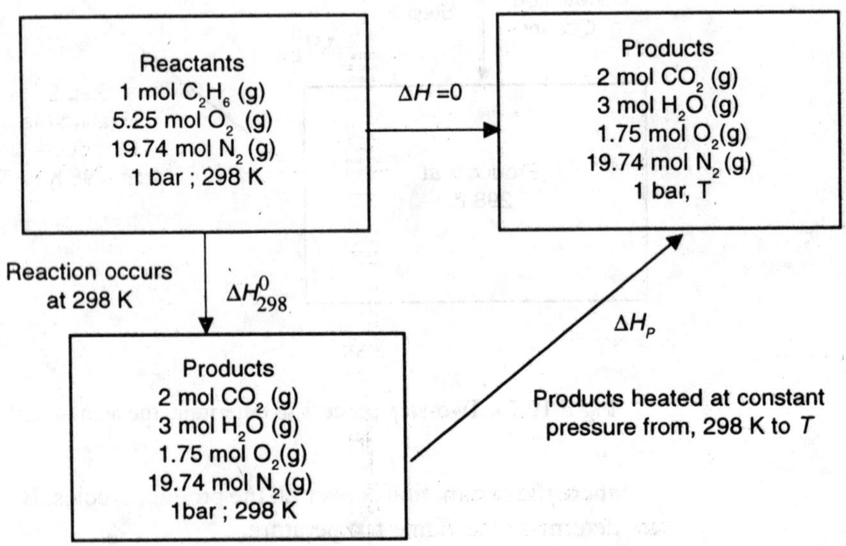


Fig.E 10.54. Sketch for Example 10.54.

Then $$\Delta H = 0 = \Delta H^0_{298} + \Delta H_P = \Delta H^0_{298} + \int_{298}^{T} (\Sigma \upsilon_i C^o_{P_i} dT)_P$$

The standard enthalpy change for the combustion reaction at 298 K is given by

$$\Delta H^0_{298} = 2\Delta H^0_{f298}\,[CO_2(g)] + 3\Delta H^0_{f298}\,[H_2O(g)] - \Delta H^0_{f298}\,[C_2H_6(g)]$$

$$= 2 \times (-393.978) + 3 \times (-241.997) - (-84.573)$$

$$= -1429.374 \text{ kJ}$$

$$\Delta C_P^0 = (\Sigma \upsilon_i C_{P_i}^0)_P = 2C_P^0 (CO_2) + 3C_P^0 (H_2O) + 1.75C_P^0 (O_2)$$

$$+ 19.74C_P^0 (N_2)$$

$$= 2 \times 53.1 + 3 \times 41 + 1.75 \times 34.3 + 19.74 \times 32.2$$

$$= 924.853 \text{ J/K} = 0.9249 \text{ kJ / K}$$

We know that

$$\Delta H_{298}^0 + \int_{298}^{T} (\Sigma \upsilon_i C_{P_i}^0 dT)_P = 0$$

or $-1429.374 + 0.9249 (T - 298) = 0$

or $T = 1843.4$ K

Therefore, the adiabatic flame temperature = 1843.4 K.

### Example 10.55

Estimate the maximum flame temperature that can be attained in a burner if hydrogen gas at 1 bar and 298 K is used as a fuel. The oxidant also enters the burner at 1 bar and 298 K and the combustion products leave the burner at 1 bar. Assume that the molar heat capacity of $H_2O$ is constant and is equal to 41 J / mol K.

*Solution* :

The flame temperature attains a maximum value when pure oxygen in stoichiometric quantity is supplied as the oxidant and the burner is adiabatic. The combustion also should be complete to attain maximum flame temperature. Then the combustion reaction is given by

$$H_2 (g) + \frac{1}{2} O_2(g) \rightarrow H_2O (g)$$

$$\Delta H_{298}^0 = -241.997 \text{ kJ}$$

We know that $\Delta H_{298}^0 + \int_{298}^{T} (C_P^0 \, dT)_{H_2O} = 0$

or $= -241.997 + 41 \times 10^{-3} (T - 298) = 0$

or $T = 6200$ K

**Example 10.56**

Rework Example 10.54, taking into account the dependence of molar heat capacity of gases on temperature.

*Solution* :

The constants in the molar heat capacity expression for the product species are

| Species | $a$ | $b \times 10^3$ | $e \times 10^{-5}$ |
|---|---|---|---|
| $CO_2(g)$ | 45.369 | 8.688 | - 9.619 |
| $H_2O(g)$ | 28.850 | 12.055 | 1.006 |
| $O_2(g)$ | 30.255 | 4.207 | - 1.887 |
| $N_2(g)$ | 27.270 | 4.930 | 0.333 |

The combustion reaction is given by

$$C_2H_6\,(g) + 1.5 \times 3.5\, O_2(g) + 1.5 \times 3.5 \times 3.76\, N_2\,(g)$$
$$\rightarrow 2\, CO_2\,(g) + 3\, H_2O\,(g) + 1.75\, O_2(g) + 19.74\, N_2\,(g)$$

$$\Delta H^0_{298} = -\,1429.374 \text{ kJ} \qquad \text{[see Example 10.54]}$$

We know that $$\Delta H^0_{298} + \int_{298}^{T} (\Delta C^0_p dT)_P = 0$$

where $$\Delta C^0_P = \Delta a + \Delta b T + \frac{\Delta e}{T^2}$$

$$\Delta a = 2 \times 45.369 + 3 \times 28.850 + 1.75 \times 30.255 + 19.74 \times 27.270 = 768.54$$

$$\Delta b = (2 \times 8.688 + 3 \times 12.055 + 1.75 \times 4.207 + 19.74 \times 4.930) \times 10^{-3}$$
$$= 158.22 \times 10^3$$

$$\Delta e = [2 \times (-9.619) + 3 \times 1.006 + 1.75 \times (-1.887) + 19.74 \times 0.333] \times 10^5$$
$$= -12.949 \times 10^5$$

Then

$$\Delta H^0_{298} + \int_{298}^{T} (\Delta C^0_P\, dT)_P = \Delta H^0_{298} + \int_{298}^{T} (\Delta a + \Delta b T + \frac{\Delta e}{T^2}) dT = 0$$

$$= \Delta H^0_{298} + \Delta a\,(T - 298) + \frac{\Delta b}{2}(T^2 - 298^2) - \Delta e\left(\frac{1}{T} - \frac{1}{298}\right) = 0$$

$$\text{or}\quad = -1\,429\,374 + 768.54\,(T-298) + \frac{158.22\times10^{-3}}{2}\,(T^2-298^2) + 12.949\times10^5\left(\frac{1}{T}-\frac{1}{298}\right) = 0$$

$$\text{or}\quad -1\,669\,769 + 768.54T + \frac{158.22\times10^{-3}}{2}T^2 + \frac{12.949\times10^5}{T} = 0$$

$$\text{or}\quad T = 2172.65 - 1.0294\times10^{-4}T^2 - \frac{1684.88}{T} \qquad (A)$$

Eqn. ($A$) can be solved by trial and error to determine the value of $T$. Let us guess a value for $T$, say $T$ = 1850 K and subsitute this value on the RHS of Eqn.($A$) to obtain the revised value for $T$. Then

$$T = 2172.65 - 1.0294\times10^{-4}(1850)^2 - \frac{1684.88}{1850}$$

or $T$ = 1819.4 K

Now, use this value $T$ = 1819.4 K as the guess value on the RHS of Eqn.($A$) to obtain another revised value for $T$. Then

$$T = 2172.65 - 1.0294\times10^{-4}(1819.4)^2 - \frac{1684.88}{1819.4}$$

or $T$ = 1831 K

Continue the inerative calculations till convergence is achieved

$$T = 2172.65 - 1.0294\times10^{-4}(1831)^2 - \frac{1684.88}{1831}$$

or $T$ = 1826.6 K

$$T = 2172.65 - 1.0294\times10^{-4}(1826.6)^2 - \frac{1684.88}{1826.6}$$

or $T$ = 1828.3 K

$$T = 2172.65 - 1.0294\times10^{-4}(1828.3)^2 - \frac{1684.88}{1828.3}$$

or $T$ = 1827.6 K

The difference in the calculated value of $T$ in the last two iterations is 0.7 K. Hence, we may assume that convergence has achieved at $T$ = 1827 K.

Therefore, the adiabatic flame temperature = 1827 K.

**Example 10.57**

What is a partial molar property ?

*Solution* :

The partial molar property is the change in the property of a system per mole of component $i$ at constant temperature, pressure and composition (expressed in moles) of all other constituents of the system. We know that the state of a multicomponent system can be expressed in terms of temperature, pressure and composition. That is, the property enthalpy of a multicomponent mixture is given by

$$H = H(T,P,n_1 n_2 ... n_c)$$

Where $n_i$ represents the moles of component $i$. Then, the partial molar enthalpy of component $i$ is given by

$$\bar{h}_i = \left(\frac{\partial H}{\partial n_i}\right)_{T,P,n_J} \quad \text{Where } J \neq i$$

**Example 10.58**

What is a chemical potential ?

*Solution* :

The chemical potential $\mu_i$ of component $i$ in a multicomponent system is defined as

$$\mu_i = \bar{g}_i = \left(\frac{\partial G}{\partial n_i}\right)_{T,P,n_J} \quad \text{where } J \neq i$$

That is, the chemical potential $\mu_i$ of component $i$ is the partial molar Gibbs free energy of component $i$. The Gibbs free energy of a multicomponent mixture is given by

$$G = G(T,P,n_1,n_2..n_C)$$

Then the differential change in the Gibbs free energy of the solution is given by

$$dG = \left(\frac{\partial G}{\partial T}\right)_{P,n_i} dT + \left(\frac{\partial G}{\partial P}\right)_{T,n_i} dP + \sum_{i=1}^{c} \left(\frac{\partial G}{\partial n_i}\right)_{T,P,n_J} dn_i$$

$$= -SdT + VdP + \Sigma \mu_i dn_i$$

From this we find that $\mu_i$ is also an intensive property like $T$ and $P$. The Gibbs free energy of a system can be expressed in terms of the chemical potentials of the constituent components as

$$G = \Sigma n_i \bar{g}_i = \Sigma n_i \mu_i$$

**Example 10.59**

What is fugacity?

*Solution* :

We know that for a pure substance, the change in the Gibbs free energy is given by

$$dg = -sdT + vdP$$

At constant temperature, the above relation reduces to

$$dg = vdP \qquad \text{(at constant } T) \ (A)$$

For an ideal gas, Eqn.($A$) gives

$$dg^0 = vdp = \frac{RT}{P}dP = RTd\ln P \qquad (B)$$

For real gases, Eqn. ($B$) is not correct. However, the simplicity of Eqn. ($B$) is quite appealing and one can define a new parameter called fugacity such that the form of Eqn.($B$) is retained for a real gas. Hence fugacity is defined as

$$dg = vdP = RTd\ln f \qquad \text{(at constant } T) \qquad (C)$$

Suppose a real gas is changed from a low pressure $P \to 0$, where it behaves like an ideal gas to the state at $P$ while temperature is held constant. Then the change in the Gibbs free energy is given by

$$g - g^0 = \int_{P\to 0}^{P} dg = \int RTd\ln f = RT\ln\frac{f}{P}$$

For an ideal gas $f = P$ or $f/P = 1$. Thus fugacity is defined by the relations

$$dg = vdP = RT\,d\ln f \text{ and } f/P = 1 \text{ as } P \to 0.$$

The fugacity has the units of pressure.

**Example 10.60**

What is meant by extent of a reaction ?

*Solution* :

Suppose a chemical reaction

$$aA + bB \to rR + sS$$

is occurring. Then the changes in the moles of the reactants and products are not independent of each other but they are governed by the stoichiometry of the chemical reaction. Let $dn_A$, $dn_B$, $dn_R$ and $dn_S$ denote the change in the moles of the constituents. Then the stoichiometry of the reaction gives

$$\frac{-dn_A}{a} = \frac{-dn_B}{b} = \frac{dn_S}{s} = \frac{dn_R}{r}$$

In general, if a chemical reaction is expressed as $\Sigma \upsilon_i A_i = 0$, then the stoichiometry of the reaction gives the following relation for the changes in the moles of reactants and products.

$$\frac{dn_{A_1}}{\upsilon_1} = \frac{dn_{A_2}}{\upsilon_2} = \frac{dn_{A_3}}{\upsilon_3} = d\varepsilon$$

where $\varepsilon$ is the extent of the reaction or the degree to which the reaction progresses.

**Example 10.61**

Suppose a system is initially charged with 3 mol $CH_4$, 1mol $H_2$, 1 mol $CO_2$ and 5 mol $H_2O$ and the following reaction takes place between the species

$$CH_4\ (g) + 2\ H_2O\ (g) \rightarrow CO_2\ (g) + 4H_2\ (g)$$

Express the composition of the mixture (in terms of mole fractions) as a function of the extent of reaction.

*Solution* :

Total moles of initial mixture = 3 + 1 + 1 + 5 = 10

Let $\varepsilon$ = extent of reaction.

Then moles of $CH_4 = 3 - \varepsilon$

moles of $H_2O = 5 - 2\varepsilon$

moles of $CO_2 = 1 + \varepsilon$

moles of $H_2 = 1 + 4\varepsilon$

Total moles of mixture = $10 + 2\varepsilon$

Then $y_{CH_4} = \frac{3-\varepsilon}{10+2\varepsilon}$; $y_{H_2O} = \frac{5-2\varepsilon}{10+2\varepsilon}$; $y_{CO_2} = \frac{1+\varepsilon}{10+2\varepsilon}$ and $y_{H_2} = \frac{1+4\varepsilon}{10+2\varepsilon}$

**Example 10.62**

Methanol can be produced according to the reaction

$$CO\ (g) + 2H_2O\ (g) \rightarrow CH_3OH(g)$$

Suppose a reactor is initially fed with 1 mol CO and 4 mol $H_2$ and the mixture is allowed to react. Express the composition of the mixture in terms of mole fractions as a function of the extent of reaction.

*Solution* :

Let $\varepsilon$ = extent of reaction.

Then the moles of the constituents in the reactor are as given below,

| Constituent | Initial moles | Moles as reaction proceeds | Mole fraction $y_i$ |
|---|---|---|---|
| CO | 1 | $1-\varepsilon$ | $(1-\varepsilon)/(5-2\varepsilon)$ |
| $H_2$ | 4 | $4-2\varepsilon$ | $(4-2\varepsilon)/(5-2\varepsilon)$ |
| $CH_3OH$ | 0 | $\varepsilon$ | $\varepsilon/(5-2\varepsilon)$ |
| Total | 5 | $(5-2\varepsilon)$ | |

**Example 10.63**

What is the criterion of equilibrium for a chemically reacting system ?

*Solution* :

We know that the criterion of equilibrium at constant temperature and pressure is given by $dG = 0$. That is, at equilibrium the virtual changes that occur in a system, which is held at constant temperature and pressure, produce no change in the Gibbs free energy of the system and the system is in a state of minimum Gibbs free energy. We know that

$$dG = -SdT + VdP + \Sigma \mu_i dn_i = 0$$

At constant temperature and pressure, this reduces to

$$dG = \Sigma \mu_i dn_i = 0 \qquad (A)$$

Suppose a chemical reaction given by

$aA + bB \rightarrow rR + sS$ or $\Sigma \upsilon_i A_i = 0$

occurs in the system. Then the changes in the moles $dn_i$ of the species are not independent, but they are related by the expression

$$\frac{-dn_A}{a} = \frac{-dn_B}{b} = \frac{dn_R}{r} = \frac{dn_S}{s} = d\varepsilon$$

or $$\frac{dn_i}{\upsilon_i} = d\varepsilon \text{ or } dn_i = \upsilon_i d\varepsilon \qquad (B)$$

Substituting for $dn_i$ in Eqn.($A$) from Eqn. ($B$), we get

$$\Sigma \mu_i dn_i = \Sigma \mu_i \upsilon_i d\varepsilon_i = 0$$

or $\Sigma \upsilon_i \mu_i = 0$ (C)

which is the criterion of equilibrium for a chemically reacting system.

**Example 10.64**

What is meant by standard Gibbs free energy change of a chemical reaction ?

*Solution* :

The standard Gibbs free energy change of a chemical reaction is defined in an analogous way to standard enthalpy change of a chemical reaction. We know that

$$\Delta H^0_{298} = \Sigma\, \upsilon_i\, \Delta H^0_{f298}\,(A_i)$$

In a similar way, the standard Gibbs free energy charge for the reaction $\Sigma \upsilon_i A_i = 0$ is given by $\Delta G^0_{298} = \Sigma \upsilon_i\ \Delta G^0_{f298}\,(A_i)$ where $\Delta G^0_{f298}$ is the standard Gibbs free energy of formation of $A_i$. By convention, the standard Gibbs free energy of formation of elements in their standard states have been assigned a value of zero.

The standard Gibbs free energy change $\Delta G^0_{298}$ for the reaction

$$aA + bB \rightarrow rR + sS$$

is given by

$$\Delta G^0_{298} = r\,\Delta G^0_{f298}\,(R) + s\Delta G^0_{f298}\,(S) - a\Delta G^0_{f298}\,(A) - b\Delta G^0_{f298}\,(B)$$

The values of the standard free energy of formation at 298 K for a few selected compounds are given in Appendix –10

**Example 10.65**

Derive a relation to estimate the equilibrium constant from a knowledge of the standard Gibbs free energy change of a chemical reaction.

*Solution* :

We know that the criterion of equilibrium for a chemically reacting system is given by

$$\Sigma\, \upsilon_i\, \mu_i = 0 \qquad (A)$$

and we also know that the chemical potential $\mu_i$ of component $i$ is defined as

$$d\,\mu_i = d\,\bar{g}_i = RT\,d\ln f_i \qquad \text{(at constant temperature)} \quad (B)$$

where $f_i$ = fugacity of component $i$

Integrating Eqn. $(B)$ at constant temperature from the standard state of the component $i$ to the state in the multicomponent mixture gives

$$\mu_i - g_i^0 = RT\ln\frac{f_i}{f_i^0} = RT\ln a_i$$

or $\mu_i = g_i^0 + RT \ln a_i$ (*C*)

where $f_i^0$ = standard state fugacity of component *i*

$a_i$ = activity of component *i*

The activity of component *i*, is defined as the ratio of the fugacity of component *i* in the mixture to the fugacity of the component *i* in the standard state. That is, the activity $a_i$ of component *i* is given by

$$a_i = \frac{f_i}{f_i^0}$$

Substituting Eqn. (*C*) in Eqn. (*A*), we get

$$\Sigma\upsilon_i\mu_i = \Sigma\upsilon_i(g_i^0 + RT\ln a_i) = 0$$

or $\Sigma\upsilon_i g_i^0 + RT\,\Sigma\upsilon_i \ln a_i = 0$

or $\Sigma\upsilon_i g_i^0 = \Delta G^0 = -RT\,\Sigma\upsilon_i \ln a_i = -RT\,\Sigma \ln a_i^{\upsilon_i}$

or $\Delta G^0 = -RT\ln\left(\Pi a_i^{\upsilon_i}\right) = -RT\ln K$ or $K = \exp\left(\frac{-\Delta G^0}{RT}\right)$

where $K = \prod_i a_i^{\upsilon_i}$ = equilibirium constant

$\Delta G^0 = \Sigma\upsilon_i g_i^0$ = standard Gibbs free energy change.

**Example 10.66**

Calculate the standard Gibbs free energy change $\Delta G_{298}^0$ and the equilibrium constant for the ammonia synthesis reaction $N_2\,(g) + 3H_2\,(g) \rightarrow 2NH_3\,(g)$ at 1 bar and 298.15 K.

*Solution* :

For the reaction

$$N_2\,(g) + 3H_2\,(g) \rightarrow 2NH_3(g)$$

The standard Gibbs free energy change $\Delta G_{298}^0$ is given by

$$\Delta G_{298}^0 = 2\Delta G_{f298}^0\,[NH_3(g)] - \Delta G_{f298}^0[N_2(g)] - 3\Delta G_{f298}^0\,[H_2(g)]$$
$$= 2 \times (-16.747) - 0 - 0 = -\,33.494 \text{ kJ}$$

We know that the equilibrium constant K is given by

$$K = \exp\left(-\frac{\Delta G^0}{RT}\right) = \exp\left(\frac{33\,494}{8.314 \times 298}\right) = 7.433 \times 10^5$$

**Example 10.67**

Is it possible to judge whether a given reaction is feasible or not from a knowledge of the standard Gibbs free energy change of the reaction ?

*Solution* :

The criterion for equilibrium at constant temperature and pressure is given by $dG = O$. That is, the Gibbs free energy of a system reaches a minimum value in a state of equilibrium. Thus the change in the Gibbs free energy indicates the feasibility of a given reaction. This can be summarized as

$\Delta G < O$; the reaction is feasible and spontaneous.

$\Delta G = O$; the system attained a state of equilibrium.

$\Delta G > O$; the reaction proceeds spontaneously in the reverse direction.

Thus, one can identify whether a given chemical reaction proceeds in the specified direction or not, based on a knowledge of the standard Gibbs free energy change of the reaction.

**Example 10.68**

It is proposed to carry out the reaction $CO\,(g) + H_2O\,(g) \rightarrow CO_2\,(g) + H_2\,(g)$ at 1 bar and 298 K. Calculate the standard Gibbs free energy change at 298 K for the above reaction and specify whether it is feasible to carry out the above reaction at 298 K or not. If possible, estimate the equilibrium constant.

*Solution* :

The proposed reaction is

$$CO\,(g) + H_2O\,(g) \rightarrow CO_2\,(g) + H_2\,(g)$$

$$\Delta G^0_{298} = \Delta G^0_{f298}\,[CO_2(g)] - \Delta G^0_{f298}\,[H_2O(g)] - \Delta G^0_{f298}[CO(g)]$$

Read the values of $\Delta G^0_{f298}$ from Appendix – 10. Then

$$\Delta G^0_{298} = -\,394.815 - (-\,137.327) - (-\,228.6) = -\,28.888 \text{ kJ.}$$

Since $\Delta G^0_{298} < 0$, the proposed reaction is feasible at 298 K.

The equilibrium constant $K$ is given by

$$K = \exp\left(-\frac{\Delta G^0}{RT}\right) = \exp\left(\frac{28\,888}{8.314 \times 298}\right) = 1.1582 \times 10^5$$

**Example 10.69**

What are the dimensions of the equilibrium constant $K$ ?

*Solution* :

Consider the reaction

$$aA + bB \rightarrow rR + sS$$

The equilibrium constant $K$ is given by

$$K = \prod_i a_i^{\upsilon_i} = \frac{a_R^r \, a_S^s}{a_A^a \, a_B^b}$$

The activity $a_i$ is given by $a_i = f_i/f^0_i$. Hence the activity is dimensionless. Therefore, the equilibrium constant $K$ is dimensionless.

**Example 10.70**

In a test the candidates were asked to calculate the equilibrium constant $K$ at 298 K for the synthesis of ammonia. Three different candidates have expressed the reaction in three different ways. Candidate $X$ writes the reaction as

$$N_2(g) + 3H_2(g) \rightarrow 2NH_3(g)$$

and calculates $K_x$. The candidate $Y$ writes the reaction as

$$\frac{1}{2}N_2(g) + \frac{3}{2}H_2(g) \rightarrow NH_3(g)$$

and calculates the value $K_y$, where as the candidate $Z$ expresses the reaction as

$$2NH_3(g) \rightarrow N_2(g) + 3H_2(g)$$

and determines the value of $K_z$, where $K_x$, $K_y$ and $K_z$ denote the equilibrium constants, What is the relation between $K_x$, $K_y$ and $K_z$ ?

*Solution* :

The candidate $X$ expresses the reaction as

$$N_2(g) + 3H_2(g) \rightarrow 2NH_3(g)$$

Then $$K_x = \prod_i a_i^{\upsilon_i} = \frac{a_{NH_3}^2}{a_{N_2} \, a_{H_2}^3} \qquad (A)$$

Then candidate $Y$ writes the reaction as

$$\frac{1}{2}N_2(g) + \frac{3}{2}H_2(g) \rightarrow NH_3(g)$$

Then $$K_y = \frac{a_{NH_3}}{a_{N_2}^{1/2} \, a_{H_2}^{3/2}} \qquad (B)$$

The candidate $Z$ expresses the reaction as

$$2\,NH_3(g) \rightarrow N_2(g) + 3H_2(g)$$

Then $$K_z = \frac{a_{N_2} a_{H_2}^3}{a_{NH_3}^2} \qquad (C)$$

From Eqn (*A*) – (*C*), we find that

$$K_x = K_y^2 = \frac{1}{K_z}$$

**Example 10.71**

The standard Gibbs free energy of formation of compounds which normally exist as liquid at ordinary temperatures are tabulated for the liquid state only. However, some times while analyzing a practical problem it is necessary to know the standard Gibbs free energy of formation in the gaseous state. Develop a method to estimate $\Delta G_f^0$ of a compound in gaseous state from a knowledge of $\Delta G_f^0$ of the same compound in the liquid state.

*Solution* :

We are given $\Delta G_{f298}^0$ for the compound in the liquid state and we would like to find the value of $\Delta G_{f298}^0$ for the compound in the gaseous state.

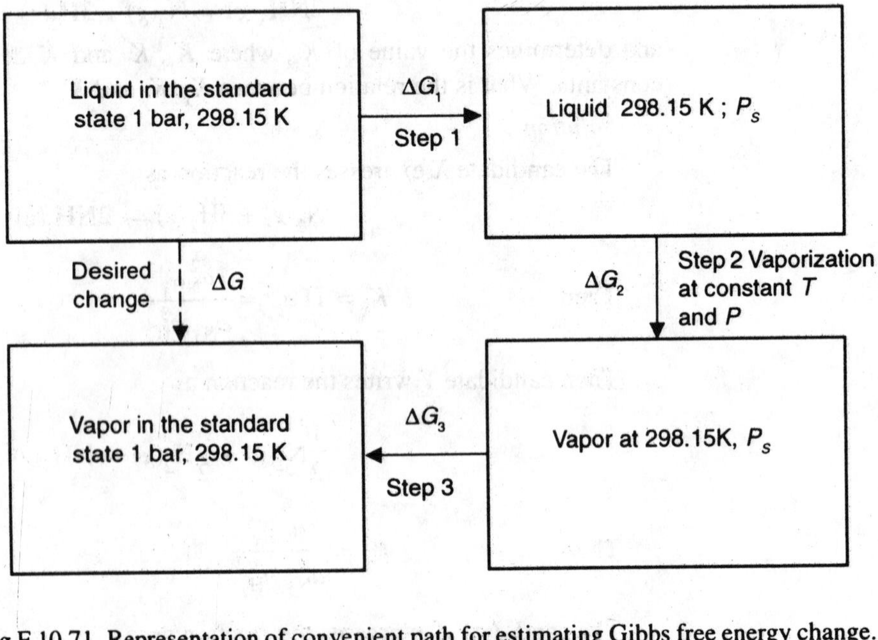


Fig.E 10.71. Representation of convenient path for estimating Gibbs free energy change.

The standard state for the liquid is pure liquid at 1 bar pressure and 298.15 K, while the standard state for the gas is the pure gas in the ideal gas state at the same pressure and temperature. The $\Delta G^0_{f298}$ of gas state is equal to the sum of $\Delta G^0_{f298}$ of the liquid state and the change in the Gibbs free energy for the change of state-liquid to gas-at 1 bar and 298.15 K. To determine the change in the Gibbs free energy for the change of state, we can devise the path as shown in FigE. 10.71. The desired change of state is shown by the broken line and the imaginary process to achieve the same change of state is shown by solid line in FigE. 10.71 which consists of the following three steps.

**Step 1:** Liquid in the standard state at 1 bar and 298.15 K is changed to $P_S$ the saturation pressure at 298.15 K. We know that for a pure substance, the change in the Gibbs free energy is given by

$$dg = -sdT + vdP$$

At constant temperature it reduces to $dg = vdP$.

$$\text{or} \quad \Delta G_1 = \int_{1\text{ bar}}^{P_S} vdP = v\Delta P \text{ (since liquid is incompressible)}$$

The molar volume of liquid is usually very small and hence it is reasonable to approximate $\Delta G_1 = 0$

**Step 2**: Vaporization at constant temperature and pressure for which $\Delta G = 0$

**Step 3**: Vapor at low pressure $P_S$ changes to $P = 1$ bar in the ideal gas state. At low pressure $P_S$ the vapor behaves like an ideal gas. For an ideal gas undergoing a change of state at constant temperature

$$dg = vdP = \frac{RT}{P}dP = RTd\ln P$$

$$\text{or} \quad \Delta G_3 = RT\ln\frac{P}{P_S}$$

Then the required change in Gibbs free energy is given by

$$\Delta G = RT\ln\frac{P}{P_S}$$

Therefore

$$\Delta G^0_{f298}(g) = \Delta G^0_{f298}(l) + RT\ln\frac{P}{P_S}$$

**Example 10.72**

Calculate the standard Gibbs free energy of formation of $CH_3OH$ ($g$) at 298.15 K from $\Delta G^0_{f298}$ of $CH_3OH(l)$, given that the saturation pressure of methanol at 298.15 K is 0.167 16 bar.

*Solution*:

We know that

$$\Delta G^0_{f298}\,(g) = \Delta G^0_{f298}(l) + RT\ln\frac{P}{P_S} \quad \{\text{see Example 10.71}\}$$

$$\Delta G^0_{f298} \text{ of } CH_3OH(l) = -166.215 \text{ kJ} \qquad \{\text{see Appendix -10}\}$$

Therefore,

$\Delta G^0_{f298}\,[CH_3OH(g)]$

$$= -166.215\times10^3 + 8.314\times298\ln\frac{1}{0.16716} = -161.783 \text{ kJ}$$

**Example 10.73**

Given that $\Delta G^0_{f298}$ for $H_2O$ (*g*) is – 228.600 kJ, estimate the value of $\Delta G^0_{f298}$ for $H_2O$ (*l*) and compare it with the value given in Appendix–10

*Solution* :

At 25°C, the saturation pressure of water = 0.031 66 bar.

We know that

$$\Delta G^0_{f298}\,(g) = \Delta G^0_{f298}\,(l) + RT\ln\frac{P}{P_S}$$

or $$-228.600\times10^3 = \Delta G^o_{f298}\,(l) + 8.314\times298\ln\frac{1}{0.03166}$$

or $\Delta G^0_{f298}\left[H_2O(l)\right] = = -237.154$ kJ

The value given in Appendix –10 is – 237.392 kJ

**Example 10.74**

What is Vant Hoff equation and what does it predict ?

*Solution* :

We know that $dg = -sdT + vdp$ or $\left(\dfrac{\partial g}{\partial T}\right)_P = -s$

Then we can obtain

$$\left(\frac{\partial \Delta G^0}{\partial T}\right)_P = \left[\frac{\partial}{\partial T}(\Sigma\upsilon_i g_i^0)\right]_P = -\Sigma\upsilon_i s_i^0 = -\Delta S^0$$

Where $\Delta S^0$ = standard entropy change of the reaction

$$\left[\frac{\partial(\Delta G^0 / T)}{\partial T}\right]_P = \frac{1}{T}\left(\frac{\partial \Delta G^0}{\partial T}\right)_P - \frac{\Delta G^0}{T^2} = -\frac{\Delta S^0}{T} - \frac{\Delta G^0}{T^2}$$

Since $\Delta G^0 = \Delta H^0 - T\Delta S^0$, we obtain

$$\left[\frac{\partial(\Delta G^0 / T)}{\partial T}\right]_P = -\frac{\Delta S^0}{T} - \frac{\Delta G^0}{T^2} = \frac{-\Delta H^0}{T^2}$$

We also know that $\Delta G^0 = -RT \ln K$ or $\dfrac{\Delta G^0}{T} = -R \ln K$

Therefore

$$\left[\frac{\partial(\Delta G^0 / T)}{\partial T}\right]_P = -R\left(\frac{\partial \ln K}{\partial T}\right) = -\frac{\Delta H^0}{T^2}$$

or

$$\left[\frac{\partial \ln K}{\partial T}\right]_P = \frac{\Delta H^0}{RT^2}$$

Which is known as vant Hoff equation.

For an endothermic reaction $\Delta H^0 > 0$. Therefore, the vant Hoff equation predicts that the equilibrium constant $K$ increases with increasing temperature for an endothermic reaction. It also says that the equilibrium constant $K$ decreases with increasing temperature for an exothermic reaction ($\Delta H^0 < 0$). That is the vant Hoff equation predicts the effect of temperature on the equilibrium constant. If $\Delta H^0$ is assumed to be constant in the temperature range $T_1$ to $T_2$, the vant Hoff equation can be integrated to obtain

$$\ln\left(\frac{K_2}{K_1}\right) = \frac{\Delta H^0}{R}\left(\frac{1}{T_1} - \frac{1}{T_2}\right)$$

That is a plot of ln $K$ versus $1/T$ yields a straight line.

**Example 10.75**

The equilibrium constant $K$ for the reaction CO $(g)$ + $H_2O$ $(g) \rightarrow CO_2$ $(g)$ + $H_2$ $(g)$ at 1 bar and 298 K is $1.1582 \times 10^5$. Assuming that $\Delta H^0$ remains constant in the temperature range 298 K to 1000 K, estimate the value of $K$ at 1 bar and 1000 K.

*Solution* :

Then given reaction is

$$CO\,(g) + H_2O\,(g) \longrightarrow CO_2\,(g) + H_2\,(g)$$

$$\Delta H^0_{298} = -393.978 - (-110.532) - (-241.997)$$

$$= -41.449 \text{ kJ}$$

The integrated form of the Vant Hoff equation is

$$\ln\left(\frac{K_2}{K_1}\right) = \frac{\Delta H^0}{R}\left(\frac{1}{T_1} - \frac{1}{T_2}\right)$$

Therefore

$$\ln\left(\frac{K_{1000}}{K_{298}}\right) = \frac{-414\,49}{8.314}\left(\frac{1}{298} - \frac{1}{1000}\right) = -11.7442$$

or
$$\frac{K_{1000}}{K_{298}} = 7.934 \times 10^{-6}$$

or $K_{1000} = 7.934 \times 10^{-6} \times 1.1582 \times 10^5 = 0.9189$

**Example 10.76**

The ammonia synthesis reaction $N_2(g) + 3H_2(g) \rightarrow 2NH_3(g)$ at 1 bar and 298 K has the equilibrium constant $K = 7.433 \times 10^5$. Though this reaction is highly favourable at 298 K, the rate of the reaction is very low. The rate of a chemical reaction increases with temperature. Hence, in the industry this advantage is made use of by carrying out the synthesis reaction at elevated temperatures. Assuming that $\Delta H^0$ remains constant, estimate the equilibrium constant at 800 K and 1 bar.

*Solution* :

The ammonia synthesis reaction is given by

$$N_2\,(g) + 3H_2\,(g) \rightarrow 2NH_3(g)$$

$$\Delta H^0_{298} = 2 \times (-46.055) = -92.11 \text{ kJ}$$

That Vant Hoff equation gives

$$\ln\left(\frac{K_2}{K_1}\right) = \frac{\Delta H^0}{R}\left(\frac{1}{T_1} - \frac{1}{T_2}\right)$$

Therefore,

$$\ln\left(\frac{K_{800}}{K_{298}}\right) = \frac{-92\,110}{8.314}\left(\frac{1}{298} - \frac{1}{800}\right) = -23.3289$$

or $\left(\dfrac{K_{800}}{K_{298}}\right) = 7.3856 \times 10^{-11}$

Therefore, $K_{800} = 7.3856 \times 10^{-11} \times 7.433 \times 10^5 = 5.4897 \times 10^{-5}$

### Example 10.77

Derive a general relation to estimate the standard Gibbs free energy change of a reaction as a function of temperature, if the constant pressure molar heat capacities of the gaseous species are expressed as $C_P^0 = a + bT + cT^2 + dT^3 + eT^{-2}$.

*Solution* :

We know that the standard enthalpy change of a reaction is given by

$$\Delta H_T^0 = \Delta H_0 + \Delta aT + \frac{\Delta b}{2}T^2 + \frac{\Delta c}{3}T^3 + \frac{\Delta d}{4}T^4 - \frac{\Delta e}{T} \qquad (A)$$

and the vant Hoff equation is given by

$$\left(\frac{\partial \ln K}{\partial T}\right)_P = \frac{\Delta H^0}{RT^2} \text{ or } \ln K = \int \frac{\Delta H^0}{RT^2} dT \qquad (B)$$

Substitute Eqn.($B$) in Eqn.($A$) to obtain

$$\ln K = \frac{1}{R}\int\left(\frac{\Delta H_0}{T^2} + \frac{\Delta a}{T} + \frac{\Delta b}{2} + \frac{\Delta c}{3}T + \frac{\Delta d}{4}T^2 - \frac{\Delta e}{T^3}\right)dT$$

$$\text{or} \quad \ln K = \frac{1}{R}\left(-\frac{\Delta H_0}{T} + \Delta a \ln T + \frac{\Delta bT}{2} + \frac{\Delta cT^2}{6} + \frac{\Delta d}{12}T^3 + \frac{\Delta e}{2T^2}\right) + I \qquad (C)$$

We know that $\Delta G^0 = -RT \ln K$ $\qquad (D)$

Substitute Eqn ($C$) in Eqn. ($D$), to obtain

$$\Delta G_T^0 = \Delta H_0 - \Delta aT \ln T - \frac{\Delta b}{2}T^2 - \frac{\Delta c}{6}T^3 - \frac{\Delta d}{12}T^4 - \frac{\Delta e}{2T} - IRT$$

where $I$ is a constant, the value of which can be determined from a knowledge of $\Delta G^0$ at one temperature.

### Example 10.78

Determine the standard Gibbs free energy change and the equilibrium constant at 1000 K and 1 bar for the reaction $CO\,(g) + H_2O\,(g) \longrightarrow CO_2\,(g) + H_2\,(g)$.

*Solution*:

The given reaction is

$$CO\,(g) + H_2O\,(g) \longrightarrow CO_2\,(g) + H_2\,(g)$$

$$\Delta H^0_{298} = -393.978 - (-110.532) - (-241.997)$$
$$= -41.449 \text{ kJ}$$

$$\Delta G^0_{298} = -394.815 - (-137.327) - (-228.600)$$
$$= -28.888 \text{ kJ}$$

The constants of the isobaric molar heat capacity equation are given by

| Species | *a* | $b \times 10^3$ | $e \times 10^{-5}$ |
|---|---|---|---|
| $CO_2\,(g)$ | 45.369 | 8.688 | – 9.619 |
| $H_2\,(g)$ | 27.012 | 3.509 | 0.690 |
| $CO\,(g)$ | 28.068 | 4.631 | – 0.258 |
| $H_2O\,(g)$ | 28.850 | 12.055 | 1.006 |

$$\Delta a = 45.369 + 27.012 - 28.068 - 28.850 = 15.463$$
$$\Delta b = (8.688 + 3.509 - 4.631 - 12.055) \times 10^{-3} = -4.489 \times 10^{-3}$$
$$\Delta e = \{-9.619 + 0.690 - (-0.258) - 1.006\} \times 10^5 = -9.677 \times 10^5$$

We know that $\Delta H^0_T = \Delta H_0 + \Delta a T + \dfrac{\Delta b}{2} T^2 + \dfrac{\Delta c}{3} T^3 + \dfrac{\Delta d}{4} T^4 - \dfrac{\Delta e}{T}$

The isobaric molar heat capacities are expressed as $C^0_P = a + b\,T + eT^{-2}$
Then we get

$$\Delta H^0_T = \Delta H_0 + \Delta a T + \frac{\Delta b}{2} T^2 - \frac{\Delta e}{T} \qquad (A)$$

Substitute $T = 298$ K and $\Delta H^0{}_{298}$ in Eqn. (*A*), to obtain

$$-414\,49 = \Delta H_0 + 15.463 \times 298 - \frac{4.489 \times 10^{-3}}{2}(298)^2 + \frac{9.677 \times 10^5}{298}$$

or $\Delta H_0 = -49.105$ kJ

We know that $\Delta G^0_T = \Delta H_0 - \Delta a T \ln T - \dfrac{\Delta b}{2} T^2 - \dfrac{\Delta c}{6} T^3 - \dfrac{\Delta d}{12} T^4 - \dfrac{\Delta e}{2T} - IRT$

When $\Delta c = 0$ ; $\Delta d = 0$, this equation reduces to

$$\Delta G^0_T = \Delta H_0 - \Delta a T \ln T - \frac{\Delta b}{2} T^2 - \frac{\Delta e}{2T} - IRT \qquad (B)$$

Substitute $T = 298$ K and $\Delta G^0_{298}$ in Eqn. (*B*), to obtain

$$-28\,888 = -49\,105 - 15.463 \times 298 \ln 298 + \frac{4.489 \times 10^{-3}}{2}(298)^2$$

$$+\frac{9.677 \times 10^5}{2 \times 298} - IR \times 298 \text{ or } IR = -149.819$$

Now, substitute $T = 1000$ K and the value of $IR$ in Eqn. ($B$) to obtain

$$\Delta G^0_{1000} = -49\,105 - 15.463 \times 1000 \times \ln 1000 + \frac{4.489 \times 10^{-3}}{2}(1000)^2$$

$$+\frac{9.677 \times 10^5}{2 \times 1000} + 149.819 \times 1000$$

or $\Delta G^0_{1000} = -3.372$ kJ

Then $$K_{1000} = \exp\left(\frac{-\Delta G^0}{RT}\right) = \exp\left(\frac{3372}{8.314 \times 1000}\right) = 1.5$$

***Note*** *The equilibrium constant for this reaction at the same temperature was estimated as 0.9189 by assuming that $\Delta H^0$ remains constant (see Example 10.75). The value of $\Delta H^0_{1000}$ is given by*

$$\Delta H^0_{1000} = -49\,105 + 15.463 \times 1000 - \frac{4.489 \times 10^{-3}}{2}(1000)^2 + \frac{9.677 \times 10^5}{1000}$$

or $\Delta H^0_{1000} = -34.919$ kJ

We find that ($\Delta H^0_{298} = -41.449$ kJ and $\Delta H^0_{1000} = -34.919$ kJ). $\Delta H^0$ varies with temperature. Hence, the assumption that $\Delta H^0$ is independent of temperature leads to incorrect value of $K$. Therefore, the assumption that $\Delta H^0$ is constant provides only an approximate value of $K$.

**Example 10.79**

Calculate the standard Gibbs free energy change and the equilibrium constant at 1 bar and 800 K for the reaction $N_2\,(g) + 3\,H_2\,(g) \rightarrow 2NH_3\,(g)$ taking into account the variation of isobaric molar heat capacities of gases with temperature.

*Solution* :

The given reaction is $N_2\,(g) + 3\,H_2\,(g) \rightarrow 2\,NH_3\,(g)$

$$\Delta H^0_{298} = 2 \times (-46.055) = -92.11 \text{ kJ}$$

$$\Delta G^0_{298} = 2 \times (-16.747) = -33.494 \text{ kJ}$$

The constants of the isobaric molar heat capacities are

| **Species** | ***a*** | ***b* × 10³** | ***e* × 10⁻⁵** |
|---|---|---|---|
| $NH_3$ (*g*) | 29.747 | 25.108 | – 1.546 |
| $N_2$ (*g*) | 27.270 | 4.930 | 0.333 |
| $H_2$ (*g*) | 27.012 | 3.509 | 0.690 |

$$\Delta a = 2 \times 29.747 - 27.270 - 3 \times 27.012 = -48.812$$

$$\Delta b = (2 \times 25.108 - 4.930 - 3 \times 3.509) \times 10^{-3}$$
$$= 34.759 \times 10^{-3}$$

$$\Delta e = \{2 \times (-1.546) - 0.333 - 3 \times 0.690\} \times 10^5 = -5.495 \times 10^5$$

If the isobaric molar heat capacities are expressed as $C_P^0 = a + b\,T + e\,T^{-2}$

The expressions for $\Delta H_T^0$ and $\Delta G_T^0$ are given by

$$\Delta H_T^0 = \Delta H_0\ \Delta aT + \frac{\Delta b}{2}T^2 - \frac{\Delta e}{T} \qquad (A)$$

$$\Delta G_T^0 = \Delta H_0 - \Delta aT \ln T - \frac{\Delta b}{2}T^2 - \frac{\Delta e}{2T} - IRT \qquad (B)$$

Substitute $\Delta H^0_{298}$ and $T = 298$ in Eqn. (*A*) to determine the value of the constant $\Delta H_0$. Then using the values of $\Delta H_0$, $\Delta G^0_{298}$ and $T = 298$ in Eqn. (*B*), determine the value of the constant *I*.

Substitute $\Delta H^0_{298}$ for $\Delta H_T^0$ and $T = 298$ in Eqn. (*A*) to obtain

$$-92\,110 = \Delta H_0 - 48.812 \times 298 + \frac{34.759 \times 10^{-3}}{2}(298)^2 + \frac{5.495 \times 10^5}{298}$$

or $\Delta H_0 = -80.951$ kJ

Substituting $\Delta G^0_{298}$ and $\Delta H_0$ in Eqn. (*B*) at $T = 298$ K, we obtain

$$-33\,494 = -809\,51 + 48.812 \times 298 \ln 298$$

$$-\frac{34.759 \times 10^{-3}}{2}(298)^2 + \frac{5.495 \times 10^5}{2 \times 298} - IR \times 298$$

or $IR = 116.75$

Now, substitute the values of $\Delta H_0$, $IR$ and $T = 800$ K in Eqn. ($B$) to obtain

$$\Delta G^0_{800} = -80\ 951 + 48.812 \times 800 \ln 800$$

$$-\frac{34.759 \times 10^{-3}}{2}(800)^2 + \frac{5.495 \times 10^5}{2 \times 800} - 116.75 \times 800 \text{ or } \Delta G^0_{800} = 75.9 \text{ kJ}$$

$$K_{800} = \exp\left(-\frac{\Delta G^0}{RT}\right) = \exp\left(\frac{-75\ 900}{8.314 \times 800}\right) = 1.1067 \times 10^{-5}$$

***Note*** *Compare the value of $K_{800} = 1.1067 \times 10^{-5}$ which is obtained by taking into account the variation of $\Delta H^0$ with temperature with that obtained $K_{800} = 5.4897 \times 10^{-5}$, obtained by assuming that $\Delta H^o$ is constant in Example 10.76.*

**Example 10.80**

The equilibrium constant $K$ is expressed in terms of the activities of products and reactants. Is it possible to express the equilibrium constant in terms of measurable quantities like partial pressures and mole fractions of the species ? What is the relation between $K_P$, the equilibrium constant in terms of partial pressures, $K_y$, the equilibrium constant in terms of mole fractions and $K$ ?

*Solution* :

The equilibrium constant $K$ is defined as $K = \prod_i a_i^{\upsilon_i}$

We know that the activity $a_i$ of species $i$ is given by $a_i = f_i / f_i^0$

Let us restrict to homogeneous gas phase reactions and assume that the reaction mixture behaves like an ideal gas. Then, for an ideal gas, the fugacity is equal to pressure. Therefore, the fugacity of component $i$ in a mixture is equal to the partial pressure $p_i$ of the component. For an ideal gas, the partial pressure $p_i = y_i P$, where $P$ is the total pressure. The standard state for gases is pure component gas in the ideal gas state at 1 bar. For convenience, the unit of pressure is taken as 1 bar for the analysis of chemically reacting systems. Then, the standard state fugacity $f_i^0$ is 1 bar. Now, we can write the equilibrium constant $K$ as

$$K = \prod_i a_i^{\upsilon_i} = \prod_i \left(\frac{f_i}{f_i^0}\right)^{\upsilon_i} = \prod_i f_i^{\upsilon_i} = \prod_i p_i^{\upsilon_i} = K_P$$

and $K = K_P = \prod_i (y_i P)^{\upsilon_i} = \prod_i y_i^{\upsilon_i} P^{\Sigma\upsilon_i} = K_y P^{\Sigma\upsilon_i}$

where $K_P = \prod_i p_i^{\upsilon_i}$ and $K_y = \prod_i y_i^{\upsilon_i}$

Therefore $K = K_P = K_y\, P^{\Sigma\upsilon_i}$

Note that the unit of pressure is bar. For a chemical reaction of the type

$$aA + b\,B \rightarrow r\,R + s\,S$$

We have

$$K = \prod_i a_i^{\upsilon_i} = \frac{a_R^r\, a_S^s}{a_A^a\, a_B^b}$$

$$K_p = \prod_i p_i^{\upsilon_i} = \frac{p_R^r\, p_S^s}{p_A^a\, p_B^b}$$

$$K_y = \prod_i y_i^{\upsilon_i} = \frac{y_R^r\, y_S^s}{y_A^a\, y_B^b}$$

### Example 10.81

On what factors does the equilibrium constant $K$ of a reaction depend ?

*Solution* :

We know that the equilibrium constant $K$ is given by

$$K = \exp\left(-\frac{\Delta G^0}{RT}\right)$$

where $\Delta G^0 = \Sigma\, \upsilon_i\, g_i^0,$ and

$g_i^0$ = standard Gibbs free energy of component $i$ which is a function of temperature only.

Therefore, the equilibrium constant $K$ is a function of temperature only.

### Example 10.82

In a particular plant employing the Haber process for the synthesis of ammonia, the reactor is maintained at 800 K and 25 MPa. If a stoichiometric mixture of nitrogen and hydrogen is fed to the reactor, estimate the percent of nitrogen that can be converted into ammonia and the composition of the reactor effluent. The equilibrium constant for the reaction $N_2\,(g) + 3\,H_2\,(g) \rightarrow 2\,NH_3\,(g)$ at 800 K is $1.1067 \times 10^{-5}$. Assume that the reaction mixture behaves like an ideal gas.

*Solution* :

The synthesis reaction is given by $N_2\,(g) + 3\,H_2\,(g) \rightarrow 2\,NH_3\,(g)$

The initial feed to the reactor consists of $N_2(g)$ and $H_2(g)$ in the mole ratio 1 : 3.

Let $\varepsilon_e$ = degree of conversion or extent of reaction at equilibrium with respect to $N_2\,(g)$. Then

| Component | $N_2\,(g)$ + | $3\,H_2(g) \longrightarrow$ | $2\,NH_3\,(g)$ |
|---|---|---|---|
| Initial moles in feed | 1 | 3 | 0 |
| Moles at equilibrium | $1-\varepsilon_e$ | $3\,(1-\varepsilon_e)$ | $2\,\varepsilon_e$ |
| Mole fraction $y_i$ | $\dfrac{1-\varepsilon_e}{2(2-\varepsilon_e)}$ | $\dfrac{3(1-\varepsilon_e)}{2(2-\varepsilon_e)}$ | $\dfrac{2\varepsilon_e}{2(2-\varepsilon_e)}$ |

Total moles of equilibrium $= (1-\varepsilon_e) + 3\,(1-\varepsilon_e) + 2\,\varepsilon_e = 4 - 2\,\varepsilon_e = 2\,(2-\varepsilon_e)$

$$\Sigma \upsilon_i = 2 + (-1) + (-3) = -2$$

$$K_y = \prod_i y_i^{\upsilon_i} = \frac{\{2\varepsilon_e/2(2-\varepsilon_e)\}^2}{\dfrac{1-\varepsilon_e}{2(2-\varepsilon_e)}\left\{\dfrac{3(1-\varepsilon_e)}{2(2-\varepsilon_e)}\right\}^3} = \frac{\{4\varepsilon_e(2-\varepsilon_e)\}^2}{27(1-\varepsilon_e)^4}$$

$$P = 25 \text{ MPa} = 250 \text{ bar}$$

We know that $K = K_y\, P^{\Sigma \upsilon_i}$

Therefore,

$$1.1067 \times 10^{-5} = \frac{\{4\varepsilon_e(2-\varepsilon_e)\}^2}{27(1-\varepsilon_e)^4} \times (250)^{-2}$$

or $$\frac{4\varepsilon_e(2-\varepsilon_e)}{\sqrt{27}\,(1-\varepsilon_e)^2} = 0.8317$$

or $\varepsilon_e = 1.6933$ and $0.3067$

The value of $\varepsilon_e = 1.6933$ is inadmissible, because the degree of conversion cannot be greater than one. Hence $\varepsilon_e = 0.3067$.

Therefore, $\varepsilon_e = 0.3067$. That is 30.67% of $N_2(g)$ is converted into ammonia. Then, the composition of the reactor effluent is given by

$$y_{N_2} = \frac{1-\varepsilon_e}{2(2-\varepsilon_e)} = 0.2047$$

$$y_{H_2} = \frac{3(1-\varepsilon_e)}{2(2-\varepsilon_e)} = 0.6142$$

$$y_{NH_3} = \frac{2\varepsilon_e}{2(2-\varepsilon_e)} = 0.1811$$

**Example 10.83**

An equimolar mixture of $CO(g)$ and $H_2O(g)$ enters a reactor which is maintained at 10 bar and 1000 K. Given that the equilibrium constant for the reaction

$$CO\,(g) + H_2O\,(g) \rightarrow CO_2\,(g) + H_2\,(g)$$

at 1000 K is 1.5, estimate the composition of the gas mixture that leaves the reactor. Assume that the reaction mixture behaves like an ideal gas.

*Solution* :

Let $\varepsilon_e$ = degree of conversion at equilibrium.

| Component | $CO\,(g)$ | $H_2O\,(g)$ | $CO_2\,(g)$ | $H_2\,(g)$ |
|---|---|---|---|---|
| Moles in feed | 1 | 1 | 0 | 0 |
| Moles at equilibrium | $1-\varepsilon_e$ | $1-\varepsilon_e$ | $\varepsilon_e$ | $\varepsilon_e$ |
| Mole fraction $y_i$ | $(1-\varepsilon_e)/2$ | $(1-\varepsilon_e)/2$ | $\varepsilon_e/2$ | $\varepsilon_e/2$ |

Total moles at equilibrium = $(1-\varepsilon_e) + (1-\varepsilon_e) + \varepsilon_e + \varepsilon_e = 2$

$$\Sigma\, \upsilon_i = 1 + 1 - 1 - 1 = 0$$

$$K_y = \prod_i y_i^{\upsilon_i} = \frac{\left(\frac{\varepsilon_e}{2}\right)\left(\frac{\varepsilon_e}{2}\right)}{\left(\frac{1-\varepsilon_e}{2}\right)\left(\frac{1-\varepsilon_e}{2}\right)} = \left(\frac{\varepsilon_e}{1-\varepsilon_e}\right)^2$$

$$P = 10 \text{ bar}$$

we know that $K = K_y P^{\Sigma \upsilon i} = K_y\,(10)^0 = K_y$

Therefore, $1.5 = \left(\dfrac{\varepsilon_e}{1-\varepsilon_e}\right)^2$ or $\varepsilon_e = 0.5505$

Therefore, the degree of conversion at equilibrium = 0.5505.

The composition of the gas mixture leaving the reactor is given by

$$y_{CO} = y_{H_2O} = \frac{1-\varepsilon_e}{2} = 0.224\,75;$$

$$y_{CO_2} = y_{H_2} = \frac{\varepsilon_e}{2} = 0.275\,25$$

**Example 10.84**

An equimolar mixture of $CH_4(g)$ and $H_2O(g)$ enters a reactor which is maintained at 1000 K and 5 bar. The following reaction is likely to occur between $CH_4$ and $H_2O$,

$$CH_4\,(g) + H_2O\,(g) \rightarrow CO(g) + 3H_2\,(g)$$

Calculate the equilibrium constant at 1000 K and estimate the degree of conversion of $CH_4$ into products. Also estimate the composition of the reactor effluent assuming that the reaction mixture behaves like an ideal gas.

*Solution* :

The Chemical reaction which occurs is given as

$$CH_4\,(g) + H_2O\,(g) \rightarrow CO(g) + 3H_2\,(g)$$

$$\Delta H^0_{298} = -110.532 - (-74.943) - (-241.997) = 206.408 \text{ kJ}$$

$$\Delta G^0_{298} = -137.372 - (-50.660) - (-228.600) = 141.933 \text{ kJ}$$

The constants of the isobaric molar heat capacities of the gases are

| Component | $a$ | $b \times 10^3$ | $c \times 10^6$ | $d \times 10^9$ | $e \times 10^{-5}$ |
|---|---|---|---|---|---|
| CO (g) | 28.068 | 4.631 | — | — | – 0.258 |
| $H_2$ (g) | 27.012 | 3.509 | — | — | 0.690 |
| $CH_4$ (*g*) | 17.449 | 60.449 | 1.117 | – 7.204 | — |
| $H_2O$ (g) | 28.850 | 12.055 | — | — | 1.006 |

$$\Delta a = 28.068 + 3 \times 27.012 - 17.449 - 28.850 = 62.805$$

$$\Delta b = (4.631 + 3 \times 3.509 - 60.449 - 12.055) \times 10^{-3}$$

$$= -57.346 \times 10^{-3}$$

$$\Delta c = -1.117 \times 10^{-6}$$

$$\Delta d = 7.204 \times 10^{-9}$$

$$\Delta e = (-0.258 + 3 \times 0.690 - 1.006) \times 10^5 = 0.806 \times 10^5$$

If the molar heat capacity at constant pressure is expressed as

$$C_P^0 = a + bT + cT^2 + dT^3 + eT^{-2},$$

We know that

$$\Delta H_T^0 = \Delta H_0 + \Delta aT + \frac{\Delta b}{2}T^2 + \frac{\Delta c}{3}T^3 + \frac{\Delta d}{4}T^4 - \frac{\Delta e}{T} \quad (A)$$

$$\Delta G_T^0 = \Delta H_0 - \Delta aT \ln T - \frac{\Delta b}{2}T^2 - \frac{\Delta c}{6}T^3 - \frac{\Delta d}{12}T^4 - \frac{\Delta e}{2T} - IRT \quad (B)$$

We first determine the constants $\Delta H_0$ by using the value of $\Delta H^0_{298}$ from Eqn. (*A*).

$$206\,408 = \Delta H_0 + 62.805 \times 298 - \frac{57.346 \times 10^{-3}}{2}(298)^2$$

$$-\frac{1.117 \times 10^{-6}}{3}(298)^3 + \frac{7.204 \times 10^{-9}}{4}(298)^4 - \frac{0.806 \times 10^5}{298}$$

or $\Delta H_0 = 190.504$ kJ

Now we use this value of $\Delta H_0$ along with $\Delta G^0_{298}$ in Eqn (*B*) to determine the constant *I*.

$$141\,933 = 190\,504 - 62.805 \times 298 \ln (298) + \frac{57.346 \times 10^{-3}}{2}(298)^2$$

$$+\frac{1.117 \times 10^{-6}}{6}(298)^3 - \frac{7.204 \times 10^{-9}}{12}(298)^4 - \frac{0.806 \times 10^5}{2 \times 298} - IR \times 298$$

or $IR = -186.725$

We use Eqn (*B*) again with the known values of $\Delta H_0$ and *I* to estimate $\Delta G^0_{1000}$.

$$\Delta G^o_{1000} = 190\,504 - 62.805 \times 1000 \ln (1000) + \frac{57.346 \times 10^{-3}}{2}(1000)^2$$

$$+\frac{1.117 \times 10^{-6}}{6}(1000)^3 - \frac{7.204 \times 10^{-9}}{12}(1000)^4 - \frac{0.806 \times 10^5}{2 \times 1000} + 186.725 \times 1000$$

or $\Delta G^0_{1000} = -\,28.394$ kJ

$$K_{1000} = \exp\left(-\frac{\Delta G^0_{1000}}{RT}\right) = \exp\left(\frac{28\,394}{8.314 \times 1000}\right) = 30.423$$

Let $\varepsilon_e$ = degree of conversion of $CH_4$ at equilibrium

| Component | $CH_4(g)$ | $H_2O\,(g)$ | $CO\,(g)$ | $H_2\,(g)$ |
|---|---|---|---|---|
| Moles of in feed | 1 | 1 | 0 | 0 |
| Moles at equilibrium | $1-\varepsilon_e$ | $1-\varepsilon_e$ | $\varepsilon_e$ | $3\,\varepsilon_e$ |
| Mole fraction $y_i$ | $\frac{1-\varepsilon_e}{2(1+\varepsilon_e)}$ | $\frac{1-\varepsilon_e}{2(1+\varepsilon_e)}$ | $\frac{\varepsilon_e}{2(1+\varepsilon_e)}$ | $\frac{3\varepsilon_e}{2(1+\varepsilon_e)}$ |

Total moles at equilibrium $= (1-\varepsilon_e) + (1-\varepsilon_e) + \varepsilon + 3\varepsilon_e = 2 + 2\varepsilon_e = 2\,(1+\varepsilon_e)$

$$K_y = \prod_i y_i^{\upsilon_i} = \left\{\frac{\varepsilon_e}{2(1+\varepsilon_e)}\right\}\left\{\frac{3\varepsilon_e}{2(1+\varepsilon_e)}\right\}^3 / \left\{\frac{1-\varepsilon_e}{2(1+\varepsilon_e)}\right\}\left\{\frac{1-\varepsilon_e}{2(1+\varepsilon_e}\right\}$$

or
$$K_y = \frac{27}{4}\left\{\frac{\varepsilon_e^2}{(1-\varepsilon_e)(1+\varepsilon_e)}\right\}^2$$

We know that $\quad P = 5$ bar and $\Sigma\upsilon_i = 1 + 3 - 1 - 1 = 2$

$$K = K_y\, P^{\Sigma\upsilon_i}$$

Therefore,

$$30.423 = \frac{27}{4}\left\{\frac{\varepsilon_e^2}{(1-\varepsilon_e^2)}\right\}^2 (5)^2$$

or
$$\frac{\varepsilon_e^2}{1-\varepsilon_e^2} = 0.4246 \text{ or } \varepsilon_e = 0.5459$$

Therefore, the degree of conversion of $CH_4$ at equilibrium = 0.5459. The composition of the reactor effluent is given by

$$y_{CH_4} = y_{H_2O} = \frac{1-\varepsilon_e}{2(1+\varepsilon_e)} = 0.146\,85$$

$$y_{CO} = \frac{\varepsilon_e}{2(1+\varepsilon_e)} = 0.1766$$

$$y_{H_2} = \frac{3\varepsilon_e}{2(1+\varepsilon_e)} = 0.5297$$

## Example 10.85

Rework Example 10.82 if the reactor is maintained at 10 MPa instead of 25 MPa.

*Solution* :

We know that $K = 1.1067 \times 10^{-5}$

$$K_y = \frac{\{4\varepsilon_e(2-\varepsilon_e)\}^2}{27(1-\varepsilon_e)^4} \text{ and } \Sigma\upsilon_i = -2 \quad \text{[see Example 10.82]}$$

Given $P = 10$ MPa = 100 bar

$$K = K_y\, P^{\Sigma\upsilon_i}$$

Therefore,

$$1.1067 \times 10^{-5} = \frac{\{4\varepsilon_e(2-\varepsilon_e)\}^2}{27(1-\varepsilon_e)^4} \times (100)^{-2}$$

or
$$\frac{4\,\varepsilon_e\,(2-\varepsilon_e)}{\sqrt{27}\,(1-\varepsilon_e)^2} = 0.110\,67$$

or $\varepsilon_e = 0.065$ and 1.935

The value $\varepsilon_e = 1.935 > 1$ is inadmissible. Hence $\varepsilon_e = 0.065$. Therefore 6.5% of $N_2$ (*g*) is converted into $NH_3$ (*g*). The composition of the reactor effluent is given by

$$y_{N_2} = \frac{1-\varepsilon_e}{2(2-\varepsilon_e)} = 0.2416$$

$$y_{H_2} = \frac{3(1-\varepsilon_e)}{2(2-\varepsilon_e)} = 0.7248$$

$$y_{NH_3} = \frac{2\varepsilon_e}{2(2-\varepsilon_e)} = 0.0336$$

***Note*** *For the ammonia synthesis reaction ($\Sigma\upsilon_i < 0$) a decrease in total pressure decreases the degree of conversion.*

**Example 10.86**

Rework Example 10.83 if the reactor is operated at 1 bar instead of 10 bar.

*Solution*:

We have $K = 1.5;\ K_y = \left(\frac{\varepsilon_e}{1-\varepsilon_e}\right)^2;\ \Sigma\upsilon_i = 0$ [See Example 10.83];

$$K = K_y\, P^{\Sigma\upsilon_i} = K_y\,(1)^0 = K_y$$

or
$$1.5 = \left(\frac{\varepsilon_e}{1-\varepsilon_e}\right)^2 \text{ or } \varepsilon_e = 0.5505$$

The values of $y_{CO}$; $y_{H_2O}$; $y_{CO_2}$ and $y_{H_2}$ will be identical to those of Example 10.83.

*Note* *For the reaction CO (g) + $H_2O$ (g) → $CO_2$ (g) + $H_2$ (g) where $\Sigma \upsilon_i = 0$, a decrease in pressure does not affect the degree of conversion.*

**Example 10.87**

Rework Example 10.84 if the reactor is maintained at 1 bar instead of 5 bar.

*Solution* :

We have $K = 30.423$ ;

$$K_y = \frac{27}{4}\left\{\frac{\varepsilon_e^2}{(1-\varepsilon_e)(1+\varepsilon_e)}\right\}^2 \text{ and } \Sigma\upsilon_i = 2$$

$$K = K_y\, p^{\Sigma\upsilon_i} = K_y\,(1)^2 = K_y$$

or
$$30.423 = \frac{27}{4}\left\{\frac{\varepsilon_e^2}{(1-\varepsilon^2{}_e)}\right\}^2 \text{ or } \varepsilon_e = 0.8245$$

Therefore, the degree of conversion of $CH_4 = 0.8245$

$$y_{CH_4} = y_{H_2O} = \frac{1-\varepsilon_e}{2\,(1+\varepsilon_e)} = 0.0481$$

$$y_{CO} = \frac{\varepsilon_e}{2\,(1+\varepsilon_e)} = 0.2259$$

$$y_{H_2} = \frac{3\varepsilon_e}{2(1+\varepsilon_e)} = 0.6779$$

*Note* *For the reaction $CH_4$ (g) + $H_2O$ (g) → CO (g) + 3 $H_2$ (g) where $\Sigma\,\upsilon_i = 2 > 0$, a decrease in pressure increases the degree of conversion.*

**Example 10.88**

It is known that the equilibrium constant $K$ of a reaction depends only on temperature. How does the pressure effect the degree of conversion of the reactants into products when all other variables like temperature, ratio of reactants etc. are held constant ?

*Solution* :

At a given temperature, the equilibrium constant $K$ is a constant and it is given by

$$K = K_y\, P^{\Sigma \upsilon_i} \text{ or } K_y = \prod_i y_i^{\upsilon_i} = \frac{K}{P^{\Sigma \upsilon_i}}$$

If the reaction is such that there is no change in the number of moles ($\Sigma\ \upsilon_i = 0$) due to reaction, then $P^{\Sigma \upsilon_i} = 1$ and $K_y = K =$ constant. That is, there is no change in the degree of conversion. Thus we find that for a reaction in which there is no change in the number of moles due to reaction ($\Sigma \upsilon_i = 0$), pressure has no effect on the degree of conversion. [See Examples 10.83 and 10.86].

If the reaction is such that there is a decrease in the number of moles ($\Sigma \upsilon_i < 0$) due to reaction, an increase in pressure increases the degree of conversion. [See Examples 10.82 and 10.85]. On the other hand, if there is an increase in the number of moles ($\Sigma \upsilon_i > 0$) due to chemical reaction, an increase in pressure decreases the degree of conversion [see Examples 10.84 and 10.87].

**Example 10.89**

If a mixture of $N_2(g)$ and $H_2(g)$ in the mole ratio 1 : 5 is fed to a reactor in the Haber process for sythesis of ammonia and the reactor is maintained at 800 K and 25 MPa, calculate the percentage of nitrogen that can be converted into ammonia and the molar composition of the reactor effluent assuming that the reaction mixture behaves like an ideal gas. The equilibrium constant $K$ at 800 K is $1.1067 \times 10^{-5}$.

*Solution* :

The ammonia synthesis reaction is given by

$$N_2\,(g) + 3H_2\,(g) \rightarrow 2\,NH_3\,(g).$$

Let $\varepsilon_e$ = degree of conversion of $N_2\,(g)$ into ammonia.

| **Component** | **$N_2$ (g)** | **$H_2$ (g)** | **$NH_3$ (g)** |
|---|---|---|---|
| Moles in feed | 1 | 5 | 0 |
| Moles at equilibrium | $1 - \varepsilon_e$ | $5 - 3\,\varepsilon_e$ | $2\,\varepsilon_e$ |
| Mole fraction $y_i$ | $\frac{1-\varepsilon_e}{2(3-\varepsilon_e)}$ | $\frac{5-3\varepsilon_e}{2(3-\varepsilon_e)}$ | $\frac{2\varepsilon_e}{2(3-\varepsilon_e)}$ |

$$\text{Total Moles} = (1-\varepsilon_e) + (5-3\varepsilon_e) + 2\varepsilon_e = 6 - 2\varepsilon_e = 2(3-\varepsilon_e)$$

$$P = 25 \text{ MPa} = 250 \text{ bar}$$

$$\Sigma \upsilon_i = 2 + (-1) + (-3) = -2$$

$$K_y = \prod_i y_i^{\upsilon_i} = \frac{\{2\varepsilon_e / 2(3-\varepsilon_e)\}^2}{\dfrac{1-\varepsilon_e}{2(3-\varepsilon_e)}\left\{\dfrac{5-3\varepsilon_e}{2(3-\varepsilon_e)}\right\}^3} = \left\{\frac{\{4\varepsilon_e(3-\varepsilon_e)\}^2}{(1-\varepsilon_e)(5-3\varepsilon_e)^3}\right\}$$

$$K = K_y P^{\Sigma \upsilon_i}$$

or $$1.1067 \times 10^{-5} = \left\{\frac{\{4\varepsilon_e(3-\varepsilon_e)\}^2}{(1-\varepsilon_e)(5-3\varepsilon_e)^3}(250)^{-2}\right\}$$

or $$\varepsilon_e = 0.4337$$

Therefore, 43.37% of $N_2$ ($g$) is converted into $NH_3$.

The composition of the reactor effluent is given by

$$y_{N_2} = \frac{1-\varepsilon_e}{2(3-\varepsilon_e)} = 0.1103$$

$$y_{H_2} = \frac{5-3\varepsilon_e}{2(3-\varepsilon_e)} = 0.7207$$

$$y_{NH_3} = \frac{2\varepsilon_e}{2(3-\varepsilon_e)} = 0.1690$$

***Note*** *Compare the degree of conversion of $N_2$ (g) into $NH_3$ (g) which is 0.3067 in Example 10.82 when stoichiometric mixture of $N_2$ (g) and $H_2$ (g) is used. In Example 10.89 when excess $H_2$ (g) is used, the degree of conversion of $N_2$ (g) increased to 0.4337 under identical conditions of temperature and pressure.*

**Example 10.90**

A mixture of CO($g$) and $H_2O$ ($g$) in the mole ratio 1:2 enters a reactor which is maintained at 10 bar and 1000 K. The CO ($g$)and $H_2O$ ($g$) react according to the equation

$$CO(g) + H_2O(g) \rightarrow CO_2(g) + H_2(g)$$

The equilibrium constant for this reaction at 1000 K is 1.5. Assuming that the reaction mixture behaves like an ideal gas, determine the degree of conversion of CO ($g$).

*Solution* :

The given reaction is

$$CO\,(g) + H_2O\,(g) \rightarrow CO_2\,(g) + H_2\,(g)$$

Let $\varepsilon_e$ = degree of conversion of CO ($g$) at equilibrium.

| Component | CO (g) | $H_2O$ (g) | $CO_2$ (g) | $H_2$ (g) |
|---|---|---|---|---|
| Moles in feed | 1 | 2 | 0 | 0 |
| Moles at equilibrium | $1-\varepsilon_e$ | $2-\varepsilon_e$ | $\varepsilon_e$ | $\varepsilon_e$ |
| Mole fraction $y_i$ | $\frac{1-\varepsilon_e}{3}$ | $\frac{2-\varepsilon_e}{3}$ | $\frac{\varepsilon_e}{3}$ | $\frac{\varepsilon_e}{3}$ |

Total moles = $(1-\varepsilon_e) + (2-\varepsilon_e) + \varepsilon_e + \varepsilon_e = 3$

$$K_y = \prod_i y_i^{\upsilon_i} = \frac{\left(\frac{\varepsilon_e}{3}\right)\left(\frac{\varepsilon_e}{3}\right)}{\left(\frac{1-\varepsilon_e}{3}\right)\left(\frac{2-\varepsilon_e}{3}\right)} = \frac{\varepsilon_e^2}{(1-\varepsilon_e)(2-\varepsilon_e)}$$

$$P = 10 \text{ bar} ; \Sigma\upsilon_i = 0.$$

We know that $\quad K = K_y\, P^{\Sigma\upsilon_i}$

or $\quad 1.5 = \dfrac{\varepsilon_e^2}{(1-\varepsilon_e)(2-\varepsilon_e)}.(10)^0 = \dfrac{\varepsilon_e^2}{(1-\varepsilon_e)(2-\varepsilon_e)}$

or $\quad \varepsilon_e = 0.7251$

Therefore, 72.51 % of CO ($g$) in the feed will be converted into $CO_2$ ($g$).

**Note** *Compare the value of $\varepsilon_e = 0.5505$ when stoichiometric mixture of CO (g) and $H_2O$ (g) are used in Example 10.83, with the value $\varepsilon_e = 0.7251$ when excess $H_2O$ (g) is used in Example 10.90, while other parameteres are held constant.*

**Example 10.91**

Some of the reactants used in a chemical reaction may be expensive and in such cases it is necessary to increase the degree of conversion with respect to that reactant. Since the equlibrium constant depends only on temperature, a change in temperature effects the equilibrium constant. Is it possible to obtain a higher degree of conversion by changing the ratio of reactants fed to the reactor while temperature is held constant?

*Solution*:

We know that the equilibrium constant $K$ is given by

$$K = \prod_i a_i^{\upsilon_i} = K_y \; P^{\Sigma \upsilon_i}$$

and $$y_i = \frac{n_i}{n}$$

where $n_i$ = moles of component $i$

$n = \Sigma n_i$ = total moles of the reaction mixture.

Then $$K = \prod_i y_i^{\upsilon_i} \; (P)^{\Sigma \upsilon_i}$$

$$= \prod_i n_i^{\upsilon_i} \left(\frac{P}{n}\right)^{\Sigma \upsilon_i}$$

That is if the chemical reaction is expressed as

$$aA + b\,B \longrightarrow r\,R + s\,S$$

Then $$K = \frac{n_R^r \, n_S^s}{n_A^a \, n_B^b} \left(\frac{P}{n}\right)^{\Sigma \upsilon_i}$$

where $$\Sigma \upsilon_i = r + s - a - b$$

or $$\frac{n_R^r \, n_S^s}{n_A^a \, n_B^b} = K \left(\frac{n}{P}\right)^{\Sigma \upsilon_i}$$

The RHS of the equation is a constant at constant temperature and pressure. That is the LHS of the above equation is a constant.

This clearly shows that if the number of moles of reactant $B$ are increased, the moles of $R$ and $S$ also increase. If one is interested in increasing the degree of conversion with respect to a particular reactant, say limiting reactant, then it is better to use the other reactants in excess of the stoichiometric requirement. In Example 10.89. $N_2(g)$ is the limiting reactant and use of excess $H_2(g)$ enhanced the degree of conversion. In Example 10.90, $CO(g)$ is the limiting reactant and hence the use of excess $H_2O(g)$ increased the degree of conversion of $CO(g)$.

**Example 10.92**

Suppose a mixture of $N_2\,(g)$, $H_2\,(g)$ and $He(g)$ in the mole ratio 1:3:2 is fed to a reactor which is maintained at 800 K and 25 MPa for the production of $NH_3$. Assuming that the reaction mixture behaves like an ideal gas, estimate the degree of conversion of $N_2(g)$ into $NH_3(g)$. The equilibrium constant $K$ for the reaction

$$N_2(g) + 3H_2(g) \rightarrow 2NH_3(g)$$

at 800 K is $1.1067 \times 10^{-5}$.

*Solution* :

The synthesis reaction for $NH_3$ is given by

$$N_2(g) + 3H_2(g) \rightarrow 2NH_3(g)$$

Let $\varepsilon_e$ = degree of conversion of $N_2(g)$ into $NH_3(g)$ at equilibrium

| **Component** | $N_2(g)$ | $H_2(g)$ | $NH_3(g)$ | $He(g)$ |
|---|---|---|---|---|
| Moles in feed | 1 | 3 | 0 | 2 |
| Moles of equilibrium | $(1-\varepsilon_e)$ | $3(1-\varepsilon_e)$ | $2\varepsilon_e$ | 2 |
| Mole fraction $y_i$ | $\dfrac{1-\varepsilon_e}{2(3-\varepsilon_e)}$ | $\dfrac{3(1-\varepsilon_e)}{2(3-\varepsilon_e)}$ | $\dfrac{2-\varepsilon_e}{2(3-\varepsilon_e)}$ | $\dfrac{2}{2(3-\varepsilon_e)}$ |

$$\text{Total Moles} = (1-\varepsilon_e) + 3(1-\varepsilon_e) + 2\varepsilon_e + 2 = 6 - 2\varepsilon_e = 2(3-\varepsilon_e)$$

$$K_y = \prod_i y_i^{\upsilon_i} = \frac{\left\{\dfrac{2\varepsilon_e}{2(3-\varepsilon_e)}\right\}^2}{\left\{\dfrac{1-\varepsilon_e}{2(3-\varepsilon_e)}\right\}\left\{\dfrac{3(1-\varepsilon_e)}{2(3-\varepsilon_e)}\right\}^3} = \frac{\{4\varepsilon_e(3-\varepsilon_e)\}^2}{27(1-\varepsilon_e)^4}$$

$$P = 25 \text{ MPa} = 250 \text{ bar}; \; \Sigma\,\upsilon_i = 2 - 1 - 3 = -2.$$

We know that $K = K_y\, P^{\Sigma\upsilon_i}$

Therefore,

$$1.1067 \times 10^{-5} = \frac{\{4\varepsilon_e(3-\varepsilon_e)\}^2}{27(1-\varepsilon_e)^4} \times (250)^{-2}$$

or $\quad \dfrac{\varepsilon_e(3-\varepsilon_e)}{(1-\varepsilon_e)^2} = 1.0804$

or $\quad \varepsilon_e = 0.2308$

Therefore, the degree of conversion of $N_2(g)$ into $NH_3(g)$ is 0.2308.

**Note** *The degree of conversion at the same temperature and pressure in absence of the inert gas helium is 0.3067 [see Example 10.82]. We find that for a reaction in which $\Sigma\upsilon_i < 0$, addition of inert gas decreases the degree of conversion.*

**Example 10.93**

If a mixture of CO ($g$), $H_2O$ ($g$) and He($g$) in the mole ratio 1:1:2 enters a reactor which is maintained at 10 bar and 1000 K the following reaction occurs between CO and $H_2O$ with an equilibrium constant $K = 1.5$.

$$CO\,(g) + H_2O\,(g) \rightarrow CO_2\,(g) + H_2\,(g)$$

Assuming that the reaction mixture behaves like an ideal gas estimate the degree of conversion.

*Solution* :

The reaction is given by

$$CO(g) + H_2O(g) \rightarrow CO_2(g) + H_2(g)$$

Let $\varepsilon_e$ = degree of conversion at equilibrium.

| **Component** | **CO(g)** | **$H_2O$ (g)** | **$CO_2$ (g)** | **$H_2$ (g)** | **He (g)** |
|---|---|---|---|---|---|
| Moles in feed | 1 | 1 | 0 | 0 | 2 |
| Moles at equilibrium | $(1-\varepsilon_e)$ | $(1-\varepsilon_e)$ | $\varepsilon_e$ | $\varepsilon_e$ | 2 |
| Mole fraction $y_i$ | $\frac{1-\varepsilon_e}{4}$ | $\frac{1-\varepsilon_e}{4}$ | $\frac{\varepsilon_e}{4}$ | $\frac{\varepsilon_e}{4}$ | $\frac{2}{4}$ |

Total Moles = $(1-\varepsilon_e) + (1-\varepsilon_e) + \varepsilon_e + \varepsilon_e + 2 = 4$

$$K_y = \frac{(\varepsilon_e/4)^2}{\left[\frac{(1-\varepsilon_e)}{4}\right]^2} = \left(\frac{\varepsilon_e}{1-\varepsilon_e}\right)^2$$

$$P = 10 \text{ bar};\ \Sigma\upsilon_i = 0$$

We know that $K = K_y\, P^{\Sigma\upsilon_i} = K_y\,(10)^0 = K_y$

Therefore,

$$1.5 = \left(\frac{\varepsilon_e}{1-\varepsilon_e}\right)^2 \quad \text{or } \varepsilon_e = 0.5505$$

Therefore the degree of conversion = 0.5505

***Note*** *The degree conversion at the same temperature and pressure in absence of the inert gas is also 0.5505 [see Example 10.83]. For a reaction in which $\Sigma\upsilon_i = 0$, that is there is no change in the number of moles due to chemical reaction, the addition of inert gases does not effect the degree of conversion,*

**Example 10.94**

How does the addition of inert gases effect the degree of conversion of a chemical reaction?

*Solution* :

The equilibrium constant $K$ given by

$$K = K_y\, P^{\Sigma\upsilon_i}$$

is a function of temperature only. The equilibrium constant can be expressed as

$$K = K_y\, P^{\Sigma\upsilon_i} = \prod_i y_i^{\upsilon_i}\, P^{\Sigma\upsilon_i} = P^{\Sigma\upsilon_i} \prod_i \left(\frac{n_i}{n}\right)^{\upsilon_i}$$

or

$$K = K_n \left(\frac{P}{n}\right)^{\Sigma\upsilon_i}$$

where $K_n = \prod_i n_i^{\upsilon_i}$ = equilibrium constant expressed in moles of components participating in the reaction.

$n = \Sigma n_i$ = total moles of the reaction mixture.

Therefore,

$$K_n = K\left(\frac{n}{P}\right)^{\Sigma\upsilon_i} \qquad (A)$$

Suppose a chemical reaction given by

$$aA + bB \rightarrow rR + sS$$

is carried out at a given temperature and pressure. Then

$$K_n = \frac{n_R^r\, n_S^s}{n_A^a\, n_B^b}$$

If $n_I$ moles of an inert gas is added to the reaction mixture, then

$$n = n_A + n_B + n_R + n_s + n_I$$

Now, Eqn. (*A*) gives

$$K_n = \frac{n_R^r\, n_S^s}{n_A^a\, n_B^b} = K\left(\frac{n_A + n_B + n_R + n_S + n_I}{P}\right)^{\Sigma \upsilon_i} \qquad (B)$$

At a specified temperature $K$ is constant and at the specified pressure $P$ is constant. Therefore, the RHS of Eqn. ($B$) depends on $(n_A + n_B + n_R + n_s + n_I)^{\Sigma\upsilon_i}$ and hence $K_n$ also depends on the same factor. For a chemical reaction in which $(\Sigma\upsilon_i = 0)$ there is no change in the number of moles due to chemical reaction $K_n$ does not change. In other words, the addition of inert gases has no effect on the degree of conversion for a chemical reaction in which $\Sigma\upsilon_i = 0$ [See Example 10.93]. If there is a decrease in the number of moles $(\Sigma\upsilon_i < 0)$ due to chemical reaction, $K_n$ decrease even though $K$ is constant. That is less products are formed or the degree of conversion decreases on addition of inert gases [see Example 10.92]. On the other hand, if the number of moles increases $(\Sigma\upsilon_i > 0)$ due to chemical reaction, the addition of inert gases increases the degree of conversion. In a chemical reaction the effect of addition of inert gases on the degree of conversion is equivalent to a decrease in pressure.

**Example 10.95**

A mixture of $CO(g)$, $H_2O(g)$ and $H_2(g)$ in the mole ratio 1:1:0.5 is fed to a reactor which is maintained at 10 bar and 1000 K, in which the following reaction takes place

$$CO(g) + H_2O(g) \rightarrow CO_2(g) + H_2(g)$$

Assuming that the gas mixture behaves like an ideal gas and K = 1.5 for the above reaction at 1000 K, calculate the degree conversion.

*Solution*:

Let $\varepsilon_e$ = degree of conversion at equilibrium.

| **Component** | **$CO(g)$** | **$H_2O(g)$** | **$CO_2(g)$** | **$H_2(g)$** |
|---|---|---|---|---|
| Moles in feed | 1 | 1 | 0 | 0.5 |
| Moles at equilibrium | $(1-\varepsilon_e)$ | $(1-\varepsilon_e)$ | $\varepsilon_e$ | $\varepsilon_e + 0.5$ |
| Mole fraction $y_i$ | $\frac{1-\varepsilon_e}{2.5}$ | $\frac{1-\varepsilon_e}{2.5}$ | $\frac{\varepsilon_e}{2.5}$ | $\frac{\varepsilon_e + 0.5}{2.5}$ |

Total moles = $(1-\varepsilon_e) + (1-\varepsilon_e) + \varepsilon_e + \varepsilon_e + 0.5 = 2.5$

$$K_y = \prod_i y_i^{\upsilon_i} = \frac{\varepsilon_e(\varepsilon_e + 0.5)}{(1-\varepsilon_e)^2}$$

$$P = 10 \text{ bar}; \Sigma \upsilon_i = 0$$

We know that $K = K_y \, P^{\Sigma \upsilon_i}$

Therefore, $$1.5 = \frac{\varepsilon_e (\varepsilon_e + 0.5)}{(1 - \varepsilon_e)^2} (10)^0 = \frac{\varepsilon_e (\varepsilon_e + 0.5)}{(1 - \varepsilon_e)^2}$$

or $\varepsilon_e = 0.4586$

***Note*** *The degree of conversion without adding one of the products ($H_2$) under identical conditions of temperature and pressure is 0.5505 [ see Example 10.83]. Thus addition of products to the initial reaction mixture decreases the degree of conversion.*

**Example 10.96**

What is the effect of the presence of the products in the initial feed to a reactor, on the degree of conversion of reactants into products?

*Solution* :

Suppose a chemical reaction given by

$$aA + bB \rightarrow rR + sS$$

occurs at a specified temperature and pressure. Then we know that

$$K = K_y \, P^{\Sigma \upsilon_i} = \prod_i y_i^{\upsilon_i} \, (P^{\Sigma \upsilon_i}) = P^{\Sigma \upsilon_i} \prod_i \left( \frac{n_i}{n} \right)^{\upsilon_i}$$

$$= \left( \frac{P}{n} \right)^{\Sigma \upsilon_i} \prod_i n_i^{\upsilon_i} = K_n \left( \frac{P}{n} \right)^{\Sigma \upsilon_i}$$

or $$K_n = K \left( \frac{n}{P} \right)^{\Sigma \upsilon_i}$$

or $$\frac{n_R^r \, n_S^s}{n_A^a \, n_B^b} = K \left( \frac{n}{P} \right)^{\Sigma \upsilon_i}$$

From the above equation it can be easily seen that if the products $R$ or $S$ or both present in the initial feed, then the moles formed due to reaction are $(n_R - n_R^0)$ or $(n_S - n_S^0)$ where $n_R^0$ and $n_S^0$ denote the moles of $R$ and $S$ in the initial feed. That is, the presence or addition of products in the initial feed decreases the degree of conversion of the reactants into products.

**Example 10.97**

It is proposed to produce ethyl alcohol ($C_2H_5OH$) according to the reaction

$$C_2H_4\,(g) + H_2O\,(g) \rightarrow C_2H_5OH\,(g)$$

It is known that $\Delta H_{298}^0 = -45.792$ kJ and $\Delta G_{298}^0 = -8.378$ kJ for the above reaction.

(*a*) Is the proposed reaction feasible or not ?

(*b*) Is it better to carry out the reaction at low temperature or at high temperature if the degree of conversion of reactants into products is to be sufficiently good ?

(*c*) If the degree of conversion is to be sufficiently large, is it advisable to carry out the reaction at low pressures or high pressures ?

(*d*) Does the degree of conversion improve by using $H_2O(g)$ in excess of the stoichiometric requirement ?

(*e*) Is it possible to improve the degree of conversion by diluting the reactant gases with an inert gas like nitrogen ?

*Solution* :

(*a*) The reaction is definitely feasible at 298 K, since $\Delta G_{298}^0 = -8.378$ kJ $< 0$.

(*b*) The given reaction is exothermic ($\Delta H^0 < 0$) and hence the equilibrium constant decreases with increasing temperature. Therefore. it is advisable to carry out the reaction at low temperature.

(*c*) There is a decrease in the number of moles ($\Sigma\upsilon_i = -1$) due to chemical reaction and hence high pressures are desirable to improve the degree of conversion.

(*d*) The use of excess of the reactant $H_2O(g)$ definitely increases the conversion of the limiting reactant $C_2H_4\,(g)$.

(*e*) The addition of inert gases decreases the degree of conversion of the reactants, since $\Sigma\upsilon_i < 0$. Hence it is not desirable to dilute reactants with inert gases.

**Example 10.98**

In general more than one reaction is possible in a reacting mixture. Develop a method to determine the composition of the reaction mixture in a state of equilibrium if two or more simultaneous chemical reactions occur in a mixture.

*Solution* :

Suppose the following two chemical reactions occur simultaneously in a reacting mixture.

$$\Sigma \upsilon_{1i} A_{1i} = 0 \qquad (A)$$

$$\Sigma \upsilon_{2i} A_{2i} = 0 \qquad (B)$$

where the first subscripts 1 and 2 designate the reaction number. Then the criterion for equilibrium at constant temperature and pressure is given by

$$\Sigma \mu_i \, dn_i = 0 \qquad (C)$$

The change in the moles of $dn_i$ for the reactions ($A$) and ($B$) are not independent but they are given by

$$dn_{1i} = \upsilon_{1i} \, d\varepsilon_1 \qquad (D)$$

$$dn_{2i} = \upsilon_{2i} \, d\varepsilon_2 \qquad (E)$$

where $\varepsilon_1$ and $\varepsilon_2$ denote the extent of reactions ($A$) and ($B$), respectively. Substituting Eqns.($D$) and ($E$) in Eqn. ($C$), we get

$$\Sigma \mu_i dn_i = \Sigma(\mu_{1i} \, dn_{1i} + \mu_{2i} \, dn_{2i}) = 0$$

or

$$\Sigma \mu_{1i} \upsilon_{ii} d\varepsilon_1 + \Sigma \mu_{2i} \upsilon_{2i} . d\varepsilon_2 = 0 \qquad (F)$$

The two simultaneous reactions are independent and hence $\varepsilon_1$ and $\varepsilon_2$ are independent. The Eqn.($F$) is true for arbitrary and independent values of $\varepsilon_1$ and $\varepsilon_2$. Therefore we get

$$\Sigma \upsilon_{1i} \mu_{1i} = 0, \text{ and} \qquad (G)$$

$$\Sigma \upsilon_{2i} \mu_{2i} = 0, \qquad (H)$$

The chemical potential $\mu_i$ of a component $i$ is given by

$$\mu_i = RT \ln a_i + g_i^0 \qquad (I)$$

Substituting Eqn($I$) in Eqns. ($G$) and ($H$), we get

$$\Sigma \upsilon_{1i} (RT \ln a_{1i} + g_i^0) = 0 \text{ and} \qquad (J)$$

$$\Sigma \upsilon_{2i} (RT \ln a_{2i} + g_{2i}^0) = 0 \qquad (K)$$

or

$$\Sigma \upsilon_{1i} \, g_{1i}^0 = \Delta G^0{}_1 = -RT \ln \prod_i a_{1i}^{\upsilon_{1i}} = -RT \ln K_1 \qquad (L)$$

$$\Sigma \upsilon_{2i} \, g_{1i}^0 = \Delta G^o{}_2 = -RT \ln \prod_i a_{2i}^{\upsilon_{2i}} = -RT \ln K_2 \qquad (M)$$

The equilibrium constants $K_1$ and $K_2$ can be expressed in terms of the mole fractions of the components. Then a simultaneous solution of the Eqns.($L$) and ($M$) gives the composition of the reaction mixture at equilibrium.

**Example 10.99**

An equimolar mixture of $CH_4(g)$ and $H_2O(g)$ enters a reactor which is maintained at 1000 K and 1 bar. The following two reactions are expected to occur simultaneously in the reactor.

$$C_2H_4(g) + H_2O(g) \rightarrow CO(g) + 3H_2(g) \qquad (A)$$

$$CO(g) + H_2O(g) \rightarrow CO_2(g) + H_2(g) \qquad (B)$$

At 1000 K, the equilibrium constants for the reactions (*A*) and (*B*) are given by $K_A = 30$ and $K_B = 1.5$. Assuming that the reaction mixture behaves like an ideal gas, calculate the composition of the reactor effluent.

*Solution*:

Let $\varepsilon_A$ = degree of conversion of $H_2O$ (*g*) in reaction (*A*)

$\varepsilon_B$ = degree of conversion of $H_2O$ (*g*) in reaction (*B*)

| Component | $CH_4(g)$ | $H_2O(g)$ | $CO(g)$ | $CO_2(g)$ | $H_2(g)$ |
|---|---|---|---|---|---|
| Moles in feed | 1 | 1 | 0 | 0 | 0 |
| Moles at equilibrium | $1-\varepsilon_A$ | $1-\varepsilon_A-\varepsilon_B$ | $\varepsilon_A-\varepsilon_B$ | $\varepsilon_B$ | $3\varepsilon_A+\varepsilon_B$ |
| Mole fraction $y_i$ | $\frac{1-\varepsilon_A}{2(1+\varepsilon_A)}$ | $\frac{1-\varepsilon_A-\varepsilon_B}{2(1+\varepsilon_A)}$ | $\frac{\varepsilon_A-\varepsilon_B}{2(1+\varepsilon_A)}$ | $\frac{\varepsilon_B}{2(1+\varepsilon_A)}$ | $\frac{3\varepsilon_A+\varepsilon_B}{2(1+\varepsilon_A)}$ |

$$\text{Total Moles} = (1-\varepsilon_A) + (1-\varepsilon_A-\varepsilon_B) + (\varepsilon_A-\varepsilon_B) + \varepsilon_B + (3\varepsilon_A+\varepsilon_B)$$

$$= 2 + 2\varepsilon_A = 2(1+\varepsilon_A)$$

$$K_{yA} = \frac{\dfrac{\varepsilon_A-\varepsilon_B}{2(1+\varepsilon_A)}\left\{\dfrac{3\varepsilon_A+\varepsilon_B}{2(1+\varepsilon_A)}\right\}^3}{\dfrac{(1-\varepsilon_A)}{2(1+\varepsilon_A)}\dfrac{(1-\varepsilon_A-\varepsilon_B)}{2(1+\varepsilon_A)}} = \frac{(\varepsilon_A-\varepsilon_B)(3\varepsilon_A+\varepsilon_B)^3}{4(1+\varepsilon_A)^2(1-\varepsilon_A)(1-\varepsilon_A-\varepsilon_B)}$$

$$K_{yB} = \frac{\dfrac{\varepsilon_B}{2(1+\varepsilon_A)}\dfrac{(3\varepsilon_A+\varepsilon_B)}{2(1+\varepsilon_A)}}{\dfrac{(\varepsilon_A-\varepsilon_B)}{2(1+\varepsilon_A)}\dfrac{(1-\varepsilon_A-\varepsilon_B)}{2(1+\varepsilon_A)}} = \frac{\varepsilon_B(3\varepsilon_A+\varepsilon_B)}{(\varepsilon_A-\varepsilon_B)(1-\varepsilon_A-\varepsilon_B)}$$

We know that

$$K = K_y P^{\Sigma\upsilon_i}$$

$$P = 1 \text{ bar};\ \Sigma\upsilon_{1i} = 2;\ \Sigma\upsilon_{2i} = 0.$$

Therefore, $K_A = K_{yA}(1)^2 = K_{yA}$ and

$$K_B = K_{yB}(1)^0 = K_{yB}$$

Then

$$\frac{(\varepsilon_A - \varepsilon_B)(3\varepsilon_A + \varepsilon_B)^3}{4(1+\varepsilon_A)^2(1-\varepsilon_A)(1-\varepsilon_A-\varepsilon_B)} = 30 \tag{C}$$

$$\frac{\varepsilon_B(3\varepsilon_A + \varepsilon_B)}{(\varepsilon_A - \varepsilon_B)(1-\varepsilon_A-\varepsilon_B)} = 1.5 \tag{D}$$

Eqns. (*C*) and (*D*) can be solved simultaneously to obtain the values of $\varepsilon_A$ and $\varepsilon_B$. Let us assume that $\varepsilon_B = 0.0626$ and substitute this in Eqn. (*D*) to obtain

$$\frac{0.0626(3\varepsilon_A + 0.0626)}{(\varepsilon_A - 0.0626)(1-\varepsilon_A - 0.0626)} = 1.5$$

or $\varepsilon_A = 0.7980$ and $0.0768$

Now substitute the value $\varepsilon_A = 0.7980$ and $\varepsilon_B = 0.0626$ in Eqn (*C*) to verify whether it is satisfied or not. Then we find LHS of Eqn.(*C*) = 29.9415 ≅ 30 = RHS. Hence, our guess value $\varepsilon_B = 0.0626$ is correct.

Therefore, we have $\varepsilon_A = 0.7980$ and $\varepsilon_B = 0.0626$.

The composition of the reaction mixture at equilibrium is given by

$$y_{CH_4} = \frac{1-\varepsilon_A}{2(1+\varepsilon_A)} = 0.0562$$

$$y_{H_2O} = \frac{1-\varepsilon_A-\varepsilon_B}{2(1+\varepsilon_A)} = 0.0388$$

$$y_{CO} = \frac{\varepsilon_A-\varepsilon_B}{2(1+\varepsilon_A)} = 0.2045$$

$$y_{CO_2} = \frac{\varepsilon_B}{2(1+\varepsilon_A)} = 0.0174$$

$$y_{H_2} = \frac{3\varepsilon_A+\varepsilon_B}{2(1+\varepsilon_A)} = 0.6831$$

# Appendix – 1

*Saturated steam : Pressure table**

| Pressure | Temperature | Specific volume | | Specific enthalpy | | Specific entropy | |
|---|---|---|---|---|---|---|---|
| $P$ (bar) | $t$ (°C) | $v_f$ (m³/kg) | $v_g$ (m³/kg) | $h_f$ (kJ/kg) | $h_g$ (kJ/kg) | $s_f$ (kJ/kg K) | $s_g$ (kJ/kg K) |
| 0.01 | 6.9828 | 0.001 000 1 | 129.20 | 29.34 | 2514.4 | 0.1060 | 8.9767 |
| 0.05 | 32.8980 | 0.001 005 2 | 28.19 | 137.77 | 2561.6 | 0.4763 | 8.3960 |
| 0.10 | 45.8330 | 0.001 010 2 | 14.67 | 191.83 | 2584.8 | 0.6493 | 8.1511 |
| 0.15 | 53.9970 | 0.001 014 0 | 10.02 | 225.97 | 2599.2 | 0.7549 | 8.0093 |
| 0.20 | 60.0860 | 0.001 017 2 | 7.650 | 251.45 | 2609.9 | 0.8321 | 7.9094 |
| 0.3 | 69.1240 | 0.001 022 3 | 5.229 | 289.30 | 2625.4 | 0.9441 | 7.7695 |
| 0.4 | 75.8860 | 0.001 026 5 | 3.993 | 317.65 | 2636.9 | 1.0261 | 7.6709 |
| 0.5 | 81.3450 | 0.001 030 1 | 3.240 | 340.56 | 2646.0 | 1.0912 | 7.5947 |
| 0.6 | 85.9540 | 0.001 033 3 | 2.732 | 359.93 | 2653.6 | 1.1454 | 7.5327 |
| 0.7 | 89.9590 | 0.001 036 1 | 2.365 | 376.77 | 2660.1 | 1.1921 | 7.4804 |
| 0.8 | 93.5120 | 0.001 038 7 | 2.087 | 391.72 | 2665.8 | 1.2330 | 7.4352 |
| 0.9 | 96.7130 | 0.001 041 2 | 1.869 | 405.21 | 2670.9 | 1.2696 | 7.3954 |
| 1.0 | 99.6320 | 0.001 043 4 | 1.694 | 417.54 | 2675.4 | 1.3027 | 7.3598 |
| 1.5 | 111.3700 | 0.001 053 0 | 1.159 | 467.13 | 2693.4 | 1.4336 | 7.2234 |
| 2.0 | 120.2300 | 0.001 060 8 | 0.8854 | 504.70 | 2706.3 | 1.5301 | 7.1268 |
| 2.5 | 127.4300 | 0.001 067 5 | 0.7184 | 535.34 | 2716.4 | 1.6071 | 7.0520 |
| 3.0 | 133.5400 | 0.001 073 5 | 0.6056 | 561.43 | 2724.7 | 1.6716 | 6.9909 |
| 3.5 | 138.8700 | 0.001 078 9 | 0.5240 | 584.27 | 2731.6 | 1.7273 | 6.9392 |
| 4.0 | 143.6200 | 0.001 083 9 | 0.4622 | 604.67 | 2737.6 | 1.7764 | 6.8943 |
| 5.0 | 151.8400 | 0.001 092 8 | 0.3747 | 640.12 | 2747.5 | 1.8604 | 6.8192 |
| 6.0 | 158.8400 | 0.001 100 9 | 0.3155 | 670.42 | 2755.5 | 1.9308 | 6.7575 |
| 7.0 | 164.9600 | 0.001 108 9 | 0.2727 | 697.06 | 2762.0 | 1.9918 | 6.7052 |
| 8.0 | 170.4100 | 0.001 115 0 | 0.2403 | 720.94 | 2767.5 | 2.0457 | 6.6596 |
| 9.0 | 175.3600 | 0.001 121 3 | 0.2148 | 742.64 | 2772.1 | 2.0941 | 6.6192 |
| 10.0 | 179.8800 | 0.001 127 4 | 0.1943 | 762.61 | 2776.2 | 2.1382 | 6.5862 |

| Pressure | Temperature | Specific volume | | Specific enthalpy | | Specific entropy | |
|---|---|---|---|---|---|---|---|
| $P$ (bar) | $t$ (°C) | $v_f$ (m³/kg) | $v_g$ (m³/kg) | $h_f$ (kJ/kg) | $h_g$ (kJ/kg) | $s_f$ (kJ/kg K) | $s_g$ (kJ/kg K) |
| 11.0 | 184.0700 | 0.001 133 1 | 0.1774 | 781.13 | 2779.7 | 2.1786 | 6.5497 |
| 12.0 | 187.9600 | 0.001 138 6 | 0.1632 | 798.43 | 2782.7 | 2.2161 | 6.5194 |
| 13.0 | 191.6100 | 0.001 143 8 | 0.1511 | 814.70 | 2785.4 | 2.2510 | 6.4913 |
| 14.0 | 195.0400 | 0.001 148 9 | 0.1407 | 830.08 | 2787.8 | 2.2837 | 6.4651 |
| 15.0 | 198.2900 | 0.001 153 9 | 0.1317 | 844.67 | 2789.8 | 2.3145 | 6.4406 |
| 16.0 | 201.3700 | 0.001 158 6 | 0.1237 | 858.56 | 2791.7 | 2.3436 | 6.4175 |
| 17.0 | 204.3100 | 0.001 163 3 | 0.1166 | 871.84 | 2793.4 | 2.3713 | 6.3957 |
| 18.0 | 207.1100 | 0.001 167 8 | 0.1103 | 884.58 | 2794.8 | 2.3976 | 6.3751 |
| 19.0 | 209.8000 | 0.001 172 3 | 0.1047 | 896.81 | 2796.1 | 2.4228 | 6.3554 |
| 20.0 | 212.3700 | 0.001 176 6 | 0.099 54 | 908.59 | 2797.2 | 2.4469 | 6.3367 |
| 25 | 223.9400 | 0.001 197 2 | 0.079 91 | 961.96 | 2800.9 | 2.5543 | 6.2536 |
| 30 | 233.8400 | 0.001 216 3 | 0.066 63 | 1008.40 | 2802.3 | 2.6455 | 6.1837 |
| 35 | 242.5400 | 0.001 234 5 | 0.057 03 | 1049.80 | 2802.0 | 2.7253 | 6.1228 |
| 40 | 250.3300 | 0.001 252 1 | 0.049 75 | 1087.40 | 2800.3 | 2.7965 | 6.0685 |
| 45 | 257.4100 | 0.001269 1 | 0.044 04 | 1122.10 | 2797.7 | 2.8612 | 6.0191 |
| 50 | 263.9100 | 0.001 285 8 | 0.039 43 | 1154.50 | 2794.2 | 2.9206 | 5.9735 |
| 60 | 275.5500 | 0.001 318 7 | 0.032 44 | 1213.70 | 2785.0 | 3.0273 | 5.8908 |
| 70 | 285.7900 | 0.001 351 3 | 0.027 37 | 1267.40 | 2773.5 | 3.1219 | 5.8162 |
| 80 | 294.9700 | 0.001 384 2 | 0.023 53 | 1317.10 | 2759.9 | 3.2076 | 5.7471 |
| 90 | 303.3100 | 0.001 417 9 | 0.020 50 | 1363.70 | 2744.6 | 3.2867 | 5.6820 |
| 100 | 310.9600 | 0.001 452 6 | 0.018 04 | 1408.00 | 2727.7 | 3.3605 | 5.6198 |
| 120 | 324.6500 | 0.001 526 8 | 0.014 28 | 1491.80 | 2689.2 | 3.4972 | 5.5002 |
| 140 | 336.6400 | 0.001 610 6 | 0.011 50 | 1571.60 | 2642.4 | 3.6242 | 5.3803 |
| 160 | 347.3300 | 0.001 710 3 | 0.009 308 | 1650.50 | 2584.9 | 3.7471 | 5.2531 |
| 180 | 356.9600 | 0.001 839 9 | 0.007 498 | 1734.80 | 2513.9 | 3.8765 | 5.1128 |
| 200 | 365.7000 | 0.002 037 0 | 0.005 877 | 1826.50 | 2418.4 | 4.0149 | 4.9412 |
| 220 | 373.6900 | 0.002 671 4 | 0.003 728 | 2011.10 | 2195.6 | 4.2947 | 4.5799 |
| 221.20 | 374.1500 | 0.003 17 | 0.003 17 | 2107.40 | 2170.4 | 4.4429 | 4.4429 |

* Reprinted from Chandrasekaran, K.D. and Venkateswarlu, D., *SI Units in Chemical Engineering and Technology,* Chemical Engineering Education Development Centre, IIT-Madras (1974), with permission from Chairman, Centre for Continuing Education, IIT-Madras.

# Appendix – 2

*Saturated steam : Temperature table**

| Temperature | Pressure | Specific volume | | Specific enthalpy | | Specific entropy | |
|---|---|---|---|---|---|---|---|
| $t$ (°C) | $P$ (bar) | $v_f$ (m³/kg) | $v_g$ (m³/kg) | $h_f$ (kJ/kg) | $h_g$ (kJ/kg) | $s_f$ (kJ/kg K) | $s_g$ (kJ/kg K) |
| 0 | 0.006 108 | 0.001 000 2 | 206.3000 | −0.04 | 2501.6 | −0.0002 | 9.1577 |
| 2 | 0.007 055 | 0.001 000 1 | 179.9000 | 8.39 | 2505.2 | 0.0306 | 9.1047 |
| 4 | 0.008 129 | 0.001 000 0 | 157.3000 | 16.80 | 2508.9 | 0.0611 | 9.0526 |
| 6 | 0.009 345 | 0.001 000 0 | 137.8000 | 25.21 | 2512.6 | 0.0913 | 9.0015 |
| 8 | 0.010 720 | 0.001 000 1 | 121.0000 | 33.60 | 2516.2 | 0.1213 | 8.9513 |
| 10 | 0.012 270 | 0.001 000 3 | 106.4000 | 41.99 | 2519.9 | 0.1510 | 8.9020 |
| 15 | 0.017 039 | 0.001 000 8 | 77.9800 | 62.94 | 2529.1 | 0.2243 | 8.7826 |
| 20 | 0.023 37 | 0.001 001 7 | 57.8400 | 83.86 | 2538.2 | 0.2963 | 8.6684 |
| 25 | 0.031 66 | 0.001 002 9 | 43.4000 | 104.77 | 2547.3 | 0.3670 | 8.5592 |
| 30 | 0.042 41 | 0.001 004 3 | 32.9300 | 125.66 | 2556.4 | 0.4365 | 8.4546 |
| 35 | 0.056 22 | 0.001 006 0 | 25.2400 | 146.56 | 2565.4 | 0.5049 | 8.3543 |
| 40 | 0.073 75 | 0.001 007 8 | 19.5500 | 167.45 | 2574.4 | 0.5721 | 8.2583 |
| 45 | 0.095 82 | 0.001 009 9 | 15.2800 | 188.35 | 2583.3 | 0.6383 | 8.1661 |
| 50 | 0.123 35 | 0.001 012 1 | 12.0500 | 209.26 | 2592.2 | 0.7035 | 8.0776 |
| 55 | 0.157 41 | 0.001 014 5 | 9.5790 | 230.17 | 2601.0 | 0.7677 | 7.9926 |
| 60 | 0.199 20 | 0.001 0171 | 7.6790 | 251.09 | 2609.7 | 0.8310 | 7.9108 |
| 65 | 0.2501 | 0.001 019 9 | 6.2020 | 272.02 | 2618.4 | 0.8933 | 7.8322 |
| 70 | 0.3116 | 0.001 022 8 | 5.0460 | 292.97 | 2626.9 | 0.9548 | 7.7565 |
| 75 | 0.3855 | 0.001 025 9 | 4.1340 | 313.94 | 2635.4 | 1.0154 | 7.6835 |
| 80 | 0.4736 | 0.001 029 2 | 3.4090 | 334.92 | 2643.8 | 1.0753 | 7.6132 |
| 85 | 0.5780 | 0.001 032 6 | 2.8290 | 355.92 | 2652.0 | 1.1343 | 7.5454 |
| 90 | 0.7011 | 0.001 036 1 | 2.3610 | 376.94 | 2660.1 | 1.1925 | 7.4799 |
| 95 | 0.8453 | 0.001 039 9 | 1.9820 | 397.99 | 2668.1 | 1.2501 | 7.4166 |
| 100 | 1.0133 | 0.001 043 7 | 1.6730 | 419.06 | 2676.0 | 1.3069 | 7.3554 |
| 105 | 1.2080 | 0.001 047 7 | 1.4190 | 440.17 | 2683.7 | 1.3630 | 7.2962 |
| 110 | 1.4327 | 0.001 051 9 | 1.2100 | 461.32 | 2691.3 | 1.4185 | 7.2388 |
| 115 | 1.6906 | 0.001 056 2 | 1.0360 | 482.50 | 2698.7 | 1.4733 | 7.1832 |
| 120 | 1.9854 | 0.001 060 6 | 0.8915 | 503.72 | 2706.0 | 1.5276 | 7.1293 |
| 125 | 2.3210 | 0.001 065 2 | 0.7702 | 524.99 | 2713.0 | 1.5813 | 7.0769 |
| 130 | 2.7013 | 0.001 070 0 | 0.6681 | 546.31 | 2719.9 | 1.6344 | 7.0261 |

| Temperature | Pressure | Specific volume | | Specific enthalpy | | Specific entropy | |
|---|---|---|---|---|---|---|---|
| $t$ (°C) | $P$ (bar) | $v_f$ (m³/kg) | $v_g$ (m³/kg) | $h_f$ (kJ/kg) | $h_g$ (kJ/kg) | $s_f$ (kJ/kg K) | $s_g$ (kJ/kg K) |
| 135 | 3.131 | 0.001 075 0 | 0.5818 | 567.68 | 2726.6 | 1.6869 | 6.9766 |
| 140 | 3.614 | 0.001 080 1 | 0.5085 | 589.10 | 2733.1 | 1.7390 | 6.9284 |
| 145 | 4.155 | 0.001 085 3 | 0.4460 | 610.60 | 2739.3 | 1.7906 | 6.8815 |
| 150 | 4.760 | 0.001 090 8 | 0.3924 | 632.15 | 2745.4 | 1.8416 | 6.8358 |
| 155 | 5.433 | 0.001 096 4 | 0.3464 | 653.78 | 2751.2 | 1.8923 | 6.7911 |
| 160 | 6.181 | 0.001 102 2 | 0.3068 | 675.47 | 2756.7 | 1.9425 | 6.7475 |
| 165 | 7.008 | 0.001 108 2 | 0.2724 | 697.25 | 2762.0 | 1.9923 | 6.7048 |
| 170 | 7.920 | 0.001 114 5 | 0.2426 | 719.12 | 2767.1 | 2.0416 | 6.6630 |
| 175 | 8.924 | 0.001 120 9 | 0.2165 | 741.07 | 2771.8 | 2.0906 | 6.6221 |
| 180 | 10.027 | 0.001 127 5 | 0.1938 | 763.12 | 2776.3 | 2.1393 | 6.5819 |
| 185 | 11.233 | 0.001 134 4 | 0.1739 | 785.26 | 2780.4 | 2.1876 | 6.5424 |
| 190 | 12.551 | 0.001 141 5 | 0.1563 | 807.52 | 2784.3 | 2.2356 | 6.5036 |
| 195 | 13.987 | 0.001 148 9 | 0.1408 | 829.88 | 2787.8 | 2.2830 | 6.4654 |
| 200 | 15.549 | 0.001 156 5 | 0.1272 | 852.37 | 2790.9 | 2.3307 | 6.4278 |
| 205 | 17.243 | 0.001 164 4 | 0.1150 | 874.99 | 2793.8 | 2.3778 | 6.3906 |
| 210 | 19.077 | 0.001 172 6 | 0.1042 | 897.74 | 2796.2 | 2.4247 | 6.3539 |
| 215 | 21.060 | 0.001 181 1 | 0.094 63 | 920.63 | 2798.3 | 2.4713 | 6.3176 |
| 220 | 23.198 | 0.001 190 0 | 0.086 04 | 943.67 | 2799.9 | 2.5178 | 6.2817 |
| 225 | 25.501 | 0.001 199 2 | 0.078 35 | 966.89 | 2801.2 | 2.5641 | 6.2461 |
| 230 | 27.976 | 0.001 208 7 | 0.071 45 | 990.26 | 2802.0 | 2.6102 | 6.2107 |
| 235 | 30.632 | 0.001 218 7 | 0.065 25 | 1013.80 | 2802.3 | 2.6562 | 6.1756 |
| 240 | 33.478 | 0.001 229 1 | 0.059 65 | 1037.60 | 2802.2 | 2.7020 | 6.1406 |
| 245 | 36.523 | 0.001 239 9 | 0.054 61 | 1061.60 | 2801.6 | 2.7478 | 6.1057 |
| 250 | 39.776 | 0.001 251 3 | 0.050 04 | 1085.80 | 2800.4 | 2.7935 | 6.0708 |
| 255 | 43.246 | 0.001 263 2 | 0.045 90 | 1110.20 | 2798.7 | 2.8392 | 6.0359 |
| 260 | 46.943 | 0.001 275 6 | 0.042 13 | 1134.90 | 2796.4 | 2.8848 | 6.0010 |
| 270 | 55.058 | 0.001 302 5 | 0.035 59 | 1185.20 | 2789.9 | 2.9763 | 5.9304 |
| 280 | 64.202 | 0.001 332 4 | 0.030 13 | 1236.80 | 2780.4 | 3.0683 | 5.8586 |
| 290 | 74.461 | 0.001 365 9 | 0.025 54 | 1290.00 | 2767.6 | 3.1611 | 5.7848 |
| 300 | 85.927 | 0.001 404 1 | 0.021 65 | 1345.00 | 2751.0 | 3.2552 | 5.7081 |
| 310 | 98.700 | 0.001 448 0 | 0.018 33 | 1402.40 | 2730.0 | 3.3512 | 5.6278 |
| 320 | 112.890 | 0.001 499 5 | 0.015 48 | 1462.60 | 2703.7 | 3.4500 | 5.5423 |
| 330 | 128.630 | 0.001 561 5 | 0.012 99 | 1526.50 | 2670.2 | 3.5528 | 5.4490 |
| 340 | 146.050 | 0.001 638 7 | 0.010 78 | 1595.50 | 2626.2 | 3.6616 | 5.3427 |
| 350 | 165.350 | 0.001 741 1 | 0.008 799 | 1671.90 | 2567.7 | 3.7800 | 5.2177 |
| 360 | 186.750 | 0.001895 9 | 0.006 940 | 1764.20 | 2485.4 | 3.9210 | 5.0600 |
| 370 | 210.540 | 0.002213 6 | 0.004 973 | 1890.20 | 2342.8 | 4.1108 | 4.8114 |
| 374.15 | 221.200 | 0.003 17 | 0.003 17 | 2107.40 | 2107.4 | 4.4429 | 4.4429 |

* Reprinted from Chandrasekaran, K.D. and Venkateswarlu, D., *SI Units in Chemical Engineering and Technology,* Chemical Engineering Education Development Centre, IIT-Madras (1974), with permission from Chairman, Centre for Continuing Education, IIT-Madras.

# Appendix – 3

*Superheated steam**

$v$ (m$^3$ / kg); $h$ (kJ/kg; $s$ (kJ/kg K)

| Pressure (bar) (saturation temperature) | | Temperature (°C) | | | | | | | |
|---|---|---|---|---|---|---|---|---|---|
| | | 100 | 200 | 300 | 400 | 500 | 600 | 700 | 800 |
| 0.2 | $v$ | 8.585 | 10.507 | 13.219 | 15.529 | 17.838 | 20.146 | 22.455 | 24.762 |
| (60.09°C) | $h$ | 2686.3 | 2879.2 | 3076.4 | 3279.4 | 3489.0 | 3705.4 | 3928.7 | 4158.7 |
| | $s$ | 8.1261 | 8.5839 | 8.9618 | 9.2882 | 9.5784 | 9.8416 | 10.0836 | 10.3085 |
| 0.4 | $v$ | 4.2792 | 5.4478 | 6.6065 | 7.7625 | 8.9176 | 10.072 | 11.227 | 12.381 |
| (75.89°C) | $h$ | 2683.8 | 2878.2 | 3075.9 | 3279.1 | 3488.8 | 3705.3 | 3928.6 | 4158.6 |
| | $s$ | 7.8009 | 8.2625 | 8.6413 | 8.9680 | 9.2583 | 9.5216 | 9.7636 | 9.9885 |
| 0.6 | $v$ | 2.8440 | 3.6280 | 4.4020 | 5.1740 | 5.9440 | 6.7140 | 7.4840 | 8.2540 |
| (85.95°C) | $h$ | 2681.3 | 2877.3 | 3075.4 | 3278.8 | 3488.6 | 3705.1 | 3928.5 | 4158.5 |
| | $s$ | 7.6085 | 8.0738 | 8.4536 | 8.7806 | 9.0710 | 9.3343 | 9.5764 | 9.8013 |
| 0.8 | $v$ | 2.1260 | 2.7180 | 3.3000 | 3.8790 | 4.4570 | 5.0350 | 5.6130 | 6.1900 |
| (93.51°C) | $h$ | 2678.8 | 2876.3 | 3075.0 | 3278.5 | 3488.4 | 3705.0 | 3928.4 | 4158.4 |
| | $s$ | 7.4703 | 7.9395 | 8.3202 | 8.6475 | 8.9380 | 9.2014 | 9.4436 | 9.6685 |
| 1.0 | $v$ | 1.6960 | 2.1720 | 2.6390 | 3.1020 | 3.5650 | 4.0280 | 4.4900 | 4.9520 |
| (99.63°C) | $h$ | 2676.2 | 2875.4 | 3074.5 | 3278.2 | 3448.1 | 3704.8 | 3928.2 | 4158.3 |
| | $s$ | 7.3618 | 7.8449 | 8.2166 | 8.5442 | 8.8348 | 9.0982 | 9.3405 | 9.5654 |
| 2 | $v$ | | 1.0804 | 1.3162 | 1.5492 | 1.7812 | 2.0129 | 2.2420 | 2.4754 |
| (120.23°C) | $h$ | | 2870.5 | 3072.1 | 3276.7 | 3487.0 | 3704.0 | 3927.6 | 4157.8 |
| | $s$ | | 7.5072 | 7.8937 | 8.2226 | 8.5139 | 8.7776 | 9.0201 | 9.2452 |
| 3 | $v$ | | 0.7164 | 0.8753 | 1.0314 | 1.1865 | 1.3412 | 1.4957 | 1.6499 |
| (133.54°C) | $h$ | | 2865.5 | 3069.7 | 3275.2 | 3486.0 | 3703.2 | 3927.0 | 4157.3 |
| | $s$ | | 7.3119 | 7.7034 | 8.0338 | 8.3257 | 8.5868 | 8.8325 | 9.0577 |
| 4 | $v$ | | 0.5343 | 0.6549 | 0.7725 | 0.8892 | 1.0054 | 1.1214 | 1.2372 |
| (143.62°C) | $h$ | | 2860.4 | 3067.2 | 3273.6 | 3484.9 | 3702.3 | 3926.4 | 4156.9 |
| | $s$ | | 7.1708 | 7.5675 | 7.8994 | 8.1919 | 8.4563 | 8.6992 | 8.9246 |
| 5 | $v$ | | 0.4250 | 0.5226 | 0.6172 | 0.7108 | 0.8039 | 0.8968 | 0.9896 |
| (151.84°C) | $h$ | | 2855.1 | 3064.8 | 3272.1 | 3483.8 | 3701.5 | 3925.8 | 4156.4 |
| | $s$ | | 7.0592 | 7.4614 | 7.7948 | 8.0879 | 8.3526 | 8.5957 | 8.8213 |
| 6 | $v$ | | 0.3520 | 0.4344 | 0.5136 | 0.5918 | 0.6696 | 0.7471 | 0.8245 |
| (158.84°C) | $h$ | | 2849.7 | 3062.3 | 3270.6 | 3482.7 | 3700.7 | 3925.1 | 4155.9 |
| | $s$ | | 6.9662 | 7.3740 | 7.7090 | 8.0027 | 8.2678 | 8.5111 | 8.7368 |

| Pressure (bar) (saturation temperature) | | Temperature (°C) | | | | | | |
|---|---|---|---|---|---|---|---|---|
| | | 200 | 300 | 400 | 500 | 600 | 700 | 800 |
| 7 | *v* | 0.2999 | 0.3714 | 0.4396 | 0.5069 | 0.5737 | 0.6402 | 0.7066 |
| (164.96°C) | *h* | 2844.2 | 3059.8 | 3269.0 | 3481.6 | 3699.9 | 3924.5 | 4155.5 |
| | *s* | 6.8859 | 7.2997 | 7.6362 | 7.9305 | 8.1959 | 8.4395 | 8.6653 |
| 8 | *v* | 0.2608 | 0.3241 | 0.3842 | 0.4432 | 0.5017 | 0.5600 | 0.6181 |
| (170.41"C) | *h* | 2838.6 | 3057.3 | 3267.5 | 3480.5 | 3699.1 | 3923.9 | 4115.0 |
| | *s* | 6.8148 | 7.2348 | 7.5729 | 7.8678 | 8.1336 | 8.3773 | 8.6033 |
| 9 | *v* | 0.2303 | 0.2874 | 0.3410 | 0.3936 | 0.4458 | 0.4976 | 0.5493 |
| (175.36°C) | *h* | 2832.7 | 3054.7 | 3266.0 | 3479.4 | 3698.2 | 3923.3 | 4154.8 |
| | s | 6.7508 | 7.1771 | 7.5169 | 7.8124 | 8.0785 | 8.3225 | 8.5486 |
| 10 | *v* | 0.2059 | 0.2580 | 0.3065 | 0.3540 | 0.4010 | 0.4477 | 0.4943 |
| (179.88°C) | *h* | 2826.8 | 3052.1 | 3264.4 | 3478.3 | 3697.4 | 3922.7 | 4154.1 |
| | *s* | 6.6922 | 7.1251 | 7.4665 | 7.7627 | 8.0292 | 8.2734 | 8.4997 |
| 11 | *v* | 0.1859 | 0.2339 | 0.2782 | 0.3215 | 0.3643 | 0.4069 | 0.4493 |
| (184.07°C) | *h* | 2820.7 | 3049.6 | 3262.9 | 3477.2 | 3696.6 | 3922.0 | 4153.6 |
| | *s* | 6.6379 | 7.0778 | 7.4209 | 7.7177 | 7.9845 | 8.2289 | 8.4553 |
| 12 | *v* | 0.1692 | 0.2139 | 0.2547 | 0.2945 | 0.3338 | 0.3729 | 0.4118 |
| (187.96°C) | *h* | 2814.4 | 3046.9 | 3261.3 | 3476.1 | 3695.8 | 3921.4 | 4153.1 |
| | *s* | 6.5872 | 7.0342 | 7.3790 | 7.6765 | 7.9436 | 8.1882 | 8.4148 |
| 13 | *v* | 0.1551 | 0.1969 | 0.2348 | 0.2716 | 0.3080 | 0.3441 | 0.3800 |
| (191.61°C) | *h* | 2808.0 | 3044.3 | 3259.7 | 3475.0 | 3695.0 | 3920.8 | 4152.7 |
| | *s* | 6.5394 | 6.9938 | 7.3404 | 7.6385 | 7.9060 | 8.1508 | 8.3773 |
| 14 | *v* | 0.1429 | 0.1823 | 0.2177 | 0.2520 | 0.2859 | 0.3194 | 0.3528 |
| (195.04°C) | *h* | 2801.4 | 3041.6 | 3258.2 | 3473.9 | 3694.1 | 3920.2 | 4152.2 |
| | *s* | 6.4941 | 6.9561 | 7.3045 | 7.6032 | 7.8710 | 8.1161 | 8.3430 |
| 15 | *v* | 0.1324 | 0.1697 | 0.2029 | 0.2350 | 0.2667 | 0.2980 | 0.3292 |
| (198.29°C) | *h* | 2794.7 | 3038.9 | 3256.6 | 3472.8 | 3693.3 | 3919.6 | 4151.7 |
| | *s* | 6.4508 | 6.9207 | 7.2709 | 7.5703 | 7.8385 | 8.0838 | 8.3108 |
| 16 | *v* | | 0.1587 | 0.1900 | 0.2202 | 0.2499 | 0.2793 | 0.3086 |
| (201.37°C) | *h* | | 3036.2 | 3255.0 | 3471.7 | 3692.5 | 3918.9 | 4151.3 |
| | *s* | | 6.8873 | 7.2394 | 7.5395 | 7.8080 | 8.0535 | 8.2807 |
| 17 | *v* | | 0.1489 | 0.1785 | 0.2070 | 0.2351 | 0.2628 | 0.2904 |
| (204.31"C) | *h* | | 3033.5 | 3253.5 | 3470.6 | 3691.7 | 3918.3 | 4150.8 |
| | *s* | | 6.8557 | 7.2098 | 7.5105 | 7.7793 | 8.0250 | 8.2523 |

| Pressure (bar) (saturation temperature) | | Temperature (°C) 300 | 400 | 500 | 600 | 700 | 800 |
|---|---|---|---|---|---|---|---|
| 18 | v | 0.1402 | 0.1684 | 0.1954 | 0.2219 | 0.2481 | 0.2742 |
| (207.11°C) | h | 3030.7 | 3251.9 | 3469.5 | 3690.9 | 3917.7 | 4150.3 |
| | s | 6.8257 | 7.1816 | 7.4830 | 7.7522 | 7.9981 | 8.2256 |
| 19 | v | 0.1325 | 0.1593 | 0.1849 | 0.2101 | 0.2350 | 0.2597 |
| (209.80°C) | h | 3027.9 | 3250.3 | 3468.4 | 3690.0 | 3917.1 | 4149.9 |
| | s | 6.7970 | 7.1550 | 7.4570 | 7.7265 | 7.9727 | 8.2003 |
| 20 | v | 0.1255 | 0.1511 | 0.1756 | 0.1995 | 0.2232 | 0.2467 |
| (212.37°C) | h | 3025.0 | 3248.7 | 3467.3 | 3689.2 | 3916.5 | 4149.4 |
| | s | 6.7696 | 7.1296 | 7.4323 | 7.7022 | 7.9485 | 8.1763 |
| 22 | v | 0.113 43 | 0.137 00 | 0.159 34 | 0.181 19 | 0.202 76 | 0.224 15 |
| (217.24°C) | h | 3019.3 | 3245.5 | 3465.1 | 3687.6 | 3915.2 | 4148.4 |
| | s | 6.7179 | 7.0821 | 7.3862 | 7.6568 | 7.9035 | 8.1316 |
| 24 | v | 0.103 36 | 0.125 22 | 0.145 82 | 0.165 92 | 0.185 75 | 0.205 39 |
| (221.78°C) | h | 3013.4 | 3242.3 | 3462.9 | 3685.9 | 3914.0 | 4147.5 |
| | s | 6.6699 | 7.0384 | 7.3439 | 7.6152 | 7.8624 | 8.0908 |
| 26 | v | 0.094 83 | 0.115 26 | 0.134 38 | 0.153 01 | 0.171 35 | 0.189 52 |
| (226.04°C) | h | 3007.4 | 3239.0 | 3460.6 | 3684.3 | 3912.7 | 4146.6 |
| | s | 6.6249 | 6.9979 | 7.3048 | 7.5768 | 7.8245 | 8.0531 |
| 28 | v | 0.087 51 | 0.106 71 | 0.124 58 | 0.141 94 | 0.159 02 | 0.175 91 |
| (230.05°C) | h | 3001.3 | 3235.8 | 3458.4 | 3682.6 | 3911.5 | 4145.6 |
| | s | 6.5824 | 6.9601 | 7.2685 | 7.5412 | 7.7893 | 8.0183 |
| 30 | v | 0.081 16 | 0.099 31 | 0.116 08 | 0.132 34 | 0.148 32 | 0.164 12 |
| (233.84°C) | h | 2995.1 | 3232.5 | 3456.2 | 3681.0 | 3910.3 | 4144.7 |
| | s | 6.5422 | 6.9246 | 7.2345 | 7.5079 | 7.7564 | 7.9857 |
| 32 | v | 0.075 59 | 0.092 83 | 0.108 65 | 0.123 95 | 0.138 97 | 0.153 81 |
| (237.45°C) | h | 2988.7 | 3229.2 | 3454.0 | 3679.3 | 3909.0 | 4143.8 |
| | s | 6.5037 | 6.8912 | 7.2026 | 7.4767 | 7.7257 | 7.9552 |
| 34 | v | 0.070 68 | 0.087 11 | 0.102 09 | 0.11657 | 0.130 71 | 0.144 70 |
| (240.88°C) | h | 2982.2 | 3225.9 | 3451.7 | 3677.7 | 3907.8 | 4142.8 |
| | s | 6.4669 | 6.8595 | 7.1724 | 7.4473 | 7.6967 | 7.9266 |
| 36 | v | 0.066 30 | 0.082 02 | 0.096 26 | 0.109 96 | 0.123 37 | 0.136 61 |
| (244.16°C) | h | 2975.6 | 3222.5 | 3449.5 | 3676.1 | 3906.5 | 4141.9 |
| | s | 6.4315 | 6.8294 | 7.1439 | 7.4195 | 7.6693 | 7.8995 |

| Pressure (bar) (saturation temperature) | | Temperature (°C) | | | | | |
|---|---|---|---|---|---|---|---|
| | | 300 | 400 | 500 | 600 | 700 | 800 |
| 38 | $v$ | 0.062 37 | 0.077 47 | 0.091 04 | 0.104 06 | 0.116 81 | 0.129 37 |
| (247.31°C) | $h$ | 2968.9 | 3219.1 | 3447.2 | 3674.4 | 3905.3 | 4141.0 |
| | $s$ | 6.3973 | 6.8007 | 7.1168 | 7.3931 | 7.6434 | 7.8739 |
| 40 | $v$ | 0.058 83 | 0.073 38 | 0.086 34 | 0.098 76 | 0.110 90 | 0.122 85 |
| (250.33°C) | $h$ | 2962.0 | 3215.7 | 3445.0 | 3672.8 | 3904.1 | 4140.0 |
| | $s$ | 6.3642 | 6.7733 | 7.0909 | 7.3680 | 7.6187 | 7.8495 |
| 45 | $v$ | 0.051 34 | 0.064 72 | 0.076 43 | 0.087 57 | 0.098 43 | 0.109 10 |
| (257.41°C) | $h$ | 2944.2 | 3207.1 | 3439.3 | 3668.6 | 3901 .0 | 4137.7 |
| | $s$ | 6.2852 | 6.7093 | 7.0311 | 7.3100 | 7.5619 | 7.7934 |
| 50 | $v$ | 0.045 30 | 0.057 79 | 0.068 49 | 0.078 62 | 0.088 45 | 0.098 09 |
| (263.91°C) | $h$ | 2925.5 | 3198 3 | 3433.7 | 3664.5 | 3897.9 | 4135.3 |
| | $s$ | 6.2105 | 6.6508 | 6.9770 | 7.2578 | 7.5108 | 7.7431 |
| 60 | $v$ | 0.036 14 | 0.047 38 | 0.056 59 | 0.065 18 | 0.073 48 | 0.081 59 |
| (275.55°C) | $h$ | 2885.0 | 3180.1 | 3422.2 | 3656.2 | 3891.7 | 4130.7 |
| | $s$ | 6.0692 | 6.5462 | 6.8818 | 7.1664 | 7.4217 | 7.6554 |
| 70 | $v$ | 0.029 46 | 0.039 92 | 0.048 09 | 0.055 59 | 0.062 79 | 0.069 80 |
| (285.79°C) | $h$ | 2839.4 | 3161.2 | 3410.6 | 3647.9 | 3885.4 | 4126.0 |
| | $s$ | 5.9327 | 6.4536 | 6.7993 | 7.0880 | 7.3456 | 7.5808 |
| 80 | $v$ | 0.024 26 | 0.034 31 | 0.041 70 | 0.048 39 | 0.054 77 | 0.060 96 |
| (294.97°C) | $h$ | 2786.8 | 3141.6 | 3398.8 | 3639.5 | 3879.2 | 4121.3 |
| | $s$ | 5.7942 | 6.3694 | 6.7262 | 7.0191 | 7.2790 | 7.5158 |
| 90 | $v$ | | 0.030 73 | 0.037 64 | 0.042 80 | 0.048 53 | 0.054 08 |
| (303.31°C) | $h$ | | 3125.3 | 3389.2 | 3631.1 | 3873.0 | 4116.7 |
| | $s$ | | 6.3067 | 6.6728 | 6.9574 | 7.2196 | 7.4579 |
| 100 | $v$ | | 0.026 41 | 0.032 76 | 0.038 32 | 0.043 55 | 0.048 58 |
| (310.96°C) | $h$ | | 3099.9 | 3374.6 | 3622.7 | 3866.8 | 4112.0 |
| | $s$ | | 6.2182 | 6.5994 | 6.9013 | 7.1660 | 7.4058 |
| 110 | $v$ | | 0.023 51 | 0.029 50 | 0.034 66 | 0.039 47 | 0.048 58 |
| (318.05°C) | $h$ | | 3077.8 | 3362.2 | 3614.2 | 3860.5 | 4107.3 |
| | $s$ | | 6.1483 | 6.5432 | 6.8499 | 7.1170 | 7.3584 |
| 120 | $v$ | | 0.021 08 | 0.026 73 | 0.031 60 | 0.036 07 | 0.040 33 |
| (324.65°C) | $h$ | | 3054.8 | 3349.6 | 3605.7 | 3854.3 | 4102.7 |
| | $s$ | | 6.0810 | 6.4906 | 6.8022 | 7.0718 | 7.3147 |
| 130 | $v$ | | 0.019 02 | 0.024 40 | 0.029 02 | 0.033 19 | 0.037 16 |
| (330.83°C) | $h$ | | 3030.7 | 3336.8 | 3597.1 | 3848.0 | 4098.0 |
| | $s$ | | 6.0155 | 6.4409 | 6.7577 | 7.0298 | 7.2743 |

| Pressure (bar) (saturation temperature) | | Temperature (°C) | | | | |
|---|---|---|---|---|---|---|
| | | 400 | 500 | 600 | 700 | 800 |
| 140 | $v$ | 0.017 23 | 0.022 51 | 0.026 90 | 0.030 72 | 0.034 44 |
| (336.64°C) | $h$ | 3005.6 | 3323.8 | 3588.5 | 3841.7 | 4093.3 |
| | $s$ | 5.9513 | 6.3937 | 6.7159 | 6.9906 | 7.2367 |
| 150 | $v$ | 0.015 66 | 0.020 80 | 0.024 88 | 0.028 59 | 0.032 09 |
| (342.13°C) | $h$ | 2979.1 | 3310.6 | 3579.8 | 3835.4 | 4088.6 |
| | $s$ | 5.8876 | 6.3487 | 6.6764 | 6.9536 | 7.2013 |
| 175 | $v$ | 0.012 46 | 0.017 36 | 0.021 72 | 0.025 07 | 0.028 21 |
| (354.64°C) | $h$ | 2906.3 | 3276.5 | 3562.2 | 3822.8 | 4079.3 |
| | $s$ | 5.7274 | 6.2432 | 6.6031 | 6.8857 | 7.1366 |
| 200 | $v$ | 0.009 947 | 0.014 77 | 0.018 16 | 0.021 11 | 0.023 85 |
| (365.70°C) | $h$ | 2820.5 | 3241.1 | 3535.5 | 3803.8 | 4065.3 |
| | $s$ | 5.5585 | 6.1456 | 6.5043 | 6.7953 | 7.0511 |
| 220 | $v$ | 0.008 251 | 0.013 12 | 0.016 33 | 0.019 07 | 0.021 60 |
| (373.69°C) | $h$ | 2738.8 | 3211.7 | 3517.4 | 3791.1 | 4055.9 |
| | $s$ | 5.4102 | 6.0716 | 6.4441 | 6.7410 | 7.0001 |
| 250 | $v$ | 0.006 014 | 0.011 13 | 0.014 13 | 0.016 63 | 0.018 91 |
| | $h$ | 2582.0 | 3165.9 | 3489.9 | 3771.9 | 4041.9 |
| | $s$ | 5.1455 | 5.9655 | 6.3604 | 6.6664 | 6.9306 |
| 300 | $v$ | 0.002 831 | 0.008 681 | 0.011 44 | 0.013 65 | 0.015 62 |
| | $h$ | 2161.8 | 3085.0 | 3443.0 | 3739.7 | 4018.5 |
| | $s$ | 4.4896 | 5.7975 | 6.2340 | 6.5560 | 6.8288 |
| 400 | $v$ | 0.001 909 | 0.005 616 | 0.008 088 | 0.009 930 | 0.011 52 |
| | $h$ | 1934.1 | 2906.8 | 3346.4 | 3674.8 | 3971.7 |
| | $s$ | 4.1190 | 5.4762 | 6.0135 | 6.3701 | 6.6606 |
| 500 | $v$ | 0.001 729 | 0.003 882 | 0.006 111 | 0.007 720 | 0.009 076 |
| | $h$ | 1877.7 | 2723.0 | 3248.3 | 3610.2 | 3925.3 |
| | $s$ | 4.0083 | 5.1782 | 5.8207 | 6.2138 | 6.5222 |
| 1000 | $v$ | 0.001 446 | 0.001 893 | 0.002 668 | 0.003 536 | 0.004 341 |
| | $h$ | 1797.6 | 2316.1 | 2857.5 | 3324.4 | 3714.3 |
| | $s$ | 3.7738 | 4.4913 | 5.1505 | 5.6579 | 6.0397 |

* Adapted from Chandrasekaran, K.D. and Venkateswarlu, D., *SI Units in Chemical Engineering and Technology,* Chemical Engineering Education Development Centre, IIT-Madras (1974), with permission from Chairman, Centre for Continuing Education, IIT-Madras.

# Appendix – 4

*Critical constants of some selected substances* *

| Compound | | Critical temp $T_c$ (K) | Critical pressure $P_c$ (bar) | Critical volume $v_c$ ($m^3$/mol) | Critical compressibility $z_c$ |
|---|---|---|---|---|---|
| **Inorganic compounds** | | | | | |
| Ammonia | $NH_3$ | 405.50 | 112.77 | 0.0725 | 0.243 |
| Argon | A | 151.20 | 48.64 | 0.075 | 0.290 |
| Bromine | $Br_2$ | 584.00 | 103.35 | 0.114 | 0.306 |
| Carbon dioxide | $CO_2$ | 304.20 | 73.87 | 0 094 | 0.275 |
| Carbon disulphide | $CS_2$ | 552.00 | 79.03 | 0.170 | 0.293 |
| Carbon monoxide | CO | 133.00 | 34.96 | 0.093 | 0.294 |
| Chlorine | $Cl_2$ | 417.00 | 77.11 | 0.124 | 0.276 |
| Helium | He | 5.26 | 22.90 | 0.058 | 0.304 |
| Hydrogen | $H_2$ | 33.30 | 129.70 | 0.065 | 0.304 |
| Hydrogen chloride | HCl | 324.60 | 82.58 | 0.087 | 0.266 |
| Hydrogen sulphide | $H_2S$ | 373.60 | 90.08 | 0.098 | 0.284 |
| Neon | Ne | 44.50 | 27.26 | 0.0417 | 0.307 |
| Nitrogen | $N_2$ | 126.20 | 33.94 | 0.090 | 0.291 |
| Nitric oxide | NO | 179.20 | 65.86 | 0.058 | 0.256 |
| Nitrous oxide | $N_2O$ | 309.70 | 72.65 | 0.0963 | 0.272 |
| Oxygen | $O_2$ | 154.40 | 50.36 | 0.074 | 0.290 |
| Sulphur dioxide | $SO_2$ | 430.70 | 78.83 | 0.122 | 0.269 |
| Sulphur trioxide | $SO_3$ | 491.40 | 84.91 | 0.126 | 0.262 |
| Water | $H_2O$ | 647.30 | 221.20 | 0.056 | 0.230 |
| **Organic compounds** | | | | | |
| Acetic acid | $CH_3COOH$ | 594.80 | 57.86 | 0.171 | 0.200 |
| Acetone | $CH_3COCH_3$ | 508.70 | 47.22 | 0.213 | 0.238 |
| Acetylene | $C_2H_2$ | 309.50 | 62.42 | 0.113 | 0.274 |
| Aniline | $C_6H_7N$ | 698.80 | 52.99 | 0.274 | 0.250 |
| Benzene | $C_6H_6$ | 562.10 | 49.24 | 0.260 | 0.274 |
| Butane (*iso*) | $C_4H_{10}$ | 408.00 | 36.48 | 0.263 | 0.262 |
| Butane (*n*) | $C_4H_{10}$ | 425.20 | 37.97 | 0.255 | 0.274 |
| Butene (1) | $C_4H_8$ | 419.60 | 40.23 | 0.240 | 0.277 |
| Carbon tetrachloride | $CCl_4$ | 556.40 | 45.60 | 0.276 | 0.272 |
| Chlorobenzene | $C_6H_5Cl$ | 632.40 | 45.19 | 0.308 | 0.265 |

| Compound | | Critical temp $T_c$ (K) | Critical pressure $P_c$ (bar) | Critical volume $v_c$ (m³/mol) | Critical compressi–bility $z_c$ |
|---|---|---|---|---|---|
| Chlorodifluor omethane | $CHClF_2$ | 369.60 | 49.14 | 0.165 | 0.264 |
| Chloroform | $CHCl_3$ | 536.60 | 54.72 | 0.240 | 0.294 |
| Cyclopentane | $C_5H_{10}$ | 511.76 | 21.08 | 0.260 | 0.276 |
| Diethyl ether | $C_2H_5OC_2H_5$ | 467.00 | 39.62 | 0.281 | 0.261 |
| Ethane | $C_2H_6$ | 305.43 | 48.84 | 0.148 | 0.285 |
| Ethylamine | $C_2H_5NH_2$ | 456.40 | 56.24 | 0.181 | 0.268 |
| Ethyl acetate | $C_2H_5COOCH_3$ | 523.30 | 38.30 | 0.286 | 0.252 |
| Ethyl alcohol | $C_2H_5OH$ | 516.30 | 63.83 | 0.167 | 0.248 |
| Ethylene | $C_2H_4$ | 283.10 | 51.17 | 0.124 | 0.270 |
| Ethylene oxide | $C_2H_4O$ | 468.00 | 71.94 | 0.138 | 0.255 |
| Freon 11 | $CCl_3F$ | 471.20 | 43.77 | 0.248 | 0.277 |
| Freon 12 | $CCl_2F_2$ | 384.70 | 40.12 | 0.218 | 0.273 |
| Freon 13 | $CClF_3$ | 302.00 | 38.71 | 0.180 | 0.278 |
| Freon 21 | $CHCl_2F$ | 451.70 | 51.68 | 0.197 | 0.271 |
| Freon 113 | $C_2Cl_3F_3$ | 487.30 | 34.15 | 0.325 | 0.274 |
| Heptane (*n*) | $C_7H_{16}$ | 540.16 | 27.37 | 0.426 | 0.260 |
| Hexane (*n*) | $C_6H_{14}$ | 507.90 | 30.32 | 0.368 | 0.264 |
| Methane | $CH_4$ | 190.70 | 46.41 | 0.099 | 0.290 |
| Methyl acetate | $CH_3COOCH_3$ | 506.90 | 46.91 | 0.228 | 0.254 |
| Methyl chloride | $CH_3Cl$ | 416.30 | 66.77 | 0.143 | 0.276 |
| Methyl alcohol | $CH_3OH$ | 513.20 | 75.54 | 0.118 | 0.220 |
| Octane (*n*) | $C_8H_{18}$ | 569.40 | 24.97 | 0.486 | 0.256 |
| Pentane (*n*) | $C_5H_{12}$ | 469.80 | 33.75 | 0.311 | 0.269 |
| Propane | $C_3H_8$ | 369.90 | 42.57 | 0.200 | 0.277 |
| Propyl alcohol (*n*) | $C_3H_7OH$ | 537.30 | 50.87 | 0.220 | 0.251 |
| Propyl alcohol (*iso*) | $C_3H_7OH$ | 508.80 | 53.70 | 0.219 | 0.278 |
| Propylene | $C_3H_6$ | 365.10 | 46.00 | 0.181 | 0.274 |
| Toluene | $C_6H_5CH_3$ | 594.00 | 42.15 | 0.320 | 0.273 |

* Adapted from Chandrasekaran, K.D. and Venkteswarlu, D., *SI Units in Chemical Engineering and Technology*, Chemical Engineering Education Development Centre, IIT-Madras (1974), with permission from Chairman, Centre for Continuing Education, IIT-Madras.

# Appendix – 5

*Generalized compressibility chart*

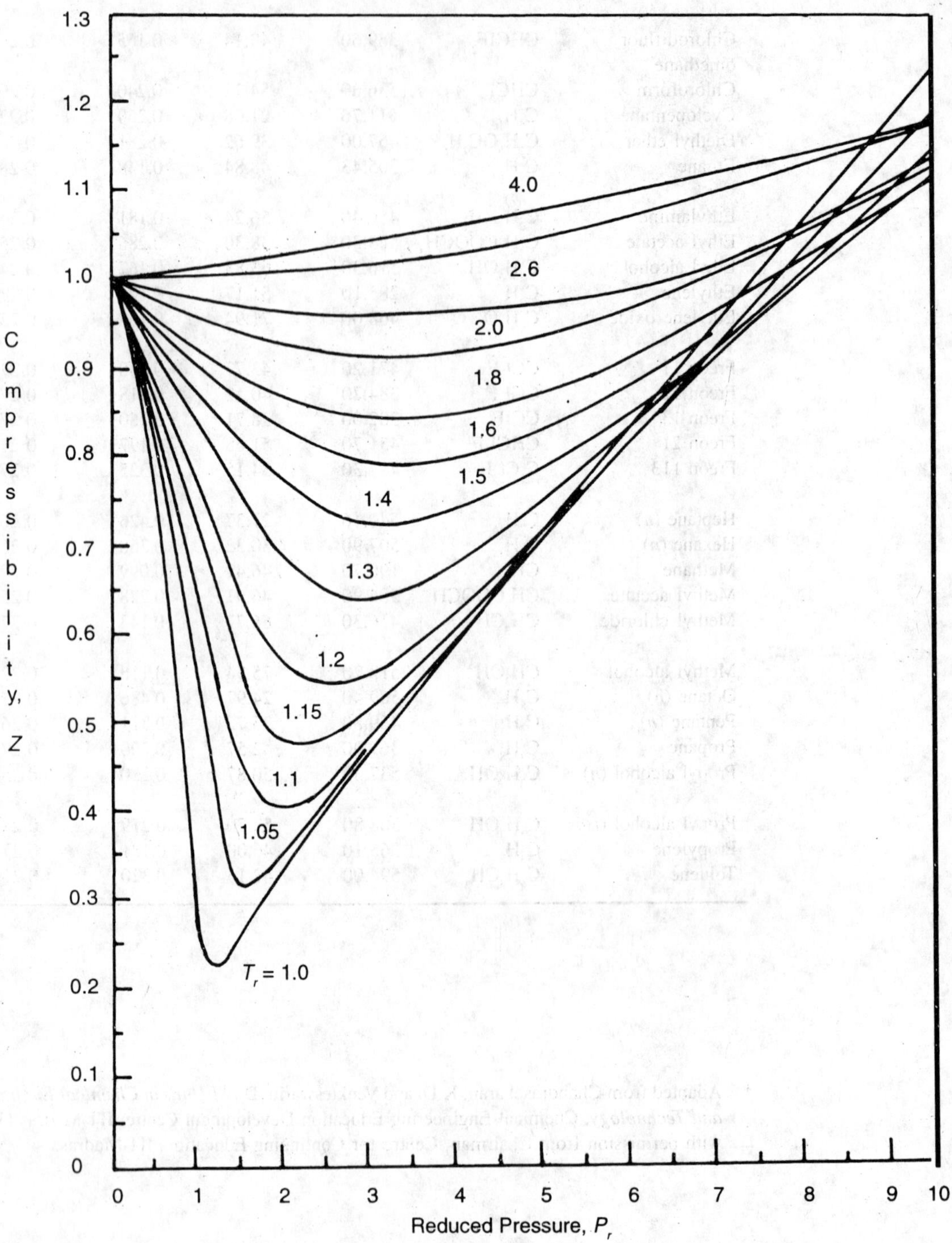

# Appendix – 6

*Saturated Freon–12: Temperature Table**

| $T$ (°C) | $P^s$ kPa | $v_f$ | $v_{fg}$ (m³/ kg) | $v_g$ | $h_f$ | $h_{fg}$ (kJ / kg) | $h_g$ | $s_f$ | $s_{fg}$ (kJ / kg K) | $s_g$ |
|---|---|---|---|---|---|---|---|---|---|---|
| - 50.0 | 39.115 | 0.000 648 | 0.383 004 | 0.383 652 | - 8.8202 | 173.8424 | 165.0222 | -0.0386 | 0.7790 | 0.7404 |
| - 48.0 | 43.355 | 0.000.650 | 0.347 990 | 0.348 640 | - 7.0638 | 173.0237 | 165.9599 | -0.0308 | 0.7685 | 0.7377 |
| - 46.0 | 47.955 | 0.000652 | 0.316 780 | 0.317 432 | - 5.3037 | 172.2005 | 166.8968 | - 0.0230 | 0.7581 | 0.7351 |
| - 44.0 | 52.935 | 0.000655 | 0.288 903 | 0.289 558 | - 3.5398 | 171.3724 | 167.8326 | - 0.0153 | 0.7479 | 0.7326 |
| - 42.0 | 58.317 | 0.000657 | 0.263 951 | 0.264 608 | - 1.7718 | 170.5391 | 168.7672 | - 0.0076 | 0.7378 | 0.7302 |
| - 40.0 | 64.123 | 0.000 659 | 0.241 573 | 0.242 233 | 0.0000 | 169.7004 | 169.7004 | 0.0000 | 0.7279 | 0.7279 |
| - 38.0 | 70.376 | 0.000 662 | 0.221 463 | 0.222 125 | 1.7759 | 168.8559 | 170.6318 | 0.0076 | 0.7181 | 0.7256 |
| - 36.0 | 77.100 | 0.000 664 | 0.203 352 | 0.204 016 | 3.5560 | 168.0053 | 171.5613 | 0.0151 | 0.7084 | 0.7235 |
| - 34.0 | 84.319 | 0.000 667 | 0.187 016 | 0.187 683 | 5.3402 | 167.1485 | 172.4887 | 0.0226 | 0.6989 | 0.7215 |
| - 32.0 | 92.057 | 0.000 669 | 0.172 251 | 0.172 920 | 7.1287 | 166.2850 | 173.4137 | 0.0300 | 0.6895 | 0.7195 |
| - 30.0 | 100.341 | 0.000 672 | 0.158 881 | 0.159 553 | 8.9215 | 165.4146 | 174.3361 | 0.0374 | 0.6803 | 0.7177 |
| - 28.0 | 109.193 | 0.000 675 | 0.146 758 | 0.147 433 | 10.7185 | 164.5372 | 175.2557 | 0.0447 | 0.6712 | 0.7159 |
| - 26.0 | 118.643 | 0.000 677 | 0.135 743 | 0.136 420 | 12.5201 | 163.6522 | 176.1723 | 0.0520 | 0.6622 | 0.7142 |
| - 24.0 | 128.715 | 0.000 680 | 0.125 720 | 0.126 400 | 14.3261 | 162.7595 | 177.0856 | 0.0592 | 0.6533 | 0.7125 |
| - 22.0 | 139.436 | 0.000 683 | 0.116 586 | 0.117 268 | 16.1366 | 161.8590 | 177.9955 | 0.0665 | 0.6445 | 0.7109 |
| - 20.0 | 150.836 | 0.000 685 | 0.108 246 | 0.108 931 | 17.9517 | 160.9500 | 178.9017 | 0.0736 | 0.6358 | 0.7094 |
| - 18.0 | 162.940 | 0.000 688 | 0.100 623 | 0.101 311 | 19.7713 | 160.0327 | 179.8040 | 0.0807 | 0.6272 | 0.7080 |
| - 16.0 | 175.778 | 0.000 691 | 0.093 644 | 0.094 335 | 21.5957 | 159.1065 | 180.7022 | 0.0878 | 0.6187 | 0.7066 |
| - 14.0 | 189.379 | 0.000 694 | 0.087 244 | 0.087 938 | 23.4249 | 158.1710 | 181.5959 | 0.0949 | 0.6103 | 0.7052 |
| - 12.0 | 203. 771 | 0.000 697 | 0.081 369 | 0.082 065 | 25.2589 | 157.2262 | 182.4851 | 0.1019 | 0.6021 | 0.7039 |
| - 10.0 | 218.985 | 0.000 700 | 0.075 967 | 0.076 667 | 27.0978 | 156.2717 | 183.3695 | 0.1089 | 0.5939 | 0.7027 |
| - 8.0 | 235.050 | 0.000 703 | 0.070 994 | 0.071 697 | 28.9417 | 155.3070 | 184.2487 | 0.1158 | 0.5857 | 0.7015 |
| - 6.0 | 251.998 | 0.000 706 | 0.066 409 | 0.067 115 | 30.7908 | 154.3320 | 185.1227 | 0.1227 | 0.5777 | 0.7004 |
| - 4.0 | 269.859 | 0.000 709 | 0.062 178 | 0.062 887 | 32.6450 | 153.3461 | 185.9911 | 0.1296 | 0.5697 | 0.6993 |
| - 2.0 | 288.661 | 0.000 712 | 0.058 268 | 00. 58 981 | 34.5045 | 152.3493 | 186.8538 | 0.1364 | 0.5619 | 0.6983 |
| 0.0 | 308.440 | 0.000 716 | 0.054 650 | 0.055 366 | 36.3695 | 151.3409 | 187.7104 | 0 1432 | 0.5541 | 0.6973 |
| 2.0 | 329.222 | 0.000 719 | 0.051 300 | 0.052 019 | 38.2398 | 149.2884 | 188.5608 | 0.1500 | 0.5463 | 0.6963 |
| 4.0 | 351.047 | 0.000 723 | 0.048 192 | 0. 048 914 | 40.1161 | 149.2884 | 189.4095 | 0.1567 | 0.5387 | 0.6954 |
| 6.0 | 373.944 | 0.000 726 | 0.045 307 | 0.046 033 | 41.9981 | 148.24.33 | 190.2414 | 0.1634 | 0.5311 | 0.6945 |
| 8.0 | 397.942 | 0.000 730 | 0.042 625 | 0.043 355 | 43.8861 | 147.1851 | 191.0712 | 0.1701 | 0.5235 | 0.6936 |
| 10.0 | 423.083 | 0.000 733 | 0.040 130 | 0.040 863 | 45.7805 | 146.1131 | 191.8935 | 0.1768 | 0.5160 | 0.6928 |
| 12.0 | 449.392 | 0.000 737 | 0.037 806 | 0.038 542 | 47.6811 | 145.0271 | 192.7082 | 0.1834 | 0.5086 | 0.6920 |
| 14.0 | 476.905 | 0.000 741 | 0.035 639 | 0.036 379 | 49.5883 | 143.9266 | 193.5149 | 0.1900 | 0.5012 | 0.6912 |
| 16.0 | 505.060 | 0.000 744 | 0.033 616 | 0.034 361 | 51.5024 | 142.8109 | 194.3132 | 0.1965 | 0.4939 | 0.6904 |
| 18.0 | 535.684 | 0.000 748 | 0.031 727 | 0.032 476 | 53.4234 | 141.6796 | 195.1030 | 0.2031 | 0.4866 | 0.6897 |
| 20.0 | 567.019 | 0.000 752 | 0.029 960 | 0.030 713 | 55.3518 | 140.5318 | 195.8836 | 0.2096 | 0.4794 | 0.6890 |
| 22.0 | 599.698 | 0.000 756 | 0.028 307 | 0.029 063 | 57.2878 | 139.3671 | 196.6550 | 0.2161 | 0.4722 | 0.6883 |
| 24.0 | 633.756 | 0.000 761 | 0.026 757 | 0.027 518 | 59.2318 | 138.1847 | 197.4165 | 0.2226 | 0.4650 | 0.6876 |
| 26.0 | 669.233 | 0.000 765 | 0 025 304 | 0.026 069 | 61.1841 | 136.9837 | 198.1679 | 0.2290 | 0.4579 | 0.6869 |

| $T$ (°C) | $P^s$ kPa | $v_f$ (m³/ kg) | $v_{fg}$ | $v_g$ | $h_f$ (kJ / kg) | $h_{fg}$ | $h_g$ | $s_f$ (kJ / kg K) | $s_{fg}$ | $s_g$ |
|---|---|---|---|---|---|---|---|---|---|---|
| 28.0 | 706.156 | 0.000 769 | 0.023 940 | 0.024 709 | 63.1449 | 135.7638 | 198.9088 | 0.2355 | 0.4508 | 0.6863 |
| 30.0 | 744.567 | 0.000 774 | 0.022 659 | 0.023 432 | 65.1147 | 134.5240 | 199.6387 | 0.2419 | 0.4438 | 0.6856 |
| 32.0 | 784.508 | 0.000 778 | 0.021 454 | 0.022 232 | 67.0939 | 133.2632 | 200.3571 | 0.2483 | 0.4367 | 0.6850 |
| 34.0 | 826.003 | 0.000 783 | 0.020 320 | 0.021 103 | 69.0827 | 131.9810 | 201.0637 | 0.2547 | 0.4297 | 0.6844 |
| 36.0 | 869.108 | 0.000 788 | 0.019 252 | 0.020 040 | 71.0821 | 130.6755 | 201.7577 | 0.2611 | 0.4227 | 0.6838 |
| 38.0 | 913.842 | 0.000 793 | 0.018 245 | 0.019 038 | 73.0921 | 129.3468 | 202.4388 | 0.2674 | 0.4157 | 0.6831 |
| 40.0 | 960.255 | 0.000 798 | 0.017 295 | 0.018 093 | 75.1134 | 127.9930 | 203.1063 | 0.2738 | 0.4087 | 0.6825 |
| 42.0 | 1008.379 | 0.000 803 | 0.016 399 | 0.017 202 | 77.1464 | 126.6134 | 203.7598 | 0.2801 | 0.4018 | 0.6819 |
| 44.0 | 1058.266 | 0.000 808 | 0.015 551 | 0.016 359 | 79.1921 | 125.2061 | 204.3982 | 0.2865 | 0.3498 | 0.6813 |
| 46.0 | 1109.948 | 0.000 814 | 0.014 749 | 0.015 563 | 81.2510 | 123.7702 | 205.0212 | 0.2928 | 0.3878 | 0.6806 |
| 48.0 | 1163.453 | 0.000 820 | 0.013 990 | 0.014 810 | 83.3234 | 122.3046 | 205.6281 | 0.2992 | 0.3808 | 0.6800 |
| 50.0 | 1218.838 | 0.000 826 | 0.013 272 | 0.014 097 | 85.4106 | 120.8071 | 206.2177 | 0.3055 | 0.3738 | 0.6793 |
| 52.0 | 1276.143 | 0.000 832 | 0.012 590 | 0.013 422 | 87.5132 | 119.2762 | 206.7894 | 0.3118 | 0.3668 | 0.6787 |
| 54.0 | 1335.408 | 0.000 838 | 0.011 943 | 0.012 781 | 89.6322 | 117.7099 | 207.3421 | 0.3182 | 0.3598 | 0.6780 |
| 56.0 | 1396.661 | 0.000 844 | 0.011 329 | 0.012 173 | 91.7681 | 116.1071 | 207.8752 | 0.3245 | 0.3527 | 0.6773 |
| 58.0 | 1459.963 | 0.000 851 | 0.010 745 | 0.011 596 | 93.9224 | 114.4648 | 208.3872 | 0.3309 | 0.3457 | 0.6766 |
| 60.0 | 1525.354 | 0.000 858 | 0.010 190 | 0.011 048 | 96.0960 | 112.7809 | 208.8769 | 0.3373 | 0.3385 | 0.6758 |

# Appendix – 7

*Saturated Freon–12: Pressure Table*

| $P^s$ | $T$ (°C) | $v_f$ | $v_{fg}$ | $v_g$ | $h_f$ | $h_{fg}$ | $h_g$ | $s_f$ | $s_{fg}$ | $s_g$ |
|---|---|---|---|---|---|---|---|---|---|---|
| (kPa) | | (m³ / kg) | | | (kJ / kg) | | | (kJ / kg K) | | |
| 45.000 | – 47.3 | 0.000 651 | 0.336 121 | 0.336 772 | – 6.4181 | 172.7220 | 166.3038 | – 0.0279 | 0.7646 | 0.7367 |
| 50.000 | – 45.2 | 0.000 653 | 0.304 683 | 0.305 336 | – 4.5641 | 171.8544 | 167.2903 | – 0.0198 | 0.7538 | 0.7340 |
| 55.000 | – 43.2 | 0.000 656 | 0.278 771 | 0.279 427 | –2.8457 | 171.0457 | 168.2000 | – 0.0123 | 0.7439 | 0.7316 |
| 60.000 | – 41.4 | 0.000 658 | 0.257 034 | 0.257 692 | –1.2435 | 170.2887 | 169.0452 | – 0.0053 | 0.7348 | 0.7295 |
| 65.000 | –39.7 | 0.000 660 | 0.238 527 | 0.239 187 | 0.2570 | 169.5783 | 169.8353 | 0.0011 | 0.7264 | 0.7275 |
| 70.000 | –38.1 | 0.000 662 | 0.222 573 | 0.223 235 | 1.6738 | 168.9037 | 170.5776 | 0.0071 | 0.7186 | 0.7258 |
| 75.000 | –36.6 | 0.000 664 | 0.208 673 | 0.209 336 | 3.0126 | 168.2656 | 171.2783 | 0.0128 | 0.7114 | 0.7242 |
| 80.000 | –35.2 | 0.000 665 | 0 196 449 | 0.197 114 | 4.2870 | 167.6549 | 171.9419 | 0.0182 | 0.7045 | 0.7227 |
| 85.000 | –33.8 | 0.000 667 | 0.185 613 | 0.186 280 | 5.5033 | 167.0694 | 172.5727 | 0.0232 | 0.6981 | 0.7213 |
| 90.000 | –32.5 | 0.000 669 | 0.175 938 | 0.176 606 | 6.6651 | 166.5090 | 173.1741 | 0.0281 | 0.6920 | 0.7200 |
| 95.000 | –31.3 | 0.000 670 | 0.167 245 | 0.167 915 | 7.7781 | 165.9707 | 173.7488 | 0.0327 | 0.6862 | 0.7188 |
| 100.000 | –30.1 | 0.000 672 | 0.159 390 | 0.160 061 | 8.8499 | 165.4495 | 174.2994 | 0.0371 | 0.6807 | 0.7177 |
| 110.000 | –27.8 | 0.000 675 | 0.145 747 | 0.146 421 | 10.8768 | 164.4597 | 175.3365 | 0.0453 | 0.6704 | 0.7157 |
| 120.000 | –25.7 | 0.000 678 | 0.134 298 | 0.134 975 | 12.7702 | 163.5288 | 176.2990 | 0.0530 | 0.6609 | 0.7139 |
| 130.000 | –23.8 | 0.000 680 | 0.124 548 | 0.125 229 | 14.5484 | 162.6496 | 177.1980 | 0.0601 | 0.6522 | 0.7123 |
| 140.000 | –21.9 | 0.000 683 | 0.116 142 | 0.116 825 | 16.2280 | 161.8138 | 178.0418 | 0.0668 | 0.6440 | 0.7108 |
| 150.000 | –20.1 | 0.000 685 | 0.108 816 | 0.109 501 | 17.8234 | 161.0139 | 178.8372 | 0.0731 | 0.6364 | 0.7095 |
| 160.000 | –18.5 | 0.000 688 | 0.102 372 | 0.103 060 | 19.3387 | 160.2513 | 179.5901 | 0.0791 | 0.6292 | 0.7083 |
| 170.000 | –16.9 | 0.000 690 | 0.096 659 | 0.097 349 | 20.7888 | 159.5160 | 180.3049 | 0.0847 | 0.6225 | 0.7072 |
| 180.000 | –15.4 | 0.000 692 | 0.091 557 | 0.092 249 | 22.1743 | 158.8112 | 180.9855 | 0.0901 | 0.6161 | 0.7061 |
| 190.000 | –13.9 | 0.000 694 | 0..086 973 | 0.087 667 | 23.5064 | 158.1290 | 181.6354 | 0.0952 | 0.6100 | 0.7052 |
| 200.000 | –12.5 | 0.000 696 | 0.082 830 | 0.083 526 | 24.7868 | 157.4704 | 182.2572 | 0.1001 | 0.6042 | 0.7043 |
| 215.000 | –10.5 | 0.000 699 | 0.077 311 | 0.078 010 | 26.6246 | 156.5184 | 183.1430 | 0.1071 | 0.5960 | 0.7030 |
| 230.000 | –8.6 | 0.000 702 | 0.072 485 | 0.073 187 | 28.3730 | 155.6049 | 183.9779 | 0.1137 | 0.5882 | 0.7019 |
| 245.000 | –6.8 | 0.000 705 | 0.068 228 | 0.068 933 | 30.0393 | 154.7288 | 184.7681 | 0.1199 | 0.5810 | 0.7009 |
| 260.000 | –5.1 | 0.000 708 | 0.064 445 | 0.065 152 | 31.6329 | 153.8853 | 185.5182 | 0.1258 | 0.5741 | 0.6999 |
| 275.000 | –3.4 | 0.000 710 | 0.016 058 | 0.061 768 | 33.1636 | 153.0685 | 186.2321 | 0.1315 | 0.5675 | 0.6990 |
| 290.000 | –1.9 | 0.058 713 | 0.058 008 | 0.058 721 | 34.6338 | 152.2796 | 186.9134 | 0.1369 | 0.5613 | 0.6982 |
| 305.000 | –0.3 | 0.000 715 | 0.055 247 | 0.055 962 | 36.0525 | 151.5124 | 187.5649 | 0.1421 | 0.5554 | 0.6974 |
| 320.000 | 1.1 | 0.000 718 | 0.052 735 | 0.053 453 | 37.4205 | 150.7690 | 188.1895 | 0.1470 | 0.5497 | 0.6967 |
| 340.000 | 3.0 | 0.000 721 | 0.049 717 | 0.050 438 | 39.1775 | 149.8058 | 188.9833 | 0.1534 | 0.5425 | 0.6958 |
| 360.000 | 4.8 | 0.000 724 | 0.047 022 | 0.037 746 | 40.8629 | 148.8744 | 189.7373 | 0.1594 | 0.5356 | 0.6950 |
| 380.000 | 6.5 | 0.000 727 | 0.044 599 | 0.045 326 | 42.4823 | 147.9729 | 190.4552 | 0.1651 | 0.5291 | 0.6943 |
| 400.000 | 8.2 | 0.000 730 | 0.042 410 | 0.043 139 | 44.0433 | 147.0969 | 191.1402 | 0.1707 | 0.5229 | 0.6935 |
| 420.000 | 9.8 | 0.000 733 | 0.040 420 | 0.041 153 | 45.5527 | 146.2424 | 191.7952 | 0.1760 | 0.5169 | 0.6929 |
| 440.000 | 11.3 | 0.000 736 | 0.038 605 | 0.039 340 | 47.0118 | 145.4107 | 192.4225 | 0.1810 | 0.5112 | 0.6923 |
| 460.000 | 12.8 | 0.000 738 | 0.036 941 | 0.037 679 | 48.4253 | 144.5993 | 193.0246 | 0.1859 | 0.5057 | 0.6917 |
| 480.000 | 14.2 | 0.000 741 | 0.035 410 | 0.036 151 | 49.7983 | 143.8046 | 193.6028 | 0.1907 | 0.5004 | 0.6911 |
| 500.000 | 15.6 | 0.000 744 | 0.033 997 | 0.034 741 | 51.1321 | 143.0275 | 194.1595 | 0.1953 | 0.4953 | 0.6906 |
| 530.000 | 17.6 | 0.000 748 | 0.032 069 | 0.032 817 | 53.0672 | 141.8893 | 194.9566 | 0.2019 | 0.4480 | 0.6898 |

| $P^s$ | $T$ (°C) | $v_f$ | $v_{fg}$ | $v_g$ | $h_f$ | $h_{fg}$ | $h_g$ | $s_f$ | $s_{fg}$ | $s_g$ |
|---|---|---|---|---|---|---|---|---|---|---|
| (kPa) | | | (m³ / kg) | | | (kJ / kg) | | | (kJ / kg K) | |
| 560.000 | 19.6 | 0.000 751 | 0.030 340 | 0.031 091 | 54.9270 | 140.7854 | 195.7124 | 0.2082 | 0.4810 | 0.6891 |
| 590.000 | 21.4 | 0.000 755 | 0.028 779 | 0.029 534 | 56.7215 | 139.7088 | 196.4303 | 0.2142 | 0.4743 | 0.6885 |
| 620.000 | 23.2 | 0.000 759 | 0.027 363 | 0.028 122 | 58.4559 | 138.6579 | 197.1138 | 0.2200 | 0.4679 | 0.6879 |
| 650.000 | 24.9 | 0.000 763 | 0.026 073 | 0.026 835 | 60.1346 | 137.6309 | 197.7655 | 0.2256 | 0.4617 | 0.6873 |
| 680.000 | 26.6 | 0.000 766 | 0.024 891 | 0.025 658 | 61.7636 | 136.6247 | 198.3883 | 0.2309 | 0.4558 | 0.6867 |
| 710.000 | 28.2 | 0.000 770 | 0.023 806 | 0.024 575 | 63.3452 | 135.6384 | 198.9836 | 0.2361 | 0.4501 | 0.6862 |
| 740.000 | 29.8 | 0.000 773 | 0.022 804 | 0.023 578 | 64.8837 | 134.6702 | 199.5539 | 0.2411 | 0.4446 | 0.6857 |
| 770.000 | 31.3 | 0.000 777 | 0.021 878 | 0.022 654 | 66.3828 | 133.7182 | 200.1010 | 0.2460 | 0.4392 | 0.6852 |
| 800.000 | 32.8 | 0.000 780 | 0.021 017 | 0.021 797 | 67.8449 | 132.7806 | 200.6255 | 0.2507 | 0.4341 | 0.6848 |
| 840.000 | 34.7 | 0.000 785 | 0.019 962 | 0.020 746 | 69.7396 | 131.5535 | 201.2931 | 0.2568 | 0.4274 | 0.6842 |
| 880.000 | 36.5 | 0.000 789 | 0.018 998 | 0.019 787 | 71.5775 | 130.3496 | 201.9270 | 0.2626 | 0.4210 | 0.6836 |
| 920.000 | 38.3 | 0.000 793 | 0.018 114 | 0.018 907 | 73.3639 | 129.1656 | 202.5294 | 0.2683 | 0.4148 | 0.6831 |
| 960.000 | 40.0 | 0.000 798 | 0.017 300 | 0.018 098 | 75.1013 | 128.0017 | 203.1030 | 0.2737 | 0.4088 | 0.6825 |
| 1000.00 | 41.7 | 0.000 802 | 0.016 549 | 0.017 351 | 76.7967 | 126.8519 | 203.6486 | 0.2790 | 0.4030 | 0.6820 |
| 1050.00 | 43.7 | 0.000 808 | 0.015 686 | 0.016 494 | 78.8585 | 125.4364 | 204.2498 | 0.2854 | 0.3959 | 0.6814 |
| 1100.000 | 45.6 | 0.000 813 | 0.014 898 | 0.015 711 | 80.8591 | 124.0452 | 204.9043 | 0.2916 | 0.3891 | 0.6808 |
| 1150.000 | 47.5 | 0.000 818 | 0.014 175 | 0.014 993 | 82.8068 | 122.6723 | 205.4791 | 0.2976 | 0.3826 | 0.6802 |
| 1200.000 | 49.3 | 0.000 824 | 0.013 509 | 0.014 333 | 84.7076 | 121.3135 | 206.0212 | 0.3034 | 0.3762 | 0.6796 |
| 1250.000 | 51.1 | 0.000 829 | 0.012 894 | 0.013 722 | 86.5615 | 119.9718 | 206.5333 | 0.3090 | 0.3700 | 0.6790 |
| 1300.000 | 52.8 | 0.000 834 | 0.012 323 | 0.013 157 | 88.3725 | 118.6441 | 207.0166 | 0.3144 | 0.3640 | 0.6784 |
| 1350.000 | 54.5 | 0.000 839 | 0.011 792 | 0.012 632 | 90.1455 | 117.3273 | 207.4728 | 0.3197 | 0.3581 | 0.6778 |
| 1400.000 | 56.1 | 0.000 845 | 0.011 297 | 0.012 142 | 91.8828 | 116.0205 | 207.9033 | 0.3249 | 0.3524 | 0.6772 |
| 1450.000 | 57.7 | 0.000 850 | 0.010 834 | 0.011 684 | 93.5870 | 114.7221 | 208.3091 | 0.3299 | 0.3468 | 0.6767 |
| 1500.000 | 59.2 | 0.000 855 | 0.010 400 | 0.011 255 | 95.2582 | 113.4337 | 208.6919 | 0.3348 | 0.3413 | 0.6761 |
| 1575.000 | 61.5 | 0.000 863 | 0.009 797 | 0.010 660 | 97.7141 | 111.5097 | 209.2238 | 0.3420 | 0.3332 | 0.6752 |

# Appendix – 8

*Super heated Freon – 12*

| | 0.05 MPa ( – 45.2°C) | | | | 0.10 MPa (– 30.1°C) | | | |
|---|---|---|---|---|---|---|---|---|
| *T* (°C) | *v* | *u* | *h* | *s* | *v* | *u* | *h* | *s* |
| | (m³/kg) | (kJ/kg) | (kJ/kg) | (kJ/kg K) | (m³/kg) | (kJ/kg) | (kJ/kg) | (kJ/kg K) |
| – 40 | 0.312 889 | 154.4476 | 170.0920 | 0.7462 | | | | |
| – 30 | 0.327 425 | 159.2255 | 175.5968 | 0.7693 | 0.160 122 | 158.3326 | 174.3448 | 0.7179 |
| – 20 | 0.341 847 | 164.1082 | 181.2006 | 0.7919 | 0.167 701 | 163.3116 | 180.0817 | 0.7410 |
| – 10 | 0.356 176 | 169.0946 | 186.9034 | 0.8140 | 0.175 178 | 168.3792 | 185.8970 | 0.7636 |
| – 0 | 0.370 429 | 174.1829 | 192.7043 | 0.8356 | 0.182 573 | 173.5367 | 191.7939 | 0.7856 |
| 10 | 0.384 619 | 179.3710 | 198.6020 | 0.8568 | 0.189 901 | 178.7842 | 197.7742 | 0.8071 |
| 20 | 0.398 756 | 184.6568 | 204.5946 | 0.8776 | 0.197 172 | 184.1213 | 203.8385 | 0.8281 |
| 30 | 0.412 848 | 190.0375 | 210.6799 | 0.8980 | 0.204 397 | 189.5467 | 209.9864 | 0.8487 |
| 40 | 0.426 903 | 195.5106 | 216.8557 | 0.9181 | 0.211 582 | 195.0589 | 216.2171 | 0.8690 |
| 50 | 0.440 925 | 201.0733 | 223.1195 | 0.9377 | 0.218 734 | 200.6560 | 222.5294 | 0.8888 |
| 60 | 0.454 919 | 206.7227 | 229.4687 | 0.9571 | 0.225 857 | 206.3359 | 228.9216 | 0.9083 |
| 70 | 0.468 889 | 212.4562 | 235.9007 | 0.9761 | 0.232 954 | 212.0965 | 235.3920 | 0.9274 |
| 80 | 0.482 838 | 218.2709 | 242.4128 | 0.9948 | 0.240 031 | 217.9355 | 241.9386 | 0.9462 |
| 90 | 0.496 760 | 224.1641 | 249.0026 | 1.0132 | 0.247 089 | 223.8503 | 248.5592 | 0.9647 |
| 100 | 0.510 684 | 230.1331 | 255.6673 | 1.0313 | 0.254 130 | 229.8388 | 255.2518 | 0.9829 |

| | 0.30 MPa (– 0.8°C) | | | |
|---|---|---|---|---|
| *T*(°C) | *v* | *u* | *h* | *s* |
| | (*m*³/kg) | (kJ/kg) | (kJ/kg) | (kJ/kg K) |
| 0 | 0.057 093 | 170.7585 | 187.8862 | 0.6997 |
| 10 | 0.059 901 | 176.2873 | 194.2576 | 0.7226 |
| 20 | 0.062 637 | 181.8618 | 200.6529 | 0.7448 |
| 30 | 0.065 314 | 187.4901 | 207.0843 | 0.7663 |
| 40 | 0.067 943 | 193.1771 | 213.5601 | 0.7873 |
| 50 | 0.070 533 | 198.9262 | 220.0860 | 0.8079 |
| 60 | 0.073 089 | 204.7391 | 226.6657 | 0.8279 |
| 70 | 0.075 617 | 210.6168 | 233.3018 | 0.8475 |
| 80 | 0.078 120 | 216.5595 | 239.9955 | 0.8668 |
| 90 | 0080 603 | 222.5668 | 246.7476 | 0.8856 |
| 100 | 0.083 068 | 228.6379 | 253.5582 | 0.9041 |
| 110 | 0.085 517 | 234.7718 | 260.4268 | 0.9223 |
| 120 | 0.087 952 | 240.9672 | 267.3529 | 0.9401 |
| 130 | 0.090 376 | 247.2228 | 274.3355 | 0.9577 |
| 140 | 0.092 788 | 253.5371 | 281.3735 | 0.9749 |
| 150 | 0.095 191 | 259.9084 | 288.4658 | 0.9919 |

| | 0.50 MPa (15.6°C) | | | |
|---|---|---|---|---|
| *T*(°C) | *v* | *u* | *h* | *s* |
| | (m³ / kg) | (kJ / kg) | (kJ / kg) | (kJ / kg K) |
| 20 | 0.035 557 | 179.3701 | 197.1487 | 0.7009 |
| 30 | 0.037 365 | 185.2545 | 203.9372 | 0.7236 |
| 40 | 0.039 113 | 191.1551 | 210.7116 | 0.7456 |
| 50 | 0.040 812 | 197.0849 | 217.4909 | 0.7669 |
| 60 | 0.042 471 | 203.0527 | 224.2883 | 0.7876 |
| 70 | 0.044 098 | 209.0643 | 231.1131 | 0.8078 |
| 80 | 0.045 697 | 215.1239 | 237.9721 | 0.8275 |
| 90 | 0.047 272 | 221.2339 | 244.8699 | 0.8468 |
| 100 | 0.048 828 | 227.3958 | 251.8097 | 0.8656 |
| 110 | 0.050 366 | 233.6106 | 258.7938 | 0.8841 |
| 120 | 0.051 890 | 239.8784 | 265.8235 | 0.9022 |
| 130 | 0.053 401 | 246.1990 | 272.8994 | 0.9200 |
| 140 | 0.054 900 | 252.5718 | 280.0219 | 0.9374 |
| 150 | 0.056 389 | 258.9963 | 287.1910 | 0.9546 |

1.00 MPa (41.7°C)

| T(°C) | v (kJ/kg) | u (kJ/kg) | h (kJ/kg) | s (kJ/ kg K) |
|---|---|---|---|---|
| 50 | 0.018 278 | 191.8051 | 210.0827 | 0.7022 |
| 60 | 0.019 324 | 198.3186 | 217.6425 | 0.7522 |
| 70 | 0.020 317 | 204.7784 | 225.0952 | 0.7473 |
| 80 | 0.021 268 | 211.2134 | 232.4817 | 0.7685 |
| 90 | 0.022 187 | 217.6428 | 239.8297 | 0.7890 |
| 100 | 0.023 079 | 224.0799 | 247.1586 | 0.8089 |
| 110 | 0.023 948 | 230.5342 | 254.4824 | 0.8283 |
| 120 | 0.024 799 | 237.0123 | 261.8113 | 0.8472 |
| 130 | 0.025 634 | 243.5190 | 269.1528 | 0.8656 |
| 140 | 0.026 455 | 250.0574 | 276.5126 | 0.8836 |
| 150 | 0.027 265 | 256.6302 | 283.8948 | 0.9013 |
| 160 | 0.028 064 | 261.2388 | 291.3025 | 0.9186 |
| 170 | 0.028 854 | 269.8842 | 298.7380 | 0.9356 |
| 180 | 0.029 636 | 276.5670 | 306.2029 | 0.9522 |
| 190 | 0.030 411 | 283.2872 | 313.6981 | 0.9686 |
| 200 | 0.031 180 | 290.0450 | 321.2246 | 0.9846 |

3.00 MPa (94.0°C)

| T(°C) | v ($m^3$/kg) | u (kJ/kg) | h (kg / kg) | s (kJ/kg K) |
|---|---|---|---|---|
| 100 | 0.005 228 | 204.2362 | 219.9197 | 0.6757 |
| 110 | 0.006 873 | 213.6597 | 231.2796 | 0.7058 |
| 120 | 0.006 407 | 222.1632 | 241.3833 | 0.7318 |
| 130 | 0.006 878 | 230.1959 | 250.8285 | 0.7556 |
| 140 | 0.007 307 | 237.9470 | 259.8688 | 0.7777 |
| 150 | 0.007 708 | 245.5165 | 268.6391 | 0.7987 |
| 160 | 0.008 086 | 252.9638 | 277.2203 | 0.8187 |
| 170 | 0.008 446 | 260.3276 | 285.6653 | 0.8380 |
| 180 | 0.008 792 | 267.6341 | 294.0105 | 0.8566 |
| 190 | 0.009 127 | 274.9020 | 302.2818 | 0.8747 |
| 200 | 0.009 451 | 282.1450 | 310.4985 | 0.8922 |
| 210 | 0.009 767 | 289.3733 | 318.6754 | 0.9093 |
| 220 | 0.010 076 | 296.5496 | 326.8235 | 0.9260 |
| 230 | 0.010 379 | 303.8150 | 334.9518 | 0.9424 |
| 240 | 0.010 676 | 311.0390 | 343.0672 | 0.9583 |
| 250 | 0.010 968 | 318.2704 | 351 1754 | 0.9740 |

4.00 MPa (110.3°C)

| T(°C) | v ($m^3$/kg) | u (kJ/kg) | h (kg / kg) | s (kJ/kg K) |
|---|---|---|---|---|
| 120 | 0.003 763 | 209.5098 | 224.5612 | 0.6763 |
| 130 | 0.004 334 | 220.3976 | 237.7338 | 0.7094 |
| 140 | 0.004 786 | 229.7668 | 248.9104 | 0.7368 |
| 150 | 0.005 176 | 238.4249 | 259.1302 | 0.7613 |
| 160 | 0.005 528 | 246.6717 | 268.7838 | 0.7838 |
| 170 | 0.005 853 | 254.6553 | 278.0656 | 0.8050 |
| 180 | 0.006 157 | 262.4605 | 287.0880 | 0.8251 |
| 190 | 0.006 445 | 270.1406 | 295.9222 | 0.8444 |
| 200 | 0.006 721 | 277.7313 | 304.6164 | 0.8630 |
| 210 | 0.006 987 | 285.2578 | 313.2047 | 0.8810 |
| 220 | 0.007 243 | 292.7381 | 321.7120 | 0.8984 |
| 230 | 0.007 493 | 300.1858 | 330.1572 | 0.9153 |
| 240 | 0.007 736 | 307.6115 | 338.5546 | 0.9319 |
| 250 | 0.007 973 | 315.0228 | 346.9157 | 0.9480 |

2.00 MPa (72.9°C)

| T(°C) | v ($m^3$ / kg) | u (kJ / kg) | h (kJ / kg) | s (kJ / kgK) |
|---|---|---|---|---|
| 80 | 0.008 654 | 201.0172 | 218.3260 | 0.6898 |
| 90 | 0.009 259 | 208.7290 | 227.4472 | 0.7153 |
| 100 | 0.009 994 | 216.1377 | 236.1250 | 0.7389 |
| 110 | 0.010 580 | 223.3633 | 244.5238 | 0.7611 |
| 120 | 0.011 132 | 230.4731 | 252.7364 | 0.7823 |
| 130 | 0.011 656 | 237.5089 | 260.8211 | 0.8026 |
| 140 | 0.012 159 | 244.4985 | 268.8165 | 0.8222 |
| 150 | 0.012 645 | 251.4609 | 276.7499 | 0.8411 |
| 160 | 0.013 115 | 228.4099 | 284.6408 | 0.8596 |
| 170 | 0.013 574 | 265.3553 | 292.5038 | 0.8775 |
| 180 | 0.014 023 | 272.3047 | 300.3499 | 0.8950 |
| 190 | 0.014 462 | 279.2635 | 308.1874 | 0.9121 |
| 200 | 0.014 894 | 286.2359 | 316.0230 | 0.9289 |

# Appendix – 9

*Psychrometric chart*

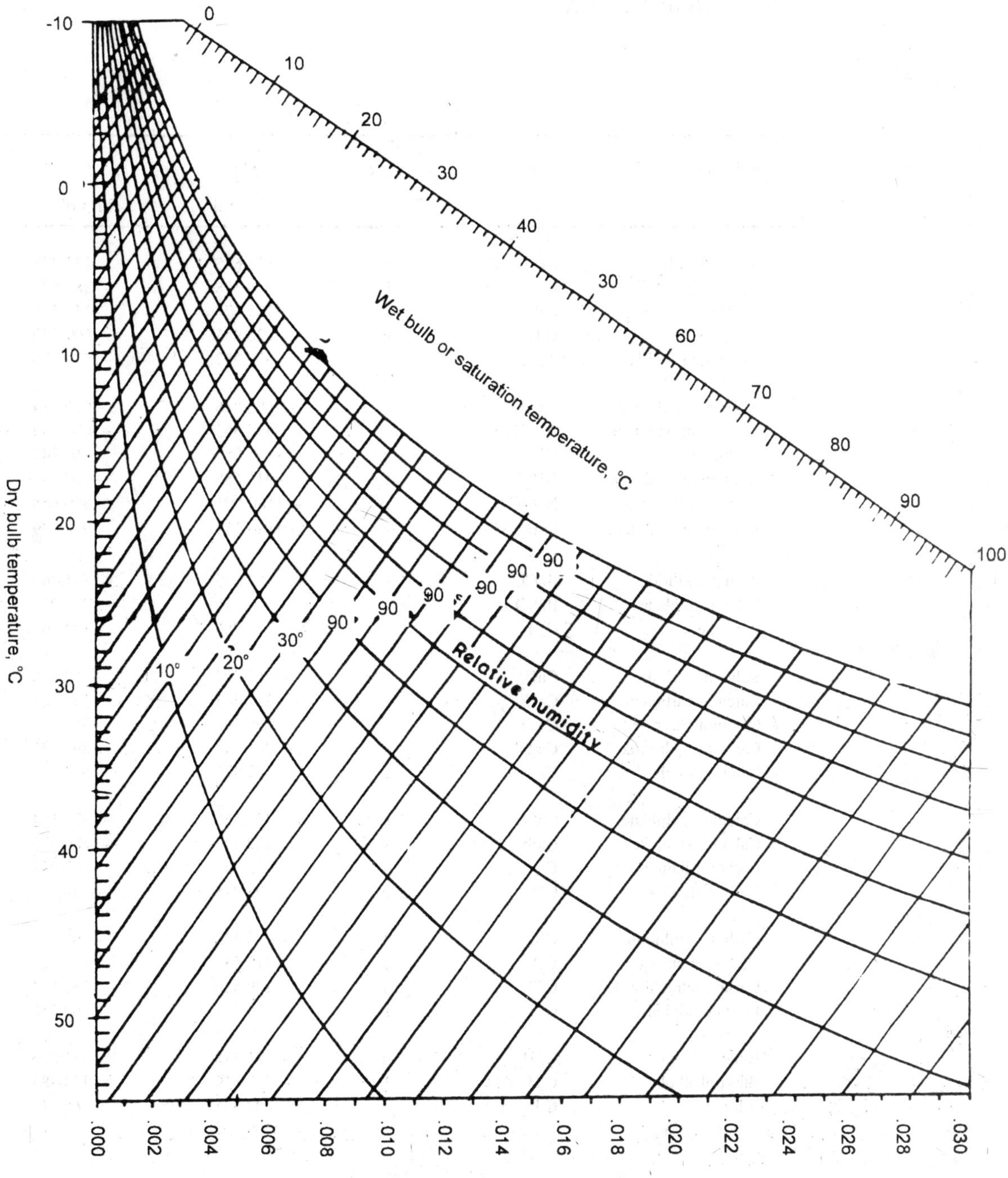

# Appendix – 10

*Standard enthalpies and Gibbs free energies of formation of some selected compounds at 298.15 K**

| Substance | Formula | State | $\Delta H_f^0$ kJ / mol | $\Delta G_c^0$ kJ / mol |
|---|---|---|---|---|
| Acetaldehyde | $CH_3CHO$ | g | —166.635 | —133.978 |
| Acetic acid | $CH_3COOH$ | l | — 438.358 | —381.836 |
| Acetone | $CH_3COCH_3$ | g | —185.475 | —144.300 |
| Acetylene | $C_2H_2$ | g | — 226.925 | —209.340 |
| Aluminium oxide | $Al_2O_3$ | s | —1670.952 | —1577.586 |
| Aluminium chloride | $AlCl_3$ | s | — 695.846 | —620.484 |
| Aluminium sulphate | $Al_2(SO_4)_3$ | s | — 3437.363 | —3094.045 |
| Ammonia | $NH_3$ | g | — 46.055 | —16.747 |
| Ammonium chloride | $NH_4CI$ | s | — 315.685 | — 20.390 |
| Ammonium nitrate | $NH_4NO_3$ | s | —1180.091 | — 900.958 |
| Ammonium sulphate | $(NH_4)_2SO_4$ | s | —1180.259 | — 900.999 |
| Barium chloride | $BaCl_2$ | s | — 860.639 | — 811.402 |
| Barium sulphate | $BaSO_4$ | s | —1466.217 | — 1354.011 |
| Benzene | $C_6H_6$ | l | 48.986 | 124.767 |
| Calcium carbide | $CaC_2$ | s | — 62.802 | — 67.826 |
| Calcium carbonate | $CaCO_2$ | s | — 1207.683 | — 1129.515 |
| Calcium chloride | $CaCI_2$ | s | — 795.492 | — 750.693 |
| Calcium hydroxide | $Ca(OH)_2$ | s | — 987.247 | — 897.231 |
| Calcium oxide | CaO | s | — 635.975 | — 604.574 |
| Calcium sulphlate | $CaSO_4$ | s | — 1433.560 | —1321.354 |
| Calcium sulphate | $CaSO_4.2H_2O$ | s | — 2022.643 | —1796.975 |
| Carbon (diamond) | C | s | 2.093 | 2.931 |
| Carbon dioxide | $CO_2$ | g | — 393.978 | —394.815 |
| Carbon disulphide | $CS_2$ | l | — 88.342 | — 63.639 |
| Carbon monoxide | CO | g | — 110.53 | — 137.327 |
| Carbon tetrachloride | $CCI_4$ | l | — 139.420 | — 68.664 |
| Copper sulphate | $CuSO_4$ | s | —770.371 | — 662.352 |
| Ethane | $C_2H_6$ | g | — 84.573 | —33.076 |
| Ethyl alcohol | $C_2H_5OH$ | I | — 277.819 | — 174.883 |
| Ethylene | $C_2H_4$ | g | 52.335 | 68.245 |

| Subtance | Formula | State | $\Delta H_f^o$ kJ / mol | $\Delta G_f^o$ kJ / mol |
|---|---|---|---|---|
| Ferric chloride | $FeCl_3$ | s | — 405.282 | — 337.456 |
| Ferric oxide | $Fe_2O_3$ | s (haematite) | — 882.206 | — 741.482 |
| Ferroso-ferric oxide | $Fe_3O_4$ | s (magnetite) | — 1117.876 | — 1014.880 |
| Ferrous sulphate | $FeSO_4$ | s | — 923.189 | — 815.588 |
| Formic acid | HCOOH | g | — 362.995 | — 335.781 |
| Formaldehyde | HCHO | g | — 115.974 | — 110.113 |
| Hydrogen chloride | HCl | g | — 92.528 | — 95.459 |
| Hydrogen peroxide | $H_2O_2$ | g | — 135.234 | — 102.577 |
| Hydrogen sulphide | $H_2S$ | g | — 20.097 | — 33.076 |
| Lead monoxide | PbO | s (red) | — 219.388 | — 189.243 |
| Lead monoxide | PbO | s (yellow) | — 218.132 | — 188.406 |
| Lead oxide, red | $Pb_3O_4$ | s | — 735.202 | —617.912 |
| Magnesium carbonate | $MgCO_3$ | s | —1113.689 | —1029.953 |
| Magnesium chloride | $MgCl_2$ | s | — 642.255 | —592.851 |
| Magnesium chloride | $MgCl_2$ $6H_2O$ | s | — 2501.194 | —2116.846 |
| Magnesium hydroxide | $Mg(OH)_2$ | s | — 925.283 | —834.429 |
| Magnesium oxide | MgO | s | — 602.061 | —569.823 |
| Magnesium sulphate | $MgSO_4$ | s | — 1279.067 | — 1174.397 |
| Manganese dioxide | $MnO_2$ | s | —521.257 | —466.410 |
| Mercury | Hg | g | 60.709 | 31.820 |
| Mercury oxide | HgO | s (red) | —90.854 | —58.615 |
| Mercury oxide | HgO | s (yellow) | —90.435 | —59.034 |
| Methane | $CH_4$ | g | —74.943 | —50.660 |
| Methyl alcohol | $CH_3OH$ | l | —238.648 | —166.215 |
| Nitric acid | $HNO_3$ | l | —173.333 | —79.968 |
| Nitric oxide | NO | g | 90.435 | 86.667 |
| Nitrous oxide | $N_2O$ | g | 81.643 | 103.833 |
| Potassium chloride | KCl | s | —436.265 | —408.632 |
| Potassium hydroxide | KOH | (aq) $m$ =1 | —481.482 | —439.869 |
| Potnssium nitrate | $KNO_3$ | s | —493.205 | —393.559 |
| Potassium permanganate | $KMnO_4$ | s | —813.914 | —714.268 |
| Propane | $C_3H_8$ | g | —103.833 | —23.446 |
| Sodium bicarbonate | $NaHCO_3$ | s | —948.310 | —852.432 |
| Sodium carbonate | $Na_2CO_3$ | s | —1131.692 | —1048.375 |
| Sodium chloride | NaCl | s | —412.818 | —386.023 |
| Sodium hydroxide | NaOH | s | —427.054 | —380.999 |
| Sodium hydroxide | NaOH | (aq) $m$ =1 | —469.910 | —419.450 |
| Sodium nitrate | $NaNO_3$ | s | —466.828 | —366.345 |

| Substance | Formula | State | $\Delta H_f^o$ kJ / mol | $\Delta G_f^o$ kJ / mol |
|---|---|---|---|---|
| Sodium sulphate | $Na_2SO_4$ | s | —1385.412 | —1267.763 |
| Sodium sulphate | $Na_2SO_4$ $10H_2O$ | s | — 4327.058 | —3646.284 |
| Sulphur | S | s (monoclinic) | 0.417 | 0 |
| Sulphur | S | g | 219.388 | 182.544 |
| Sulphur dioxide | $SO_2$ | g | — 297.263 | —300.612 |
| Sulphur trioxide | $SO_3$ | g | — 395.234 | —370.532 |
| Water | $H_2O$ | l | — 285.958 | —237.392 |
| Water | $H_2O$ | g | —241.997 | —228. 600 |
| Zinc chloride | $ZnCl_2$ | s | —416.168 | —369.694 |
| Zinc Oxide | ZnO | s | —348.342 | —318.187 |
| Zinc sulphate | $ZnSO_4$ | s | —979.293 | —872.110 |
| Zinc sulphate | $ZnSO_4 . 7H_20$ | s | —3077.717 | —2561.903 |

* Adapted from Chandrasekaran, K.D. and Venkteswarlu, D., SI Units in *Chemical Engineering and Technology*, Chemical Engineering Education Development Centre, IIT-Madras (1974), with permission from Chairman, Centre for Continuing Education, IIT-Madras.

# Appendix – 11

*Standard enthalpies of combustion of some selected compounds at 298.15 K**

| Compound | Formula (state) | $-\Delta H_c^0$ kJ/mol |
|---|---|---|
| **Hydrocarbons** | | |
| Final products : $CO_2$ (g), $H_2O$ (1) | | |
| Carbon (graphite) | C (s) | 393.78 |
| Carbon monoxide | CO (g) | 283.18 |
| Hydrogen | $H_2$ (g) | 286.03 |
| Methane | $CH_4$ (g) | 890.94 |
| Ethyne (acetylene) | $C_2H_2$ (g) | 1300.48 |
| Ethene (ethylene) | $C_2H_4$ (g) | 1411.93 |
| Ethane | $C_2H_6$ (g) | 1560.92 |
| Propyne (allylene, methylacetylene) | $C_3H_4$ (g) | 1938.94 |
| Propene (propylene) | $C_3H_6$ (g) | 2059.85 |
| Propane | $C_3H_8$ (g) | 2221.54 |
| n-Butane | $C_4H_{10}$ (g) | 2880.44 |
| n-Pentane | $C_5H_{12}$ (g) | 3538.52 |
| Benzene | $C_6H_6$ (g) | 3303.72 |
| Cyclohexane | $C_6H_{12}$ (l) | 3922.53 |
| n-Hexane | $C_6H_{14}$ (1) | 4165.91 |
| Toluene (methylbenzene) | $C_7H_8$ (l) | 3912.56 |
| n-Heptane | $C_7H_{14}$ (1) | 4820.14 |
| n-Octane | $C_8H_{18}$ (l) | 5474.37 |
| n-Decane | $C_{10}H_{22}$ (l) | 6782.87 |
| **Alcohols** | | |
| Final products: $CO_2$ (g). $H_2O$ (l) | | |
| Methyl alcohol | $CH_4O$ (g) | 764.47 |
| Ethyl alcohol | $C_2H_6O$ (g) | 1410.20 |
| n-Propyl alcohol | $C_3H_8O$ (g) | 2069.37 |
| n-Butyl alcohol | $C_4H_{10}O$ (g) | 2721.34 |

| Compound | Formula (state) | $-\Delta H_c^0$ kJ/mol |
|---|---|---|
| **CHO Compounds** | | |
| Final products: $CO_2$ (g), $H_2O$ (l) | | |
| Formaldehyde | $CH_2O$ (g) | 563.84 |
| Acetaldehyde | $C_2H_4O$ (g) | 1193.15 |
| Acetone | $C_3H_6O$ (g) | 1822.60 |
| Ethyl acetate | $C_4H_8O_2$ (g) | 2292.11 |
| Phenol | $C_6H_6O$ (g) | 3129.84 |
| **Sulphur compounds** | | |
| Final products: $CO_2$ (g), $SO_2$ (g), $H_2O$ (l) | | |
| Carbonyl sulphide | COS (g) | 553.54 |
| Carbon disulphide | $CS_2$ (g) | 1103.31 |

* Adapted from Chandrasekaran, K.D. and Venkteswarlu, D., SI Units in *Chemical Engineering and Technology*, Chemical Engineering Education Development Centre, IIT-Madras (1974), with permission from Chairman, Centre for Continuing Education, IIT-Madras.

# Appendix – 12

*Isobaric molar heat capacities of some selected gases in the ideal gas state.*

$$C_P^0 = a + bT + cT^2 + dT^3 + eT^{-2};\ C_P^0 \text{ is in J / mol K and } T \text{ is in } K$$

(from 298.16 $K$ to $T_m$)

| Compound | Formula | $T_m$ | $a$ | $b \times 10^3$ | $c \times 10^6$ | $d \times 10^9$ | $e \times 10^{-5}$ |
|---|---|---|---|---|---|---|---|
| Methane | $CH_4$ | 1000 | 17.449 | 60.449 | 1.117 | —7.204 | — |
| Ethane | $C_2H_6$ | 1000 | 4.493 | 182.229 | —74.843 | 10.797 | — |
| Propane | $C_3H_8$ | 1000 | —4.798 | 307.259 | —160.132 | 32.743 | — |
| n-Butane | $C_4H_{10}$ | 1000 | 0.469 | 385.311 | —198.849 | 39.959 | — |
| n-Pentane | $C_5H_{12}$ | 1000 | 1.443 | 476.394 | —250.366 | 51.229 | — |
| n-Hexane | $C_6H_{14}$ | 1000 | 3.083 | 565.691 | —300.319 | 62.051 | — |
| n-Heptane | $C_7H_{16}$ | 1000 | 5.020 | 653.649 | —348.678 | 72.313 | — |
| n-Octane | $C_8H_{18}$ | 1000 | 6.907 | 741.770 | —397.204 | 82.629 | — |
| **Alkenes** | | | | | | | |
| Ethylene | $C_2H_4$ | 1000 | 4.196 | 154.565 | —81.076 | 16.813 | — |
| Propylene | $C_3H_6$ | 1000 | 3.305 | 235.821 | —117.580 | 22.673 | — |
| l-Butene | $C_4H_8$ | 1000 | 2.539 | 344.871 | —191.252 | 41.657 | — |
| l-Pentene | $C_5H_{10}$ | 1000 | 7.480 | 424.413 | —231.868 | 49.593 | — |
| l-Hexene | $C_6H_{12}$ | 1000 | 8.630 | 514.563 | —281.959 | 60.256 | — |
| l-Heptene | $C_7H_{14}$ | 1000 | 9.806 | 605.750 | —334.095 | 71.869 | — |
| l-Octene | $C_8H_{16}$ | 1000 | 12.157 | 692.181 | —380.986 | 81.629 | — |
| **Acetylenes** | | | | | | | |
| Acetylene | $C_2H_2$ | 1500 | 50.969 | 16.227 | — | — | -10.797 |
| Propyne | $C_3H_4$ | 1500 | 23.456 | 85.753 | —58.332 | 15.867 | — |
| l-Butyne | $C_4H_6$ | 1500 | 13.458 | 269.990 | —149.172 | 32.538 | — |
| l-Pentyne | $C_5H_8$ | 1500 | 18.591 | 348.394 | —188.165 | 39.838 | — |
| l-Hexyne | $C_6H_{10}$ | 1500 | 19.017 | 442.250 | —243.460 | 52.555 | — |
| l-Heptyne | $C_7H_{12}$ | 1500 | 21.402 | 528.087 | —288.710 | 61.507 | — |
| l-Ocyne | $C_8H_{14}$ | 1500 | 23.431 | 615.811 | —337.270 | 72.032 | — |
| **Miscellaneous** | | | | | | | |
| Acetaldehyde | $CH_3CHO$ | 1000 | 21.879 | 161.191 | —85.883 | 18.281 | — |
| Ammonia | $NH_3$ | 1800 | 29.747 | 25.108 | — | — | -1.546 |
| Benzene | $C_6H_6$ | 1500 | —33.893 | 471.793 | —298.294 | 70.823 | — |
| 1,3-Butadiene | $C_4H_6$ | 1000 | —2.958 | 340.027 | —223.652 | 56.521 | — |
| Carbon dioxide | $CO_2$ | 2000 | 45.369 | 8.688 | — | — | -9.619 |
| Carbon monoxide | CO | 2500 | 28.068 | 4.631 | — | — | —0.258 |
| Carbon disulfide | $CS_2$ | 1800 | 52.470 | 6.693 | — | — | —7.532 |
| Chlorine | $Cl_2$ | 3000 | 36.931 | 0.740 | — | — | — 2 860 |
| Dinitrogen tetroxide | $N_2O_4$ | 2000 | 96.941 | 18.765 | — | — | — 23.171 |

| Compound | Formula | $T_m$ | $a$ | $b \times 10^3$ | $c \times 10^6$ | $d \times 10^9$ | $e \times 10^{-5}$ |
|---|---|---|---|---|---|---|---|
| Ethanol | $C_2H_5OH$ | 1000 | 20.691 | 205.346 | —99.793 | 18.825 | — |
| Elhyl benzene | $C_8H_{10}$ | 1500 | -34.579 | 661.083 | —409.809 | 96.664 | — |
| Hydrogen | $H_2$ | 3000 | 27.012 | 3.509 | — | — | 0.690 |
| Hydrogen sulfide | $H_2S$ | 2300 | 32.682 | 12.388 | — | — | —1.929 |
| Methanol | $CH_3OH$ | 1500 | 18.382 | 101.564 | —28.683 | — | — |
| Nitric oxide | $NO$ | 1000 | 29.367 | —1.548 | 10.651 | —4.547 | — |
| Nitrogen | $N_2$ | 2000 | 27.270 | 4.930 | — | — | 0.333 |
| Nitrogen dioxide | $NO_2$ | 2000 | 41.420 | 9.935 | — | — | —6.585 |
| Oxygen | $O_2$ | 2000 | 30.255 | 4.207 | — | — | —1.887 |
| n–Propyl benzene | $C_9H_{12}$ | 1500 | -29.944 | 741.946 | —452.206 | 105.289 | — |
| Sulfur dioxide | $SO_2$ | 2000 | 47.381 | 6.660 | — | — | —8.439 |
| Sulfur trioxide | $SO_3$ | 2000 | 67.011 | 8.780 | — |  | —16.861 |
| Toluene | $C_7H_8$ | 1500 | —33.876 | 556.952 | —342.315 | 79.859 | — |
| Water | $H_2O$ | 2000 | 28.850 | 12.055 | — | — | 1.006 |

# Nomenclature

## A

| | | |
|---|---|---|
| A | — | Ampere ; Area |
| A | — | Helmholtz free energy |
| A/F | — | Air-Fuel ratio |
| $a$ | — | Molar (or specific) Helmholtz free energy |
| $a$ | — | Acceleration |
| $a$ | — | constant in expression for molar heat capacity |
| $a$ | — | constant in equations of state |
| $a_i$ | — | activity of component $i$ |

## B

| | | |
|---|---|---|
| $B$ | — | Darrius function or Availability for a flow process |
| $B$ | — | Second virial coefficient |
| $B$ | — | Magnetic field |
| $B'$ | — | Second virial coefficient |
| $b$ | — | Constant in expression for molar heat capacity |
| $b$ | — | Constant in equations of state |

# C

| | | |
|---|---|---|
| $C$ | — | Number of independent components |
| $C$ | — | Capacitance |
| $C$ | — | Third virial coefficient |
| $C'$ | — | Third virial coefficient |
| $C_P$ | — | Molar (or specefic) heat capacity at constant pressure) |
| $C_v$ | — | Molar (or specific) heat capacity at constant volume |
| $Cd$ | — | Candela |
| COP | — | Coefficient of performance |
| $c$ | — | Constant in the expression for molar heat capacity |

# D

| | | |
|---|---|---|
| $D, D'$ | — | Fourth virial coefficient |
| $D_e$ | — | Bond energy |
| $D_0$ | — | Dissociation energy |
| $d$ | — | Constant in the expression for molar heat capacity |

# E

| | | |
|---|---|---|
| $E$ | — | Energy |
| $e$ | — | Specific (or molar) energy |
| $e$ | — | Constant in the expression for molar heat capacity |

# F

| | | |
|---|---|---|
| $F$ | — | Force |
| $F$ | — | Thermodynamic degrees of freedom |
| $f$ | — | Fugacity |
| $f^0$ | — | Standard state fugacity |

## G

| | | |
|---|---|---|
| $G$ | — | Gibb's Free Energy |
| $G$ | — | Universal gravitational constant |
| $\Delta G_f^0$ | — | Standard Gibbs free energy of formation |
| $\Delta G_T^0$ | — | Standard Gibbs free energy change at temperature $T$ |
| $g$ | — | Acceleration due to gravity |
| $g$ | — | Molar (or specific) Gibbs free energy |
| $\bar{g}$ | — | Partial molar Gibbs free energy |

## H

| | | |
|---|---|---|
| $H$ | — | Enthalpy |
| $\boldsymbol{H}$ | — | Magnetic field intensity |
| $\Delta H_c^0$ | — | Standard enthalpy (heat) of combustion |
| $\Delta H_f^0$ | — | Standard enthalpy of formation |
| $\Delta H_T^0$ | — | Standard enthalpy change of a reaction |
| $h$ | — | Molar (or specific) enthalpy |
| $h$ | — | Planck's constant |
| $h$ | — | Height |
| $h^*$ | — | Specific enthalpy of moist air |
| $\bar{h}$ | — | Partial molar enthalpy |

## I

| | | |
|---|---|---|
| $i$ | — | Current |

## J

| | | |
|---|---|---|
| $J$ | — | Joule |

## K

$K$ — Equilibrium constant
$K$ — Spring constant
$K_P$ — Equilibrium constant expressed in partial pressure
$K_y$ — Equilibrium constant expressed in mole fraction
$K_n$ — Equilibrium constant expressed in moles
kg — Kilogram

## L

$L$ — Length

## M

$M$ — Molar mass
$\boldsymbol{M}$ — Magnetization
MEP — Mean effective pressure
m — metre
$m$ — mass
mol — Mole

## N

N — newton
$N_0$ — Avagadro's number
$n$ — Number of moles

## P

$P$ — Pressure
$P$ — Number of phases
Pa — Pascal
$P_s$ — Saturation pressure
$p$ — Partial pressure

## Q

| | | |
|---|---|---|
| $Q$ | — | Energy transferred as heat to a system |
| $Q_c$ | — | Electric charge |
| $Q_H$ | — | Energy rejected as heat to a high temperature reservoir |
| $Q_L$ | — | Energy absorbed as heat from a low temperature reservoir |
| $q$ | — | Energy transferred as heat per unit mass or mole |

## R

| | | |
|---|---|---|
| $R$ | — | Electrical resistance |
| $R$ | — | Universal gas constant |
| $r_c$ | — | Cut-off ratio |
| $r_e$ | — | Equilibrium interatomic distance |
| $r_0$ | — | Compression ratio |
| $r_p$ | — | Pressure ratio |
| $r_{pv}$ | — | Pressure ratio for constant volume combustion |

## S

| | | |
|---|---|---|
| $S$ | — | Entropy |
| $\Delta S^0$ | — | Standard entropy change of a reaction |
| $s$ | — | Molar (or specific) entropy |
| s | — | second |

## T

| | | |
|---|---|---|
| $T$ | — | Temperature |
| $t$ | — | Time |

## U

| | | |
|---|---|---|
| $U$ | — | Internal energy |
| $u$ | — | Molar (or specific) internal energy |

## V

| | | |
|---|---|---|
| $V$ | — | Volume |
| $V$ | — | Potential difference |
| $\mathbf{V}$ | — | Velocity |
| $v$ | — | Molar (or specific) volume |

## W

| | | |
|---|---|---|
| W | — | Watt |
| $W$ | — | Work done |
| $W_s$ | — | Shaft work |

## X

| | | |
|---|---|---|
| $X$ | — | Quality of a mixture |
| $X$ | — | Mole fraction |
| $X$ | — | Compression (or extension) of a linear spring |

## Y

| | | |
|---|---|---|
| $Y_T$ | — | Isothermal Young's modulus |
| $y$ | — | Mole fraction |

## Z

| | | |
|---|---|---|
| $Z$ | — | Compressibility |
| $Z$ | — | Elevation |

## Greek Letters

| | | |
|---|---|---|
| $\beta$ | — | Coefficient of volume expansion |
| $\gamma$ | — | ratio of specific heats |
| $\varepsilon$ | — | Strain |
| $\varepsilon$ | — | Extent of reaction or degree of conversion |
| $\varepsilon_e$ | — | Extent of reaction at equilibrium |
| $\varepsilon_0$ | — | Permittivity constant |
| $\xi$ | — | Eucker coefficient |
| $\eta_i$ | — | Thermal efficiency |
| $\eta_i$ | — | Isentropic efficiency of a compressor |
| $\eta_p$ | — | Isentropic efficiency of a pump |
| $\eta_T$ | — | Isentropic efficiency of a turbine |
| $\eta_2$ | — | Second law of efficiency |
| $\theta_e$ | — | Angle |
| $\kappa$ | — | Isothermal compressibility |
| $\kappa_s$ | — | Isentropic compressibility |
| $\mu$ | — | Chemical potential |
| $\mu_0$ | — | Permeability |
| $\mu_{JT}$ | — | Joule-Thomson coefficient |
| $v$ | — | Vibrational frequency |
| $v_i$ | — | Stoichiometric coefficient associated with species $i$ |
| $\rho$ | — | Density |
| $\sigma$ | — | Stress |
| $\tau$ | — | Surface tension |
| $\tau$ | — | Torque |
| $\phi$ | — | Availability function |
| $\phi$ | — | Relative humidity |
| $\phi_i$ | — | Mass fraction of component $i$ |
| $\omega$ | — | Angular velocity |
| $\omega$ | — | Acentric factor |
| $\omega$ | — | Humidity ratio or specific humidity |

# Subscripts

| | | |
|---|---|---|
| *c* | — | Critical point |
| *c* | — | Compressor |
| *e* | — | Exit of a control volume |
| *f* | — | Final state |
| *f* | — | Saturated liquid |
| *fg* | — | Change in property due to vaporization |
| *g* | — | saturated vapor |
| *G* | — | Generation |
| *i* | — | Initial state |
| *i* | — | Inlet to a control volume |
| *N* | — | Nozzle |
| *P* | — | Pump |
| *r* | — | Reduced parameter |
| *Sat* | — | Saturation |
| *Sys* | — | System |
| *Sur* | — | Surroundings |
| *T* | — | Turbine |
| *tp* | — | Triple point |
| *Uni* | — | Universe |

# Index